Pathways to

ASTRONOMY

VOLUME 2   CUSTOM EDITION
STARS & GALAXIES

Stephen E. Schneider
Professor of Astronomy
University of Massachusetts, Amherst

Thomas T. Arny
Professor of Astronomy, Emeritus
University of Massachusetts, Amherst

Boston   Burr Ridge, IL   Dubuque, IA   New York   San Francisco   St. Louis
Bangkok   Bogotá   Caracas   Lisbon   London   Madrid
Mexico City   Milan   New Delhi   Seoul   Singapore   Sydney   Taipei   Toronto

The McGraw-Hill Companies

Pathways to ASTRONOMY
Volume 2, Custom Edition
Stars & Galaxies

This book is a McGraw-Hill Custom Publishing textbook and contains select material from *Pathways to Astronomy* by Stephen E. Schneider and Thomas T. Arny. Copyright © 2007 by The McGraw-Hill Companies, Inc. Reprinted with permission of the publisher. Many custom published texts are modified versions or adaptations of our best-selling textbooks. Some adaptations are printed in black and white to keep prices at a minimum, while others are in color.

2 3 4 5 6 7 8 9 0 QSR QSR 0 9 8 7

P/N ISBN-13: 978-0-07-327955-8 of set ISBN-13: 978-0-07-327966-4
P/N ISBN-10: 0-07-327955-2 of set ISBN-10: 0-07-327966-8

Editor: Shirley Grall
Production Editor: Tina Hermsen
Printer/Binder: Quebecor World

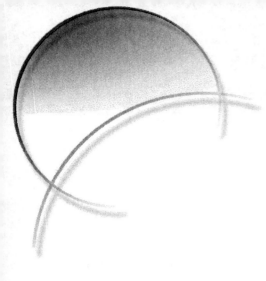

To my father, who taught me the night sky when I was little.

—Steve

BRIEF CONTENTS

CONTENTS

PART TWO

PROBING MATTER, LIGHT, AND THEIR INTERACTIONS   119

PART FIVE

GALAXIES AND THE UNIVERSE 569

PREFACE

APPROACH

There are many astronomy textbooks available today, but *Pathways to Astronomy* offers something different . . .

Created by two veteran teachers of astronomy, both recipients of outstanding teaching awards, *Pathways* breaks down introductory astronomy into its component parts. The huge and fascinating field of astronomy is divided into 84 units from which you can selectively chose topics according to your interests, while maintaining a natural flow of presentation.

One of the frustrations created by other current astronomy textbooks is that each chapter covers such a wide array of topics and is wed to such a specific order of presentation that it is difficult for students to absorb the large amount of material, and it is difficult for the professor to link the chapter readings and review questions to his or her own particular approach to teaching the subject. Whether you are learning astronomy for the first time or teaching it for the tenth, *Pathways* offers greater flexibility for exploring astronomy in the way you want.

The unit structure allows the new learner and the veteran professor to relate the text more clearly to college lectures. Each unit is small enough to be easily tackled on its own or read as an adjunct to the classroom lecture. For the faculty member who is designing a course to relate to current events in astronomy or a particular theme, the structure of *Pathways* makes it easier to assign reading and worked problems that are relevant to each topic. For the student of astronomy, *Pathways* makes it easier to digest each topic and to clearly relate each unit to lecture material.

Each unit of *Pathways to Astronomy* is like a mini-lecture on a single topic or closely related set of ideas. The same material covered in other introductory astronomy texts is included, but it is broken up into smaller self-contained units. This gives greater flexibility in selecting topics than is possible with the wide-ranging chapter in a traditional text that covers the same material as four or five *Pathways* units.

Even though the units are written to be as independent as possible, they still flow naturally from one to the next or even in alternative orders—different *Pathways*—through the book. Professors can select units to fit their course needs and cover the units in the order they prefer. They can choose individual units that will be explored in lecture while assigning other units for self-study. Or they can cover all the units in full depth in a content-rich course. With the short length of units, students can more easily digest the material covered in an individual unit before moving on to the next unit. And because the questions and problems are focused on a single topic, it is much easier to determine mastery of each topic.

The unit format also provides an opportunity to take some extra steps beyond the ordinary text. The authors have included some material of special interest that introduces topics most introductory texts do not offer—for example, units on calendar systems and special relativity. More advanced material within a particular unit topic is also organized toward the end of the unit so that the essentials are covered first—also providing flexibility for assigning readings.

Pathways to Astronomy makes it easy to tailor readings and exercises so they fit best within a course's structure. It also provides opportunities to travel down some fascinating paths to enhance a course or to provide outside reading for advanced students.

FEATURES

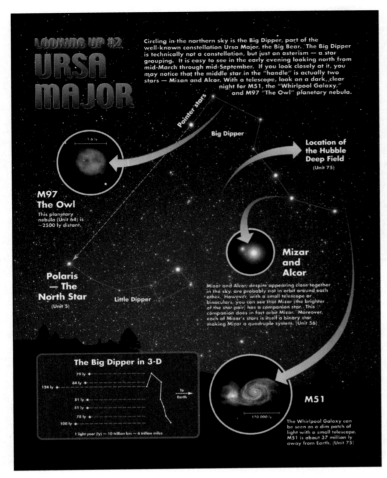

Looking Up Illustrations It can be challenging to link introductory astronomy to the sky around us. The eight "Looking Up" full-page art pieces provide another pathway to astronomy, connecting what we actually see when "looking up" at the night sky with the more academic side of astronomy. Each illustration displays a large-scale photograph of one or more constellations in the night sky. Each also contains close-up photographs and illustrations of some of the most interesting telescopic objects with cross-references to the text. Details are also given regarding the objects' distances from Earth, along with three-dimensional illustrations of some of the stars or other objects within the field of view. The Looking Up Illustrations begin on page xxi, following the Acknowledgments.

Star Chart and Planetarium Activities A good star chart and a planetarium program also help to link the study of astronomy to the night sky. *Pathways to Astronomy* offers a fold-out star chart and end-of-unit planetarium exercises designed for use with the remarkable *Starry Night* software. These will help students to take that next step beyond the book—exploring the night sky.

Detailed Art *Pathways to Astronomy* has taken each illustration a step further than the norm. Each illustration in *Pathways* is annotated to describe the processes that are actually happening within the illustration. Photos are often inserted next to the illustration for comparison so students can see the process in reality.

Writing Style *Pathways to Astronomy* provides coverage of technically complex ideas without confusing students. The authors give the students a reason to read every sentence. They relate many astronomical concepts to common, familiar experiences, and they convey with ease the drama inherent in the processes at work in our universe. This engaging style helps students to learn from what they read, allowing the professor more class time to explore more concepts.

End of Unit Material The end-of-unit sections include hundreds of review questions, mathematical problems, and test yourself questions to help students master the unit material. These elements allow students to apply what they have learned before moving on to another unit. The answers for all the test yourself questions are provided at the back of the text. Students can also access the answer to the first mathematical problem in every unit via ARIS. (See the description about ARIS in the following section about the Supplements.)

Thought Questions Dozens of thought questions are scattered throughout the margins of the units, giving students the opportunity to explore and extend their understanding of the material they are reading.

Detailed Math Steps For some mathematical material appearing in the text, additional steps and explanations have been provided in the margins rather than including these within the text. This gives instructors flexibility in the depth of mathematical emphasis they wish to include.

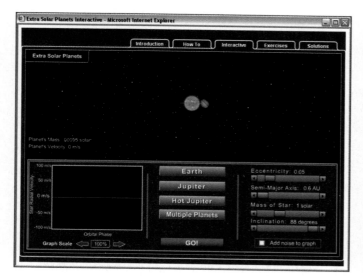

Electronic Media Integration with the Text Interactive and Animation icons have been placed where additional understanding can be gained through an animation or interactive experience on ARIS (see the description of the ARIS supplement).

Interactives McGraw-Hill is proud to bring you an assortment of 23 outstanding Interactives like no other. Each Interactive is programmed in Flash for a stronger visual appeal. These Interactives offer a fresh, dynamic method to teach astronomy basics. Each Interactive allows users to manipulate parameters and gain a better understanding of topics such as blackbody radiation, the Bohr model, a "solar system builder," retrograde motion, cosmology, and the H-R diagram by watching the effect of these manipulations. Each Interactive includes an analysis tool (interactive model), a tutorial describing its function, content describing its principal themes, related exercises, and solutions to the exercises. Users can jump between these exercises and analysis tools with just a mouse click.

SUPPLEMENTS

McGraw-Hill offers various tools and technology products to support *Pathways to Astronomy*. Instructors can obtain teaching aids by calling the Customer Service Department at 800-338-3987 or contacting their local McGraw-Hill sales representative.

Starry Night CD This planetarium software is now available free with every text. This software lets users investigate the inner workings and features of the universe at varying levels of detail using a variety of tools, including the ability to examine the night sky on a date they enter from a location they enter.

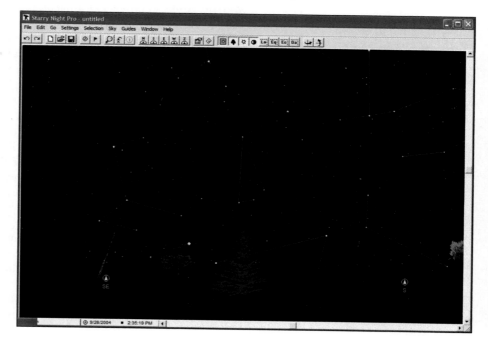

www.mhhe.com/schneider

McGraw-Hill's ARIS The Assessment, Review, and Instruction System (ARIS) for *Pathways to Astronomy* is a complete online tutorial, electronic homework, and course management system, designed for greater ease of use than any other system available. Instructors can create and share course materials and assignments with colleagues with a few mouse clicks.

The design of ARIS makes it easy for students to take full advantage of the following tools:

- **Interactive student technology:** Includes 23 outstanding Astronomy Interactives, Animations, and Online Quizzes.
- **Text-specific features:** Includes Multiple-Choice Quizzes and Test Yourself Questions.
- **General astronomy features:** Includes Astronomy Time Line, Astronomy Picture of the Day, and Constellation Quizzes.
- **Additional instructor resources:** Includes Instructor's Manual, introductory materials, sample syllabi, and CPS (Classroom Performance System) questions.

All assignments, quizzes, tutorials, and Interactives are directly tied to text-specific materials in *Pathways to Astronomy;* but instructors can also edit questions, import their own content, and create announcements and due dates for assignments. ARIS features automatic grading and reporting of easy-to-assign homework, quizzing, and testing. All student activity within McGraw-Hill's ARIS is automatically recorded and available to the instructor through a fully integrated grade book that can be downloaded to Excel.

Instructor's Testing and Resource CD McGraw-Hill's electronic testing program allows instructors to create tests from specific book items. It accommodates a wide range of question types, and instructors may add their own questions. Multiple versions of the tests can be created, and any test can be exported for use with course management systems such as WebCT, BlackBoard, or PageOut. The program is available for Windows and Macintosh environments.

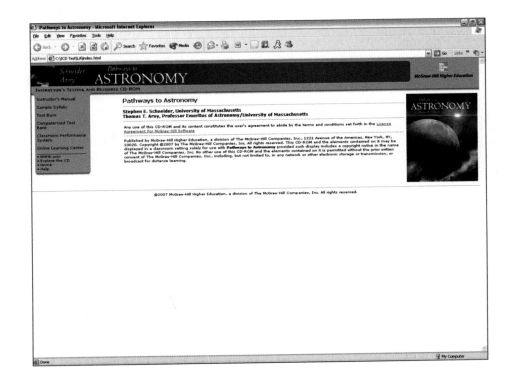

Instructor's Manual The Instructor's Manual is found on ARIS and on the Instructor's Testing and Resource CD, and can be accessed only by instructors. This manual includes hints for teaching with this text, additional thought and discussion questions, answers to end-of-unit material, and sample syllabi.

Digital Content Manager This CD contains illustrations and photographs from the text. It also includes 23 astronomy Interactives. The software makes customizing a multimedia presentation easy. The instructor can organize figures in any order desired; add labels, lines, and artwork; integrate material from other sources; edit and annotate lecture notes; and have the option of placing a multimedia lecture into another presentation program such as PowerPoint.

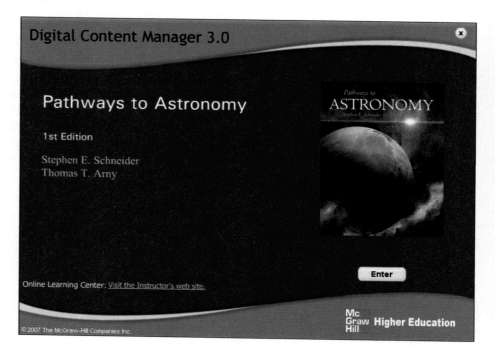

Transparencies This collection contains 90 overhead transparencies of conceptually based artwork from *Pathways to Astronomy*.

Classroom Performance System (CPS) and Questions McGraw-Hill has partnered with eInstruction to provide the revolutionary Classroom Performance System (CPS) and to bring interactivity into the classroom. CPS is a wireless response system that gives the instructor and students immediate feedback from the entire class. The wireless response pads are essentially remotes that are easy to use and engage students. CPS allows the instructor to motivate student preparation, interactivity, and active learning to receive immediate feedback and know what students understand. A text-specific set of questions, formatted for both CPS and Powerpoint, is available via download from the Instructor area of the Online Learning Center.

ACKNOWLEDGMENTS

Several people were deeply involved in the development of this book. Sharon Swihart offered helpful comments and suggestions on the entire manuscript from the perspective of a student of the subject. Meg McDonald clarified the wording and helped to check that the many pieces of this book fit together correctly. The publishing team at McGraw-Hill (Liz Recker, Daryl Bruflodt, Gloria Schiesl, Rick Noel, and many others) worked closely with us throughout the book's development, always finding a way to solve problems and keep the project moving forward.

Reviewers

Special thanks and appreciation go out to all reviewers. This first edition (through several stages of manuscript development) has enjoyed many constructive suggestions, new ideas, and invaluable advice provided by these reviewers:

Jennifer E. Bachman *Whatcom Community College*
William G. Bagnuolo, Jr. *Georgia State University*
Nadine Barlow *Northern Arizona University*
Cecilia Barnbaum *Valdosta State University*
Peter A. Becker *George Mason University*
Debra L. Burris *Oklahoma City Community College*
Eugene R. Capriotti *The Ohio State University*
George L. Cassiday *University of Utah*
Stan Celestian *Glendale Community College*
Thomas M. Corwin *UNC Charlotte*
John J. Cowan *University of Oklahoma*
Christopher J. Crow *Indiana/Purdue University, Fort Wayne*
Deborah M. Dann *Corning Community College*
Robert N. DeWitt *Penn State Altoona*
Ryan E. Droste *Trident Technical College, Charleston*
Robert J. Dukes, Jr. *College of Charleston*
Robert A. Egler *North Carolina State University*
Paul B. Eskridge *Minnesota State University*
Rica Sirbaugh French *MiraCosta College*
Robert B. Friedfeld *Stephen F. Austin State University*
Donna H. Gifford *Pima Community College*
Terry L. Goforth *Southwestern Oklahoma State University*
Alec Habig *University of Minnesota, Duluth*
Harold M. Hastings *Hofstra University*
Scott Hildreth *Chabot College*
Paul Hintzen *California State University, Long Beach*
N. Brian Hopkins *Rio Salado College*
William Hussong *College of DuPage*
James N. Imamura *University of Oregon*
Douglas R. Ingram *Texas Christian University*
Charles Kerton *Iowa State University*
Dave Kriegler *University of Nebraska*
Claud Lacy *University of Arkansas*
Henry J. Leckenby *University of Wisconsin, Oshkosh*
Loris Magnani *University of Georgia*
Phil Matheson *Utah Valley State College*
Roy C. McCord *Irvine Valley College*
José Mena-Werth *University of Nebraska at Kearney*
Zdzislaw E. Musielak *The University of Texas at Arlington*
Arnold L. O'Brien *University of Massachusetts, Lowell*
Cynthia Peterson *University of Connecticut*
Bob Powell *State University of West Georgia*
Mike Reed *Missouri State University*
Richard Rees *Westfield State College*

R. S. Rubins *University of Texas at Arlington*
M. Alper Sahiner *Seton Hall University*
Ann Schmiedekamp *Penn State University, Abington College*
Larry C. Sessions *Metropolitan State College of Denver*
James R. Sowell *Georgia Tech*
Michael E. Summers *George Mason University*
Lisa M. Will *Mesa Community College*
J. Wayne Wooten *Pensacola Junior College/University of West Florida*

Special thanks and appreciation also are due to those who contributed to the production of the ancillaries that accompany *Pathways to Astronomy:*

Cecilia Barnbaum *Valdosta State University*
Debra L. Burris *Oklahoma City Community College*
John J. Cowan *University of Oklahoma*
David Kriegler *University of Nebraska*
Henry J. Leckenby *University of Wisconsin/Oshkosh*
Roy S. Rubins *University of Texas at Arlington*
Larry C. Sessions *Metropolitan State College of Denver*
J. Wayne Wooten *Pensacola Junior College/University of West Florida*

Acknowledgments for Interactives

McGraw-Hill would like to thank Adam Frank, professor at the University of Rochester and president of Truth-N-Beauty, as well as the other employees of Truth-N-Beauty, especially Ted Pawlicki and Carol Latta.

LOOKING UP #1
NORTHERN CIRCUMPOLAR CONSTELLATIONS

For observers over most of the northern hemisphere, there are 5 constellations that are circumpolar (Unit 5), remaining visible all night long: Ursa Major (the Big Bear), Ursa Minor (the Little Bear), Cepheus (the King), Cassiopeia (the Queen), and Draco (the Dragon). The brightest stars in Ursa Major and Ursa Minor form two well-known asterisms: the Big and Little Dippers.

Delta Cepheus
A pulsating variable star (Unit 63) at a distance of ~300 ly.

Cassiopeia

Cepheus

~12 ly

M52
This is an open star cluster (Unit 69). Its distance is uncertain — perhaps 3000–5000 ly.

Draco

Polaris (The North Star)

Little Dipper

M101

~170,000 ly

This spiral galaxy is ~27 million light years from us (Unit 75).

Thuban
This was the pole star when the pyramids were built in ancient Egypt (Unit 6).

M81 and M82
Gravitational interactions between M81 and M82 have triggered star formation in both galaxies (Unit 75).

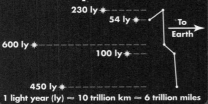

Cassiopeia in 3-D
230 ly ✳----
54 ly ✳---
To Earth
600 ly ✳- - - - -
100 ly ✳
450 ly ✳- - - - -
1 light year (ly) ≈ 10 trillion km ≈ 6 trillion miles

Big Dipper

NORTH

URSA MAJOR

Circling in the northern sky is the Big Dipper, part of the well-known constellation Ursa Major, the Big Bear. The Big Dipper is technically not a constellation, but just an asterism — a star grouping. It is easy to see in the early evening looking north from mid-March through mid-September. If you look closely at it, you may notice that the middle star in the "handle" is actually two stars — Mizan and Alcor. With a telescope, look on a dark, clear night for M51, the "Whirlpool Galaxy," and M97 "The Owl" planetary nebula.

Pointer stars

Big Dipper

~1.6 ly

M97 The Owl

This planetary nebula (Unit 64) is ~2500 ly distant.

Location of the Hubble Deep Field

(Unit 75)

Polaris — The North Star

(Unit 5)

Little Dipper

Mizar and Alcor

Mizar and Alcor, despite appearing close together in the sky, are probably not in orbit around each other. However, with a small telescope or binoculars, you can see that Mizar (the brighter of the star-pair) has a companion star. This companion does in fact orbit Mizar. Moreover, each of Mizar's stars is itself a binary star, making Mizar a quadruple system. (Unit 56)

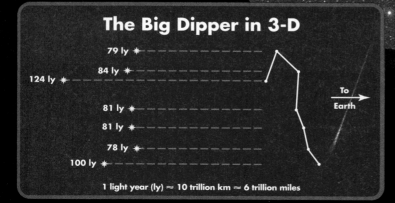

The Big Dipper in 3-D

79 ly

84 ly

124 ly

81 ly

81 ly

78 ly

100 ly

To Earth

1 light year (ly) ≈ 10 trillion km ≈ 6 trillion miles

170,000 ly

M51

The Whirlpool Galaxy can be seen as a dim patch of light with a small telescope. M51 is about 37 million ly away from Earth. (Unit 75)

M31 &
PERSEUS

The galaxy M31 lies in the constellation Andromeda, near the constellations Perseus and Cassiopeia. Northern hemisphere viewers can see M31 dimly with the naked eye, but easily with binoculars. It is about 2.9 million ly from us. The best times to see it are August through December.

Andromeda

~150,000 ly

M31
Andromeda
Galaxy (Unit 76)

~200 ly

The Double Cluster

If you scan with binoculars from M31 toward the space between Perseus and Cassiopeia, you will see the Double Cluster — two groups of massive, luminous but very distant stars. The Double Cluster is best seen with binoculars. The two clusters are about 7000 ly away and a few hundred light years apart. (Unit 69)

Perseus

Perseus is easy to identify: it looks a little like the Eiffel tower.

California
Nebula

An emission nebula (Unit 72).

M45
Pleiades
(Unit 69)

Capella

The brightest star in the constellation Auriga, the Charioteer. A binary star (Unit 56)

Auriga

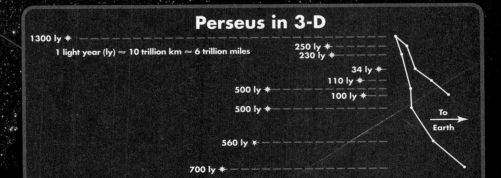

Perseus in 3-D

1300 ly ✳
1 light year (ly) ≈ 10 trillion km ≈ 6 trillion miles
250 ly ✳
230 ly ✳
34 ly ✳
110 ly ✳
500 ly ✳
100 ly ✳
500 ly ✳
To Earth
560 ly ✳
700 ly ✳

SUMMER TRIANGLE

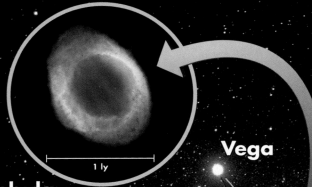

M57
Ring Nebula

This planetary nebula (Unit 64) is about 2000 ly away from us. It is 7000 years old and has a white dwarf at its center.

Vega

Lyra

The Summer Triangle consists of the three bright stars Deneb, Vega, and Altair. (Unit 58) These stars, the brightest to our eyes in the constellations Cygnus, Lyra, and Aquila, respectively, rise in the east shortly after sunset in late June and are visible throughout the northern summer and into late October (when they set in the west in the early evening). Deneb is intrinsically the brightest of the three, although Vega looks the brightest to us. Deneb looks dim only because it is so much farther from us than Vega and Altair.

Cygnus

Epsilon Lyra

A double, double star (Unit 56)

Deneb

Deneb is a Blue Supergiant (Unit 66), one of the most luminous stars we can see, Deneb emits about 250,000 times more light than the Sun.

Altair

Alberio

This star pair (easily seen in a small telescope) shows a strong color contrast (orange and blue). Astronomers disagree about whether they orbit each other or just happen to lie in the same direction in the sky. (Unit 56)

M27
Dumbbell
Nebula

~2.5 ly

Another planetary nebula (Unit 64), The Dumbbell is about 900 ly distant and is about 2.5 ly in diameter.

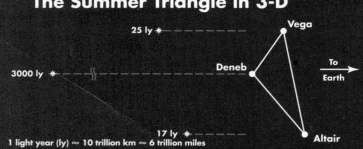

The Summer Triangle in 3-D

25 ly ✳

Vega

3000 ly ✳

Deneb

To Earth →

17 ly ✳

Altair

1 light year (ly) ≈ 10 trillion km ≈ 6 trillion miles

M1
The Crab Nebula

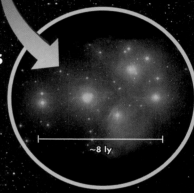

~7 ly

The Crab Nebula is the remnant of a star that blew up in the year AD 1054 as a supernova. It is about 5000 ly away from us. (Units 26 and 66)

Taurus — The Bull
One of the constellations of the zodiac and one of the creatures hunted by Orion. Taurus is visible in the evening sky from November through March. The brightest star in Taurus is Aldebaran, the eye of the bull.

Zeta Tauri

This open star cluster is easy to see with the naked eye and looks like a tiny dipper. It is about 400 ly from Earth. (Unit 69)

M45 Pleiades

~8 ly

Aldebaran

Aldebaran is a red giant star. (Unit 62) It is about 65 ly from us and its diameter is about 45 times the diameter of the Sun. It lies between us and the Hyades.

Taurus in 3-D

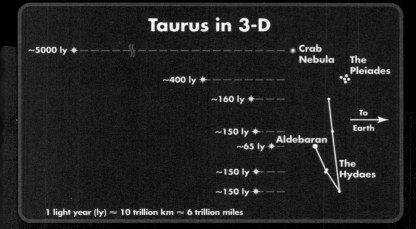

~5000 ly — Crab Nebula
The Pleiades
~400 ly
~160 ly
~150 ly
~65 ly — Aldebaran
~150 ly — The Hydaes
~150 ly
To Earth

1 light year (ly) ≈ 10 trillion km ≈ 6 trillion miles

Hyades

The "V" in Taurus is a nearby star cluster, about 137 ly away. It is easy to see its many stars with binoculars. (Unit 69)

LOOKING UP #6
ORION

Orion is easy to identify because of the three bright stars of his "belt." You can see Orion in the evening sky from November to April, and before dawn from August through September.

Betelgeuse is a Red Supergiant star (Unit 62) that has swelled to a size that is as large as our Solar System out to Mars. Its red color indicates that it is relatively cool for a star, about 3000 Kelvin.

Betelgeuse

Sun

Mars' orbit

1 ly

3 ly

Horsehead Nebula

The horsehead shape is caused by dust in an interstellar cloud blocking background light. (Unit 72)

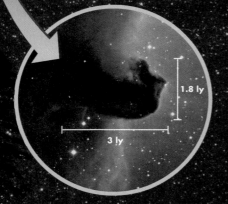

1.8 ly

3 ly

M42 Orion Nebula

The Orion Nebula is an active star-forming region rich with dust and gas. (Units 60, 72)

Rigel

Rigel is a Blue Supergiant star (Unit 66). Its blue color indicates a surface temperature of about 10,000 Kelvin

1 ly

Orion in 3-D

Betelgeuse
430 ly
240 ly

920 ly
1300 ly
820 ly
1500 ly

To
Earth

770 ly
720 ly
Rigel

1 light year (ly) ≈ 10 trillion km ≈ 6 trillion miles

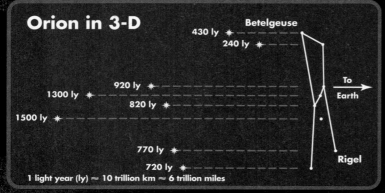

Sun

Pluto's orbit

A protoplanetary disk

This is the beginning of a star; our early Solar System may have looked like this! (Unit 34)

SAGITTARIU

Sagittarius marks the direction to the center of th
Milky Way. It is best located by its "teapot" sha
with the Milky Way seeming to rise like steam fr
the spout. For northern latitude viewers, the
constellation is best seen in the evening, July to
September. For such viewers, it is low on the
southern horizon. Many star-forming nebulae a
visible in this region. (Units 60,72)

M17
Swan Nebula

M20
Trifid Nebula

The distance to the Trifid
Nebula (so named because
of the black streaks that
divide it into thirds) is very
uncertain. It lies between
2000 and 8000 ly away.
This makes its size very
uncertain, too.

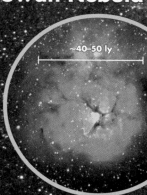

~40–50 ly

~100 ly

M22

M22 is one of the many
globular clusters that are
concentrated toward the
core of our galaxy. Easy to
see with binoculars, it is just
barely visible to the naked
eye. It is about 11,000 ly
away from us. (Unit 69)

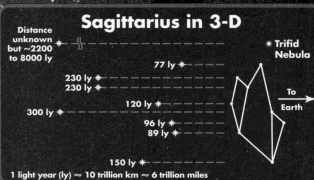

Sagittarius in 3-D

Distance
unknown
but ~2200
to 8000 ly

Trifid
Nebula

77 ly ✳

230 ly ✳
230 ly ✳

120 ly ✳

To
Earth

300 ly ✳

96 ly ✳
89 ly ✳

150 ly ✳

1 light year (ly) ≈ 10 trillion km ≈ 6 trillion miles

CENTAURUS AND CRUX, THE SOUTHERN CROSS

These constellations are best observed from the southern hemisphere. Northern hemisphere viewers can see Centaurus low in the southern sky in May – July. Crux may be seen just above the southern horizon in May and June from the extreme southern US (Key West and South Texas).

Centaurus A

This unusual galaxy, ~11 million ly distant, is one of the brightest radio sources in the sky (Unit 77)

Alpha Centauri

~50,000 ly

Proxima Centauri

This dim star is the nearest star to the Sun, 4.22 ly distant (Unit 52)

Crux The Southern Cross

~50 ly

The Jewel Box

NGC 4755, an open star cluster (Unit 69) ~500 ly from us.

The Coal Sack

An interstellar dust cloud (Unit 72)

Eta Carinae

A very high-mass star doomed to die young (Unit 60) ~8000 ly distant

~200 ly

Omega Centauri

The largest globular cluster in the Milky Way (Unit 69), ~17,000 ly distant and containing millions of stars

Southern Cross in 3-D

352 ly
321 ly
228 ly
363 ly
88 ly

To Earth

1 light year (ly) ≈ 10 trillion km ≈ 6 trillion miles

Pathways to Astronomy

Part One

The content of some Units will be enhanced if you have previously studied some earlier Units in this textbook. These Background Pathways are listed below each Unit image.

Unit 5 The Night Sky

Background Pathways: Units 1 and 2

Unit 1 Our Planetary Neighborhood

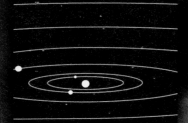

Unit 3 Astronomical Numbers

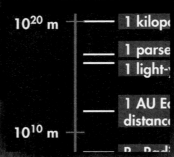

10^{20} m

10^{10} m

1 kilop

1 parse

1 light-y

1 AU Ea
distanc

Unit 6 The Year

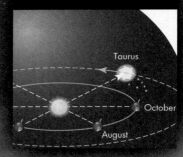

Taurus

October

August

Background Pathways: Unit 5

Unit 2 Beyond the Solar System

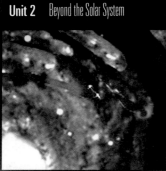

Unit 4 Foundations of Astronomy

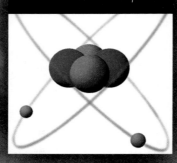

Unit 7 The Time of Day

+3h30m

Eastern

+3h45m

+3h30m

+0h44m

Background Pathways: Unit 6

The Cosmic Landscape

Unit 10 Geometry of the Earth, Moon, and Sun

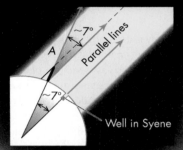

~7°
A
~7°
Parallel lines
Well in Syene

Background Pathways: Unit 8

Unit 8 Lunar Cycles

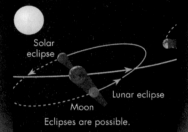

Solar eclipse
Lunar eclipse
Moon
Eclipses are possible.

Background Pathways: Unit 6

Unit 11 Planets: The Wandering Stars

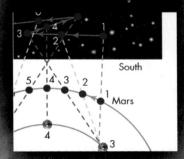

3
4
2
1
South
5 4 3 2
1 Mars
4
3

Background Pathways: Units 1 and 6

Unit 13 Observing the Sky

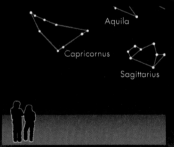

Aquila
Capricornus
Sagittarius

Background Pathways: Units 7, 8, and 11

Unit 9 Calendars

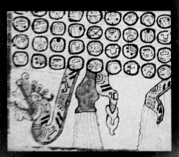

Background Pathways: Units 6 and 8

Unit 12 The Beginnings of Modern Astronomy

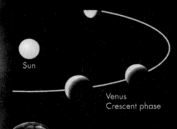

Sun
Venus
Crescent phase

Background Pathways: Units 11 and 4.1

I Our Planetary Neighborhood

Astronomy is the study of the universe: from the Earth itself to the most distant galaxy, from deciphering its nature at the beginning of time to predicting its eventual fate in the remote future. Within its vast space lie planets with dead volcanoes whose summits dwarf Mount Everest. There are stars a thousand times the size of the Sun—so large that in the Sun's place, they would swallow up the Earth. And there are galaxies—slowly whirling systems of billions of stars—so vast that the Earth is smaller by comparison than a single grain of sand is to the Earth itself. On this small planet within this immense cosmic landscape, life has evolved over billions of years, giving rise to creatures who seek to understand the nature of the universe.

Through astronomy we gain the ability to study places so remote that there is no possibility of ever visiting them. We gain insights into alien environments unlike anything found on Earth. And our insights provide new perspectives on our home planet and about ourselves.

The vast variety of planets, their distances, and their sizes are almost unimaginable. We will explore the Earth and its neighboring planets in detail in Part III; but to gain some sense of scale, we begin with a brief look at the Earth's neighborhood—the Solar System.

1.1 THE EARTH

We begin with the Earth, our home **planet** (Figure 1.1). This spinning ball of rock and metal, coated with a thin layer of gas and liquid, is huge by human standards, but it is one of the smaller bodies in the cosmic landscape. Nevertheless, it is an appropriate place to start because, as the base from which we view the universe, it influences what we can see. We cannot travel to any but the nearest objects in our quest to understand the universe. Instead, we are like children who know their neighborhood well but for whom the larger world is still a mystery, known only from books and television.

But just as children use knowledge of their neighborhood to build their image of the world, so astronomers use their

FIGURE 1.1
The planet Earth, our home, with blue oceans, white clouds, and multihued continents.

4

knowledge of Earth as a guide to distant worlds. The size of the Earth and features on it, for example, are useful reference points for appreciating the sizes of other objects. We will often refer to other planets in terms of their radii relative to the Earth's own radius of about 6371 kilometers (3909 miles). Similarly, it is convenient to refer to other planets' masses in terms of the Earth's own mass, which is itself so large that it is difficult to imagine: 5,970,000,000,000,000,000,000,000 kilograms, or about 6 billion trillion tons.

Although few people realize it, we use the size of the Earth for defining the fundamental unit of the **metric system** (Unit 3): the meter. The meter was originally defined to be one 10 millionth of the distance from the equator to the North Pole—so that the Earth's circumference equals 40,000 kilometers (about 25,000 miles). This was a convenient system for marking distances that ships traveled on the Earth's oceans when the system was introduced, and it offered a less arbitrary standard than the many other measurement systems used at the time.

The geological processes that occur on Earth provide another kind of measure—one for interpreting the processes that shape the other planets. When a volcano spews molten lava, it provides a hint that below the surface, our planet is extremely hot. During the last century geologists discovered that this internal heat drives slow but powerful currents that shake our planet's crust, move continents, build mountains, and heave up volcanoes. We can carry over our understanding of such geological processes here on Earth to help us make hypotheses about the processes that create similar features on other planets. And when we discover different features on other worlds, we can use them to help us think about the Earth in new ways.

The calculations made in the original determination of the meter were slightly off, so that the equator-to-pole distance is today found to be 10,002 kilometers. More significantly, though, the Earth is not a perfect sphere, so the circumference around the equator is larger than around the poles—about 40,075 kilometers. The definition of the meter today is based on the wavelength of light (Unit 22) produced by a particular kind of laser.

I.2   THE MOON

Our nearest neighbor, the Moon, is a profoundly different world from the Earth. The Moon is our **satellite,** orbiting the Earth nearly four hundred thousand kilometers (about a quarter million miles) away. A string stretched from the Earth to the Moon could wrap around the Earth 10 times. The Moon is much smaller, only about one-quarter our planet's diameter. The Moon also has symbolic significance for us—it marks the present limit of direct human exploration of space.

With the naked eye, the Moon appears to be a quiet glowing orb (Figure 1.2A); but with a small telescope or binoculars, we can see that the Moon's surface exhibits a landscape that reminds us that it is a rocky world itself—somewhat like the Earth, yet utterly unlike the Earth as well (Figure 1.2B). Instead of white whirling clouds, green-covered hills, and blue oceans, we see an airless, pitted ball of rock. Instead of crumpled mountain ranges and volcanoes, the Moon's surface is peppered with craters blasted into the surface when bodies—ranging from the microscopic to the size of mountains—crashed into its surface. This surface records a history of a steady pounding by objects impacting at speeds more than 10 times faster than any rifle bullet. Some of these larger collisions carved out craters more than 100 kilometers (60 miles) in diameter, while innumerable smaller impacts pulverized the surface rock to rubble and dust.

The Earth, so near to the Moon in space, must have suffered a similar pounding, yet looks utterly different. Why do these two worlds bear so little resemblance? Much of the explanation lies in their greatly different mass. The Moon's mass is only about 1/80th the Earth's, and its smaller bulk made the Moon less able to retain internal heat or an atmosphere after its formation. With less internal heat, the

What impressions of the character of the Moon do you have based on observing it with the naked eye? How does your impression change after viewing it through a telescope or examining high-resolution photographs? How do you imagine people's perceptions changed after the invention of the telescope in the early 1600s?

The internal heat of planets appears to come from two sources—heat left over from their formation and the decay of radioactive elements in their interior (Unit 35). Their ability to retain this heat depends mostly on their size.

A   **B**

FIGURE 1.2
(A) The Moon as we see it with unaided eyes. (B) The Moon's surface as seen through a small telescope.

Moon's interior is quiet. Heat-driven motion, so important in shaping and changing the Earth's surface over the eons, is nearly absent in the Moon. Without an atmosphere, the Moon is not protected from small impacting objects, which are vaporized through the heat of friction before reaching the ground on Earth. And without erosion caused by an atmosphere or modification of the surface caused by geological activity, the Moon's surface exhibits all of its old scars. Thus, the Moon can help us understand events in the Earth's past that have been largely erased from its surface.

1.3   THE PLANETS

Beyond the Moon, circling the Sun as the Earth does, are eight other planets, sister bodies of Earth. To the unaided eye, the other planets are mere points of light whose positions shift slowly from night to night. But by observing them, first with Earth-based telescopes, then ultimately by remotely piloted spacecraft, we have learned that they are truly other worlds.

In order of increasing distance from the Sun, the nine planets are Mercury, Venus, Earth, Mars, Jupiter, Saturn, Uranus, Neptune, and Pluto. These worlds have dramatically different sizes and landscapes. For example:

- Craters scar the airless surface of Mercury.
- Dense clouds of sulfuric acid droplets completely shroud Venus.
- White clouds, blue oceans, green jungles, and red deserts tint Earth.
- Huge canyons and deserts spread across the ruddy face of Mars.
- Immense atmospheric storms with lightning sweep across Jupiter.
- Trillions of icy fragments orbit Saturn, forming its bright rings.
- Dark rings girdle Uranus, the axis of its spin tipped by some catastrophic impact long ago.
- Icy methane clouds whirl in the deep blue atmosphere of Neptune.
- Perpetual ice glazes dim Pluto, the only planet that we have not yet visited with a spacecraft.

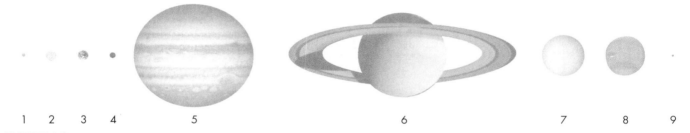

FIGURE 1.3

Pictures of the nine planets along with silhouettes showing their correct relative size. All pictures except Pluto's are images taken by spacecraft. At the scale shown, Pluto would be located about 2.5 km (1.5 miles) away.

Pluto's status as a planet has been questioned in recent years, especially with the discovery of many more objects nearly as large or even larger in the outer Solar System (Unit 46).

Figure 1.3 shows pictures of these nine distinctive bodies and reveals something of their relative size and appearance. We can indicate the sizes of the planets by comparing them to Earth's radius, which we abbreviate as $R_\oplus$ using the international symbol $\oplus$ for the Earth. The planets range in size from Jupiter, with a radius of 11 Earth radii (11 $R_\oplus$), down to Pluto with a radius of just one-fifth the Earth's radius (0.2 $R_\oplus$).

Our understanding of the Earth as a planet is made deeper through our examination of other worlds. Our two nearest neighboring planets, Venus and Mars, are also most similar to Earth in size, with radii of 0.95 and 0.53 times $R_\oplus$ respectively. Their landscapes include features that resemble many seen on Earth. For example, they both have volcanoes and mountain ranges, so we conclude that geologic processes like those that shaped the Earth must have occurred on both.

Despite their similarities in size and in distance from the Sun, Venus and Mars have dramatically different atmospheres. On Venus we would be crushed and cooked by its intensely hot, dense atmosphere, whereas on Mars we would suffocate and freeze. And the other worlds we have studied are even more alien. Every other planet and moon that we have examined also would be hostile to human life. By studying the factors that make other planets so different from Earth, we may also begin to understand the potential consequences of our own impact on the Earth.

1.4   THE SUN

Q The Sun and all the planets are hotter in their interiors than in their surfaces. What different sources of energy might produce their internal heat?

All of the planets are dwarfed by the Sun, whose immense gravity holds them in orbit. The Sun is a **star,** a huge ball of gas over 100 times the diameter of the Earth and over 300,000 times more massive. The differences in size between the Sun, Jupiter, and Earth are illustrated in Figure 1.4.

The Sun, of course, differs from the planets in more than just size: It generates energy in its core by nuclear reactions that convert hydrogen into helium. The Sun is producing energy at a furious rate—more in just one second than all of the bombs and the energy ever generated on Earth during human history. From the Sun's intensely hot core, its energy flows to the surface, which is more than a thousand times cooler but is still hot enough to vaporize iron. From there the Sun's energy streams into space, illuminating and warming the planet.

The energy the Sun can produce is enormous but nonetheless limited. Its stream of energy has already lasted more than 4 billion years—enough time for life to form and evolve on Earth and for intelligent creatures who can marvel at these

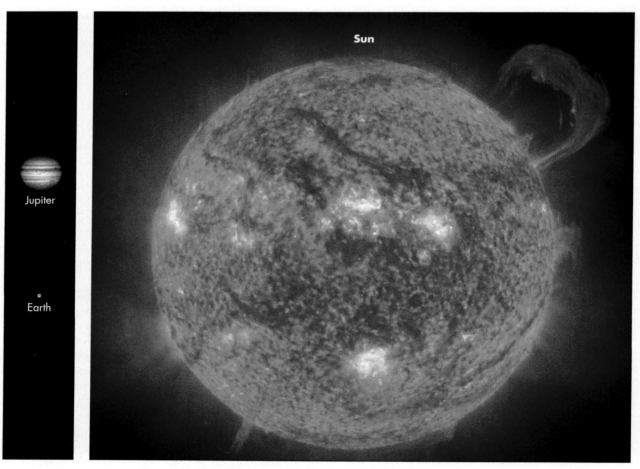

FIGURE 1.4

The Sun as viewed through a filter that allows its hot outer gases to be seen. The Earth and Jupiter are shown to scale beside it for comparison. (The filter allows astronomers to see extremely hot helium gas.)

things to have come into being. However, much evidence indicates that the Sun will run out of fuel eventually, in about another 5 or 6 billion years. It will then fade away like a cooling ember. Thus astronomy helps us to look deep into the past and far into the future as we consider phenomena of an enormous range in sizes and distances.

1.5   THE SOLAR SYSTEM

The Sun and the bodies orbiting it form the **Solar System,** bound together by the enormous gravity of the Sun. In addition to the nine planets, the Solar System is filled with a vast number of smaller bodies—satellites (moons) orbiting the planets and asteroids and comets orbiting the Sun (Figure 1.5).

If we could make visible the paths that the planets follow around the Sun, we would see that the Solar System is like a huge set of approximately circular rings, one inside the other, with the Sun at their center. The orbits of the planets lie in nearly the same plane, so that the whole system is disk-shaped. The outermost planet—Pluto—lies about 6 billion kilometers (about 4 billion miles) from the Sun. Its orbit is less circular and more tilted than that of any other planet.

If we constructed a scale model of the Solar System with the Sun represented by a grapefruit, Earth would be the size of a grain of sand or a poppy seed, orbiting about 12 meters (about 39 feet) away. This scale is illustrated in Figure 1.6, where the sizes and distances of the planets are illustrated on a U.S. football field. In this scale model, Pluto would be a spec of dust orbiting about 0.5 kilometer (about 0.3 mile) away.

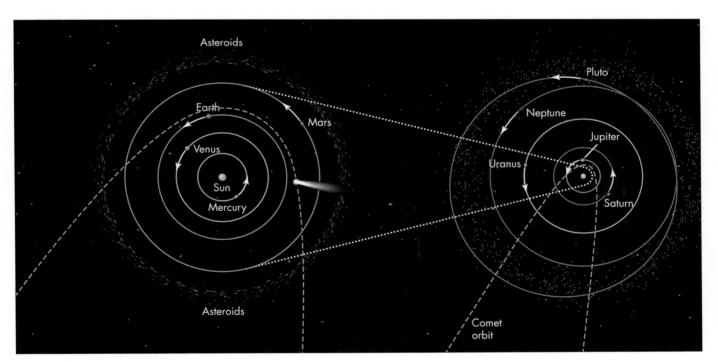

FIGURE 1.5
Sketch of the orbits of the planets in our Solar System in 2006. To show the orbits to scale, the inner and outer Solar System are shown separately.

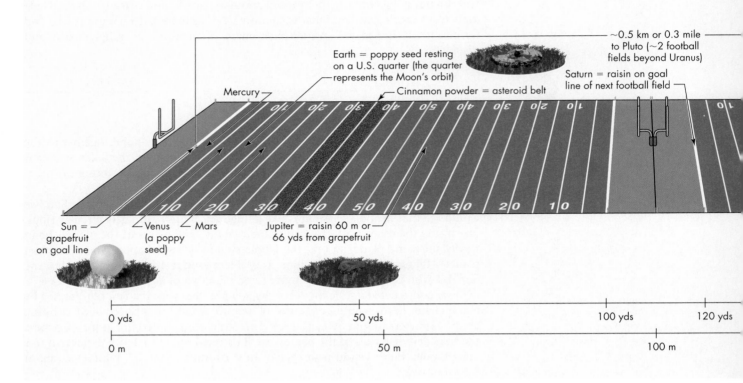

Earth = poppy seed resting
on a U.S. quarter (the quarter
represents the Moon's orbit)

Mercury

Cinnamon powder = asteroid belt

~0.5 km or 0.3 mile
to Pluto (~2 football
fields beyond Uranus)

Saturn = raisin on goal
line of next football field

Sun =
grapefruit
on goal line

Venus
(a poppy
seed)

Mars

Jupiter = raisin 60 m or
66 yds from grapefruit

0 yds   50 yds   100 yds   120 yds

0 m   50 m   100 m

FIGURE 1.6
A scale model of the Solar System with the Sun the size of a grapefruit.

I.6   THE ASTRONOMICAL UNIT

It is as difficult to comprehend distances in the Solar System reported in kilometers as it is to comprehend the distance between New York and Tokyo in millimeters. Whenever possible, we try to use units of measurement—like millimeters or kilometers—appropriate to the size of what we seek to measure.

In earlier times people used units that were quite literally at hand, such as finger widths or the spread of a hand to measure a piece of cloth, and paces to measure the size of a field. In the same tradition, although on a different scale, astronomers have adopted **units**—quantities used for measurement—related to familiar objects, such as the Earth. For example, the Earth's distance from the Sun makes a good unit for measuring the size of the Solar System.

As the Earth orbits the Sun, its distance varies from 147.1 to 152.1 million kilometers (91.4 to 94.5 million miles). The **astronomical unit,** abbreviated as **AU,** is approximately the average of these extremes, or about 150 million kilometers (93 million miles). If we use the AU to measure the scale of the Solar System, Mercury turns out to be 0.4 AU from the Sun, while Pluto is about 40 AU.

The Solar System remains the limit to our exploration of the universe with spacecraft. Most of our probes have explored only the inner part of the Solar System, although we have sent a few to explore the outer planets. The *Voyager I* and *II* spacecraft launched in 1977 represent our most distant explorers. In 1990 *Voyager I*, having passed Pluto's orbit at more than 40 AU, looked back and made the image of the Solar System shown in Figure 1.7. From this distance the planets were just points of

Q What do you suppose are the effects of the varying distance of the Earth from the Sun? How could we observe that?

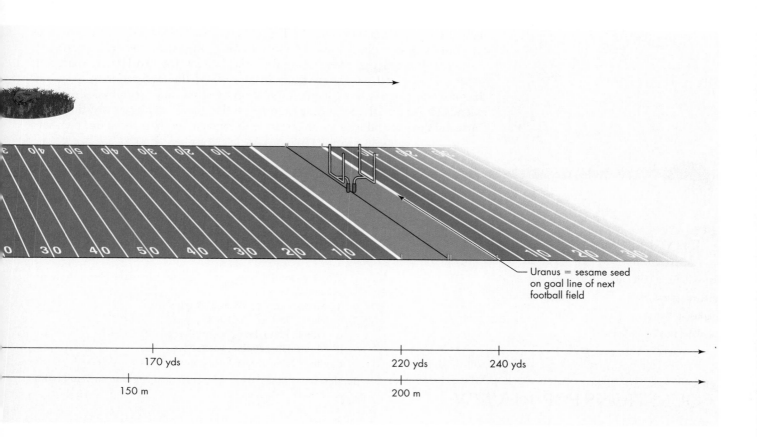

Uranus = sesame seed on goal line of next football field

170 yds

150 m

220 yds

240 yds

200 m

FIGURE 1.7

(A) This view of the Solar System is based on a series of real images made by the *Voyager I* spacecraft. The craft was about 40 AU from the Sun and about 20 AU above Neptune's orbit. The images of the planets (mere dots because of their immense distance) and the Sun have been made a little bigger and brighter in this view for clarity. Mercury is lost in the Sun's glare; Mars happened to lie nearly in front of the Sun at the time the image was made, so it too is invisible. Pluto lies outside the field of view. (B) A sketch of the orbits of the planets, showing approximately where each object was located when the image was made (February 1990). Notice the flatness of the system and the immense, empty spaces between the planets.

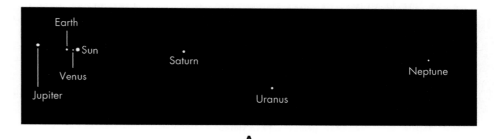

Earth
Sun
Venus
Jupiter
Saturn
Uranus
Neptune

A

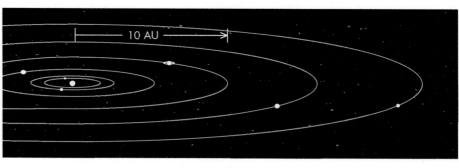

10 AU

B

light and the Sun a bright star. The *Voyager* spacecraft are now more than twice as far from the Sun as Pluto, traveling about 10 AU farther away every three years.

Even at their high speeds leaving the Solar System, the two *Voyager* spacecraft would take about 60,000 years to reach the nearest star. But exploration by spacecraft does not represent the limit of our vision. With telescopes and astrophysics, our view extends far beyond the Solar System to reveal that there are planets orbiting other stars, and there are billions of other stars with much to teach us about our own star.

KEY TERMS

astronomical unit (AU), 10   Solar System, 9

metric system, 5   star, 8

planet, 4   unit, 10

satellite, 5

QUESTIONS FOR REVIEW

1. About how many times bigger in radius is the Sun than the Earth? How many times bigger in mass?
2. How does Pluto's size compare to the Moon's?
3. What is an astronomical unit?

PROBLEMS

1. If you use a volleyball as a model of the Earth, how big would 1 kilometer be on it? The volleyball has a circumference of 68 cm.
2. What would be the circumference and diameter (circumference = $\pi \times$ diameter) of a ball that would represent the Moon if the Earth were a volleyball? What ball or object matches this size?
3. If the Earth were a volleyball, what would be the diameter of the Sun? What object matches this size?
4. If the Earth were a volleyball, how large would an astronomical unit be?

TEST YOURSELF

1. Which of the following lists gives the sizes of the objects from smallest to largest?
   a. Moon, Earth, Pluto, Mars, Jupiter
   b. Pluto, Mars, Moon, Earth, Jupiter
   c. Jupiter, Pluto, Mars, Moon, Earth
   d. Moon, Mars, Jupiter, Earth, Pluto
   e. Pluto, Moon, Mars, Earth, Jupiter
2. How many times larger is the Sun's diameter compared to Earth's?
   a. 2 times
   b. 5 times
   c. 10 times
   d. 25 times
   e. 100 times
3. Venus and Mars have features that resemble the sand dunes seen in Earth deserts, but the Moon does not. What feature is the Moon *lacking* that most likely explains this difference?
   a. Stronger gravity
   b. Life
   c. Liquid water
   d. An atmosphere
   e. Sand (or other fine particles)

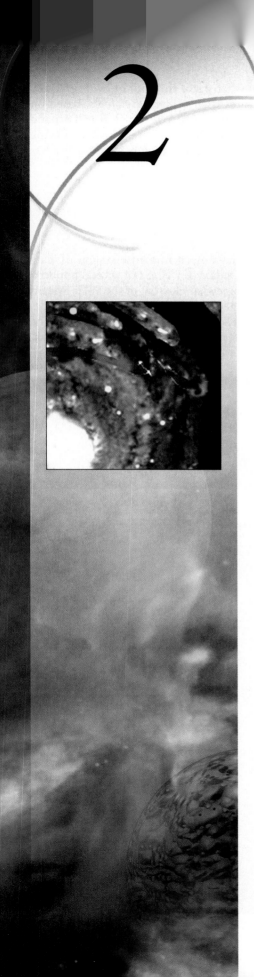

2 Beyond the Solar System

Beyond the Solar System lie billions of other stars. On a clear night, even with just your eyes, you can see that the stars are not scattered randomly across the sky but are often clumped into small groups. That clumping is the work of **gravity**, a force that attracts every object toward every other object across the farthest reaches of space. You can see that attraction in everyday life—if you let go of a book, the Earth's gravitational force makes the book fall. The same force holds the whole Earth together and gives the planets and the Sun their round shapes.

Gravity holds a star together against the enormous forces generated by the fusion of matter in the star's core. Indeed, without that furious outpouring of energy, the gravitational attraction pulling the matter together in a star is so powerful that the star must necessarily collapse in on itself. Moreover, the force of gravity spans the empty space between objects and can link separate objects together into gravitationally bound systems. It keeps the Moon in its orbit around the Earth and holds our planet in its orbit around the Sun. The Solar System is an example of one of the *smaller* systems of objects bound together by their mutual gravitational attraction.

The pull of gravity creates enormous structures—clusters of thousands of stars, galaxies of hundreds of billions of stars, or even clusters of galaxies pulled together and orbiting each other at millions of kilometers per hour. The intense gravity around exotic objects like black holes and quasars can pull in matter at immensely greater speeds—close to the speed of light—leading to titanic collisions and enormous energy bursts.

2.1 STELLAR EVOLUTION

The night sky is filled with myriad stars, some much like the Sun, but others thousands of times larger or smaller. The brightest stars produce over a billion times more light than the dimmest stars. Some stars are much hotter than the Sun and shine a dazzling blue-white, whereas others are cooler and glow a deep red. The range of masses is smaller, though, from about one-tenth the Sun's mass to about 100 times more. The mass of a star appears to be the driving factor behind the diverse properties of stars. The stronger gravity of a more massive star drives up the rate of nuclear fusion dramatically, while smaller stars consume their fuel at a more leisurely pace.

Astronomers have pieced together evidence that allows us to understand how stars are born, how they change as they age, and the dire fates they face when they run out of fuel. These changes, called **stellar evolution** by astronomers, are driven by the inexorable pull of gravity. Discovering the story of stars' lives has been one of the great triumphs of astrophysics and the scientific method during the last few centuries.

The life stories of other stars tell us that when a star runs out of hydrogen fuel, it undergoes drastic changes as it restructures itself to use other fuels. Based on this, we can predict that the Sun will go through a phase in which it will expand and nearly swallow up the Earth, vaporizing our planet entirely, before blowing away its own atmosphere and then fading like a cooling ember after a fire has burned out.

2.2   THE MILKY WAY GALAXY

Our Sun and the stars we see at night are part of an immense system of stars called the Milky Way galaxy. The **Milky Way galaxy** is a cloud of several hundred billion stars with a flattened disklike shape somewhat like that of the Solar System (Figure 2.1), but roughly 100 million times the diameter. The Sun and other stars orbit around the center of the Milky Way, but so vast is our galaxy that it takes the Sun several hundred million years to complete one trip around this immense disk.

A

100,000 light-years

B

FIGURE 2.1
The Milky Way galaxy. The top picture is a side view made by plotting stars in the 2MASS star catalog. The bottom picture is an artist's depiction of how our galaxy might look to an observer seeing it from above.

A   **B**

FIGURE 2.2

Photographs of interstellar clouds in the Milky Way taken with an earthbound telescope. On the scale of these pictures, the orbit of Pluto is about 1000 times smaller than the period ending this sentence. (A) A cold, dark cloud and a star cluster beside it. Dust in the cloud blocks our view of the stars behind it. (B) A group of clouds heated by young stars. Glowing hydrogen in the clouds creates their red color.

In the Milky Way, as in many other **galaxies,** stars intermingle with immense clouds of gas and dust. These mark the sites of stellar birth and death. Within cold, dark clouds, gravity may draw the gas into dense clumps that eventually turn into new stars, lighting the gas and dust around them (Figure 2.2). Stars eventually burn themselves out, but they do not disappear quietly. Stars like the Sun last billions of years before tearing themselves apart in their final phases—driving most of their matter back into space before fading away. The most massive stars die after just a few million years in titanic explosions, spraying matter outward to mix with the surrounding clouds. This matter from exploded stars is ultimately recycled into new stars.

In this huge swarm of stars and clouds, the Solar System is all but lost—a single grain of sand on a vast beach—forcing us again to grapple with the problem of scale. Stars are almost unimaginably remote: even the nearest one to the Sun is about 40 trillion kilometers (25 trillion miles) away—about 7000 times farther than Pluto. Such distances are so immense that analogy is often the only way to grasp them. For example, if we think of the Sun as a pinhead, the nearest star would be another pinhead about 60 kilometers (35 miles) away, and the space between them would be nearly empty. On this scale, the Sun is to the size of the Milky Way as a pinhead is to the size of the Sun itself.

2.3 THE LIGHT-YEAR

While the Astronomical Unit works well for describing distances in our Solar System, distances between stars are so immense that the AU is an inappropriately small unit. The second nearest star (after the Sun!) is hundreds of thousands of AU away. A convenient way of describing such distances is the light-year.

Measuring a distance in terms of a unit of time may at first sound peculiar, but we do it all the time in everyday life. For example, we say that our town is a 2-hour drive from the city, or our dorm is a 5-minute walk from the library. In making such statements, however, we imply that we are traveling at a standard speed: freeway driving speed or a walking pace.

Astronomers are fortunate to have a superb speed standard: the speed of light in empty space, which is a constant of nature and equal to 299,792,458 meters per second. We usually round this off to 300,000 kilometers per second (about 186,000 miles per second). Moving at this constant, universal speed, light in 1 year travels a distance defined to be 1 **light-year,** abbreviated as 1 ly. This works out to be about 10 trillion kilometers (6 trillion miles). As an example of the use of light-years, the star nearest our Sun is 4.2 light-years away. Although we achieve a major convenience in adopting such a huge distance for our scale unit, we should not lose sight of how truly immense such distances are. For example, if we were to count off the kilometers in a light-year, one every second, it would take us about 300,000 years!

We can use the light-year for setting the scale of the Milky Way galaxy. In light-years, the visible disk is over 100,000 light-years across, with the Sun orbiting roughly 28,000 light-years from the center. In the Sun's vicinity, stars are separated by typically a few light-years, but they are much more crowded toward the center and gradually thin out in the outer regions, so that there is no precisely defined outer edge. Throughout the disk of the Milky Way are scattered gas clouds with sizes of up to hundreds of light-years. These clouds sometimes contain more than a million times the Sun's mass, but they are so diffuse that their overall density is less than a billionth-trillionth of the density of the air we breathe.

2.4   GALAXY CLUSTERS AND BEYOND

M31 is visible to the naked eye in the constellation Andromeda. See Looking Up #3: M31 & Perseus.

Having gained some sense of scale for the Milky Way, we now resume our exploration of the cosmic landscape, pushing out to the realm of other galaxies. Here we find that just as stars are gravitationally bound into star clusters and galaxies, so galaxies are themselves bound into **galaxy groups** and **clusters.**

The Milky Way is part of a collection of galaxies called the **Local Group.** It is "local," of course, because it is the one we inhabit. The Local Group is a relatively small concentration of galaxies, containing only about three dozen galaxies as members, but it is still about 3 million light-years in diameter. The Milky Way is the second largest galaxy in the Local Group after the Andromeda Galaxy, also known as M31, the catalog number of a galaxy about 2.5 million light-years away from us (Figure 2.3).

To gain some perspective about the size of the Milky Way and Local Group, imagine that we built a model in which a light-year was scaled down to just 1 meter. At this scale Pluto's orbit would fit inside the head of a pin, and the Sun would be the size of a virus. The Milky Way galaxy would be the size of a large city—about 100 km (60 miles) across, while M31 would be another city about 2500 km away (Figure 2.4). Given the enormous scale of this model, light crawls along about as fast as a plant might grow—just a meter per year; so that even at light speed, a trip from the Milky Way to M31 would take 2.5 million years!

We have to adjust the scale of our thinking again to imagine even larger scales. The Local Group is one of dozens of galaxy groups surrounding a much larger assemblage of galaxies called the **Virgo Cluster,** like suburbs surrounding a major city. The Virgo Cluster is centered about 50 million light-years away and itself

Q Suppose we extended the scale model in Figure 2.4. How far away would the Virgo Cluster be? How large would the observable universe (out to 13.7 billion light-years) be?

FIGURE 2.3
Photograph of M31, the Andromeda galaxy. This is the largest galaxy in our Local Group.

contains thousands of galaxies. The central region of the Virgo Cluster is shown in Figure 2.5.

The Local Group is currently moving at over a thousand kilometers per second (about 4 million kilometers per hour or 2.5 million miles per hour) away from the Virgo Cluster. However, the gravitational pull of all the galaxies in the Virgo Cluster and its environs is so immense that it is predicted that the Local Group and other surrounding galaxies will eventually slow down, stop, and ultimately fall into the cluster. Far in the future, as the Milky Way careens into the heart of the Virgo Cluster, the stars and gas in its disk will be flung into wild new orbits, perhaps flying off into intergalactic space. This fate for the Milky Way will occur long after the Sun dies, perhaps in 100 billion years.

The term *Local Supercluster* has sometimes been used to describe the enormous "metropolitan area" of the Virgo Cluster with its surrounding galaxy groups, but this collection of groups and clusters is just a modest example of an even larger scale of clustering. The term **supercluster** is now generally reserved for collections of many galaxy clusters (and their associated surrounding regions), gravitationally bound to one another in structures that span hundreds of millions of light-years.

The mind boggles at these enormous gravitational structures, and perhaps you have begun to wonder whether this hierarchy of structures extends ever upward. But structures of such vast size are about the largest objects we can see before we take the final jump in scale to the **universe** itself. Although for centuries our knowledge of the visible universe was confined by the limits of our telescopes and instruments, today we are reaching a fundamental limit: We can see only as far away as light has had time to travel in the age of the universe. The best evidence today indicates that the universe is 13.7 billion years old, and therefore we cannot see any farther than light can have traveled in that time. The largest superclusters we see span nearly 1% of the visible universe, but as best we can determine the universe is relatively uniform over larger spans than this.

Some astronomers hypothesize that the universe extends limitlessly, whereas according to others it gradually curves back on itself to form a closed finite system like the surface of the Earth but on scales far larger than the visible universe. Regardless of our uncertainty about the known universe's overall shape and size, we can observe that it has a well-ordered hierarchy of smaller structures. Small objects are clustered into larger systems, which are themselves clustered: satellites around planets, planets around stars, stars in galaxies, galaxies in clusters, clusters in superclusters as illustrated in Figure 2.6 and Table 2.1. Although astronomers do not yet understand completely how this orderly structure originated, they do know that gravity plays a crucial role.

TABLE 2.1	Scales of the Universe
Object	**Approximate Radius**
Earth	6400 km (4000 miles) = $R_\oplus$
Moon's orbit	380,000 km $\approx$ 30 $\times$ $R_\oplus$
Sun	700,000 km = $R_\odot$ $\approx$ 110 $\times$ $R_\oplus$
Earth's orbit	150 million km = 1 AU $\approx$ 210 $\times$ $R_\odot$
Solar System to Pluto	40 AU $\approx$ 8500 $\times$ $R_\odot$
Nearest star	270,000 AU $\approx$ 4.2 ly
Milky Way galaxy	50,000 ly $\approx$ 100 million $\times$ radius of the Solar System
Local Group	1.5 million ly $\approx$ 30 $\times$ radius of the Milky Way
Local Supercluster	40 million ly $\approx$ 30 $\times$ radius of the Local Group
Visible universe	13.7 billion ly $\approx$ 300 $\times$ radius of the Local Supercluster

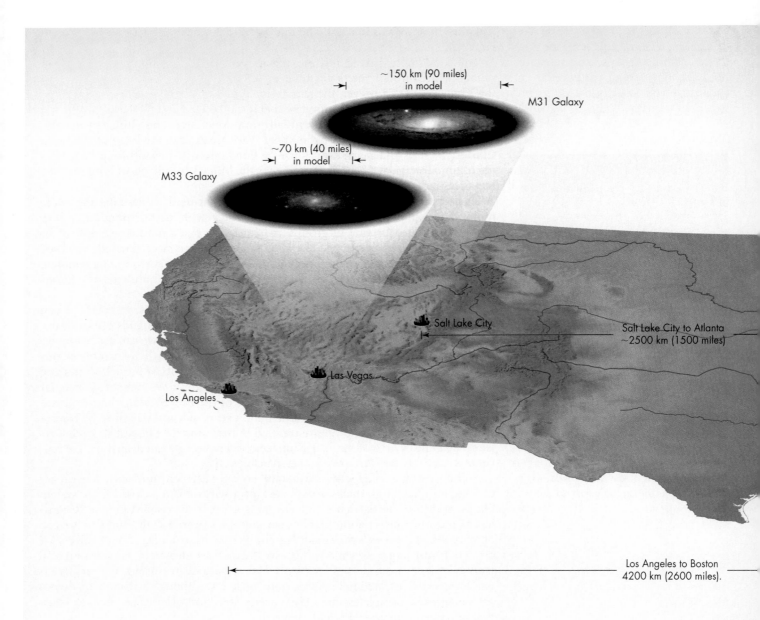

~150 km (90 miles)
in model

M31 Galaxy

~70 km (40 miles)
in model

M33 Galaxy

Salt Lake City

Salt Lake City to Atlanta
~2500 km (1500 miles)

Las Vegas

Los Angeles

Los Angeles to Boston
4200 km (2600 miles).

FIGURE 2.4

A scale model of the Local Group with a scale of 1 meter representing 1 light-year. At this scale the Local Group is nearly the size of the United States, and the three largest galaxies in the group are the size of metropolitan areas. The entire Solar System is the size of a pinhead on this scale.

FIGURE 2.5

Photograph of the central region of the Virgo Cluster. The Milky Way and entire Local Group are gravitationally bound to this galaxy cluster.

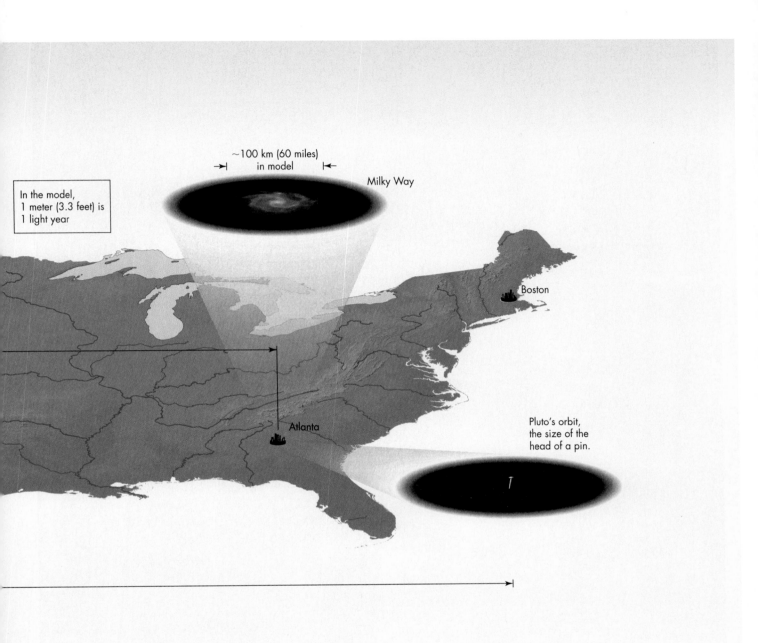

~100 km (60 miles)
in model

Milky Way

In the model,
1 meter (3.3 feet) is
1 light year

Boston

Atlanta

Pluto's orbit,
the size of the
head of a pin.

2.5   THE STILL UNKNOWN UNIVERSE

Q The idea of dark matter may seem quite unusual, but can you think of any situations in which you have been able to find evidence that something was present without actually seeing it? Were you later proved correct?

Over the last century astronomers have uncovered evidence that the entire universe behaves in ways that were utterly unexpected a hundred years ago. Measurements show that galaxies are flying away from each other as a result of what *resembles* an all-encompassing explosion, called the **Big Bang.** Albert Einstein's general theory of relativity showed that this is the wrong way to interpret what we see, though—space itself is expanding, carrying the galaxies along with it. Einstein's theories, now cornerstones of modern physics, have forced us to reevaluate our most basic notions of space and time, and we are discovering that there is much more to the universe than we can directly see through our telescopes.

Far more matter is found to fill the universe than what we can see in planets, stars, gas clouds, galaxies, and so on. Even after we correct for the amount of matter

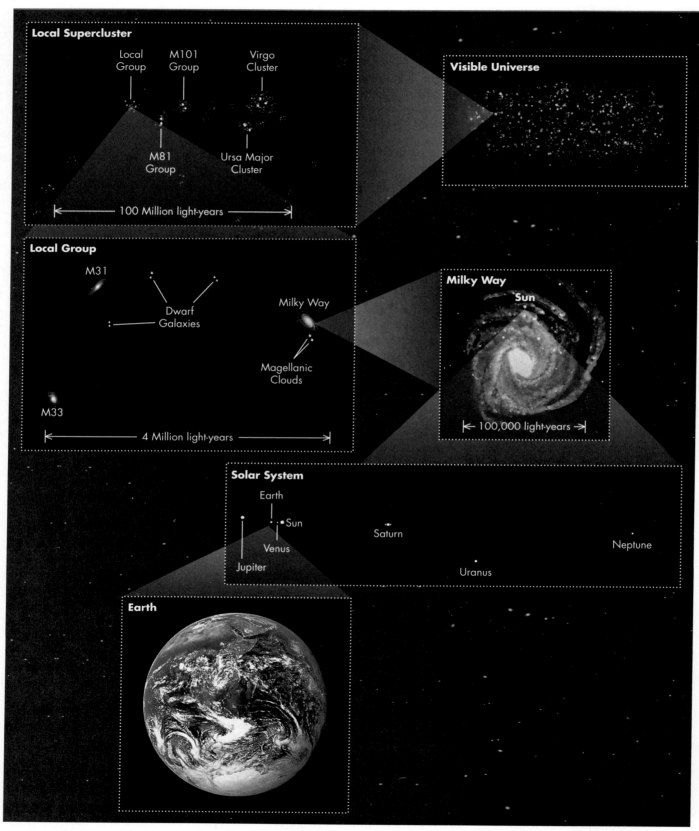

FIGURE 2.6
How the Earth fits into the universe.

that is hard to detect, there is strong evidence that there is about 10 times more utterly invisible **dark matter** that is made of strange substances that *cannot* form stars or planets. Yet this matter appears to be a critical component of every system the size of a galaxy or larger. The evidence that such matter exists comes from a variety of sources. The motion of stars in the outer parts of galaxies is so fast that they would fly off into intergalactic space unless much more mass is present than we can see. The same has been found true for galaxies in galaxy clusters. That is, the gravitational force that would be exerted by the observable stars and galaxies is far too small to hold the systems together. This is the subject of intense ongoing investigations, and it may be that particle physicists, exploring the submicroscopic world at scales far smaller than atoms, will uncover the nature of this strange kind of matter.

Over the last decade, astronomers have also accumulated evidence for an even stranger substance that has been dubbed **dark energy.** This is a strange kind of energy that pervades every corner of space—even the emptiest vacuum. Einstein predicted the possibility of such an all-pervasive energy, but based on the evidence of his day, he abandoned the idea. New evidence suggests that not only does dark energy exist, it "outweighs" all of the visible and dark matter in the universe. If the latest measurements are correct, dark energy will drive the expansion of the universe faster and faster, perhaps in the extremely remote future causing space to expand so rapidly that it will tear apart the visible universe!

KEY TERMS

Big Bang, 19

dark energy, 21

dark matter, 21

galaxy, 15

galaxy cluster, 16

galaxy group, 16

gravity, 13

light-year, 16

Local Group, 16

Milky Way galaxy, 14

stellar evolution, 13

supercluster, 17

universe, 17

Virgo Cluster, 16

QUESTIONS FOR REVIEW

1. To what systems (in increasing order of size) does the Earth belong?
2. How is a light-year defined?
3. Roughly how big across is the Milky Way galaxy?

PROBLEMS

1. If the Milky Way were the size of a nickel (about 2 centimeters):
    a. How big would the Local Group be?
    b. How big would the Local Supercluster be?
    c. How big would the visible universe be?
    (The data in Table 2.1 may help you here.)
2. If the Milky Way is moving away from the Virgo Cluster at 1000 kilometers per second, how long does it take for the distance between them to increase by 1 light-year?

3. The Milky Way is moving toward the larger galaxy M31 at about 100 kilometers per second. M31 is about 2 million light-years away. How long will it take before the Milky Way collides with M31 if it continues at this speed?

TEST YOURSELF

1. You write your home address in the order of street, town, state, and so on. Suppose you were writing your cosmic address in a similar manner. Which of the following is the correct order?
    a. Earth, Milky Way, Solar System, Local Group
    b. Earth, Solar System, Local Group, Milky Way
    c. Earth, Solar System, Milky Way, Local Group
    d. Solar System, Earth, Local Group, Milky Way
    e. Solar System, Local Group, Milky Way, Earth
2. Which of the following astronomical systems is/are held together by gravity?
    a. The Sun
    b. The Solar System
    c. The Milky Way
    d. The Local Group
    e. All of them
3. The light-year is a unit of
    a. time
    b. distance
    c. speed
    d. age
    e. All of the above

3

Astronomical Numbers

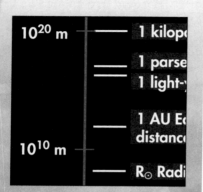

Astronomy deals with a greater range of numbers than any other science. It is challenging to grasp, even figuratively, the vast range of the measurements needed to study the universe (Figure 3.1). To understand a planet that has a diameter of 13,000,000 meters, we also need to understand the interactions between atoms that are only 0.0000000001 meters in diameter. And as we discuss other aspects of the universe, we will be considering sizes, masses, brightness, energy, and times that span even greater ranges.

To deal with this vast array of numbers, astronomers use four different strategies: (1) the metric system, (2) scientific notation, (3) special units, and (4) approximation. We will look at each of these approaches in this unit, but we can summarize these procedures briefly as follows:

The *metric system*, as opposed to the English system, allows easy conversions between larger and smaller **units** of measure, such as between meters and kilometers. To carry out

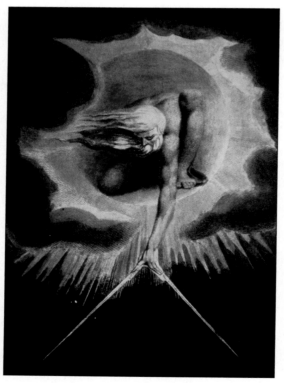

FIGURE 3.1
The "Ancient of Days" taking the measure of the universe. Etching by William Blake (c.1794).

many calculations, we must express measurements in fundamental units like meters, seconds, and kilograms. In this case we are faced with numbers like the 13,000,000 meters and .0000000001 meter just mentioned. Astronomers use a way of abbreviating these numbers called *scientific notation* that keeps track of the order of magnitude of a number separately from its precise value.

Metric units do not always prove convenient for interpreting the sizes of things, so sometimes we create special units with a clear physical meaning, such as light-years (Unit 2.3). Often the essential information we need about a number is its approximate size, so astronomers round off values to avoid focusing on a level of precision that is not important—so we say, for example, that the Astronomical Unit, describing the Earth's distance from the Sun, is 150 million kilometers even though it is known to be 149,597,870 kilometers.

This mathematical background is important for understanding much of the material in this book, but it does not necessarily have to be mastered all at once. Return to this unit whenever the numbers seem overwhelming.

3.1   THE METRIC SYSTEM

Before the metric system was introduced in 1791, widely varying systems of measurement were used in different countries, or even in different regions within a single country. For example, the Paris foot was 6.63% longer than the English foot, while the Spanish *vara* was 8.67% shorter than the English yard. Similarly, the French pound (*livre*) was 10.2% larger than the English pound, and many other weight measures were in common use throughout even just Europe. Confusion about measurements was widespread and required frequent conversion between different units.

That same confusion caused the loss of a Mars *Orbiter* in 1999, when the company building navigation jets provided its thrust data in English units while NASA mission controllers believed the numbers were in metric units. Firing the thrusters sent the craft too close to the planet, where it burned up in the atmosphere. Today the metric system has widespread adoption, while the complex system of English units is still the primary choice in only the United States, Liberia, and Myanmar. In the end, the units used should not interfere with appreciating the important ideas of astronomy, but it is helpful to be familiar with the metric system for the calculations we will carry out.

The great advantage of the **metric system** over the English system is that different units are related by factors of 10, making it much simpler to convert between different units—1 kilometer is 1000 meters, whereas 1 mile is 5280 feet. We can usually convert between metric units by just "moving the decimal point" instead of multiplying or dividing by a conversion factor like 5280 feet per mile.

Each time we move the decimal point we are performing the equivalent of multiplying or dividing by factors of 10. Mathematically, we can use **exponents** to indicate the *number* of times we multiply by 10. An exponent indicates how many times we multiply a quantity by itself. For example, there are 1000 meters in a kilometer. We can write 1000 as "10 to the third power" or "10 to the exponent three" as follows:

$$1000 = 10 \times 10 \times 10 = 10^3.$$

When we convert meters to kilometers, we also move the decimal point over three places:

$$1 \text{ km} = 1.\text{km} \times \frac{10^3 \text{ m}}{\text{km}} = 1000.\text{m}$$

The trick to converting from one unit to another is to "multiply by 1." *One* can be written in many ways. If a and b are equal then $a/b = 1$. In the conversion carried out here, because $1 \text{ km} = 10^3 \text{ m}$, we have multiplied by $\frac{10^3 \text{ m}}{1 \text{ km}}$. This lets us cancel out the units of km because they appear in both the numerator and the denominator.

where we have used the abbreviation m for meters and km for kilometers, and we have explicitly included the decimal point for clarity.

When we multiply by 10^3, we move the decimal point to the right by 3 places. If we were to divide by 10^3, we would move the decimal point to the *left* by 3 places. This second case can also be written as multiplying by 10^{-3}. In other words "10 to the negative third power" is the equivalent of one thousandth (1/1000). When we take 10 (or any number) to the "zeroth power" the result is always one: $10^0 = 1$.

Similarly, we can write 1 billion as

$$1,000,000,000. = 10 \times 10 \times 10 \times 10 \times 10 \times 10 \times 10 \times 10 \times 10 = 10^9$$

or one millionth as

$$0.000001 = \frac{1}{1,000,000.} = \frac{1}{10 \times 10 \times 10 \times 10 \times 10 \times 10} = \frac{1}{10^6} = 10^{-6}.$$

In brief, rather than writing out all the zeros, we use the exponent to tell us the number of zeros.

In the metric system, **metric prefixes** identify various possible **powers of 10.** The prefix *kilo,* for example, indicates 1000, while *milli* indicates one thousandth, and *mega* indicates 1 million. These prefixes can be added to any unit of measure to create a new unit that is closer to sizes we are interested in discussing: millimeter, kilogram, or megabyte, for example. The last term, *megabyte,* indicates a million pieces of information (such as letters or digits) stored in a computer and demonstrates how we can use metric prefixes in front of any unit of measure we like.

Table 3.1 shows the standard metric prefixes along with their meanings in words and exponential notation. This is a complete listing of metric prefixes; in this book we use only the seven shown in boldface in the table. A nanometer, for example, is a unit we will use when discussing light waves. It is a billionth of a meter or 10^{-9} m, and can be abbreviated "nm." To convert 5 nanometers to meters, we would move

It is often forgotten today that the English system also offers many intermediate size scales, but their relationships are complex. The inch is divided into 12 "lines," the foot into 12 inches, the yard into 3 feet, the "rod" into 5.5 yards, the "chain" into 4 rods, the "furlong" into 10 chains, the mile into 8 furlongs, and the "league" into 3 miles. The acre, still commonly used as a measure of area, is one furlong long by one chain wide. As if this is not enough to give one a headache, weights and volumes were divided in completely different ways!

	TABLE 3.1	Metric Prefixes	
Power of 10	Exponential Notation	Metric Prefix	Abbreviation
septillion (trillion trillion)	10^{24}	yotta	Y
sextillion (billion trillion)	10^{21}	zetta	Z
quintillion (million trillion)	10^{18}	exa	E
quadrillion (thousand trillion)	10^{15}	peta	P
trillion	10^{12}	tera	T
billion	10^9	**giga**	**G**
million	10^6	**mega**	**M**
thousand	10^3	**kilo**	**k**
hundred	10^2	hecto	h
ten	10^1	deca	da
tenth	10^{-1}	deci	d
hundredth	10^{-2}	**centi**	**c**
thousandth	10^{-3}	**milli**	**m**
millionth	10^{-6}	**micro**	**μ**
billionth	10^{-9}	**nano**	**n**
trillionth	10^{-12}	pico	p
quadrillionth	10^{-15}	femto	f
quintillionth	10^{-18}	atto	a
sextillionth	10^{-21}	zepto	z
septillionth	10^{-24}	yocto	y

TABLE 3.2		MKS Units	
Quantity	Metric Unit	MKS Equivalent	English Equivalent
Length	meter	m	3.28 feet
Mass	kilogram	kg	2.2 pounds (of mass)
Time	second	sec	(same)
Area	square meter	m^2	10.76 square feet
Volume	liter (L)	$10^{-3}\,m^3$	1.06 U.S. quarts
Speed	meters per second	m/sec	2.24 miles per hour
Acceleration	meters per square second	m/sec^2	3.28 feet per square second
Density	kilograms per liter	$10^3\,kg/m^3$	0.036 pounds per cubic inch
Force	newton (N)	$kg \cdot m/sec^2$	0.225 pounds (of force)
Energy	joule (J)	$kg \cdot m^2/sec^2$	0.000948 BTUs
Power	watt (W)	$kg \cdot m^2/sec^3$	0.00134 horsepower

the decimal point to the left by 9 places: 5. nm = 0.000000005 m. Thus starting from a fundamental unit of measure like the meter, the metric prefixes give us a wide range of units that are appropriate in different contexts.

Any system of physical measurements requires three fundamental units—those describing length, mass, and time. In the metric system these are the meter (which is about 10% longer than the yard), the kilogram (which is about 2.2 pounds), and the second. This set of units defines the **MKS system,** which stands for <u>m</u>eter, <u>k</u>ilogram, and <u>s</u>econd. Units for measuring other quantities, such as force, energy, and power, can all be written in terms of these fundamental units. Some of the more common kinds of metric units we use in this book, and their nearest equivalent in the English system, are listed in Table 3.2.

One unit commonly used in this book, the liter—a unit of volume—is not quite as simple a combination of MKS units as the others. Because we will often be comparing the **densities** (mass per volume) of materials, the density in kilograms per liter is a convenient reference. The liter is equivalent to one thousandth of a cubic meter. This in turn leads to the original definition of the kilogram, which is the mass of one liter of water. Water has a density of 1 kilogram per liter, but we will encounter centers of dying stars where the density is enormously larger, and regions of interstellar space where the density is a minuscule fraction of this.

The units in the Table 3.2 can all employ metric prefixes, so the specifications on the power of a car would be likely listed as 200 kilowatts. To convert this to the English unit of horsepower, we could carry out the following calculation:

$$200 \text{ kW} = 200{,}000 \times 0.00134 \text{ horsepower} = 268 \text{ horsepower}$$

Also be aware that in the English system, the term *pound* is used interchangeably for a mass (a measure of the amount of matter) as well as for the gravitational force with which the Earth pulls on that mass (see Unit 14.1 for details). As we will discover, the same mass weighs a different amount on different planets.

Because of difficulties in reproducing a precisely consistent mass from a liter of water, the kilogram has been redefined today in terms of a platinum-alloy cylinder stored at the International Bureau of Weights and Measures in France.

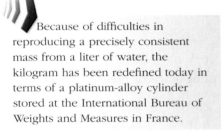

The conversion here is accomplished first by converting kilowatts to watts, multiplying by $1 = \dfrac{1000 \text{ W}}{1 \text{ kW}}$, and then again multiplying by

$1 = \dfrac{0.00134 \text{ horsepower}}{1 \text{ W}}$.

3.2   SCIENTIFIC NOTATION

When we make a calculation in scientific notation, we usually begin by converting all of the values into MKS units. For example, the mass of the Sun is 1,989,000,000,000,000,000,000,000,000,000 kilograms. Partly because there is no metric prefix close to the size of such an enormous number, we use a more concise way to express such numbers, called *scientific notation.*

Scientific notation combines the powers-of-10 notation just described with the particular value of the number. We divide the number into a value between 1 and 10 and a power of 10 that when multiplied together yield the original number. Thus we can write 600 (six hundred) as

$$600 = 6 \times 10 \times 10 = 6 \times 10^2$$

We write 543,000 (five hundred forty-three thousand) as

$$543{,}000 = 5.43 \times 10 \times 10 \times 10 \times 10 \times 10 = 5.43 \times 10^5$$

and 21 millionths becomes

$$\frac{21}{1{,}000{,}000} = 0.000021 = \frac{2.1}{100{,}000} = \frac{2.1}{10^5} = 2.1 \times 10^{-5}.$$

Once again the exponent indicates how we have moved the decimal point. The Sun's mass expressed this way is 1.989×10^{30} kg.

With scientific notation, multiplying and dividing very large numbers becomes easier. This is because when we multiply two powers of 10 we just add the exponents, whereas to divide we subtract them. For example,

$$10^2 \times 10^5 = (10 \times 10) \times (10 \times 10 \times 10 \times 10 \times 10) = 10^{2+5} = 10^7$$

or as an example of division,

$$\frac{10^8}{10^3} = \frac{10 \times 10 \times 10 \times 10 \times 10 \times 10 \times 10 \times 10}{10 \times 10 \times 10}$$

$$= 10 \times 10 \times 10 \times 10 \times 10 = 10^5.$$

We can write this as a pair of general rules:

$$10^A \times 10^B = 10^{A+B} \quad \text{and} \quad 10^A/10^B = 10^{A-B}.$$

An important thing to remember is that $10^A + 10^B$ does *not* equal 10^{A+B}. To add or subtract numbers, both should be converted to the *same* power of 10.

As an illustration of the use of scientific notation, we can calculate the number of kilometers in a light-year. To do this, we multiply light's speed by the number of seconds in a year. A year is 365¼ days, each day having 24 hours, each hour 60 minutes, and each minute 60 seconds, so the total number of seconds in a year is

The conversion here is accomplished by multiplying by 1 written in three different ways. Make sure you can identify them.

$$365.25 \text{ days} \times \frac{24 \text{ hours}}{1 \text{ day}} \times \frac{60 \text{ minutes}}{1 \text{ hour}} \times \frac{60 \text{ seconds}}{1 \text{ minute}}$$

$$= 31{,}557{,}600 \text{ seconds} \approx 3.156 \times 10^7 \text{ sec.}$$

The speed of light is 299,793 km/sec $\approx 2.998 \times 10^5$ km/sec. Multiplying the speed by the time gives us the distance:

$$\begin{aligned}
1 \text{ ly} &= \text{Speed of light} \times 1 \text{ year} \\
&= 2.998 \times 10^5 \text{ km/sec} \times 3.156 \times 10^7 \text{ sec} \\
&= 2.998 \times 3.156 \times 10^{5+7} \text{ km} \\
&\approx 9.46 \times 10^{12} \text{ km}
\end{aligned}$$

or nearly 10 trillion kilometers.

3.3   SPECIAL UNITS

For good or bad, astronomers have invented several new units to describe objects and phenomena that are far more immense than what we encounter on Earth. Usually these units make it easier to gain physical intuition, but they add to the amount of information to learn.

Quantity	Special Unit	Abbreviation	Metric Equivalent
Length	Earth's radius	$R_\oplus$	6.37×10^6 m
	Sun's radius	$R_\odot$	6.97×10^8 m
	Astronomical Unit (Earth–Sun distance)	AU	1.50×10^{11} m
	Light-year (distance light travels in one year)	ly	9.46×10^{15} m
	parsec (distance calculated by special geometric technique)	pc	3.09×10^{16} m
Mass	Earth's mass	$M_\oplus$	5.97×10^{24} kg
	Sun's mass	$M_\odot$	1.99×10^{30} kg
Time	Year (orbital period of Earth)	yr	3.16×10^7 sec
	Sun's lifetime (estimated total time Sun will generate energy)	$t_\odot$	3.16×10^{17} sec
Speed	Speed of light (through empty space)	c	3.00×10^{10} m/sec
Acceleration	Earth's surface gravity (the rate at which falling objects accelerate)	g	9.81 m/sec^2
Energy	Kiloton of TNT (energy released by a standard 1000-ton bomb)	kt	4.18×10^{12} J
Power	Sun's luminosity (energy the Sun generates per second)	$L_\odot$	3.86×10^{26} W

TABLE 3.3   Special Units

The **light-year,** the distance light travels in a year, is a good example. In principle, we could use metric prefixes and use a unit like "petameters" (Table 3.1) and write a light-year as 9.46 petameters or 9.46 Pm. However, the light-year has such a useful interpretation that we prefer to introduce it as another unit. Specifically, when we see an object 10 million light-years away, the light has taken 10 million years to reach us. That means that we are seeing the object *as it was 10 million years ago.* This becomes even more interesting as we look out billions of light-years, when the universe was just a fraction of its current age. We are literally able to look back toward the beginning of time, and the light-year unit helps us to understand what, or rather "when," we are seeing.

A list of the special units we will use in this book is given in Table 3.3. Many of these units are used to relate other objects to the more familiar Sun and Earth. These special units add a level of complication to a calculation whenever we need to convert to the MKS system first, but some of the units were invented to make our calculations easier. For example, the Astronomical Unit simplifies the calculation of orbital parameters of objects in the Solar System as well as for stars that orbit each other. The **parsec** was invented as a unit to simplify the formula for the primary method we have for finding stars' distances (see Unit 52.2 for details). The relative sizes of the parsec and other special units for describing sizes are illustrated in Figure 3.2. These additional units are intended to clarify rather than confuse.

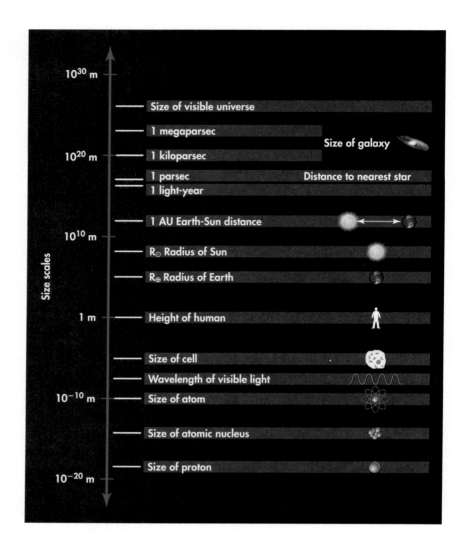

FIGURE 3.2

Sizes of special units and various objects in the universe—ranging from the submicroscopic to the astronomical on a scale of powers of 10.

3.4   APPROXIMATION

Astronomy is a science in which we often have to make uncertain estimates. Even though we can measure some numbers precisely, many others are subject to such uncertainty that there is no point in keeping track of more than the first couple of digits of a number.

For example, we know that there are 3.261633 light-years in a parsec. However, we rarely know the distance of a star more precisely than 10%. If a star's distance is listed as 10 parsecs, it is *not* correct to say it is at 32.61633 light-years. The latter number gives the appearance of precision that does not exist. It would be more appropriate to convert 10 parsecs to 33 light-years or even 30 light-years—these better reflect the precision with which such distances are known.

Generally, measured values are reported showing only the number of digits that are accurately known. Values written as 374,000 or 0.00512 or 1.90×10^7, for example, are each reported to three **significant digits.** Having three significant digits means that the first two digits (37, 51, 19 in these three examples) are well-determined, and the last one is fairly accurate, although it may have some uncertainty. Scientific notation here is useful for indicating when a trailing 0 is significant. This would be unclear in the example if 1.90×10^7 was written 19,000,000.

In the technical literature, astronomers often report numbers with a "plus or minus" range to indicate the uncertainty in a value, like 71 ± 6. Then if we multiplied the number by two, we would find 142 ± 12, with both the value and uncertainty doubling. This notation becomes cumbersome, especially when we start multiplying a set of numbers that each has its own precision.

Any result from multiplying or dividing numbers in a formula can be only as accurate as the *least* accurate measurement used. For example, the nearest star, Proxima Centauri, is measured to be 1.3 parsecs away—to two digits of precision. We can multiply this by the number of light-years per parsec to get the distance in light-years:

$$1.3 \text{ pc} \times \frac{3.261633 \text{ ly}}{\text{pc}} = 4.2401229 \text{ ly} \approx 4.2 \text{ ly}$$

We **round off** the result to 4.2 ly, with two digits to reflect the level of precision for this calculation.

Notice that because we are limited by the least precise number, there was really no need to use all of the digits 3.261633 in our calculation. If we rounded off our ly-to-pc conversion factor to 3.26, we would have gotten the same result:

$$1.3 \text{ pc} \times \frac{3.26 \text{ ly}}{\text{pc}} = 4.23 \text{ ly} \approx 4.2 \text{ ly}$$

On the other hand, if we rounded off our conversion factor to just one digit, 3, we get this:

$$1.3 \text{ pc} \times \frac{3 \text{ ly}}{\text{pc}} = 3.9 \text{ ly} \approx 4 \text{ ly}$$

The result should be reported to only one digit because one of the numbers we used had only one significant digit.

Finally, note that if we rounded the conversion factor to 3.3, the result is

$$1.3 \text{ pc} \times \frac{3.3 \text{ ly}}{\text{pc}} = 4.29 \text{ ly} \approx 4.3 \text{ ly}$$

which is rounded *up* because the digit (9) after the last significant digit 2 is greater than or equal to 5. Unfortunately, when we carry out a calculation with several numbers that are approximate, we multiply the error, and so the result has a *greater* uncertainty than the least accurate of the numbers.

Good use of approximation is an art as well as science. For example, someone might ask whether the distance of 4.2 light-years to Proxima Centauri is its distance from the Earth or the Sun. We know the Earth–Sun separation very accurately: 1 AU is 149,597,900 km. Should we worry about this when working out the correct distance? A little reflection will show that as big as an AU is, it is utterly insignificant compared to 4.2 light-years. The first time you think about this may require some calculations to convince yourself. Calculations show that light traverses 1 AU in just 8½ minutes—so just as you would not worry about whether 8½ minutes made a difference to the date of the last day of classes, you need not worry about it for the distance to Proxima Centauri.

KEY TERMS

QUESTIONS FOR REVIEW

1. What are the advantages of the metric system?

2. What does a positive exponent mean? A negative exponent? An exponent of zero?

3. What are significant digits?

PROBLEMS

1. The radius of the Sun is 7×10^5 kilometers, and that of the Earth is about 6.4×10^3 kilometers. Show that the Sun's radius is about 100 times the Earth's radius.

2. A typical bacterium has a diameter of about 10^{-6} meters. A hydrogen atom has a diameter of about 10^{-10} meters. How many times smaller than a bacterium is a hydrogen atom?

3. Using scientific notation, show that it takes sunlight about 8½ minutes to reach Earth from the Sun.

4. Calculate approximately how long it takes light to travel from the Sun to Pluto.

5. Suppose two galaxies move away from each other at 6000 km/sec and are 300 million (3×10^8) light-years apart. If their speed has remained constant, how long has it taken them to move that far apart? Express your answer in years.

6. Using scientific notation, evaluate $(4 \times 10^8)^3/(5 \times 10^{-6})^2$.

7. Using scientific notation, evaluate $(3 \times 10^4)^2/\sqrt{4 \times 10^{-6}}$

TEST YOURSELF

1. An orbiting spacecraft travels about 4×10^4 kilometers in about 5×10^3 seconds. What is its speed?
   a. 8×10^0 kilometers/second
   b. 2×10^8 kilometers/second
   c. 8×10^6 kilometers/second
   d. 9×10^7 kilometers/second
   e. 1×10^1 kilometers/second

2. Which of the following is *not* equivalent to 30 kilometers?
   a. 30,000,000 millimeters   d. 3×10^3 meters
   b. 3×10^6 cm   e. 0.03 megameters
   c. 30,000 meters

3. From the following, choose the best approximation for the sum of 3.14×10^{-1} and 6.86×10^4
   a. 10.00×10^3 d. 1.00×10^3
   b. 6.86×10^4 e. 3.72×10^3
   c. 3.14×10^{-1}

4

Foundations of Astronomy

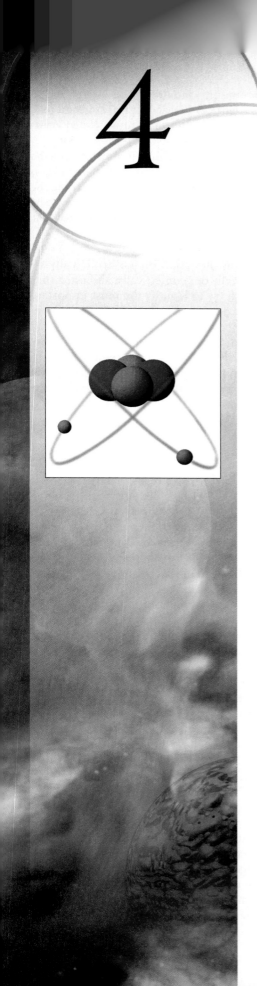

It seems that whenever we try to understand *larger* objects, we need to know more about ever smaller elements of the *submicroscopic* world. To study planets we must understand atoms and molecules; to study stars we have to understand the nuclei of atoms; to study the universe we need to understand subatomic particles.

The discoveries we have made about the submicroscopic properties of matter in laboratories here on Earth appear to apply to the rest of the universe. This is remarkable in itself, given how tiny a piece of the universe the Earth represents. In studying the cosmos, we encounter extremes of temperature, size, gravity, and energy that no one could dream of testing in Earth-based laboratories.

Our understanding of the universe has not come easily. Astronomy builds on many ideas from physics, chemistry, geology, mathematics, and biology. Learning how to extend our understanding from our experiments here on Earth to the far reaches of the universe is part of the science of astronomy. And astronomers sometimes encounter surprises when comparing what we have learned from our Earth-based experiments to more distant realms. These surprises may lead to new ideas for revising or extending our theories. This understanding represents the intellectual work of thousands of men and women over thousands of years, through the testing of ideas. That testing is part of what we generally call the *scientific method*.

4.1 THE SCIENTIFIC METHOD

In its simplest form, the **scientific method** is the procedure by which scientists construct their ideas about the universe and its contents, regardless of whether those ideas concern stars, planets, living things, or matter itself. In the scientific method, a scientist proposes an idea—a **hypothesis**—about some property of the universe and then tests the hypothesis by experiment. Ideally the experiment's results either support the hypothesis or refute it. Astronomers face a special challenge in applying the scientific method because they cannot experiment with their subject matter directly. Astronomy is primarily an observational science, like geology or the study of human behavior; so the strength of astronomers' theories is tied to their ability to predict phenomena or circumstances not yet observed.

One of the requirements for a hypothesis to be regarded as scientific is that it must be clear how it can be disproved. In experimental sciences this may be the prediction of the outcome of a laboratory experiment. For an observational science like astronomy, it might be a prediction that, for example, whenever we find a star with two particular properties, we will find that it also has a third property. To disprove the hypothesis, we need only demonstrate that the prediction is incorrect.

By requiring hypotheses to be disprovable, science not only provides a way of introducing new ideas; it also encourages scientists to test and retest older ideas from new perspectives. This openness to retesting should not be interpreted as vagueness or uncertainty—far from it. For an idea to stand up to this constant testing, it must be robust, providing us with much of the most rigorously tested information we possess.

Once a hypothesis has been thoroughly tested and verified, it may be termed a *theory* or *law*. When we use the word *theory* here we do *not* mean that the idea is unproved or tentative, and the word *law* should not be interpreted as meaning the idea is beyond testing. Rather, theories and laws have achieved wide acceptance by successful testing over a long time. **Laws** are generally mathematical statements, whereas **theories** are generally expressed in words.

Sometimes more complex ideas are described as **models.** Models express relationships between different quantities and how they affect each other. Usually astronomical models are expressed mathematically or geometrically, and often they oversimplify a complex set of relationships to try to identify the most important elements. For example, contrary models were developed to explain the motions of planets, one assuming that the Earth was the center of their motion, another that the Sun was. Both models can accurately predict the motions we see, but today we recognize the Sun-centered model as the more nearly correct of the two models. However, even it is not quite right because, in fact, the bodies in the Solar System all interact with each other through the force of gravity in a more complex way than simply circling the Sun. And in some situations the Earth-centered model is more convenient for describing the motions we see—as when we speak of the "Sun setting" as opposed to saying that the Earth rotated so that the Sun became hidden below our local horizon.

When an idea has achieved the status of a theory or law, its ability to make accurate predictions does not end even when a new idea overtakes it. For example, in the late 1600s Isaac Newton proposed several mathematical relationships describing motion and gravity, known as Newton's Laws. They are extremely successful mathematical models that are still used today despite the fact that they are now known to make small errors in the prediction of motions under extreme conditions beyond our common experience, as shown in the early 1900s by Albert Einstein. He proposed revisions to these laws, known as the theories of special and general relativity, which explain how space and time are altered by motion and gravity. Scientists have subjected Einstein's theories to an enormous number of tests, and they have passed all such tests with the highest precision. Einstein's theories may someday also need revision; but it should be understood that, as with Newton's Laws, their validity will not suddenly vanish. The many tests of the theories' predictions show that they work with a high degree of precision over a great range of circumstances.

Application of the scientific method is no guarantee that its results will be believed. For example, even before 300 B.C., Greek philosophers pursued several lines of inquiry demonstrating that the Earth is a sphere (Unit 10). Yet despite the evidence supporting this hypothesis, many people continued to believe the Earth to be flat for thousands of years. Today many people believe that it is possible to make a spaceship move faster than the speed of light (this belief is depicted in many science fiction movies), even though Einstein's theory of relativity has provided strong evidence that this cannot happen.

We can be quite confident of the longest-standing theories because so many scientists have attempted to overturn them. Nevertheless, an exciting aspect of science is that it invites skepticism. At the forefront of the sciences, debate is very lively, with new hypotheses being proposed to explain new measurements, and challenges to accepted theories always encouraged. Individual scientists often dispute particular pieces of evidence. One astronomer may find the evidence supporting a hypothesis convincing, while another astronomer may think that the experiment was done incorrectly or the data were analyzed improperly. This is all part of the scientific process.

We need therefore to keep in mind that when we discuss ideas, they are never "proved," and sometimes they are not even widely accepted. This is especially true of ideas at the frontiers of our knowledge—for example, those dealing with the origin and structure of the universe or those dealing with black holes. However, the tentativeness of such ideas does not always stop astronomers from being positive about them, leading the Russian physicist Lev Landau to joke that astrophysicists are "often in error, but never in doubt." Therefore, keep in mind that some of the ideas we discuss in this book will be vastly improved on or perhaps proved wrong in the future. That is not a failing of science, however. It is its strength.

4.2   THE NATURE OF MATTER

One of the most powerful ideas in astronomy is that we can apply what we learn about nature here on Earth to the universe at large. This is sometimes called the hypothesis of **universality.**

Starting from this hypothesis, the step-by-step application of physical laws to the nature of matter drives us to many extraordinary conclusions. We will conclude that there are exotic things in the universe such as black holes, quasars, neutron stars, and dark matter, as discussed in later units. Yet the volume of space we have explored directly with space probes is just a thousandth-trillionth-trillionth-trillionth (10^{-39}) of the volume of the visible universe. How can we actually be sure that the properties of matter and physical laws do not change elsewhere in the universe?

Our studies of matter on Earth show that it is made up of tiny particles called **atoms,** and 92 kinds of these occur naturally. The simplest type, a hydrogen atom, is about one 10-billionth of a meter (10^{-10} m) in diameter, so that 10 million (10^7) hydrogen atoms could be put in a line across the diameter of the period at the end of this sentence. Despite this tiny size, atoms themselves have structure. Every atom has a small (about 1/10,000th the atom's diameter) central core called the **nucleus** that is surrounded by one or more even smaller lightweight particles called **electrons** (Figure 4.1A). The nucleus is composed of two kinds of heavy particles called **protons** and **neutrons.**

Superficially, the atom resembles a miniature solar system, but different physical forces and behaviors are at play than the **gravitational force** between planets and the Sun. At subatomic scales, particles do not behave in the ways familiar to us—their motions are more wavelike (Unit 21). It is impossible to predict their precise position at any moment, so the location of the electron is described in terms of an "orbital cloud" (Figure 4.1B).

Experiments show that electrons repel each other, and protons repel each other, but electrons and protons *attract* each other. This situation is different from

FIGURE 4.1

An atom consists of a nucleus around which electrons orbit. The nucleus is itself composed of particles called protons and neutrons. (A) A classical view of an atom. (B) A more modern view with the electrons in "orbital clouds."

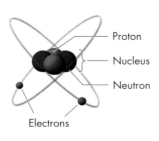

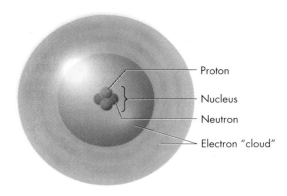

gravitational force, where all kinds of matter exert only an attraction on each other. You can get some sense of these effects by rubbing a balloon on your hair in dry weather. The balloon collects electrons from your hair, while your hair is left with an excess of protons. The electric charge on the balloon attracts strands of your hair, while your hair stands on end because the strands of like-charged hair repel each other. This is the same force at play on an atomic level.

To model this idea, we have designated that the proton has what we call a "positive" **electric charge,** and an electron has a "negative" electric charge. A neutron, as its name suggests, is neutral and has no charge. With this description, we can say that electric charges of the same sign repel each other ($+$ repels $+$; $-$ repels $-$) and opposites attract ($+$ attracts $-$). That attraction is what holds the electrons in their orbits around the nucleus of an atom.

The light we see is also made up of elementary particles, called **photons,** which can be generated or affected by electrically charged particles such as electrons and protons. The characteristics of the light emitted by each kind of atom are distinct, and a careful analysis of photon energies allows us to determine the kind of matter that generated the light. Astronomers have used this fact to test the hypothesis of universality. In fact, by examining the characteristics of light emitted by very distant objects, we are also testing whether the universe was the same in the past when the light left those objects. We find that everywhere we look in the universe, light has the same characteristics that we find in matter on Earth. The same kinds of atoms made of the same kinds of electrons, protons, and neutrons undergo the same kinds of chemical reactions we observe here on Earth as they did "a long time ago in a galaxy far, far away."

4.3   THE FOUR FUNDAMENTAL FORCES

The world of matter at submicroscopic scale is a whole universe unto itself. But we need to understand the universe at this scale to understand the larger universe. For example, reactions occurring in the nuclei of atoms are the source of power for stars, and the characteristics of the fundamental forces in nature appear to be linked to how the universe itself formed.

At this most basic level, there are only a few kinds of elementary particles interacting through just four forces. Two of the four forces are relatively common to our everyday experience: gravitational force and electric force. Actually, the electric force that holds the atom together is also fundamental to the magnetic force that makes a compass work or holds magnets on the door of your refrigerator. Moving electric charges generate fields of magnetic force, and moving magnets can generate electric fields. Scientists refer to them jointly as **electromagnetic force.** For subatomic particles, electromagnetic force is far more important than gravitational force. The strength of the electromagnetic force between two protons is 10^{36} (a trillion trillion trillion) times stronger than the gravitational force.

Two other forces play critical roles inside atoms. One of these, called the strong nuclear force or simply the **strong force,** binds protons and neutrons together to form an atom's nucleus. It is strong in the sense that, for two protons in a nucleus, it is about 100 times stronger than the electric repulsion they have for each other. Although the effects of the strong force cannot be seen directly in everyday life, without it, the nuclei of atoms would fly apart, and so would our familiar world. The last of the four forces is called the **weak force** because it is about a billion times weaker than the strong force between protons. The weak force can cause one kind of particle to decay into another through a process described further in the following section. Both the strong and weak forces are short-range forces that have virtually no influence outside an atomic nucleus. The strength of the weak force can

TABLE 4.1   Fundamental Forces

Force	Associated Property	Effect	Range	Carrier Particle	Relative Strength
Gravitational	Mass	All masses attract each other.	Infinite but weakens with distance	Graviton	10^{-36}
Electromagnetic	Electric charge	Opposites attract, likes repel.	Infinite but weakens with distance	Photon	1
Strong	Color charge	Three colors combine to make neutral combinations.	$\approx 10^{-15}$ meters (distance between protons in atomic nucleus)	Gluon	10^2
Weak	Weak charge	Massive particles decay to lower mass particles.	$\approx 10^{-18}$ meters (1/1000th proton diameter)	W and Z	10^{-7}

become more nearly comparable to the other forces at very, very small distances—less than 10^{-18} (a millionth-trillionth) of a meter—only a thousandth of the size of a proton.

The strong and weak forces each have their own associated "charges." These are not electric charges, but different properties of the subatomic particles that respond to the particular force. The strong force has three kinds of charge that must always occur in balanced combinations, analogous to the way red, blue, and green light can be combined to make white light. This analogy has led the charge associated with the strong force to be called "color charge." Associated with the weak force is a "weak charge," which has a complex representation that depends on a pair of properties in different combinations.

If the different kinds of charges are neutralized in a particle, then it will not respond to the associated force. This is why gravity is the dominant force at astronomical size scales despite being so extremely weak compared to the other forces. The great strength of the electromagnetic force keeps pulling electrons into an atom until their negative charges cancel out the positive charge exerted by the protons of the nucleus. As a result, the net electromagnetic force of the Sun on our planet is negligible compared to the gravitational force. The same occurs with both the strong and weak forces. Gravity remains because its "charge"—mass—does not come in opposites that can cancel each other's effects.

In the language of particle physics, each of the four forces is "mediated," or carried, by a particle. We just discussed the photon's interaction with electric charges; it is the carrier of the electromagnetic force. A particle called a **gluon** (the name invented to suggest its strong gluelike properties) likewise carries the strong force, while particles called W and Z carry the weak force. Physicists hypothesize that another particle—the **graviton**—carries the gravitational force, although this particle has not been detected. Some of the basic properties of the forces are summarized in Table 4.1.

4.4   THE ELEMENTARY PARTICLES

The inner space of an atom is a universe that is very foreign to our everyday experience. Many kinds of subatomic particles populate this universe and respond to the four forces in unfamiliar ways. In interactions, a number of properties, like charge, are always conserved; but some particles can be converted from one type to another, and mass and energy can sometimes be converted into each other.

Protons and neutrons are built of elementary particles called **quarks.** Two types of quarks, called the *up quark* (*u*) and *down quark* (*d*), combine in sets of three to make the proton (*uud*) and neutron (*ddu*). The strong force acting between quarks grows stronger with distance, making it impossible to remove a quark from nuclear particles to study in isolation. However, particle physicists can "see" the internal lumpy structure of protons or neutrons when they bombard them with high-energy electrons, and from this kind of experiment, they can determine much about the characteristics of quarks.

The weak force can convert quarks from one type into another. In fact, the weak force, despite its name, is important because it is the only force that can change one particle type into another. It can cause a particle to spontaneously decay into a less massive particle—for example, causing a neutron to decay into a proton. In this process a down quark turns into an up quark and emits two other particles, an electron and a **neutrino,** which balance the electric and weak charges in the interaction.

The discovery of the neutrino is a good example of the scientific method in action. Physicists studying the decay of the neutron in the 1930s could see that the proton and electron produced in the decay did not have as much energy as predicted by existing theories of energy and mass conservation. Various hypotheses were put forward, including the idea that there was a new particle that interacted so weakly that it was difficult to detect. This new particle had to have no electric charge and extremely little mass compared to any other known particle, and many scientists doubted the possibility of such a "ghost" particle. However, the neutrino was finally detected more than 20 years later, confirming the hypothesis and reinforcing the earlier conservation theories.

Neutrinos remain one of the most mysterious of the known particles today. For many years they were thought to have no mass, but it is now suspected that they do have a tiny (but still unmeasured) mass. Besides gravity, they interact only through the weak force, whereas electrons interact through the weak and electromagnetic forces and quarks interact through all the forces. Because the neutrino's strongest interaction is through such a weak force, it truly can seem "ghostly." The weak force requires that neutrinos must be *extremely* close to interact with another elementary particle. They might have to pass within 10^{-24} m of a quark—a millionth the diameter of a proton—for an interaction to be likely, so neutrinos can travel through an atomic nucleus without much chance of interacting with any of the particles. As a result, normal matter looks nearly transparent to them. Neutrinos pass through the Earth more readily than photons pass through a window! On the rare occasion when they interact with other particles, though, they have the ability to change their character, turning a down quark into an up quark, for example (Figure 4.2).

FIGURE 4.2

Illustration of the interaction of a neutrino (*v*) with the quarks in a neutron. (A) The neutrino passes very close to a down (*d*) quark, and they exchange a weak force (W) particle. (B) The down quark becomes an up (*u*) quark and an electron is emitted. The strong force between quarks is illustrated by gluon exchange, whereas the electron responds to electromagnetic force through photons.

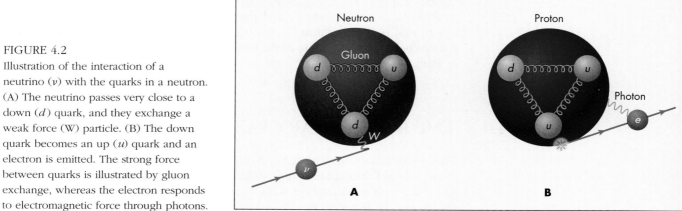

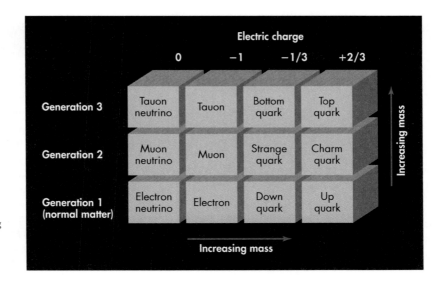

FIGURE 4.3

The elementary particles of the standard model. The general direction of increasing mass is shown, although the down quark appears to be more massive than the up quark.

Each particle also has a corresponding "antiparticle" with opposite properties. When an antiparticle meets its partner, they annihilate each other, leaving nothing but energy. Antiparticles, and their role at the beginning of the universe, are discussed further in Unit 81.

Up quarks, down quarks, electrons, and neutrinos make up nearly all of the matter in the known universe. There are two sets of more massive particles with parallel properties to each of these four particles, making "generations" of particles in the terminology of physicists, as illustrated in Figure 4.3. These alternate generations of matter are created at high temperatures and during high-energy collisions, but the weak force causes them to decay down to the first generation of particles. The other generations of quarks have been given fanciful names—*strange, charm, bottom,* and *top*—because they exhibit properties that we do not see in normal matter. Corresponding to the electron, the two other generations of particles are called the *muon* and *tauon,* and they each have an associated neutrino, called the *muon neutrino* and the *tauon neutrino.*

These particles, along with the carrier particles of the forces, are the basis of the **standard model** of particle physics. This model was developed before all of the particles were discovered and it predicted their properties before they were detected, so today it is regarded as a robust theory of the fundamental nature of matter and the forces or interactions that occur between its particles.

In principle, then, the story of the universe has just a few unique actors, and if we understood physics and chemistry perfectly, we should be able to predict the stunning array of phenomena that we find throughout the visible universe. The complexity that grows out of the interactions of these elementary particles produces many surprises, however. It seems impossible that any theory of these particles could have predicted that 13.7 billion years into its history the universe would create beings who would contemplate these questions!

The science of astronomy uses what we have learned from particle physics and other sciences to interpret what we see. Occasionally astronomers have had to hypothesize some piece of "missing physics," or a previously unknown force or particle, because of contradictions between what is observed and what was predicted. For example, our understanding of the neutrino's mass has been changed by studies of the neutrinos emitted by the Sun. In this way astronomy can sometimes drive forward new understandings in particle physics—or chemistry, geology, and biology. Astronomy is detective work in which we piece together what is known and guess at what must be missing to build a more complete, consistent explanation of the phenomena we observe.

KEY TERMS

atom, 33	neutron, 33
electric charge, 34	nucleus, 33
electromagnetic force, 34	photon, 34
electron, 33	proton, 33
gluon, 35	quark, 36
gravitational force, 33	scientific method, 31
graviton, 35	standard model, 37
hypothesis, 31	strong force, 34
law, 32	theory, 32
model, 32	universality, 33
neutrino, 36	weak force, 34

QUESTIONS FOR REVIEW

1. What is meant by the scientific method?
2. What force holds the Sun together?
3. What particles make up an atom?
4. What force holds the protons in an atom's nucleus?

PROBLEMS

1. Make a hypothesis about the cause of some weather phenomenon you have observed. For example, what produces lightning or hail? Describe the different observations you are familiar with that support your hypothesis. What experiment could be conducted that could disprove your hypothesis?
2. Explain at a fundamental level what is happening when you take clothes out of a drier and find them clinging to each other.

What do you think is happening when you hear a crackling sound as you pull a sock and shirt apart?

3. If the strong force grows larger with distance, explain why the force is limited in its range to acting within the atomic nucleus.
4. It is estimated that a neutrino has to pass within a distance of about 10^{-24} meters of a quark for a high likelihood of interaction. If an atom has a radius of 10^{-10} meters and 12 nuclear particles in its nucleus, what is the probability that a neutrino passing through an atom will undergo an interaction? (Hint: Think of the atom as a target and the area around each quark as a bull's-eye.)

TEST YOURSELF

1. According to the standard model, which of the following is not an elementary particle?
    a. Photon
    b. Proton
    c. Electron
    d. Neutrino
    e. Up quark
2. The best measure of the validity of an astronomical hypothesis is
    a. Its level of acceptance by scientists.
    b. The ease with which it can be understood by nonscientists.
    c. The sophistication of the mathematics it requires.
    d. Its ability to make predictions about future observations.
    e. How well it describes what has been previously observed.
3. The best model among the following of the physical structure of an atom is
    a. electrons orbiting a dense nucleus of protons and neutrons.
    b. electrons and positrons orbiting a microscopic black hole.
    c. protons and electrons orbiting around a nucleus of neutrons.
    d. protons orbiting a low-density nucleus of electrons and neutrons.
    e. neutrons, protons, and electrons all in orbit around each other in an "atomic cloud."

5 The Night Sky

Background Pathways: Units 1 and 2

One of nature's spectacles is the night sky seen from a clear, dark location with the stars scattered across the vault of the heavens. From ancient records we know that the pattern of stars has changed little over the last several thousand years. Thus, the night sky affords us a direct link with our remote ancestors as they tried to understand the nature of the heavens. When you look up at the stars, you might imagine a shepherd in ancient Egypt, a hunter–gatherer on the African plains, a trader sailing along the coast of Persia, or even an airplane navigator in the early twentieth century.

Astronomical observations are part of virtually every culture and include the obvious events that anyone who watches the sky can see without the need of any equipment, such as the setting of the Sun in the western sky and the moving pattern of stars seen in the night sky. Sadly, many of the astronomical phenomena well known to ancient people are not familiar to us today because the bright lights of cities make it hard to see the sky and its rhythms. Therefore, if we are to appreciate the context of astronomical ideas, we need to first understand what our distant ancestors knew and what we ourselves can learn by watching the sky throughout a night.

5.1 THE CELESTIAL SPHERE

From a dark location on a clear night, thousands of stars are visible to the naked eye. The stars appear to be sprinkled on the inside of a huge dome that arches overhead. At any moment as you look at the stars, another half of this dome lies below the **horizon,** hidden by the solid Earth beneath your feet. If we took a spaceship far enough away from the Earth, we would see the whole panorama of stars surrounding us.

Although the stars look as if they form a dome over us, the stars are at vastly different distances from us. For example, the nearest star is about four light-years away, while others we can see are thousands of times more distant. And even the planets, the Sun, and the Moon, which is 100 million times closer than the nearest star, are still so distant that we are unable to get a direct sense of their true three-dimensional arrangement in space without precise measurements and careful deduction.

Nevertheless, for the purposes of studying the patterns of the night sky with our naked eyes, we can treat all stars as if they are at the same distance from us—a distant starscape on the interior of a giant **celestial sphere,** with the Earth at its center, as depicted in Figure 5.1. The celestial sphere is a model (Unit 4.1). It has no physical reality, but it serves as a useful way to visualize the arrangement and motions of celestial bodies.

5.2 CONSTELLATIONS

The patterns formed by brighter stars sometimes suggest the shapes of animals, personages, or objects of cultural relevance. For example, the pattern of stars shown in Figure 5.2A looks a little like a lion, although other cultures saw a dragon

Stars are scattered throughout space in
different directions and at different distances.

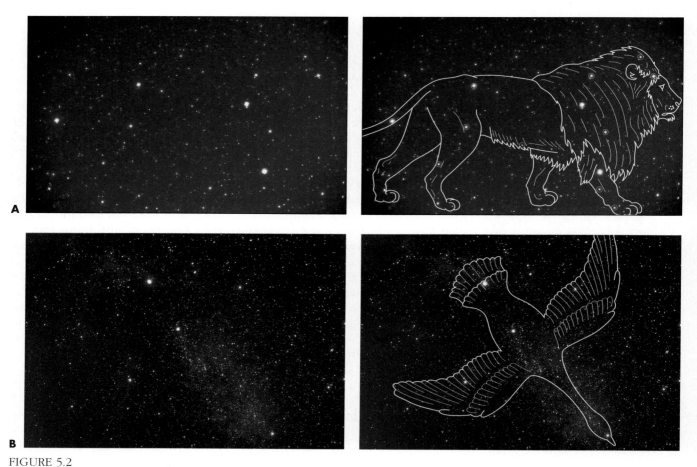

Model

A

Stars *appear*
to all lie at
same distance
on what we
call the
celestial sphere.

Our Experience of the Celestial Sphere.

B

FIGURE 5.1

(A) Although the stars are scattered through space at very different distances, they appear to lie at the same distance from us on what we
call the *celestial sphere*. Note: Sizes and distances are drastically exaggerated. For Earth at the size shown, the nearest star would really be
10,000 km (6000 miles) away and would be about 100 times larger than the Earth. (B) The celestial sphere meets the ground at the horizon.

FIGURE 5.2

The two constellations Leo (A) and Cygnus (B) with figures sketched in to help you visualize the animals they resemble.

In Figure 5.1 and many others throughout the book, the sizes of astronomical bodies are exaggerated compared to the distances between them for clarity.

or a sphinx in this pattern of stars. Today we call this **constellation**—a recognized grouping of stars—Leo. This name has been carried down through the centuries from the Latin word for lion, and the connection with a lion dates back millennia earlier.

Today the International Astronomical Union recognizes 88 official constellations that cover the entire celestial sphere. About half of these have ancient roots, particularly the larger constellations. Some smaller constellations were named more recently by cartographers making maps of the skies and filling in areas of the sky between the better-known constellations. Some groupings of stars remain in common use even though they are not part of the 88 recognized constellations. These "unofficial" groupings of stars are called **asterisms.** For example, the pattern of stars that makes up the head of Leo is sometimes called the "Sickle." Similarly, the "Big Dipper" is an asterism within the constellation Ursa Major (Latin for "large bear").

Many constellations bear no obvious resemblance to their namesakes. It takes some imagination to see a swan in the constellation Cygnus (Figure 5.2B), which is one of the more recognizable constellations. In some cases, factors other than their shape may have played a role. The location of constellations like Ursa Major and Ursa Minor (the large and small bear), always in the northern part of the sky, may have made early sky watchers think of real or legendary bears in northern lands. In some cases the names are connected with legends regarding other constellations, like Canis Major and Canis Minor, the large and small dogs who were hunting companions of Orion—one of the more recognizable constellations. Some areas of the southern sky that are not visible from Europe were given names by early European navigators and explorers. This yielded names like Telescopium, Microscopium, and Antlia, named respectively for the telescope, microscope, and the air pump!

Although most of the constellation names come to us from the classical world of ancient Greece and Rome, many other astronomical terms show the influence of Islam and other Middle Eastern cultures. For example, the **zenith** (the point of the sky that is straight overhead) and the **nadir** (the point on the celestial sphere directly below you and opposite the zenith) come from Arabic words. So do the names of nearly all the bright stars—Aldebaran, Betelgeuse, and Zubeneschamali, to name a few. Many of these star names describe the parts of a constellation; for example, the second brightest star in Leo is called Denebola, from Arabic words meaning "the lion's tail."

Keep in mind that stars in a constellation generally have no physical relation to one another. They simply happen to be in more or less the same direction in the sky. Also, all stars move through space; but as seen from Earth, their positions change very slowly, usually taking tens of thousands of years to make any noticeable shift. Thus we see today virtually the same pattern of stars that was seen by ancient peoples.

Q What point on the Earth is directly opposite where you live? Where is the zenith for someone there?

5.3   DAILY MOTION

If you watch a star near the horizon, in as little as 10 minutes you will notice a shift in its position. If you locate a constellation near your zenith and then come back to look at it after few hours, you will find it has changed position dramatically. Almost everyone has noticed these phenomena for one star—the Sun. Our cycle of day and night and the motion of the Sun across the sky occur because, as the Earth spins, we face different parts of the celestial sphere.

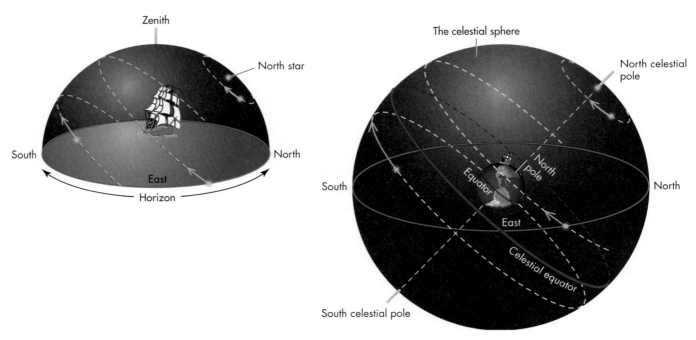

FIGURE 5.3
Stars appear to rise and set as the celestial sphere rotates overhead. Also shown are the celestial equator and poles and where they lie on the celestial sphere with respect to the Earth's equator and poles.

Star rise and star set

From our perspective standing on Earth, it seems as if the celestial sphere rotates around a stationary Earth (Figure 5.3). Objects within the Solar System, including the nearest star to us, the Sun, move more perceptibly on the celestial sphere, generally over the course of days or weeks. These motions are gradual enough that in 24 hours the objects remain at nearly the same position on the celestial sphere.

As with the Sun, the stars, the planets, and the Moon all rise in the east, move across the sky, and set in the west. Ancient peoples had no compelling reasons to believe that the Earth spun, so they attributed all daily motion—that of the Sun, Moon, stars, and planets—to the turning of the vast celestial sphere overhead. Today, of course, we know that it is not the celestial sphere that spins but the Earth; however, we still speak of the Sun "setting."

As the celestial sphere turns overhead, two points on it do not move, as you can see in Figure 5.3. These points are defined as the north and south **celestial poles.** The celestial poles lie exactly above the North and South Poles of the Earth, and just as our planet turns about a line running from its North to South Poles, so the celestial sphere rotates around the celestial poles from our perspective on Earth. The star **Polaris** in the constellation Ursa Minor is very close to the north celestial pole, making it an important navigation aid. Because the north celestial pole lies directly above the Earth's North Pole, it always marks the direction of true north, and it is frequently called the "North Star." There is no comparably bright star anywhere near the south celestial pole, so finding true south from the stars usually requires some triangulation between brighter constellations fairly far from the pole.

Another useful sky marker frequently used by astronomers is the **celestial equator.** The celestial equator lies directly above the Earth's equator, just as the celestial poles lie above the Earth's poles, as Figure 5.3 shows. Stars on the celestial equator rise due east and set due west.

5.4   LATITUDE AND LONGITUDE

The part of the sky we can see depends on where we are located on the Earth. Some parts of the celestial sphere rotate into view during the night, but some parts may remain forever hidden unless we move to a different location on the Earth. Our location is determined by a **longitude,** defining our east–west position, and a **latitude,** defining our north–south position.

We can divide a map of the Earth into a grid by drawing lines of latitude, each at a fixed distance from the equator, and lines of longitude running from pole to pole. The positions of these lines are measured not in a linear unit like kilometers, but instead by an angle in degrees. Ninety degrees (90°) makes up a right angle, and 360° makes up a full circle. Latitudes are measured north and south relative to the equator; for example, Boston, Massachusetts, and Rome, Italy, are both about 42°N, while Buenos Aires, Argentina, and Sydney, Australia, are both about 34°S.

The position east or west has no such natural reference line as the equator. By international agreement, the line of constant longitude that runs through the Royal Observatory in Greenwich, England, is used to mark 0° longitude. Using our sample cities above, Boston is 71°W, Rome is 12°E, Buenos Aires is 58°W, and Sydney is 151°E of this line.

Any two locations at the same latitude will see the same parts of the celestial sphere during a night, so Boston's sky is the same as Rome's. The difference in longitude means they do not see it at the same time. For anyone observing the sky north of the equator, there will be some stars near the north celestial pole that

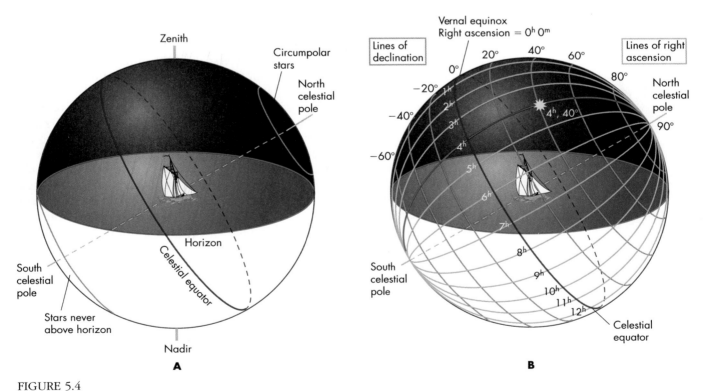

FIGURE 5.4

(A) Some stars rise and set, whereas circumpolar stars near the north celestial pole remain above the horizon continuously. Others near the south celestial pole never rise for a northern observer. (B) Locating a star according to right ascension and declination.

always remain above the horizon, said to be **circumpolar,** and some stars near the south celestial pole that always remain below the horizon (Figure 5.4A). The reverse is true south of the equator.

The most extreme case is at the poles. If you were standing at the North Pole, Polaris, the North Star, would remain always straight overhead. The Earth under your feet would permanently block the view of the southern half of the sky, while all the stars north of the celestial equator would be circumpolar, appearing to move around you parallel to the horizon. Again, the reverse would be true if you were at the South Pole, but there happens to be no stellar equivalent of a "South Star" located almost directly overhead.

The location of the celestial poles is particularly important for navigation by the stars because the celestial pole is at an angle above the horizon equal to your latitude. Observing from the Earth's North Pole, 90°N, the north celestial pole is 90° above the horizon. From the equator, the north celestial pole is just on the horizon—we could say 0° above it. Thus, if you can find the celestial pole, you can determine your latitude.

Q If you were standing on the Earth's equator, where would you look to see the north celestial pole? Could you see this pole from Australia?

Q Can you think of a way to build a simple device to measure the angle of a star above the horizon?

5.5   CELESTIAL COORDINATES

The coordinate grid used by astronomers is similar to the longitude and latitude system used to describe positions on Earth. A star's location in the sky is described by a **right ascension,** defining its east–west position, and a **declination,** defining its north–south position. Right ascension (or **RA** for short) plays the same role as longitude, while declination (or **dec**) plays the same role as latitude.

The celestial sphere can thus be divided into a grid consisting of east–west lines parallel to the celestial equator, and north–south lines connecting one celestial pole to the other (Figure 5.4B). Declination values run from +90° to −90° (at the north and south celestial poles), with 0° at the celestial equator. Right ascension values can likewise be recorded in degrees, but they are commonly listed in "hours." Just as we can divide a circle into 360°, we can divide it into 24 "hours." Each hour of RA equals 15°; that is, 360° ÷ 24 = 15°. The convenience of this system is that if a star at RA = 2^h is overhead now, a star at RA = 5^h will be overhead three hours from now. The right ascension of an object can be further refined to minutes (m) and seconds (s), just as we divide time intervals.

With a set of coordinate lines established, we can now locate astronomical objects in the sky the same way we can locate places on the Earth. Astronomers use star charts for this purpose, much as navigators use maps to find places on Earth. Part of a detailed star chart is shown in Figure 5.5. It shows the location of the constellations, the stars, and other objects. It also gives some indication of the relative brightness of the stars by marking their positions with larger or smaller dots. Many charts also have information about the season and time of night at which the stars are visible. The foldout star chart in the back of the book labels the dates when different stars will be overhead at 8 P.M. in the evening—a common time to be outside viewing the stars.

In Figure 5.5 an oval-shaped object is located at right ascension 0 hours 42.7 minutes ($0^h 42.7^m$), declination +41°. This is a nearby galaxy called M31, located in the Local Group. Because its declination very nearly matches the latitude of Boston and Rome, M31 passes nearly through the zenith, the point directly overhead, as observed from both cities. This holds true anywhere: If a star's declination matches your latitude, it will pass through your zenith.

The location of the chart in Figure 5.5 can be found on the foldout chart at the back of the book. Compare these charts to the photograph of the sky in Looking Up #3: Perseus & M31.

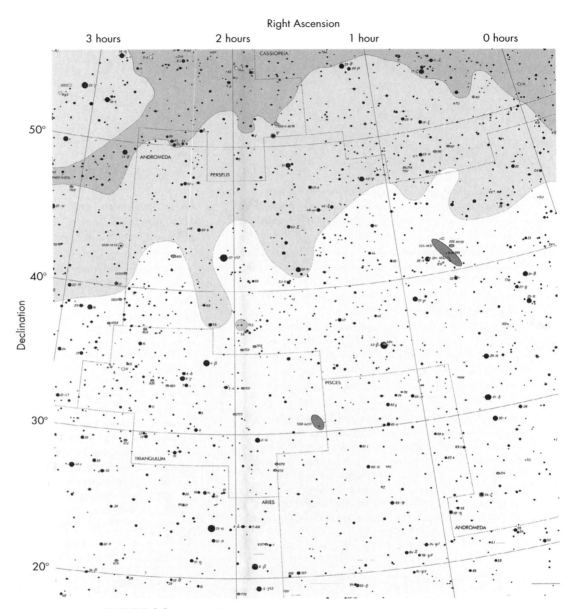

FIGURE 5.5

A modern star chart showing stars, galaxies, and coordinates. Black circles are stars. Their size indicates their brightness—larger circles are brighter stars. Red ellipses are galaxies. The shaded blue area is the Milky Way.

KEY TERMS

asterism, 41	celestial sphere, 39	declination (dec), 44	nadir, 41
celestial equator, 42	circumpolar, 44	horizon, 39	Polaris, 42
celestial pole, 42	constellation, 41	latitude, 43	right ascension (RA), 44
		longitude, 43	zenith, 41

QUESTIONS FOR REVIEW

1. What is the celestial sphere?

2. What is a constellation?

3. What is right ascension?

PROBLEMS

1. What are your latitude and longitude? What are the latitude and longitude at the point on the Earth exactly opposite you (in the direction of your nadir)? What geographic features are located there?

2. From different locations on the Earth, stars will appear to rise at different angles with respect to the horizon. If you were located on the equator, what would be the angle to the horizon of stars rising in the east? What if you were at the North Pole? What about your own latitude?

3. Boston and Rome are at the same latitude, but Boston is at a longitude of 71°W, while Rome is at 12°E. If M31 passes through the zenith at a particular time in Rome, how many hours later will it pass through the zenith in Boston?

4. How far is the celestial equator from your zenith if your latitude is 42°N? 34°S?

5. Using a protractor, draw a diagram of the Earth. Mark the poles, the equator, and your own latitude. With a ruler, find the line that is "tangent" to your location—that is, the line should be parallel to the Earth's surface and just touching this point. This tangent line represents your horizon. Draw a line extending from pole to pole to indicate the Earth's axis, and extend both this line and your horizon line until they cross. Show that your latitude is the same as the angle of the celestial pole with respect to your horizon.

TEST YOURSELF

1. What makes the star Polaris special?
   a. It is the brightest star in the sky.
   b. The Earth's axis points nearly at it.
   c. It sits a few degrees above the northern horizon from any place on Earth.
   d. It is part of one of the most easily recognized constellations in the sky.
   e. All of the above

2. Where must you be located if a star with a declination of 10° passes through your zenith?
   a. The North Pole
   b. A longitude of 10°W
   c. A latitude of 10°N
   d. A longitude of 10°E
   e. A latitude of 10°S

3. If you are standing at the Earth's North Pole, which of the following will be at the zenith?
   a. The celestial equator
   b. The Moon
   c. M31
   d. The north celestial pole
   e. The Sun

PLANETARIUM EXERCISES

All the planetarium exercises use the Starry Night planetarium software provided with your text. You should install the program and familiarize yourself with the various parameters and controls before trying these exercises. When the program asks you to enter a password, simply click "OK" to run Starry Night.

Set the location from which you are viewing to the nearest large city or your latitude and longitude. Accomplish this by clicking on the Location box on the toolbar. Verify that the local time is correct.

1. Set the field to 60 degrees on the left toolbar, the magnitude limits from −2.0 to 5.0, by going under the heading Settings and finding Star Magnitudes. Next set the time step to 30 minutes with the Time option on the toolbar. Set the program so that it displays stars and illustrations for the constellations by using the heading Settings, then Constellations, Illustrations. Set your view to due East. Now step the program forward in 30-minute increments using the Time option in the toolbar. Try the same thing facing due west (Azimuth 270), due south (Azimuth 180), and due north (Azimuth 0). What does this star motion tell you about the rotation of the Earth on its axis? If stars rise in the east and set in the west just like the Sun, is the Earth moving west to east or east to west?

2. Pick a well-known constellation that is easily visible, such as Orion. Center on it using the heading Settings and then using the Find option. Set the time step to half-hour intervals and step ahead a few hours. Find that same constellation in your sky and follow its motion in half-hour intervals just as in the program. Does this motion follow what you envisioned after seeing it in the planetarium program? What seems different?

3. Set the field to 60 degrees on the left toolbar, the magnitude limits from −2.0 to 5.0 under the Settings heading, and the time step to 30 minutes. Look north and center on Polaris. Step the program ahead for six hours. Do the stars near Polaris appear to rise and set? In what direction do they rotate around Polaris during the evening? What does this suggest about the direction in which the Earth rotates on its axis as seen from above? What do you see near the south celestial pole?

6 The Year

For people long ago, observations of the heavens had more than just curiosity value. Because so many astronomical phenomena are cyclic—that is, they repeat at a regular interval—they can serve as timekeepers. The most basic of these cycles is the rhythm of day and night, as the celestial sphere appears to rotate about the Earth as discussed in Unit 5. This cycle is not completely uniform—days and nights alternately lengthen and shorten over the year. This slow rhythm is tied to a gradual shift of the Sun's position in the celestial sphere, which leads to changes in the weather and temperature.

The motion of the Sun in the celestial sphere provides a means for tracking these changes predictably. For example, when is it time to plant crops? Or move to the next location to ensure a ready supply of water? Or prepare for winter? Some of the impetus for studying the heavens probably came from the desire to plan for future events, and it may have motivated early cultures to build monumental stone structures such as Stonehenge (Figure 6.1).

Background Pathways: Unit 5

A

B

FIGURE 6.1
(A) Stonehenge, a stone monument built by the ancient Britons on Salisbury Plain, England. Its orientation marks the seasonal rising and setting points of the Sun. (B) Sketch showing how the stones are aligned with respect to the solstice.

6.1   ANNUAL MOTION OF THE SUN

As the Earth orbits the Sun, the stars that are visible each night change. The shift is so slow that it is difficult to appreciate from one night to the next, but in the span of a month, the changes are obvious. Many ancient peoples built monuments to keep track of these motions. Because these movements repeat yearly, they are called *annual motions.*

If you watch the sky each evening over several months, you will discover that new constellations appear in the eastern sky and old ones disappear from the western sky. Across most of North America, Europe, and Asia on an early July evening, the constellation Scorpius will be visible in the southern half of the sky. However, on December evenings, the brilliant constellation Orion, the hunter, is visible instead.

The realization that different stars are visible at different times of the year was extremely important to early peoples because it provided a way to measure the passage of time other than by maintaining accurate records over several hundred days. For example, if you live in the Northern Hemisphere and the evening sky shows Leo in the south instead of Scorpius or Orion, it will soon be time to plant. Even where the seasons show strong temperature variations, farmers may be tricked into planting too early by a short spell of warmer-than-normal weather in the spring, so they are better off waiting for a date that is reliably later than any past experience of frost. Even in semitropical climates, where seasonal temperature differences are much smaller, planting with accuracy is also necessary to avoid crop damage due to annual flooding or dry periods.

ANIMATIONS

Constellations by season

The changing of constellations throughout the year is caused by the Earth's motion around the Sun. As the Earth **revolves** (orbits) around the Sun, the Sun's glare blocks our view of the part of the celestial sphere that lies in the direction of the Sun, making the stars that lie beyond the Sun invisible, as Figure 6.2 shows. For example, in early June, a line from the Earth to the Sun points toward the constellation Taurus, and its stars are completely lost in the Sun's glare. In the dusk after sunset, however, it is possible to see the neighboring constellation, Gemini, just above the western horizon. A month later, from the Earth's new position, the Sun lies in the direction of Gemini, causing this constellation to disappear in the Sun's glare. Looking to the west just after sunset, it is possible to see, just barely, the dim stars of the constellation Cancer above the horizon. A month after that, the Sun is in Cancer, and the constellation Leo is visible just above the horizon after sunset.

Month by month, the Sun hides one constellation after another. It is like sitting around a campfire and not being able to see the faces of the people on the far side. But if we get up and walk around the fire, we can see faces that were previously hidden. Similarly, the Earth's motion allows us to see stars previously hidden in the Sun's glare.

6.2   THE ECLIPTIC AND THE ZODIAC

The name *ecliptic* arises because only when the new or full moon crosses this line can an eclipse occur. (See Unit 8.)

If we could mark on the celestial sphere the path traced by the Sun as it moves through the constellations, we would see a line that runs around the celestial sphere, as illustrated in Figure 6.3. Astronomers call the line that the Sun traces across the celestial sphere the **ecliptic.** If you look at Figure 6.3, you can see that the ecliptic is the extension of the Earth's orbit onto the celestial sphere, just as the celestial equator is the extension of the Earth's equator onto the celestial sphere.

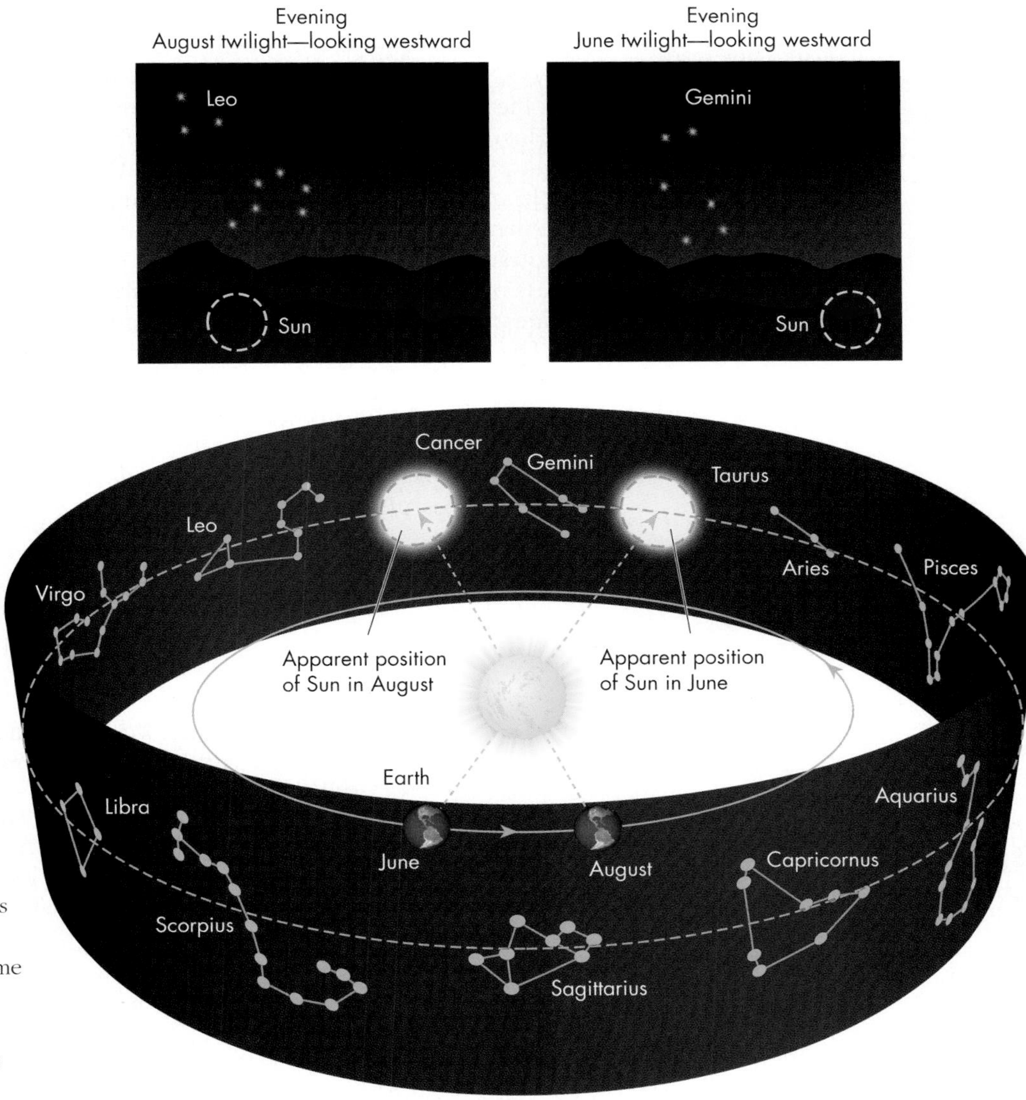

Evening
August twilight—looking westward

Leo

Sun

Evening
June twilight—looking westward

Gemini

Sun

Cancer
Gemini
Taurus
Leo
Virgo
Aries
Pisces

Apparent position
of Sun in August

Apparent position
of Sun in June

Earth

June   August

Libra

Aquarius

Capricornus

Scorpius

Sagittarius

FIGURE 6.2
The Sun hides from our view stars that lie beyond it. As we move around the Sun, those stars become visible, and the ones previously seen are hidden. Thus, the constellations that we see change with the seasons.

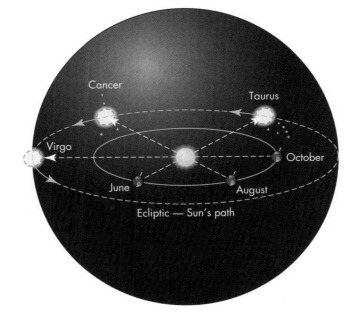

Cancer

Taurus

Virgo

October

June

August

Ecliptic — Sun's path

FIGURE 6.3
As the Earth orbits the Sun, the Sun appears to move around the celestial sphere through the background stars. The Sun's path is called the *ecliptic*. The Sun appears to lie in Taurus in June, in Cancer during August, in Virgo during October, and so forth. Note that the ecliptic is the extension of the Earth's orbital plane out to the celestial sphere. (Sizes and distances of objects are not to scale.)

Q Can you think of an astronomical reason why the zodiac may have been divided into 12 signs rather than 8 or 16?

The ecliptic passes through a dozen constellations, which are collectively called the **zodiac.** The word *zodiac* comes from Greek roots meaning animals (as in **zo**ology) and circle (as in **dia**meter). That is, zodiac refers to a circle of animals, which for the most part its constellations represent: Aries (ram), Taurus (bull), Gemini (twins), Cancer (crab), Leo (lion), Virgo (virgin), Libra (scale), Scorpius (scorpion), Sagittarius (archer), Capricornus (goat), Aquarius (water bearer), and Pisces (fish). Actually, the ecliptic passes through a thirteenth constellation, Ophiuchus (serpent holder), during the first half of December (between Scorpius and Sagittarius); but this constellation was not included in the zodiac, probably because of some uncertainty in ancient times about the precise path of the Sun and some vagueness about the boundaries of the constellations.

The names of some of the constellations of the zodiac may have originated in the seasons when the Sun passed through them. For example, rainy weather in much of Europe during winter was foretold by the Sun's appearance in the constellation Aquarius (the water bearer). Likewise, the harvest time was indicated by the Sun's appearance in Virgo (the virgin), a constellation often depicted as the goddess Proserpine, holding a sheaf of grain.

6.3   THE SEASONS

INTERACTIVE

Seasons

Q When it is summer in the Northern Hemisphere, what is the season in the Southern Hemisphere? What does this demonstrate about possible causes of the seasons?

ANIMATIONS

Earth's rotation axis

The constancy of Earth's tilt is a consequence of the *conservation of angular momentum*. (See Unit 20.)

Many people believe that we have seasons because the Earth's distance from the Sun changes. They assume that summer occurs when we are closest to the Sun and winter when we are farthest away. It turns out, however, that the Earth is several million kilometers closer to the Sun in early January, when the Northern Hemisphere is coldest, than it is in July. Thus, seasons must have some other cause.

To see what causes our seasons, we need to look at how our planet is oriented in space. As the Earth orbits the Sun, our planet also spins or **rotates.** That spin is around a line—the **rotation axis**—that we might imagine running through the Earth from its North Pole to its South Pole. The Earth's rotation axis is not perpendicular to its orbit around the Sun. Rather, it is tipped by 23.5° from the vertical, as shown in Figure 6.4A.

As our planet moves along its orbit, its rotation axis maintains nearly exactly the same tilt and direction, as Figure 6.4B shows. That is, the Earth behaves much like a giant spinning top. In fact, this tendency to maintain its orientation is shown by every spinning object. It is what keeps a rolling coin from tipping over and it is why a quarterback puts "spin" on a football.

The constancy of our planet's tilt as we move around the Sun causes sunlight to fall more directly on the Northern Hemisphere for half of the year and more directly on the Southern Hemisphere for the other half of the year, as illustrated in Figure 6.5. This in turn changes the amount of heat each hemisphere receives from the Sun.

A surface directly facing the Sun receives the most concentrated sunlight (Figure 6.6). If the surface receives the sunlight at an angle, the light is spread out over a larger area and therefore is less concentrated. An astronomer might express this in terms of the energy received per square meter. A portion of the Earth directly facing the Sun receives about 1300 watts on every square meter. Where the surface is tilted at an angle to the sun's light, the same 1300 watts are spread out over a larger area on the ground, and each square meter of the Earth's surface receives only a fraction as much energy. You take advantage of this effect instinctively when you warm your hands at a fire by holding your palms flat toward the fire. You also may have experienced the high temperature of pavement around noon, when the Sun is shining most directly on it, whereas the same surface will be cooler in the

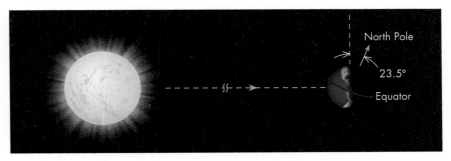

A

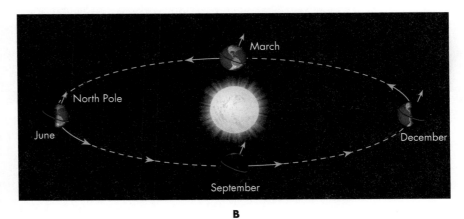

B

FIGURE 6.4

(A) The Earth's rotation axis is tilted 23.5° to its orbit around the Sun. (B) The Earth's rotation axis keeps nearly the same tilt and direction as it revolves (orbits) around the Sun. (Sizes and distances are not to scale.)

FIGURE 6.5

Because the Earth's rotation axis keeps the same tilt as we orbit the Sun, sunlight falls more directly on the Northern Hemisphere during part of the year and on the Southern Hemisphere during the other part of the year. (Sizes and distances are not to scale.)

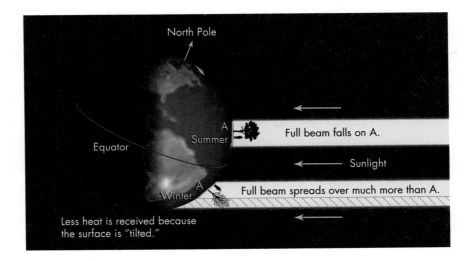

FIGURE 6.6

A flat surface directly facing the Sun receives more light (and thus more heat) than a tilted surface.

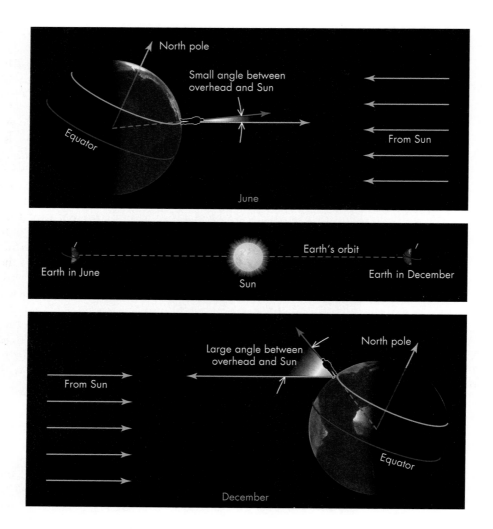

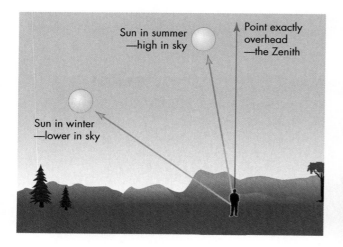

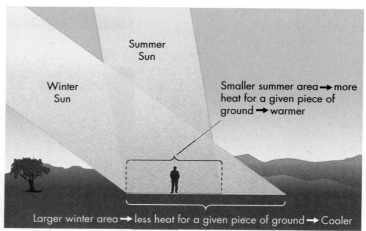

The summer and winter beams carry the same amount of energy,
but spread that energy over very different amounts of ground.

FIGURE 6.7
Why the Sun at noon is high in the sky in summer and low in the sky in winter.

late afternoon when sunlight strikes it more obliquely, even though the pavement is not shaded.

As the Earth orbits the Sun, the same region will face the Sun either more—or less—directly depending on the time of year. For example, a region in the Northern Hemisphere receives sunlight most directly in late June and most obliquely in late December, as illustrated in Figure 6.7. The direct sunlight produces the strongest heating, whereas the large angle in December produces the least. This heating difference is enhanced because the Earth's tilt also leads to more hours of daylight in the spring and summer than in the fall and winter (Unit 7.2). As a result, we not only receive the Sun's light more directly, we receive it for a longer time. Thus the seasons are caused by the tilt of the Earth's rotation axis. From Figure 6.7 you can also see that this makes the seasons reversed between the Northern and Southern Hemispheres; when it is summer in one, it is winter in the other.

An important point here is not to confuse the "directness" of the Sun's light with one hemisphere being closer to the Sun. It is true that the Northern Hemisphere of the Earth is a few thousand kilometers closer to the Sun than the Southern Hemisphere during the northern summer. However, the effect of this difference in distance is tiny. Compared to the millions of kilometers of distance to the Sun, this difference in distance between the two hemispheres does not produce a change of even one-tenth of a degree in the temperature. By contrast, the differing angle at which the Sun shines on higher latitudes during the year changes the solar energy absorbed by the ground by a factor of two or more. The same size bundle of sunlight is spread out over a much larger area in winter, as shown in the last panel of Figure 6.7; this is the source of seasonal heating changes.

6.4 THE ECLIPTIC'S TILT

ANIMATIONS

The Sun's seasonal motion

The tilt of the Earth's rotation axis not only causes seasons, it makes the ecliptic tilted with respect to the celestial equator. The Sun lies north of the celestial equator for half of the year and south of the celestial equator for the other half of the year. As you can see in Figure 6.5, in June the Northern Hemisphere is tipped toward the Sun. As a result, the Sun lies north of the celestial equator. But in December, when the Earth is on the other side of the Sun, the Sun lies south of the celestial equator.

The consequence of such north–south motion is that the Sun's path—the ecliptic—must cross the celestial equator, and therefore the ecliptic must be tilted with respect to that line, as the sequence of sketches in Figure 6.8 shows. This motion of the Sun north and south in the sky during the year is also why at noon the Sun is so high in the summer sky and so low in the winter sky. For example, at mid-northern latitudes (40°), the noon Sun is about 74° above the horizon in June, but it is only about 27° above the horizon in December.

6.5 SOLSTICES AND EQUINOXES

The tilt of the ecliptic with respect to the celestial equator means that during the year, the points on the horizon where we see the Sun rise and set are not due east and west, except when the Sun is crossing the celestial equator. This occurs on two days of the year called the **equinoxes,** from the Latin for "equal night," so named because the length of the night is approximately equal to the length of the day on those dates. The spring or **vernal equinox** occurs near March 21, when the Sun is

Earth's position in its orbit
at different times of year

Sun's position on celestial
sphere at different seasons

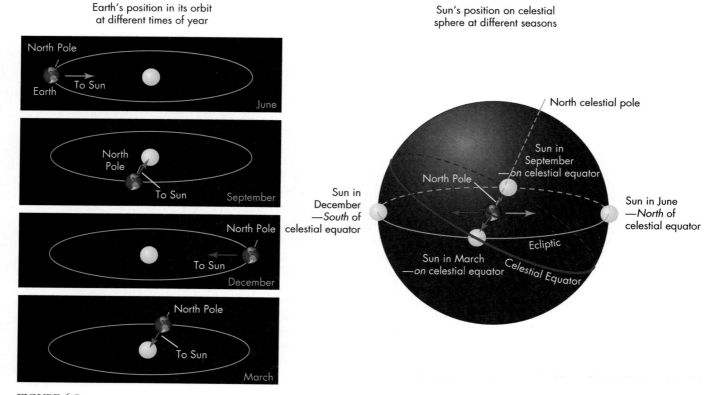

FIGURE 6.8

As the Earth orbits the Sun, the Sun's position with respect to the celestial equator changes. The Sun lies *north* of the celestial equator in June but *south* of the celestial equator in December. The Sun crosses the celestial equator in March and September. The times of these crossing define the equinoxes.

The exact dates of the equinoxes and solstices vary slightly from year to year, mainly because of differences in the calendar due to leap years, but also because of slight variations in the Earth's orbit.

The Sun's position in the celestial sphere is shown in star charts by the curving line of the ecliptic—see the foldout star chart in the back of the book.

moving from the Southern Hemisphere of the celestial sphere into the Northern Hemisphere. Six months later the **autumnal equinox** occurs near September 23 as the Sun crosses the celestial equator on its way south.

On every other day of the year the Sun rises either to the north or south of due east in a regular, predictable fashion as illustrated in Figure 6.9A. From the vernal equinox to the autumnal equinox (during the Northern Hemisphere spring and summer, and the Southern Hemisphere fall and winter) the Sun rises in the northeast and sets in the northwest. During the rest of the year, the Sun rises in the southeast and sets in the southwest.

Midway between the equinoxes, the Sun reaches its furthest point north or south on the celestial sphere, 23.5° from the celestial equator. At these times of year, the Sun pauses in its steady north–south motion and changes direction. Accordingly, these times are called the **solstices,** meaning the Sun (*sol*) stops its northward or southward motion and begins to reverse direction. The dates of the solstices are close to June 21 and December 21.

In terms of celestial coordinates (Unit 5.5), the Sun is at a declination of 0° on the equinoxes, while it is at plus or minus 23.5° on June 21 and December 21 respectively. The right ascension of the Sun is, by definition, zero on the vernal equinox, around March 21. The Sun's position at the moment it crosses the celestial equator on this date is used to define 0° (or 0 hours) for the right ascension system. On June 21 the Sun moves to a right ascension of about 6 hours, then 12 hours on September 23, then 18 hours on December 21, before returning to 0 hours of right ascension a year later (see Figure 6.8).

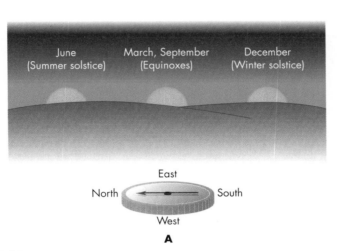

June
(Summer solstice)

March, September
(Equinoxes)

December
(Winter solstice)

East

North ← → South

West

A

B

FIGURE 6.9

(A) The direction of the rising and setting Sun changes throughout the year. At the equinoxes, the rising and setting points are due east and west. The sunrise direction shifts slowly north from March until the summer solstice, after which it shifts back, reaching due east at the autumn equinox. The sunrise direction continues moving south until the winter solstice, then reverses direction again back to the north. The sunset point similarly shifts north from approximately December 21 until June 21 and south from June 21 until December 21.
(B) Sunrise on the summer solstice at Stonehenge.

	TABLE 6.1		Monthly Average Temperatures in Four Cities (in Degrees Celsius)										
City	Latitude	Jan.	Feb.	Mar.	April	May	June	July	Aug.	Sept.	Oct.	Nov.	Dec.
Buenos Aires	34°S	23.5	22.7	20.6	16.7	13.3	10.4	10.0	11.1	13.2	16.0	19.3	22.0
Boston	42°N	−2.2	−1.6	2.5	8.2	14.1	19.4	22.5	21.5	17.3	11.5	5.5	0.0
Rome	42°N	7.1	8.2	10.5	13.7	17.8	21.7	24.4	24.1	20.9	16.5	11.7	8.3
Sydney	34°S	22.1	22.1	21.0	18.4	15.3	12.9	12.0	13.2	15.3	17.7	19.5	21.2

The seasonal shifts of the Sun also define three kinds of regions on the Earth: the tropics, the polar regions, and the temperate latitudes, which lie in between. The tropics lie between latitudes 23.5°S and 23.5°N, where the Sun passes directly overhead at some time during the year. The polar regions mark the latitudes where the Sun does not rise during some portion of the year. This occurs within 23.5° of the poles—north of 66.5°N, the **Arctic Circle,** and south of 66.5°S, the **Antarctic Circle.** The northern limit of tropical latitudes, 23.5°N, is called the **Tropic of Cancer** because the Sun is in the zodiac constellation Cancer when it is overhead at this latitude on the summer solstice. The southern limit is the called the **Tropic of Capricorn,** the zodiac constellation where the Sun is at the winter solstice.

The seasons "officially" begin on the solstices and equinoxes, with northern spring running from the vernal equinox to the solstice in June. Even though the longest day is on the first day of summer, the hottest period of the year occurs roughly six weeks later, as shown for four cities in Table 6.1. The delay, known as the *lag of the seasons,* results from the oceans and land being slow to warm up in summer. Similarly, there is about a six-week lag after the shortest day of the year until the coldest period of the year.

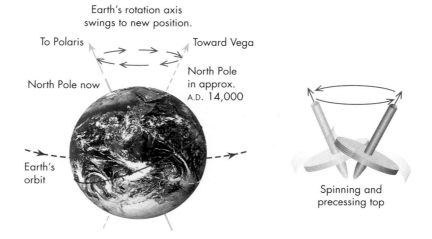

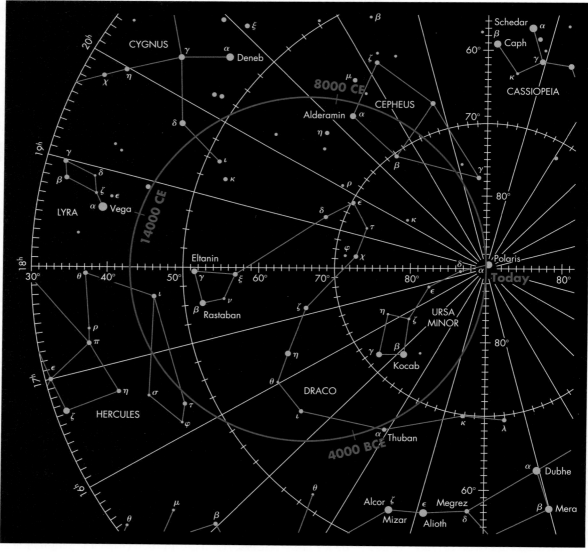

FIGURE 6.10
Precession makes the Earth's rotation axis swing slowly in a circle.

Q Why does the position of sunrise along the eastern horizon change during the year? Is this shift the same at every latitude?

Just as the changing position of the Sun against the constellations could be used as an indicator of the seasons to ancient peoples, so too could the northward and southward journeys of the Sun. One well-known example of a structure to mark these journeys is Stonehenge, the ancient stone circle in England, shown in Figure 6.1. Although its exact use in ancient times is lost to us, it appears that it was laid out so that seasonal changes in the Sun's position could be observed by noting through which stone arches the Sun rose or set. For example, on the summer solstice an observer standing at the center of this immense circle of standing stones would see the rising Sun framed by an arch (Figure 6.9B). A variety of structures were built by cultures all over the world to detect the extreme limits of the Sun's motion in the sky because of the importance of knowing when the changes of seasons occur.

6.6   PRECESSION

If you watch a spinning top, you will see that it "wobbles," often more and more extremely as it slows down. That it wobbles is another way of saying that its rotation axis slowly shifts direction (Figure 6.10). The spinning Earth wobbles too in a motion called **precession.** Precession occurs very slowly for the Earth. A single "wobble" takes about 26,000 years, but it has both interesting and important consequences.

At this time our planet's northern rotation axis points at the star Polaris. But this is only temporary. When the Egyptian pyramids were built 4000 years ago, the "North Star" was Thuban (meaning "the star") in the constellation Draco (see Looking Up #1: Northern circumpolar constellations). In the future the north axis will continue shifting past Polaris, and it will not point near any bright stars for thousands of years. In about 7000 years the south celestial pole will be very close to a star slightly brighter than Polaris in the constellation Vela. In that future time there will be a "South Star." In 12,000 years the rotation axis will have shifted so that the north celestial pole points fairly close to the bright star Vega. Then we will have a new, much brighter "North Star."

Precession also slowly alters Earth's climate. At this time we are closest to the Sun during the northern winter. In 13,000 years we will be farthest from the Sun during the northern winter. This will make seasons in the Northern Hemisphere more severe at that time. Precession is suspected to be one of the important components that affect long-term changes in climate, which may have triggered past ice ages.

KEY TERMS

Antarctic Circle, 55
Arctic Circle, 55
autumnal equinox, 54
ecliptic, 48
equinox, 53
precession, 57
revolve, 48

rotate, 50
rotation axis, 50
solstice, 54
Tropic of Cancer, 55
Tropic of Capricorn, 55
vernal equinox, 53
zodiac, 50

QUESTIONS FOR REVIEW

1. What is the ecliptic?
2. What is the zodiac?
3. What causes the seasons?
4. When it is winter in Australia, what season is it in the United States?

PROBLEMS

1. If the shape of the Earth's orbit was unaltered but its rotation axis was shifted so that it had no tilt with respect to the orbit, how would seasons be affected?

2. Suppose the Earth's axis was tilted by 10° instead of 23.5°. Where would the tropics and arctic regions be? How would seasons be different?

3. Suppose the Earth's axis was tilted by 50° instead of 23.5°. Where would the tropics and arctic regions be? How would seasons be different?

4. If you were observing the Sun from the Arctic Circle, at a latitude of 66.5°N, describe the motion you would see on the solstices and the equinoxes.

TEST YOURSELF

1. From what location on Earth will the Sun always rise due east and set due west?
   a. The North Pole
   b. The South Pole
   c. The equator
   d. A latitude of 23.5°N
   e. Nowhere

2. On what day(s) of the year are nights longest at the equator?
   a. They are the same length throughout the year there.
   b. The solstices
   c. The equinoxes
   d. Approximately June 21
   e. Approximately December 21

3. During winter in either hemisphere the temperature is lower because the Sun
   a. stops moving.
   b. is furthest south.
   c. doesn't rise as high in the sky.
   d. has a lower temperature.
   e. is farther away due to the Earth's eccentric orbit.

PLANETARIUM EXERCISES

All the planetarium exercises use the Starry Night planetarium software provided with your text. You should install the program and familiarize yourself with the various parameters and controls before trying these exercises. When the program asks you to enter a password, simply click "OK" to run Starry Night.

Set the location from which you are viewing to the nearest large city or your latitude and longitude. Accomplish this by clicking on the Location box on the toolbar. Verify that the local time is correct.

1. Set the field to 120 degrees on the left toolbar, and "turn off" the stars by clicking on the star icon on the top toolbar. Look south (north if you live in the Southern Hemisphere) and set the time step to 1 hour. Set the program to run forward. In what direction does the Sun move across the sky? Stop the program with the Sun in the south. Change the time step to 24 hours. Note the time and date, and step forward. Which way does the Sun move? Run the program forward to see when the Sun changes direction—what date is this?

2. Set up the program as in the previous problem, adjust the time for the current date, and look toward the western horizon. Adjust the time until the Sun is setting. Note this time and location of the Sun on the horizon. Move the date forward a week. When is the Sun setting? Where is it on the horizon? Try a month in the future and find the sunset again. Look for the next date that the Sun starts to move its sunset position on the horizon in the other direction. When is that, and what is its direction?

7

The Time of Day

From before recorded history, people have used events in the heavens to mark the passage of time. The day was the time interval from sunrise to sunrise, and the time of day could be determined from how high the Sun was in the sky. As our ability to measure time independently has become more accurate, we find that the apparent motions of the Sun across the sky are not as uniform as we once thought, and in this age of rapid travel and communications it no longer makes sense for each town to set its own time according to the Sun. With high-precision modern clocks we have even detected a gradual slowing of the Earth's spin!

Each of these adjustments to our understanding of how to keep time provides an insight into the workings of astronomy.

Background Pathways: Unit 6

7.1 THE DAY

The length of the day is set by the Earth's rotation speed on its axis. One day is defined to be one rotation. However, we must be careful how we measure our planet's rotation. For example, we might use the time from one sunrise to the next to define a day. That, after all, is what sets the day–night cycle around which we structure our activities. However, we would soon discover that the time from sunrise to sunrise changes steadily throughout the year as a result of the seasonal change in the number of daylight hours. A better time marker might be the time it takes the Sun to move from its highest point in the sky on one day (what we technically call **apparent noon**) to its highest point in the sky on the next day—a time interval that we call the **solar day.**

We often divide a day into "**A.M.**" and "**P.M.,**" which stand for **ante meridian** and **post meridian,** respectively. The **meridian** is a line that divides the eastern and western halves of the sky. The meridian extends from the point on the horizon due north to the point due south and passes directly through the zenith, the point exactly overhead. As the Sun moves across the sky (Figure 7.1), it crosses the meridian at the time called *apparent noon.* Before (*ante*) noon is thus A.M., while after (*post*) noon is P.M.

Apparent solar time is what a sundial measures, and during the year this time may be ahead or behind clock time by up to about a quarter of an hour. This variation arises from the Earth's orbital characteristics (Unit 13). Thus, although the Sun's motion across the sky determines the day–night cycle, the Sun's motion does *not* make a stable reference for measuring Earth's spin.

We can avoid most of the variation in the day's length if, instead of using the Sun, we use a star as our reference. For example, if we pick a star that crosses our meridian at a given moment and measure the time it takes for that same star to return to the meridian again, we will find that this time interval repeats itself precisely. However, this interval is not 24 hours, but about 23 hours, 56 minutes, and 4.0905 seconds. This day length, measured with respect to the stars, is called a **sidereal day.**

Astronomers find that a clock set to run at this speed is extremely useful. It is not just that the sidereal day is much more stable. Another reason is that, at a given

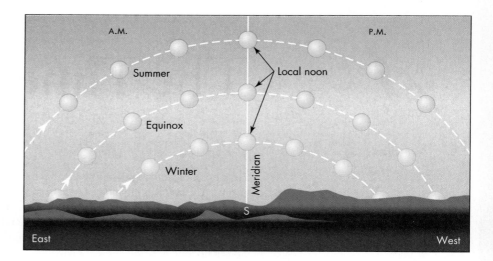

FIGURE 7.1

The Sun rises in the east, crosses the meridian at local noon, then sets in the west. This figure depicts the path of the Sun seen from the Northern Hemisphere at the equinoxes and summer and winter solstices.

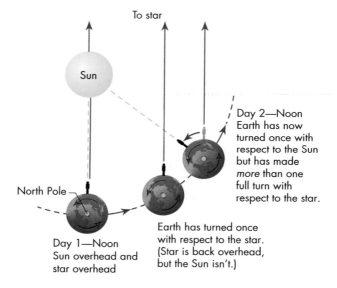

FIGURE 7.2

The length of the day measured with respect to the stars is not the same as the length measured with respect to the Sun. The Earth's orbital motion around the Sun makes it necessary for the Earth to rotate a tiny bit more before the Sun will be back overhead. (Motion is exaggerated for clarity.)

location, any particular star will always rise at the same sidereal time. To avoid the nuisance of A.M. and P.M., sidereal time is measured on a 24-hour basis. For example, the bright star Procyon in the constellation Canis Minor ("small dog") rises at about 10 P.M. in November but at about 8 P.M. in December and 6 P.M. in January by solar time. However, on a clock keeping sidereal time, it always rises at the same time at a given location: about 01:30 by the sidereal clock.

Why is the sidereal day shorter than the solar day? We can see the reason by looking at Figure 7.2, where we measure the interval between successive apparent noons—a solar day. Let us imagine that at the same time we are watching the Sun, we can also watch a star and that we measure the time interval between its passages across the meridian—a sidereal day.

As we wait for the Sun and star to move back across the meridian, the Earth moves along its orbit. The distance the Earth moves in one day is so small compared with the star's distance that we see the star in essentially the same direction as on the previous day. However, we see the Sun in a measurably different direction, as Figure 7.2 shows. The Earth must therefore rotate a bit more before the Sun is again on the meridian. That extra rotation, needed to compensate for the Earth's orbital motion, makes the solar day slightly longer than the sidereal day.

It is easy to figure out how much longer on the average the solar day must be. Because it takes us 365¼ days to orbit the Sun and because there are by definition 360 degrees in a circle, the Earth moves approximately 1 degree per day in its orbit around the Sun. That means that for the Sun to reach its noon position, the Earth must rotate approximately 1 degree past its position on the previous day. Another way of thinking about this is that the Sun is slowly moving eastward across the sky through the stars at the same time the Earth is rotating. Thus, in a given "day," the Earth must rotate a bit more to keep pace with the Sun than it would to keep pace with the stars.

In 24 hours there are 1440 (24 × 60) minutes, and during this time the Earth rotates 360 degrees. Therefore, for the Earth to rotate about 1 degree extra so that the same side is facing the Sun again, takes about 1440/360 = 4 minutes. The solar day is therefore about 4 minutes longer than the sidereal day.

7.2   LENGTH OF DAYLIGHT HOURS

ANIMATIONS
Hours of daylight

Although each day lasts 24 hours, the number of hours of daylight, or the amount of time the Sun is above the horizon, changes dramatically throughout the year unless you are close to the equator. For example, in northern middle latitudes, including most of the United States, southern Canada, and Europe, summer has about 15 hours of daylight and only 9 hours of night. In the winter, the reverse is true. Just north of the **Arctic Circle** at a latitude of 66.5° (90° − 23.5°), the Sun remains above the horizon for 24 hours on the summer solstice and below the horizon the entire day on the winter solstice. On the equator, day and night are each 12 hours every day.

This variation in the number of daylight hours is caused by the Earth's tilted rotation axis. Remember that as the Earth moves around the Sun, its rotation axis points in very nearly a fixed direction in space. Thus the Sun shines more directly on the Northern Hemisphere during its summer and less directly during its winter. The result (as you can see in Figure 7.3) is that a large fraction of the Northern Hemisphere is illuminated by sunlight at any time in the summer, but a small fraction is illuminated in the winter. So as rotation carries us around the Earth's axis, only a relatively few hours of a summer day are unlit, but a relatively large number of winter hours are dark. Figure 7.3 also shows that on the first days of spring and autumn (the equinoxes), the hemispheres are equally lit, so that day and night are of equal length everywhere on Earth.

If we change our perspective and look out from the Earth, we see that during the summer the Sun's path is high in the sky, so that the Sun spends a larger portion of the day above the horizon. This gives us not only more heat but also more hours of daylight. On the other hand, in winter the Sun's path across the sky is much shorter, giving us less heat and fewer hours of light.

7.3   TIME ZONES

Because the Sun is our basic timekeeping reference, most people like to measure time so that the Sun is highest in the sky at about noon. Technically this is unnecessary now that we have good electronic clocks that can keep time independent of the Sun. Nevertheless, it is a tradition that is hard to break, and as a result, clocks in

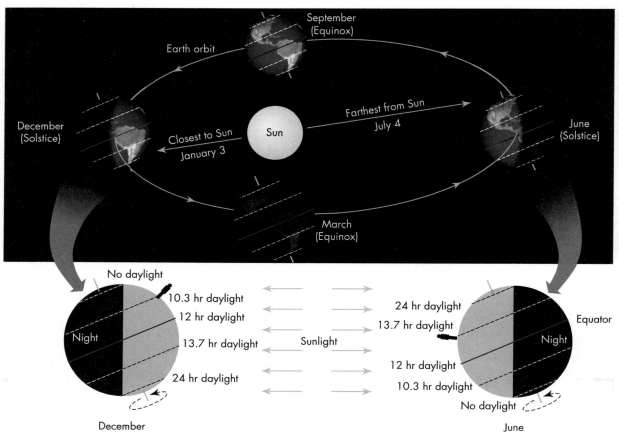

FIGURE 7.3

The tilt of the Earth affects the number of daylight hours. Locations near the equator always receive about 12 hours of daylight, but locations toward the poles have more hours of darkness in winter than in summer. In fact, above latitudes 66.5°, the Sun never sets for part of the year and never rises for another part of the year (the *midnight sun* phenomena). At the equinoxes, all parts of the Earth receive the same number of hours of light and dark. (Sizes and separation of the Earth and Sun are not to scale.)

different parts of the world are set to different times so that the local clock time approximately reflects the position of the Sun in the sky. Because the Earth is round, the Sun can't be "overhead" everywhere at the same time, so it can't be noon everywhere at the same time.

By the late 1800s, with the increasing speed of travel and communications, it became confusing for each city to maintain its own time according to the position of the Sun in the sky. By international agreement, the Earth was therefore divided into 24 major **time zones,** centered every 15° of longitude, in which the time differs by one hour from one zone to the next. With this system, clocks in a time zone all read the same, and they are at most a half hour ahead or behind what they would be if the time was measured locally. Many regions use local geographic or political borders to define the boundaries between time zones rather than strictly following the longitude limits. Authorities in a few countries and regions did not adopt this agreement, choosing instead to maintain a time standard that was closer to local time. For example, Newfoundland, India, Nepal, and portions of Australia are offset by 30- or 15-minute differences from the international standard.

Across the lower 48 United States, the time zones are, from east to west, Eastern, Central, Mountain, and Pacific (see Figure 7.4). The time within each zone is the

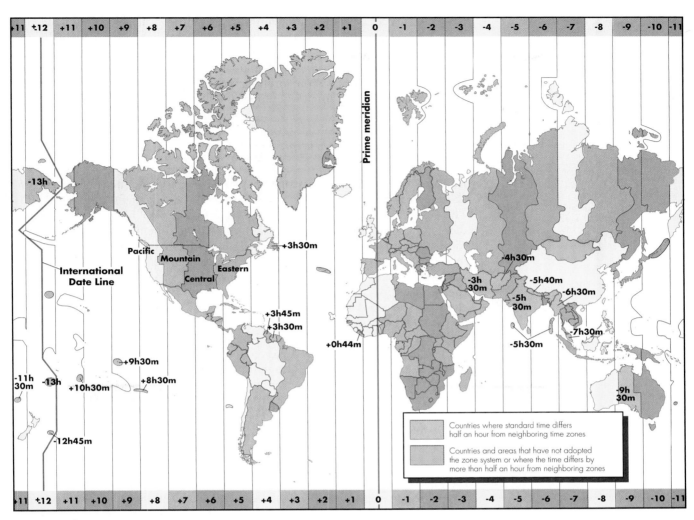

FIGURE 7.4
Time zones of the world and the international date line. Local time = universal time − numbers on top or bottom of chart.

Q In the story *Around the World in 80 Days* by Jules Verne, the travelers discover that although they have experienced 80 days and nights, they still have one more day before 80 days have passed in London. How can this be?

same and is called **standard time.** Thus in the eastern zone the time is denoted Eastern Standard Time (EST), whereas in the central zone it is denoted Central Standard Time (CST). As you travel across the country it is necessary to reset your watch if you cross from one time zone to another, adding one hour for each time zone as you move from west to east and subtracting one hour when you move from east to west.

If you travel through many time zones, you may need to make such a large time correction that you shift your watch past midnight. For example, if you could travel westward quickly enough that little time elapsed, setting your watch back each time you crossed a time zone, you could end up at your starting point with your watch turned backward by 24 hours. But you would not have traveled back in time!

A traveler does not actually "gain" a day: when you cross longitude 180° (roughly down the middle of the Pacific Ocean), you add a day to the calendar if you are traveling west and subtract a day if you are traveling east. For example, you could celebrate the New Year in Japan and take a flight after midnight to Hawaii, where it would still be the day before, so you could celebrate the New Year that night too! The precise location where the day shifts is called the **international date line** (Figure 7.4). It generally follows 180° longitude but bends around extreme eastern Siberia and some island groups to ensure they keep the same calendar time as their neighbors.

The nuisance of having different times at different locations can be avoided by using **universal time,** abbreviated as UT. Universal time is the time kept in the time zone containing the longitude zero, which passes through Greenwich, England. By using UT, which is based on a 24-hour system to avoid confusion between A.M. and P.M., two people at remote locations can decide to do something at the same time without worrying about what time zones they are in.

7.4   DAYLIGHT SAVING TIME

In many parts of the world, people set clocks ahead of standard time during the summer months and then back again to standard time during the winter months. This has the effect of shifting sunrise and sunset to later hours during the day, thereby creating more hours of daylight during the time most people are awake. Time kept in this fashion is called **daylight saving time** in the United States. In some other parts of the world it is called "Summer Time."

Daylight saving time was originally established during World War I as a way to save energy. By setting clocks ahead, less artificial light was needed in the evening hours, In effect, it is a method to get everyone to wake up and go to work an hour earlier than they would normally and take advantage of the earlier rising of the Sun. Clocks are not set ahead permanently because, in some parts of the country, winter sunrise is so late that daylight saving time would require people to wake and go to work in the dark. Starting in 2007, daylight savings time in the U.S. runs from the second Sunday of March to the first Sunday in November. In Europe it runs from the last Sunday in March to the last Sunday in October, whereas in Australia it runs in exact reverse. Many other countries, and even some states within the United States, follow different rules, while most tropical countries keep their clocks fixed on standard time.

Q If daylight saving time saves energy, why not just have people come to work earlier when the Sun rises earlier and later when it rises later?

7.5   LEAP SECONDS

Highly accurate time measurements have now allowed us to detect a gradual slowing of the Earth's spin. The second was defined as 1 part in 86,400 ($24 \times 60 \times 60$) of a day, based on astronomical measurements made more than a century ago. With modern atomic clocks it has been determined that the length of the mean solar day is now about 86,400.002 seconds, and it is increasing by about 0.0014 second each century.

Over a year the 0.002 seconds add up to most of a second: $365 \times 0.002 \approx 0.7$ seconds each year. This means that an accurate clock set at midnight on New Year's Eve would signal the beginning of the next new year almost 1 second too early. By the year 2100, an accurate clock will be off by 1.2 seconds after a year.

This time change might sound insignificant, but over long periods its effects accumulate. Fossil records from corals that put down visible layers each day indicate that the Earth had about 400 days in a year 400 million years ago. The orbital year wasn't any longer, but the Earth took only about 22 hours to spin on its axis.

With the precise timing needed today for technological uses such as the Global Positioning System (GPS), even millisecond errors are critical. To keep our clocks in agreement with the Sun and stars, we now adjust atomic clocks with a **leap second** every year or two. This is coordinated worldwide so that clocks everywhere remain in agreement.

The slowing of the Earth is a consequence of the interaction of the spinning Earth with ocean tides, which are held relatively stationary by the Moon's gravity. In effect, the eastern coasts of the continents "run into" high tide as the solid Earth spins beneath the tide, and this creates a **tidal friction** force that is slowing the Earth. Tides are discussed further in Unit 19.

Note that atomic clocks are not radioactive. They use an internal vibration frequency of cesium atoms to determine precise time intervals.

KEY TERMS

A.M. (ante meridian), 59	P.M. (post meridian), 59
apparent noon, 59	sidereal day, 59
Arctic Circle, 61	solar day, 59
daylight saving time, 64	standard time, 63
international date line, 63	tidal friction, 64
leap second, 64	time zones, 62
meridian, 59	universal time (UT), 64

QUESTIONS FOR REVIEW

1. How is the sidereal day defined?
2. Why do the sidereal and solar days differ in length?
3. What is universal time?

PROBLEMS

1. Suppose the Earth's spin slowed down until there were just 180 days in a year. Compare the length of a sidereal and a solar day in this new situation. (Do not redefine units of time—just express them in terms of our current hours, minutes, and seconds.)

2. It is thought that the angle of the Earth's axis varies by a few degrees over tens of thousands of years. If the angle was 20° instead of 23.5°, what would be the angle of the noontime Sun from the horizon at your own latitude on the solstices?

3. What is your longitude, and what is the longitude of the center of the nearest time zone? Calculate what time the Sun should, on average, cross your own meridian, assuming that the Sun crosses the meridian at exactly 12:00 at the center of your time zone.

TEST YOURSELF

1. In which of the following locations can the length of daylight range from zero to 24 hours?
   a. Only on the equator
   b. At latitudes closer than 23.5° to the equator
   c. At latitudes between 23.5° and 66.5° north or south
   d. At latitudes greater than 66.5° north or south
   e. Nowhere on Earth

2. In which of the following locations is the length of daylight 12 hours throughout the year?
   a. Only on the equator
   b. At latitudes closer than 23.5° to the equator
   c. At latitudes between 23.5° and 66.5° north or south
   d. At latitudes greater than 66.5° north or south
   e. Nowhere on Earth

3. Daylight saving time
   a. reduces the amount of daylight in summer and shifts it to winter.
   b. corrects clocks for errors caused by the Earth's tilted axis.
   c. results in the Sun crossing the meridian around 1 P.M.
   d. keeps clocks in closer agreement with the position of the Sun in the sky.
   e. All of the above

PLANETARIUM EXERCISES

All the planetarium exercises use the Starry Night planetarium software provided with your text. You should install the program and familiarize yourself with the various parameters and controls before trying these exercises. When the program asks you to enter a password, simply click "OK" to run Starry Night.

Set the location from which you are viewing to the nearest large city or your latitude and longitude. Accomplish this by clicking on the Location box on the toolbar. Verify that the local time is correct.

1. Set the local time to the summer solstice, June 21, the longest day of the year in the Northern Hemisphere. Set the view to the east at sunrise. Note the time. Step the day through and note the sunset time. Using sunrise and sunset times, determine the length of the day for your latitude on this longest day. Using what you know about the summer and winter solstices, calculate the length of day for the shortest day of the year, December 21. Verify your results by setting the program to that date and once again calculating the length of the day.

2. The length of day on the summer solstice should get longer the further north you are and shorter as you move south. Verify this by choosing two additional locations to compare to your own and calculating the length of day as you have done previously. Choose one to the north and a second in the same hemisphere to the south. As in the previous exercise, verify the relationship between the length of the longest and shortest days at these two sites. Use the Location option in the toolbar to change viewing positions.

8 Lunar Cycles

The Moon is one of the loveliest of astronomical sights; it has appeared countless times in literature, poetry, song, and art. It is forever changing shape, brightness, position, and the times when it is visible. Sometimes it is so bright that it can illuminate the night well enough to hike or even read by its light, whereas at other times it provides little light or is altogether absent. The pattern of these changes may not be obvious unless you spend many successive nights studying the Moon.

Like all celestial objects, the Moon rises in the east and sets in the west. Also, like the Sun, the Moon shifts its position across the background stars from west to east, but about 12 times more rapidly. You can easily detect this motion yourself if the Moon happens to lie close to a bright star. In as little as 10 minutes you can see the Moon shift eastward with respect to the star. This rapid motion makes the Moon the most quickly changing of the astronomical bodies that we can see with the naked eye. Its changes follow a regular, clocklike rhythm over about 29½ days. This is the origin of our time period of a **month**—the term coming from the word *Moon.*

Solar eclipse

Lunar eclipse

Moon

Eclipses are possible.

Background Pathways: Unit 6

8.1 PHASES OF THE MOON

The shifting position of the Moon is quite apparent from night to night. If you spot the Moon in the west shortly after sunset one evening, on subsequent nights it will have shifted progressively farther east, and in about two weeks it will be in the eastern sky as the Sun sets. These changes occur as the Moon orbits the Earth, causing the Moon to rise and set about 50 minutes later each day. If the Earth did not spin, the Moon would rise in the west and set in the east about two weeks later.

One of the most striking features of the Moon is that its shape seems to change from night to night in what is called the cycle of lunar **phases.** That is, starting from when the Moon appears as a thin **crescent** just after sunset, more of it becomes illuminated each night—it **waxes**—until after about two weeks it appears as a fully illuminated disk. Then it steadily decreases—it **wanes**—back to a thin crescent again, and then disappears for a day or two before beginning the cycle again. In addition to *crescent,* several descriptive names are used for the phases seen during the period of approximately 29½ days of the **lunar month.** When more than half of the Moon is lit, it is **gibbous;** when it is completely illuminated, it is **full;** and when it is completely dark, it is **new.**

The half-lit Moon is also given a special name. Because this phase occurs one-quarter or three-quarters of the way through the lunar cycle, when the phase is waxing or waning, it is called a **first-quarter moon** or **third-quarter moon.** The **quarter,** new, and full phases refer to particular moments in the orbit of the Moon around the Earth—although we often refer to the Moon being full, for example, if it is within about half a day of that phase. A new moon occurs when the Moon lies as near as possible to the line between us and the Sun. A full moon occurs when the Moon is on the other side of the Earth from the Sun, opposite it in the sky. The quarter moons occur when the Moon is 90° from the Sun. All of these terms can become confusing, so they are summarized in Table 8.1 and illustrated in Figure 8.1.

Position in Orbit (Figure 8.1)	Day	Phase	Location and Time for Viewing
TABLE 8.1		The Lunar Cycle	
1	0.0	New	In the same direction as the Sun
2	0.0–7.4	Waxing crescent	Near the Sun in the evening sky
3	7.4	First quarter	90° from the Sun in the evening sky
4	7.4–14.7	Waxing gibbous	Far from the Sun in the evening sky
5	14.7	Full	Opposite the Sun, rising at sunset
6	14.7–22.1	Waning gibbous	Far from the Sun in the morning sky
7	22.1	Third quarter	90° from the Sun in the morning sky
8	22.1–29.5	Waning crescent	Near the Sun in the morning sky
1	29.5	New	In the same direction as the Sun

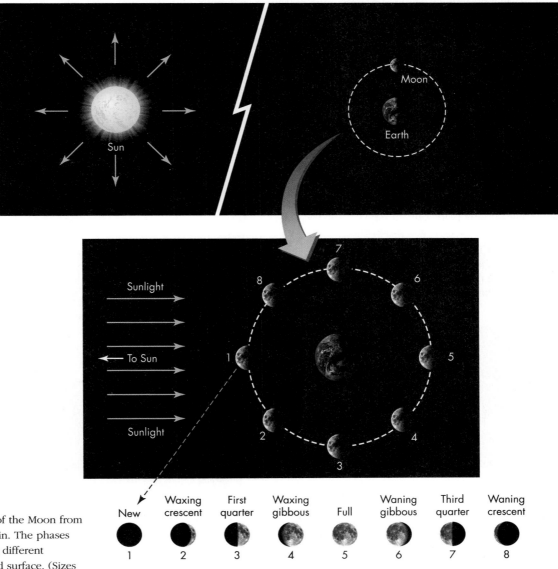

FIGURE 8.1

The cycle of the phases of the Moon from new to full and back again. The phases are caused by our seeing different amounts of its illuminated surface. (Sizes and distances of objects are not to scale.)

Appearance of the Moon from Earth

The times of day and night when you can see the Moon also change throughout the lunar cycle. This is indicated in Table 8.1 and shown graphically in Figure 8.2.

Many people mistakenly assume the changes in shape are caused by the Earth's shadow falling on the Moon. However, you can deduce that this cannot be the explanation because crescent phases occur when the Moon and Sun lie approximately in the same direction in the sky, and the Earth's shadow must therefore point away from the Moon. In fact, we see the Moon's shape change because as it moves around us, we see different amounts of its illuminated half. For example, when the Moon lies approximately opposite the Sun in the sky, the side of the Moon toward the Earth is fully lit. On the other hand, when the Moon lies approximately between us and the Sun, its fully lit side is turned nearly completely away from us, and therefore we glimpse at most a sliver of its illuminated side, as illustrated in Figure 8.1.

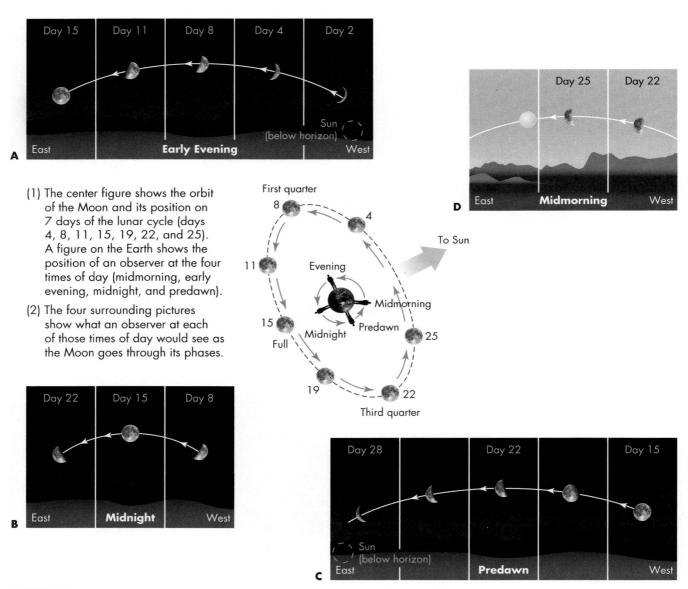

(1) The center figure shows the orbit of the Moon and its position on 7 days of the lunar cycle (days 4, 8, 11, 15, 19, 22, and 25). A figure on the Earth shows the position of an observer at the four times of day (midmorning, early evening, midnight, and predawn).

(2) The four surrounding pictures show what an observer at each of those times of day would see as the Moon goes through its phases.

FIGURE 8.2

When the Moon is visible. This figure shows where to look for the Moon and how it appears at different times of day as it goes through its monthly cycle of phases. The central sketch shows the Earth and Moon and where the Moon is in its orbit at different phases, indicated here by the number of days since the new moon. The yellow arrow points toward the Sun.

Another confusing point is that the Moon's orbit around the Earth is actually about 27.3 days, not the 29.5 days of the lunar cycle. This is also called the **sidereal month** (from the Latin *siderea* for starry) because it is the time the Moon takes to return to the same position among the stars. The difference is a result of the Earth's orbital motion around the Sun—after a month, the Sun shifts into a new constellation of the zodiac, and so the position of the new moon shifts as well.

8.2   ECLIPSES

INTERACTIVE

Eclipse

An **eclipse** occurs when the Moon lies exactly between the Earth and the Sun, or when the Earth lies exactly between the Sun and the Moon, so that all three bodies lie on a straight line. When this happens the Moon will cast a shadow on the Earth or vice versa. A **solar eclipse** occurs whenever the Moon passes directly between the Sun and the Earth and blocks our view of the Sun, as depicted in Figure 8.3. A **lunar eclipse** occurs when the Moon passes through the Earth's shadow as it moves through the portion of its orbit opposite the Sun, as shown in Figure 8.4. Thus solar eclipses can occur *only* when the Moon is new, and lunar eclipses *only* when the Moon is full.

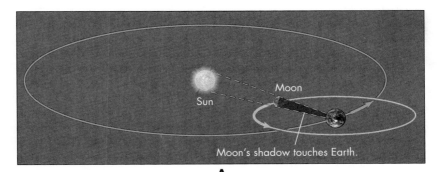

A

B

C

FIGURE 8.3
(A) A solar eclipse occurs when the Moon passes between the Sun and the Earth so that the Moon's shadow touches the Earth. (B) A photograph of the Earth from space during a solar eclipse, showing the Moon's shadow on Earth. (C) This photograph shows what the solar eclipse looks like from Earth, from the center of the Moon's shadow, where the Sun is completely covered.

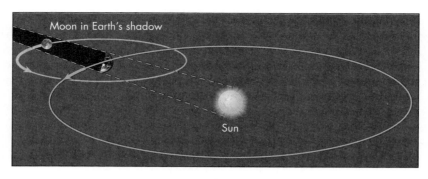

Moon in Earth's shadow

Sun

A

What you see from Earth

B

FIGURE 8.4
(A) A lunar eclipse occurs when the Earth passes between the Sun and Moon, causing the Earth's shadow to fall on the Moon. Some sunlight leaks through the Earth's atmosphere, casting a deep reddish light on the Moon. (B) The photo inset shows what the eclipse looks like from Earth.

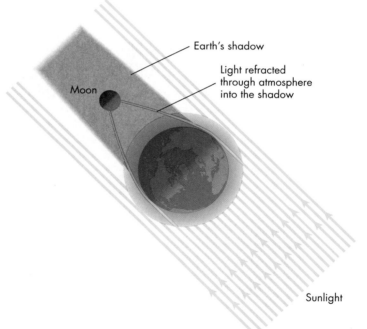

Earth's shadow

Light refracted through atmosphere into the shadow

Moon

Sunlight

FIGURE 8.5
As sunlight falls on the Earth, some passes through the Earth's atmosphere and is slightly bent so that it ends up in the Earth's shadow. In its passage through our atmosphere, most of the blue light is removed, leaving only the red. That red light then falls on the Moon, giving it its ruddy color at totality.

Total eclipses are amazing sights. During a total *solar* eclipse, the Moon completely covers the Sun. The midday sky may become dark as night for several minutes. Birds and other animals may react as if it were nightfall, and the glow from the outer atmospheric layers of the Sun becomes visible. During a total *lunar* eclipse, the Moon usually turns dark orange or red for over an hour. It does not usually become completely black in the Earth's shadow because some sunlight is bent by the Earth's atmosphere into the shadow (Figure 8.5). During a lunar eclipse, if you were looking at the Earth from the Moon, the Earth would appear as a fiery ring of red—in effect you would be seeing sunsets from all over the Earth simultaneously!

Given the remarkable nature of these events, it is not surprising that early people recorded them and sought (successfully) to predict them. Ancient Babylonian and Chinese astronomers kept records stretching back centuries and, based on the patterns they found, were able to predict future eclipses. In more recent times

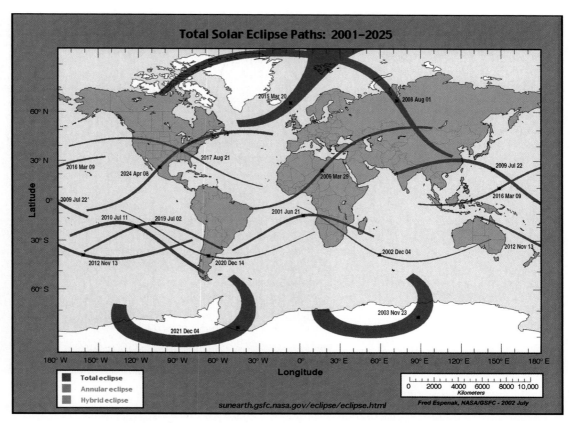

FIGURE 8.6
Location of some recent and upcoming total solar eclipses.

astronomers have used total solar eclipses to study the outer layers of the Sun, which are normally hidden from sight by the enormously brighter central region of the Sun.

Because the Moon is so small compared with the Earth, its shadow is small, and therefore you can see a total solar eclipse only from within a narrow band, as illustrated in Figure 8.6. On the other hand, when the Moon passes through the Earth's shadow in a lunar eclipse, the eclipse is visible from half of the Earth—anywhere that the Moon is above the horizon at the time of the eclipse.

Most solar eclipses pass almost unnoticed because the Moon only partially blocks the Sun's light. The shadow cast by the Moon or by the Earth contains regions where the Sun's light is blocked completely, called the **umbra,** and an outer region where the Sun's light is blocked only partially, called the **penumbra** (Figure 8.7). From the Earth, when we are in the Moon's penumbra, we call it a **partial eclipse.** These occur about twice as often as total eclipses, in which the entire Sun is covered by the Moon from a vantage point somewhere on Earth.

The Moon's orbit around the Earth is not perfectly circular, so sometimes it is far enough that the umbra does not quite reach the Earth. When this happens, we may be in a position where the Sun and Moon are exactly aligned, but the Moon will appear smaller than the Sun, allowing us to see the Sun's surface in a ring or *annulus* around the Moon. This is a type of partial eclipse called an **annular eclipse.**

Similarly, some portion of the Moon passes through the penumbra of the Earth's shadow about twice as often as it completely enters the umbra, for a total lunar eclipse in which the Earth completely blocks the light of the Sun from directly striking any part of the Moon. In partial lunar eclipses, only part of the

Q The Moon's distance from the Earth is measured to be gradually increasing. When the Moon is much farther away than it is now, what will happen to the frequency of the various kinds of eclipses?

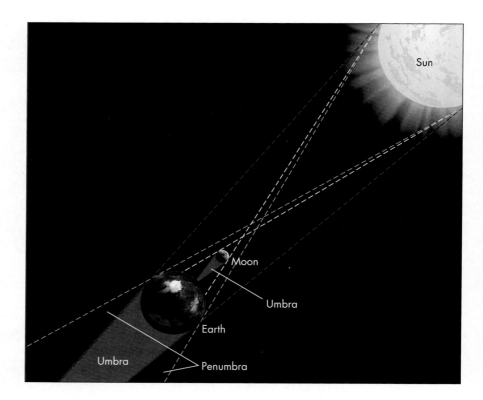

FIGURE 8.7
The shadow of the Earth contains regions where the Sun is completely blocked (the umbra) and regions where sunlight is only partially blocked.

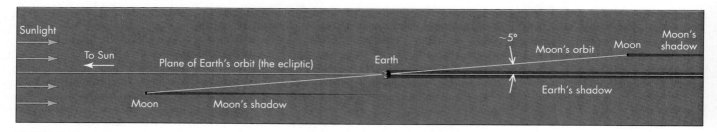

FIGURE 8.8
The Moon's orbit (seen here edge on) is tilted with respect to the Earth's orbit by about 5°. The Earth and Moon are drawn to correct relative size and separation. Note how thin their shadows are.

Moon enters the umbra. The weakest types of partial eclipses are called **penumbral eclipses** because no part of the Moon enters the umbra. They are often so minor that they are not marked on astronomical calendars, although someone living on the Moon would observe the Earth partially eclipsing the Sun.

You may wonder why we do not have eclipses every month. The answer is that the Moon's orbit is tipped by about 5 degrees with respect to the Earth's orbit around the Sun (Figure 8.8). Because of this tilt, even if the Moon is new, its shadow may pass above or below Earth. As a result, no eclipse occurs. Similarly, when the Moon is full, it may pass above or below the Earth's shadow so that again no eclipse occurs.

When the Earth and Moon and their separation are drawn accurately to scale (Figure 8.8) it becomes clearer that a nearly exact alignment of the Earth, Moon, and Sun is required for an eclipse to occur. Eclipses can occur only when the Moon is within about 1 degree of crossing the ecliptic—the plane of the Earth's orbit around the Sun. This usually occurs only twice during the year (Figure 8.9).

For example, if a solar eclipse occurs on a given day in May, a lunar eclipse is likely to occur roughly two weeks later or earlier (the interval between full and new

The name *ecliptic* for the Earth's orbital plane around the Sun arises because only when the new or full moon crosses this line can an eclipse occur.

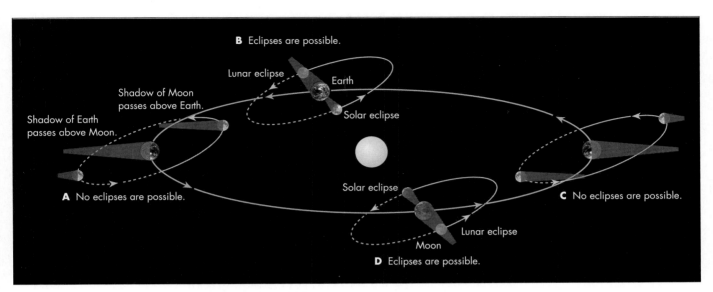

FIGURE 8.9

The Moon's orbit keeps approximately the same orientation as the Earth orbits the Sun. Because of its orbital tilt, the Moon generally is either above or below the Earth's orbit. Thus the Moon's shadow rarely hits the Earth, and the Earth's shadow rarely hits the Moon, as you can see in (A) and (C). Eclipse seasons occur when the Moon's orbital plane, if extended, intersects the Sun. A solar eclipse will then occur at new moon and a lunar eclipse at full moon, as you can see in (B) and (D).

Astronomers can calculate when and what type of eclipses occurred in the distant past and will occur into the distant future. Counting all kinds of partial and total eclipses visible from somewhere on Earth, there are always at least two solar and two lunar eclipses each year, and there can be as many as five of either. Total eclipses are less common, although some years have as many as two total solar eclipses or three total lunar eclipses.

moons), and it is even possible to have two partial solar eclipses on two new moons in a row. Half a year later in November, the Moon will again be close to the ecliptic when it is new or full, and another set of eclipses may occur.

The times of year at which eclipses can occur gradually shift from year to year because the plane of the Moon's orbit "wobbles." This wobble (called *precession*; see also Unit 6.6) swings the Moon's orbital plane back to the same orientation every 18.6 years, so patterns of eclipses tend to be similar every 18 years or so.

8.3   MOON LORE

The Moon figures prominently in folklore around the world. Most stories concerning its powers are false. For example, people often claim that the full moon triggers antisocial behavior—hence the term *lunatic*. However, all carefully controlled studies to look for such effects have found nothing. Automobile accidents, murders, admissions to clinics, and so forth show no increase when the Moon is full. It is true that the extra light at night around the time of the full moon makes it easier to see and do things outside; a variety of links to animal behavior have been found, with predators and prey using the extra light or the cover of darkness to advantage.

The 18 year, 11 day, 8 hour period, called the *saros,* between similar eclipses was discovered by ancient astronomers. For a solar eclipse, the eight additional hours shift the location where the eclipse can be seen about one-third of the way around the globe. However after three of these periods (about 54 years, 31 days), a similar eclipse occurs near the original location.

The position of the full moon in the sky changes during the year. The full moon is almost opposite the Sun, so in winter the full moon is located nearly where the Sun is on the celestial sphere in summer, and vice versa. Thus the full moon traces a path through the sky similar to that of the Sun six months earlier or later, so the full moon reaches a much higher point in the sky in the winter than in the summer (compare Figure 8.10 to Figure 7.1).

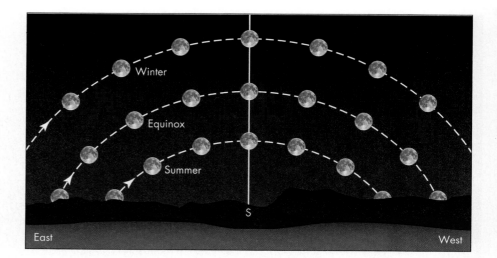

FIGURE 8.10

Path of the full moon through the sky during different seasons. It reaches its highest point in the sky at midnight, but this is lower in the sky in summer than in winter.

Month	Name of Full Moon	Month	Name of Full Moon
January	Old moon	July	Thunder or hay moon
February	Hunger moon	August	Grain or green corn moon
March	Sap or crow moon	September	Harvest moon
April	Egg or grass moon	October	Hunter's moon
May	Planting moon	November	Frost or beaver moon
June	Rose or flower moon	December	Long night moon

TABLE 8.2   Names Used for Full Moons

Q What natural phenomena that occur on Earth do you think might be linked to the lunar cycle? How could you test your idea?

Special names are sometimes given to the full moon at different times of the year. Some examples from North American folklore are listed in Table 8.2. The full moon nearest the time of the autumn equinox is often called the harvest moon because, as it rises in the east at sunset, the light from the harvest moon helps farmers see to get the crops in. Full moons in other months also have special names, but only the harvest and hunter's moon in October are widely used.

The phrase "once in a blue moon," indicating a rare event, has some basis in fact because, on rare occasions, the Moon may look blue. This odd coloration comes from particles in the Earth's atmosphere. Normally our atmosphere filters out blue colors and allows red to pass through, which is why the Sun looks red when it is low in the sky. However, if the atmosphere contains particles whose size falls within a narrow range, the reverse may occur. Dust from volcanic eruptions or smoke from forest fires occasionally has just the right size particles to filter out the red light, allowing mainly the blue colors of the spectrum to pass through. Under these unusual circumstances, we may therefore see a "blue moon."

A different meaning for "blue moon" has appeared more recently. This new meaning applies to months with two full moons: On some calendars the symbol for a second full moon in a month is printed in blue ink. Two full moons are unusual in one month because the cycle of phases is 29.5 days; therefore, unless the Moon is full on the first day of the month, the next full moon will fall in the following month. The odds of the Moon being full on the first of the month are about 1 in 30, so two full moons in a given month happen about every 2½ years. This printing practice appears to be more recent than the earliest uses of the phrase, however.

FIGURE 8.11
Circles beside converging rails illustrate how your perception may be fooled. The bottom circle looks smaller than the circle on the horizon but is in fact the same size. Similarly, the circle high in the sky looks smaller than the circle on the horizon.

8.4   THE MOON ILLUSION

To most people the Moon appears to be much larger when they see it rising or setting than when it is high in the sky. But if you measure the Moon's apparent diameter carefully, you will find it to be slightly *smaller* when it is near the horizon than when it is nearly overhead. In fact, when the Moon is nearest its zenith, we are slightly closer to it. The difference in distance occurs because, as the Earth spins, we are carried closer to and farther from the Moon by a distance equal to the Earth's radius, and when we see the Moon overhead, we are on the part of the Earth closest to the Moon.

This misperception of sizes, known as the *Moon illusion,* is still not well understood. We know that it is an optical illusion caused, at least in part, by the observer's comparing the Moon with objects seen near it on the horizon, such as distant hills and buildings. You know those objects are big even though their distance makes them appear small. Therefore, you unconsciously magnify both them and the Moon, making the Moon seem larger. You can verify this sense of illusory magnification by looking at the Moon through a narrow tube that blocks out objects near it on the skyline. You can also hold up a small coin at arm's length and compare its size to the Moon's. Observed these ways, you will find that the Moon appears roughly the same size regardless of where it is on the sky.

Figure 8.11 shows a similar effect. Because you know that the rails are really parallel, your brain ignores the apparent convergence of the railroad tracks and mentally spreads the rails apart. That is, your brain provides the same kind of enlargement to the circle near the rails' convergence point as it does to the rails, causing you to perceive the middle circle as larger than the lower one, even though they are both the same size.

KEY TERMS

QUESTIONS FOR REVIEW

1. How long does it take the Moon to go through a cycle of phases?

2. Why is phase related to the angle between the Moon and Sun?

3. Describe the various kinds of eclipses.

PROBLEMS

1. What time would a first-quarter moon rise? What time would it set? Draw a diagram of the Earth, Moon, and Sun to explain your answers.

2. The Moon takes about 27.3 days to complete a trip around the ecliptic and return to the same position relative to the stars. This is called its sidereal period. Show that the extra 2.2 days needed to complete a cycle of lunar phases requires the Moon to move through the same angle that the Earth has moved around the Sun in 29.5 days.

3. A total solar eclipse occurred over Europe on August 11, 1999. Based on the 18 year, 11 day, 8 hour period between repetitions of similar eclipses, a similar total eclipse will occur somewhere on Earth at a similar latitude. When and where will it be?

TEST YOURSELF

1. Suppose you observe a solar eclipse shortly before sunset. The phase of the Moon must be
   a. full.
   b. new.
   c. first quarter.
   d. third quarter.

2. During a solar eclipse,
   a. the Earth's shadow falls on the Sun.
   b. the Moon's shadow falls on the Earth.
   c. the Sun's shadow falls on the Moon.
   d. the Earth's shadow falls on the Moon.
   e. the Earth stops turning.

3. Each day, the Moon rises about
   a. the same time.
   b. an hour later.
   c. It depends on the year.
   d. an hour earlier.
   e. It depends on the season.

PLANETARIUM EXERCISES

All the planetarium exercises use the Starry Night planetarium software provided with your text. You should install the program and familiarize yourself with the various parameters and controls before trying these exercises. When the program asks you to enter a password, simply click "OK" to run Starry Night.

Set the location from which you are viewing to the nearest large city or your latitude and longitude. Accomplish this by clicking on the Location box on the toolbar. Verify that the local time is correct.

1. The text discusses the lunar cycle to signify a full cycle of phases for our satellite. Use the planetarium program to estimate what phase and percentage of illumination the Moon is at this evening. Set the time step to one day. Step the time forward and name and list the phases at each step. Keep track of the phase and illumination, and mark how many days it takes for the Moon to return to its original illumination. This should correspond to the length of the lunar cycle. Test this for at least two lunar cycles.

9 Calendars

Different cultures record the passage of time in many ways. The array of approaches may seem bewildering at first (Figure 9.1), but they are based on just a few astronomical cycles.

We group days into weeks, months, and years to track time periods, which reflect natural cycles that shape our lives. A problem arises, though. These various time periods are not simple multiples of one another. The lunar cycle is about 29.53 days, and the year (from vernal equinox to vernal equinox) is about 365.24 days; so there is not a whole number of days in the month or year, nor a whole number of weeks in a month or months in a year. As a result, there is no obvious or best way to record the passage of time.

For example, is it more important that spring arrives on the same date, that the Moon is new on the first day of the month, or that the method of counting days is simple? This question has been answered in different ways by different cultures. Some developed complex systems for keeping their calendars connected with the astronomical cycles, while others ignored the incompatibilities and let their calendars lose that connection.

Background Pathways: Units 6 and 8

9.1 THE WEEK

The origin of the seven-day week is uncertain. It may reflect an approximate measure of quarters of the Moon's phases, although, because the lunar cycle is about 29.5 days, there is no regular interval of days that will keep alignment with the Moon's phases. The seven-day week was used by the Babylonians and Sumerians, although other cultures used anywhere from six to ten days. In ancient Rome an eight-day week (marking the interval between market days) was used; in more recent history, after the French Revolution, a ten-day week was briefly adopted in France.

FIGURE 9.1
Salvador Dali. *The Persistence of Memory (Persistance de la Mémoire).* 1931. Oil on canvas, 9½ × 13 inches (24 × 33.0 cm).
(Copyright © The Museum of Modern Art/Licensed by SCALA/Art Resource, NY.)

Another possibility is that there are seven days in the week to recognize the seven visible objects that move across the sky with respect to the stars: the Sun, the Moon, and the planets Mercury, Venus, Mars, Jupiter, and Saturn. We can see the names of some of these bodies in our English day names (Sunday, Monday, and Saturday). The influence is even clearer in the romance languages, which have their roots in Latin. An example is Spanish, with the day names *lunes, martes, miércoles, jueves,* and *viernes* relating to the Moon, Mars, Mercury, Jupiter, and Venus.

Some English day names come to us through the names of Germanic gods (many of whom have a direct parallel with the Greco–Roman gods after whom the planets are named). For example, Tuesday is from Tiw, a god of war like Mars (matching Spanish *martes*). Wednesday is named for Woden, the chief god of Germanic peoples and identified with the Roman Mercury (*miércoles*). Thursday is named for Thor, the thunder god. He had powers like the Roman god Jupiter, who was also called Jove (*jueves*). Friday is named for Freya, a love goddess like Venus (*viernes*).

Despite the astronomical connections to days' names, the week may simply reflect a human cycle: reasonable periods for work and rest, for purchasing goods and social gatherings. The seven-day week became dominant worldwide with the spread of Jewish, Christian, and Islamic cultures, which all base their week on the biblical account of the creation of the world in seven days, as well as several other cultures that also used seven days without a biblical origin. Longer time periods such as the month and year have a clearer connection to astronomical cycles.

Q Where else do the names of ancient gods show up in everyday life? What qualities do the names convey?

9.2   THE MONTH

The month is the next largest unit that is used by nearly all cultures. This time interval, and its name, derives from the Moon's cycle of phases. The interval between new moons is 29.53059. . . days, and because the year has about 365¼ days in it, there are about 12 lunar cycles per year.

However, 12 lunar cycles end up 11 days short of a full year, so there is no way to build a simple calendar that reflects both cycles. This has led to several different calendar systems used by different cultures and religions.

The commonly used calendar system maintains the "month," but it has no relationship to the phases of the Moon. It is an interval of 28 to 31 days, 12 of which add up to a year. This system is really just a solar calendar, maintaining the shorter period we call a month only as a timekeeping convenience. The history of this system is quite interesting, and we will return to it soon.

Other calendar systems have retained a much closer connection to the Moon. For example, the Islamic calendar system is purely lunar, defining a year as 12 lunar cycles (Figure 9.2). However, this means dates shift relative to the seasonal position of the Sun by almost 11 days each year. Thus the holy month of Ramadan begins on September 24 in 2006 but on September 13 in 2007 and September 2 in 2008. This system is a logical development for a culture that arose in the Middle East, where seasonal differences are not especially pronounced and where night travel by moonlight through the desert was commonplace.

The Chinese and Jewish calendars are called **lunisolar** because they maintain a connection to the cycles of both the Sun and Moon by adding a 13th month about every three years. The extra month is added whenever the calendar gets too far out of alignment with the seasonal year, so some years have about 354 ($\approx$ 12 $\times$ 29.5) days while others have about 384 ($\approx$ 13 $\times$ 29.5) days. Thus the Jewish New Year, Rosh Hashanah, occurs on September 30, 2008, then shifts to September 19 and September 9 in the subsequent two years. It shifts back to September 29 in 2011 after a year in which a 13th month is added between the sixth and seventh named months.

The Chinese calendar is similar, but the month names are based on the segment of the ecliptic where the Sun is located when the Moon is new (Figure 9.3). This is

1426 A.H.

| 1 MUHARRAM مُحَرَّم Feb-Mar 2005 |||||||
SAT	SUN	MON	TUE	WED	THU	FRI
					1 / 10	2 / 11
3 / 12	4 / 13	5 / 14	6 / 15	7 / 16	8 / 17	9 / 18
10 / 19	11 / 20	12 / 21	13 / 22	14 / 23	15 / 24	16 / 25
17 / 26	18 / 27	19 / 28	20 / 1	21 / 2	22 / 3	23 / 4
24 / 5	25 / 6	26 / 7	27 / 8	28 / 9	29 / 10	30 / 11

| 2 SAFAR صَفَر Mar-Apr 2005 |||||||
SAT	SUN	MON	TUE	WED	THU	FRI
1 / 12	2 / 13	3 / 14	4 / 15	5 / 16	6 / 17	7 / 18
8 / 19	9 / 20	10 / 21	11 / 22	12 / 23	13 / 24	14 / 25
15 / 26	16 / 27	17 / 28	18 / 29	19 / 30	20 / 31	21 / 1
22 / 2	23 / 3	24 / 4	25 / 5	26 / 6	27 / 7	28 / 8
29 / 9						

⋮

| 11 DHUL QA'DAH ذُوالقَعْدَة Dec-Jan 2005-06 |||||||
SAT	SUN	MON	TUE	WED	THU	FRI
	1 / 4	2 / 5	3 / 6	4 / 7	5 / 8	6 / 9
7 / 10	8 / 11	9 / 12	10 / 13	11 / 14	12 / 15	13 / 16
14 / 17	15 / 18	16 / 19	17 / 20	18 / 21	19 / 22	20 / 23
21 / 24	22 / 25	23 / 26	24 / 27	25 / 28	26 / 29	27 / 30
28 / 31	29 / 1					

| 12 DHUL HIJJAH ذُوالحِجَّة Jan 2006 |||||||
SAT	SUN	MON	TUE	WED	THU	FRI
		1 / 2	2 / 3	3 / 4	4 / 5	5 / 6
6 / 7	7 / 8	8 / 9	9 / 10	10 / 11	11 / 12	12 / 13
13 / 14	14 / 15	15 / 16	16 / 17	17 / 18	18 / 19	19 / 20
20 / 21	21 / 22	22 / 23	23 / 24	24 / 25	25 / 26	26 / 27
27 / 28	28 / 29	29 / 30				

FIGURE 9.2
An Islamic calendar, showing the first two and last two months of the year 1426. Corresponding dates in the common calendar are shown in green (in 2005–2006 C.E.).

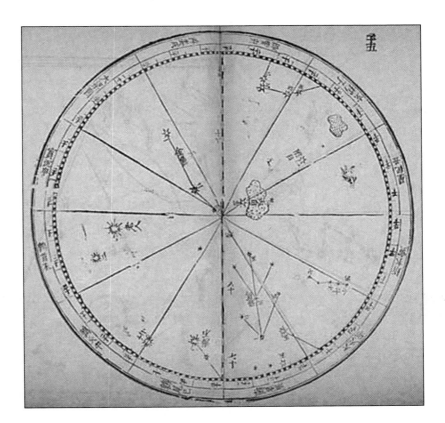

FIGURE 9.3
A Chinese calendar dividing up the year by days and position of the Sun along its path through the stars.

something like naming the months according to the constellation of the zodiac that the Sun is in at each new moon. When two new moons occur while the Sun is in the same segment of the ecliptic, there can be two of the same named month. This can happen, for example, when a new moon occurs just as the Sun enters the segment of the ecliptic designated for "August," and then another new moon occurs just before the Sun leaves that segment. In such a case, there will be a second August that year. Using this procedure adds an extra month once every three or so years at varying times during the year. This shifts the date of the Chinese New Year in a pattern similar to that of Rosh Hashanah.

9.3   THE ROMAN CALENDAR

Our commonly used calendar is based on one developed by the Romans about 200 B.C.E (Before Common Era—see Section 9.5). In fact, the word *calendar* is itself of Roman origin. There is some controversy about how the original Roman calendar was organized. It may have had only 10 months, and it probably began on the first day of spring (the vernal equinox) rather than in January.

Many of the names of our months reflect that early calendar system. For example, if the year began in March, then September, October, November, and December would be the 7th (Sept.), 8th (Oct.), 9th (Nov.), and 10th (Dec.) months, each word's root reflecting the Latin name for the number of the month. Origins of the names of the months, some of which are less certain, are listed in Table 9.1.

Because the Roman calendar was not required to conform to the cycles of astronomical phenomena, it became a form of political patronage. The priests who regulated the calendar would add days and even months to please one group, and take days off to punish another. Think about the possibilities for changing when rent and bills are due! Such confusion resulted from these abuses that in 46 B.C.E Julius Caesar asked the astronomer Sosigenes to design a calendar that would fit the astronomical events better and give less room for the priests and politicians to manipulate it. The resulting calendar, known as the **Julian calendar,** consisted of 12 months, alternating in order between 31 or 30 days, except February, which is discussed further in the next section.

The Julian calendar barely survived Caesar before the politicians were at it again. First the name of the month *Quintilis* (originally the fifth month of the Roman year) was changed to *Julio* to honor Julius Caesar—hence our July. Next, on the death of Julius Caesar's successor, Augustus Caesar, an able and highly respected leader, the following month, *Sextilis,* was renamed in his honor—hence the name August. However, the order of months originally made August 30 days long, and it would have been impolitic to have Augustus's month a day shorter than Julius's; so August was changed to have 31 days, and all the following months had the number of their days changed to reestablish the 30/31 alternation. Unfortunately, this used up one more day than the year allowed. Thus February, already one day short, was trimmed by a second day, leaving it with only 28 days. With only minor modifications, this is the calendar we use today.

TABLE 9.1   Origin of the Names of the Months

Month	Origin of Name
January	Janus (gate), the two-faced god looking to the past and future; hence, beginnings.
February	*Februa* ("expiatory offerings").
March	The god Mars.
April	Etruscan *apru* (April), probably shortened from the Greek Aphrodite, goddess of love and earlier of the underworld.
May	Maia's month; Maia ("she who is great"), the eldest of the Pleiades and the mother of Hermes by Zeus.
June	Junius, an old Roman noble family (from Juno, wife and sister of Jupiter, equal to Greek Hera).
July	Julius Caesar (*Julius* means "descended from Jupiter"; the *Ju* of June and July are the same: Jupiter, "Sky-father").
August	Augustus Caesar (*augustus* means "sacred" or "grand").
September–December	"Seventh month" to "tenth month." The -*ember* may come from the same root as *month*.

9.4   THE LEAP YEAR

The ancient Egyptians knew that the year is not exactly 365 days long. It takes about 365¼ days for the Earth to complete an orbit around the Sun. Because we can't have fractions of a day in the calendar, a calendar based on a year of 365 days will come up 1 day short every 4 years.

Does a quarter of a day really matter? Consider that the seasons are set by the orientation of the Earth's rotation axis with respect to the Sun, not by how many days have elapsed. We therefore want to make sure that we start each year when the Earth has the same orientation. Otherwise the seasons get out of step with the calendar. For example, because in 4 years you will lose 1 day, in 120 years you will lose a month, and in 360 years you will lose an entire season. With a 365-day year, in a little over three centuries, April would begin in what is now January.

This problem is corrected by the **leap year,** a device used by the ancient Romans to keep the calendar in step with the seasons. The leap year corrects for the quarter day by adding a day to the calendar every fourth year. The extra day is added to February, which alternates between 28 and 29 days. The tradition of making adjustments to February may date back to when February was the last month, and year-end adjustments were made to keep the calendar in agreement with the vernal equinox. The civil calendar of India today has a similar system of leap years. It has a set of 12 months of 30 or 31 days that begins at the vernal equinox, and the first month of the year increases from 30 to 31 days during a leap year.

Unfortunately, the year is actually a bit shorter than 365¼ days, so the leap year corrects by a tiny bit too much. To address this problem, centuries that are not divisible by 400 are not leap years. Thus 1900 was not a leap year, but 2000 was. This modification of omitting the leap year for all century years not divisible by 400 was not part of the Julian calendar but was added in 1582 at the direction of Pope Gregory XIII. The calendar we use today is thus known as the **Gregorian calendar** (Figure 9.4).

The inauguration of the Gregorian calendar in 1582 was not a smooth transition. Because it was adopted roughly 1600 years after the Julian calendar, the error in the relationship between the calendar and the seasons had accumulated to about 10 days. To bring the calendar back into synchrony with the seasons, Pope Gregory XIII simply eliminated 10 days from the year 1582 so that the day after October 4 became October 15.

Although the changeover was accepted in much of Europe, non-Catholic countries such as Protestant England refused to abide by the Pope's edict. The change was not made in England until 1752. In Russia the change was not made until the revolution in the early 1900s. Other countries (Greece and Turkey, for example) changed in the 1920s.

Q If you were to create a calendar completely on your own, what would it look like?

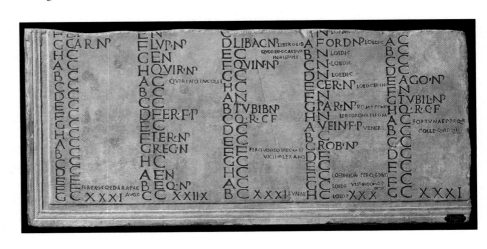

FIGURE 9.4

A portion of an ancient Roman calendar showing the first five months of the year. (Note the Roman numerals indicating the number of days in each month at the bottom of each column.) The letters A through H indicate the 8 days of the week.

9.5   THE CHRONICLING OF YEARS

The common scheme for numbering years began over a millennium ago in the year 532. But why "532"? The numbering of years by different cultures often relates to important religious dates, or the estimated starting date of a nation, or the years since an important ruler came to power. Sometimes these numbering systems were developed decades or even centuries after the event whose anniversary they are intended to mark, so the numbering is somewhat uncertain.

For example, until 532, throughout Europe the year was still commonly counted from the supposed founding of Rome (in 753 B.C.E., according to the common calendar), which made 532 the year 1285. Others, though, marked 532 as the year 248, reckoning from the year that Diocletian became emperor of Rome. Dionysius Exiguus, a monk of this era, carried out calculations for determining the date of the Christian holiday of Easter, which is related to the date of the full moon after the vernal equinox. The monk published a table of Easter dates beginning in the year 532 "Anno Domini" (or A.D., meaning "in the year of the Lord," *not* "after death") to honor Jesus's life instead of Rome or its emperors. However, the monk was a better astronomer than historian. Christian scholars even at the time pointed out that this could not mark the years since the birth of Jesus because of obvious inconsistencies with other historical dates. But the table enjoyed wide circulation, and the idea of marking years this way became popular in Christian countries. So despite its historical inaccuracy, it is the calendar widely used today.

For the years that preceded A.D.1, the convention became to write them as 1, 2, 3 . . . B.C., standing for "before Christ." There is no year 0, which complicates mathematical treatment of years and introduces confusion about when centuries and millennia begin. In fact, the 21st century and the third millennium began on January 1, *2001*, because a decade, century, or millennium is complete only after its 10th, 100th, or 1000th year is completed. December 31, 2000, marked such a completion. Hence, starting from A.D. 1, the transition from A.D. 1999 to A.D. 2000 marked the completion of only 1999 years, not two full millennia.

Recently two different abbreviations have begun to replace A.D. and B.C. They are C.E. and B.C.E., which stand, respectively, for "common era" and "before common era." "Common era" refers to our present calendar, which is used nearly worldwide for most business purposes and thereby avoids reference to a particular religion. Yet another abbreviation—B.P., which stands for "before present (era)"—is used, especially in archaeological, paleontological, and geological works. B.P., used for dates determined by analyzing radioactive carbon, takes 1950 C.E. as its base year.

In the Jewish calendar, dated from the biblical creation of the world, Rosh Hashanah in 2000 began the year 5761. The Islamic calendar is dated from Mohammed's emigration to Medina in 622 C.E. Because the Islamic year is shorter than the solar year, its numbering increases by one year relative to the common system every 33 or 34 years. The beginning of the Islamic year 1421 was April 6, 2000.

The Chinese year-numbering system is based on cycles of 60 years, broken into five sequences of 12 years, each year of which is associated with a particular animal—Rat, Ox, Tiger, Rabbit, Dragon, Snake, Horse, Sheep, Monkey, Chicken, Dog, Pig. Every two years are associated with a particular element—Wood, Fire, Earth, Metal, Water—and the combination of animal and element characteristics repeats after 60 years. We are currently in the 79th of these 60-year cycles, year 1 of which was in 1984, associated with the Rat and Wood. According to tradition this system originated in 2697 B.C.E. Continuous chronologies have been maintained since about the 9th century B.C.E.

These are only a few examples of calendar systems that have survived in relatively common use. Other cultures, like the Maya for example, had systems that folded in additional astronomical cycles, making them even more complex (Figure 9.5).

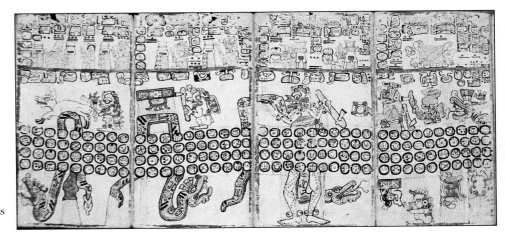

FIGURE 9.5
A portion of a Mayan calendar, which is broken up into 20-day "months."

KEY TERMS

A.D. (*anno domini*), 82

B.C. (before Christ), 82

B.C.E. (before common era), 82

B.P. (before present), 82

C.E. (common era), 82

Gregorian calendar, 81

Julian calendar, 80

leap year, 81

lunisolar calendar, 78

QUESTIONS FOR REVIEW

1. Explain how the lunisolar calendar works.
2. Why do we need a leap year?
3. What establishes the initial year in the various calendar systems?

PROBLEMS

1. What year will 2050 C.E. be in the Chinese, Islamic, and Jewish calendar systems?
2. The first day of the month in the Islamic calendar begins when the crescent moon is first spotted. The earliest this is usually possible is about 16 hours after the new moon. What is the angular distance between the Moon and the Sun at this time? Approximately how long after the Sun sets will the Moon set if it is just 16 hours past new?
3. Under the Julian calendar system, the year averages to 365.25 days after 4 years. In the Gregorian system, determine the average length of the year after 400 years.

4. If the Gregorian calendar system had not been adopted, on what date would the Sun currently reach its northernmost point (according to the Julian calendar system)?
5. Suppose the year was 365.170 days long. How would you design a pattern of leap years to make the average year length exactly match this value within a span of 1000 years?

TEST YOURSELF

1. Suppose that the length of the year was 365.2 days instead of 365.25 days. How often would we have leap year? Every
   a. 2 years
   b. 5 years
   c. 10 years
   d. 20 years
   e. 50 years
2. Why does the Chinese New Year occur on different dates in the Western calendar each year?
   a. The Chinese year always starts when the Moon is new.
   b. The precession of the Earth's axis shifts it a little each year.
   c. The date is based on when Mars undergoes retrograde motion.
   d. The date is chosen each year for when a positive horoscope is cast.
   e. The Chinese calendar slips by a quarter-day each year because it has no leap days.
3. Why was the Roman calendar system of a leap year every 4 years changed in the 1500s?
   a. The year is not exactly 365.25 days long.
   b. The Earth had slowed down its spin since ancient Roman times.
   c. Precession of the Earth had changed the length of the year.
   d. The lunar phases had gone out of phase with our months.

IO Geometry of the Earth, Moon, and Sun

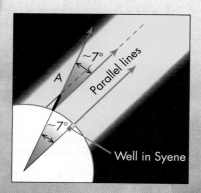

~7°
A
Parallel lines
~7°
Well in Syene

Background Pathways: Unit 8

It is possible to draw several important conclusions about the sizes of and distances between the Earth, Moon, and Sun based on careful but simple measurements. As far as we know, the ancient Greek astronomers of classical times were the first to do this. The values they determined were not always highly accurate, but they were surprisingly good. They demonstrated that you do not need telescopes, high-speed communications, computers, and satellites to determine, for example, that the Earth is a sphere with a circumference of about 40,000 kilometers.

We study ancient ideas of the heavens not so much for what they tell us about the heavens but to learn how observation, geometry, and careful reasoning can lead us to a deeper understanding of the universe. Moreover, these same kinds of geometric ideas can be applied today to find the distances to the stars.

10.1 THE SHAPE OF THE EARTH

The ancient Greeks knew that the Earth was round. As long ago as about 500 B.C.E., the mathematician **Pythagoras** (about 560–480 B.C.E.) was teaching that the Earth was spherical, but the reason for his support of this idea was as much mystical as rational. Like many of the ancient philosophers, he believed that the sphere was the perfect shape and that the gods would therefore have utilized that perfect form in creating the Earth.

By 300 B.C.E., however, **Aristotle** (384–322 B.C.E.) was presenting arguments for the Earth's spherical shape that were based on simple naked-eye observations that anyone could make. Such reliance on careful, firsthand observation was the first step toward generating scientifically testable hypotheses about the contents and workings of the universe. For instance, Aristotle noted that if you look at a lunar eclipse when the Earth's shadow falls on the Moon, the shadow can be clearly seen as curved, as Figure 10.1A shows. As he wrote in his treatise "On the Heavens,"

The shapes that the Moon itself each month shows are of every kind—straight, gibbous, and concave—but in eclipses the outline is always curved; and, since it is the interposition of the Earth that makes the eclipse, the form of this line will be caused by the form of the Earth's surface, which is therefore spherical.

Another of Aristotle's arguments that the Earth is spherical was based on the observation that a traveler who moves south will see stars that were previously hidden below the southern horizon, as illustrated in Figure 10.1B. For example, the bright star Canopus is easily seen in Miami but is below the horizon in Boston. This could not happen on a flat Earth.

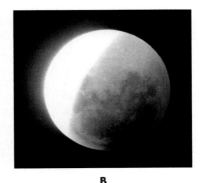

A

B

C

Star is invisible from A, always below the horizon. Star is visible from B.

FIGURE 10.1

(A) During a lunar eclipse, we see that the Earth's shadow on the Moon is curved. Thus the Earth must be round. (B) Photograph of a partial lunar eclipse. (C) As a traveler moves from north to south on the Earth, the stars that are visible change. Some disappear below the northern horizon, whereas others, previously hidden, become visible above the southern horizon. This variation would not occur on a flat Earth.

10.2   DISTANCE AND SIZE OF THE SUN AND MOON

One of the most remarkable ancient Greek astronomers was **Aristarchus** of Samos (about 310–230 B.C.E). He was able to estimate the size and distance of the Moon and Sun. Even though his values were not highly accurate, they gave at least the correct sense of which was larger and which smaller than the Earth, as well as approximate indications of their distances from Earth. Few of his writings survive intact, but references to his works by later astronomers allow us to reconstruct many of his findings. For example, by comparing the size of the Earth's shadow on the Moon during a lunar eclipse to the size of the Moon's disk, illustrated in Figure 10.2, Aristarchus calculated that the Moon's diameter was about one-third of the Earth's. His estimate that the Moon's diameter is about 0.33 that of the Earth's is not far from the correct value of about 0.27.

Aristarchus then estimated the distance to the Moon by noting that when the Moon passes through the middle of the Earth's shadow during a lunar eclipse, it takes about three hours to complete its passage through the shadow. On the other hand, the Moon takes about 660 hours to complete its passage around the zodiac. If

If the circumference of the Moon's orbit is 220 times Earth's diameter, it is $440 \times R_\oplus$. The circumference of a circle is 2π times its radius, so the Moon's orbital radius is $440 \times R_\oplus/2\pi \approx 70 \times R_\oplus$.

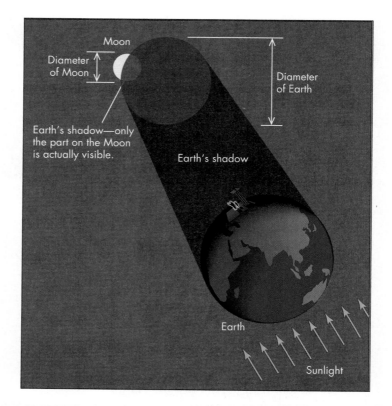

FIGURE 10.2
Aristarchus used the size of the Earth's shadow and the time the Moon took to pass through it during a lunar eclipse to estimate the relative sizes of the Earth and Moon.

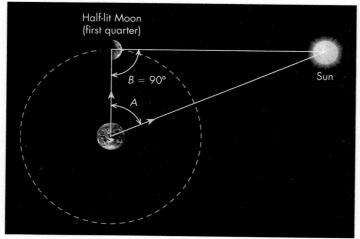

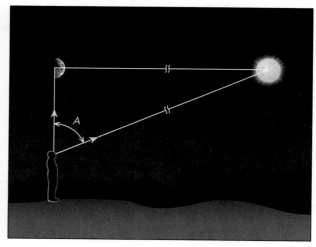

FIGURE 10.3
Aristarchus estimated the relative distance of the Sun and Moon by observing the angle *A* between the Sun and the Moon when the Moon is exactly half lit. Angle *B* must be 90° for the Moon to be half lit. Knowing the angle *A*, he could then set the scale of the triangle and thus the relative lengths of the sides. (Sizes and distances not to scale.)

Even with modern instruments, it is impossible to measure the angle between the quarter Moon and Sun well enough to obtain the Sun's distance accurately. If the Sun was as close as about 10 times the Moon's distance, though, there would be about a day less between new Moon and first quarter than between first quarter and full Moon, which would be fairly easy to detect.

the Earth's shadow is about the same size as the Earth itself, then the entire circumference of the Moon's orbit is about 220 (= 660/3) times bigger than the Earth's diameter, and its distance is about 70 times the Earth's radius. This is not far from the modern estimate that the Moon is about 60 Earth radii away from the planet.

Aristarchus also calculated the Sun to be about 20 times farther away from the Earth than the Moon is. He carried out this calculation by measuring the angle between the Sun and the Moon when the Moon was exactly half lit at the first and third quarter phases (Figure 10.3). From the apparent size in the sky of the Sun and

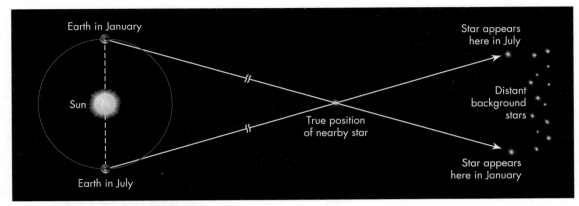

FIGURE 10.4

Motion of the Earth around the Sun causes *stellar parallax*—a shift in the position of nearby stars. Because the stars are so remote, this shift is too small to be seen by the naked eye. This lack of observable parallax led the ancient Greeks to deduce incorrectly that the Sun could not be the center of the Solar System. (Sizes and distances are not to scale.)

Moon and their relative distances, and his calculation of the Moon's diameter relative to the Earth's, Aristarchus deduced that the Sun's diameter was about seven times that of the Earth's (see Section 10.4). This is far smaller than we know it is today—the Sun is more than 100 times larger than the Earth. Nevertheless, it was the first demonstration that the Sun is much bigger than the Earth and much more distant than the Moon.

It was perhaps his recognition of the vast size of the Sun that led Aristarchus to the revolutionary idea that the Earth revolved around the Sun rather than the reverse. Aristarchus was, of course, correct; but his idea was too radical, and another 2000 years passed before scientists became convinced of its correctness. However, this was not mere stubbornness. There was a good reason for not concluding that the Earth moves around the Sun. If it did, the positions of stars should change during the year as the Earth moved from one side to the other within the celestial sphere. Looking at Figure 10.4, you can see that the nearby star should appear to lie at a different angle from and in a different position with respect to each of the more distant stars in January and in July.

This shift in a star's apparent position resulting from the Earth's motion around the Sun is called **parallax,** and Aristarchus's critics were right in supposing that it should occur. Actually, many astronomers of the time assumed the stars were all at the same distance, but even so there would be a parallax effect. If the Earth moved around inside this hypothesized celestial sphere, we would see changes in the stars' apparent separation—just as we can tell how we are moving in the dark by looking at nearby lights. Because no parallax was seen, astronomers rejected Aristarchus's idea. What they failed to realize was that because of the immense distance to the stars, the parallax is so tiny that we need a powerful telescope and very precise equipment to measure it. In fact, parallax of stars was not successfully measured until 1838 (see Unit 52). Thus an idea may be rejected for reasons that are logically correct but are flawed because of limited data accuracy.

10.3   THE SIZE OF THE EARTH

Even though few astronomers accepted Aristarchus's idea that the Earth moved around the Sun, they did accept his estimates of the sizes and distances of the Sun and Moon. However, all of these measurements were scaled to the size of the Earth,

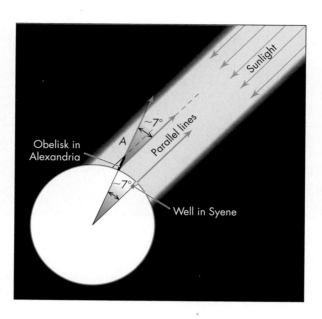

FIGURE 10.5

Eratosthenes's calculation of the circumference of the Earth. The Sun is directly overhead on the summer solstice at Syene in southern Egypt. On that same day, Eratosthenes found the Sun to be 1/50th of a circle (about 7°) from the vertical in Alexandria, in northern Egypt. Eratosthenes deduced that the angle between two verticals placed in northern and southern Egypt must be 1/50th of the circumference of the Earth.

With our modern perspective, we would say that this occurred because Syene was at a latitude close to 23.5°N, whereas Alexandria lies farther north where the Sun never passes through the zenith.

The "stadium" is an ancient unit of measure used by many cultures, and with different lengths, so the exact value used by Eratosthenes is uncertain and may have ranged from about 0.15 to about 0.21 km. The modern use of the word *stadium* comes from the use of the term to refer to a racetrack one stadium long.

whose size was not known. Thus the Moon might be one-third the size of the Earth; but was that 1000 km or 100,000 km?

Eratosthenes (276–195 B.C.E), head of the famous library at Alexandria in Egypt, made the first measurement of the Earth's size. He obtained a value for its circumference that very nearly matches the presently measured value. Eratosthenes's demonstration is one of the most elegant ever performed because it so superbly illustrates how scientific inference links observation and logic.

Eratosthenes, a geographer as well as an astronomer, heard that to the south, in the Egyptian town of Syene (the present city of Aswân), the Sun would be directly overhead at noon on the summer solstice and thus it cast no shadow. Proof of this was that on the solstice the Sun lit the bottom of a deep well there. However, on that same day, the Sun did cast a shadow in Alexandria, where he lived. Knowing the distance between Alexandria and Syene and appreciating the power of geometry, Eratosthenes realized he could deduce the size (circumference) of the Earth. He analyzed the problem as follows: Because the Sun is far away from the Earth, its light travels in parallel rays toward the Earth. Thus two rays of sunlight, one hitting Alexandria and the other shining down the well, are parallel lines, as depicted in Figure 10.5.

At noon on the summer solstice, the Sun's rays passing through Syene are aimed straight at the center of the Earth. However, because of the curved surface of the Earth, parallel rays passing through Alexandria would miss the Earth's center. We can calculate the angle between the two cities with a geometric construction. Draw a straight line from the center of the Earth outward so that it passes vertically through the Earth's surface in Alexandria. The line makes an angle, *A*, with all of the Sun's parallel rays, so it also indicates the angle of latitude between the two cities (see Figure 10.5). The reason is that the corresponding angles formed where a single line crosses two parallel lines are equal (a geometric theorem).

The angle *A* can be measured with sticks and a protractor (or their ancient equivalent) and is the angle between the direction to the Sun and a straight stick pointing vertically upward (see Figure 10.5). Eratosthenes found this angle to be about 1/50th of a circle, or roughly 7°. Therefore the angle formed by a line from Alexandria to the Earth's center and a line from the well to the Earth's center must also be 1/50th of a circle, as deduced from the parallel line–equal angle theorem.

Eratosthenes reasoned that if the angle between Alexandria and Syene was 1/50th of a circle, the distance between these cities must be 1/50th the circumference of the Earth. Because he knew the distance between Alexandria and the well to be 5000 stadia (where a stadium is about 0.16 kilometer), the distance around the entire Earth must be 5000 stadia × 50, or 250,000, stadia. When expressed in modern units, this is roughly 40,000 kilometers, or a diameter of about 13,000 kilometers, which is approximately the size of the Earth as we know it today.

By modern standards, there is absolutely nothing wrong with this technique. You can use it yourself to measure the size of the Earth. Eratosthenes's success was a triumph of logic and the scientific technique, but it made one assumption that we have not yet justified. The method required that he assume the Sun was so far away that its light reached Earth along parallel lines. That assumption, however, was supported by the earlier set of measurements made by Aristarchus that the Sun was much more distant than the Moon (Section 10.2).

10.4   MEASURING THE DIAMETER OF ASTRONOMICAL OBJECTS

One of the most basic properties of an astronomical object is its size. Because we cannot stretch a tape measure across the disk of the Sun or Moon, how do we know the size of such heavenly bodies?

The basic method for measuring the size of a distant object was worked out long ago and is still used today. This method involves first measuring how big an object *looks,* a quantity called its **angular size.** We can measure an object's angular size by drawing imaginary lines to each side of it, as shown in Figure 10.6, and then measuring the angle between the lines using a protractor.

As an astronomical example, we can measure the angular size of the Moon, or equivalently, because it is a round object, its *angular diameter.* The Moon's angular size is small and it moves on the sky, so it is not practical to aim lines on each side. Instead, we can determine its angular size by the following procedure: Hold up a ruler at arm's length in front of the Moon. If you do this, you might find, for example, that the Moon is about 6 mm across on the ruler. Next measure the distance from your eye to the ruler; about 70 cm in this example. Draw this carefully to scale on a sheet of paper (divide all of the sizes by 3 to fit it on a standard sheet of paper).

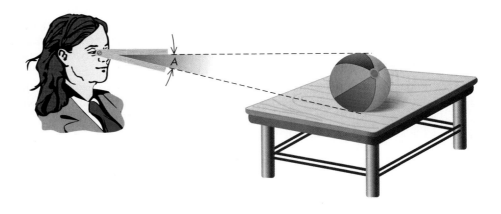

FIGURE 10.6
The beach ball has an angular size *A* for the observer.

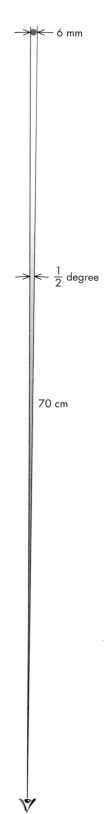

FIGURE 10.7
Scale drawing to find the angular diameter of the Moon. It is the equivalent size of a 6 mm object held 70 cm from your eye, or about one-half degree.

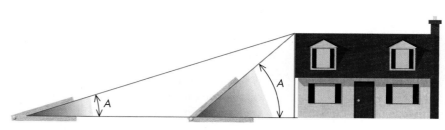

FIGURE 10.8
How angular size varies with distance.

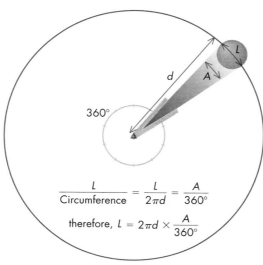

$$\frac{L}{\text{Circumference}} = \frac{L}{2\pi d} = \frac{A}{360°}$$

$$\text{therefore, } L = 2\pi d \times \frac{A}{360°}$$

FIGURE 10.9
How to determine linear size from angular size.

This yields a picture something like Figure 10.7. Measuring the angle between the lines with a protractor, we find that the Moon's angular size is about one-half degree.

Notice that an object's angular size depends on its distance (Figure 10.8). For example, if you move closer to an object, its angular size increases. Therefore,

The angular size of an object depends inversely on its distance.

From its angular size and distance, we can find an object's *linear size*—that is, its size in units of length such as meters. To work out a mathematical formula relating angular size and linear size, imagine that you are at the center of a circle whose radius is d, the distance to the body. The circle passes through the object, as illustrated in Figure 10.9. Let L be the linear size of the body. Draw lines from the center to each end of L, letting the angle between the lines be A, the object's angular size.

We now determine the object's linear size, L, by forming the following proportion: L is to the circumference of the circle as A is to the total number of degrees around the circle, which we know is 360. Thus

$$\frac{L}{\text{Circumference}} = \frac{A}{360°}.$$

However, we know from geometry that the circle's circumference is $2\pi \times d$. Thus we can solve for L by multiplying both sides by the circumference:

$$L = \text{Circumference} \times \frac{A}{360°} = 2\pi \times d \times \frac{A}{360°}.$$

L = linear size

A = angular size

d = distance

Dividing 360 by 2π gives an important formula relating angular size, linear size, and distance:

$$L = d \times \frac{A}{57.3°}.$$

Thus, given a body's angular size and distance, we can calculate its linear size. For example, suppose we apply this method to measure the Moon's diameter. We stated previously that the Moon's angular size is about 1/2°. Its distance is about 384,000 kilometers. Thus its diameter is

$$L = 384{,}000 \text{ km} \times \frac{0.5°}{57.3°} = 3400 \text{ km}.$$

So the Moon's diameter is a little more than one-quarter of the Earth's.

KEY TERMS

angular size, 89   Eratosthenes, 88

Aristarchus, 85   parallax, 87

Aristotle, 84   Pythagoras, 84

QUESTIONS FOR REVIEW

1. What were the major astronomical contributions of Aristotle, Aristarchus, and Eratosthenes?

2. What is meant by the phrase *angular size*?

3. If you triple your distance from an object, what happens to its angular size?

PROBLEMS

1. Some people still believe the Earth is flat. What "proof" would you offer them that it is round? Could you persuade them?

2. Suppose you were an alien living on the fictitious warlike planet Myrmidon and you wanted to measure its size. The Sun is shining directly down a missile silo 1000 miles to your south, while at your location, the Sun is 36° from straight overhead. What is the circumference of Myrmidon? What is its radius?

3. The great galaxy in Andromeda has an angular diameter along its long axis of about 5°. Its distance is about 2.2 million light-years. What is its linear diameter?

TEST YOURSELF

1. The circular shape of the Earth's shadow on the Moon led early astronomers to conclude that
   a. the Earth is a sphere.
   b. the Earth is at the center of the Solar System.
   c. the Earth must be at rest.
   d. the Moon must orbit the Sun.
   e. the Moon is a sphere.

2. The ancient Greeks are *not* credited with
   a. measuring the size of the Earth.
   b. finding the relative sizes of the Earth and Moon.
   c. determining that the Earth is round.
   d. detecting the parallax of stars.
   e. suggesting a heliocentric model of the Solar System.

3. If the Moon were half as far away from the Earth, its angular size would be
   a. twice as big.
   b. half as big.
   c. four times as big.
   d. one-quarter as big.
   e. the same as it is now.

11

Planets: The Wandering Stars

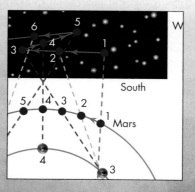

Background Pathways: Units 1 and 6

Five of the brightest "stars" in the sky do not remain in the same position on the celestial sphere, but instead gradually shift from night to night. The motions of these objects are quite complex and difficult to predict. The Greeks gave them the name *planetai,* meaning "wanderers," from which our word **planet** comes.

Until the invention of telescopes, they were thought of as special stars, free to move in the sky, unlike the thousands of other stars visible on the celestial sphere. This freedom of movement suggested they had special powers, and perhaps this inspired their association with the gods Mercury, Venus, Mars, Jupiter, and Saturn. When it was eventually proposed that the motions of these objects could be more easily understood if Earth was a "wandering star" too, is it any wonder that this was a difficult idea for most to accept?

11.1 MOTIONS OF THE PLANETS

Planets move against the background of stars because of a combination of the Earth's and their own orbital motion around the Sun. One of the more important features of this motion is that the planets always remain within a narrow band on either side of the ecliptic, within the constellations of the zodiac, as the Sun does (Unit 6). The planets remain close to the ecliptic because their orbits, including that of the Earth, all lie in nearly the same plane (Figure 11.1).

Like the Sun and the Moon, the planets usually move from west to east through the stars. This does not mean that the planets rise in the west and set in the east. As seen from a spinning Earth, planets always rise in the east and set in the west because their movements on the celestial sphere are far slower than the Earth's rotation. The motion of a planet along the zodiac can be seen by marking off its position on a star chart over a period of several months. Figure 11.2 illustrates such a plot and shows that planets normally move eastward through the stars as a result of their orbital motion around the Sun.

However, unlike for the Sun and Moon, this simple pattern of movement is sometimes interrupted. Sometimes a planet will move west with respect to the stars, a condition known as **retrograde motion** and shown in Figure 11.3. The word *retrograde* means "backward moving," and when a planet is in retrograde motion, its path through the stars may turn or loop backward for a month or more. All planets undergo retrograde motion when they lie in the same direction as the Earth from the Sun. This extraordinary behavior proved difficult to explain and became one of the major challenges of astronomers for almost 2000 years.

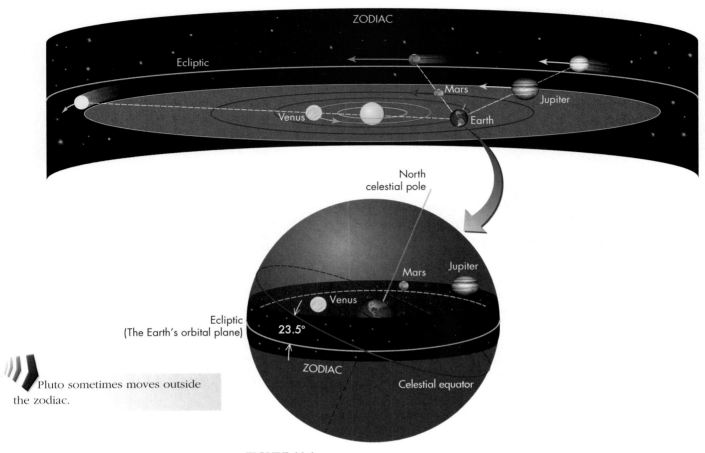

FIGURE 11.1
The planets move through the zodiac, which is tilted 23.5° with respect to the celestial equator because the Earth's rotation axis is tilted by that amount with respect to the plane in which the planets orbit.

Pluto sometimes moves outside the zodiac.

Q If you see a bright "star" in the sky, how can you tell whether it is a star or a planet (such as Venus) without using a telescope?

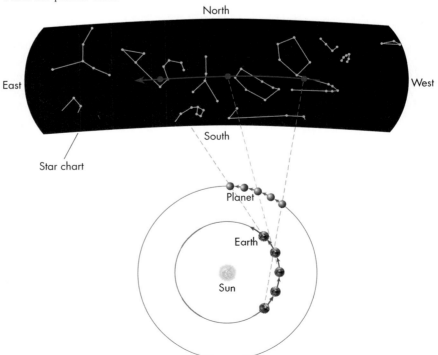

FIGURE 11.2
A planet's eastward drift against the background stars plotted on the celestial sphere. Note: Star maps usually have east on the left and west on the right, so that they depict the sky when looking south.

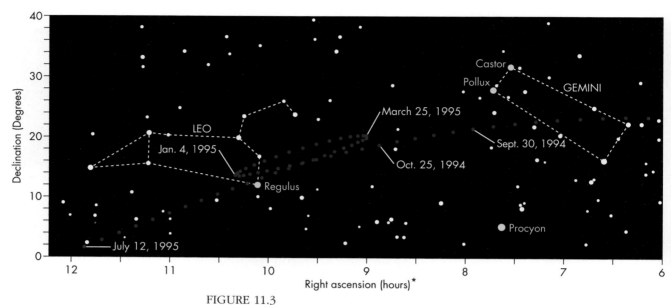

FIGURE 11.3

The position of Mars marked out on the background stars and showing its retrograde motion.

11.2 EARLY THEORIES OF RETROGRADE MOTION

As one observes the sky, the most basic of observations is that everything seems to move around the Earth from east to west. From earliest Greek times, this led to the model that the Earth was at the center of the universe and the planets and stars moved around it. Models of the universe of this type are called Earth-centered or **geocentric models.**

Figure 11.4 shows a typical geocentric model based on the work of the Greek astronomer Eudoxus, who lived about 400–347 B.C.E. The Sun, Moon, and planets all revolve around the Earth. The bodies that move fastest across the sky are those that are nearest to the Earth. Thus the Moon completes its orbit most rapidly and is nearest, whereas Saturn, whose path through the stars takes roughly 29 years to complete, is located the farthest out (of the planets then known). By assuming that each body was mounted on its own revolving transparent (crystalline) sphere and by tipping the rotation axis of each sphere slightly, Eudoxus was able to give a satisfactory explanation of the normal motions of the heavenly bodies. However, this model does not explain retrograde motion, unless one believes that the crystalline spheres sometimes stop turning, reverse direction, pause, and then resume their original motion. This idea is clumsy and unappealing.

Eudoxus was able to develop a more complex model to explain retrograde motion by requiring that each planet's crystalline sphere should itself be attached to another crystalline sphere that rotated at a different angle. By combining steady rotations of both spheres, he was able to make zigzag paths that roughly resembled retrograde motion. Over the next five centuries, astronomers proposed variants of this complex model, leading to a model that could predict the planets' positions with good accuracy, developed by Claudius **Ptolemy,** the great astronomer of Greco–Roman times, who lived about 100 to 170 C.E.

Ptolemy lived in Alexandria, Egypt, which at that time was one of the intellectual centers of the world, in part because of its magnificent library. Ptolemy fashioned a model of planetary motions in which each planet moved on one small circle, which in turn moved on a larger one (Figure 11.5). The small circle, called an

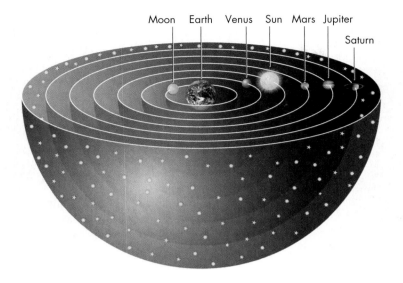

FIGURE 11.4

A cutaway view of the geocentric model of the Solar System according to Eudoxus. (Some spheres are omitted for clarity.)

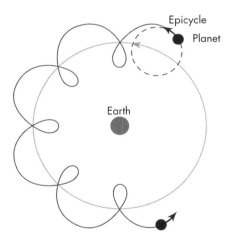

FIGURE 11.5

Epicycles are a bit like a bicycle wheel with a Frisbee bolted onto its rim.

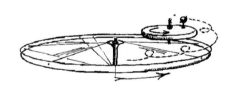

ANIMATIONS

Ptolemy's model

epicycle, was supposed to be carried along on the large circle like a Frisbee spinning on the rim of a bicycle wheel or the children's toy called a "Spirograph."

According to Ptolemy's model, the motion of the planet from east to west across the night sky is caused by the rotation of the large circle (the bicycle wheel in our analogy). Retrograde motion occurs when the epicycle carries the planet in a reverse direction (caused by the rotation of the Frisbee in our analogy). Thus, with epicycles, it is possible to account for retrograde motion, and Ptolemy's model was able to predict planetary motions with fair precision.

However, discrepancies remained between the predicted and true positions of the planets. This led to further modifications of the model, each of which led to slightly better agreement but at the cost of adding further complexity. Ptolemy's model survived until the 1500s thanks to its success at predicting the positions of the planets. In the end it was rejected because it was too complex to be plausible: Simplicity is an important element of scientific theory. As the medieval British philosopher William of Occam wrote in the 1300s, "Entities must not be unnecessarily multiplied." This expresses the idea that a scientific model that requires many parts or parameters to explain a phenomenon is less desirable than a simpler model that achieves the same result, a principle known as **Occam's razor.** It expresses a common situation encountered in science: If we begin with incorrect assumptions, as new evidence accumulates, we will often have to add more and more complexity to our models to explain discrepancies between the evidence and the model.

Although we call our numerals "Arabic," they were borrowed by Arab traders from the civilizations farther east in what is now India.

Ptolemy's era was one of decay and general political instability for the Greco–Roman civilization, and much of what we know of his work (and of Greek and Roman civilization in general) we owe to Islamic scholars around the southern edge of the Mediterranean, who preserved and expanded on his work from about 700 to 1200. In addition, Islamic scholars revolutionized mathematical techniques through innovations such as algebra (another Arabic word) and Arabic numerals. These greatly aided in the development of a mathematical understanding of the universe.

11.3   THE HELIOCENTRIC MODEL

The man who began the demolition of the geocentric model and started the revolution in astronomical ideas that continues to this day was a Polish physician and lawyer by the name of Nicolas **Copernicus** (1473–1543). Copernicus (Figure 11.6) tried many modifications of the geocentric model to explain the centuries of data on planetary positions that had been collected since Ptolemy's geocentric model was developed. All of his attempts failed. Thus he reconsidered the idea that the Earth moves around the Sun.

Heliocentric models in which the Sun (*helios* in Greek) was the center of a system of planets, of which the Earth was one, had been proposed nearly 2000 years earlier by Aristarchus (Unit 10.2), but had been rejected at the time. Nevertheless, Copernicus showed that such a model offered a far simpler explanation of retrograde motion. In fact, if the planets orbit the Sun, retrograde motion becomes a simple consequence of one planet on a smaller, faster orbit overtaking and passing another on a larger, slower orbit.

To see why retrograde motion occurs, look at Figure 11.7. Here we see the Earth and Mars moving around the Sun. The Earth completes its orbit, circling the Sun, in 1 year, whereas Mars takes 1.88 years to complete an orbit, with the Earth overtaking and passing Mars every 780 days.

If we draw lines from the Earth to Mars, we see that Mars will appear to change its direction of motion against the background stars each time the Earth overtakes it. This occurs around the time Mars is directly opposite the Sun in the sky, or at what seen from Earth is called **opposition.** When Mars is at opposition, it is also nearest Earth, as well as being brightest and easiest to see. A similar phenomenon happens when you drive on a highway and pass a slower car. Both cars are, of course, moving in the same direction. However, as you pass the slower car, it looks as if it is moving backward against the stationary objects behind it.

Jupiter and Saturn (as well as Uranus, Neptune, and Pluto) exhibit retrograde motion around the time of opposition, but Mercury and Venus are different. Because they are nearer the Sun than the Earth, they can never be in opposition. Instead they appear to reverse direction against the stars as they pass between the Earth and the Sun. The term **conjunction** refers to when a planet lies in the same direction as the Sun. This occurs for all of the planets when they are on the far side of the Sun. Venus and Mercury, however, can be in conjunction either when they are farther than the Sun (often called **superior conjunction**) or when they are nearer than the Sun (**inferior conjunction**). Venus and Mercury undergo retrograde motion around the time they are in inferior conjunction. These planetary configurations are shown in Figure 11.8.

Copernicus found that he could determine the relative sizes of the planets' orbits from information about the planets' configurations. For example, Venus and Mercury never get very far from the Sun, as shown in Figure 11.9. Mercury can never be more than 28° from the Sun and Venus never more than 47° as seen on the sky. The size of this largest angular separation from the Sun, or **greatest elongation,** is determined by the sizes of these planets' orbits relative to the Earth's orbit. By

INTERACTIVE

Retrograde motion

ANIMATIONS

Retrograde motion of Mars

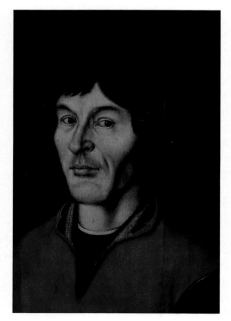

FIGURE 11.6
Nicolas Copernicus (1473–1543).

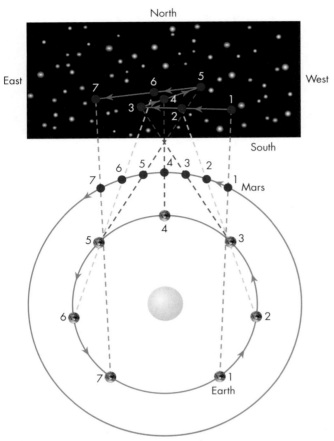

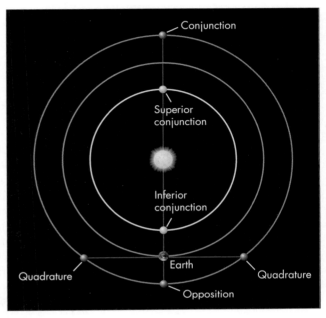

FIGURE 11.7
Why we see retrograde motion. (Object sizes, positions, and distances are exaggerated for clarity.)

FIGURE 11.8
Planetary configurations: opposition, superior conjunction, inferior conjunction, and quadratures.

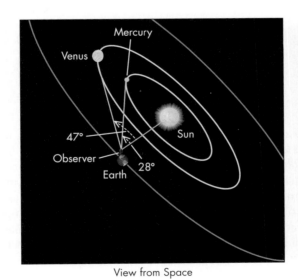
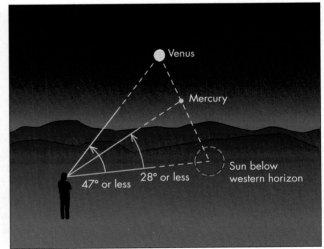

FIGURE 11.9
The greatest elongations of Mercury and Venus and the Evening Star phenomenon. The left-hand diagram also shows that Mercury and Venus can *never* appear more than 28° and 47°, respectively, from the Sun.

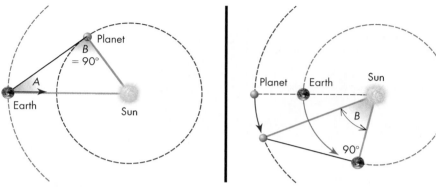

FIGURE 11.10

How Copernicus calculated the distance to the planets. (A) When an inner planet appears in the sky at its farthest point from the Sun, the planet's angle on the sky away from the Sun, A, can be measured. You can see from the figure that at the same time angle B is 90°. The planet's distance from the Sun can then be calculated with geometry, if one knows the measured value of angle A and the fact that the Earth–Sun distance is 1 AU. (B) Finding the distance to an outer planet requires determining how long it takes the planet to move from being opposite the Sun in the sky (the planet rises at sunset) to when the Sun–Earth–planet angle is 90° (the planet rises or sets at midnight). Knowing that time interval, one then calculates what fraction of their orbits the Earth and planet moved in that time. Multiplying those fractions by 360° gives the angles the planet and Earth moved. The difference between those angles gives angle B. Finally, using geometry and the value of angle B as just determined, the planet's distance from the Sun can be calculated.

Planet	TABLE 11.1 Planetary Distances from the Sun According to Copernicus	
	Distance in AU According to Copernicus	Actual Distance in AU
Mercury	0.38	0.39
Venus	0.72	0.72
Earth	1.00	1.00
Mars	1.52	1.52
Jupiter	5.22	5.20
Saturn	9.17	9.54

assuming each planet's orbit is a concentric circle around the Sun, we can make a geometric construction (or use trigonometry) to find their sizes (Figure 11.10).

Determining the sizes of the other planets' orbits is a little more complicated. For this, Copernicus could examine how long the planets took to move from opposition (180° from the Sun) to **quadrature** (90° from the Sun), as illustrated in Figure 11.10. For a distant planet like Saturn, the time between opposition and quadrature is nearly a quarter of the orbit. However, for a planet like Mars, whose orbit is not much bigger than the Earth's, the portion of the orbit between opposition and quadrature is a fairly small fraction of the total orbit. Again, a geometric construction can be used to establish the relative size of the planets' orbits.

Thus, with his heliocentric model, Copernicus could provide not only a simple explanation of retrograde motion, but he could also calculate the relative sizes of all of the planets' orbits. He did not know just how far away the Earth was from the Sun, but he could calculate that Jupiter was about five times farther from the Sun than the Earth and that Mercury was about one-third as far. The distances found in this manner must be expressed in terms of the Earth's distance from the Sun—the astronomical unit—whose value was not accurately known until several hundred years later. Table 11.1 illustrates that Copernicus's calculations agree remarkably well with modern values. These distances became an important starting place for those who came later.

11.4   THE COPERNICAN REVOLUTION

Copernicus described his model of a Sun-centered universe in one of the most influential scientific books of all time, *De revolutionibus orbium coelestium (On the Revolutions of the Celestial Orbs)*. Because his ideas were counter to the beliefs of the day, they were met with hostility and skepticism. Copernicus quietly circulated a manuscript of his ideas for about 30 years before allowing the book to be published in 1543. Reportedly, he saw the first copy while on his deathbed.

Some criticism of Copernicus's work was justified because his model could not predict the positions of the planets any better than Ptolemy's. This lack of success arose in part because Copernicus assumed that the planetary orbits were circles, an idea that was disproved only in the next century. To make his model more accurate, Copernicus even resorted to the use of epicycles to adjust his predictions of the planets' positions. Furthermore, his model again raised the question of why no stellar parallax (Unit 10) could be seen.

There were counterarguments to the objections, however. The lack of parallax could be explained if the stars were very distant. On a technical level, Copernicus's model did not require large epicycles to correct his positions—because the circular orbits were not bad approximations to the planets' orbits. Thus it could be said that Copernicus's model was no more successful than Ptolemy's at predicting the positions of planets as observed from Earth, but it offered a simpler explanation of retrograde motion.

The main objection to Copernicus's views of planetary motion was that they ran counter to the widely accepted teachings of the ancient Greek philosopher Aristotle, views supported both by "common sense" and by religious belief at that time. After all, when we observe the sky, it looks as if it moves around us. Moreover, we do not detect any sensations caused by the Earth's motion—it feels at rest. And if the Earth is just one of the planets, it is not the center of the universe. This mixture of rational and irrational objections made even many scientists slow to accept the Copernican view.

However, conditions were favorable for such new ideas. The cultural Renaissance in Europe was at its height; the Protestant Reformation had just begun; Europeans were sailing to the new world; and there was a growing acceptance of the immensity of the universe. In such an intellectually stimulating environment, new ideas flourished and found a more receptive climate than in earlier times. In the late 1500s the English astronomer Thomas Digges argued in favor of the Copernican model, proposing that the stars were other suns at enormous and various distances. The Italian philosopher and monk Giordano Bruno went so far as to claim that the planets around these other stars were inhabited.

The aesthetic appeal of the heliocentric system led to a growing acceptance of the model despite church efforts to suppress it. Bruno was burned at the stake in 1600 for his heretical ideas, but others continued their quest for a deeper understanding of the workings of the cosmos. Are the stars other suns? How can we reconcile motion of the Earth with our everyday experience that it seems stationary? What keeps the planets in their orbits? The steady work of science was on the verge of answering these and many more challenging questions.

Q. Imagine a planet far beyond Saturn. What would its retrograde motion look like? How does the idea of parallax compare to the idea of retrograde motion?

KEY TERMS

conjunction, 96

Copernicus, Nicolas, 96

epicycle, 95

geocentric model, 94

greatest elongation, 96

heliocentric model, 96

inferior conjunction, 96

Occam's razor, 95

opposition, 96

planet, 92

Ptolemy, Claudius, 94

quadrature, 98

retrograde motion, 92

superior conjunction, 96

QUESTIONS FOR REVIEW

1. What is retrograde motion?

2. Compare geocentric and heliocentric models.

3. Describe the different planetary configurations.

PROBLEMS

1. If you lived on Mars, what would the greatest elongation of the Earth be?

2. Make a geometric construction (or use trigonometry) to show that Venus's orbit is about 0.72 times the size of Earth's based on the fact that Venus's greatest elongation is 47°.

3. Mars spends, on average, about 11.5% of its orbit moving from opposition to quadrature. Make a geometric construction (or use trigonometry) to show that Mars's orbit is about 1.52 times the size of Earth's.

TEST YOURSELF

1. A planet in retrograde motion
   a. rises in the west and sets in the east.
   b. shifts westward with respect to the stars.
   c. shifts eastward with respect to the stars.
   d. will be at the north celestial pole.
   e. will be exactly overhead no matter where you are on Earth.

2. Which are the only planets that can be seen in opposition?
   a. All planets with orbits larger than the Earth's
   b. All planets that undergo retrograde motion
   c. All planets with epicycles
   d. All planets that are larger than the Earth
   e. All of the above

3. Occam's razor is used to
   a. discriminate between models based on their simplicity.
   b. measure the angle between the Sun and planets.
   c. explain why planets change the direction of their orbits.
   d. describe the path that the planets follow across the ecliptic.
   e. execute heretics.

PLANETARIUM EXERCISES

All the planetarium exercises use the Starry Night planetarium software provided with your text. You should install the program and familiarize yourself with the various parameters and controls before trying these exercises. When the program asks you to enter a password, simply click "OK" to run Starry Night.

Set the location from which you are viewing to the nearest large city or your latitude and longitude. Accomplish this by clicking on the Location box on the toolbar. Verify that the local time is correct.

1. Use the planetarium program to determine what planets are currently visible in your night sky as well as where they are and when they can best be viewed. After setting the program for the proper time and location, center on each of the visible planets and Uranus (visible with a good pair of binoculars on a dark night). In each case, find the planets visible in the night sky for your location by clicking the Find option on the horizontal toolbar (it looks like a magnifying glass). Move the time forward and back to determine the planets' rising and setting times as well as the constellations they are closest to (click on Guides in the Toolbar, then select Constellations). Record your data. Go outside and verify your data by locating at least one of the planets.

2. Set the time step to 7 days. Use the Selection heading and the Find option to select Mars. Set the field of view to 60 degrees using the toolbar. Step by step, advance the time forward and observe the position of Mars with respect to the background stars and constellations. Advance the steps faster and note the path of Mars. Note that there are times when it appears to move backward (retrograde). About how often does this happen? Consult the text to determine why this happens.

3. Following the same formula used in the previous exercise, follow the paths of Jupiter and Saturn over a year or more. Describe your results in comparison to Mars. How often does retrograde motion take place for them? Discuss your results in relation to their orbital periods and that of the Earth.

12
The Beginnings of Modern Astronomy

Venus
Gibbous phase

Sun

Venus
Crescent phase

Background Pathways: Unit 11

A story is told that in medieval Europe, a gathering of scholars debated how many teeth a horse has—according to Aristotle versus the Bible. The debate went on for hours with no apparent resolution. A young novitiate proposed going out to the stable to look in a horse's mouth and count. After being castigated for his unscholarly proposal, he was ejected from the assembly!

This story may be no more than legend, but it illustrates a difference in scientific approach before about 1600. Scholars used to believe that one should look for answers in the writings of great authorities of the past. Scientists today take it for granted that if they have questions or disagreements, they should conduct experiments and make new observations to settle the issue. In the century following Copernicus's proposal that the Earth orbited the Sun, new observations finally settled the issue.

12.1 PRECISION ASTRONOMICAL MEASUREMENTS

Early first steps in solving the problem of astronomical motion were made by the sixteenth-century astronomer Tycho **Brahe** (1546–1601), pronounced *TEE-koh BRAH-hee.* Born into the Danish nobility, Brahe (Figure 12.1) used his position and wealth to indulge his passion for study of the heavens, a passion based in part on his professed belief that God placed the planets in the heavens to be used as signs to mankind of events on Earth. Driven by this interest in the skies, he designed, had built, and used instruments that permitted the most accurate pretelescopic measurements ever made (Figure 12.2). These were something like giant protractors that he could turn to measure positions in the sky with high precision. His instruments allowed him to measure the positions of planets to about 1 **arc minute**, or 1/60th of a degree. This is 30 times smaller than the angular size of the Moon, and about the limit of human vision.

Brahe was more than just a recorder of planetary positions; he studied other astronomical events as well. For example, in 1572, when he saw an exploding star (what we would now call a supernova—Unit 64), he showed that this bright new object maintained a fixed position in the celestial sphere and therefore had to be far beyond the supposed spheres on which the planets move. Likewise, when a bright comet appeared in 1577, he showed that it lay far beyond the Moon, not within the Earth's atmosphere as taught by the ancients. These observations suggested that

FIGURE 12.1
Tycho Brahe (1546–1601).

FIGURE 12.2
Picture of Tycho Brahe's astronomical instruments at Uraniborg.

Q Because the sky moves overhead, Tycho Brahe's time measurements had to be precise within a few seconds to achieve an accuracy of 1 arc minute for positions relative to the horizon. Can you think of ways to measure positions on the celestial sphere that might not require such accurate timing?

the heavens were both more changeable and more complex than previously accepted.

Although Brahe could see the virtues of the simplicity offered by the Copernican model, he was unconvinced of its validity because even his instruments were unable to detect stellar parallax (Unit 10)—the apparent shift in star positions expected if the Earth moved around the Sun. Therefore, he offered a compromise model in which all of the planets except the Earth went around the Sun, while the Sun, as in earlier models, circled the Earth. Brahe was the last of the great astronomers to hold that the Earth was at the center of the universe. Brahe's model won favor with those who sought to maintain the idea of a geocentric universe. It was difficult to disprove because it was essentially the same as the Copernican model, except for the change of perspective about whether the Earth moved around the Sun or vice versa. Others, though, began to find additional reasons for doubting that the Earth was so central to the rest of the universe.

12.2   THE NATURE OF PLANETARY ORBITS

A few years before he died, Tycho Brahe hired a younger assistant, Johannes **Kepler** (1571–1630). They did not get along well, and Kepler (Figure 12.3) was fired and rehired during their brief association. Kepler was a clever and hardworking man with a strong grasp of geometry and unusual ideas. For example, he sought some

FIGURE 12.3
Johannes Kepler (1571–1630).

INTERACTIVE

Kepler's second law

INTERACTIVE

Kepler's third law

fundamental relationship between the spacing between the planets and how tightly various geometrical figures could nest inside each other—such as a sphere inside a cube. Kepler did not, however, think highly of Brahe's planetary model, which he described as a "pretzel."

Upon Brahe's death, his observational data passed into Kepler's hands. Brahe's family accused Kepler of stealing them, but their real value was apparent only after years of Kepler's painstaking work. He was able to derive from this huge set of precise data a detailed picture of the path of the planet Mars. Whereas all previous investigators had struggled to fit the planetary paths to circles, Kepler showed—with Brahe's superb data—that Mars moved not in a circular path but rather in an **ellipse.**

An ellipse can be drawn with a pencil inserted in a loop of string that is hooked around two thumbtacks. If you move the pencil while keeping it tight against the string, as shown in Figure 12.4A , you will draw an ellipse. Each point marked by a tack is called a **focus** of the ellipse. The shape of an ellipse is described by its long and short dimensions, called its *major axis* and *minor axis* (Figure 12.4B) respectively; for most calculations the most important value is half the major axis length, or the **semimajor axis** (Figure 12.4C). If the two foci are brought closer together, the ellipse becomes more nearly circular, and a circle's radius is the same as its semimajor axis.

Kepler's careful measurements of Mars's orbit showed that the Sun is located *not* at the center of the ellipse but off-center, at one focus of the ellipse (Figure 12.5A). Kepler found further that Mars moves faster in its orbit when it is closer to the Sun in a way he could describe geometrically. Imagining a line joining a planet to the Sun, Kepler found that the planet's speed changes so that the line sweeps out equal areas in equal time intervals, as illustrated by the shaded areas in Figure 12.5B. For the areas to be equal, the distance traveled along the orbit in a given time must be larger when the planet is near the Sun. Thus, as a planet moves along its elliptical orbit, its speed changes, increasing as it nears the Sun and decreasing as it moves away from the Sun.

With the elliptical shape and speed variations of the orbit now established, Kepler was able to obtain excellent agreement between the calculated and the observed positions not only of Mars but also of the other planets.

Along with discovering the characteristics of each planet's orbit, Kepler also measured the relative sizes of the orbits. When he compared a planet's orbital size with how long it takes to orbit the Sun—the orbital **period**—Kepler determined that a mathematical relationship described exactly how the period increased with

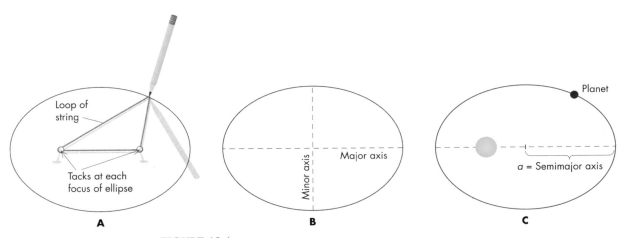

FIGURE 12.4
(A) Drawing an ellipse. (B) The major and minor axes. (C) The Sun lies at one focus of the ellipse.

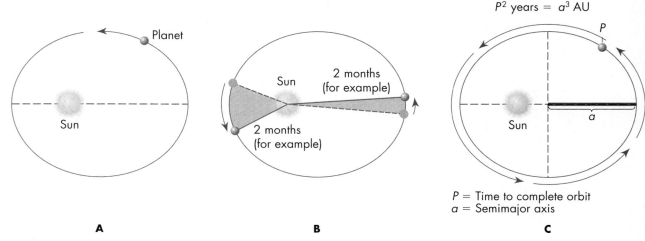

FIGURE 12.5

Kepler's three laws. (A) A planet moves in an elliptical orbit with the Sun at one focus. (B) A planet moves so that a line from it to the Sun sweeps out equal areas in equal times. Thus the planet moves fastest when nearest the Sun. For the drawing a two-month interval is chosen. (C) The square of a planet's orbital period (in years) equals the cube of the semimajor axis or its orbit (in AU), the planet's distance from the Sun if the orbit is a circle.

TABLE 12.1　　Table Illustrating Kepler's Third Law

Planet	Distance from Sun (a) (in Astronomical Units)	Orbital Period (P) (in Years)	a^3	P^2
Mercury	0.387	0.241	0.058	0.058
Venus	0.723	0.615	0.378	0.378
Earth	1.0	1.0	1.0	1.0
Mars	1.524	1.881	3.54	3.54
Jupiter	5.20	11.86	141	141
Saturn	9.54	29.46	868	868

Kepler spent much of his life searching for connections between pure geometry and nature. Where have you encountered geometric patterns in nature?

ANIMATIONS

Kepler's laws

P = orbital period, measured in years

a = semimajor axis, measured in AU

the size of the orbit. The size of an ellipse cannot be described by a single number, but Kepler found that the average of the largest and smallest distances of the planet from the Sun—in other words, the semimajor axis—obeyed his mathematical relationship, as illustrated in Table 12.1. He found that the square of the orbital period measured in years equaled the cube of the semimajor axis measured in astronomical units (Unit 1.6).

Kepler's discoveries of the nature of planetary motions are expressed in what are known today as **Kepler's laws:**

I. **Planets move in elliptical orbits with the Sun at one focus of the ellipse (Figure 12.5A).**

II. **The orbital speed of a planet varies so that a line joining the Sun and the planet will sweep over equal areas in equal time intervals (Figure 12.5B).**

III. **The amount of time a planet takes to orbit the Sun, P, is related to its orbit's size, a (Figure 12.5C):**

$$P^2 = a^3$$

Kepler's three laws describe the essential features of planetary motion around the Sun. They describe not only the shape of a planet's path but also its speed and distance from the Sun.

Kepler's laws are *empirical* findings—that is, they are determined directly from measurements rather than from a theoretical idea about what causes the planets to move. They remain valid today, even though we no longer agree with many of Kepler's own interpretations of their causes. Later scientists discovered that these laws can all be derived from more fundamental laws about the nature of motion and gravity.

Describing the heavens with mathematical laws revolutionized our way of thinking about the universe. Without such mathematical formulations of physical laws, much of our technological society would be impossible. Kepler's laws were a major breakthrough in our quest to understand the world around us.

It is perhaps ironic that such mathematical laws should come from Kepler because so much of his work is tinged with mysticism. For example, Kepler's third law evolved from his attempts to link planetary motion to music, using the mathematical relations known to exist between different notes of the musical scale. Kepler even attempted to compose "music of the spheres" based on such a supposed link. Nevertheless, despite such excursions into these nonastronomical matters, Kepler's discoveries remain the foundation for our understanding of how planets move.

The third law has implications for the relative speeds of planets whose orbits are at different distances from the Sun. Because the third law states that $P^2 = a^3$, a planet far from the Sun (larger a) has a longer orbital period (P) than one near the Sun. If the third law had stated that $P = a$ then the speeds of the planets would all have been the same because the distance traveled in the orbit is proportional to the semimajor axis. However, the third law implies that planets nearer the Sun move faster than outer planets. For example, the Earth takes 1 year to complete its orbit; but Jupiter, whose orbit is about 5 times larger than the Earth's, takes about 12 years. Thus the Earth moves 2.4 ($=12/5$) times faster than Jupiter; and as Earth overtakes and passes Jupiter, we perceive it moving backward in retrograde motion, as discussed in Unit 11.

Kepler's third law has other important applications. It allows us to calculate the distance from the Sun of any body orbiting it if we measure the body's orbital period. Or if we measure the semimajor axis, we can predict the period.

Suppose we wish to determine how far Pluto is from the Sun compared with the Earth's distance from the Sun. Although Pluto has not yet completed an orbit since its discovery, we have measured its motion against the background stars to determine how long Pluto takes to circle the Sun, finding from such observations that its orbital period, P, is 248 years. Kepler's third law tells us then that

$$(248)^2 = 61{,}504 = a^3.$$

Taking the cube root of 61,504 gives us $a \approx 39.5$ AU. That is, Pluto is about 40 times farther from the Sun than the Earth is.

For example, the second law is a consequence of the conservation of angular momentum, which we will discuss more fully in Unit 20.

12.3   THE FIRST TELESCOPIC OBSERVATIONS

By longstanding tradition, Galilei is usually referred to by his first name, Galileo.

At about the same time that Tycho Brahe and Johannes Kepler were striving to understand the motion of heavenly bodies, the Italian scientist **Galileo** Galilei (1564–1642) was likewise trying to understand the heavens. However, his approach was dramatically different.

FIGURE 12.6
Galileo Galilei (1564–1642).

Kepler introduced the word *satellite*. When he saw the moons of Jupiter with a small telescope, their motion around the planet made him think of attendants or bodyguards—*satelles* in Latin.

ANIMATIONS
Phases of Venus

Q Brahe argued that the Sun orbited the Earth but that the other planets orbited the Sun. Could Brahe's model explain the phases of Venus as observed by Galileo? Why or why not?

Galileo (Figure 12.6) was interested not just in celestial motion but in all aspects of motion. He studied falling bodies, swinging weights hung on strings, and so on. In addition, he used the newly invented telescope to study the heavens and interpret his findings, and what he found was astonishing.

In looking at the Moon, Galileo saw that its surface had mountains and was in that sense similar to the surface of the Earth. Therefore, he concluded that the Moon was not some mysterious ethereal body but a ball of rock. He discovered that the Sun had dark spots (now known as **sunspots**) on its surface. He noticed that the position of the spots changed from day to day. This showed that the Sun not only had blemishes and was not a perfect celestial orb, but also that it changed. Both these observations conflicted with previously held conceptions of the heavens as perfect and unchanging. In fact, by observing the changing position of the spots from day to day, Galileo deduced that the Sun rotated.

These observations gave evidence that the Sun and Moon are physical bodies. This was a problem for geocentric theories. To explain the rapid motion of all the objects in the sky around the Earth (roughly once every 24 hours), early astronomers supposed the heavenly bodies were made of some light substance that could move at extremely high speeds—perhaps like the beam of a searchlight. It was harder to conceive of mountains of matter swinging around the Earth at such speeds.

Galileo looked at Jupiter and saw four smaller objects orbiting it, which he concluded were moons of the planet. These bodies, known today in his honor as the **Galilean satellites,** proved that at least some bodies in the heavens do not orbit the Earth. Seeing these satellites remain in orbit around Jupiter also refuted the notion that if the Earth moved, the Moon would fall behind it. But perhaps even more important, the motion of these bodies raised the crucial problem of what held them in orbit.

Galileo also looked at Saturn and discovered that it did not appear as a perfectly round disk but that it appeared to have "satellites" that remained stationary on either side of the planet—these were actually Saturn's rings, but his telescope was too small and too crudely made (inferior to today's inexpensive binoculars) to show them clearly. When Galileo examined the faint band of light on the celestial sphere that we call the Milky Way, he saw that it was populated with an uncountable number of stars. This single observation, demonstrating that there were far more stars than previously thought, shook the complacency of those who believed in a simple Earth-centered universe.

Perhaps more than any other observation, Galileo's observations of Venus proved the most decisive in ruling out the geocentric model. Galileo observed that Venus went through a cycle of phases like the Moon, as illustrated in Figure 12.7. Consider what the geocentric model predicts: Venus should be on a crystalline sphere inside the Sun's own crystalline sphere. Because it never reaches an angle larger than 47° from the Sun, sunlight shining on Venus would always illuminate it from behind, like a crescent moon.

By contrast, the heliocentric model predicted that Venus would cross both in front of and behind the Sun from our perspective. When Venus is on the far side of the Sun, its angular size is small (because of the large distance), but we see the side of the planet illuminated by the Sun. Therefore its phase is nearly full. On the other hand, when it is passing between the Earth and the Sun, it will be angularly large, but we are facing mostly the dark side of the planet, with only a thin crescent illuminated. This behavior was exactly what Galileo found, leaving no doubt that Venus orbited the Sun.

Galileo's probings into the laws of nature led him into trouble with religious "law." He was a vocal supporter of the Copernican view of a Sun-centered universe and wrote and circulated his views widely and somewhat tactlessly. He presented his arguments as a dialog between a wise teacher and a nonbeliever in the

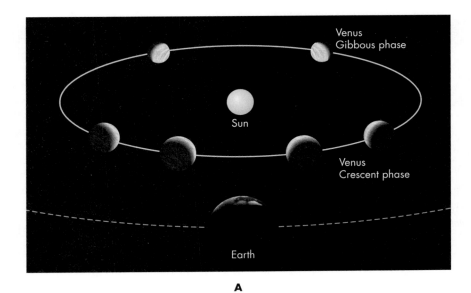

A

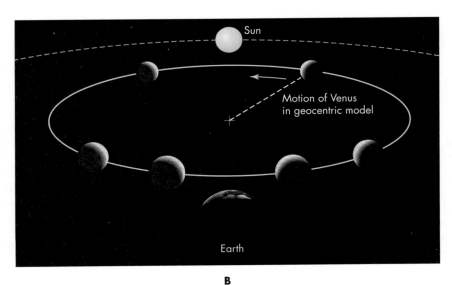

B

FIGURE 12.7

As Venus orbits the Sun, it goes through a cycle of phases (A). The phase and its position with respect to the Sun show conclusively that Venus cannot be orbiting the Earth. The gibbous phases Galileo observed occur for the heliocentric model but cannot happen in the Earth-centered Ptolemaic model, as illustrated in (B).

Copernican system whom he called *Simplicio* and who, according to his detractors, was patterned after the pope. Although the pope was actually a friend of Galileo, conservative churchmen urged that Galileo should be brought before the Inquisition because his views that the Earth moved were counter to the teachings of the Catholic Church. Considering that his trial took place when the papacy was attempting to stamp out heresy, Galileo escaped relatively lightly. He was forced to recant his "heresy" and was put under house arrest for the remainder of his life. Only in 1992 did the Catholic Church admit it had erred in condemning Galileo for his ideas.

Galileo's contributions to science would be honored even had he not made so many important observational discoveries. Beyond his revolutionary work in astronomy, he carried out fundamental experiments in physics that provided the beginnings of our understanding of the nature of motion and gravity. He was one of the principal founders of the experimental method for studying scientific problems, and his methods provided the direction for a new approach toward studying the cosmos.

KEY TERMS

arc minute, 101

Brahe, Tycho, 101

ellipse, 103

focus, 103

Galilean satellite, 106

Galilei, Galileo, 105

Kepler, Johannes, 102

Kepler's three laws, 104

period, 103

semimajor axis, 103

sunspot, 106

QUESTIONS FOR REVIEW

1. Describe the major astronomical contributions of Brahe, Kepler, and Galileo.

2. What are Kepler's three laws?

3. How did Galileo's observations help to rule out a geocentric model of the solar system?

PROBLEMS

1. Suppose a planet was found with an orbital period of 64 years. How might you estimate its distance from the Sun? If its orbit is circular, what is its radius?

2. A planet is discovered orbiting a nearby star once every 125 years. If the star is identical to the Sun, how could you find the planet's distance from its star? If the planet's orbit is a perfect circle, how far from the star is the planet in AUs?

3. All objects orbiting the Sun obey Kepler's laws. If a comet orbits the Sun and reaches 1 AU at its closest approach to the Sun, and its orbital period is 27 years, what is its maximum distance from the Sun?

TEST YOURSELF

1. Galileo used his observations of the changing phases of Venus to demonstrate that
   a. the Sun moves around the Earth.
   b. the universe is infinite in size.
   c. the Earth is a sphere.
   d. the Moon orbits the Earth.
   e. Venus follows an orbit around the Sun rather than around the Earth.

2. Kepler's third law
   a. relates a planet's orbital period to the size of its orbit around the Sun.
   b. relates a body's mass to its gravitational attraction.
   c. allowed him to predict when eclipses occur.
   d. allowed him to measure the distance to nearby stars.
   e. showed that the Sun is much farther away than the Moon.

3. Which of the following did the models of Copernicus and Kepler have in common?
   a. Planets move in elliptical orbits.
   b. The inner planets move faster than the outer planets.
   c. The motions of the planets are uniform.
   d. All of the above

13 Observing the Sky

Experiencing the astronomical changes that occur each night, or over several days or months, will give you a much deeper appreciation of the cosmos. This unit offers a number of projects for exploring the sky and connecting your observations to the models and theories discussed in earlier units.

There are many important aspects of the universe you can learn by simply watching the sky. You do not need fancy equipment. Just a star atlas or a computer with planetarium software can help you get oriented as you look at the sky. Best of all, stargazing is fun and connects you to cultural traditions reaching into prehistory as well as current scientific research. If you find astronomical observation rewarding, there are opportunities to engage in more serious pursuits, such as searching for new comets or monitoring variable stars. These possibilities are discussed further in Unit 31.

Background Pathways: Units 7, 8, and 11

13.1 LEARNING THE CONSTELLATIONS

One of the best ways to get started in your studies of the sky is to learn the constellations. All you need is a star chart, a dim flashlight, and a place that is dark and has an unobstructed view of the night sky. A foldout star chart is located in the back of this book. The star chart gives directions for how to hold it so that you can begin matching it to the sky for the date and time at which you are observing. The planetarium software program included with this book can provide star charts tailored to your location and time.

Start by determining which way is north, using a compass if necessary. Then try to locate a few of the brighter stars, matching them up with the chart. For example, a large asterism called the Summer Triangle spans three constellations. It consists of the three bright stars conspicuous in the summer evening: Deneb (in Cygnus, the Swan), Altair (in Aquila, the Eagle), and Vega (in Lyra, the Harp), shown in Figure 13.1. This will give you some sense of how big a piece of the sky the chart corresponds to. A photograph of this region of the sky is shown in Looking Up #4: Summer Triangle.

Next, try to identify a few star patterns. Focus at first on just some brighter ones. For example, if you live at midlatitude in the Northern Hemisphere (in most of the United States, Canada, Europe, and Asia), the Big Dipper—the asterism that is part of the constellation Ursa Major—is a good group to start with because it is circumpolar for anyone living at a latitude north of about 35°N (see Looking Up #2: Ursa Major).

As you attempt to find and identify stars, your spread hand held at arm's length makes a useful scale. For most people, a fully spread hand at arm's length covers an angular size of about 20° of sky, or about the length of the Big Dipper from tip of handle to bowl, as shown in Figure 13.2. For smaller distances, you can use finger widths: held at arm's length, your thumbnail is about 2° wide, and your little fingernail is about 1° wide.

This scaling of sky distances with your hand makes it easy to point out stars to other people. For example, you can say that a star is two hands away from the Moon and at the 4 o'clock position, as illustrated in Figure 13.3.

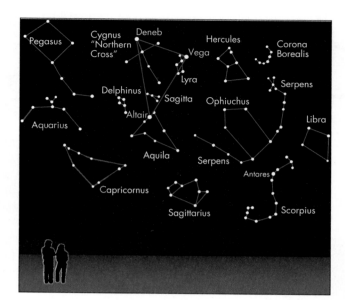

FIGURE 13.1

Dominating the night sky in July, August, and September are the three bright stars Vega, Altair, and Deneb, which form the Summer Triangle. This sketch shows almost half of the whole sky looking south (from mid-northern latitudes) at about 8 P.M. in early September.

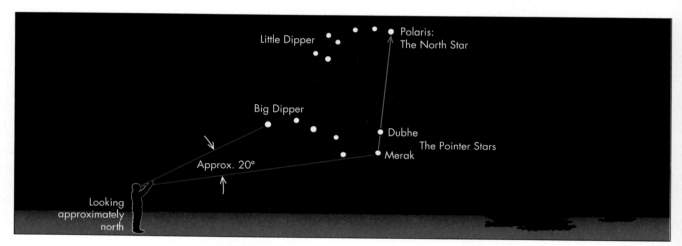

FIGURE 13.2

The Big Dipper, part of the constellation Ursa Major, the Great Bear. A line through the two pointer stars points toward Polaris. The Big Dipper spans about 20° of the sky, about a hand width at arm's length for most people. The sky is shown approximately as it looks in early October at about 8 P.M.

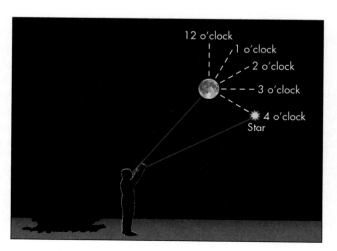

FIGURE 13.3

Describing the location of stars by clock position. The star is two hands from the Moon and at the 4 o'clock position.

Once you recognize a few constellations, you may find that learning the stories behind them will help you find and remember their shapes and locations. Star lore is part of virtually all cultures. The ancient Greeks, the Pawnee tribes of the American Midwest, and the Australian aborigines, for example, all created stories about the star groupings they saw in the sky. Because the star groupings do not change except on time scales of tens of thousands of years, the night sky we see is essentially the same night sky that ancient peoples saw. Star lore can therefore link us to our ancestors in the remote past.

It has been suggested that many such stories were created as aids to memory, especially important when familiarity with the stars could be literally a matter of life or death to a farmer or a navigator. Scientists have even shown that baby birds learn to recognize star patterns and movements and use them to navigate safely—unguided by their parents—across thousands of miles of ocean to their winter homes. Perhaps we too have such instinctive faculties that help us learn the stars.

Probably the most familiar star grouping in the Northern Hemisphere is the Big Dipper (Figure 13.4). The Big Dipper is not only easy to spot, but it is also an excellent signpost to other asterisms and stars. For example, the two stars at the end of its "bowl" away from the "handle" (see Figure 13.2) are called the "pointers" because they point, roughly, to the North Star, Polaris, about 30° (1½ handspreads) away. If you extend the arc formed by the stars in the handle of the Big Dipper, you will find a path that curves to the bright star Arcturus ("follow the arc to Arcturus") in the constellation Boötes (*boh-OH-teez*). Photographs of this part of the sky are also shown in Looking Up #1: Northern Circumpolar Constellations and #2: Ursa Major.

Polaris lies almost exactly above the Earth's North Pole, and because of its position there, it is the only star in the northern sky that shows, to the naked eye, no obvious motion during the night. Its relatively fixed position is illustrated by the time

FIGURE 13.4

Perseus, Andromeda, Cassiopeia, Cepheus, Cetus, and Pegasus. The sky is drawn as it looks in early November at about 8 P.M., looking approximately northeast (from mid-northern latitudes).

Polaris

FIGURE 13.5

A time exposure showing how Polaris remains essentially fixed while the sky appears to pivot around it.

exposure in Figure 13.5, showing the other stars rotating around it. Because Polaris always lies nearly true north, it is useful in orienting yourself to compass directions. Polaris marks the end of the handle of the Little Dipper, another asterism and part of the constellation Ursa Minor, the Little Bear.

The native inhabitants of North America had a story about the Big Dipper. Its bowl represented a huge bear, and the handle represented three warriors in pursuit of the bear. They had wounded it, and it was bleeding. The red color of the leaves in autumn was said to be caused by the bear's blood dripping on them when the constellation lies low in the sky during the evening hours of the autumn months.

Stories are also told about stars in other parts of the sky. For example, in the winter sky you can see the sad legend of the Hunter, Orion (see Looking Up #6: Orion), and the maiden who refused to fall in love with him. The story also involves Orion's hunting dogs (Canis Major and Canis Minor), a bull (Taurus), a rabbit (Lepus), the maiden's sisters (the Pleiades—a cluster of stars in the constellation Taurus—see Looking Up #5: Taurus), and a scorpion (Scorpius).

The king of the island Chios had a lovely daughter, Merope. His island was filled with savage beasts, and to rid his kingdom of these dangerous animals, the king called on Orion to kill the beasts and make his kingdom safe. When the task was done, Orion met Merope and made unwelcome advances. In punishment, he was blinded by the king, but after doing penance, he had his sight restored. After reaching an old age, however, Orion one day stepped on a scorpion, which stung and killed him. On his death, the gods placed him in the sky with his faithful dogs (one of whom chases Lepus, the rabbit), forever attacking the wild bull, Taurus. Beyond the bull, Merope and her sisters (the Pleiades) run from the hunter, who pursues them each night across the sky. The scorpion was also placed in the sky, but on the other side of the heavens so that Orion would never again be threatened by it. (Orion is visible in the evening only in the winter, whereas Scorpius is visible in the evening only in the summer.) Like many sky myths, the Orion story has several versions, but the one described here fits together many of the astronomical references.

There are many other stories about constellations, but the one just described may give you some sense of those that have been handed down over thousands of

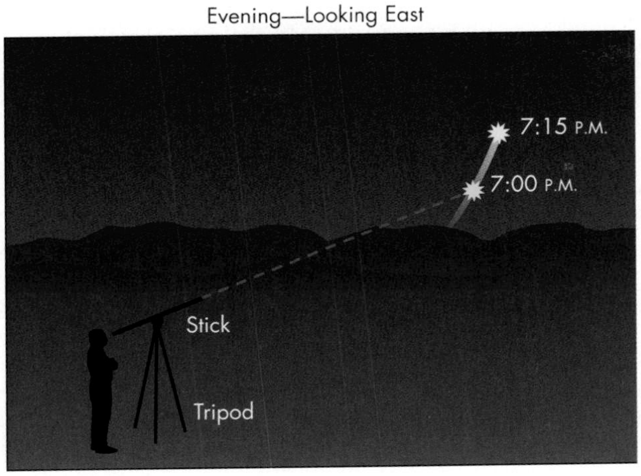

Evening—Looking East

7:15 P.M.

7:00 P.M.

Stick

Tripod

FIGURE 13.6
A sketch illustrating how to observe the motion of the stars across the sky by sighting along a stick.

years of written and oral history. Explore these stories as you learn the constellations: They will help you remember the relative locations in the sky of the various constellations.

13.2   MOTIONS OF THE STARS

Many people are surprised when they are told that the stars rise and set and move across the sky in much the same way that the Sun does. However, it is easy to show that they do.

Use a tripod or a stick that you can poke into the ground so it will stand upright. Get a second straight stick that you can tape or affix to the upright in some manner. A ruler taped to a camera tripod would be ideal, as sketched in Figure 13.6.

Find a bright star and sight along the top stick toward the star. If you now wait five or so minutes and again sight along the stick, you will see that it no longer points to the star. That is, the star has moved so that it now lies west of where the stick is pointing. You can do this experiment indoors if you have a window on which you can put a small mark. Set up a chair by the window so you can watch a star through the glass. While you remain seated, place a mark on the glass where the star appears to be with a grease pencil or piece of tape. Again, wait a few minutes. The star's motion will be clearly visible.

13.3   MOTION OF THE SUN

Observing the motions of stars teaches us about the Earth's daily rotation; but by observing the nearest star to us, the Sun, we can detect the effects of the Earth's revolution and tilted axis of rotation. You can observe the Sun's shift north and south on the celestial sphere based on its location at sunset. If you observe the Sun setting from the same location for even a few nights, you can begin to trace the patterns that led ancient peoples to build remarkable structures like Stonehenge.

Find a spot, perhaps on a hill or out a window facing west, where you can see the western horizon in the evening. Sketch the horizon, noting hills, buildings, or trees that might serve as reference marks. Use your hand and fingers to estimate the angular size of features on the horizon, as discussed previously. From your chosen

Although the Sun is often dimmed enough at sunset for safe viewing, it is not always. The only way to be certain it is safe photographing the Sun at sunset is to use a camera that shows you the image of the Sun indirectly on a display screen, such as most digital cameras.

viewing spot, watch the sunset and mark on your sketch where the Sun goes down. Label the date and time.

Make observations for as many consecutive days as you can. You will discover that the Sun's position changes by an obvious amount in a single night near the equinoxes, but much more slowly near the solstices (Unit 6). Observe how the times of sunset change as its position changes. If you enjoy photography, you might try taking a photograph or videotaping the sunset or sunrise. It is extremely important, however, never to look directly at the Sun, especially through any kind of magnifying lenses, because doing so can damage your eyes.

Observe the angle at which the Sun approaches the horizon: Unless you live on the equator, it will not go straight down.

13.4   MOTIONS OF THE MOON AND PLANETS

Determine from the calendar when the Moon is a few days past new so it will be visible in the early evening. Go outside shortly after sunset and look for the Moon in the west near where the Sun went down.

On the foldout star chart in the back of the book, you will find a list of dates along the top. At 8 P.M. the stars below this date lie along the meridian—the north–south line running overhead. The stars below the next date to the left are overhead an hour later, and to the right an hour earlier. Once the sky is dark, you will be able to see most of the stars six hours to the right and six hours to the left of the ones that are overhead. As the sky darkens, locate the brighter stars near the Moon, and then mark on the chart where the Moon is with respect to those stars. Finally, sketch the Moon's shape.

To keep your chart clean for future use, you may want to apply removable stickers or write lightly with a pencil.

Repeat this process for the next four or five nights. The Moon will set about 50 minutes later each evening; note these times and adjust your observing time accordingly. After watching for a few nights, mark out the Moon's path on the star chart. Ideally, you might want to follow the Moon's track for about two weeks, although as the Moon reaches its third quarter (three weeks after new moon), you'll have to stay up late because the third-quarter moon does not rise until about midnight. Early risers who get up before dawn can watch the crescent moon shrink as it approaches the new phase.

You can also use a star chart to study the motion of the planets. If it is visible, Venus is a good choice because it is bright and moves rapidly across the sky. It is often called the **Evening Star** because it stands out as the brightest "star" in the evening sky. However, Venus spends half its time as the **Morning Star** and is sometimes too close to the Sun to be easily seen, so it is not always a convenient target for observation.

ANIMATIONS

Morning and Evening Stars

To locate a suitable planet, you can look in an almanac, which will tell you what constellation the planet is in on the day you are observing. You may also find this information in the local newspaper or online. Because the outer planets move relatively slowly across the sky, you should space out your observations, perhaps marking positions once a week rather than every night.

As you progress in this project, you should examine how closely the planets follow the ecliptic—shown in the star chart by a curving line that crosses both sides of the equator—and whether they are making direct or retrograde motion against the background stars. Retrograde motion occurs for a month or more around the time the planets are closest to and passing each other in their orbits. This is centered on the time of opposition for the outer planets and inferior conjunction for the inner planets (Unit 11).

The interval between these periods of retrograde motion (or any other successive planetary configurations such as opposition or conjunction) is called the

synodic period. The synodic period differs from the planet's orbital period because both the Earth and the other planets move around the Sun. Thus the interval between oppositions is neither an Earth year nor the other planet's orbital period. For example, the Earth takes about two years to catch up to and overtake Mars after an opposition. The Earth overtakes the slower-moving, more distant planets more quickly, and the interval between oppositions is close to a year. Thus the Martian synodic period is about 780 days, whereas the Saturnian synodic period is 378 days.

13.5   A SUNDIAL: ORBITAL EFFECTS ON THE DAY

The preceding projects reveal many basic features of the sky and planetary motion. A surprising amount of more sophisticated information about the Earth's orbit can be learned from careful observations of a sundial. The Sun's motion each day reveals many aspects of the Earth's orbital motion that are not at first apparent.

One of the best kinds of sundial is a flagpole or any other fixed tall pole where you can mark the shadow cast by the top of the pole. For example, if we were to measure the length of the solar day from noon to noon with a stopwatch, we would discover that it is in general *not* exactly 24 hours. We can do this with a flagpole by marking when the shadow lies exactly along a north–south line. The time between successive noons varies by as much as half a minute at different times during the year. Our clocks, of course, do not change speed during the year, but instead use the average length of a day over the year. That average day length is called the **mean solar day,** which has, by definition, 24 hours of clock time.

The difference in length between the mean solar day and the true solar day accumulates to a difference of over 16 minutes between clock time and time based on the position of the Sun at different times of year. This difference is described by the **equation of time,** which is shown graphically in Figure 13.7. The equation of time gives the correction needed on a sundial if it is to give the same time as your watch. The difference may seem just a curiosity today; but for a navigator using the Sun to determine a ship's longitude, it could cause an error of more than 300 km (200 miles) if no corrections were made.

The variation in the solar day arises because of two effects: The Earth's orbit is not circular, and the Earth's axis is tilted with respect to the orbit. Both of these have similarly sized effects on the length of the day; but they follow different patterns that are offset in time, which makes the equation of time quite complicated.

The effect of the Earth's elliptical orbit on the day arises because the Earth moves faster in its orbit when it is near the Sun and slower when it is farther away

FIGURE 13.7
The equation of time is the correction that must be applied to sundial time to determine mean solar time.

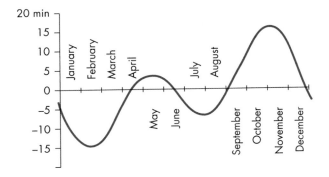

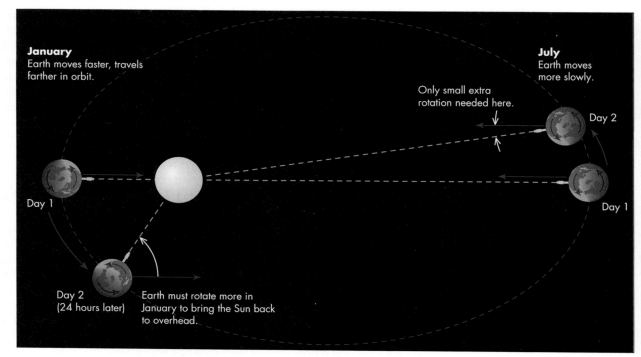

FIGURE 13.8

As the Earth moves around the Sun, its orbital speed changes as a result of Kepler's second law of motion. For example, the Earth moves faster in January when it is near the Sun than in July when it is far from the Sun. Thus in 24 hours the Earth moves farther along its orbit in January than in July. As a result, the Earth must turn slightly more in January to bring the Sun back to overhead. This makes the interval between successive noons longer in January than in July and means they are not exactly 24 hours. For that reason, time is kept using a "mean Sun" that moves across the sky at the real Sun's average rate. (The extremes of the Earth's distance from the Sun have been greatly exaggerated in this figure.)

(Kepler's second law), as illustrated in Figure 13.8. In January, when the Earth is closest to the Sun, the Earth sweeps through a larger angle in its orbit, and the Sun appears to shift farther eastward on the celestial sphere than it does on average. This in turn means the Earth has to rotate a bit farther to face the Sun again, with the interval between successive noons becoming about 10 seconds longer than average. This effect is in the same direction for about half the year, and consequently solar time falls further and further behind clock time. The reverse applies in July and surrounding months, when solar time gets ahead of clock time.

The effect of the Earth's tilted axis on the day is easiest to understand if we think of the Sun as moving around the celestial sphere. Suppose the Earth's orbit was perfectly circular and the Sun shifted by the same distance along the ecliptic each day. In this case the Sun would make its most rapid progress in an eastward direction among the stars when it was moving due east rather than when part of its motion was northward or southward. (This may seem more obvious when we speak of a car driving at a fixed speed; clearly the more directly eastward it is driven, the more progress to the east it will make in one day.) Because the ecliptic is tilted, the Sun moves due east among the stars, parallel to the celestial equator, only during the two solstices. It makes the least eastward progress when its motion north or south is largest: during the equinoxes. As we discussed, the Earth has to turn farther when the Sun's apparent position has shifted farther eastward, and this effect can lengthen the day by up to about 20 seconds at the solstices, shortening it by the same amount near the equinoxes.

Because both effects reach maxima near the end of the year (winter solstice and closest approach to the Sun), the longest day of the year is actually a couple of days after the winter solstice—about 30 seconds longer than average. This is not the same as the length of daylight hours, which are shortest (in the Northern Hemisphere) at this time of year as discussed in Unit 7.

KEY TERMS

equation of time, 115   Morning/Evening Star, 114

mean solar day, 115   synodic period, 115

QUESTIONS FOR REVIEW

1. What methods can you use for describing positions in the sky?
2. Of the Sun, Moon, and planets, which move most quickly and slowly against the stars?
3. Why does sundial time differ from clock time?

PROBLEMS

1. Use your hand and fingers to estimate the angular size of at least three constellations. Sketch the constellations to scale with your measurements.
2. Sailors have "handy" rules for estimating the time until sunset. For example, how many minutes before sunset is the Sun "one finger" above the horizon at the equator? At 45° N or S latitude?
3. If you mark the position of the shadow cast by the top of a flagpole at the same clock time every day throughout the year, you will find that the marks trace out a figure eight pattern. Explain why this happens by referring to the equation of time.

The content of some Units will be enhanced if you have previously studied some earlier Units in this textbook. These Background Pathways are listed below each Unit image.

Unit 17 Measuring a Body's Mass Using Orbital Motion

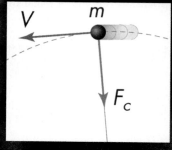

Background Pathways: Unit 16

Unit 14 Astronomical Motion: Inertia, Mass, and Force

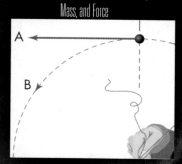

Background Pathways: Unit 12

Unit 18 Orbital and Escape Velocities

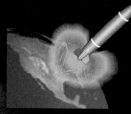

Background Pathways: Unit 16

Unit 21 Light, Matter, and Energy

Background Pathways: Unit 4

Unit 15 Force, Acceleration, and Interaction

Background Pathways: Unit 14

Unit 19 Tides

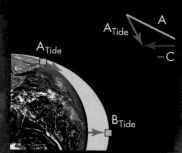

Background Pathways: Units 8 and 16

Unit 22 The Electromagnetic Spectrum

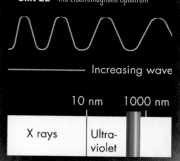

Background Pathways: Unit 21

Unit 16 The Universal Law of Gravity

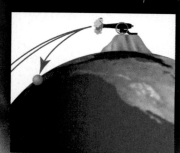

Background Pathways: Unit 15

Unit 20 Conservation Laws

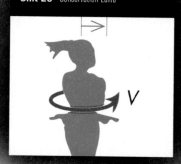

Background Pathways: Unit 16

Probing Matter, Light, and Their Interactions

Unit 23 Thermal Radiation

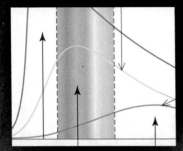

Background Pathways: Unit 22

Unit 24 Atomic Spectra: Identifying Atoms by Their Light

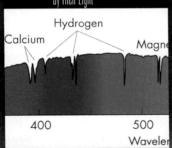

Background Pathways: Unit 23

Unit 25 The Doppler Shift

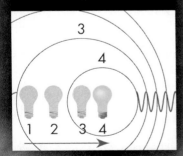

Background Pathways: Unit 24

Unit 26 Detecting Light

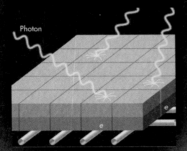

Photon

Background Pathways: Unit 22

Unit 27 Collecting Light

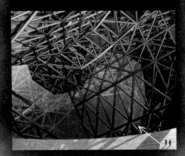

Background Pathways: Unit 26

Unit 28 Focusing Light

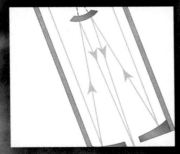

Background Pathways: Unit 26

Unit 29 Telescope Resolution

Background Pathways: Unit 26

Unit 30 The Earth's Atmosphere and Space Observatories

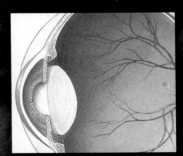

Background Pathways: Unit 22

Unit 31 Amateur Astronomy

Background Pathways: Unit 28

14 Astronomical Motion: Inertia, Mass, and Force

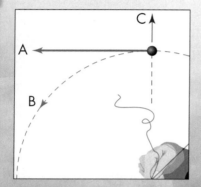

Background Pathways: Unit 12

Astronomers of antiquity did not make the connection between gravity and astronomical motion that we recognize today. People puzzled over why, if the Earth moved, they did not simply fly off into space; and they were also mystified about what kept the planets moving in their orbits.

The solutions to these mysteries began with a series of careful experiments conducted by Galileo Galilei in the early 1600s. Apart from his famous—and perhaps fictitious—demonstration of weights dropped from the Leaning Tower of Pisa, Galileo experimented with projectiles and with balls rolling down planks. The behavior of balls rolling down planks sounds far removed from the behavior of planets, but these basic experiments led him to recognize several properties of motion. A new understanding of forces and motion was essential to make Copernicus's heliocentric model of the Solar System plausible.

14.1 INERTIA AND MASS

Central to Galileo's laws of motion is the concept of **inertia.** Inertia is the tendency of a body at rest to remain at rest and a body in motion to keep moving in a straight line at a constant speed. Galileo's contemporary, Johannes Kepler, introduced the term, but Galileo demonstrated it by real experiment.

In one such experiment, Galileo rolled a ball down a sloping board repeatedly and noticed that it always sped up as it rolled down the slope (Figure 14.1). He next rolled the ball up a sloping board and noticed that it always slowed down as it approached the top. He hypothesized that if a ball rolled on a flat surface and no forces—such as friction—acted on it, its speed would neither increase nor decrease but remain constant. That is, in the absence of forces, inertia keeps an object already in motion moving at a fixed speed.

Inertia is familiar to us all in everyday life. Apply the brakes of your car suddenly, and the inertia of the bag of groceries beside you keeps the bag moving forward at its previous speed until it hits the dashboard or spills onto the floor. We commonly think about the amount of inertia an object has in terms of how heavy it is, but our senses can be fooled—your own body or an object you are carrying may feel lighter underwater or heavier on an amusement park ride.

In scientific terms we measure inertia by an object's **mass.** Mass can be described as the amount of matter an object contains. It is generally measured in kilograms. One kilogram—abbreviated 1 kg—equals 1000 grams. For example, a liter of water (1.1 quarts) has a mass of 1 kg. Different substances may have the same mass in a larger or smaller volume—for example, about one-third liter of rock has a mass of a

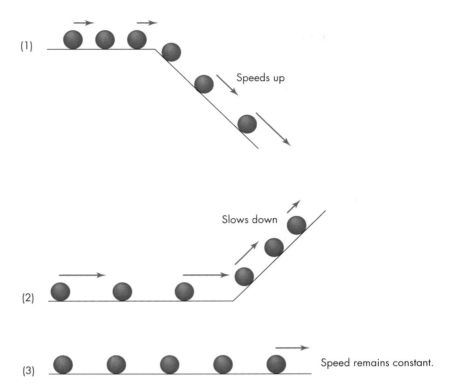

FIGURE 14.1
A ball rolling down a slope speeds up. A ball rolling up a slope slows down. A ball rolling on a flat surface rolls at a constant speed if no forces (including friction) act on it.

Q Suppose you were in an elevator standing on a bathroom scale, and the cable broke, allowing the elevator to fall. What would your weight register on the scale? Why?

 Some scientists prefer to say that *weight* refers only to the gravitational pull on you, and other effects (like buoyancy or orbiting in a spacecraft) change only your *apparent weight*.

kilogram, and about 800 liters of air have this mass. The kilogram standard is a cylinder of an alloy of platinum, copies of which are kept by governments around the world. All other mass measurements are made relative to these official standards.

It is important to remember that mass is not the same as **weight.** Because an object's mass describes the amount of matter in it, its mass in kilograms is a fixed quantity. An object's weight, however, is a measure of the net force on it. Your weight will differ on the Moon because the strength of its gravitational pull on you is different. Your weight can also vary on Earth because of other forces acting on you. For example, the buoyancy of water may make you feel lighter or even weightless. In an elevator, as it first starts rising, you may sense a greater weight—and in fact if you took a bathroom scale into the elevator, you would discover that your weight does increase momentarily. On the other hand, when an elevator starts downward, you will feel momentarily lighter, and in an orbiting space capsule, astronauts feel weightless. We experience changes in weight as a result of the gravitational force on us, other forces acting on us, or the way our surroundings move; but no matter where we are, we have the same mass.

14.2   THE LAW OF INERTIA

From his experiments on the manner in which bodies move and fall, Galileo deduced the first correct "laws of motion." But a more complete understanding was achieved by another scientist, arguably the greatest of all time, who was born the year Galileo died. Isaac Newton (1642–1727) made astounding contributions to mathematics, physics, and astronomy. Moreover, Newton (Figure 14.2) pioneered the modern studies of motion, optics, and gravity. Many of these ideas were conceived when he was 23, forced to stay home from college because the plague was ravaging England. Newton was a fascinating individual. He was a deeply religious man and wrote prolifically on theological matters as well as science.

In his attempts to understand the motion of the Moon, Newton not only deduced the law of gravity; realizing the mathematical methods he needed did not

FIGURE 14.2
Isaac Newton (1642–1727).

How is the term *inertia* used in everyday conversation? How similar is it to the physical meaning?

exist, he invented calculus! What is especially remarkable about Newton's work is that the discoveries he made in the seventeenth century form the basis for calculating the trajectories of spacecraft today.

Newton recognized the special importance of inertia and helped clarify various aspects of it. He described it in what is now called **Newton's first law of motion** (sometimes referred to simply as the *law of inertia*). The law can be stated as follows:

I. A body continues in a state of rest, or in uniform motion in a straight line at a constant speed, unless made to change that state by forces acting on it.

An important point here is that inertia causes an object to resist changes in either speed or direction. This is again exemplified by groceries on a car seat. If the car turns a corner at a constant speed, the grocery bag will slide on the seat and tip over. Its inertia keeps it going in the same direction as before unless you apply a force to it. To keep the bag from falling over as it continues with its former speed and direction, whether you are stopping or turning, you need to apply a force on the bag by, for example, reaching over and holding it.

Because the speed and direction of motion are both important, scientists use a quantity that incorporates both: **velocity.** A velocity might be written as 100 km/hr (or 60 mph) to the northeast. In space we have to define a velocity in three dimensions. A body's velocity changes if either its speed or direction changes. This lets us simplify Newton's first law:

I. A body maintains the same velocity unless forces act on it.

14.3   FORCES

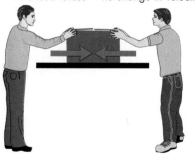

FIGURE 14.3
Balanced forces lead to no change in velocity.

In effect, Newton's first law defines what a **force** is—anything that can cause a body to change velocity. We need to note that when we use the term *force,* we are talking about net force—that is, the total of all forces acting on a body. For example, if a brick is at rest but pushed equally by two opposing forces, the forces are balanced. Therefore the brick experiences no net force and does not move (Figure 14.3).

Newton's first law may not sound impressive at first, but it carries an idea that is crucial in astronomy: If a body is changing speed or moving along a curved path, some net force must be acting on it.

Actually, Newton was preceded in stating this law by the seventeenth-century Dutch scientist Christiaan Huygens. However, Newton went on to develop additional physical laws and—more important for astronomy—showed how to apply them to the universe. For example, if we swing a mass tied to a string in a circle, Newton's first law tells us that the mass's inertia will carry it in a straight line if no forces act. What force, then, is acting on the circling mass? The force is the one exerted by the string, preventing the mass from moving in a straight line and keeping it in a circle. We can feel that force as a tug on our hand from the string, and we can see its importance if we suddenly let go of the string. With the force no longer acting on it, the mass flies off in a straight line, demonstrating the first law, as illustrated in Figure 14.4.

Actually, if we watch the mass in the last example a little longer, we will notice that its path is not completely straight even after we let go of the string. If we release it parallel to the ground, we will notice that the trajectory of the mass begins curving down toward the ground. Newton's great insight was that the Earth must therefore be exerting a force on the object—the force of gravity.

We can translate this example to an astronomical setting and apply it to the orbit of the Moon around the Earth, or the Earth around the Sun, or the Sun around the Milky Way. Each of these bodies follows a curved path. Therefore each must have a force acting on it, a force that can be transmitted through space like an invisible string.

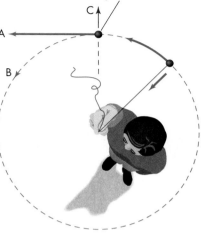

If string is released when ball is here, ball goes straight toward A, *not* toward B, *nor* toward C.

FIGURE 14.4
For a mass on a string to travel in a circle, a force (green arrow) must act along the string to overcome inertia. Without that force, inertia makes the mass move in a straight line.

Side view

Top view

KEY TERMS

force, 122

inertia, 120

mass, 120

Newton's first law of motion, 122

velocity, 122

weight, 121

QUESTIONS FOR REVIEW

1. What is meant by *inertia?*
2. What is the difference between mass and weight?
3. What does Newton's first law of motion tell you about the difference between motion in a straight line and motion along a curve?

PROBLEMS

1. Which do you think has more inertia: a small, inflated balloon or a bowling ball? If each were moving toward you at 1 meter/second, which would be easier to catch? Why?
2. In some amusement park rides, you are spun in a cylinder and are pressed against the wall as a result of the spin. People sometimes describe that effect as being due to "centrifugal force." What is really holding you against the wall of the spinning cylinder? Drawing a sketch and using Newton's first law may help you answer the question.
3. A cinder block can be weightless in space. Would you be hurt if you kicked it with your bare foot? Explain your answer using Newton's first law.

TEST YOURSELF

1. Which of the following demonstrate(s) the property of inertia?
   a. A car skidding on a slippery road
   b. The oil tanker Exxon Valdez running aground
   c. A brick sitting on a tabletop
   d. Whipping a tablecloth out from under the dishes on a table
   e. All of the above
2. If an object moves along a curved path at a constant speed, you can infer that
   a. a force is acting on it.
   b. it is accelerating.
   c. it is in uniform motion.
   d. Both (a) and (b) are true.
   e. Neither (a) nor (b) is true.
3. The mass of a 5 kg bowling ball would be _____ if it were located in deep space, far from any star or planet.
   a. zero
   b. much smaller
   c. slightly smaller
   d. the same

15

Force, Acceleration, and Interaction

Background Pathways: Unit 14

Suppose a force acts on an object. How much deviation from straight-line or, as it is sometimes called, uniform motion will the force produce? To answer that question, we need to define carefully what we mean by *motion*.

Motion of an object is a change in its position, which we can characterize in two ways: by the direction of the object and by its speed. For example, a car is moving east at 40 miles per hour. If the car's speed and direction remain constant, we say it has a constant velocity. If the car changes either its speed or direction, it is no longer moving uniformly, as depicted in Figure 15.1. Any change in velocity is defined as an **acceleration**.

15.1 ACCELERATION

We are all familiar with acceleration as a change in speed. For example, when we step on the accelerator in a car and it speeds up, we say the car is accelerating. Although in everyday usage *acceleration* implies an increase in speed, scientifically *any* change in speed is defined as acceleration. So in scientific terms a car "accelerates" when we apply the brakes and it slows down.

In the previous example we produced acceleration by changing the car's speed. We can also produce acceleration by changing the car's direction of motion. For example, suppose we drive a car around a circular track at a steady speed. At each moment, the car's direction of travel is changing, and therefore its velocity is changing. Similarly, a mass swung on a string or a planet orbiting the Sun is experiencing a change in velocity and is therefore accelerating. In fact, a body moving in a circular orbit constantly accelerates, even if its speed is not changing.

The acceleration of an object is defined as its change in velocity divided by the time taken to change it. This can be written mathematically as

$$\text{Acceleration} = \frac{\text{Change in velocity}}{\text{Change in time}}$$

or using symbols,

$$a = \frac{\Delta V}{\Delta t}.$$

Suppose a car is traveling at 10 meters per second (this is the same as 36 km per hour or about 22 mph) along a straight road. If the car increases its speed to 16 m/sec over 3 seconds, we would say that its change of velocity is $\Delta V = 16 \text{ m/sec} - 10 \text{ m/sec} = 6 \text{ m/sec}$, while the time change is $\Delta t = 3$ sec. The acceleration is therefore

$$a = \frac{6 \text{ m/sec}}{3 \text{ sec}} = 2 \frac{\text{m}}{\text{sec}^2}.$$

Uniform motion
(Same speed (*V*), same direction)

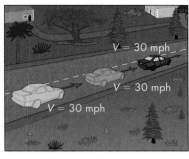

A

Acceleration
(A change in speed)

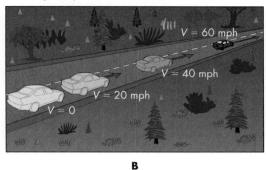

B

Acceleration
(A change in direction)

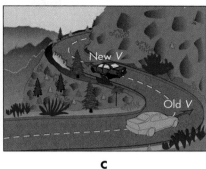

C

FIGURE 15.1

Views of a car in uniform motion and accelerating. (A) Uniform motion implies no change in speed or direction. The car moves in a straight line at a constant speed. If an object's speed (B) or direction (C) changes, the object undergoes an acceleration.

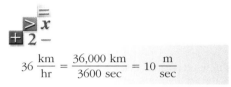

$$36 \, \frac{\text{km}}{\text{hr}} = \frac{36,000 \text{ km}}{3600 \text{ sec}} = 10 \, \frac{\text{m}}{\text{sec}}$$

To describe velocities in two dimensions, we can use "vector addition." The speed eastward could be represented by an arrow to the right 10 units long: ———→. The speed northward could be represented by a vertical arrow 1 unit long: ↑. The length of the two added together can be found from the Pythagorean formula: $\sqrt{10^2 + 1^2} = \sqrt{101} = 10.05$. Note that the overall speed of the car barely changes, even though we have changed the component of velocity northward by 1 km/sec.

Acceleration is usually written in units of "m/sec²," and we might say that the car's acceleration during this three-second interval is "two meters per second squared" or "two meters per second per second." What this means is that the speed changes, on average, by 2 m/sec during each second. After 1 second of this acceleration, the car sped up by 2 m/sec and was traveling at 12 m/sec. After 2 seconds, it was traveling at 14 m/sec. After 3 seconds, it was traveling at 16 m/sec.

Suppose the driver turns the car slightly to avoid an obstacle in the right lane. The driver maintains the same speed eastward (10 m/sec), but at the end of the turn, the car now has a component of motion 1 m/sec northward. Depending on how quickly the turn is completed, the acceleration we would feel inside the car might be large or small. If the direction change is done gradually over 2 seconds, the acceleration is small: $a = 1 \text{ m/sec}/2 \text{ sec} = 0.5 \text{ m/sec}^2$—a gentle push to the north. But if it is made quickly, in just 0.2 seconds, the acceleration is much larger: $a = 1 \text{ m/sec}/0.2 \text{ sec} = 5 \text{ m/sec}^2$—an acceleration that might topple a bag of groceries.

How do we produce acceleration? Newton realized that for a body to accelerate, a force must act on it. By pressing the accelerator pedal, we make the engine run faster and transmit a force to the car by turning the tires faster. By turning the steering wheel, we transmit a sideways force to the car that changes its direction of motion. For example, to accelerate—change the direction of—a mass whirling on a string, we must constantly exert a pull on the string. Similarly, to accelerate a shopping cart, we must exert a force on it. In addition, experiments show that the acceleration we get is proportional to the force we apply. That is, greater force produces a larger acceleration. For example, if we push a shopping cart gently, its acceleration is slight. If we push harder, its acceleration is greater. But experience shows us that more than just force is at work here. For a given push, the amount of acceleration also depends on how full the cart is. A lightly loaded cart may scoot away under a slight push, but a heavily loaded cart hardly budges given the same push, as illustrated in Figure 15.2. Thus the acceleration produced by a given force also depends on the amount of matter being accelerated.

15.2   NEWTON'S SECOND LAW OF MOTION

With an understanding of how to measure the acceleration of an object, we are now prepared to write an equation that can describe how forces affect the motions of any object in the universe. This is **Newton's second law of motion,** and

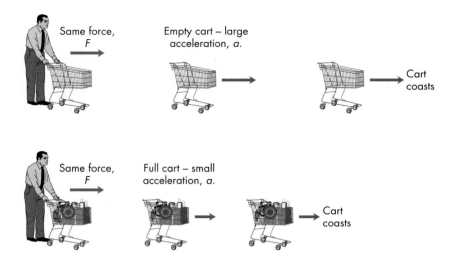

FIGURE 15.2
A loaded cart will not accelerate as easily as an empty cart.

In MKS units (Unit 3), the unit of force is appropriately enough called the *newton,* which is the force needed to accelerate a 1 kg mass to a speed of 1 m/sec in 1 second (1 newton = 1 kg · m/sec²). This is one-tenth the downward force of a 1 kg mass in Earth's gravitational field, and about a quarter pound of force in English units.

a = Acceleration (in m/sec²)

F = Force (in newtons or kg · m/sec²)

m = Mass (in kg)

Q In what situations in everyday life have you experienced Newton's second law?

it is surprisingly simple. Mathematically, in its most familiar form, the law states this:

$$F = m \times a.$$

In words,

I. The force (F) acting on an object equals the product of its acceleration (a) and its mass (m).

The way the equation is commonly written, as it is here, would be useful if we wanted to determine what force must be acting on an object of known mass that is undergoing a measured acceleration. Forces are expressed in the MKS unit **newtons.**

Much more often we are interested in predicting how we will change the velocity of an object (accelerate it) when applying a known force. Thus a useful way of thinking about Newton's formula is this:

$$a = F/m.$$

In words,

II. The amount of acceleration (a) that a body will experience is equal to the force (F) applied divided by the body's mass (m).

Another way of saying this is that the acceleration is proportional to the force and inversely proportional to the mass of the object.

Incidentally, we can also write Newton's second law as $m = F/a$. This form is useful for measuring the mass of an object, independent of the gravitational pull it is experiencing—or even if it is floating, weightless, in space. This is the way astronauts measure their mass while in orbit. A known force is applied, and the resulting acceleration is measured; these numbers are "plugged into" the equation, and the mass is calculated.

Astonishingly, this simple equation allows scientists to predict virtually all features of a body's motion. With $a = F/m$ and with knowledge of the masses and the forces in action, engineers and scientists can, for example, target a spacecraft safely between Saturn and its rings by setting off its thrusters to produce forces on the spacecraft to change its velocity (speed and direction) by a predictable amount even though it is millions of miles away from its target and hundreds of millions of miles away from the Earth.

15.3   ACTION AND REACTION: NEWTON'S THIRD LAW OF MOTION

Newton's studies of motion led him to yet another critical law, which relates the forces that bodies exert on each other. This additional relation, **Newton's third law of motion,** is sometimes called the *law of action–reaction:*

> **III. When two bodies interact, they create equal and opposite forces on each other.**

This law is sometimes counterintuitive because we usually think of one object supplying a force and another receiving it; but the force is felt by both with equal intensity. Two skateboarders side by side may serve as a simple example of the third law (Figure 15.3). If X pushes on Y, both move. According to Newton's law, when X exerts a force on Y, Y exerts a force on X so that both accelerate.

Suppose, though, that one of the skateboarders has a much larger mass than the other. No matter which of the skateboarders pushes on the other, the one with the smaller mass will move away faster. The force F acting on each is the same, but because of Newton's second law the resulting accelerations differ: $a = F/m$. The skateboarder with the smaller mass m will have the larger acceleration because m appears in the denominator. A larger force would have to be applied to the heavier skateboarder to make him or her accelerate as much as the lighter skateboarder.

The gravitational force between the Earth and the Sun affords an astronomical example of Newton's second and third laws and at the same time leads us a step closer to understanding gravity. According to Newton's third law, the gravitational force of the Earth on the Sun must be exactly the same as the gravitational force of the Sun on the Earth. Why, then, does the Earth orbit the Sun and not the other way around? The answer is the same as for the skateboarders: Because of Newton's second law, $a = F/m$. Thus, even though the forces acting on the Earth and Sun are precisely equal, the Sun accelerates 300,000 times less because it is 300,000 times more massive than the Earth. Because the Earth's acceleration is so much larger

Q Suppose the skateboarder pushed off a wall instead of another skateboarder. How does Newton's third law apply?

FIGURE 15.3

(A) Skateboarders illustrate Newton's third law of motion. When X pushes on Y, an equal push is given to X by Y. (B) When X pushes on a much heavier person Z, an equal and opposite push is given to each, but Z moves off at a smaller speed.

A   **B**

FIGURE 15.4
A hammer thrower feels an equal and opposite force matching the centripetal force on the swinging hammer.

than the Sun's, the Earth does most of the moving. In fact, however, the Sun does move a little bit as the Earth (and every other planet) orbits it, much as you must lean back and move in a circle yourself if you swing a heavy weight around you (Figure 15.4). However, because the Sun is so much more massive than all of the planets, it wobbles by only a fraction of its own radius.

KEY TERMS

acceleration, 124

newton, 126

Newton's second law of motion, 125

Newton's third law of motion, 127

QUESTIONS FOR REVIEW

1. How does acceleration differ from velocity? From force?

2. What is Newton's second law?

3. How can objects move if every force is associated with an equal and opposite force?

PROBLEMS

1. A car driving at 90 kph brakes and comes to a stop in 3 seconds. What was its acceleration during this time in units of meters/second^2?

2. If a 10 kg · m/sec^2 force is applied to an initially stationary 1 kg mass, how fast will it be moving after 1 second? After 2 seconds? Assume no other forces are acting on the mass.

3. If the same force in the previous problem is applied to a 2 kg mass, how fast is it moving after 1 second? What force would be needed to accelerate it at the same rate as the 1 kg mass?

TEST YOURSELF

1. A rocket blasts propellant out of its thrusters and "lifts off," heading into space. What provided the force to lift the rocket?
   a. The propellant pushing against air molecules in the atmosphere
   b. The propellant heating and expanding the air beneath the rocket, and so pushing the rocket up
   c. The action of the propellant accelerating down, giving a reaction force to the rocket
   d. The propellant reversing direction as it strikes the ground below the rocket, then bouncing back and pushing the rocket up

2. Which of the following cases does *not* describe an acceleration?
   a. A car rounding a curve at a steady 50 kph
   b. A car changing its speed from 100 kph to 90 kph
   c. A car falling off a cliff
   d. A race car driving at 200 kph on a straight highway

3. If you apply the same force to two carts, the first with a mass of 100 kg, the second 10 kg, the acceleration of the 100 kg cart will be _____ the acceleration of the 10 kg cart.
   a. 10 times larger than
   b. 10 times smaller than
   c. the same as

16

The Universal Law of Gravity

According to one story Newton realized gravity's role when he saw an apple falling from a tree. The apple falling down to the Earth's surface made him speculate whether Earth's gravity might extend to the Moon. Newton realized that if the Earth's gravitational pull reached all the way to the Moon, it could provide the force that keeps the Moon circling the Earth—like a string pulling on a twirling mass.

16.1 ORBITAL MOTION AND GRAVITY

Much of Newton's work is highly mathematical, but as part of his discussion of orbital motion, he described a thought experiment to demonstrate how a body can move in orbit. Thought experiments are not actually performed; rather, they serve as a way to think about problems. In Newton's thought experiment, we imagine a cannon on a mountain peak firing a projectile (Figure 16.1A). From our everyday experience, we know that whenever a body is thrown horizontally, gravity pulls it downward so that its path is an arc. Moreover, the faster we throw the body, the farther it travels before striking the ground.

Now let us imagine increasing the projectile's speed more and more, allowing it to travel ever farther. However, as the distance traveled by the ball becomes very large, we see that the Earth's surface curves away below the projectile (Figure 16.1B). Therefore, if the projectile moves at the right speed, its curvature downward will

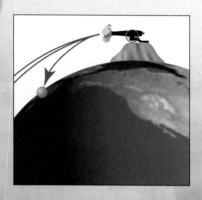

Background Pathways: Unit 15

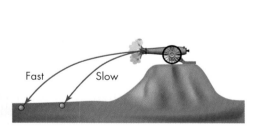

FIGURE 16.1

(A) A cannon on a mountain peak fires a projectile. If the projectile is fired faster, it travels farther before hitting the ground. (B) At a sufficiently high speed, the projectile travels so far that the Earth's surface curves out from under it, and the projectile is in orbit.

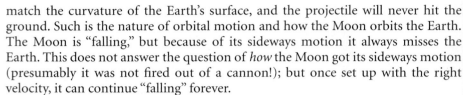

If we continue increasing the speed, the cannonball will begin to swing farther and farther from the Earth in an elliptical path. If its speed is great enough, however, it will escape from the Earth and never return (Unit 18).

ANIMATIONS

Newton's cannon

INTERACTIVE

Gravity variations

Q Why do you suppose most spacecraft are launched to the east? Why are they generally launched from near the equator?

match the curvature of the Earth's surface, and the projectile will never hit the ground. Such is the nature of orbital motion and how the Moon orbits the Earth. The Moon is "falling," but because of its sideways motion it always misses the Earth. This does not answer the question of *how* the Moon got its sideways motion (presumably it was not fired out of a cannon!); but once set up with the right velocity, it can continue "falling" forever.

We can phrase this thought experiment more specifically using Newton's first law of motion. According to that law, in the absence of forces, the projectile would travel in a straight line at constant speed. But because a force—gravity—is acting on the projectile, its path is not straight but curved. Moreover, the law helps us understand that the projectile does not stop because it has inertia.

Notice that in this discussion we used no formulas. All we needed was Newton's first law and the idea that gravity supplies the force. To make further progress—for example, to determine how rapidly the projectile must move to be in orbit—we need laws that have a mathematical formulation.

16.2 NEWTON'S UNIVERSAL LAW OF GRAVITATION

Newton was not the first person to attempt to discover and define the force that holds planets in orbit around the Sun. Nearly 100 years before Newton, Kepler recognized that some force must hold the planets in their orbits and proposed that something similar to magnetism might be responsible. Newton was not even the first person to suggest that gravity was responsible. Other members of the Royal Society in England speculated about gravity's role, but none was able to present a convincing case. Years after developing his initial ideas, Newton published his law of gravity in 1687 in one of the milestones in the history of science, his *Philosophiae Naturalis Principia Mathematica* (Mathematical Principles of Natural Philosophy). He demonstrated the properties that gravity must have if it is to control planetary motion. Moreover, Newton went on to derive the **law of gravity** in mathematical form, allowing astronomers to predict the position and motion of the planets and other astronomical bodies.

The law of gravity must describe the force that acts in many circumstances—a falling apple, the Moon, or the planets—with a single equation. We can begin to see what form this equation must have if we consider all of these different situations. First, it seems clear that gravity must depend on mass because larger bodies, like the Sun, produce a stronger force than smaller bodies like the Earth or an apple. And the pull an object exerts must also depend on the mass of the object being pulled—for example, the Earth must pull on the Moon with a far larger force than it does on the apple in order to overcome the Moon's larger inertia. Newton determined that the only way to explain these diverse situations and make the law of gravity consistent with his laws of motion was that the gravitational force between two bodies must depend on the product of their masses. Finally, Newton determined that the force must grow weaker with distance to explain the slower speeds of the outer planets in their orbits about the Sun (Kepler's third law, Unit 12.2).

Newton analyzed these issues and concluded the following:

> **Every mass exerts a force of attraction on every other mass. The strength of the force is directly proportional to the product of the masses divided by the square of the distance between them.**

An important note about the distances used in this calculation: it is the distance between the centers, or technically the **centers of mass,** of the two objects. For a spherical object the center of mass is at the center, but for more complicated shapes

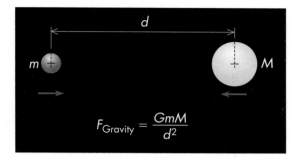

FIGURE 16.2
Gravity produces a force of attraction (green arrows) between bodies. The strength of the force depends on the product of their masses, *m* and *M*, and the square of their separation, *r*. *G* is the universal gravitational constant.

ANIMATIONS

Force of gravity

F_G = Force of gravity

G = Newton's gravitational constant

M, m = Masses of objects attracting each other

d = Distance between their centers

Because the unit of force, the newton, is defined as 1 kg · m/sec², we can also write this as $G = 6.67 \times 10^{-11}$ m³/sec² · kg.

Q Does the saying "The bigger they are, the harder they fall" make sense given Newton's laws and the law of gravity?

it is essentially a balance point for the mass distribution—for your body, this is roughly at the center of your pelvis. Thus, if you are standing on the Earth, the distance used to calculate the gravitational force exerted on your body is not zero but approximately the radius of the Earth, or about 6380 km.

We can write this extremely important result in a shorthand mathematical manner by defining several algebraic variables. Let *m* and *M* be the masses of the two bodies (Figure 16.2) and let the separation between their centers be *d*. Then the strength of the gravitational force between them, F_G, is

$$F_G = G\frac{M \times m}{d^2}.$$

The factor **G** is a constant, a conversion factor, that lets us translate from the units of mass and distance on the right side of the equation to units of force for F_G. The value of *G* is found by measuring the force between two bodies of known mass and separation—for example, two large lead masses in a laboratory. The resulting number for *G* depends on the units chosen to measure *M*, *m*, *d*, and F_G. For example, if *M* and *m* are measured in kilograms, *d* in meters, and F_G in newtons, then

$$G = 6.67 \times 10^{-11} \text{ newtons} \cdot \text{meters}^2/\text{kilogram}^2.$$

As long as we make measurements using the same units, *G* is the same whether we are dealing with stars, planets, or apples.

Writing the law of gravity as an equation helps us see several important points. If either *M* or *m* increases, and the other factors remain the same, the force increases by the same amount. We call this a *direct proportionality*. On the other hand, if *d* (the distance between the objects) increases, the force gets weaker. Moreover, the force weakens as the square of the distance. That is, if the distance between two masses is doubled, the gravitational force between them decreases by a factor of four, not two. We call this an *inverse-square proportionality*.

Finally, the law of gravity shows us that even though one body's gravitational force on another weakens with increasing distance, the gravitational force never completely disappears. Thus the gravitational attraction of a body reaches across the entire universe. The Earth's gravity not only holds you onto its surface but also extends to the Moon and exerts the force that holds the Moon in orbit around the Earth. Earth's pull extends even to distant stars and galaxies, although its pull is minuscule and just one among the forces from countless other objects.

16.3 SURFACE GRAVITY AND WEIGHT

Recall Galileo's observation that balls with different weights dropped from a height all strike the ground simultaneously. This seems counterintuitive at first because we feel a greater force downward when we heft a large mass than a light one. In fact a feather does fall more slowly than a cannonball if there is air resistance, but in a vacuum they fall at the same speed.

The Earth *does* pull with a larger force on a massive object, but this is counter-balanced by the object's greater inertia. It requires a greater force to accelerate an object with larger inertia, and gravity provides just the right force to accelerate all objects at the same rate on our planet's surface. We can show this mathematically using Newton's laws.

Consider the gravitational acceleration on an object with mass m dropped near the surface of Earth. The Earth has a mass $M_{\oplus}$ and a radius $R_{\oplus}$. The distance between the center of the object and the center of the Earth is approximately $R_{\oplus}$, so we can use $R_{\oplus}$ as a close approximation to the actual distance. We can use Newton's second law, $a = F/m$, to calculate the acceleration due to the Earth's gravity. This acceleration is usually written as g. Using Newton's universal law of gravity, we find

$$g = \frac{F_G}{m} = G\frac{m \times M_{\oplus}}{m \times R_{\oplus}^2} = G\frac{M_{\oplus}}{R_{\oplus}^2}.$$

This is often simply called the **surface gravity.**

Note that the mass of the falling object does not matter. The surface gravity, g, depends only on the mass and radius of the Earth and the gravitational constant G. If we "plug in" the values for these quantities, we can find the acceleration all objects experience at the Earth's surface:

$$g = 6.67 \times 10^{-11}\, \frac{\text{newton} \cdot \text{m}^2}{\text{kg}^2} \times \frac{5.97 \times 10^{24}\,\text{kg}}{(6.37 \times 10^6\,\text{m})^2}$$

$$= 9.81\frac{\text{newton}}{\text{kg}}$$

$$= 9.81\frac{\text{m}}{\text{sec}^2}$$

In the last step we have used the definition of a **newton:** 1 newton = $1\,\text{kg} \cdot \text{m/sec}^2$. Thus all objects at the Earth's surface accelerate at this same rate. After falling for one second, they have a velocity of 9.81 m/sec (about 35 kph or 22 mph) downward; after two seconds, 19.62 m/sec (70 kph, 44 mph); after three seconds, 29.43 m/sec (105 kph, 66 mph)—increasing by 9.81 m/sec each second.

The "g" is often used in describing other rates of acceleration. For example, you may experience up to about 5 g's on an amusement park ride or race car, and automobile airbags are triggered by a negative acceleration (*deceleration*) of about 10 g's.

We can also compare the rates of acceleration experienced at the surface of other astronomical bodies. The Moon's surface gravity is about 0.17 g—that is, you would weigh about 17% as much on the Moon as on the Earth. Figure 16.3

Q In Figure 16.3, why do you suppose the flag appears to be waving? Can you locate the astronaut's shadow? How might you use the shadow to determine how high he is jumping?

FIGURE 16.3
Astronaut John Young making a jumping salute at the *Apollo 16* lunar landing site near Descartes Crater. Despite a total mass of over 170 kg (370 pounds) in his space suit, he easily jumps up because of the Moon's weak gravity.

shows astronaut John Young jumping about a meter in the air as he salutes while wearing a spacesuit with a mass of about 130 kg (about 280 pounds). On the Moon your mass remains the same; but your weight depends on the forces you are experiencing, and the Moon's force on you at its surface is only 17% of the Earth's force.

KEY TERMS

center of mass, 130

law of gravity, 130

g (acceleration due to
    gravity at Earth's surface), 132

newton (unit of force), 132

surface gravity, 132

G (a constant), 131

QUESTIONS FOR REVIEW

1. Describe the variables that determine the force of gravity between two objects.

2. What is the function of the gravitational constant G?

3. What is surface gravity?

PROBLEMS

1. Suppose you were standing at the top of a *very* tall tower, 6370 km tall. What would be the gravitational force of the Earth on you at the top of this tower compared to the force you feel standing on Earth's surface? (Hint: You do not have to do as much arithmetic if you work this out with proportions.)

2. The Moon's radius is 1.74×10^6 m, while its mass is 7.35×10^{22} kg. Find the surface gravity on the Moon from these values.

3. The Sun's radius is 6.97×10^8 m, while its mass is 1.99×10^{30} kg. Find the surface gravity on the Sun from these values. How much would you weigh if you could stand on the Sun's surface?

TEST YOURSELF

1. The Earth's mass is about 80 times larger than the Moon's. What is the ratio of the gravitational force of the Earth on the Moon to the gravitational force of the Moon on the Earth?
    a. 80 to 1
    b. 1 to 80
    c. 1 to 1
    d. 80^2 to 1
    e. 1 to 80^2

2. If the distance between two bodies is quadrupled, the gravitational force between them is
    a. increased by a factor of 4.
    b. decreased by a factor of 4.
    c. decreased by a factor of 8.
    d. decreased by a factor of 16.
    e. decreased by a factor of 64.

3. Gravity
    a. is the result of the pressure of the atmosphere on us.
    b. occurs between objects that are touching each other (or that are both touching the atmosphere).
    c. is the force larger objects exert on smaller ones.
    d. is the attraction between all objects that have mass.
    e. is caused only by planets and the Sun.

17

Measuring a Body's Mass Using Orbital Motion

Background Pathways: Unit 16

Knowledge of orbital motion is important for more than simply understanding the paths of astronomical objects. From properties such as the size and period of an orbit, astronomers can deduce the masses of one or both of the orbiting objects.

The method for determining an astronomical object's mass was first worked out by Newton using his laws of motion and gravity. The underlying idea is simple: The masses of the orbiting bodies determine the gravitational force between them. The gravitational force in turn sets the properties of the orbit. Thus, from knowledge of the orbit, astronomers can work backward to find the masses of the objects.

17.1 MASSES FROM ORBITAL SPEEDS

We first consider a relatively uncomplicated case: the circular orbit of a body having a mass so small it shifts the position only negligibly of the more massive body. We can therefore treat the more massive object as essentially stationary. These restrictions are met to high precision in many astronomical systems, such as the Earth's motion around the Sun and the Sun's motion around the Milky Way. These assumptions simplify the mathematics, but the results we will find are essentially the same as if we were to consider more complex cases.

From Newton's first law, we know that there must be a force acting on a body that moves along a circular path. This **centripetal force** must be applied to any body moving in a circle, whether it is a car rounding a curve, a mass swung on a string, or the Earth orbiting the Sun.

Newton used calculus (which he invented to help solve this type of problem) to determine the acceleration a body undergoes when it travels in a circle, as illustrated in Figure 17.1. He derived the following equation for the centripetal force, F_C, on a mass m moving with a velocity V at a distance d from the center of the circle:

$$F_C = \frac{m \times V^2}{d}.$$

Without going into the details of the derivation of this formula, we can still understand why it has these dependencies. The *size* of the change in velocity at any moment and the *rate* at which it changes are both proportional to the velocity. Hence the result depends on V^2. The turns become tighter and the acceleration is greater if the circle is smaller, which gives an inverse dependence on the distance from the center. Finally, the mass term comes from Newton's second law—a larger mass takes a greater force to turn.

F_C = Centripetal force

m = Mass moving on circular path

V = Velocity of circular motion

d = Distance from center of circular motion

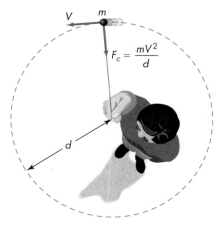

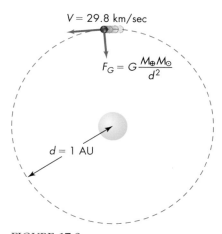

FIGURE 17.1

The centripetal force, F_C, depends on the mass and speed at which an object swings in a circle as well as its distance from the center.

FIGURE 17.2

The gravitational force between the Sun and the Earth holds the Earth in its nearly circular orbit.

Q If you are rounding a corner in a car, what does the centripetal force equation tell you about what conditions might make the car skid?

Step-by-step derivation:

$$F_G = F_C$$

$$G\frac{m \times M}{d^2} = \frac{m \times V^2}{d}$$

$$G\frac{M}{d} = V^2$$

$$\frac{d}{G} \times G\frac{M}{d} = \frac{d}{G} \times V^2$$

$$M = \frac{d \times V^2}{G}$$

Step-by-step derivation of the Earth's orbital speed: $V = 2\pi \times d/P = (6.28 \times 1.50 \times 10^{11} \text{ m})/3.16 \times 10^7 \text{ sec} = 2.98 \times 10^4 \text{ m/sec} = 29.8 \text{ km/sec}$.

Using this equation, we can find the mass of the Sun if we know the speed and radius of a planet's orbit around it. Let the Sun's mass be M and the planet's mass be m, the latter of which is assumed to be much smaller than M. Assume the planet moves in a circular orbit at a distance d from the Sun with a velocity V. The gravitational attraction between the Sun and the planet provides the force that deflects the planet from its tendency to move in a straight line, creating the force needed to produce the observed centripetal acceleration.

For an object in a circular orbit, the gravitational force F_G must equal the centripetal force F_C. If we set $F_G = F_C$ and carry out the algebra, we find that the mass is related to the size and speed of the orbit as follows:

$$M = \frac{d \times V^2}{G}.$$

Therefore, we can determine the mass of an object if we know the speed and distance of another body orbiting it.

For example, the orbit of the Earth around the Sun (Figure 17.2) allows us to determine the Sun's mass. The Earth's orbital velocity V is 29.8 km/sec, or 2.98×10^4 m/sec. Using this value together with the Earth–Sun distance (1 AU = 1.50×10^{11} m), we find that the Sun's mass must be

$$M_\odot = \frac{1.50 \times 10^{11} \text{ m} \times (2.98 \times 10^4 \text{ m/sec})^2}{6.67 \times 10^{-11} \text{ m}^3/\text{sec}^2 \cdot \text{kg}} = 2.0 \times 10^{30} \text{ kg}.$$

This same method can be used to calculate, for example, the mass of the Earth from an orbiting satellite, or the mass of a galaxy from an orbiting star. This is an especially convenient way of finding masses because astronomers have methods for directly finding the speed of an orbiting object from the way motions affect light (Unit 25).

17.2   KEPLER'S THIRD LAW REVISITED

In the previous section we could have written the expression for M in a slightly different way: by expressing the velocity, V, in terms of the orbital circumference ($2\pi d$) and the period, P. If we were to do that, we would end up with

$$M = \frac{4\pi^2}{G} \times \frac{d^3}{P^2}.$$

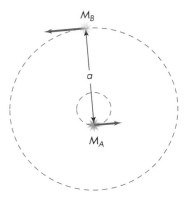

FIGURE 17.3

A pair of stars in orbit around each other. The sum of their masses can be determined by the modified form of Kepler's third law.

Astronomers and other scientists often use subscripts as reminders about the meaning or use of the terms in an equation. The subscripts do not otherwise have any importance for carrying out the arithmetic.

This expression bears a certain resemblance to Kepler's third law that $P^2 = a^3$, where P is measured in years and a in astronomical units. As a reminder that these values need to be measured in these units, we sometimes write P_{yr} and a_{AU}.

The resemblance to Kepler's third law is a little clearer if we rewrite it as

$$1 = \frac{a_{AU}^3}{P_{yr}^2}.$$

The difference from Newton's formula is that instead of measuring distances in meters, periods in seconds, and masses in kilograms, Kepler could just use the Earth's orbital parameters as a reference, ignoring the Sun's mass and the constants (4, π, and G). Kepler's version also applies to any elliptical orbit for which we can measure the semimajor axis a, not just the distance in a circular orbit.

Newton was able to derive a **modified form of Kepler's third law,** which applies not just to the Solar System but to any two objects orbiting each other (Figure 17.3). Even if the mass of the smaller body is not negligibly small, and even if the orbit is elliptical, Newton found that the sum of the two bodies' masses obeys the following law:

$$M_A + M_B = \frac{a_{AU}^3}{P_{yr}^2}$$

where the masses of the two orbiting bodies, M_A and M_B, are in units of the Sun's mass $M_\odot$.

Newton's version of the law matches Kepler's version within the Solar System because the Sun's mass is so much larger than anything else. As a result, the sum of the Sun's mass and any other object's mass is not much different from the Sun's mass alone. For example, the Earth's mass is just 0.000003 $M_\odot$, so when it is added to the Sun's mass, the left side of the equation becomes 1.000003. It requires very precise measurements to detect the small difference of this value from 1. The only planet for which it is not too difficult to detect the difference is Jupiter, which has a mass of about 0.001 $M_\odot$.

The modified form of Kepler's law is especially important because we can use it to measure the masses of stars that are orbiting each other (Unit 56). Thus if we can measure the orbital period, P, of two stars in years, along with the semimajor axis of their orbit, a, in AU, we can determine the sum of the two stars' masses.

With these equations, gravity becomes a tool for determining the mass of astronomical bodies, and we shall use this method many times throughout our study of the universe.

KEY TERMS

centripetal force, 134   modified form of Kepler's third law, 136

QUESTIONS FOR REVIEW

1. Describe a situation in which you have experienced centripetal force.

2. Might alien astronomers living in a different system of planets orbiting a star have derived Kepler's third law too?

PROBLEMS

1. Given that Jupiter is about five times farther from the Sun than the Earth, calculate its orbital velocity. How many years does it take Jupiter to complete an orbit around the Sun?

2. The Sun orbits the center of the Milky Way galaxy at about 220 km/sec at a distance from the center of about 2.6 × 10^{20} meters (about 28,000 light-years). What is the mass of the Milky Way inside the Sun's orbit? (Note: In this problem, we assume that the galaxy can be treated as a sphere of matter. Strictly speaking, this is not precisely correct, but the far more elaborate math needed to calculate the problem properly ends up giving almost the same answer.)

3. Derive the mass of the Earth from the Moon's orbital period of 27.3 days and orbital distance of 3.84×10^8 m.

TEST YOURSELF

1. If you are riding a merry-go-round and experience a centripetal acceleration of $0.1g$, and the merry-go-round starts spinning twice as fast, how big will your acceleration be?
    a. $0.05g$
    b. $0.1g$
    c. $0.2g$
    d. $0.3g$
    e. $0.4g$

2. You determine the mass of a galaxy from the observation that it is rotating at 200 km/sec at a distance 100,000 ly from its center. Later you make a measurement and find that it is rotating at the same speed two times farther from the center. How many times bigger will the new mass you calculate be compared to the old mass you calculated?
    a. 100,000 times bigger
    b. 200 times bigger
    c. 2 times bigger
    d. 4 times bigger
    e. 200,000 times bigger

3. In the distant past, it was thought that the Moon orbited four times closer to the Earth. What would its speed have been in its orbit at this distance compared to its currently measured speed?
    a. 4 times slower
    b. 4 times faster
    c. 2 times slower
    d. 2 times faster
    e. The same

18

Orbital and Escape Velocities

What goes up does not always come down. Without friction to slow it down, a satellite around a planet, or a planet around the Sun, can remain in orbit essentially forever. Newton's image of a cannon ball circling the Earth (Unit 16) is not far removed from the satellites of today.

If we were to continue Newton's thought experiment with a cannon beyond the point where the ball circled the Earth, the ball would begin to travel farther and farther out. It would follow an elliptical path, as found by Kepler, and return to its starting point; at a high enough speed, the ball would continue to travel outward forever.

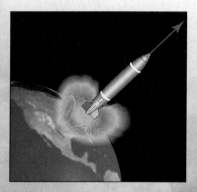

Background Pathways: Unit 16

18.1 ORBITS

Most spacecraft orbit the Earth only as high as necessary to avoid friction from the Earth's atmosphere. The international space station, for example, orbits at about 360 km (220 miles) above the Earth's surface. Even at that height, drag from the very tenuous atmosphere there causes it to drop about 50 m closer to Earth each day. Periodic visits by the space shuttle bring not only food and supplies, but an opportunity to boost the station back to higher altitudes.

Compared to the Earth's radius of about 6400 km, the space station's, and most other satellites', orbits are less than 10% farther from the Earth's center than the Earth's own surface. If we calculate the balance between gravity and centripetal force (Unit 17), we find that an object in a circular orbit must have an **orbital velocity** of

$$V_{circ} = \sqrt{\frac{GM}{R}}$$

where M is the mass of the planet (or other object) being orbited and R is the orbital distance from the *center* of the planet. Without deriving this equation, we can see that it makes sense because the orbital speed is larger for a more massive planet, and it grows slower at larger radii, where gravity is weaker.

For an orbit just above the Earth's surface (Figure 18.1) the circular orbital speed is

$$V_{circ} = \sqrt{\frac{GM_{\oplus}}{R_{\oplus}}}$$

$$= \sqrt{\frac{6.67 \times 10^{-11} \text{ m}^3/\text{sec}^2 \cdot \text{kg} \times 5.97 \times 10^{24} \text{ kg}}{6.37 \times 10^6 \text{ m}}}$$

$$= \sqrt{6.25 \times 10^7 \text{ m}^2/\text{sec}^2}$$

$$= 7900 \text{ m/sec}$$

V_{circ} = Velocity of a circular orbit

   G = Newton's gravitational constant

   M = Mass of body being orbited

   R = Radius of orbit (from center of body being orbited)

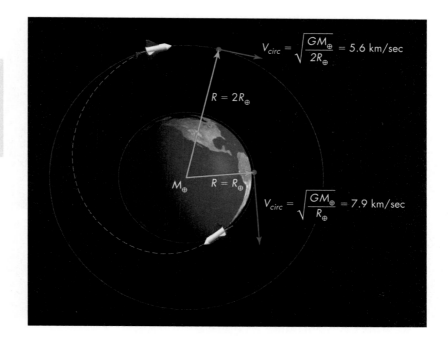

$$V_{circ} = \sqrt{\frac{GM_{\oplus}}{2R_{\oplus}}} = 5.6 \text{ km/sec}$$

$$R = 2R_{\oplus}$$

$$M_{\oplus} \quad R = R_{\oplus}$$

$$V_{circ} = \sqrt{\frac{GM_{\oplus}}{R_{\oplus}}} = 7.9 \text{ km/sec}$$

FIGURE 18.1

A circular orbit is slower at larger distances from the Earth. A rocket has to fire its thrusters to speed up to travel out to the slower orbit. At twice the Earth's radius, the circular velocity drops to 5.6 km/sec.

INTERACTIVE

Orbital velocity

or a little under 8 kilometers per second. Notice that for a satellite orbiting at larger radii, the circular velocity becomes slower.

The slower speed in higher orbits leads to an oddity—to "slow down" a rocket's orbit, it must "speed up." For example, to travel out to the larger, slower orbit in Figure 18.1, the rocket must fire its thrusters to speed itself up. That puts it into an elliptical orbit that carries it out to the larger radius. When it reaches the larger orbital radius, it is traveling even more slowly than the outer circular orbit speed, and it will fall back along its elliptical orbit unless it again fires its rockets to speed up and match the circular orbit speed at this radius.

18.2   ESCAPE VELOCITY

Q If you were commanding the space shuttle and wanted to reach the space station several thousand kilometers ahead of you, how would you maneuver to reach it?

V_{esc} = Velocity needed to escape from gravitational pull of a body

   G = Newton's gravitational constant

   M = Mass of body

   R = Starting distance from center of body

To overcome a planet's gravitational force and escape into space, a rocket must achieve a critical speed known as the **escape velocity.** Escape velocity is the speed at which an object needs to move away from a body so as not to be drawn back by its gravitational attraction. We can understand how such a speed might exist if we think about throwing an object into the air. The faster the object is tossed upward, the higher it goes and the longer it takes to fall back. Escape velocity is the speed an object needs so that it will never fall back, as depicted in Figure 18.2. Thus escape velocity is of great importance in space travel if craft are to move away from one body and not be drawn back to it. However, escape velocity is also important for understanding many astronomical phenomena, such as whether a planet has an atmosphere and the nature of black holes.

The escape velocity, V_{esc}, for a spherical body such as a planet or star can be found from the law of gravity and Newton's laws of motion. It is given by the following formula:

$$V_{esc} = \sqrt{\frac{2GM}{R}}.$$

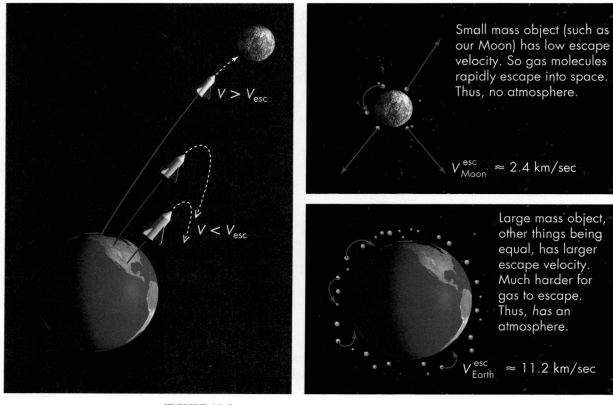

FIGURE 18.2

Escape velocity is the speed an object must have to overcome the gravitational force of a plane or star and not fall back. A low escape velocity, in general, leads to the absence of an atmosphere on a planet or satellite.

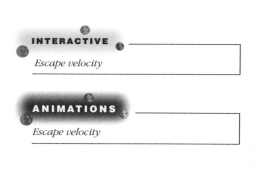

INTERACTIVE

Escape velocity

ANIMATIONS

Escape velocity

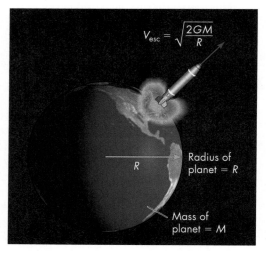

FIGURE 18.3

Calculating the escape velocity from a body.

Multiplying $\sqrt{2}$ times the speed of 7.9 km/sec calculated in the previous section for a circular orbit just above the Earth's surface gives the escape velocity from Earth's surface: 11.2 km/sec.

Here G is Newton's gravitational constant, M stands for the mass of the body from which we are attempting to escape, and R is its radius, as shown in Figure 18.3. The escape velocity is only $\sqrt{2}$ (≈ 1.414) times larger than the circular orbit speed at any radius. Thus if we know one, we can multiply or divide by $\sqrt{2}$ to find the other.

Notice in the equation for V_{esc} that if two bodies of the same radius are compared, the larger mass will have the larger escape velocity. On the other hand, if two

Q How would the escape velocity change if we were starting not from the surface of a planet but on a launch platform twice as far from the center?

bodies of the same mass are compared, the one with the smaller radius will have the greater escape velocity. We will see in Unit 68 that the huge escape velocity of a black hole arises from its abnormally small radius.

To illustrate the use of the formula, we calculate the escape velocity from the Moon. The Moon's mass is 7.35×10^{22} kilograms. Its radius is 1.74×10^6 meters. We insert these values in the formula for escape velocity and find

$$V_{esc}(Moon) = \sqrt{\frac{2 \times 6.67 \times 10^{-11} \text{ m}^3/\text{sec}^2 \cdot \text{kg} \times 7.35 \times 10^{22} \text{ kg}}{1.74 \times 10^6 \text{ m}}}$$

$$= 2.4 \times 10^3 \text{ m/sec}$$
$$= 2.4 \text{ km/sec}.$$

Q Can two bodies have the same escape velocity but different densities?

This low escape velocity compared to the Earth's 11.2 km/sec means it is much easier to blast a rocket off the Moon than the Earth.

The low escape velocity from the Moon is also one of the primary reasons why it lacks an atmosphere. The average speed of molecules in a gas at the Earth's and Moon's temperatures is typically about 0.5 km/sec. Individual molecules may randomly travel faster, though. It has been found that if the escape velocity is not at least 10 times larger than the average molecular speed of a gas, a planet does not generally retain the gas. By contrast, the Sun's escape velocity is 617 kilometers per second; thus it is able to hang onto the gas in its atmosphere even though it is heated to a high temperature.

KEY TERMS

escape velocity, 139   orbital velocity, 138

QUESTIONS FOR REVIEW

1. How do orbital speeds depend on the distance from a planet?
2. What is meant by *escape velocity*?

PROBLEMS

1. Calculate the orbital speed for a satellite 1000 km above the Earth's surface, using the fact that $M_\oplus = 5.97 \times 10^{24}$ kg and $R_\oplus = 6.37 \times 10^6$ m.

2. At what distance would a satellite orbiting the Earth be *geosynchronous* (orbiting the Earth once every 24 hours)?

3. What is the escape velocity from a galaxy at a radius of 50,000 ly if the mass of the galaxy is 10^{12} times the Sun's mass?

4. Which body has a larger escape velocity, Mars or Saturn?

   $M_{Mars} = 0.1 \, M_\oplus$

   $M_{Saturn} = 95 \, M_\oplus$

   $R_{Mars} = 0.5 \, R_\oplus$

   $R_{Saturn} = 9.4 \, R_\oplus$

TEST YOURSELF

1. The Moon's escape velocity is smaller than the Earth's because
   a. its radius is smaller.
   b. its mass is smaller.
   c. its distance from Earth is greater.
   d. it has no atmosphere.
   e. All of the above

2. Suppose the Sun suddenly collapsed in on itself, dropping to half its current radius. How big would the escape velocity from the surface be compared to its current value?
   a. ½ as big
   b. 2 times bigger
   c. $\sqrt{2}$ times bigger
   d. 4 times bigger
   e. The same

3. Suppose the Sun suddenly collapsed in on itself, dropping to half its current radius. How big would the escape velocity be at Earth's orbital distance of 1AU compared to its current value?
   a. ½ as big
   b. 2 times bigger
   c. $\sqrt{2}$ times bigger
   d. 4 times bigger
   e. The same

19 Tides

Anyone who has spent even a few hours by the sea knows that the ocean's level rises and falls during the day. A blanket set on the sand 10 feet from the water's edge may be under water an hour later, or an anchored boat may be left high and dry. This regular change in the height of the ocean is called the **tides**, which are caused primarily by the Moon.

In fact, tides occur everywhere bodies interact gravitationally: between planets and their satellites; between stars that orbit each other; and between neighboring galaxies. Tidal interactions are a direct consequence of the nature of gravitational forces. In this unit we explore Earth–Moon tidal interaction specifically, but the same analysis might be applied to a pair of stars or a pair of galaxies.

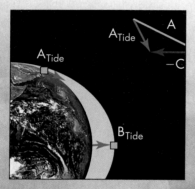

Background Pathways: Units 8 and 16

19.1 CAUSE OF TIDES

Just as the Earth exerts a gravitational pull on the Moon, so too the Moon exerts a gravitational attraction on the Earth. Until now we have considered the gravitational pull between interacting bodies as if each were a discrete whole, but in fact gravity pulls on every atom within each body. Recall, too, that the force of gravity weakens with distance because of the $1/d^2$ dependence of gravity. Hence the Moon pulls on different parts of the Earth with different strengths. For example, the attraction is stronger on the side of the Earth nearer the Moon and weaker on the far side (see Figure 19.1). The difference between a stronger force pulling on one side of a body while a weaker force is experienced on its opposite side produces a stretching effect. Astronomers call this stretching effect a "differential gravitational force" or simply a **tidal force.**

The tidal force draws water in the oceans into **tidal bulges** on the side of the Earth facing the Moon as well as on the Earth's far side, as shown in Figure 19.2. The tidal bulge on the far side of the Earth may seem counterintuitive. Consider the following analogy: Imagine that you were being lifted up by hundreds of strings tied to all parts of you and your clothing. Suppose the strings tied to the hair on your head were being pulled a bit harder than the rest—your hair would end up being pulled upward until it stood on end. On the other hand, the strings tied to your feet and shoelaces are pulled a bit less strongly, so your feet and shoelaces dangle downward, away from the direction you are being pulled. The difference of forces across your body produces a stretching effect.

How big is this effect? It turns out to be quite small. When the Moon is overhead (or on the far side of the Earth below your feet), you weigh about one tenmillionth less than when the Moon is near the horizon. This is barely measurable, so why do the oceans change in height so dramatically? The answer is that water can flow, and the tidal force can create forces that pull the water along the surface. For example, in Figure 19.2, water at point A feels a net force that causes it to flow toward the point directly beneath the Moon. On the far side of the Earth from the Moon, there are similar net tidal forces along the surface causing water to flow toward the furthest point from the Moon. Here again, the flow is not a strong current of water traveling thousands of kilometers, but mostly small shifts in the

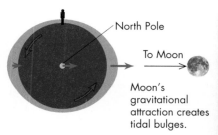

FIGURE 19.1
Tides are caused by the Moon's gravity creating tidal bulges.

ANIMATIONS

Origin of tides

To calculate the strength and direction of tidal forces, we must add and subtract *vectors,* shown as force arrows in Figure 19.2. This is done by putting each vector's tail to the head of the previous vector being summed up—or if subtracting a vector, doing this with a vector in the opposite direction. The final vector is from the tail of the first vector to the head of the last vector added together.

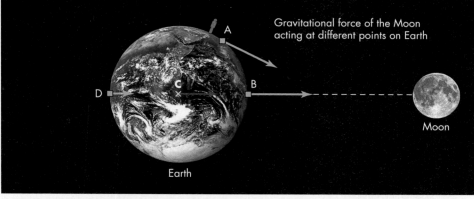

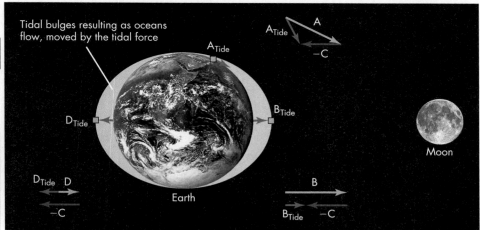

FIGURE 19.2
(Top) Arrows schematically show the Moon's gravitational force at different points on the Earth. (Bottom) Tidal forces from the point of view of an observer on the Earth. These arrows represent the difference between the Moon's gravitational force at a given point and its force at the Earth's center (C). Graphically, you can find the tidal force by "adding" the arrows. The figure shows schematically how to do this, but details are omitted.

Q The Moon does not generally lie directly above the Earth's equator, but may be as much as about 30 degrees north or south of the equator. How can this cause the two high tides to have different strengths?

Q One kind of alternative energy source is tidal power. How could you extract energy out of the tides?

position of water in oceans all over the world creating the excess we see in the tidal bulges.

In this discussion we have ignored the Earth's rotation. The tidal bulges are aligned approximately with the Moon, but the Earth spins. The Earth's rotation therefore carries us first into one bulge and then the next. As we enter the bulge, if we are on the ocean, the water level rises, and as we leave the bulge, the level falls. Because there are two bulges, we are carried into high water twice a day; these are the twice-daily high tides. Between the times of high water, as we move out of the bulge, the water level drops, making two low tides each day (Figure 19.3).

This simple picture becomes more complicated when the tidal bulge reaches shore. In most locations the tidal bulge has a depth of about 2 meters (6 feet), but it may reach 10 meters (30 feet) or more in some long narrow bays (as you can see in the photographs of high and low tides along the Maine coast—Figure 19.3) and may even rush upriver as a *tidal bore*—a cresting wave that flows upstream. On some rivers surfers ride the bore upstream on the rising tide. The rising tide water in these regions is funneled into a narrower region, "piling" up the water to greater depth.

The motion of the Moon in its orbit makes the tidal bulge shift slightly from day to day. Thus high tides come about 50 minutes later each day, the same delay as in moonrise (see Unit 7).

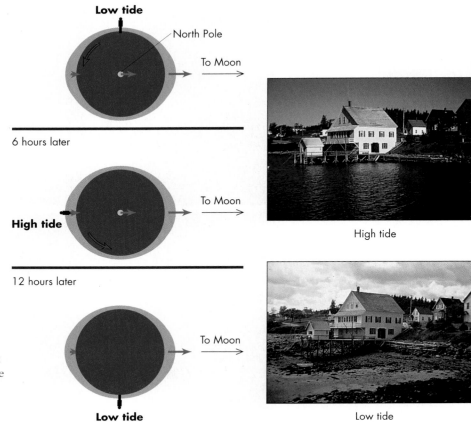

Low tide

North Pole

To Moon

6 hours later

High tide

To Moon

12 hours later

To Moon

Low tide

A

High tide

Low tide

B

- Moon's gravitational attraction creates tidal bulges.
- Earth's rotation carries a person through the tidal bulges.

FIGURE 19.3
As the Earth rotates, it carries points along the coast through the tidal bulges. Because there are two bulges where the water is high and two regions where the water is low, we get two high tides and two low tides each day at most coastal locations.

19.2   SOLAR TIDES

The Sun also creates tides on the Earth. But although the Sun is much more massive than the Moon and exerts a larger gravitational force on the Earth, it is also much farther away; so the *differential* force from one side of the Earth to the other is smaller. The result is that the Sun's tidal effect on the Earth is only about one-half the Moon's. Nevertheless, it is easy to see the effect of their tidal cooperation in **spring tides,** which are unusually large tides that occur at new and full moons. At these times the lunar and solar tidal forces work together, adding their separate tidal bulges, as illustrated in Figure 19.4A. Notice that spring tides have nothing to do with the seasons; rather they refer to the "springing up" of the water at new and full moons.

It may seem odd that spring tides occur at both new and full moons because the Moon and Sun pull together when the Moon is new but in opposite directions when it is full. However, both the Sun and Moon create two tidal bulges, and the bulges combine regardless of whether the Sun and Moon are on the same or opposite sides of the Earth. On the other hand, at first and third quarters, the Sun and Moon's tidal forces work at cross-purposes, creating tidal bulges at right angles to one another, as shown in Figure 19.4B. The **neap tides** that result are therefore not as extreme as average high and low tides.

The Sun's tidal effects are much stronger on Mercury and Venus because those planets orbit closer to the Sun. In fact, the Sun's tidal pull has probably slowed down the spin of these planets. The same is true of the tidal effects of planets on many satellites. These effects are examined further in Part III of the book.

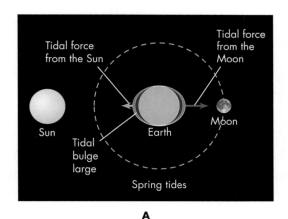
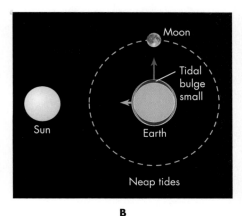

A   **B**

FIGURE 19.4

The Sun's gravity creates tides too, though its effect is only about half that of the Moon.
(A) The Sun and Moon each create tidal bulges on the Earth. When the Sun and Moon are in line, their tidal forces add together to make larger-than-average tides. (B) When the Sun and Moon are at 90° as seen from Earth, their tidal bulges are at right angles and partially nullify each other, creating smaller-than-average tidal changes.

KEY TERMS

neap tide, 144   tidal force, 142

spring tide, 144   tides, 142

tidal bulge, 142

QUESTIONS FOR REVIEW

1. What is a tidal force?

2. Explain the tidal bulge on the side of the Earth opposite the Moon.

3. What are spring and neap tides?

PROBLEMS

1. At what times would you expect high and low tides when the Moon is full?

2. At what times would you expect high and low tides when the Moon is at first quarter?

3. Using the law of gravity, calculate the Moon's pull (in newtons = kg · m/sec^2) on a kilogram of water on the near side of the Earth and on the far side of the Earth. The Moon's mass

is 7.35×10^{22} kg, the distance between the Earth and Moon is 3.84×10^8 m, and the Earth's radius is 6.37×10^6 m.

4. Carry out the same calculation as in the previous problem, but for the Sun. The Sun's mass is 1.99×10^{30} kg, and the distance between the Earth and Sun is 1.50×10^{11} m.

TEST YOURSELF

1. As a result of the Moon's gravitational pull, when would you weigh less?
   a. whenever it is high tide locally
   b. whenever it is low tide locally
   c. only when the Moon is overhead
   d. only if you were near one of the Earth's poles
   e. Your weight never changes as a result of the Moon's gravity.

2. Low tide during the new moon occurs at about
   a. midnight.
   b. 6 P.M.
   c. midnight and noon.
   d. Noon.
   e. 6 A.M. and 6 P.M.

3. Ocean tides are caused primarily by
   a. seismic pressure waves beneath the surface.
   b. sunlight reflecting off waves.
   c. the Moon's gravitational pull.
   d. tectonic motion of the spreading ocean floor.
   e. all of the above

20 Conservation Laws

Newton's laws of motion are but one set of laws important for understanding the universe. Another set of laws, the **conservation laws**, are also very powerful. Some physical properties of matter or a system remain the same—are conserved—despite whatever is done to them. For example, in cooking (and other chemical processes) the numbers of each type of atom are conserved; the atoms may combine with each other in different ways, but no atoms are created or destroyed. Knowing this allows us to make a number of predictions about what is possible when we cook a certain combination of ingredients. It may not teach us how to bake a cake, but it can help us predict the consequences of leaving out certain ingredients!

Conservation laws identify fundamental properties that do not change under almost any circumstances. We will discuss two such laws here: conservation of energy and conservation of angular momentum. The power of these conservation laws is that if we can measure the energy or the angular momentum of a system at one time, we know that all subsequent changes in the system can occur only in ways that conserve both quantities. With these laws we can predict how fast a collapsing star will spin or the explosive energy of an asteroid striking the Earth.

Background Pathways: Unit 16

20.1 CONSERVATION OF ENERGY

Energy is a familiar term from everyday conversation, but that usage is not always the same as is meant in the sciences. **Energy** can come in many forms, but in general, it is the ability to generate motion. Thus a moving ball may be able to hit a stationary ball and set it in motion. The moving ball thus has energy. Energy can also be present where there is no motion. When you lift a ball up off the ground it is stationary, but if you let go of it, it will be pulled downward by gravity and be set into motion. Thus picking it up gave it a different form of energy that would allow it to generate motion.

Energy is neither created nor destroyed—it just changes forms. This can be described by the law of conservation of energy, which states the following:

> **The energy in a closed system may change form, but the total amount of energy does not change as a result of any process.**

The idea of a *closed system* is important: It means that energy is not exchanged with anything outside the system. For example, the Earth is not a closed system because it receives energy from the Sun and radiates heat energy into space. Very few systems remain isolated at all times, but often we can examine an individual process—energy conversion during a roller coaster ride, for example—that occurs in isolation.

To appreciate the power of this law, we need to look more closely at how we measure energy as well as some of the forms it may take. Energy is interesting in part because it can have so many forms. The most obvious form is the energy of an object that is already in motion, or **kinetic energy.** A simple example of kinetic energy is the energy of a moving car or a thrown ball (Figure 20.1A). Physicists measure energy in joules, the unit of energy in the MKS system (see Unit 3) named

Kinetic energy

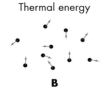

Thermal energy

A   **B**

Potential energy

Conversion of potential energy
to kinetic and thermal energy

C   **D**

FIGURE 20.1
Energy can be found in many forms.
A moving body has kinetic energy
(A), while the energy we associate with
heat is produced by the motions of
particles within the substance (B). Energy
can also be found in a "potential" form,
such as a bowling ball at the edge of a
table (C). If it falls off the edge of the
table it produces kinetic energy as it falls
to the floor (D).

E_K = Kinetic energy

m = Mass of moving object

V = Speed of moving object

after the nineteenth-century British physicist James Joule, who helped verify the
law of conservation of energy. The kinetic energy of a body of mass m moving with
a speed V can be expressed mathematically as

$$E_K = \frac{1}{2}\, m \times V^2.$$

If m is measured in kilograms and V in meters per second, the calculation will give
the energy in joules.

Another form of energy is heat or **thermal energy.** Heat is also energy of motion, but at the atomic or molecular level. That is, the hotter something is, the more
rapidly its atoms and molecules move within it. Sometimes that motion is a vibration. Other times it is random careening of particles (Figure 20.1B).

Yet another common form of energy is **potential energy.** Potential energy is energy that has not been liberated but is available. For example, water stored behind
a dam, or a bowling ball poised on the edge of a table, both have potential energy
(Figure 20.1C). If the dam bursts or the ball rolls off the table, the potential energy
can be released and converted into motion. These examples of potential energy result from the gravitational attraction between the Earth and the water or the bowling ball. To lift an object up against Earth's gravity takes energy, and that energy is
stored until the object falls back to the Earth. The **gravitational potential energy**
can be written mathematically as

E_G = Gravitational potential energy

m, M = Masses of objects

d = Distances between their centers

G = Newton's gravitational constant

$$E_G = -G\frac{m \times M}{d}$$

where G, m, M, and d are all defined as in the law of gravity (Unit 16). Again, if all
the quantities are measured in MKS units, the resulting energy will be in joules.
The form of this equation is what you might expect—potential energy has a bigger
magnitude for larger masses, and it grows weaker with distance as gravity grows
weaker. A curious aspect of this equation, though, is the minus sign. We discuss this
further a little later.

The usefulness of the law of conservation of energy is that when you add together all the forms of energy at the beginning of a process, their sum must equal
the total amount of energy in all its forms at the end. As an example, let's consider
the bowling ball on the table edge. If it is just sitting there, it initially has only
potential energy. If it falls from the table, it converts that potential energy to energy

Q What kind of energy is nuclear
energy?

of motion as the ball falls. On hitting the ground it may look as if the energy has disappeared: The ball is at rest again so that it no longer has energy of motion. Likewise, it has given up its potential energy. However, its energy has *not* disappeared. On impact, the ball shakes the ground and gives its energy of motion to the atoms in the ground. Their motion is heat and sound (vibrations of atoms in the air), so that the ball's energy has changed form yet again (Figure 20.1D). If you pick up the bowling ball and set it back on the table, you will expend the same amount of energy lifting the ball in **chemical energy**—potential energy in the electrical bonds of molecules such as fat and sugar stored in your body—as the potential energy that the bowling ball gains.

An astronomical example that illustrates the conservation of energy and its conversion to different forms is an asteroid approaching Earth. Initially it has energy of motion. During its passage through the atmosphere, a small amount of the energy of motion is converted to heat and light as it tears through the atmosphere. The asteroid's kinetic energy drives it deep into the solid crust of the Earth, transferring its kinetic energy as it slows by compressing and heating the rock at the point of impact. This violent compression is transferred to surrounding rock, expanding outward in all directions. In a fraction of a second, so much heat is generated that the asteroid and surrounding rock melt or vaporize. Some rock is blasted up and out, giving it kinetic energy, and seismic waves (Unit 35) travel through the Earth, shaking the ground thousands of kilometers away. Most of the asteroid's kinetic energy ends up as thermal energy, which is eventually radiated away into space (Unit 23). Some of the energy remains as potential energy—stored in chemical bonds created by reactions in the heat of the blast or stored as gravitational potential energy in rock pushed upward in the region surrounding the blast crater. Thus the energy changes form, but the total energy at the beginning equals the total energy at the end.

Galileo's experiments showing that a ball slows down as it goes up a ramp or speeds up as it goes down (Unit 12) can also be understood in terms of conservation of energy. The ball's kinetic energy decreases as gravitational potential energy increases and vice versa. Extending this idea, we can phrase the question of how fast a rocket must travel to leave the Earth (its *escape velocity,* Unit 18) in terms of the kinetic energy it must have to completely overcome the Earth's gravitational potential energy. The rocket slows, and as it reaches an infinite distance where the gravitational pull approaches zero, its kinetic energy also drops to zero. If a rocket has enough (positive) kinetic energy to overcome the (negative) gravitational potential energy, it will escape from the Earth's surface.

This helps explain why gravitational potential energy (or potential energy based on any attractive force) has a negative value. When two objects are very far apart, the force of gravity becomes negligible, so it is natural to describe the gravitational potential energy as near zero. If the two objects fall together, their kinetic energy increases—subtracting energy from the gravitational potential energy. If we start with nearly zero joules of gravitational potential energy and subtract thousands of joules, E_G must become more negative as gravity pulls objects closer together.

Curiously, on astronomical scales many objects in the universe, and perhaps the entire universe itself, exhibit a near balance between the gravitational potential and kinetic energies. Both may be individually large, but their sum is usually close to zero.

Q What forms of energy have you used today?

 Assigning a negative value to gravitational potential energy may seem strange, but the alternatives are perhaps stranger. If we assigned "zero energy" to when two objects were close together, we would have to say that the potential energy was very large when the objects were barely feeling each other's gravity—and the gravitational potential energy would *still* become negative if the objects got even closer together.

20.2   CONSERVATION OF ANGULAR MOMENTUM

You have perhaps seen water going down a drain and noticed that as the water moves inward toward the drain hole, it spirals faster and faster. Likewise, ice skaters who go into a spin with outstretched arms and then fold their arms inward spin even faster. These are both examples of the conservation of angular momentum.

Angular momentum is a measure of the tendency of a spinning object to keep spinning with its rotation axis pointing in the same direction unless acted on by an external "rotational force," called a **torque,** that works to alter the rotation. The distinction between a force and a torque is that some forces act in directions that do not affect rotation, while the strength of a torque is measured by its ability to increase or decrease rotation. If a mass m is moving around a rotation axis at a distance r with a velocity V, then mathematically its angular momentum is $m \times V \times r$. The law of conservation of angular momentum states this:

> **If no external force acts to change the spin rate, then $m \times V \times r$ remains constant.**

How does this explain the faster spin of the ice skater whose arms are pulled in? If $m \times V \times r$ is constant, then if the skater makes r (the length of the extended arms) shorter, then V must get larger (Figure 20.2). Notice that when skaters are pulling their own arms in, no *external* forces are involved. Likewise, as the water flowing toward a drain circles nearer the drain hole, its r decreases; therefore its V of rotation must increase.

We have already encountered an astronomical example of this conservation law in Kepler's second law of planetary motion. As a planet moves closer to the Sun, its orbital velocity must increase to keep $m \times V \times r$ constant. Because nearly all objects have some intrinsic amount of rotation, if any process causes them to change size, they will change their spin rate in response.

m = Mass of rotating object

r = Distance from axis of rotation

V = Speed of rotation around axis

Q Suppose you were in a spinning spacecraft. How could you use a jet thruster to stop your spin? If you can slow your spin, in what sense is angular momentum conserved?

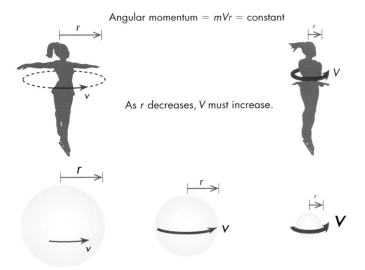

Angular momentum = mVr = constant

As r decreases, V must increase.

FIGURE 20.2

Conservation of angular momentum spins up a skater and a spinning star. Angular momentum is proportional to a body's mass, m, times its rotation speed, V, times its radius, r. Conservation of angular momentum requires that mathematically $m \times V \times r$ = constant. Thus if a given mass changes its r, its V must also change. In particular, if r decreases, V must increase, as the ice skater demonstrates by spinning faster as she draws her arms in. Similarly, a rotating star spins faster if its radius shrinks.

KEY TERMS

angular momentum, 149

chemical energy, 148

conservation laws, 146

energy, 146

gravitational potential energy, 147

kinetic energy, 146

potential energy, 147

thermal energy, 147

torque, 149

QUESTIONS FOR REVIEW

1. What is a conservation law?

2. What is potential energy?

3. What is a torque?

PROBLEMS

1. A basketball has a mass of about 0.5 kg. How fast would it have to be thrown to have 1 joule of kinetic energy? Convert your answer to kph (or mph if that is more familiar to you). Is that a hard throw or a light throw?

2. Show that if the gravitational potential energy and kinetic energy add up to zero for an object on the Earth's surface, you can derive the escape velocity formula of Unit 18.

3. Suppose that a comet with a mass of 10^9 kg struck the Earth at a velocity of 40 km/sec. How much energy would be released by the impact? Convert this energy into the equivalent of "kilotons of TNT" where 1 kt ≈ 4×10^{12} joules.

4. If a star collapses inward until it is 10 times smaller than it was originally (while losing no mass), how many times faster will its surface be moving? If it was spinning around once every 100 hours initially, how long will it take to spin around after it collapses?

TEST YOURSELF

1. A volleyball has about half the mass of a basketball. If they are both moving at the same speed, the volleyball's kinetic energy would be _____ the basketball's.
    a. 2 times more than
    b. ½ as much as
    c. 4 times more than
    d. ¼ as much as
    e. the same as

2. Suppose a comet has an elliptical orbit that brings it twice as close to the Sun at its closest approach as when it is at its largest separation. If its speed is 10 km/sec at its largest separation, what will its speed be when it is closest?
    a. 5 km/sec
    b. 10 km/sec
    c. 20 km/sec
    d. 40 km/sec

3. How many times more chemical energy does a car have to expend to reach 100 kph instead of 50 kph?
    a. 50
    b. 4
    c. 100
    d. 2
    e. 200

21

Light, Matter, and Energy

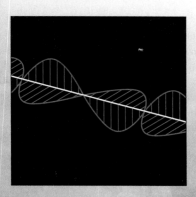

Background Pathways: Unit 4

Our home planet is so far from other astronomical bodies that, with few exceptions, we cannot bring samples from them to study in our laboratories. And even for objects within the range of our spacecraft, physical conditions may make it impossible to send probes to make direct measurements. For example, if we want to know how hot the Sun is, we cannot stick a thermometer into it. Nor can we directly sample the atmosphere of a distant star. However, we can sample such remote bodies indirectly by analyzing their light.

At first this might appear to be an enormous limitation—allowing us to say something about only the direction to an object and the quantity of light it produces. However, if we can understand the energetics of light and the way it interacts with matter, we can also learn about what the body is made of and its temperature. Light, therefore, is our key to studying the universe; but to use the key, we need to understand the nature of light and matter and how they interact.

21.1 THE NATURE OF LIGHT

Light is radiant energy; that is, it is energy that can travel through space from one point to another without the need of a direct physical link. Light is, therefore, very different from, for example, sound. Sound can reach us only if it is carried by a medium such as air or water. Light can reach us even across empty space. In empty space we can see the burst of light of an explosion, but we will hear no sound from it at all.

Light's capacity to travel through the vacuum of space is paralleled by another special property: its high speed. In fact, the speed of light is an upper limit to all motion (Unit 53). In empty space light travels at the incredible speed of 299,792,458 meters per second—a speed that would take an object seven times around the Earth in one second! The speed of light in empty space is a universal constant and is denoted by c, and although it is very precisely measured, for most purposes it can be rounded off to 300,000 km/sec.

Observation and experimentation on light throughout the last few centuries have produced two different models of what it is and how it works. According to one model, light is a wave that is a mix of electric and magnetic energy, changing together, as depicted in Figure 21.1. The ability of such a wave to travel through empty space comes from the interrelatedness of electricity and magnetism.

You can see this relationship between electricity and magnetism in everyday life. For example, when you start your car, turning the ignition key sends an electric current from the battery to the starter. There the current generates a magnetic force that turns over the engine. Similarly, when you pull the cord on a lawn mower, you spin a magnet, generating an electric current that creates the spark to ignite the gas in the engine.

151

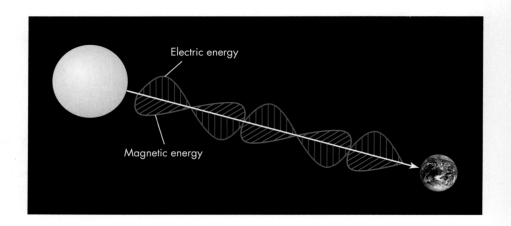

FIGURE 21.1

A wave of electromagnetic energy moves through empty space at the speed of light, 299,792.5 kilometers per second. The wave carries itself along by continually changing its electric energy into magnetic energy and vice versa.

The speed of light is slower when it travels through matter. Compared to its speed in a vacuum, light travels through glass about 60% as fast; in water, about 75%; in air, about 99.97%.

Q If you start a water wave from one side of a bathtub (or a sink or pool), does any water travel from one side of the tub to the other? How can you test your idea?

Joules are the MKS unit of energy (Unit 3).

ANIMATIONS

Photons from a light source

Light can travel through empty space because a small disturbance of an electric field creates a magnetic disturbance in the adjacent space, which in turn creates a new electric disturbance in the space adjacent to it, and so on. In this fashion, light can move wavelike through empty space, "carrying itself by its own bootstraps." Thus we say that light is an **electromagnetic wave.** When such a wave encounters the electrons in matter, it may interact with them, a little bit like the way a water wave makes a boat rock. It is from such tiny disturbances in our eyes or in an antenna that we can detect electromagnetic waves.

Many of the early experiments on light demonstrated this distinctly wavelike behavior. For example, sources of light can **interfere** with each other like the interacting waves in an ocean, reinforcing and canceling each other's effects in complex patterns. And like ocean waves passing through a narrow inlet into a bay, light passing through an opening does not travel in a straight line, but spreads out. This is an important consideration in the design of telescopes because it limits the sharpness of images a telescope can produce (Unit 29).

The wave model of light works well to explain many phenomena—for example, the length of the wave determines the color of light (Unit 22). However, the wave model fails to explain other properties of light. If light was like a water wave, for example, we could deliver any amount of electromagnetic energy by varying the intensity of the electromagnetic wave. However, we find something quite different for light. For example, the red light we see coming from a stoplight is made up of little parcels containing 3×10^{-19} joules of energy. We cannot get half this much energy or 1.7 times this much energy from red light, just whole multiples of this much energy. We can vary the energy in a water wave by making the waves taller or shorter, but the behavior of light is like saying we can make waves in a bathtub that are 3, 6, or 9 centimeters tall, but no size in between. To describe this property, we say that the energy of light is **quantized;** that is, it comes in discrete packets.

Based on this property, light is better described as being made of particles. These particles are called **photons** (Figure 21.2). You can have 1 or 2 or 3 photons, but not a half a photon or 1.7 photons. In this model the photons are packets of energy. When they enter your eye, they produce the sensation of light by striking your retina like microscopic bullets, releasing their energy and causing a chemical change in your photoreceptor cells (Unit 31). In empty space, photons move in a straight line at the speed of light.

For centuries, scientists argued over whether light consisted of waves or particles, pointing out experiments in which the opposing model could not give an adequate explanation. By the early twentieth century, however, it became clear that neither model completely describes light. Moreover, at the submicroscopic scales in which light operates, it has a more complex dual nature that we can describe only incompletely with our wave and particle models. So today scientists describe

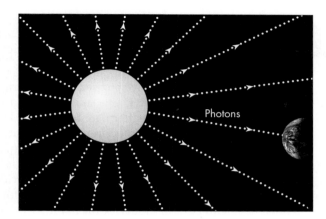

FIGURE 21.2
Photons—particles of energy—stream away from a light source at the speed of light.

light as having a **wave–particle duality,** and they use whichever model—wave or particle—best describes or explains a particular phenomenon. For example, the way light reflects off a mirror is easily understood if you imagine photons striking the mirror and bouncing back just the way a ball rebounds when thrown at a wall. On the other hand, the focusing of light by a lens is best explained by the wave model.

21.2   THE EFFECT OF DISTANCE ON LIGHT

An individual photon traveling through space maintains its speed and energy; however, we also know that a distant object appears dimmer. This dimming is the result of spreading out the light over a larger area as the photons move away from the source of light—much as the height of a wave produced by a rock tossed into a pond grows shorter as the ring expands.

This spreading out of photons explains why the planets in the inner Solar System are warmer than the outer planets (Unit 32), even though they receive photons that have the same energy. For example, photons traveling away from the Sun become spread farther apart, as illustrated in Figure 21.2. At the distance of the Earth from the Sun, the photons are spread over the area of a sphere that has a radius of 1 astronomical unit (the Earth–Sun distance). This surface area, given by the geometric formula $4\pi(1 \text{ AU})^2$, is about 12 square AU. At Jupiter's distance from the Sun—about 5 AU—the light is spread out over a sphere with a surface area of $4\pi(5 \text{ AU})^2$, which is an area 25 times larger than at the Earth's distance. Thus each square meter of the Earth's surface receives 25 times more light than each square meter of Jupiter's surface. As a result, the Sun looks 25 times dimmer at Jupiter's distance than at the Earth's distance.

Astronomers describe this change in the amount of light as a change in its **brightness**. At a distance d from the source, the light is spread out over an area of $4\pi d^2$, so we can write this relationship in mathematical form as

$$\text{Brightness} = \frac{\text{Total light output}}{4\pi d^2}.$$

ANIMATIONS

Inverse-square law

This relationship is known as the **inverse-square law** because the light grows dimmer in proportion to the square of the distance. The inverse-square law is important for determining how much energy is received at different distances from a source of light. The relationship also allows us to calculate the distances to many stars (Unit 54).

21.3 THE NATURE OF MATTER

Just as scientists have puzzled over the nature of light, they have also puzzled over the nature of matter. Like so many of our ideas about the nature of the universe, our ideas about matter date back to the ancient Greeks. For example, Leucippus and his student Democritus, who lived around the fifth century B.C.E. in Greece, taught that matter was composed of tiny indivisible particles. They called these particles *atoms,* which means "uncuttable" in Greek. They argued, for example, that the water in a tub could be subdivided into smaller and smaller pieces—droplets—only down to some finite size for the smallest particle of water.

Our current model for the nature of atoms dates back to the early 1900s and the work of the New Zealand physicist Ernest Rutherford. He showed that an atom can be thought of as a dense core called a *nucleus* around which smaller particles called *electrons* orbit (Figure 21.3), as described in Unit 4. Those orbits are generally extremely small. For example, the diameter of the smallest electron orbit in a hydrogen atom is only about 10^{-10} (1 ten-billionth) meter.

This submicroscopic size creates effects that operate at an atomic level and have no counterpart in larger systems. Our model of electrons as particles is incomplete because, like light, they too have a wave–particle duality. Their wave nature is not another kind of electromagnetic wave, and it is not related to the fact that an electron has an electric charge; rather, it concerns fluctuations of "electron-ness." In the early twentieth century it was discovered that all matter behaves like a wave in certain respects, although this is usually apparent only for the individual

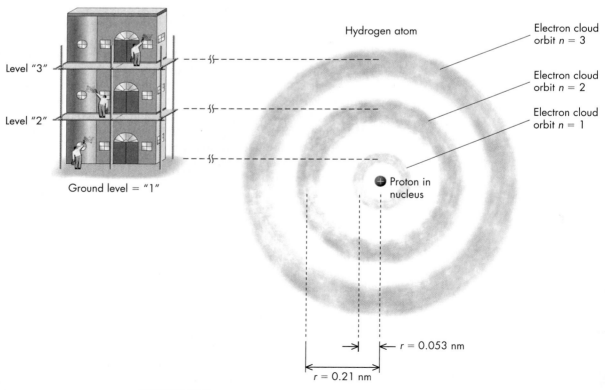

FIGURE 21.3

Just as the painters can be only at levels 1, 2, 3 . . . of the scaffold (and cannot "float in between"), so too an electron must be in orbital 1, 2, 3, ...

subatomic particles. The existence and position of such particles can fluctuate according to rules that are described by the area of physics called *quantum mechanics*.

One of the most important consequences of electrons' duality is that their orbits may have only certain quantized sizes. Although a planet may orbit a star at any distance, an electron may orbit an atomic nucleus only at certain distances—much as when you climb a set of stairs, you can gain only certain discrete heights. This peculiar steplike character to electrons' orbits was worked out by the Danish physicist Niels Bohr in the early 1900s. The electron's wave nature forces the electron to move only in orbits whose circumference is a whole number of waves. If it moved in an orbit whose circumference was ½ or 1½ times the length of a wave, the crest of a subsequent wave would land where the trough of the wave had been previously, and the electron wave would "cancel" itself out. This property prevents the electron from falling into the nucleus despite the electrical attraction because the smallest orbit, or **ground state,** is just one electron wave long. Thus in the **Bohr model** of the hydrogen atom, the electron can have an orbital radius of 0.053 n^2 nanometers (a nanometer is 10^{-9} meters; see Unit 3) where $n = 1, 2, 3$, and so on; but it cannot have intermediate values, as illustrated in Figure 21.3. This model of the atom is highly successful at explaining the properties of hydrogen as well as other atoms.

The wave nature of the electron has another important effect. It "smears out" the position of the electrons. Just as a water wave does not exist at a single point, the electron wave cannot be defined as a single point. As a result, although we have described the electrons as orbiting like tiny particles around the nucleus, a better description is that they exist as a three-dimensional electron wave or **orbital.**

Electron orbitals have another property totally unlike those of planets in orbit: They can shift from one orbital to another as they interact with light. This shifting changes their energy, as can be understood by a simple analogy. The electrical attraction between the nucleus and the electron creates a force between them like a spring. If the electron increases its distance from the nucleus, the spring must stretch. The atom requires energy to accomplish this "stretch." Similarly, if the electron moves closer to the nucleus, the spring contracts and the atom must give up, or emit, energy. We perceive that emitted energy as electromagnetic waves.

We can try to describe this interaction in two ways corresponding to the wave–particle duality of nature at submicroscopic scales:

Particles: An electron is orbiting an atomic nucleus at a particular distance. It is struck by a photon that contains the exact amount of energy needed to knock the electron into the next larger possible orbit. The electron absorbs the photon and jumps into the new, larger orbit.

Waves: An atom has an electron orbital of a particular wave configuration. An electromagnetic wave comes by with electric and magnetic fields oscillating at just the exact rate so that the orbital interacts with the wave, absorbing energy from it, and begins to vibrate in a new, more energetic mode in a larger orbital.

Neither description is exactly right because matter at these scales is both a particle *and* a wave—a duality that is foreign to our everyday experience.

In some ways the particle description is clearer. Discrete particles interact in a clear and sequential way. But it may lead one to wonder how the electron "knows" that it should not react when a photon with slightly less or slightly more energy hits it. The wave picture is more challenging in other ways, but the interaction between a photon and an electron may be clearer. The electron orbital responds to the electromagnetic oscillations, much as the string of a musical instrument may begin vibrating in response to a sound wave of the right pitch, but not to all pitches.

For our purposes, we can simplify the discussion of electron behavior by focusing on just the energies that are involved. When an electron moves between

In quantum mechanics the location of a particle is described in terms of a *wave function* that describes the probability of finding it at any position.

TABLE 21.1	Astronomically Important Elements	
Element	Number of Protons	Number of Neutrons*
Hydrogen	1	0
Helium	2	2
Carbon	6	6
Nitrogen	7	7
Oxygen	8	8
Silicon	14	14
Iron	26	30

*The number of neutrons listed here is the number found in the most abundant form of the element. Different neutron numbers can occur and lead to what are called *isotopes* of the element.

orbitals—an **electronic transition**—the energy change depends on the particular orbital and on the kind of atom. To understand why, we need to say a bit more about atomic structure.

21.4   THE CHEMICAL ELEMENTS

The structure of atoms determines both their chemical properties and their light-emitting and light-absorbing properties. For example, iron and hydrogen not only have very different atomic structures, but they also emit photons with different energies. From such differences astronomers can deduce whether an astronomical body—a star or a planet—contains iron, hydrogen, or whatever chemicals happen to be present. Therefore, understanding the structure of atoms ultimately helps us understand the nature of stars.

Iron and hydrogen are examples of what are called chemical **elements.** The characteristics of a chemical element are determined by the number of protons in its nucleus. For example, hydrogen atoms contain 1 proton; helium atoms contain 2 protons; carbon 6; oxygen 8; and iron 26. A complete list of the elements is given in the periodic table of the elements shown on the inside back cover of the book. The **atomic number** of each element indicates the number of protons in its nucleus.

Although the identity of an element is determined by the number of protons in its nucleus, the chemical properties of each element are determined by the number of electrons orbiting its nucleus. The number of electrons normally equals the number of protons. This means that each atom normally has an equal number of positive and negative electrical charges and is therefore electrically neutral.

An atom with six protons does not simply have electron orbitals that interact with photons six times more energetic than in hydrogen, or any other such simple relationship. The electron orbital clouds interact with each other in complex ways, giving each element a rich variety of unique atomic properties.

Table 21.1 lists most of the elements that will be important for our exploration of the universe, along with the number of protons and neutrons each contains.

Q The atomic mass of each element in the periodic table indicates (approximately) the sum of the number of protons (the atomic number) and the number of neutrons. Why do you suppose the number of neutrons increases relative to the number of protons at high atomic numbers?

21.5   ATOMIC ENERGIES

We have seen that when an electron moves from one orbital to another, the energy of the atom changes. If the electron is in its innermost orbital, closest to the nucleus, the atom's energy is at its lowest possible energy, called the *ground state.* If the atom's energy is increased, one or more electrons moves outward into a larger orbital. Such an atom is said to be in an **excited state.**

Although the energy of an atom may change, the energy does not just appear or disappear; rather, it changes form (see Unit 20 on the conservation of energy). According to this principle, if an atom loses energy, exactly the same amount of energy must reappear in some other form, such as an electromagnetic wave.

How is electromagnetic radiation created? When an electron drops from one orbital to another, it alters the electric energy of the atom. As we described earlier, such an electrical disturbance generates a magnetic disturbance, which in turn generates a new electrical disturbance. Thus the energy released when an electron drops from a higher to a lower orbital becomes an electromagnetic wave, a process called **emission** (Figure 21.4).

Emission plays an important role in many astronomical phenomena. The aurora (northern lights) is an example of emission by atoms in the Earth's upper atmosphere, and sunlight and starlight are examples of emission in those bodies.

The reverse process, in which light's energy is stored in an atom as energy, is called **absorption** (Figure 21.5). Absorption lifts an electron from a lower to a

Emission of electromagnetic waves

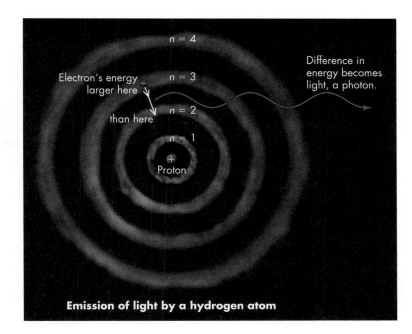

Emission of light by a hydrogen atom

FIGURE 21.4
Energy is released when an electron drops from an upper to a lower orbital, causing the atom to emit electromagnetic radiation.

Absorption of electromagnetic waves

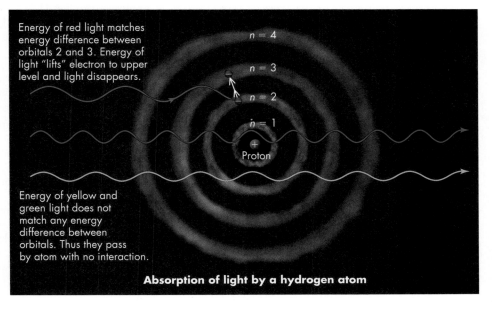

Absorption of light by a hydrogen atom

FIGURE 21.5
An atom can absorb a photon if its energy matches the energy required for an electron to jump to a higher orbital.

higher orbital and excites the atom by increasing the electron's energy. Absorption may even **ionize** the atom—this occurs if the electron gains so much energy that it leaves orbit completely. Absorption is important in understanding such diverse phenomena as the temperature of a planet and the identification of star types, as we will discover in later units.

Whether you picture emission and absorption as photons and electrons knocking about or as waves exciting vibrations, the important thing to keep track of is how much energy has been transferred. You may find it helpful in understanding emission and absorption if you think of an analogy. Absorption is a bit like drawing an arrow back, preparing to shoot it from a bow. Emission is like the arrow being shot. In one case, energy of your muscles is transferred to and stored in the flexed bow. In the other, the stored energy is released as the arrow takes flight.

KEY TERMS

absorption, 157	ground state, 155
atomic number, 156	interfere, 152
Bohr model, 155	inverse-square law, 153
brightness, 153	ionize, 158
electromagnetic wave, 152	light, 151
electronic transition, 156	orbital, 155
element, 156	photon, 152
emission, 157	quantized, 152
excited state, 156	wave–particle duality, 153

QUESTIONS FOR REVIEW

1. Why is light called an *electromagnetic wave?*
2. What is a photon? What is duality?
3. What is the inverse-square law?
4. What is the structure of an atom?
5. In what sense is matter wavelike?

PROBLEMS

1. Given that the Sun is 150 million kilometers from the Earth, calculate how long it takes light to travel from the Earth to the Sun.
2. Suppose you are operating a remote-controlled spacecraft on Mars from a station here on Earth. How long will it take the craft to respond to your command if Mars is at its nearest point to Earth? (You will need to look up a few numbers here.)

3. In the Bohr model of hydrogen, the energy associated with an electron being in level n is found to be -2.2×10^{-18} joules/n^2. (See Unit 3 for metric units.) What is the energy of an electron in level $n = 2$? In level $n = 100$?

4. Based on the formula in the previous problem, what is the energy of an electron if n goes to infinity? Explain.

5. a. At Mars's distance of 1.52 AU from the Sun, how many times less bright is the light from the Sun than at Earth's distance?

    b. What about at Mercury's distance of 0.39 AU?

TEST YOURSELF

1. Light is an electromagnetic wave in that it
    a. is made of magnetized electrons moving back and forth.
    b. can be generated by the motions of electrical charges.
    c. travels through water in microscopic oscillations.
    d. carries no energy.

2. When a photon interacts with an atom, what changes occur in the atom?
    a. The atomic number increases
    b. The nucleus begins to glow
    c. An electron changes its orbital
    d. The photon becomes trapped in orbit

3. Describing an atom's orbitals as being *quantized* refers to the fact that
    a. the atom is made of individual electrons, protons, and neutrons.
    b. the electron can oscillate only with certain discrete patterns.
    c. there are the same number of electrons as protons.
    d. there are a whole number of electrons in an atom.
    e. the electron's electric charge has a single fixed value.

22

The Electromagnetic Spectrum

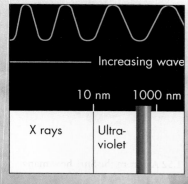

Increasing wave

10 nm   1000 nm

| X rays | Ultra-violet |

Background Pathways: Unit 21

There is more to light than what meets the eye. Our eyes can detect electromagnetic waves—known as **electromagnetic radiation**—with certain energies, but there are waves we cannot see. You are already familiar with many other forms of electromagnetic waves from many different sources. For example, radio waves, X-rays, and ultraviolet light are also electromagnetic radiation.

Other kinds of electromagnetic radiation differ from visible light only in the length of their electromagnetic waves. We can alternatively describe this in terms of the energies their photons contain. X-rays are rapidly oscillating waves; equivalently, we might call them highly energetic photons. Radio waves are slowly oscillating waves—or low-energy photons. This entire range of waves is called the **electromagnetic spectrum**.

22.1 THE COLORS OF VISIBLE LIGHT

Regardless of whether we consider light to be a wave or a stream of photon particles, our eyes perceive one of its most fundamental properties as color. Human beings can see colors ranging from deep red through orange and yellow into green, blue, and violet. The colors to which the human eye is sensitive define what is called the **visible spectrum.** But what property of photons or electromagnetic waves corresponds to light's different colors?

According to the wave interpretation of light, its color is determined by the light's **wavelength,** which is the spacing between wave crests (Figure 22.1). That is, instead of describing a quality, light's *color,* we can specify a quantity, its *wavelength,* denoted by the Greek letter lambda, λ. The wavelengths of visible light are very small—roughly the size of a bacterium. For example, the wavelength of deep red light is about 7×10^{-7} meters or 700 nanometers. The wavelength of violet light is about 4×10^{-7} meters or 400 nanometers. Intermediate colors have intermediate wavelengths.

Sometimes it is useful to describe electromagnetic waves by their frequency rather than their wavelength. **Frequency** is the number of wave crests that pass a given point in 1 second. The number of waves or "cycles" per second is indicated with the MKS unit **hertz** (abbreviated as Hz), named after Heinrich Hertz, who pioneered the broadcast and reception of radio waves in the late 1800s. In equations the frequency is usually denoted by the Greek letter nu: ν.

The frequency and wavelength of a wave have a simple relationship to each other. When the wavelength is longer, the frequency is lower (slower); when the wavelength is shorter, the frequency is higher (faster). If you float in ocean waves, as illustrated in Figure 22.2, you can experience this in terms of the rate at which you bob up and down (frequency) as waves pass by.

The relationship between the frequency and wavelength of a wave is determined by the wave speed because in one vibration, a wave must travel a distance equal to

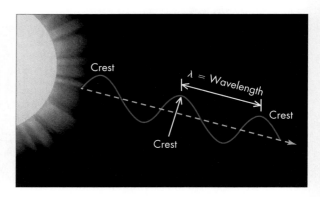

FIGURE 22.1
The distance between crests defines the wavelength, λ, for any kind of wave, be it water or electromagnetic.

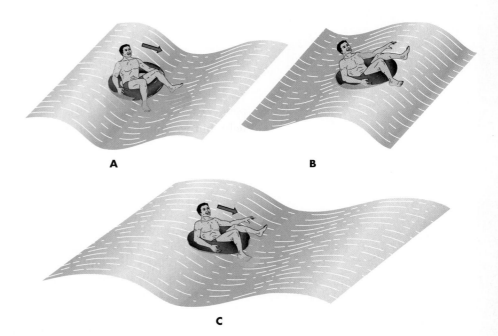

FIGURE 22.2
The frequency at which you bob up and down while floating in the water depends on the wavelength of the wave. Short-wavelength waves move past you quickly, so you bob up and down at a high frequency (A–B). A long-wavelength wave (C) takes longer to pass by even though the speed of the wave crest is the same, so the frequency is slower.

The speed of light is reduced when it travels through materials that appear transparent to us, such as glass, water, and gases. Furthermore, different colors—wavelengths—of light are slowed differently. For example, in nearly all materials, blue light travels slightly more slowly than red light.

one wavelength. Thus if we multiply the wavelength λ by the number of waves per second *ν*, the product is the speed of light:

$$\lambda \times \nu = c.$$

Because all light waves travel at the same speed *c*, the wavelength determines the frequency and vice versa. We will generally use λ to characterize electromagnetic waves, but *ν* is just as good. For example, yellow light vibrates about 500 trillion times a second (5×10^{14} Hz); red light vibrates a little slower; blue light a little faster.

Be sure to remember that *shorter wavelengths of visible light correspond to bluer colors.* Equivalently, *higher frequencies correspond to bluer colors.* We will use this information many times as we interpret the significance of different colors of light coming from astronomical objects.

22.2   WHITE LIGHT

Although wavelength is an excellent way to specify most colors of light, some light seems to have no color. For example, the Sun high in the sky and an ordinary light-bulb appear to have no dominant color. Light from such sources is called **white light.**

White light is not a special color of light; rather it is a mixture of all colors. That is, the sunlight we see is made up of all the wavelengths of visible light—a blend of red, yellow, green, blue, and so on—and our eyes perceive the combination of all these as white. Newton demonstrated this property of sunlight by a simple but elegant experiment. He passed sunlight through a prism (Figure 22.3) so that the light was spread out into the visible spectrum (which we see as a rainbow of colors). He then recombined the separated colors with a lens and reformed the beam of white light.

You can see how colors of light mix if you look at a color television screen with a magnifying glass. You will notice that the screen is covered with tiny red, green, and blue dots. In a red object, only the red spots are lit. In a blue object, only the blue spots are illuminated. In a white object, all three color spots are lit, and your brain mixes these three colors to form white. Other colors are made by appropriate blending of red, green, and blue.

You may have noticed that photos taken in artificial light look much more yellow than the scene you remember, or outdoor photographs may look bluer than you remember. That's because most lightbulbs are much more yellow than sunlight, but we adjust to the difference. We generally see a white piece of paper as white even under different color lighting conditions. Presumably because our senses have evolved to make us aware of differences in our surroundings, we ignore the ambient "color" of sunlight just as in time we learn to ignore a steady background sound or smell.

The adding of light of different colors is very different from the way that pigments of paint mix because the pigments reflect the color we see and absorb other colors. When you mix red, green, and blue paint, you will generally get a dark gray or brown color because most wavelengths of light are being absorbed. This is called "subtractive" color mixing.

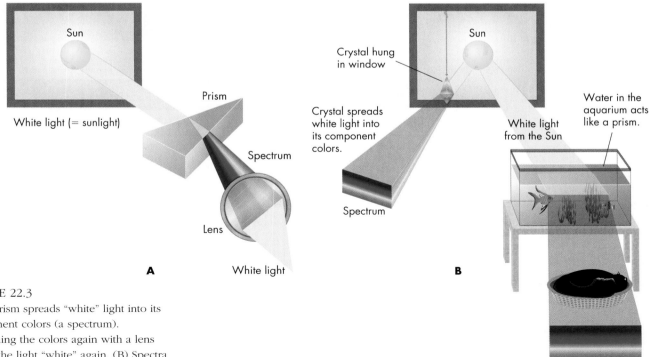

FIGURE 22.3
(A) A prism spreads "white" light into its component colors (a spectrum). Combining the colors again with a lens makes the light "white" again. (B) Spectra in everyday life.

White light and the enormous variety of colors we can see represent just some of the richness possible for electromagnetic waves ranging between 400 and 700 nm. Yet today there are instruments capable of detecting electromagnetic waves with wavelengths thousands of kilometers long down to 10^{-18} meters or less. Ordinary visible light falls in a narrow section in the middle of a broad spectral range (see Figure 22.4 and Table 22.1). Objects may emit and absorb light in every part of the

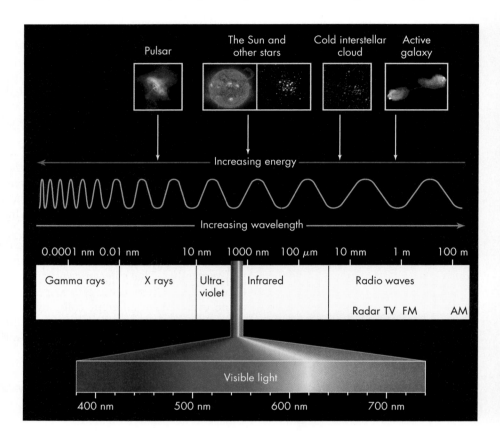

FIGURE 22.4
The electromagnetic spectrum.

TABLE 22.1	Electromagnetic Spectrum	
Wavelength	**Kind of Radiation**	**Astronomical Sources**
100–500 meters	Radio (AM broadcast)	Pulsars (remnants of exploded stars)
10–100 meters	Short-wave radio	Active galaxies
1–10 meters	TV, FM radio	
10–100 centimeters	Radar	Planets, active galaxies
1–100 millimeters	Microwaves	Interstellar clouds, cosmic background radiation
1–10^3 micrometers	Infrared	Young stars, planets, interstellar dust
400–700 nanometers	Visible light	Stars, Sun
1–300 nanometers	Ultraviolet	Stars
0.01–1 nanometers	X-rays	Collapsed stars, hot gas in galaxy clusters
10^{-7}–0.01 nanometers	Gamma rays	Active galaxies and gamma-ray bursters

electromagnetic spectrum, creating "colors" completely outside the range of our eyes. Using our eyes alone, an object may appear to be dark; but it may be the source of an assortment of different-wavelength photons when viewed with a camera that is sensitive to a different part of the electromagnetic spectrum.

22.3   INFRARED AND ULTRAVIOLET RADIATION

The exploration of the electromagnetic spectrum beyond visible light began in 1800, when Sir William Herschel (discoverer of the planet Uranus) showed that heat radiation, such as you feel from the Sun or from a warm radiator, though invisible, is related to visible light.

Although humans cannot see infrared radiation, several kinds of snakes, including the rattlesnake, have special infrared sensors located just below their eyes. These allow the snake to "see" in total darkness, helping it to find warm-blooded prey such as rats.

Herschel was trying to measure heat radiated by astronomical sources. He projected the Sun's spectrum onto a tabletop and placed a thermometer in each color to measure its energy. He was surprised that when he put a thermometer just off the red end of the visible spectrum, the thermometer registered an elevated temperature there just as it did in the red part of the spectrum. He concluded that some form of invisible energy detectable as heat existed beyond the red end of the spectrum, and he therefore called it **infrared.** Even though your eyes cannot see infrared light, when it strikes your skin it deposits energy that warms the skin, and the nerves in your skin can feel the heat.

Just a year later, in 1801, the German scientist Johann Ritter was studying the light sensitivity of chemicals that were later to become the basis for photography. When he shone light through a prism onto one of these chemicals, he discovered that there was an intense photochemical reaction beyond the last visible portion of the spectrum at its violet end—**ultraviolet** radiation.

Infrared and ultraviolet radiation differ in no physical way from visible light except in their wavelength. Infrared has longer wavelengths and ultraviolet shorter wavelengths than visible light (see Table 22.1). Both can be described as waves or photons of electromagnetic energy, like visible light. They just cannot be seen by the human eye.

22.4   ENERGY CARRIED BY PHOTONS

In MKS units, E is measured in joules and λ in meters, and c is the speed of light. Laboratory measurements of the constant h have determined its value: $h = 6.63 \times 10^{-34}$ joule-seconds, and $h \times c = 1.99 \times 10^{-25}$ joule-meters. So for a red photon with a wavelength of 670 nm

$$E = \frac{h \times c}{\lambda} = \frac{1.99 \times 10^{-25} \text{ joule} \cdot \text{m}}{670 \text{ nm}}$$

$$= \frac{1.99 \times 10^{-25} \text{ joule} \cdot \text{m}}{6.7 \times 10^{-7} \text{ m}}$$

$$= 3.0 \times 10^{-19} \text{ joule}.$$

The warmth we feel on our face from a beam of sunlight demonstrates that light carries energy, but not all wavelengths carry the same amount of energy. It turns out that the amount of energy, E, carried by a photon depends on its wavelength, λ. Each photon of wavelength λ carries an energy, E, given by

$$E = \frac{h \times c}{\lambda}.$$

The quantity h is called *Planck's constant,* a constant of nature that describes the fundamental relationship between energies and waves in the submicroscopic realm of quantum mechanics. This equation can also be written in terms of the frequency of a photon. From the equation relating wavelength and frequency, we find that $\nu = c/\lambda$, so the energy of a photon is equivalently $E = h \times \nu$.

Thus the wavelength of a photon determines how much energy it carries. Long-wavelength, low-frequency photons contain less energy. In other words,

A photon carries an amount of energy proportional to its frequency and inversely proportional to its wavelength.

For example, an ultraviolet photon with a wavelength of 300 nm carries twice as much energy as a 600 nm yellow photon and three times as much energy as a 900 nm infrared photon. In fact, ultraviolet photons of sufficiently short wavelength carry so much energy that they can break apart atomic and molecular bonds, which is why ultraviolet light gives you a sunburn but an infrared "heat lamp" does not.

22.5   RADIO WAVES

It might help you to remember the connection between energy, wavelength, and frequency to make waves on a rope (or a telephone cord). Tie one end to a doorknob and shake the loose end of it. To make short-wavelength waves along the rope, how rapidly (frequently) do you shake the rope? Does it take more of your energy to make short waves or long waves?

Q Why do you suppose that the centers of some filled pastries grow hot in a microwave oven while their outer portions do not grow as hot?

James Clerk Maxwell, a Scottish physicist, predicted the existence of **radio waves** in the mid-1800s. It was some 20 years later, however, before Heinrich Hertz produced them experimentally in 1888, and another 50 years had to pass before Karl Jansky discovered naturally occurring radio waves coming from cosmic sources.

Radio waves range in length from millimeters to hundreds of meters and longer, making them much longer than visible and infrared waves. An AM "radio" is a receiver for electromagnetic waves that are hundreds of meters long; FM radio corresponds to wavelengths of about one-third of a meter. Actually radio wavelengths are more commonly listed by their frequency. You can see this on a radio dial, where you tune in a station by its frequency in kilohertz or megahertz (thousands or millions of waves per second) rather than its wavelength.

Today we can generate radio waves and use them in many ways, ranging from communication to radar to cooking. Given the enormous span of wavelengths and uses, the wide range of radio waves is sometimes subdivided, with the shorter wavelengths being called **microwaves.** These are generally described as waves of about 20 cm or less, although there is no definite standard on what the boundaries are between infrared, microwave, and radio. Most microwave ovens generate electromagnetic waves of about 12 cm wavelength.

The strong connection between radio waves and their use for communication systems confuses many people into thinking that radio waves are a kind of sound wave. Not so! Radio waves are a kind of light, traveling at the same enormous speed as all forms of electromagnetic radiation. If radio waves traveled anywhere near as slowly as sound, you would have to wait tens of seconds for every response when calling on a cell phone to a friend just a few kilometers away, and hours when calling across the country!

Many kinds of astronomical objects such as forming stars, exploding stars, active galaxies, and interstellar gas clouds generate these low-energy radio waves by natural processes. Understanding this radiation reveals information about the temperature of the gas, the molecules present, magnetic fields in the region, and a wide variety of other factors. Astronomers detect this radiation with radio telescopes to study the kinds of astrophysical processes and conditions different from those that generate visible light.

22.6   HIGH-ENERGY RADIATION

Over the last several decades, astronomers have also probed very short-wavelength regions called the **X-ray** and **gamma-ray** parts of the spectrum. X-ray wavelengths are even shorter than those of visible and ultraviolet light, typically between 0.01 and 10 nanometers, and gamma-ray wavelengths are shorter yet (see Table 22.1).

The fact that X-rays and gamma rays have such short wavelengths implies that they carry a great deal of energy and are generated only in regions with extremely high temperatures or high-energy reactions taking place. These wavelengths therefore reveal to astronomers some of the universe's most exotic objects and processes, such as hot gas falling into black holes.

The high energy of these photons allows them to penetrate many objects that are opaque to visible light. One familiar example is the ability of X-rays to penetrate soft tissue, which allows us to make diagnostic medical X-ray images. The X-ray photons that are not blocked by bone or other dense material travel through and expose a photographic negative—much like Ritter's 1801 experiment with ultraviolet light.

Note that because X-rays are outside the range that our eyes can see, there is no such thing as a pair of "X-ray glasses" that lets you see through objects. If you looked through a pair of glasses that transmitted only X-rays, you would see . . . *nothing!*

Despite the enormous variety of electromagnetic waves, they are all the same physical phenomenon: the vibration of electric and magnetic energy traveling at the speed of light. The essential difference between kinds of electromagnetic radiation is merely their wavelength—or frequency—or energy.

Q Our eyes are also insensitive to infrared light, but there are "infrared night-vision goggles." How do you suppose we can see infrared light with these?

KEY TERMS

electromagnetic radiation, 159
electromagnetic spectrum, 159
frequency, 159
gamma ray, 164
hertz, 159
infrared, 163
microwaves, 164

radio waves, 164
ultraviolet, 163
visible spectrum, 159
wavelength, 159
white light, 161
X-ray, 164

QUESTIONS FOR REVIEW

1. How are color and wavelength related?
2. What is meant by the *electromagnetic spectrum?*
3. Name the regions of the electromagnetic spectrum from short to long wavelengths.
4. What is the difference between emission and absorption in terms of what happens to an electron in an atom?

PROBLEMS

1. a. A yellow photon has wavelength of 500 nanometers. What is its frequency?
   b. The frequency of a radio station transmitter is 90 megahertz. What is its wavelength?
2. The energies of electrons in their atomic orbitals are typically about 10^{-18} joules. What wavelength of light would have

photons with this energy? Which band of electromagnetic radiation is this in?

3. a. What is the energy (in joules) of an extremely powerful gamma ray with a wavelength of 10^{-7} nanometers?
   b. How many of these gamma-ray photons would have the same amount of energy as a housefly, which has a kinetic energy of about one millionth of a joule as it flies along?

TEST YOURSELF

1. Which type of electromagnetic radiation has the longest wavelength?
   a. Ultraviolet
   b. Visible
   c. X-ray
   d. Infrared
   e. Radio

2. Which kind of photon has the highest energy?
   a. Ultraviolet
   b. Visible
   c. X-ray
   d. Infrared
   e. Radio

3. A photon may be Doppler shifted to a lower frequency if the source of light is moving away from us at high speed. If the frequency drops to half its former value, the energy of the photons will
   a. drop by half too.
   b. double.
   c. remain the same.

23 Thermal Radiation

When a body becomes hotter, the atoms and molecules inside it vibrate more rapidly, undergo more collisions, and generally interact more energetically. It is not surprising, then, that a hot object generally emits more electromagnetic radiation and that the photons emitted tend to have higher energies. For example, when a chunk of charcoal becomes hot, it may glow with an orange color. This kind of electromagnetic emission is called **thermal radiation**.

Thermal radiation is produced by many objects that you might not think of as "glowing." Your body or an animal's body produces infrared thermal radiation, for example (Figure 23.1). On the other hand, some objects, like the collapsed remnants of stars, are so hot that their thermal radiation is mostly at X-ray wavelengths. Not all kinds of materials produce thermal radiation—in particular, low-density gas produces emission at the individual wavelengths associated with electrons changing orbitals. A wide variety of denser materials produce thermal emission that looks similar no matter what the chemical composition is of the emitting materials.

Wherever thermal radiation occurs, it follows two simple rules: (1) The photons have higher energy when the material is hotter, and (2) the total amount of radiation produced increases rapidly as the temperature climbs. These two rules have mathematical formulations that allow us to learn about the physical conditions in material from the thermal radiation it emits. These provide important tools for studying objects throughout the universe.

Background Pathways: Unit 22

23.1 BLACKBODIES

A **blackbody** is an object that absorbs all the radiation falling on it. Because such a body reflects no light, it looks black to us when it is cold, hence its name. For an object to appear black, it must be capable of absorbing all wavelengths of electromagnetic radiation. This implies that the substance is capable of undergoing energy transitions corresponding to all wavelengths. This is *not* true of a low-density gas, but it is true or nearly true for a wide range of other substances.

When blackbodies are heated, they can also radiate at all wavelengths. Once again, this is because transitions of all energies (or equivalently all wavelengths) are possible. Thus they are both excellent absorbers and excellent emitters. Moreover, the intensity of their radiation changes smoothly from one wavelength to the next with no gaps or narrow peaks of brightness. Very few objects are perfect blackbodies, but many of the objects astronomers study are near enough to being blackbodies that their radiation looks very similar. For example, a piece of charcoal, an electric stove burner, the Sun, and the Earth all produce thermal emission that closely resembles the idealized blackbody.

On the other hand, gases (unless compressed to a high density) are generally not blackbodies and do not produce thermal emission. The wavelengths emitted by a gas are determined by its composition and the electron energy-level transitions that are possible for the kinds of atom and molecules present in the gas.

What makes a solid or dense gas different? In them, the orbitals of each atom's electrons are disturbed by neighboring atoms. The possible energies of the electrons

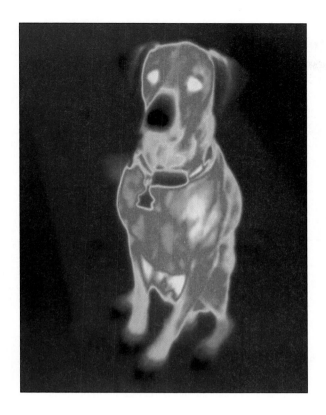

FIGURE 23.1

A picture of the infrared thermal radiation from a dog. The dog's eyes are warmest and produce the most light, while its nose is cold and emits little light.

INTERACTIVE

Blackbody radiation

are no longer precisely defined as they are for isolated atoms, so an incoming photon is likely to encounter an atom that can absorb its energy. Solid materials often retain some characteristics of their component atoms and molecules, which is why the things we see are not all black. For example, a leaf appears green because it absorbs most wavelengths *except* those with green colors.

When a solid or dense gas becomes hotter, it usually appears more and more like a blackbody. This is because the vibrations and energetic collisions associated with higher temperatures cause the electrons' orbitals to be disturbed even more, further "smearing out" the range of energies—and wavelengths—they can absorb or emit.

23.2 COLOR, LUMINOSITY, AND TEMPERATURE

One important property we can infer about an object from the light it emits is its temperature. Temperature gives rise to subtle differences in the colors of stars at night. And by observing beyond the wavelengths of visible radiation, we can detect objects both much cooler and much hotter than stars.

A hot object emits light, as you can easily see if you turn on the burner of an electric stove. You can also see from such a burner that the color of the light is related to the temperature of the burner. In particular, as the burner grows hotter, at first it emits infrared light that you might sense as heat; then it shifts to red, then orange (Figure 23.2A). This relation between color and temperature allows astronomers to measure the temperature of stars and other astronomical objects from the wavelength of the light they emit. They can make this measurement using a relation called **Wien's law.**

Wien's law states that the wavelength at which an object radiates most strongly is inversely proportional to the object's temperature:

Hotter bodies radiate more strongly at shorter wavelengths.

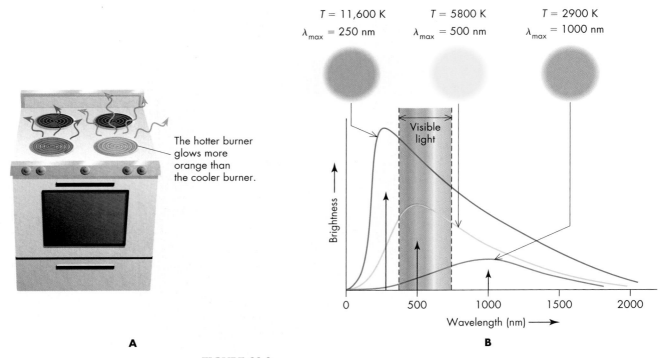

FIGURE 23.2

(A) The hotter burner glows more orange than the cooler burner. (B) As an object is heated, the wavelength at which it radiates most strongly, λ_{max}, shifts to shorter wavelengths, a relation known as *Wien's law*. Note also that as the object's temperature rises, the amount of energy radiated increases at *all* wavelengths.

We can see this effect illustrated in Figure 23.2B, where we plot the amount of energy radiated at each wavelength (color) for three bodies of different temperatures. Notice that the hotter body has its most intense emission (highest point) at a shorter wavelength than the cooler body. This is what gives it a different color.

Note also that for objects of the same size, the hotter object emits more energy at *every* wavelength than the cooler one does. This is obvious looking at a stove burner. As it grows hotter, going from dull red to orange, it also grows much brighter. This is also the basis for the dimmer switches on lights: Changing the amount of electricity going to the lamp changes the temperature of the filament in the lightbulb and thereby its brightness. The light output of an object is known as its **luminosity.** What is observed is that even if the filament rises only a little bit in temperature, its luminosity increases substantially. Thus,

The luminosity of a hot body rises rapidly with temperature.

A mathematical form of this principle is formulated as the *Stefan-Boltzmann law*.

Before we can write out these mathematical laws in a form that allows us to carry out specific calculations, we must decide what temperature scale we wish to use. To make this choice, it will be helpful if we briefly discuss what temperature measures.

23.3   MEASURING TEMPERATURE

An object's temperature is directly related to its energy content and to the speed of its molecular motion. That is, the hotter the object, the faster its atoms move and the more energy it possesses. Similarly, if the body is cooled sufficiently, molecular

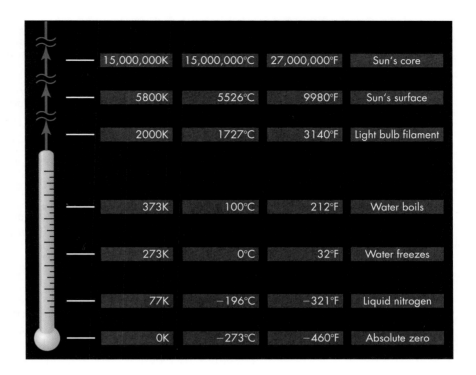

15,000,000K	15,000,000°C	27,000,000°F	Sun's core
5800K	5526°C	9980°F	Sun's surface
2000K	1727°C	3140°F	Light bulb filament
373K	100°C	212°F	Water boils
273K	0°C	32°F	Water freezes
77K	−196°C	−321°F	Liquid nitrogen
0K	−273°C	−460°F	Absolute zero

FIGURE 23.3
Temperatures in Kelvin (K) and on the Celsius and Fahrenheit scales.

Fahrenheit temperatures, °F, can be calculated by using the formula $T_{\circ F} = 9/5\ T_K - 459.4$, where T_K is the temperature in Kelvin. The Fahrenheit scale was designed to have 0°F at the temperature at which concentrated saltwater freezes and 100°F at body temperature.

Note that temperatures on the Kelvin scale are not given in "degrees" but are simply called "Kelvin."

motion within it slows to a virtual halt and its energy approaches zero. In the late 1800s the British scientist Lord Kelvin devised a temperature scale based on these properties (Figure 23.3). At zero on the **Kelvin** scale, molecular motion essentially ceases, and objects have no heat energy. This is known as **absolute zero**—the lowest possible temperature. Notice that negative temperatures therefore have no meaning on the Kelvin scale because there cannot be less motion than none.

It is simple to convert from Kelvin to the more familiar Celsius scale. The temperature in Kelvin equals 273.15 plus the Celsius temperature. For example, the freezing and boiling points of water, 0 °C and 100 °C, are approximately 273 K and 373 K, respectively. Room temperature is about 300 Kelvin.

The Kelvin scale is useful in astronomy because the temperature in Kelvin is directly proportional to an object's **thermal energy** (the energy due to the internal vibrations and motions of the atoms within the object). Thus when the thermal energy doubles, the value of the temperature in Kelvin also doubles. By contrast, when the temperature of an object doubles in the Celsius and Fahrenheit scales, it is *not* true that its thermal energy doubles.

Because of its direct relation to so many physical processes, we will use the Kelvin scale in most of the remainder of this book. With a temperature scale chosen, we can now learn how we might measure a star's temperature.

23.4   TAKING THE TEMPERATURE OF ASTRONOMICAL OBJECTS

Wien's law, named for the German physicist who discovered it around 1900, is important because we can use it to measure how hot something is simply from the color of light it radiates most strongly. To measure a distant body's temperature using Wien's law, we proceed as follows. First we measure the body's brightness at many different wavelengths to find the particular wavelength at which it is brightest (that is, its wavelength of maximum emission). Then we use the law to calculate the body's temperature. To see how this is done, however, we need a mathematical expression for the law.

If we let T be the body's temperature measured in Kelvin, and we let λ_{max} be the wavelength in nanometers at which the body radiates most strongly (Figure 23.2B), Wien's law can be written in this form:

$$T = \frac{2.9 \times 10^6 \text{ K} \cdot \text{nm}}{\lambda_{max}}.$$

The subscript max on λ is to remind us that it is the wavelength of maximum emission.

As an example, we can measure the Sun's temperature. The Sun turns out to radiate most strongly at a wavelength of about 500 nanometers. Substituting $\lambda_{max} = 500$ nanometers, we find

$$T_\odot = \frac{2.9 \times 10^6 \text{ K} \cdot \text{nm}}{500 \text{ nm}} = 5800 \text{ K}.$$

The Sun is not a perfect blackbody; nevertheless, this result is within about 20 K of the value found by much more sophisticated analyses.

You might note that the wavelength at which the Sun radiates most strongly corresponds to a blue-green color, yet the Sun looks yellow-white to us. The reason we see it as whitish is related to how our eyes perceive color. Physiologists have found that the human eye interprets sunlight (and light from all extremely hot bodies) as whitish, with only tints of color. Such hot bodies emit a significant amount of light at all visible wavelengths, not just at the color corresponding to the wavelength of maximum emission. Thus cool stars look white tinged with red, whereas very hot stars look white tinged with blue.

Wien's law works quite accurately for most stars and planets or almost any material that has atoms packed tightly together. One important thing to remember, though, is that the law tells us only about thermal radiation *emitted* by an object. An object may simultaneously be emitting thermal radiation and *reflecting* light from another source. For example, the red color of an apple and the green color of a lime have nothing to do with their temperatures. They both do emit thermal radiation, but if they are at normal room temperature, the radiation they emit will be primarily in the infrared (Figure 23.4).

Q Suppose you wanted to design a paint to keep a building cool in hot sunlight. At what wavelengths would you want it to absorb light well, and at what wavelengths would you want it to reflect light well?

65°C

10°C

FIGURE 23.4
Side-by-side infrared and visible photographs of the same scene. Most of the visible light we see on Earth is light reflected off a surface, not thermal radiation.

23.5   THE STEFAN-BOLTZMANN LAW

In the late 1800s two Austrian scientists, Josef Stefan and Ludwig Boltzmann, showed how the total amount of light emitted by an object increases as its temperature rises, as indicated in Figure 23.2B. The **Stefan-Boltzmann law,** as their discovery is now called, states that a body of temperature T (measured in Kelvin) radiates an amount of energy each second equal to σT^4 per square meter. The quantity σ is called the **Stefan-Boltzmann constant,** and its value is 5.67×10^{-8} watts per meter^2 per Kelvin^4. Like Wien's law, this law is precisely true for blackbodies, but it is also quite accurate for many different kinds of materials, both solids and gases, as long as the material is relatively dense. It works well for embers of coal, a lightbulb filament, or the surface of a star.

The Stefan-Boltzmann law shows mathematically what we observe as rapid brightening as an object gets hotter. If the temperature doubles, for example, the light output increases by a factor of 16. We can show this mathematically for material heated to one temperature that's twice another. For example, suppose a lightbulb filament is heated to 1000 K. The amount of light it generates is

$$\sigma \times (1000\ K)^4 = (5.67 \times 10^{-8}\ watt/m^2/K^4) \times (10^3)^4 \times K^4$$
$$= 5.67 \times 10^{-8} \times 10^{12}\ watt/m^2$$
$$= 5.67 \times 10^4\ watt/m^2.$$

On the other hand, if its temperature is raised to 2000 K, the luminosity is

$$\sigma \times (2000\ K)^4 = (5.67 \times 10^{-8}\ watt/m^2/K^4) \times (2 \times 10^3)^4 \times K^4$$
$$= 5.67 \times 10^{-8} \times 16 \times 10^{12}\ watt/m^2$$
$$= 90.72 \times 10^4\ watt/m^2.$$

Comparing the second line in each of these calculations, you can see that the only difference is the factor of 16, which comes from taking 2 to the fourth power: $2^4 = 2 \times 2 \times 2 \times 2 = 16$. It would not matter what pair of temperatures we put into the equation: As long as one was *two times hotter* than the other, the amount of light generated by each square meter of the hotter object will always be *16 times more luminous* than the cooler one.

The other thing to notice from these calculations is the enormous wattages generated by material at these high temperatures. A square meter (about 10 square feet) of material heated to 2000 K generates over 900,000 watts of luminous power! A lightbulb filament is typically about 2000 to 2500 K. If the total surface area of the filament is one square centimeter, that's equivalent to $10^{-4}\ m^2$. Such a filament would produce about $91 \times 10^4\ watt/m^2 \times 10^{-4}\ m^2 = 91$ watts of thermal energy.

For a 2000 K filament, Wien's law tells us that most of the emission is at infrared wavelengths we cannot see. Typically, less than 10% of the power is actually produced as visible light. By raising the square-centimeter filament's temperature to 2500 K, we would now produce 221 watts. In addition, by raising the temperature, Wien's law tells us that we would shift the peak of the emission closer to optical wavelengths, so the amount of visible light would increase even more.

The Sun's surface is about 5800 K, so compared to the 1000 K thermal emission we just calculated, the Sun's surface must generate $5.8^4 = 5.8 \times 5.8 \times 5.8 \times 5.8 = 1130$ times more power. This is over 64 million watts per square meter emitted by every square meter over the entire surface of this enormous body! When you look at photographs of the Sun's surface (Figure 23.5), you will see some regions that are darker and some lighter. These are regions with slightly different temperatures. The strong dependence of luminosity on temperature means that a region that is at 5000 K will look dimmer than the surrounding regions.

1 square centimeter = 1 cm $\times$ 1 cm = 10^{-2} m $\times$ 10^{-2} m = 10^{-4} m^2

Q Why do you suppose a lightbulb filament is usually a thin winding wire instead of simply being thicker and straight? How might this thinking apply to the thermal radiation from a rocky asteroid versus the same amount of rock broken up into small pieces?

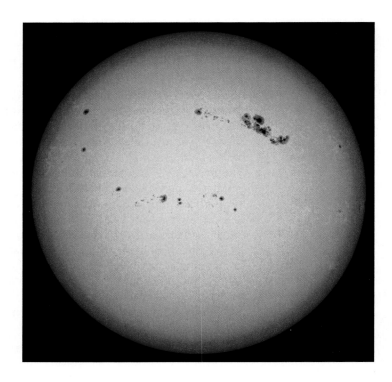

FIGURE 23.5
A portion of the Sun's surface. The darker spots in the picture are cooler than surrounding regions, so they look dark by contrast. However, they are generating light.

KEY TERMS

absolute zero, 169

blackbody, 166

Kelvin, 169

luminosity, 168

Stefan-Boltzmann constant, 171

Stefan-Boltzmann law, 171

thermal energy, 169

thermal radiation, 166

Wien's law, 167

QUESTIONS FOR REVIEW

1. How are color and temperature related?
2. What are the advantages of the Kelvin scale?
3. What is Wien's law?
4. What is the Stefan-Boltzmann law?

PROBLEMS

1. A lightbulb radiates most strongly at a wavelength of about 3000 nanometers. How hot is its filament?
2. The Earth's temperature averaged over the year is about 300 Kelvin. At what wavelength does it radiate most strongly? What part of the electromagnetic spectrum does this wavelength lie in? Can you see it?
3. If a lightbulb filament could be heated to 10,000 K, how much more thermal radiation would it produce than if it were at 2500 K?

TEST YOURSELF

1. Suppose we doubled the thermal energy of a rock that had a temperature of 7 °C = 45 °F = 280 K. What would its new temperature be?
   a. 14 °C
   b. 90 °F
   c. 560 K
   d. All of the above
2. If the temperature of an object doubles, the wavelength where it emits the most amount of light will be
   a. 4 times longer.
   b. 2 times longer.
   c. the same.
   d. 2 times shorter.
   e. 4 times shorter.
3. If the temperature of an object doubles, the total amount of its thermal radiation will be
   a. the same.
   b. 2 times larger.
   c. 4 times larger.
   d. 8 times larger.
   e. 16 times larger.

24

Atomic Spectra: Identifying Atoms by Their Light

Calcium
Hydrogen
Magne

400   500

Wavelen

Background Pathways: Unit 23

The keys to determining the composition and conditions of an astronomical body are the wavelengths of light it absorbs and emits. The technique used to capture and analyze the light from an astronomical body is called **spectroscopy**. In spectroscopy, electromagnetic radiation that is emitted or reflected by the object being studied is collected with a telescope and spread into its component colors to form a spectrum. For visible wavelengths this is done by passing the light through a prism or through a grating consisting of numerous tiny, parallel lines (Figure 24.1). Because photons of particular wavelengths are emitted and absorbed by atoms as electrons shift between orbitals, the spectrum of light will bear an imprint of the kinds of atoms that the light interacted with. This allows astronomers to search for various atoms' "signatures" by measuring how much light is present at each wavelength.

Spectroscopy is such an important tool for astronomers that we will examine it at a fairly close level of detail. Specifically, why does an atom produce a unique spectral signature? To understand that, we need to recall how light is produced.

24.1 HOW A SPECTRUM IS FORMED

When an electron moves from one orbital to another, the atom's energy changes by an amount equal to the difference in the energy between the two orbitals. Because the atom's energy is determined by what orbitals its electrons move in, orbitals are often referred to as **energy levels.**

As an example, suppose we look at light from heated hydrogen. Heating speeds up the atoms, causing more forceful and frequent collisions that can knock each atom's electron into more energetic orbitals. At the same time, the electrical attraction between the nucleus and the electron draws the electron back down to lower energy levels. As the electron shifts downward, the atom's energy decreases, and the energy that is lost appears as a photon.

Suppose we look at an electron shifting from orbital 3 to orbital 2, as shown in Figure 24.2A. The wavelength of the emitted light can be calculated from the energy difference of the levels and the relation between energy and wavelength: $E = hc/\lambda$ (Unit 22). If we evaluate the wavelength of this light, we find that it is 656 nanometers, a bright red color. An electron dropping from orbital 3 to orbital 2 in a hydrogen atom always produces light of this wavelength.

If, instead, the electron moves between orbital 4 and orbital 2 (Figure 24.2B), there will be a different change in energy because orbital 4 has higher energy than orbital 3. The larger energy change corresponds to a shorter-wavelength photon.

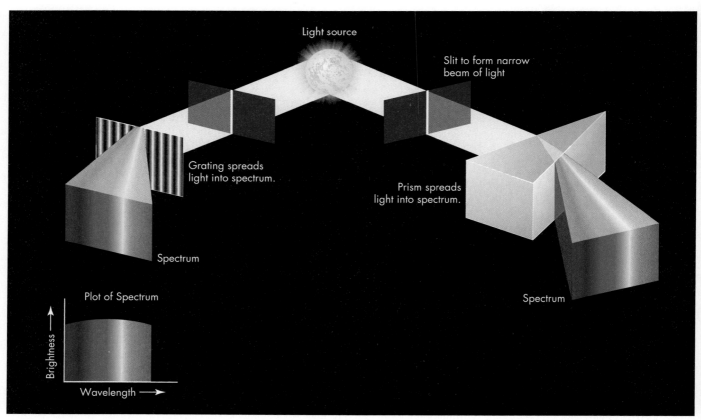

FIGURE 24.1

Sketch of a spectroscope and how it forms a spectrum. Either a prism or a grating may be used to spread the light into its component colors.

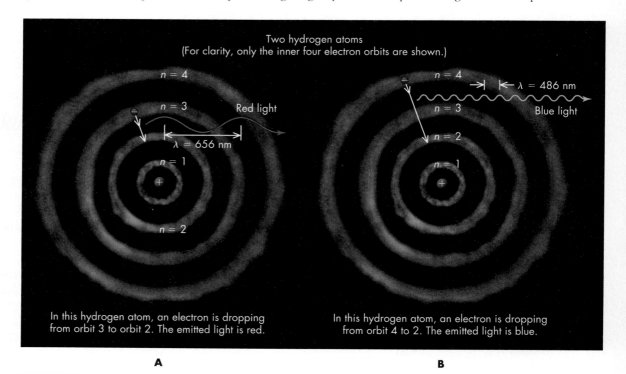

A   **B**

FIGURE 24.2

Emission of light from a hydrogen atom. The energy of an electron dropping from an upper to a lower orbital is converted to light. The light's color depends on the orbitals involved.

A calculation of its energy change leads in this case to a wavelength of 486 nanometers, a turquoise blue color.

24.2   IDENTIFYING ATOMS BY THEIR LIGHT

If we spread the light from a hot gas into a spectrum, we will see that in general the spectrum contains light at only certain wavelengths. In a gas made up of hydrogen atoms, we will see light from atoms undergoing changes between many different energy levels simultaneously, as depicted in Figure 24.3. Thus hydrogen gas produces a spectrum containing the red 656-nanometer and turquoise 486-nanometer

Hydrogen atoms in tube

Atom emits at wavelength set by the orbit its electron happens to be in. Thus, if an electron jumps from level 3→2, the atom emits red light. If the electron jumps from 5→2, it emits violet, etc. No level jump corresponds to yellow or green light so those colors do not appear in the hydrogen spectrum.

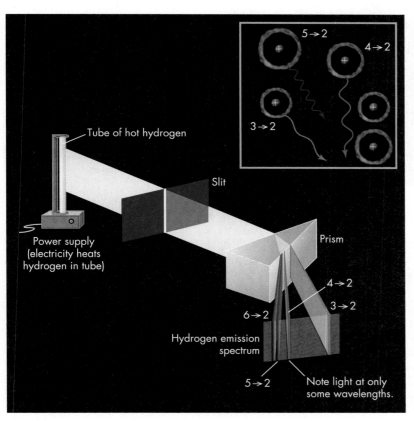

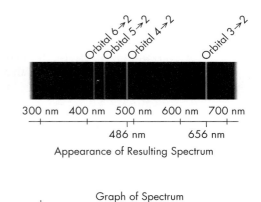

Appearance of Resulting Spectrum

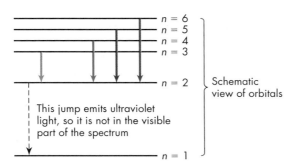

FIGURE 24.3

The spectrum of hydrogen in the visible wavelength range.

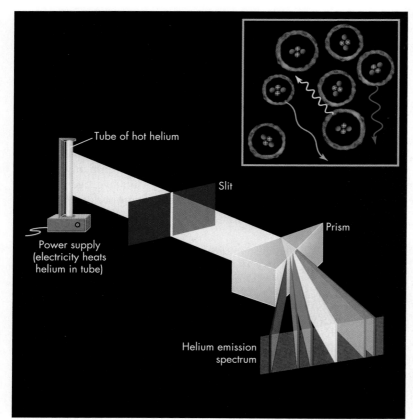

Helium atoms in tube
The electron orbits for helium atoms are different from the orbits in hydrogen. The light they emit therefore differs from that of hydrogen.

Tube of hot helium

Slit

Prism

Power supply (electricity heats helium in tube)

Helium emission spectrum

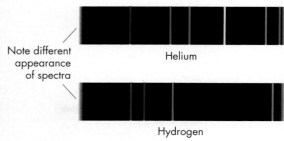

Note different appearance of spectra

Helium

Hydrogen

FIGURE 24.4
The spectrum of helium in the visible wavelength range.

lines, as well as violet light corresponding to electrons dropping from orbitals 5 to 2, and other lines that lie outside the wavelength range that our eyes can detect. This is known as an **emission-line spectrum.** This set of lines is the "signature" of hydrogen, unique to this element, and it allows astronomers to determine whether an astronomical object contains hydrogen.

If instead of hydrogen we looked at heated helium, we would see a very different spectrum. The reason is simple. Helium has two protons instead of one, and the electrons interact with each other, which leads to a different set of electron energy levels and thus a different set of spectral lines, as you can see in Figure 24.4. The spectrum thus becomes a means of identifying what atoms are present in a gas.

In fact helium was first discovered in 1868 by spectroscopic observations of the Sun during an eclipse. Astronomers recognized that the wavelength of the strong yellow line in its spectrum had no known counterpart, and they gave the name *helium* to the unknown element after the Greek word for the Sun, *helios*. Helium was not discovered by chemists on Earth until the 1890s.

For each element present in a gas we will see specific sets of spectral lines. We will see no light at most other colors because there are no electron orbital transitions corresponding to those energies. Therefore, the spectrum shows a set of brightly colored lines separated by wide, dark gaps.

It is also possible to identify atoms in a gas from the way they absorb light. Light is absorbed by an atom if the energy of its wavelength corresponds to an energy that matches the difference between two energy levels in the atom. If the wavelength does not match, the light will not be absorbed, and it will simply move past the atom, leaving itself and the atom unaffected.

Q If an atom had just five different possible energy levels for its electrons, how many different wavelengths of light might it be possible for the atom to emit?

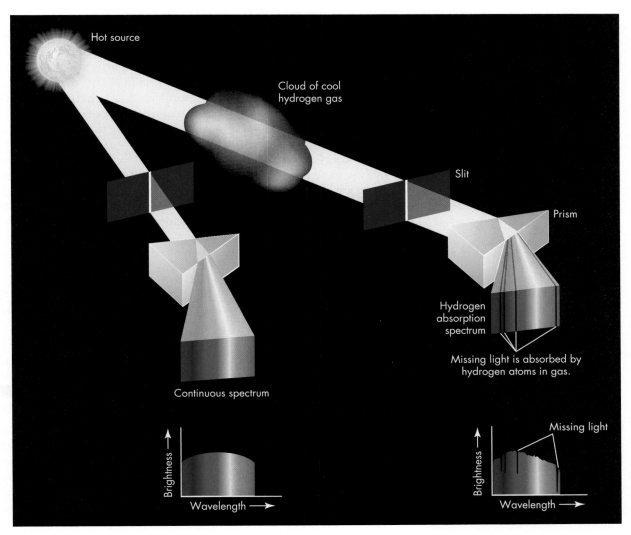

FIGURE 24.5

A hot, dense substance produces a continuous spectrum. Atoms in an intervening gas cloud absorb only those wavelengths whose energy equals the energy difference between their electron orbitals. The absorbed energy lifts the electrons to upper orbitals. The lost light makes the spectrum darker at the wavelengths where it is absorbed.

Even if an atom has a pair of energy levels separated by an amount matching a photon's energy, the atom cannot absorb the photon unless an electron is sitting in the lower of the pair of energy levels. This can limit the ability of a very cold gas or a very hot gas to absorb photons.

For example, suppose we shine a beam of light that initially contains all the colors of the visible spectrum through a box full of hydrogen atoms. If we examine the spectrum of the light after it has passed through the box, we will find that certain wavelengths of the light have been removed and are missing from the spectrum (Figure 24.5). In particular, the spectrum will contain gaps that appear as dark lines at 656 nanometers and 486 nanometers, precisely the wavelengths at which the hydrogen atoms emit. The **absorption-line spectrum** is, in effect, the opposite of the emission-line spectrum because an atom's set of possible absorption lines have exactly the same wavelengths as its possible emission lines.

The gaps in the spectrum are created when photons at 656 nanometers and 486 nanometers interact with hydrogen atoms' electrons, lifting them from orbital 2 to 3 or orbital 2 to 4, respectively. Light at other wavelengths in this range has no effect on the atoms. Thus we can tell that hydrogen is present from either its emission or its absorption spectral lines.

In our discussion we have considered light emitted and absorbed by individual atoms in a gas. If the atoms are linked to one another to form molecules, such as water or carbon dioxide, the new configuration of the electron orbitals results in an entirely different set of emission and absorption lines, so the molecules can also be identified from spectroscopy. In fact, even solid objects may imprint spectral lines on light that reflects off them. For example, when light from the Sun reflects from an asteroid, spectral features appear that were not present in the original sunlight. This gives astronomers information about the surface composition of bodies too cool to emit significant light of their own.

We conclude that in general we can identify the kinds of atoms or molecules that are present by examining either the bright or dark spectrum lines. Gaps in the spectrum at 656 nanometers and 486 nanometers imply that hydrogen is present. Similar gaps at other wavelengths would show that other elements are present. By matching the observed gaps to a directory of absorption lines, we can identify the atoms and molecules that are present. This is the fundamental way astronomers determine the chemical composition of astronomical bodies.

24.3   TYPES OF SPECTRA

The basic forms of spectra we have discussed can be categorized into three basic classes, and each implies a set of physical conditions. This classification system was first formulated by the German physicist Gustav Kirchhoff in the mid-1800s, so it is usually known as **Kirchhoff's laws**:

Continuous spectrum

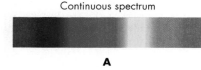

A

Emission line spectrum (hydrogen gas)

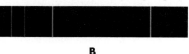

B

Absorption line spectrum (hydrogen gas)

C

FIGURE 24.6
Types of spectra: (A) continuous,
(B) emission-line, and (C) absorption-line.

1. Some sources emit light in such a way that the intensity changes smoothly with wavelength and all colors are present. We say such a light has a **continuous spectrum** (Figure 24.6A). For a source to emit a continuous spectrum, its atoms must in general be packed so closely that the electron orbitals of each atom are distorted by the presence of neighboring atoms. Isolated atoms all behave essentially identically, absorbing or emitting frequencies specific to the type of atom. However, when an atom is bumping into other atoms, the electron energy levels become altered by various amounts so that each atom can interact with photons of different energies, allowing photons of all wavelengths to be emitted or absorbed. Such conditions are typical of solid or dense materials such as the heated filament of an incandescent lightbulb, a glowing piece of charcoal, or the denser gas in the atmosphere of a star.

2. Conversely, an emission-line spectrum (Figure 24.6B) implies that the atoms or molecules are well separated and the atoms are hot enough to have excited the electrons into higher energy levels so they can emit the light at the specific frequencies associated with a drop to a lower energy level. Emission-line spectra are usually produced by hot, low-density gas, such as that in a fluorescent tube, the aurora, and many interstellar gas clouds.

3. An absorption-line spectrum (Figure 24.6C) is produced when light from a hot, dense body (as in 1 above) passes through cooler gas between it and the observer. In this case nearly all the colors are present, but light is either missing or much dimmer at the specific wavelengths corresponding to energy transitions in the gas atoms. Absorption-line spectra are seen, for example, in most star's spectra produced by cooler low-density gas above the hotter layers (that produce a continuous spectrum) beneath. Cool interstellar gas clouds between a star and us can also produce absorption lines in a star's spectrum.

Kirchhoff's laws give us a starting point for interpreting the physical conditions that need to be present to produce each kind of spectrum we have discussed.

24.4   ASTRONOMICAL SPECTRA

Let us now apply what we know about spectra to astronomical bodies. As an example, consider the spectra in Figure 24.7.

Spectra can be displayed in several ways. One method is simply to make an image of the light after it has been spread out by a prism or grating. Figure 24.7A shows two spectra, one an emission-line spectrum from a comet and the other an absorption-line spectrum from the Sun. Notice that in the former bright regions are separated by large regions where there is no light, but in the latter certain very narrow regions (the dark absorption lines) have little or no light. Another useful way to depict a spectrum is to plot the brightness of the light at each wavelength. Figure 24.7B shows the spectra of the comet and the Sun in this latter fashion.

The first step facing an astronomer is to identify the spectral features. This is done by measuring the wavelengths and then looking them up in a directory of spectral lines. By matching the wavelength of the line of interest to a line in the table, astronomers can determine what kind of atom or molecule created the line. A look at a typical spectrum will show you that some lines are faint and hard to see, whereas other lines are obvious and strong. The strength or weakness of a given line depends on the number of atoms or molecules absorbing (or emitting, if we are looking at an emission line) at that wavelength.

We can see from the spectral lines that the Sun contains hydrogen. In fact, when a detailed calculation is made of the strength of the lines, it turns out that roughly 90% of the atoms in the Sun are hydrogen. Complicating the interpretation, the number of atoms or molecules that can absorb or emit depends not just on how many of them are present but also on their temperature. Nevertheless, astronomers can deduce from the strength of emission and absorption lines the relative quantity of each atom producing a line, and thereby deduce the composition of the material in the body. Table 24.1 shows the result of such an analysis for our Sun, a typical star.

Because the 10% of atoms in the Sun besides hydrogen are heavier than hydrogen, hydrogen contributes "only" about 71% of the Sun's mass.

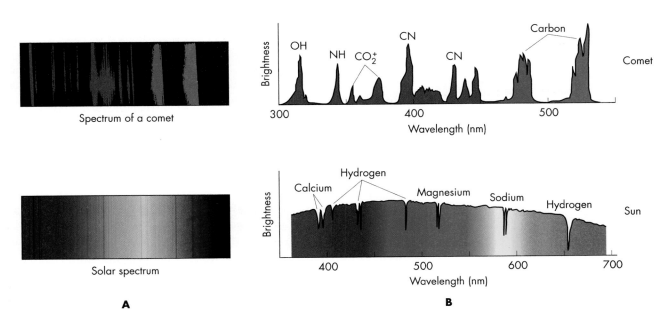

FIGURE 24.7
(A) The spectrum of a comet (an emission-line spectrum) and of the Sun (an absorption-line spectrum). (B) A graphical representation of their spectra.

TABLE 24.1	Composition of a Typical Star, Our Sun*	
Element	Relative Number of Atoms	Percentage by Mass
Hydrogen	10^{12}	70.6%
Helium	9.77×10^{10}	27.4%
Carbon	3.63×10^8	0.31%
Nitrogen	1.12×10^8	0.11%
Oxygen	8.51×10^8	0.96%
Neon	1.23×10^8	0.18%
Silicon	3.55×10^7	0.07%
Iron	4.68×10^7	0.18%
Gold	10.0	1.4×10^{-7}%
Uranium	Less than 0.3	Less than 5.7×10^{-9}%

(From Anders and Grevesse, *Geochim Cosmochim Acta* 53 [1989]:197.)

*The table lists some of the most common elements. Gold and uranium are included only to illustrate how extremely rare they are. Notice that they are a million or more times rarer than iron, which is itself thousands of times rarer than hydrogen.

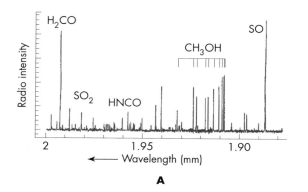

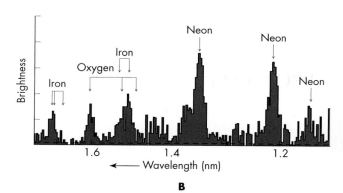

FIGURE 24.8

(A) A radio spectrum of a cold interstellar cloud. (B) An X-ray spectrum of hot gas from an exploding star.

Similar observations show that the spectrum of a comet contains emission lines from such substances as the molecules carbon dioxide and cyanogen (a molecule composed of carbon and nitrogen). Thus we know that comets contain these substances. Moreover, recalling Kirchhoff's laws, we can tell that the cyanogen and carbon dioxide must be gaseous because the spectrum consists of emission lines. There may be other gases present too, but without seeing their spectral features, we cannot tell for sure.

Although the examples we have used involve spectra of visible light, one of the most useful features of spectroscopy is that it may be used in any wavelength region. For example, Figure 24.8 shows a radio spectrum and an X-ray spectrum. Regardless of the wavelength region we use, the spectrum allows us to determine what kind of material is present.

KEY TERMS

absorption-line spectrum, 177

continuous spectrum, 178

emission-line spectrum, 176

energy level, 173

Kirchhoff's laws, 178

spectroscopy, 173

QUESTIONS FOR REVIEW

1. What are some of the things astronomers can learn about astronomical objects from their spectra?

2. Why don't atoms emit a continuous spectrum?

3. How can you tell what sort of gas is emitting light?

PROBLEMS

1. How could you use a spectrum produced from sunlight reflected off the atmosphere of Venus to determine something about Venus's atmospheric composition?

2. If you were to look at the spectrum of the gas flame of a stove or the blue part of a Bunsen burner flame, what sort of spectrum would you expect to see: absorption, emission, or continuous? Why?

3. In the Bohr model of hydrogen, the energy associated with an electron being in level n is found to be -2.2×10^{-18} joules/n^2. (See Unit 3 for metric units.) Find the energy difference between levels $n = 1$ and 2, and determine the wavelength of a photon with this energy. What part of the electromagnetic spectrum is this?

TEST YOURSELF

1. An astronomer finds that the visible spectrum of a mysterious object shows bright emission lines. What can she conclude about the source?
   a. It contains cold gas.
   b. It is an incandescent solid body.
   c. It is rotating very fast.
   d. It contains hot, relatively tenuous gas.
   e. It is moving toward Earth at high speed.

2. Most stars have spectra showing dark lines against a continuous background of color. This observation indicates that these stars
   a. are made almost entirely of hot, low-density gas.
   b. are made almost entirely of cool, low-density gas.
   c. have a warm interior that shines through hotter, high-density gas.
   d. have a hot interior that shines through cooler, low-density gas.

3. Suppose we detect the 656 nm wavelength of the $n = 3$ to 2 transition of hydrogen from a 5000 K gas cloud. If the cloud was heated to 10,000 K, what would the wavelength of the $n = 3$ to 2 transition now be?
   a. 278 nm
   b. 1312 nm
   c. 5656 nm
   d. 658 nm
   e. 656 nm

25

The Doppler Shift

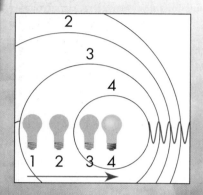

Background Pathways: Unit 24

If we observe light from a source that is moving toward or away from us, we will find that the wavelengths we receive from it are altered by its motion. If it moves toward us, its wavelengths will be shorter. If it moves away from us, its wavelengths will be longer, as illustrated in Figure 25.1. Furthermore, the faster the source moves, the greater those changes in wavelength will be. This change in wavelength caused by motion is called the **Doppler shift**, and it is a powerful tool for measuring the speed of approach or recession of astronomical objects. Motion perpendicular to the line of sight creates no Doppler shift because the source is neither approaching nor moving away from us.

The Doppler shift occurs for all kinds of waves. You have probably heard the Doppler shift of sound waves from the siren of an emergency vehicle: The siren's noise drops from a high pitch to a low pitch (corresponding to a shift from short to longer wavelengths of the sound) as the approaching vehicle passes you and then moves away (Figure 25.1A). The same thing happens with electromagnetic waves, as when, for example, the Doppler shift of a radar beam that bounces off your car reveals to a law enforcement officer how fast your car is moving (Figure 25.1B).

25.1 CALCULATING THE DOPPLER SHIFT

We can understand the Doppler shift by thinking about the spacing between wave crests. If a source is generating electromagnetic waves at a particular frequency (like those hydrogen atoms might emit, for example) and is stationary, the wave crests reach us with the normal "rest" wavelength λ. If the object is moving toward us, the object still generates the waves at the same frequency, but in the time between the generation of one wave crest and the next, the object has moved closer, so the wavelength we observe, λ', is shorter. This is illustrated in Figure 25.2, where the position of the light source is shown at the time it emits the crest of the each wave. The wave crest expands out spherically around the position from which it was emitted, but the spacing between wave crests is tighter in the direction the source is moving.

The difference between λ' and λ is usually written $\Delta\lambda$, (called "delta lambda" or "the change in λ"). For example, suppose a spectral line is generated at 656 nm (λ), but the line instead appears to be at 658 nm (λ'). The change in λ is

$$\Delta\lambda = \lambda' - \lambda = 658 \text{ nm} - 656 \text{ nm} = 2 \text{ nm}.$$

This is the distance the source has moved in the time it took to generate the wave.

The time between the generation of one wave crest and the next is, by definition, the inverse of the frequency, $1/\nu$. With the aid of the relationship between wavelength

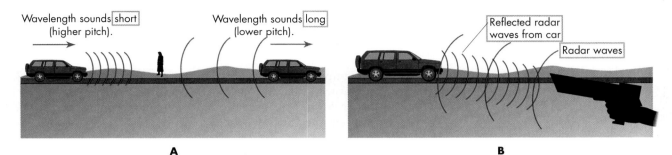

A
B

FIGURE 25.1

The Doppler shift: Waves shorten as a source approaches and lengthen as it recedes. (A) The Doppler shift of sound waves from a passing car. (B) The Doppler shift of radar waves in a speed trap.

ANIMATIONS

The Doppler effect

INTERACTIVE

Doppler shift

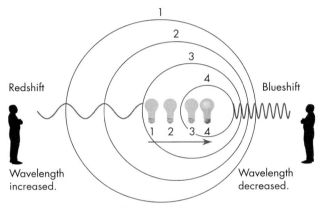

Redshift   Blueshift

Wavelength   Wavelength
increased.   decreased.

Bulb moves from 1 to 4.

FIGURE 25.2

A source of light moving toward you will shorten the distance the next wave must travel to reach you, so the time between wave crests grows shorter. If it moves away, the distance grows larger and the wavelength longer.

V_R = Radial velocity (toward or away from observer)

$\Delta\lambda$ = Change from normal wavelength

λ = Normal wavelength

c = Speed of light

Q About how fast would you have to be driving as you approached a red light for it to be Doppler shifted to a green color?

Q Wien's law indicates that the wavelength of light is longer from a cooler object. How could we tell the difference between this and an object that is moving away from us?

and frequency (Unit 22), this can be written as c/λ. So the speed of the object along our line of sight, called its **radial velocity,** V_R, is

$$V_R = \frac{\Delta\lambda}{\lambda} \times c.$$

This fairly simple formula relates the change in wavelength to the source's speed, allowing astronomers to find out how fast a source is moving and whether the motion is toward or away from us. By convention, $\Delta\lambda$ and V_R are negative if the object is approaching us (so that the observed wavelength, λ', is shorter than the normal rest wavelength) and positive if it is moving away from us. In the previous example, the speed is

$$V_R = \frac{2 \text{ nm}}{656 \text{ nm}} \times c = 0.003 \times 300,000 \text{ km/sec} = 900 \text{ km/sec}$$

and the object must be moving away from us because the wavelength has grown longer.

Doppler shift measurements can be made at any wavelength of the electromagnetic spectrum. But regardless of the wavelength region observed, astronomers refer to shifts that *increase* the measured wavelength of the radiation as **redshifts** and those that *reduce* the wavelength as **blueshifts.** This harkens back to the early days of spectroscopy when only the visible portion of the electromagnetic spectrum was observable.

 The constancy of the speed of light was one of the great mysteries resolved by Einstein's theory of special relativity (see Unit 53).

Incidentally, we can also understand the Doppler shift in the particle model of light by thinking of the motion of a source toward or away from us as adding to or subtracting from the energy of the photon. This would be like throwing a ball to a receiver while running. If you are running toward the receiver, the ball will arrive with a greater impact; but if you are running away, the impact will be weaker, even though you throw the ball equally hard in each case. However, we have to be careful with this analogy—in the case of the photon, experiments conclusively show that its energy changes, but *not* its speed.

KEY TERMS

blueshift, 183   radial velocity, 183
Doppler shift, 182   redshift, 183

QUESTIONS FOR REVIEW

1. What is the Doppler shift?
2. Do we always see a Doppler shift when an object is moving?
3. Describe the difference between a redshift and a blueshift.

PROBLEMS

1. If the $n = 3$ to 2 transition of hydrogen, normally at 656 nm, is detected from a star at 655.5 nm, what is the radial velocity of the star?
2. Hydrogen also produces spectral lines at radio wavelengths, notably at 21.1 cm. If a galaxy is moving away from us at 10% of the speed of light, at what wavelength will we detect this line? Convert this into a frequency.

TEST YOURSELF

1. If an object's spectral lines are shifted to longer wavelengths, the object is
    a. moving away from us.
    b. moving toward us.
    c. very hot.
    d. very cold.
    e. emitting X-rays.
2. What properties of stars can be determined from their spectra?
    a. Chemical composition
    b. Surface temperature
    c. Radial velocity
    d. All of the above
    e. None of the above
3. The wavelength of a radio emission line from a galaxy normally seen at 100 cm is detected at 101 cm. This implies that the galaxy is
    a. moving toward us at 1% of the speed of light.
    b. moving away from us at 1% of the speed of light.
    c. moving toward us at 1 centimeter per second.
    d. moving away from us at 1 centimeter per second.

26 Detecting Light

Like all scientists, astronomers rely heavily on observations to guide them in proposing hypotheses and in testing theories already developed. Unlike most scientists, however, astronomers usually cannot actively probe the objects they study. Instead, they must perform their observations passively, collecting whatever light and other forms of radiation have been emitted by the bodies they seek to study.

The quest to understand the cosmos ultimately depends on the quality and detail of the data we can collect. Sometimes that quality and detail are limited by the available technology. Sometimes they are constrained by the physical properties of the light we are attempting to detect. In this unit we present an overview of the instruments used for astronomical observation. In subsequent units we will explore in greater detail several of the limitations and challenges posed by these instruments.

26.1 TECHNOLOGICAL FRONTIERS

Astronomy is the most ancient of the sciences, but many of its most exciting discoveries are very recent. For example, in the long history of astronomy, only in the last 200 years have astronomers known that "light"—electromagnetic radiation—extends beyond the visible spectrum; and only in the last half century have instruments been developed to study these other bands of electromagnetic radiation. And at visual wavelengths, telescopes have progressed to achieve remarkable precision and size (Figure 26.1).

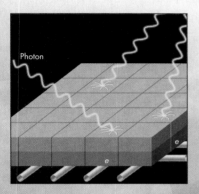

Background Pathways: Unit 22

FIGURE 26.1
One of the twin Keck telescopes on the summit of Mauna Kea, Hawaii, during a total eclipse of the Sun. These are the two largest individual optical telescopes in the world.

The design and construction of instruments to collect and detect electromagnetic radiation of all wavelengths is a branch of astronomy called **instrumentation.** The amazing images and data that have changed many of our ideas about the nature of the cosmos in recent decades are due to the remarkable strides made in instrumentation.

The most recognizable tool of astronomical technology is the **telescope.** Telescopes allow astronomers to observe things not visible to the naked eye, either because they are too dim or because they emit radiation outside the visible range of the electromagnetic spectrum. They are basically devices for collecting a large amount of electromagnetic radiation, focusing this light onto detectors, and magnifying the images to resolve fine detail. Astronomers strive to build bigger telescopes because the amount of light that can be collected depends on the size of the telescope's **aperture,** or collecting area (Unit 27). Collecting more light allows us to see dimmer and more distant objects. If we collect enough light, it is possible to examine the spectrum of radiation from a source, which reveals details of its physical and chemical nature (Unit 24).

A variety of telescope designs use different techniques to take the light that enters the telescope aperture and concentrate or **focus** it (Unit 28). Many of the principles involved in focusing light are similar whether we are dealing with radio waves or X-rays; but the construction details differ enormously because the precision of the construction is determined by the wavelength of the light being observed. This allows radio telescopes to be built with rougher surfaces; however, they have to be built much larger to achieve a comparable **resolution**—a telescope's ability to detect fine detail. This is because resolution is also a function of the size of the telescope relative to the wavelength of the radiation it is detecting (Unit 29).

The Earth's own atmosphere presents challenges for observing electromagnetic radiation. It absorbs some wavelengths and alters the path of light in ways that badly degrade the quality of images. For these reasons, a number of space telescopes have been developed in recent years and placed in orbit above the atmosphere (Unit 30). However, putting telescopes in space presents significant challenges—and expense—and one of the goals of modern instrumentation is to find ways to compensate for the atmosphere for ground-based telescopes and to improve the resolution of telescopes with new designs.

26.2 DETECTING VISIBLE LIGHT

Modern telescopes bear little resemblance to the long tubes depicted in cartoons (Figure 26.2). Moreover, professional astronomers almost never sit at the eyepiece of a telescope. And few astronomers ever wear lab coats! Not only are astronomers not needed close to a telescope, but the heat from their bodies can ruin the image quality. With the major research telescopes of today, an astronomer is more likely to be dressed in jeans and a T-shirt, operating the telescope from a computer terminal that may be thousands of miles away. She or he will examine the data being collected, along with a wide array of data about the telescope's performance and weather conditions, and attempt to gauge whether enough data of sufficient quality have been collected to move on to the next set of observations. To make the most efficient use of the instruments, it is becoming more common for telescopes to be operated and the data checked by computer programs. A team of astronomers develops computer programs to observe the targeted sources efficiently, then waits for the requested data but has no immediate control over the telescope at all.

Before astronomical photography became widespread in the late 1800s, astronomers did stare through telescope eyepieces, and they wrote down data or made sketches of the objects being observed. However, despite the eye's marvelous abilities, it has difficulty detecting very faint light (Unit 31). Many astronomical objects are too dim for their few photons to create a sensible effect on the eye. For example, even if you were to look at a galaxy through the largest telescope available today, the light is so spread out that it would appear merely as dim smudges to your eye. Only by storing galaxies' light, sometimes for many hours, can we discern the detailed structure of the galaxies. Thus to see very faint objects, astronomers must use detectors that can store light in some manner.

Until the 1980s, astronomers generally used photographic film to record the light from the objects they were studying. Astronomers today, however, almost always use an electronic detector similar to those in digital cameras called a **CCD,** which stands for "charge-coupled device." These devices use the **photoelectric effect,** in which a photon that has sufficient energy causes an electron to become unbound from the atom it is orbiting. Thus when the incoming light strikes the "semiconductor" surface of the CCD, it frees electrons to move within the material. The surface is divided into thousands of little squares, or **pixels** (Figure 26.3). Electrical voltages on the surface pull the freed electrons into the nearest pixel. Here they are temporarily stored before being "read out" by the electronics. Then the whole chip is cleared and returned to its original state. The number of freed electrons in each pixel is proportional to the number of photons hitting it (that is, proportional to the intensity of the light). An electronic device coupled to a computer then scans the CCD, counting the number of electrons in each pixel and generating a picture.

Such cameras are extremely efficient, generally recording more than 75% of the photons striking them. This allows astronomers to record images much faster than with film, which might detect only 10% of the photons. Furthermore, unlike film, CCDs are reusable, and the digital images can be processed rapidly by a computer. Astronomers work with the digital images to correct them for instrumental effects, remove extraneous light, and determine the exact amount of light striking different parts of the image.

"It's somewhere between a nova and a supernova -- probably a pretty good nova."

FIGURE 26.2
Astronomers discussing the nature of their discovery? (Copyright © 2005 by Sidney Harris)

Albert Einstein explained the photoelectric effect in 1905, providing one of the foundations for understanding the particle-like nature of light. His work showed that light is composed of discrete photons with energies that depend on their wavelength. His Nobel prize was awarded for this work, though he is even better known for his development of relativity theory, explaining the nature of space and time.

Q Many CCDs are refrigerated to very low temperatures. How do you suppose keeping the temperature low might make the image quality better? (Consider thermal effects on atoms within the CCD.)

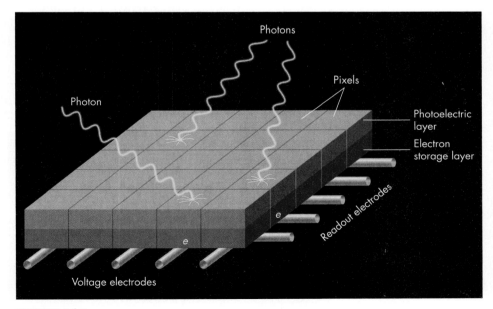

FIGURE 26.3
Simplified diagram of a CCD. Photons striking the photoelectric layer free an electron (e). A positive voltage applied to one set of electrodes attracts the electrons and holds them in place under each pixel. During the readout phase, the voltages are changed to "push" the collected electrons along the readout electrodes.

26.3   OBSERVING AT NONVISIBLE WAVELENGTHS

Essentially, the long-wavelength radio waves cause electrons in an antenna to oscillate in response, and the small electric currents of these moving electrons are detected and amplified.

Many astronomical objects radiate electromagnetic energy at wavelengths outside the range of visible light, and so astronomers have devised new ways to observe such objects. For example, cold clouds of gas in interstellar space emit little visible light but large amounts of radio energy. To observe them, astronomers use radio telescopes, which employ technologies like those used for detecting television or radio station transmissions, but which are far more sensitive.

At infrared and shorter wavelengths astronomers generally use CCDs or similar technologies to detect photons. For example, dust clouds in space are too cold to emit detectable visible light, but they do radiate infrared photons. Normal CCDs are much more sensitive to infrared photons than photographic film (with wavelengths up to about twice the wavelength of optical photons). But longer-wavelength photons do not have enough energy to free electrons in the photoelectric material. New photoelectric materials have been developed for this purpose. These detectors must be kept very cold, however, because heat from the instrument itself or any surrounding equipment will generate photons of these energies, which will overwhelm the photons from the astronomical source of interest.

Astronomers use X-ray and gamma-ray telescopes to observe, for example, the hot gas that accumulates around black holes. A single X-ray photon has sufficient energy to free many electrons within the CCD material used, and thus the number of electrons freed can be used as a measure of the energy of the incoming X-ray photon.

These technologies improved dramatically in the last few decades. An astronomer spending several years using cutting-edge infrared detectors around 1980 to collect data for a Ph.D. dissertation could with today's instruments collect those same data in five minutes! Astronomy at other wavelengths is likewise advancing rapidly, and astronomers are learning to interpret an ever-growing array of data as these other areas begin to "catch up" with the much better established field of visible wavelength astronomy.

One of the challenges in interpreting data collected at nonvisible wavelengths is how to display it. Because our eyes cannot see these other wavelengths, astronomers must devise ways to depict what such instruments record. One way is with **false-color images,** as shown in Figure 26.4. In a false-color image, colors are used to represent information from other wavelengths in a way that takes advantage of our eyes' ability to discern fine differences in color. In a common type of

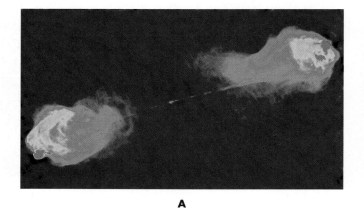

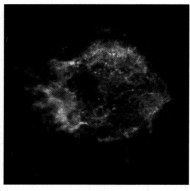

A   B

FIGURE 26.4

(A) False-color picture of a radio galaxy. We can't see radio waves, so colors are used to represent their brightness—red brightest, blue dimmest. (B) False-color X-ray picture of Cas A, an exploding star.

false-color image, colors are used to represent the intensity of radiation that we cannot see—often using colors as they are shown in a geographic map, with reds and oranges representing the high levels (like mountains) and blues and purples representing low areas (like oceans). For example, in Figure 26.4A (a radio image of a galaxy and the jet of hot gas spurting from its core), astronomers color the regions emitting the most intense radio energy red; they color areas emitting somewhat less energy yellow and the faintest areas blue.

Figure 26.4B shows a different kind of false-color image (of the gas shell ejected by an exploding star). It is an X-ray image where different energy X-ray photons have been assigned colors analogously to how we see colors in the visible spectrum—blue for high energies, green for medium, and red for lower energies.

A false-color image may be coded to depict data in many different ways. Colors may be used to represent different wavelength ranges—for example, red representing radio, yellow representing infrared, and blue representing X-ray emission. Or the colors may be coded to indicate derived quantities like the temperature of gas, the strength of magnetic fields, or Doppler shift values. These techniques of visualizing data help astronomers to "see" connections between processes going on within an astronomical object.

26.4   THE CRAB NEBULA: A CASE HISTORY

The history of observational astronomy is illustrated by the story of the discovery and gradual piecing together of the astronomical mystery of an exploding star, first seen nearly 1000 years ago. Although the story began with naked-eye observations, it continues with observations made with telescopes on the ground and in space. Moreover, the story illustrates how astronomers have come to rely on observing radiation at many wavelengths, not just visible light.

On July 4, 1054 C.E., just after sunset, astronomers in ancient China and other far eastern countries noticed a brilliant star near the crescent moon in a part of the sky where no bright star had previously been seen. They wrote of this event, "In the last year of the period *Chih-ho,* a guest star appeared. After more than a year it became invisible." The star was so bright that for several weeks it was visible during the day; but it gradually faded away, and after more than a year it became invisible and was forgotten.

Curiously, no records of this spectacular event have been found in European archives. It is difficult to imagine that such a spectacular event, visible for over a year from most of the Earth, went unnoticed by anyone in Europe. Some scholars speculate that written records of this event might have been suppressed because they ran counter to religious beliefs of the medieval church.

Nearly 700 years later, in 1731, John Bevis, a British physician and amateur astronomer, noticed with his telescope a faint dim patch of light in the constellation Taurus. In the late 1700s the French astronomer Charles Messier, who was hunting for comets, developed a catalog of similar dim patches of light that might be confused with comets. This object was first on his list, and so it became known as "Messier 1" or M1. M1 can be seen with a small telescope—Looking Up #5 shows its location in Taurus.

In 1844 William Parsons, an Irish astronomer and telescope builder (also known as Lord Rosse), noticed that M1 contained filaments that to his eye resembled a crab (Figure 26.5A). He therefore named it the "Crab Nebula." Modern photographs (Figure 26.5B) look quite different from Rosse's sketch, but perhaps still resemble a crab—for someone with a generous imagination!

In 1921 John Duncan, an American astronomer, compared two photographs of the nebula taken 12 years apart and noticed that it had increased slightly in

Messier's catalog contains more than 100 of the brightest of these *nebulae* (Latin for "clouds"). The positions of them are given in the appendix. A number of them are identified on the Looking Up Illustrations. The complete Messier catalog is given in Appendix Table 13.

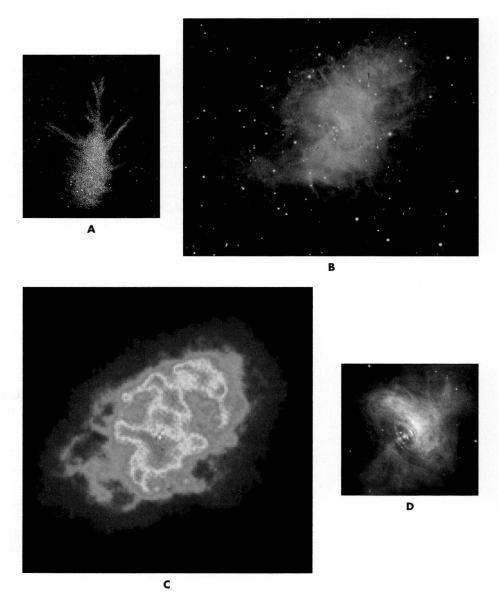

FIGURE 26.5

(A) Lord Rosse's 1844 drawing of the "Crab Nebula." (B) Visible-light photograph of the Crab Nebula. (C) Radio image of the Crab Nebula. (D) X-ray image in false color of the Crab Nebula. X-rays are shown as blue/white while optical light is shown as red. The flattened round shape of the glowing gas suggests a spinning disk.

diameter. He therefore deduced that the nebula was expanding. At the same time, several other astronomers came across the ancient Chinese records and noticed the coincidence in position of the nebula with the report of the exploding star. Then, seven years later, Edwin Hubble, at Mount Wilson Observatory in California, measured the increase of size more accurately and calculated from the rate of expansion that the nebula was about 900 years old—roughly matching the date recorded by the Chinese astronomers. The speed of the nebula's expansion calculated from these observations was astonishingly large—thousands of kilometers per second. Astronomers realized that the Chinese astronomers had witnessed an explosion that had sent gas flying violently outward at almost unimaginable speeds.

Since the second half of the twentieth century, astronomers have examined the Crab Nebula with telescopes using virtually all wavelength bands and, in doing so, have added yet more to their understanding of a star's demise. For example, the nebula is a powerful source of both radio waves (Figure 26.5C) and X-ray radiation (Figure 26.5D). These observations have led astronomers to the conclusion that this type of *supernova* explosion occurs after massive stars use up their fuel— teaching us that stars may die violent deaths (Units 65, 66).

Based on what you know about the interactions of light and matter, what kinds of information might you hope to learn by studying radiation in the different electromagnetic bands?

The assortment of modern observations at many wavelengths has given astronomers one of their most detailed understandings of the death throes of a star. From radio-wavelength observations, they deduce that the "remains" of the original star are spinning incredibly fast—about 30 times per second. This in turn provides an essential clue that the star has collapsed to such a small size that it is as dense as the nucleus of an atom—an object called a *neutron star* (Unit 67). The X-ray observations reveal details of high-energy particles being ejected at near the speed of light from the dead star and trapped in rings around the star by its strong magnetic fields. Thus, by observing at a variety of wavelengths, astronomers have deduced far more than they could have from observations at one wavelength alone.

KEY TERMS

aperture, 186

CCD, 187

false-color image, 188

focus, 186

instrumentation, 186

photoelectric effect, 187

pixel, 187

resolution, 186

telescope, 186

QUESTIONS FOR REVIEW

1. What advantages does a larger telescope have over a smaller one?

2. How does a CCD work?

3. What is a false-color image?

4. How have astronomical observing techniques changed over the last century?

PROBLEMS

1. One of the 10-m diameter Keck telescopes can observe 500 nm visible wavelength photons. If a radio telescope observing 10 cm photons was built to be proportionally as large relative to the photons it was observing, what would its diameter be? What about a gamma-ray telescope observing 0.01 nm photons?

2. Small knots of gas in the Crab Nebula are observed to be moving away from the position of the dead star at its center.

Measurements of the position from photographs taken in 1940 and 1990 show that knots originally located 150 arc seconds from the center had moved between 8 and 9 arc seconds farther out. Estimate the range of dates when the explosion might have occurred from this information.

TEST YOURSELF

1. Which of the following is *not* a reason that astronomers seek to build larger telescopes?
   a. Larger telescopes can resolve greater detail.
   b. Larger telescope collect more photons.
   c. Larger telescopes can detect fainter objects.
   d. Larger telescopes are less affected by the Earth's atmosphere.

2. Which of the following is an *advantage* of a CCD over photographic film?
   a. CCDs can collect light for a long time.
   b. CCDs record a greater percentage of the photons striking them.
   c. CCDs are not affected by blurring of the Earth's atmosphere.
   d. CCDs do not need to be corrected for instrumental effects.
   e. All of the above

3. Which of the following is an important difference between radio waves and other wavelengths of electromagnetic radiation for the design of astronomical instrumentation?
   a. Radio waves are detected as sounds, not light.
   b. Radio wavelengths travel much more slowly than other electromagnetic radiation.
   c. Radio photons have very small energies.
   d. Radio waves cannot be focused.
   e. All of the above

27 Collecting Light

Because most astronomical objects are so remote, their radiation is extremely faint by the time it reaches Earth. For example, to collect enough light to study the most remote galaxies, astronomers use telescopes with mirrors the size of swimming pools, and it may be necessary to point a telescope for hours at a spot so small that it would require thousands of years to look at every object in the sky this way. Therefore astronomers have to be selective in their approach—choosing objects to study that may help to confirm or reject a detailed hypothesis or to refine a theory.

Moreover, special instruments must often be used to extract the information desired from the radiation—instruments that can measure the brightness, the spectrum, and the position of objects to high precision. These instruments generally require much more light than is needed simply to detect that an object is present at a particular position in the sky, and this in turn requires larger telescopes to keep the required observing time within practical limits.

Background Pathways: Unit 26

27.1 MODERN OBSERVATORIES

Telescopes are essentially large "photon buckets" used to collect light over a large area and then concentrate the light for use by a high-sensitivity detector. The challenge is to build a telescope as large as possible but with every surface and separation built to a very exacting standard—to within a fraction of a wavelength of the electromagnetic radiation being observed. This high precision is required to keep the light waves **coherent:** The electromagnetic waves passing through different parts of lenses or bouncing off different parts of mirrors must all arrive at the detector with their wave crests in unison to achieve the maximum effect. If the waves arrive out of unison, they will partially cancel each other—somewhat like how the rowers in a large racing shell must row in unison for maximum effect.

The immense telescopes and the associated equipment that astronomers use are expensive. Therefore, the largest telescopes are often national or international facilities. For example, the National Optical Astronomical Observatory and the National Radio Astronomy Observatory of the United States, the European Southern Observatory, and dozens of other countries each operate many different telescopes. They are open to use by astronomers from around the world through a competitive review of observing proposals. Many colleges and universities have their own research telescopes (in addition to smaller ones near campus for instructional purposes). Altogether, several thousand observatories exist around the world on every continent. There are even telescopes at the South Pole in Antarctica. These take advantage of the long night and extreme dryness of the bitterly cold Antarctic air to view wavelengths of light that are absorbed by water vapor.

The largest individual optical telescopes are currently the twin 10-meter Keck telescopes on Mauna Kea in Hawaii (Figure 27.1). The telescope "array" with the greatest collecting area in the world is the VLT (for "Very Large Telescope"), which consists of four 8-meter telescopes that can work as a unit. The VLT is operated by

FIGURE 27.1
Photograph of the twin Keck telescopes on the summit of Mauna Kea in Hawaii. As the telescopes track the stars or galaxies being observed, all 36 mirrors in each telescope can be measured with a laser and then adjusted to keep them in precise alignment.

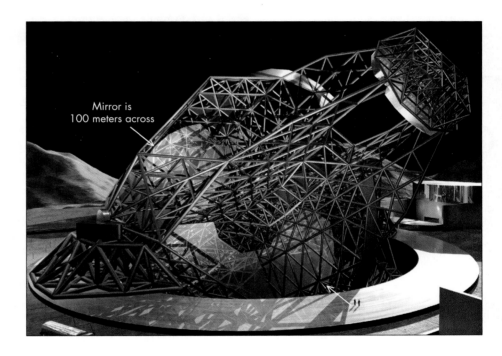

Mirror is
100 meters across

FIGURE 27.2
Proposed new OWL ("overwhelmingly large") telescope.

Q How thick is a piece of paper in nanometers? Determine this by measuring the thickness of a stack of 100 sheets. How does this compare to the precision that an optical telescope must maintain?

a consortium of European countries and Chile and is located in the extremely dry, high-altitude mountains in northern Chile.

To observe the faint trickle of light from forming galaxies billions of light-years away, astronomers are examining the possibilities for even bigger telescopes. One dubbed the "Overwhelmingly Large" Telescope (OWL for short) would be larger than a football field (Figure 27.2). Such a telescope might allow us to study the very first stages of star and galaxy formation, shortly after the universe began. However, there are enormous challenges in building a telescope so large with a structure precise to about 50 nanometers, and these need to be overcome before such a colossus can be built.

Radio telescopes are generally the largest of all. Radio astronomers must collect very large numbers of these lowest-energy photons to produce a signal powerful

FIGURE 27.3
Photograph of the 300-meter diameter
Arecibo radio telescope in Puerto Rico.

Q The "dish" of the Arecibo telescope is fastened to the ground. How do you suppose it is possible to look in any direction other than straight overhead with this telescope?

enough to be detected. Fortunately, because the wavelengths being observed are of the order of centimeters or longer, the telescopes need be built only to a precision of millimeters or longer, which is well within the capability of current manufacturing techniques.

One such huge radio telescope is located in Arecibo, Puerto Rico (Figure 27.3). The Arecibo radio telescope is more than 300 m (1000 ft) in diameter. The huge "dish" is fixed in place in a natural basin and thus has limited ability to look in different directions. It depends in large part on the rotation of the Earth to observe different parts of the sky. Fully steerable radio telescopes of 100 m diameter are located in West Virginia and in Bonn, Germany. Radio astronomers are also thinking about ways to collect signals from very faint sources, and plans for a "Square Kilometer Array," which would have a collecting area of 1 kilometer square, are currently being developed. In the more distant future, astronomers would like to build a radio telescope on the far side of the Moon, where it would be shielded from most human-made radio-wavelength transmissions.

27.2 COLLECTING POWER

For us to see an object, light or photons from it must enter our eyes. How bright the object appears to us depends on the number of its photons that enter our eye per second. Astronomers can increase this amount by collecting photons with a telescope, which then funnels the photons to our eye. The bigger the telescope's **collecting area,** the more photons it gathers, as shown in Figure 27.4. The result is a brighter image, which allows us to see dim stars that are invisible in telescopes with smaller collecting areas. The area of a circular collector of diameter D is

$$\text{Collecting area} = \frac{\pi}{4} \times D^2$$

(or equivalently, $\pi \times \text{radius}^2$). As a result, a relatively small increase in the radius of the collecting area gives a larger increase in the number of photons caught. For example, doubling the radius of a lens or mirror increases its collecting area by a

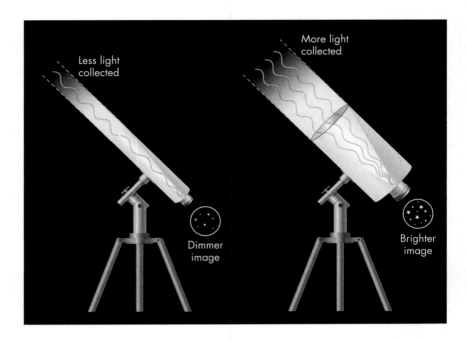

FIGURE 27.4
A large lens collects more light (photons) than a small one, leading to a brighter image.

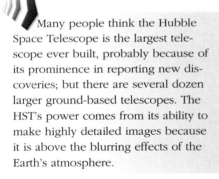
Many people think the Hubble Space Telescope is the largest telescope ever built, probably because of its prominence in reporting new discoveries; but there are several dozen larger ground-based telescopes. The HST's power comes from its ability to make highly detailed images because it is above the blurring effects of the Earth's atmosphere.

factor of 4. Because the collecting area is so important to a telescope's performance, astronomers usually describe a telescope by the diameter of its **aperture**—the area over which it collects photons. Thus the 10-meter Keck Telescope in Hawaii has mirrors spanning 10 meters (roughly 30 feet) in diameter.

When your eye's pupil is completely dilated, it has an aperture with a diameter of about 8 mm. The Hubble Space Telescope has an aperture with a 2.4 m diameter. The relative collecting areas of these are

$$\frac{\frac{\pi}{4}(2.4 \text{ m})^2}{\frac{\pi}{4}(8 \text{ mm})^2} = \left(\frac{2400 \text{ mm}}{8 \text{ mm}}\right)^2 = 300^2 = 90,000.$$

Thus the Hubble Space Telescope collects 90,000 times more light per second than your eye. Carrying out the same calculation for the Keck telescope shows that it collects more than 1.5 million times more light than your eye, and 17 times more light than the Hubble Space Telescope. The advantage of the Hubble, of course, is that it is free of the atmospheric distortions that Earth-based collected light is subject to (Unit 30).

27.3   FILTERING LIGHT

Telescopes generally collect a wide range of wavelengths. For example, most visible-wavelength telescopes also work into the infrared and ultraviolet. However, astronomers rarely collect all of the incoming light but instead **filter** out all but a narrow range of wavelengths, selected to learn about the particular objects they are studying.

Filters are often thin pieces of colored glass. The filter glass is designed to reject all wavelengths but the range of interest and to be highly transparent within that range. At visual wavelengths astronomers use an assortment of filters:

B, or blue, transparent primarily at 390–480 nm.
V, or "visual," transparent at 500–590 nm.
R, or red, transparent at 570–710 nm.

Different kinds of film, CCDs, and lenses may behave differently depending on the wavelength of the light. By narrowing the range of wavelengths observed, filters make it easier to compare and combine data gathered with different telescopes and detectors and under different atmospheric conditions.

Some specialized filters select narrow ranges of wavelength to examine the light from a single spectral line, whereas other filters extend into the ultraviolet and infrared regions.

By observing stars and other objects through filters, astronomers can characterize more precisely how much red light or blue light an object emits, which can help to determine its temperature (Unit 23). In addition, filters help to reject unwanted light—when we observe an object known to emit most strongly at one wavelength, a filter can be chosen to admit light at that wavelength and no others.

Note that a filter does not "color" the light passing through it. When you look through a blue filter, you are seeing just the blue photons from each object, not changing the wavelength of other color photons. Most objects produce or reflect some photons of every wavelength, so a yellow object will not necessarily appear black through a blue filter.

KEY TERMS

aperture, 195

coherent, 192

collecting area, 194

filter, 195

QUESTIONS FOR REVIEW

1. What is meant by the term *collecting area*?
2. How does a larger telescope make it easier to see faint objects?
3. What is a filter?

PROBLEMS

1. If you look at stars with binoculars with 8-cm (about 3-inch) diameter lenses, how many times fainter a star should you be able to see compared to what you can see with your unaided, fully dilated eye, with the pupil open to about 8 mm?

2. Show that the total collecting area of the four 8-meter telescopes of the VLT is greater than that of the two 10-meter Keck telescopes.

3. At a radio wavelength around 30 cm, the Arecibo radio telescope can collect about 3×10^{-14} watts (joules per second) of radiation from the Crab Nebula. About how many photons is it collecting each second?

4. The orbiting Chandra X-ray telescope, observing at about 0.1 nm, can collect about 2×10^{-13} watts (joules per second) of radiation from the Crab Nebula. About how many photons is it collecting each second?

TEST YOURSELF

1. The Palomar telescope is 5 meters in diameter, whereas one of the Keck telescopes is 10 meters in diameter. How many times larger is the Keck telescope's collecting area?
   a. 2 times
   b. 4 times
   c. 5 times
   d. 10 times
   e. 25 times

2. Which of the following is a good reason for building larger telescopes?
   a. They permit us to detect dimmer stars.
   b. They make stars look much bigger.
   c. They let us see much dimmer surface brightness levels.
   d. All of the above

3. The purpose of a blue filter is to
   a. alter the wavelength of light we observe to make it easier to detect.
   b. eliminate the color of the sky from photographs.
   c. let through only the blue wavelengths of light.
   d. allow a telescope to collect more light.

28 Focusing Light

In addition to a telescope's role as a large "photon bucket," telescopes are designed to distinguish the direction from which light arrives and to **focus** the light. Focusing light is the process of bringing the electromagnetic waves together to form the strongest and most detailed signal possible.

Galileo Galilei is sometimes credited with the invention of the telescope, but the invention seems to have been the work of the Dutch spectacle maker Hans Lippershey in 1608. News of the invention spread rapidly, and by the summer of 1609, Thomas Harriott, an English mathematician and scientist, had used a telescope to study the Moon. Just a few months later, Galileo, quickly grasping the idea behind the design of the telescope, built a telescope of better quality than any previously available, capable of magnifying images by about a factor of 20. With this telescope he not only observed the Moon but discovered Jupiter's moons, Venus's phases, and the vast numbers of stars in the Milky Way. Most important, he published and interpreted his findings (Unit 12.3).

The competition to build better telescopes has continued nonstop to this day, and now includes specialized telescopes for every wavelength range. Astronomers have not only sought to build bigger telescopes that can detect dimmer objects (Unit 27); they have developed improved designs for focusing the light to provide more detailed images. The methods used for focusing light fall into two broad categories, employing either a lens or mirror to concentrate the light that enters the main aperture of the telescope. Each of these methods has advantages and disadvantages, as we discuss next.

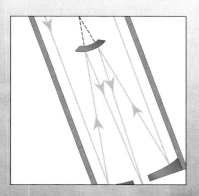

Background Pathways: Unit 26

28.1 REFRACTING TELESCOPES

The first telescopes used **lenses** to focus the light. Transparent substances like the glass of a lens can alter the path of electromagnetic waves. This effect is called **refraction.** If a piece of glass is properly shaped, it can focus light. Refraction is the principle by which eyeglasses and the lenses inside our eyes focus light.

Telescopes in which the light that enters the aperture is collected and focused by a lens are called *refracting telescopes,* or **refractors** for short. The lens of a refractor focuses the light by bending the rays, as shown in Figure 28.1. A lens that is just the right shape can collect light over the entire lens area and bend it so that it arrives at a focus behind the lens, creating an image. The figure depicts the light from two stars, a yellow star straight in front of the telescope and a red star above it. The light from each star is focused to a point on a flat region called the **focal plane.** This is where photographic film or a CCD might be placed, although there are often a variety of additional lenses in front of the detector to provide extra magnification and to correct for distortions produced by a single lens.

Lenses can bend light because when light moves from one transparent material into another at an angle, the direction in which the light travels is bent. Bending occurs when light from a distant star enters the Earth's atmosphere, or when that light travels from air into water or glass. The bending is generally stronger with a greater difference in density of the materials. For example, glass is slightly more dense than water, and light entering glass from air is bent slightly more than when

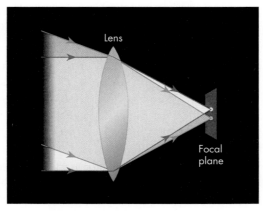

FIGURE 28.1
Light is shown arriving from a yellow star straight in front of the lens and from a red star above it. Each star's light is brought to a focus at the focal plane.

FIGURE 28.2
Refraction of light in a glass of water. Notice how the pencil appears to be bent.

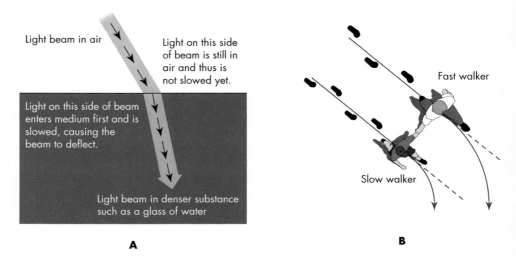

A

B

FIGURE 28.3
Cause of refraction. (A) Light entering the dense medium is slowed, while the portion still in the less dense medium proceeds at its original speed. (B) A similar effect occurs when you walk hand-in-hand with someone who walks more slowly than you do.

Q If you are scuba diving and you look up at the surface of the water overhead, you will see light from the sky in a region only out to about 45° away from the zenith. How might refraction explain this?

Ocean waves travel more slowly where the water becomes less deep, so a change in wave direction can sometimes indicate where the water grows shallow.

it enters water. However, a clear piece of glass underwater may be hard to see because of the small difference in refracting properties.

You can easily see the effects of refraction by placing a pencil in a glass of water and noticing that the pencil appears bent, as shown in Figure 28.2. The pencil in water also illustrates an important property of refraction. If you look along a pencil placed partly in the water and change the pencil's tilt, you will see that the amount of bending (refraction) changes. Exactly vertical rays are not bent at all, nearly vertical rays are bent only a little, and rays entering at a grazing angle are bent most.

Light is bent because its speed changes as it enters matter, generally slowing in denser material. This decrease in the speed of light arises from its interaction with the atoms through which the electromagnetic waves move. To understand how this reduction in light's speed makes it bend, imagine a light wave approaching a slab of material. The part of the wave that enters the material first is slowed while the part remaining outside is unaffected, as depicted in Figure 28.3A. Imagine what would happen if the right side of your car went off the road into mud. As that side of the

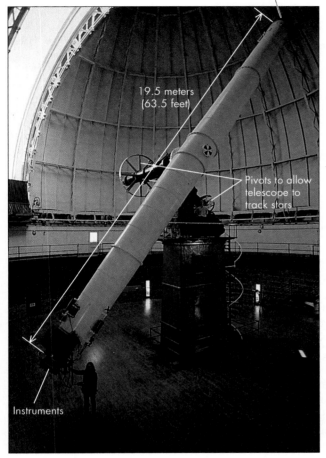

Lens 1 meter
(40 inches)
in diameter

19.5 meters
(63.5 feet)

Pivots to allow
telescope to
track stars

Instruments

FIGURE 28.4
A refracting telescope. Completed in 1897 for the University of
Chicago's Yerkes Observatory in Williams Bay, Wisconsin, this
refractor has a lens approximately 1 meter (40 inches) in
diameter, making it the world's largest refracting telescope.

car slowed, the car would swerve to the right. Or if two people are
walking hand-in-hand (as sketched in Figure 28.3B), if the one
on the right slows down, the direction of their motion turns to
the right. So, too, if one portion of a light wave moves more
slowly than another, the light's path will bend.

Figure 28.4 shows a photograph of the world's largest refractor,
the 1-meter (40-inch) diameter Yerkes telescope of the University
of Chicago. This telescope, completed in 1897, was one of the last
large-lens telescopes ever built. Building a telescope with such a
large lens presents serious structural problems. The Yerkes lens
has a mass of over 200 kilograms (over 450 pounds). This massive
piece of glass has to be supported by its edges, where the glass is
thinnest, so the lens flexes slightly. In addition, the large mass of
glass is located at the end of a long telescope tube, which must be
even more massive—about 20 tons—to keep it from flexing.
Building larger refractors would require even more massive telescopes with even greater problems of structural flexing. To build
larger-aperture telescopes, an alternative approach was needed.

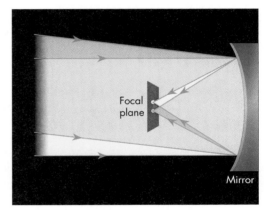

Focal
plane

Mirror

FIGURE 28.5
A curved mirror focuses light from a yellow star
straight in front of it and a red star above that.
The focal plane is in front of the mirror and
blocks some of the starlight.

28.2   REFLECTING TELESCOPES

Newton's first telescope mirror
was made from polished metal because the technique of laying a thin
reflective coat of metal on polished
glass had not been developed.

Not long after the invention of telescopes, people realized that mirrors could be
used to bring light to a focus; but a practical design to do this was not worked out
until 1670 by Isaac Newton. Technology at the time did not make it easy to build
reflecting telescopes, so it was not until the late 1800s that reflecting telescopes
began to replace refracting telescopes. Today almost all research telescopes,
whether used for visible, radio, or X-ray wavelengths, employ mirrors to collect
and focus light. They are called *reflecting telescopes* or **reflectors.**

As Figure 28.5 shows, a curved mirror can focus light rays reflected from it. The
figure depicts the light from two stars, a yellow star straight in front of the telescope
and a red star above the yellow star. Each star's light is focused at a point on the focal
plane, but this is now in front of the mirror, so the observer or detector blocks some
of the light from reaching the mirror. If the mirror is big enough, the fraction of the

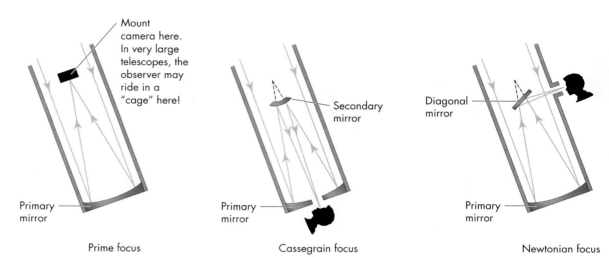

FIGURE 28.6

Sketches of different focus arrangements for reflectors.

mirror blocked by a detector (or even an astronomer!) may be only a small fraction of the aperture. Some large telescopes contain a "cage" in which an astronomer may ride inside the telescope at the focal plane.

A variety of other designs have been developed to move the focus outside the path of the incoming light. Newton's solution was to use a **secondary mirror** to deflect the light out to the side (Figure 28.6). This is today known as a **Newtonian telescope,** and it remains a popular design for smaller reflectors. Most modern research telescopes, whether optical or radio, use a secondary mirror that reflects the light back through a hole cut in the middle of the main mirror. This is called a **Cassegrain telescope,** named for the French sculptor Guillaume Cassegrain, who designed it in 1672. The secondary mirror in these designs still blocks part of the light from reaching the **primary mirror,** but only a small fraction of the total.

In optical telescopes the mirrors are usually made of glass that has been shaped to a smooth curve, polished, and then coated with a thin layer of aluminum or some other highly reflective material. The reflecting surface must be smooth to about one-tenth of the wavelength of the photons bouncing off it; otherwise the light will scatter in many directions. Glass can be polished to such a smoothness, and the electromagnetic waves therefore can be brought together in sharp focus.

A radio telescope designed to study 10-centimeter wavelength light, by contrast, needs to be smooth only at the level of 1 centimeter, so it can be built of metal plates bolted together. At the other extreme, X-ray (and gamma-ray) photons have wavelengths that may be smaller than the size of an atom. X-ray photons encountering an optical telescope mirror would be scattered in many directions or absorbed by the mirror. However, they will reflect off a surface if they encounter it at a very shallow angle. (You can demonstrate this principle to yourself with visible light by tilting a rough surface so light hits it at a glancing angle.) The mirror designs for X-ray telescopes therefore look completely different from optical designs, but they achieve the same result of funneling the photons toward the detector (Figure 28.7).

It is important that the electromagnetic waves arrive at the detector in unison, so the shape of the reflector is critical in addition to its smoothness. The mirror needs to be designed so that after electromagnetic waves bounce off the mirror, the crests of the waves come together at the focus all at the same time. By achieving this simultaneous arrival, the strength of the electromagnetic vibration at the focus will be the concentrated strength of the entire wave collected over the whole aperture. If the mirror shape is not correct, portions of the crest of the wave will arrive ahead

When the Hubble Space Telescope was first put into service, it was discovered that the surface had been polished to the wrong shape because of a manufacturing error. Fortunately, it was possible to insert a lens to correct for the shape error, much as you can wear glasses to correct for focusing problems caused by the shape of your eye.

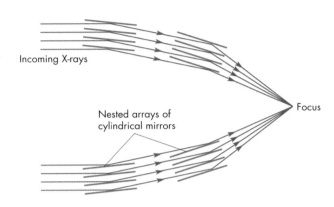

A The cylindrical mirrors of an X-ray telescope

B Cross-sectional view of the grazing-incidence mirrors showing how they focus X-rays

FIGURE 28.7

Grazing incidence optics in an X-ray telescope. X-rays pass through normal mirrors, so X-ray telescopes bring X-rays to a focus by letting them graze the surfaces of nested cylindrical mirrors at shallow angles.

or behind one another, and they will not combine to the maximum strength. As a result, the signal will be weaker at the detector, and the images will be blurry.

Moreover, as a telescope moves, its mirrors and lenses must keep their same precise shapes and relative positions if the images are to be sharp. This is one of the most technically demanding parts of building a large telescope because large pieces of glass or metal bend slightly when their positions shift. Unlike lenses, however, a mirror can be supported from behind, thereby helping to hold the glass in shape. But holding a large mirror stiff requires a massive support structure behind the mirror. Thus building ever larger aperture telescopes becomes more difficult.

Development of large reflectors proceeded rapidly during the early 1900s, greatly surpassing the apertures of the largest refractors. A 5-meter (17-foot) diameter mirror was completed in 1948 at the Palomar observatory in California—the mirror alone weighing nearly 15 tons, the whole telescope nearly 500 tons. Building larger telescopes along similar lines was found infeasible because the weight of glass and supporting structures grew too rapidly with size.

Recent innovations have worked around the problem of building extremely massive, stiff "back structures." One way is to collect and focus the light with several smaller mirrors and then align each one individually. So instead of a telescope having a single large mirror, it may have many small ones. Telescopes designed this way are called *multimirror instruments*. Currently the largest such multimirror instruments are the twin 10-meter (33-foot) Keck Telescopes (Figure 28.8A), operated by the California Institute of Technology and the University of California and located on the 4200-meter (14,000-foot) volcanic peak Mauna Kea in Hawaii. The total weight of the glass in this design is about the same as in the Palomar 5-meter telescope, but the total telescope weight is only about half as much. Each Keck telescope consists of 36 separate mirrors (visible in Figure 28.8A) that are kept aligned by lasers that measure precisely the tilt and position of each mirror. If any misalignment is detected, tiny motors shift the offending mirror segment to keep the image sharply in focus. The same principle is being used to build large new radio telescopes.

Another approach is to build mirrors much thinner and lighter than those in earlier, smaller telescopes. For many years telescope mirrors were made thicker as they were made larger to keep them stiff. By making a back structure that uses sensors and tiny motors to compensate for bending as the mirror is tilted, much as described for the Keck telescopes, a thin mirror can be kept precisely shaped. A number of thin

A

B

FIGURE 28.8

(A) Photograph of one of the twin Keck telescopes. The 36 mirrors cover an area 10 meters (about 33 feet) in diameter, making them the world's largest single optical telescopes.
(B) One of the two Gemini telescopes. One of this telescope pair is on Mauna Kea, Hawaii; the other one is in Chile. The mirror of each telescope is about 8 meters (26 feet) in diameter. The yellow lines show the light path through the telescope.

8-meter (about 26-foot) glass mirrors have been built in recent years, achieving overall weights of glass and telescope that are similar to the Palomar 5-meter telescope. Four are used in unison at the VLT (Very Large Telescope) of the European Southern Observatory in Chile, giving the combination of the four the largest total collecting area of any current visible-wavelength observatory. A photograph of the 8-meter Gemini telescope on Mauna Kea is shown in Figure 28.8B.

28.3   REFLECTORS VERSUS REFRACTORS

After 1900 almost all telescopes built for research were reflectors. This was primarily because of the expense of building larger lenses and the technological problems of supporting such massive pieces of glass, as we have discussed. Note that lenses are still frequently used as part of the optics of many research telescopes, but these telescopes continue to be called *reflectors* if a mirror does the primary work of focusing the light from the extent of the aperture.

Refracting telescopes have the advantage that the focus is behind the lens. This simplifies the design, and there is no need for mirrors or other equipment to

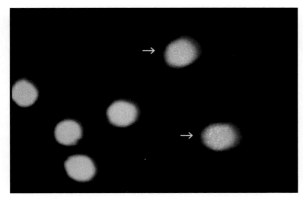

FIGURE 28.9

The dispersion of light caused by glass can result in different wavelengths focusing differently. The result can appear in images as fringes of color.

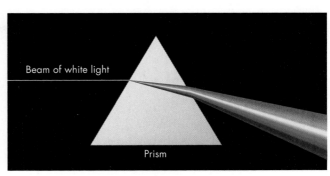

FIGURE 28.10

Glass refracts blue light more than red light, producing a spectrum.

 This effect on image quality is called *chromatic aberration* and is a problem for cameras as well as telescopes.

partially block the telescope aperture. The image quality is degraded by anything that blocks part of the aperture. For example, the "spikes" seen on stars in most astronomical images are caused by the support structures that hold the secondary mirror in place in a reflector.

Refractors have some significant disadvantages, though, because of the properties of glass. The lenses may be as opaque as a chunk of concrete to shorter-wavelength ultraviolet photons. Mirrors, by contrast, are generally much less selective in the wavelengths of electromagnetic radiation they reflect, allowing much broader wavelength coverage with a single telescope. Mirrors also reflect light at the same angle, independent of wavelength, but the same is not true of lenses. The angle by which glass bends light depends on the light's wavelength. So, for example, the red wavelengths of light may be out of focus when the blue wavelengths are sharp, creating images fringed with color (Figure 28.9). This property of refracting materials is a problem for making sharp images, but it has important benefits in other applications.

28.4   COLOR DISPERSION

Different wavelengths of light travel at slightly different speeds in most materials and are therefore refracted by different amounts. Shorter wavelengths of light are generally refracted more strongly. Thus when white light—which is a mix of all colors—passes through a refracting material at an angle, it is spread into a spectrum, or rainbow, in a process called **dispersion.** This happens to light passing through the glass of a **prism** (Figure 28.10) or through drops of water in a rainstorm.

Because of this effect of lenses, most modern research instruments avoid lenses if mirrors can achieve the same result. The dispersion caused by glass is one reason astronomical mirrors have reflective aluminum coating on top of the glass instead of underneath it (as in a conventional mirror); telescope mirrors must be recoated over time as the metal becomes tarnished by exposure to air.

Dispersion is also greatly useful for carrying out spectroscopy on the light from astronomical sources. Astronomers can use a large prism to split the light of all of the astronomical sources within the telescope's field of view into spectra all at once. Stars and other objects may display slight or large differences in the characteristics of their spectra even though their color and appearance otherwise look the same. This gives astronomers an important tool for classifying objects and discovering new kinds of sources.

KEY TERMS

Cassegrain telescope, 200   primary mirror, 200

dispersion, 203   prism, 203

focal plane, 197   reflector, 199

focus, 197   refraction, 197

lens, 197   refractor, 197

Newtonian telescope, 200   secondary mirror, 200

QUESTIONS FOR REVIEW

1. What do we mean by the phrase *to focus light*?
2. What is the difference between reflecting and refracting telescopes?
3. What are some reasons for using mirrors rather than lenses in telescopes?
4. How is dispersion of light useful in astronomy?

PROBLEMS

1. An image will appear upside down when we look through a simple refracting telescope, as can be seen from Figure 28.1. Trace the light through a similar diagram to show the orientation in which you would expect to see things through a Cassegrain telescope. Assume the secondary mirror in the Cassegrain telescope is flat.
2. After the Palomar 5-m telescope was completed, astronomers estimated that the mass of larger telescopes would increase as the (mirror diameter)^3 or (mirror diameter)^4 because of the greater mass of glass and support needed to keep the entire telescope stiff. If the 10-m Keck telescopes followed this rule,

how many times more massive would they have been than the Palomar telescope?

3. The *index of refraction* measures the factor by which light slows down in a substance. The index of refraction of water is 1.33. If you sent a pulse of light through water and a pulse of light through air (index of refraction of 1.00), how much longer would it take the light to travel 1 km through the water? (The speed of light in air is 299,800 km/sec.)

TEST YOURSELF

1. A reflector and a refractor have the same diameter aperture. Which of the following would be a disadvantage of using the refractor?
   a. Refractors bend the light at too large an angle.
   b. Refractors do not focus all colors at the same place.
   c. Refractors have less magnification.
   d. Refractors allow in less light.
2. Suppose you were examining a pulse of radio signals from a distant civilization far away across interstellar space. Knowing that the ionized gas in interstellar space causes dispersion of radio waves, what effect would you expect this to have on the signal?
   a. The signal would be slowed down—stretched out to fill a much longer time.
   b. The path would be bent so the signal would come from a different direction than it started from.
   c. The time when the pulse arrived would be different for different wavelengths.
   d. The wavelengths would all grow longer as they ran out of energy.
3. How does the speed of light in glass compare to the speed of light in empty space?
   a. It is faster in glass.
   b. It is faster in space.
   c. It is the same in both.

29

Telescope Resolution

Background Pathways: Unit 26

People often ask how high a magnification a telescope can produce. In reality, any telescope can deliver as high a magnification as we want with the right eyepiece. What is more important to understand is the **resolution** of a telescope—the level of detail it is capable of revealing. A highly magnified blurry image is of little use.

Resolution is determined in part by the quality of the optics; but even with perfectly made lenses and mirrors, there is a fundamental limit on resolution imposed by the aperture size of a telescope. The larger the aperture, the finer the detail it is capable of detecting. This is the other reason astronomers strive to build larger telescopes—so they can resolve increasingly finer structural details of the objects they observe.

29.1 RESOLUTION AND DIFFRACTION

If you mark two black dots close together on a piece of paper and look at them from the other side of the room, at a great enough distance your eyes will no longer be able to see them as separate spots. Similarly, stars that lie very close together, or markings on a planet, may not be clearly distinguishable. The smallest separation between features that a telescope is able to distinguish is a measure of the telescope's resolution.

Resolution is fundamentally limited by the wave nature of light. Whenever waves pass through an opening, they are bent at the edges of the opening by a phenomenon called **diffraction.** Figure 29.1A shows how water waves, all initially parallel to each other, are diffracted after they pass through a narrow opening. The central portion of the set of waves travels straight through the opening, but weaker waves radiate away from the edges of the opening.

Light waves are similarly diffracted at the edge of a telescope's aperture. The result is that the light waves from a star are shifted into slightly different directions, smearing out the light. Figure 29.1B shows an image of a star from the Hubble Space Telescope. The image has been enhanced to bring out the very faint **diffraction pattern** produced when a star is imaged. Diffraction spreads the star's image into a blur, even though the light is coming effectively from a single point within this image. A small percentage of the light is diffracted at larger angles and produces the complex outer parts of the diffraction pattern.

The complex set of rings and spikes seen in the diffraction pattern is caused by interference between waves coming through different parts of the aperture or diffracting from the edges of any other structures within the telescope's aperture. **Interference** occurs when light waves add to or subtract from each other. The diffracted light travels through the telescope along slightly different paths, offset from its original direction. Light diffracted by different edges of the aperture and following slightly longer or shorter paths may arrive at the detector with the wave

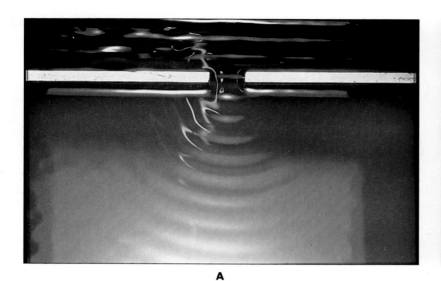

A

B

FIGURE 29.1
(A) Water ripples diffracted as they pass (moving down in image) through a narrow opening. (B) Image of a star made with the Hubble Space Telescope. Light coming from a single point is spread out by diffraction. Diffraction at the edges of the aperture and at internal support structures produces a very faint, complex diffraction pattern.

One way of describing the resolution of a telescope is to find the closest pair of stars that can still be discerned from each other. If the stars are separated by too small an angle, diffraction by the telescope aperture (or poor optics) will overlap the light from the stars to such an extent that they become indistinct from each other.

crests adding together or canceling each other out. Where they add together, we see a brighter spot in the diffraction pattern. Where the interference causes the waves to cancel one another, we see a darker spot in the diffraction pattern. The diffraction rings are produced by interactions between light waves diffracted from different parts of the edges of the aperture. The X-shaped pattern surrounding the star is caused by light diffracted by the support structure holding a secondary mirror in the telescope.

The diffraction pattern can make brighter stars appear larger in an astronomical image even though the light from all stars is essentially coming from a single point. This is especially true when images are optimized to examine fainter objects. When the image is brightened, the inner portion of the diffraction pattern around a bright star may become completely "burned out" (completely white or bright). For the brightest stars, the diffraction pattern will be "burned out" even farther from the star. Detecting a faint star near a bright star can be difficult because a dim object may become lost within the bright star's diffraction pattern even if the separation is relatively large. This makes it extremely difficult to image planets around other stars, for example.

29.2   CALCULATING THE RESOLUTION OF A TELESCOPE

The best possible resolution of a telescope depends on both the wavelength of the radiation being observed and the diameter of the lens or mirror that collects the light. Resolution can be improved by using a larger mirror or lens, and it can also be improved by observing at shorter wavelengths.

Detailed calculations show that if two points of light are separated by an angle α (measured in arc seconds) and are observed at a wavelength λ (measured in

An *arc second* is a unit of angle and is equal to 1/3600th of a degree.

α_{arcsec} = Smallest resolvable angle (in arc seconds)

λ_{nm} = Wavelength of observation (in nanometers)

D_{cm} = Diameter of telescope aperture (in centimeters)

Most people would say that they can resolve greater detail in bright light, when their pupils are very small, than in dim light. What other factors might account for this impression?

nanometers), they cannot be seen as separate sources unless they are observed through a telescope whose diameter D (in centimeters) is larger than

$$D_{cm} > 0.02\, \lambda_{nm}/\alpha_{arcsec},$$

where the subscripts cm, nm, and arcsec have been added as reminders of the units that must be used for measuring each term. Notice that the diameter needed to resolve two sources increases as the sources get closer together.

Alternatively, we can express the smallest separation angle a telescope is capable of resolving by rearranging the equation terms:

$$\alpha_{arcsec} > 0.02\, \lambda_{nm}/D_{cm}.$$

For example, a telescope of diameter 100 cm (about 40 inches) observing visible light ($\lambda \cong 500$ nanometers) will be able to resolve two stars if they are separated by at least 0.1 seconds of arc.

Radio telescopes, despite their enormous size, observe very long wavelengths of light. In consequence, their resolution is usually much poorer than optical telescopes. The Arecibo radio telescope has a diameter of 300 m ($D = 30{,}000$ cm). Observing a radio wavelength of 10 cm ($\lambda = 10^8$ nanometers), the smallest resolvable angle at Arecibo is

$$0.02 \times 10^8/30{,}000 \approx 70 \text{ arc seconds.}$$

The unaided human eye (pupil diameter 0.8 cm) observing visible wavelengths (500 nm) can resolve

$$0.02 \times 500/0.8 \approx 13 \text{ arc seconds.}$$

Thus the human eye can resolve finer detail than the largest single radio telescope in the world! Of course, if our eyes were sensitive to radio wavelengths, they would have extremely poor resolution.

29.3   INTERFEROMETERS

The limitation on resolution caused by diffraction represents a major impediment to studying distant objects, where the angular size becomes very small. Building larger and larger telescopes has structural and financial limits, so how can we achieve better resolution?

An answer to this problem begins to become apparent when we realize that a telescope aperture does not need to be continuous. If you take binoculars or a small telescope and put narrow strips of black tape in an X across the front of them—leaving gaps where light can pass through—the image you see through the binoculars does not have an X through it. The separate parts of the lens where light can still pass will focus the light perfectly well. Similarly, if you blacked out portions of the mirror on a reflecting telescope, the telescope would still function. The interesting thing is that as long as portions near the edges of the mirror remain uncovered, the resolution remains about the same. Less light may be collected, and the diffraction pattern would grow more complex, but the resolution would be little affected.

The pieces of the aperture actually do not need to be connected at all. As long as we can determine the relative positions of the waves precisely enough, we can add the light waves together so that they are in sync. More remarkable still, it is possible to record the electromagnetic wave from the separate "pieces" and later join together the recorded signals to form the equivalent of a single extended telescope

If you try putting black tape over a lens, do not stick the tape to the glass of the lens! Some lenses have special coatings or might be difficult to clean.

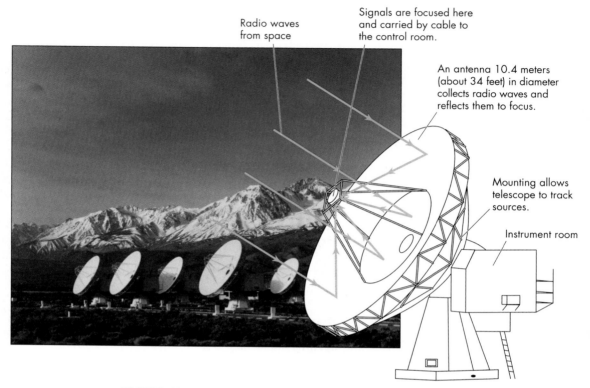

Radio waves
from space

Signals are focused here
and carried by cable to
the control room.

An antenna 10.4 meters
(about 34 feet) in diameter
collects radio waves and
reflects them to focus.

Mounting allows
telescope to track
sources.

Instrument room

FIGURE 29.2

Photograph of the Owens Valley Millimeter Wavelength radio telescope, operated by the California Institute of Technology. In the background you can see the Sierra Nevada mountain range. The "telescope" is an array of six separate dishes that collect the radio waves. The captured radiation is then combined by computer to increase the resolution of the instrument.

aperture. This process is now regularly carried out at radio wavelengths (Figure 29.2) to join the signals from many separate telescopes. Major radio telescope arrays are located in many countries, the largest in steady use being the VLBA (Very Long Baseline Array) that spans the United States from the Caribbean to Hawaii. Radio astronomers sometimes arrange to make observations with even larger arrays by linking observations from radio telescopes around the world, and even space-borne radio telescopes, to achieve a resolution the equivalent of a single telescope that is tens of thousands of kilometers across!

Astronomers call a combined set of telescopes like this an **interferometer.** The reason behind this name relates to the principle of interference discussed earlier. To achieve the equivalent of a single large aperture, the same wave crest must be combined from the individual telescopes so that the interference between waves does not cause them to subtract from each other. Because the telescopes may be widely separated, the same wave crest will arrive at the different telescopes at different times. The electromagnetic waves from each telescope must be delayed by corresponding amounts of time so that the combined wave generates the strongest possible coherent signal. This indicates that the waves are all being brought together in unison—similar to the function of properly shaped lenses and mirrors in an individual telescope.

The result of this process is the ability to produce images in which the resolution is set not by the size of the individual telescopes but rather by their separation. The

Q Interferometers can provide superb resolution, but most observations are still made with individual telescopes. What sorts of trade-offs do you think are likely to be made when using an interferometer?

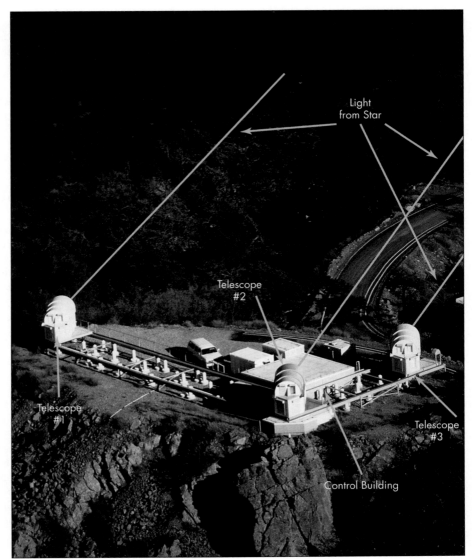

FIGURE 29.3

Photograph of an infrared and optical wavelength interferometer (IOTA). Light from the object of interest is collected by the three telescopes and sent to a control room. Computers there combine the light and reconstruct an image of the object.

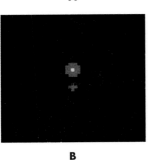

FIGURE 29.4

(A) A young star observed with an ordinary telescope. (B) The same star observed with an interferometer. The higher resolution of the interferometer reveals that the "star" is actually two stars in orbit around one another.

twin Keck telescopes and the four telescopes of the VLT are designed to be used in this way. A photograph of an interferometer using three telescopes is shown in Figure 29.3. For example, if the mirrors are 100 meters apart, the interferometer has the same resolution as a telescope 100 meters in diameter.

The high resolution of interferometers is far beyond what can be obtained in other ways. Figure 29.4A shows a view of two closely spaced stars as observed with a single telescope. Their images are blended as a result of diffraction. Figure 29.4B shows the same stars observed with an interferometer after the image has been processed by a computer. The two stars can now be easily distinguished: The two separate mirrors produce the resolution of a single mirror with a diameter equal to the spacing between them.

KEY TERMS

diffraction, 205

diffraction pattern, 205

interference, 205

interferometer, 208

resolution, 205

QUESTIONS FOR REVIEW

1. What is resolution of a telescope? What physical process limits it?

2. How is resolution affected by the size of a telescope's mirror or lens?

3. What is the purpose of an interferometer?

PROBLEMS

1. What is the smallest resolvable angle of the 2.4-m Hubble Space Telescope when it is observing 240 nm ultraviolet light? What is its resolution when it is observing 1.2 micron infrared light?

2. Find the smallest resolvable angle of the 100-m diameter Green Bank Telescope when it is observing radio waves at 20-cm wavelength. What is its resolution when observing a 3-mm wavelength?

3. What is the smallest resolvable angle of a radio interferometer that extends 10,000 km when it is observing 20-cm or 3-mm wavelength?

TEST YOURSELF

1. A telescope's resolution measures its ability to see
   a. fainter sources.
   b. more distant sources.
   c. finer details in sources.
   d. larger sources.
   e. more rapidly moving sources.

2. Astronomers use interferometers to
   a. observe extremely dim sources.
   b. measure the speed of remote objects.
   c. detect radiation that otherwise cannot pass through our atmosphere.
   d. improve the ability to see fine details in sources.
   e. measure accurately the composition of sources.

3. Suppose we made observations with a telescope, but found that we needed better resolution. Which of the following would allow us to make higher-resolution observations?
   a. Observe with a telescope that has a bigger mirror.
   b. Use a telescope with a lens instead of a mirror.
   c. Use a telescope with a mirror made of gold.
   d. Observe with the same telescope but at longer wavelengths.
   e. All of the above

30

The Earth's Atmosphere and Space Observatories

Background Pathways: Unit 22

The Earth's atmosphere presents challenges for carrying out astronomical observations. Although our atmosphere is transparent at visible wavelengths, it is not transparent at all wavelengths. Moreover, the atmosphere distorts the radiation that does pass through it. For example, sometimes it acts like an imperfect lens, distorting light as it moves from space into progressively denser layers of air. To complicate matters further, these effects vary with wavelength and weather conditions.

The atmosphere makes it challenging to build telescopes that work well. Some wavelengths cannot penetrate the Earth's atmosphere, so they cannot be observed except from space. For other wavelengths, conditions may be better in space, but observations from the ground are possible too. In these cases there are trade-offs to consider. Space telescopes are generally much more expensive to build, launch, and maintain than ground-based telescopes. Also, limitations on the weight that can be launched into space set limits on the size of a space telescope. For such reasons astronomers have explored many technological innovations, which by compensating for the distorting effects of the Earth's atmosphere will improve observations made from the ground.

30.1 ATMOSPHERIC ABSORPTION

Telescopes operating at many wavelengths face a major obstacle: Most of the radiation they seek to measure cannot penetrate the Earth's atmosphere. Gases in the Earth's atmosphere absorb electromagnetic radiation, and the amount of this absorption varies greatly with wavelength. For example, atmospheric gases affect visible light relatively little, so our atmosphere is nearly completely transparent to the wavelengths we see with our eyes. On the other hand, some of the gases strongly absorb infrared radiation while others absorb ultraviolet radiation.

The transparency of the atmosphere to visible light compared with its nontransparency to infrared and ultraviolet radiation creates what is called an **atmospheric window.** An atmospheric window is a wavelength region in which energy comes through our atmosphere easily compared with other nearby wavelengths (Figure 30.1). For example, there is a large window at radio wavelengths that makes most ground-based radio observations practical.

211

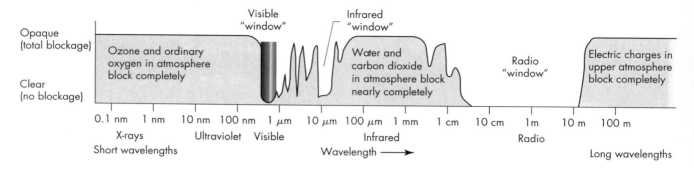

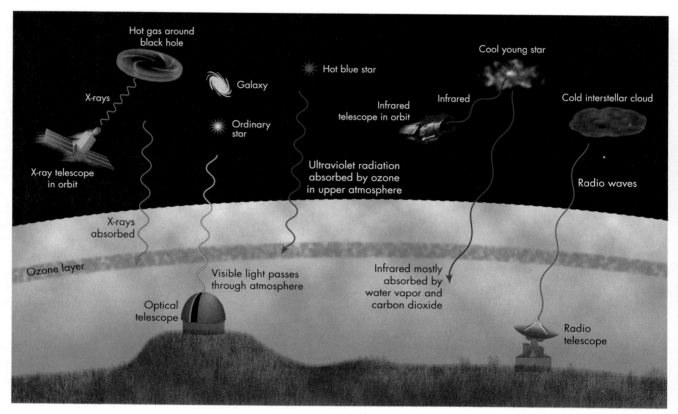

FIGURE 30.1

Atmospheric absorption. Wavelength regions where the atmosphere is essentially transparent, such as the visible spectrum, are called *atmospheric windows*. (Wavelengths and atmosphere are not drawn to scale.)

Atmospheric absorption

Other wavelength ranges present greater difficulties. Water (H_2O) and carbon dioxide (CO_2) absorb most infrared wavelengths, although there are some narrow windows within the range. This blockage by molecules in the Earth's atmosphere makes it difficult to detect emission from water or carbon dioxide located elsewhere in the cosmos. At shorter wavelengths than visible light, ozone (O_3) and ordinary oxygen (O_2) strongly absorb ultraviolet radiation, while oxygen and nitrogen absorb X-rays and gamma radiation. As a result, only from space can we observe the high-energy photons from some of the most violent processes occurring in the cosmos.

The atmospheric window at visible wavelengths permits us to observe the Sun and other stars, which allows us to do much astronomy from the ground. But the

FIGURE 30.2
Photographs illustrating light pollution.
(A) Los Angeles basin viewed from
Mount Wilson Observatory in 1908.
(B) Los Angeles at night in 1988.
(C) Notice the pattern of the interstate
highway system visible in the satellite
picture of North America at night.

Get involved in your community
by letting people know that they can
save money by using light fixtures
that have reflectors to keep the light
shining downward. They are paying
for the electricity a poorly designed
fixture uses to light up the night sky!

blockage at other wavelengths is beneficial, too. For example, because our atmosphere blocks wavelengths shorter than visible light, it protects us from high-energy photons that are dangerous to living organisms. Similarly, because our atmosphere is not transparent to infrared wavelengths, it traps infrared radiation emitted by the warmed ground at night, retaining much of its heat and preventing the oceans and us from freezing.

Recently astronomers have had to contend with another factor that limits their observations from the ground: **light pollution.** Most inhabited areas are peppered with nighttime lighting such as streetlights and outdoor advertising displays (Figure 30.2A and B). Although some such lighting may increase safety, much of it is wasted energy, illuminating unessential areas and spilling light upward into the sky, where it serves no purpose. Figure 30.2C shows a satellite view of North America at night and illustrates the wasted energy created by light pollution. Such stray light can interfere with astronomical observations. Many early observatories were built in cities to make them more accessible, but they have become essentially useless for research because of light pollution. In some places astronomers have persuaded regional planning bodies to develop lighting codes to minimize light pollution. Light pollution not only wastes energy and interferes with astronomy; it also destroys a part of our heritage—the ability to see stars at night. The night sky is a beautiful sight, and it is shameful to deprive people of it.

30.2 ATMOSPHERIC REFRACTION

Why stars twinkle

Even from a location where there is no light pollution and the air is clear, the atmosphere still causes significant problems for observations at visible wavelengths. Refraction—the bending of light's path as it travels through materials of different density—alters the path of light as it passes through the air. If you have ever noticed a star twinkling in the night sky, then you have observed atmospheric refraction. Twinkling, more properly called **scintillation,** is caused by various pockets of warmer and cooler air that have slightly different densities. Each acts as a lens on the light traveling through it. As hot air rises or as winds and turbulence stir the air, these pockets will cross in front of a star, magnifying and demagnifying, and bending the path of the light first one way and then another. The color dispersion of these "lenses" will even make the star's color appear to vary. These effects are particularly apparent when we view a star near the horizon, where the light travels a longer path through the atmosphere. Depending on atmospheric conditions, a star may flicker and shift more or less, conditions astronomers call, respectively, bad or good **seeing**.

With bad seeing, the starlight you see at any instant is a blend of light from many slightly different directions, which smears the star's image and makes it dance (Figure 30.3). You can see a similar effect if you look down through the water at something on the bottom of a swimming pool. If the surface of the water has even slight disturbances, a pebble or coin on the bottom seems to dance around. Such similar refractive twinkling in our atmosphere limits the ability of Earth-bound observers to see fine details in astronomical objects. The dancing image of a star or planet distorts its picture when recorded by a camera or other device.

Until recently, ground-based astronomers had to submit to the distortions of seeing, but now they can partially compensate for such seeing in several ways. One technique involves observing a bright star simultaneously with the object of interest. Extremely rapid measurements are made of the bright star's shifting position, then tiny motors make compensating adjustments to the tilt of a secondary

Q One way you can often tell a planet from a bright star without using a telescope is by how much they twinkle. Planets tend to shine more steadily than a star in the same position. Why do you suppose there is a difference?

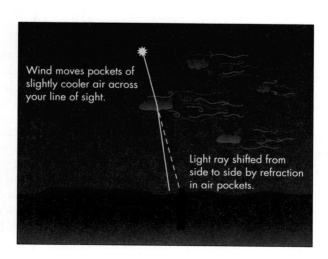

Wind moves pockets of slightly cooler air across your line of sight.

Light ray shifted from side to side by refraction in air pockets.

FIGURE 30.3
Twinkling of stars (seeing) is caused by moving atmospheric irregularities that refract light in random directions.

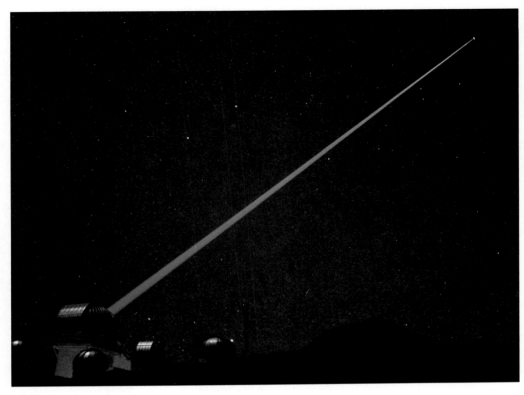

FIGURE 30.4

A laser beam creates an artificial star whose image serves as a reference to eliminate the atmosphere's distortion of real stars. This photograph was taken at the Starfire Optical Range of the Phillips Laboratory at Kirtland Air Force Base in New Mexico.

mirror to keep the bright star's position the same. Because the pockets of air in the atmosphere are fairly large, faint objects in the vicinity of the bright star are also kept stationary, and the image remains sharp. Astronomers are learning to make even more rapid adjustments for atmospheric effects, eliminating many of the effects of atmospheric seeing. This technique, called **adaptive optics,** has already given astronomers dramatically improved views through the turbulence of our atmosphere. Because bright stars are not present next to all of the objects we might want to observe, astronomers have developed a technique using a powerful laser beam to create an artificial star high in the Earth's atmosphere, as shown in Figure 30.4.

A less obvious effect is that refraction by the atmosphere as a whole also bends the light from stars near the horizon, so that we can see stars that are actually below the horizon! The bending of light is greater for objects that are closer to the horizon (Figure 30.5A and B). At the horizon, the effect of the bending is about half a degree. As a result, when the Sun or Moon appears to be just touching the horizon, it is in fact below the horizon. If Earth had no atmosphere, it would already be out of view. Thus atmospheric refraction alters the time at which the Sun seems to rise or set. It also distorts the shape of the Sun because of differences in the amount of bending from the bottom to the top of the Sun's disk (Figure 30.5C).

By "lifting" the Sun's image above the horizon, even though it has set, refraction slightly extends the length of daylight hours. As a result, the day of the year the Sun appears to be above the horizon for exactly 12 hours is not the equinox, but rather a few days before the first day of spring and a few days after the first day

Q How do the refraction and absorption of light by the atmosphere relate to the color of the Moon during a total lunar eclipse?

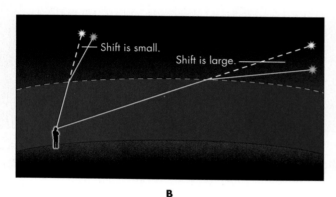

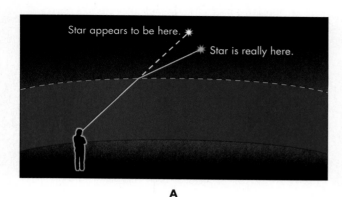

FIGURE 30.5

(A) Atmospheric refraction makes the Sun or a star look higher in the sky than it really is.
(B) Refraction is stronger for objects nearer the horizon. (C) The Sun is flattened because
refraction "lifts" its lower edge more than its upper edge. The Sun's reflection in a band
below it on the water is called the *glitter path*.

of autumn. This depends on latitude and atmospheric conditions, but it turns
out that near latitude 40° N, on St. Patrick's Day (March 17) the day usually has
12 hours between sunrise and sunset, whereas the actual equinox occurs on about
March 21.

30.3   OBSERVATORIES IN SPACE

Figure 30.6 shows several of the major orbiting telescopes presently in use. Some,
like the Hubble Space Telescope, which was launched in 1990, have operated for
many years thanks to their low orbit around Earth that allows servicing missions
with the space shuttle to repair problems and upgrade equipment. Most other
space telescopes, like the Spitzer Infrared Space Telescope, have a shorter lifetime
because the detectors require coolants such as liquid helium to keep them close to
a temperature of absolute zero. When the supply of liquid helium is used up after
several years, it is unlikely that it can be replaced because the telescope was placed
in deep space millions of kilometers from Earth. It was located far from Earth so
that the Earth's own heat would not interfere with observations. By contrast,

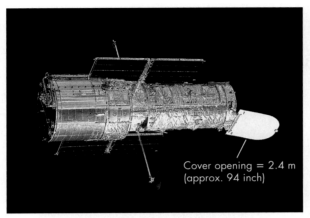

Cover opening = 2.4 m
(approx. 94 inch)

Hubble Space Telescope (13.6 m long) – HST

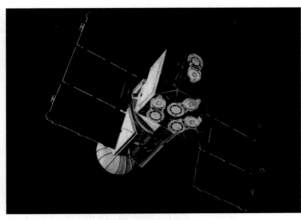

Extreme Ultraviolet Explorer – EUVE

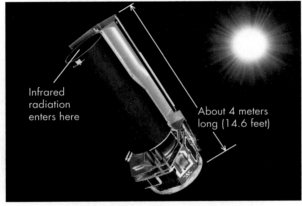

Infrared
radiation
enters here

About 4 meters
long (14.6 feet)

Spitzer Infrared Space Telescope

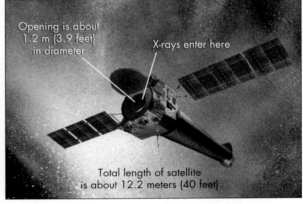

Opening is about
1.2 m (3.9 feet)
in diameter

X-rays enter here

Total length of satellite
is about 12.2 meters (40 feet)

Chandra X-ray Telescope Satellite

FIGURE 30.6
Photograph of the Hubble Space Telescope and drawings of some other orbiting
observatories: EUVE, Spitzer, and Chandra.

the Hubble Space Telescope orbits about 500 km (300 miles) above the Earth's surface.

More than 50 space telescopes have been launched by countries around the world to examine all different parts of the electromagnetic spectrum. Many of these missions were brief and were designed to give us our first clues about cosmic radiation outside of the Earth's atmospheric windows. With each mission we have developed a clearer idea of the characteristics of electromagnetic emission, leading to improved designs. Recent telescopes like the Chandra X-ray telescope, launched in 1998, have provided some of the first images of X-ray emissions, with resolutions comparable to visible-wavelength images.

Of the many orbiting telescopes used by astronomers, the Hubble Space Telescope (HST) is the most ambitious to date. The HST is designed to observe at visible, infrared, and ultraviolet wavelengths and has a mirror 2.4 meters (about 8 feet) in diameter. Its instruments allow it to take both pictures and spectra of astronomical objects. Although the HST initially had a number of problems, astronauts have repaired the major defects; and astronomers are now delighted with the clarity of its images (Figure 30.7). These images reveal details never seen by telescopes on the ground because such telescopes must peer through the blurring effects of our atmosphere.

Sombrero Galaxy
A system of billions of stars and dark dusty interstellar matter.

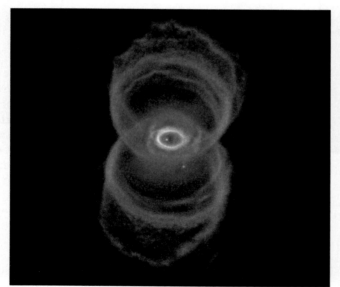

Hourglass Nebula
A dying star

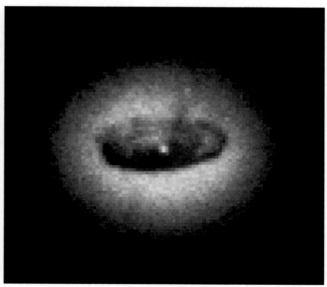

Disk of hot gas in the core of the galaxy NGC 4261;
a black hole may lie in the disk's center.

FIGURE 30.7
Some images obtained with the Hubble Space Telescope.

Despite the freedom from atmospheric blurring and absorption that space observatories enjoy, much astronomical work will be done from the ground for the
foreseeable future. For many years, ground-based telescopes will be much larger
than orbiting telescopes. Moreover, equipment problems can be corrected easily
without the expense, delay, danger, and complexity of a space shuttle launch.

Because huge telescopes in space or even on the Moon will remain dreams for
years to come, astronomers choose with care the location of ground-based observatories. Sites are picked to minimize clouds and the inevitable distortions and
absorption of even clear air. Thus nearly all observatories are in dry, relatively
cloud-free regions of the world such as the American Southwest, the Chilean

desert, Australia, and a few islands such as Hawaii and the Canary Islands off the coast of Africa. Moreover, astronomers try to locate observatories on mountain peaks to get them above the haze that often develops close to the ground in such dry locales and to improve the seeing.

30.4   OPENING A NEW DOMAIN: GAMMA RAYS

Astronomers have made many of their most exciting discoveries when new telescopes have allowed them to observe the sky at wavelengths not previously detectable. Gamma rays (electromagnetic waves with wavelengths shorter than about 0.01 nanometers) are among the last of the wavelength regions to be explored, and astronomers are still trying to interpret what they see.

Gamma-ray astronomy began in 1965 with a small and (by modern standards) primitive satellite designed to detect cosmic gamma rays. A few years later, a slightly more advanced satellite detected gamma rays coming from the center of our galaxy, the Milky Way. By the 1970s astronomers had discovered that many familiar sources, such as the Crab Nebula and the remnants of other exploded stars, emitted gamma rays. Ironically, and unknown to astronomers, perhaps the most interesting gamma-ray sources had already been discovered accidentally in the 1960s.

In 1967 the United States placed several military surveillance satellites in orbit to watch for the gamma rays produced when a nuclear bomb explodes. The satellites were designed to monitor the United States–Soviet Union ban on nuclear bomb tests in the atmosphere. Curiously, on a number of occasions, the satellites detected gamma-ray bursts coming not from the Earth but from space. This caused great concern among political and military leaders, but fortunately some scientists involved had enough background to recognize these as astronomical sources and avert a crisis. Unfortunately for astronomers, the discovery of the bursts was top secret at the time and was not made public until 1973.

Astronomers' thirst for more information about these high-energy sources was unsatisfied for many years because our atmosphere absorbs gamma rays, and ordinary telescopes cannot focus gamma rays. Nevertheless, with further development of gamma-ray detectors in satellites, astronomers discovered that gamma-ray sources—apart from the bursts—coincided with known astronomical objects such as dying stars and some peculiar galaxies. The gamma-ray bursts, on the other hand, would appear suddenly in otherwise blank areas of the sky, flare in intensity for a few seconds, and then fade to invisibility.

Neutron stars are the remnants of massive stars that have exploded. They are discussed in more detail in Unit 67.

It has taken more than 30 years of study to determine even whether they are near or far. The breakthrough came in December 1997, when astronomers detected a gamma-ray burst that coincided with a distant galaxy. This seems to have solved the mystery of the bursts' distance but leaves unanswered what they are. Hypotheses to explain the bursts abound. According to one popular hypothesis, the bursts occur when a pair of dead, collapsed objects called *neutron stars* collide in a distant galaxy. According to another hypothesis, the bursts are *hypernovas,* stellar explosions caused when massive stars run out of fuel and collapse to form black holes. This latter proposal gained support when in 2002 astronomers obtained spectra of a burst that shows emission lines suggesting the explosion of a massive star. As data accumulate, it also appears that there may be different classes of gamma-ray bursts, and perhaps some of them are not so far away. Today gamma-ray bursts are still mysterious, but that is what makes science exciting.

KEY TERMS

adaptive optics, 215

atmospheric window, 211

light pollution, 213

scintillation, 214

seeing, 214

QUESTIONS FOR REVIEW

1. What is an atmospheric window?
2. What do astronomers mean by *seeing*?
3. Describe the advantages and disadvantages of putting telescopes in space.

PROBLEMS

1. Why do stars twinkle more when they are near the horizon than when they are near the zenith? Draw a diagram to explain your answer.
2. In the polar regions, there can be entire days when the Sun remains below the horizon. Given the refraction of the atmosphere, what is the latitude farthest from the poles where you might experience a 24-hour night?
3. If the refraction of the atmosphere causes the Sun's position to shift by one degree at the horizon, calculate how long the day would be on the equinox for someone living at the equator. Would that same day be longer or shorter for someone living at latitude 45°? Explain using a diagram.

TEST YOURSELF

1. Which of the following telescopes would be most suitable for ground-based observations?
   a. Radio telescope
   b. X-ray telescope
   c. Ultraviolet telescope
   d. Gamma-ray telescope
   e. Infrared telescope
2. Which of the following explains the advantage of the Hubble Space Telescope compared with ground-based telescopes?
   a. It is the largest reflector ever made.
   b. It can detect X-rays from space.
   c. It is not affected by atmospheric scintillation.
   d. It is closer to the objects it is imaging.
   e. All of the above
3. What is the purpose of adaptive optics?
   a. To make telescope designs more flexible so they can fit in smaller buildings.
   b. To allow telescopes to look in several directions without having to turn the primary mirror.
   c. To increase the effective collecting area of a telescope by capturing more photons.
   d. To rapidly adjust for the changing distortions caused by the Earth's atmosphere.
   e. All of the above

31 Amateur Astronomy

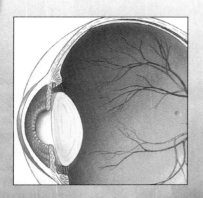

Background Pathways: Unit 28

Anyone with access to even modest equipment such as binoculars or low-powered telescopes has better equipment than Galileo ever had. With such equipment and access to a dark sky, that person can become an amateur astronomer. The pleasures of the hobby can range from the aesthetic satisfaction of taking a lovely photograph to the thrill of discovering a new comet or an exploding star. Here we offer a few suggestions for increasing your enjoyment of the night sky and taking your interest in astronomy a step further.

31.1 THE HUMAN EYE

The human eye is a remarkable device. It not only detects visible-wavelength photons, but can discriminate between different wavelengths. It has a self-adjusting aperture and can adjust the sensitivity of its receptor cells by nearly a factor of a million. Some parts of the eye allow for high resolution and color discrimination, while other portions permit us to detect extremely low light levels. Understanding how the eye works will help you get the most from your observations of the night sky.

Your eye (Figure 31.1) contains many of the same elements as a telescope. The eye's **iris** can expand or contract to adjust the aperture, or **pupil,** to allow in more

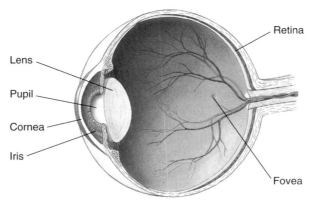

FIGURE 31.1

Structure of the human eye. The transparent cornea and lens focus light onto the retina at the back of the eye, which contains the rods and cones. The center of each eye's visual field is focused on the fovea. Focusing is accomplished by contraction and relaxation of the muscles that adjust the curvature of the lens.

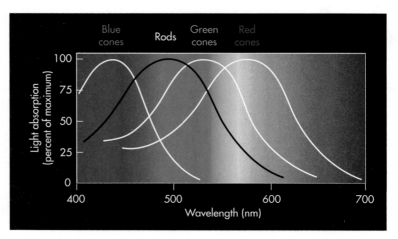

FIGURE 31.2

The spectral sensitivities of the photopigments in rods and three types of cones.

or less light. Light passing through the pupil is focused on the back surface called the **retina.** The shape of a transparent layer of your eye called the **cornea** does most of the focusing of light, but an internal lens can be adjusted to allow the eye to focus at different distances.

The light focused on the retina can trigger a response in light-sensitive nerve cells called **rods** and **cones,** so named because of their shapes. There are more than 100 million of these "detectors"—akin to the pixels of a CCD—that send signals to your brain, providing your sense of vision. The rods and cones all use **photopigments**—proteins that change form when they absorb light, causing a sequence of chemical reactions. Your eye uses four different kinds of photopigments that respond to different wavelengths of light (Figure 31.2). The rods use a photopigment that has a peak sensitivity at a blue-green wavelength. There are three kinds of cones, each using one kind of photopigment that is most sensitive to blue, green, or red light.

The cones and rods are not distributed uniformly in the eye. Instead, the cones are concentrated in a region called the **fovea** (Figure 31.1) in the center of your field of vision. By comparing the response of the three kinds of cones, your brain assigns a color. Note that a 500 nm photon can excite any of the photopigments, so your brain assigns a perceived color based on the relative strength of the reaction by different cones. Digital cameras employ a similar strategy to produce color images. Alternate pixels are behind one of three different colors of filters in front of each pixel, mimicking the way the eye works.

Your color perception is strongly concentrated in the middle of your field of vision, so your color discrimination is best when you look straight at an object, and it drops sharply away from there. There are few cones beyond 10° from the center of your vision, and almost none beyond 40°. You may think you see colors in your peripheral vision, but that is because your eye constantly glances in different directions to fill in detail, and your brain remembers the colors from when you saw an object earlier. Your brain also compares the images from both eyes to estimate the distances of objects. In addition, it tries to fill in missing information from parts of the eye where there are no cones and rods—such as the blind spot about 15° to the outside of the center of your field of vision in each eye, where the nerves come together to carry the signals to your brain.

Q To "see" your blind spot, put a large black dot near one end of a piece of paper, then draw a straight line from the dot that is about 20 cm (8 inches) long. Close one eye and hold the paper in front of you with the line horizontal and the dot at the right side for your right eye (or vice versa). Look first at the dot and then "slide" your center of vision along the line slowly. You will find that at a particular angle the dot seems to disappear. Why do you suppose we are unaware of our own blind spots?

31.2   YOUR EYES AT NIGHT

You will discover that the longer you stay outside in dim light, the more sensitive your eyes will become and the fainter the stars you will be able to see. This is the result of physiological changes in your eye referred to as **dark adaption.**

The simplest change in your eye occurring in dim light is that the pupil opens wider. This is easy to verify by looking at yourself in a mirror in a dimly lit room. In bright sunlight your pupil can shrink to a diameter of as little as 1 millimeter, but in total darkness its diameter may expand to about 7 or 8 millimeters, which allows more light to enter your eye. A change from 1 to 8 millimeters is a factor of 8 in diameter and therefore a factor of 64 in the amount of light admitted to the eye. The maximum size of the pupil tends to decrease with age to about 5 or 6 mm.

Your eyes undergo another change in the dark. The rods can build up the level of photopigments within them until they are about 1 million times more sensitive to light than they are in full daylight. The process takes about 20 minutes to get fully established but is undone by even a few seconds' exposure to bright light. Thus, once you are dark-adapted, you should stay away from bright lights for as long as you intend to observe. Smoking, alcohol, and other drugs also reduce your ability to adapt to the dark by reducing the iris's ability to expand and by interfering with the chemistry of the rods, cones, and other nerve cells.

The cones have much more limited ability to increase their sensitivity than the rods; so as light levels decrease, your vision relies more on the rods. This has several effects: Your ability to discriminate between colors declines or disappears completely; the wavelengths of light you are sensitive to shifts to blue wavelengths, a phenomenon known as the *Purkinje effect;* and your best sensitivity to light shifts outside of the perceived center of your field of vision in the fovea.

The Purkinje effect can change your perception of the relative brightness of different colored objects. In full daylight, the eye's overall response is strongest to yellow-green colors. At low light levels, you become most sensitive to the blue-green colors that the rods' photopigment responds to. This is possibly the result of natural selection because the average color of starlight is bluer than sunlight, so eyes responsive to blue will therefore aid survival. As a result of this effect, your perception of the relative brightness of two differently colored stars will change as your eyes adapt to the dark—an important consideration if you are trying to study the changes in brightness of **variable stars** (Unit 63) by comparing them to neighboring stars. It is certainly the case that night-flying insects see blue light better than yellow. That is why bug zappers use a blue light to attract insects, and also why a yellow lightbulb is often used for outdoor night lighting to be less attractive to insects.

Within a circle about 1° across in the center of your field of vision, the fovea contains no rods at all. The number of rods relative to cones climbs steadily in the surrounding area of the retina. You can apply this information practically with the following observational technique for viewing dim objects. Look at a point about 5° away from the dim object you are trying to see (about the length of your outstretched thumb). By doing so you gain sensitivity to faint objects. This technique, called **averted vision,** places the focused image of the object you are trying to see on a part of the iris with a high density of rods. This is not easy to do at first, but with practice you will be able to see some extremely faint objects that are "invisible" when you look straight at them, even though you know of their existence from star charts.

31.3   SMALL TELESCOPES

Although your eye can see remarkably faint stars and even a few galaxies, a small telescope will greatly increase the number of objects you can observe and will offer you far better and more interesting views of the Moon, planets, and even more remote objects. Such telescopes come in a wide range of styles and prices, but selecting the best one for your needs can be confusing.

Many amateur astronomers begin with about a 10-cm (4-inch) reflecting telescope. The size refers to the diameter of the mirror—the aperture size (Unit 27). Refractors are also an excellent choice, but they tend to be more expensive for the same size aperture. Diffraction (Unit 29) limits the resolution—the smallest discernible feature—of a 10-cm aperture to about 1 arc second. This is about the same as the smearing usually caused by atmospheric turbulence or "seeing" (Unit 30.2). Thus an aperture larger than that of a 10-cm telescope can offer little or no improvement in the detail it is possible to see unless you are observing from a mountaintop under excellent atmospheric conditions. The detail visible through a telescope of this size or larger is more likely to be limited by the quality of the optics, and the quality of the optics needs to be *better* to get as good an image from a larger telescope. So when choosing a telescope, it is a good idea to concentrate on the quality of the optics before spending more on a larger aperture. It is also important to read the manual and check that the optics are well aligned—procedures for adjusting telescope alignment are often explained in the telescope's manual. If your telescope is not performing up to your expectations, try to find an amateur astronomy club and bring your telescope along when you visit.

With a 10-cm telescope, you can easily see the moons of Jupiter, the rings of Saturn, and many lovely star clusters and galaxies. However, these latter objects will not look like the pictures in books because your eyes—unlike a photograph or electronic detector—cannot store up light as a camera can with a long exposure.

To directly observe dimmer objects, you will need a larger aperture. Here "the sky's the limit." Amateur telescopes can be enormous—even more than a meter in diameter. Also keep in mind that using your eyes well will have much more effect on what you can see than increasing the telescope's size. You should make sure you are getting enough Vitamin A, which is an essential part of the chemical process of the photopigments in your eye. Keep your eyes shielded from bright lights for at least half an hour, and learn to use averted vision even when looking through an eyepiece. Do not smoke; smoking renders the eye less responsive to changes in light level. With these steps you will be able to see things through the telescope that would otherwise be invisible to you.

Notice that we have said nothing about magnifying power. Some telescopes advertise their "magnifying power," but this is an unimportant number because any telescope can magnify by any amount with different eyepieces. You can determine the magnification of a telescope–eyepiece combination by dividing the **focal length** of the telescope by the focal length of the eyepiece:

$$\text{Magnification} = \frac{\text{Telescope focal length}}{\text{Eyepiece focal length}}.$$

The focal length of a refractor is usually the length of the telescope tube—the distance over which it focuses the light entering the aperture down to a point. With the multiple bounces and shaped mirrors in many reflectors, the focal length may not be obvious. However, you can calculate it by multiplying the telescope's aperture size by its *f-ratio*:

$$\text{Telescope focal length} = (\text{Aperture diameter}) \times (\text{f-ratio}).$$

For example, a 10 cm f/8 telescope has a focal length of 10 cm $\times$ 8 = 80 cm.

Eyepieces almost universally have their focal length listed on them in mm. Thus for a telescope with an 80 cm focal length, an 8 mm eyepiece will give a magnification of

$$\text{Magnification} = \frac{80 \text{ cm}}{8 \text{ mm}} = \frac{800 \text{ mm}}{8 \text{ mm}} = 100.$$

Likewise, a 4 mm eyepiece on this telescope will give a magnification of 200, whereas a 16 mm eyepiece will give a magnification of 50.

Distortions caused by the atmosphere make a magnification of about 100 to 200 the useful limit. For the clearest images, you need to get the best quality optics you can afford. Often buying a better eyepiece than came with your telescope will improve the overall optics dramatically. If you visit an astronomy club, you may get the chance to try different members' eyepieces on your own telescope to see which kinds work well for you.

The way a telescope is supported is another important choice. An **alt–az mount** is the simplest and least expensive. It allows the telescope to swing up and down from the horizon to the zenith (the **altitude**) and pivot to point toward different compass directions (the **azimuth**). This lets you see all parts of the sky, but unless you are observing from the North or South Pole, it will not be easy to follow objects on the celestial sphere as the Earth rotates. Because of the Earth's rotation, objects will move out of your field of view within a minute or two, or even seconds under high magnification.

To compensate for the Earth's rotation, an **equatorial mount** has a pivot axis that can be positioned so that it points at the celestial pole. With an equatorial mount, you can swing the telescope from the polar axis until you match an object's **declination,** and then pivot the telescope to match its **right ascension** (Unit 5.5). By rotating the telescope about the pivot axis, you can exactly counteract the effects of the Earth's rotation and keep the telescope pointed at the same celestial coordinate. Telescopes on equatorial mounts are more difficult to balance and generally have to be much heavier than alt–az mounts. However, they allow you to follow objects smoothly because they are oriented to rotate parallel to the Earth's axis—and with a motor drive, they can remain pointing at the same object for long periods.

If you set up the telescope correctly, exactly aligning it (no easy task), you can find faint astronomical objects by their right ascension and declination. Generally, though, it is easier to locate an object by pointing to the approximate position relative to bright stars as determined from a sky chart or planetarium software program. Then progressively close in on the position of the object by "star hopping" to fainter stars using the small "finder scope" mounted on the side of the telescope. Some newer telescopes offer computer-guided assistance to point the telescope, and even some alt–az mounts are able to follow celestial objects by using a computer-guided drive to adjust both the altitude and azimuth after some initial setup to establish the direction to the celestial pole.

Whatever type of telescope you choose, be sure to get a sturdy mounting for it. At a magnification of a factor of 100, tiny vibrations of the telescope caused by wind or the touch of your hand will make the image jiggle, hopelessly blurring it.

Before you actually buy a telescope, you might want to talk with your instructor or a local amateur astronomer. Such people often belong to an astronomy club, some of whose members may have secondhand telescopes they are willing to sell at reduced prices. For information about new telescopes, browse through magazines like *Astronomy* or *Sky & Telescope*—publications widely read by both amateur and professional astronomers—which contain many advertisements for small telescopes.

31.4   ASTRONOMICAL PHOTOGRAPHY

Q In the star-trail picture, how long was the exposure? In what direction was the observer looking? Were the stars rising or setting?

You can take surprisingly fine photographs, like the picture shown in Figure 31.3, with some digital cameras or many older 35-millimeter cameras. For astronomical photography it is best to have as much manual control of the camera as possible. Many modern cameras have automatic features that require much higher light levels, and you may be frustrated when you try to use them. You want to set the focus for infinity and open the diaphragm (f stop) all the way. Start with an exposure of 10 or so seconds, and then experiment with different exposure times. If you are using film, stick to one of the high-speed color films (ASA 400 or higher). For many 35-millimeter cameras, you can make long exposures by putting the camera on the B (for "flashbulb") setting and holding the shutter release down for the desired time. Better still, mount the camera on a tripod, then use a cable release to make the exposure without shaking the camera. Some digital cameras have settings for long exposure times or will sense the low light levels and stay open automatically for a longer time.

If you expose for more than about 15 seconds, the Earth's rotation will smear the star image into a streak. Deliberately allowing the smearing to occur can produce dramatic pictures of what are called "star trails" (see Figure 31.4). To make star-trail pictures, leave the shutter open for 20 minutes or so. Longer exposures are possible, but if the Moon is out or if there is much stray light or light pollution (Unit 30), the image may become foggy.

To take untrailed long exposures or to use a telephoto lens, you will need a way to compensate for the Earth's rotation. Many telescopes allow you to attach a camera to the body of the telescope. If the telescope has an equatorial mount, then you

FIGURE 31.3
Sunset view of four planets strung along the zodiac on March 1, 1999. Their straight-line arrangement results from the flatness of the Solar System. From top to bottom, you can see Saturn, Venus, Jupiter, and Mercury (nearly lost in the twilight).

Q If you did not have a telescope to compensate for the Earth's rotation, you might be able to attach your camera to a board, itself attached by hinges to a base. How would you need to orient the hinge to compensate for the Earth's rotation? What could you build to slowly tilt the board at the same rate the sky is "turning"?

FIGURE 31.4
A picture made with a fixed camera showing star trails. The constellation Orion is rising, along with the planet Saturn (the bright trail to the left).

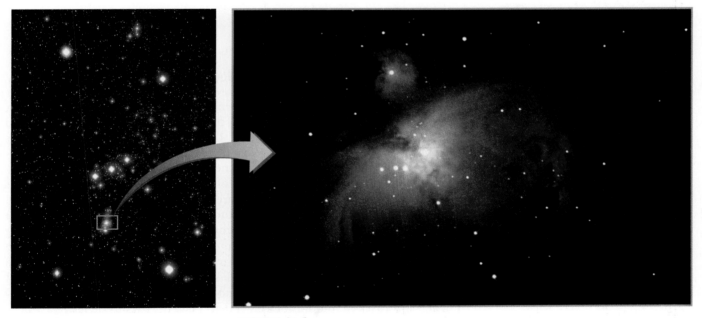

FIGURE 31.5
(A) Photograph of the constellation Orion made with an ordinary camera. (B) Picture of the Orion Nebula taken with a small backyard telescope.

can point the camera in any direction (it need not be the same direction the telescope is looking) while you follow a star with the telescope. This will keep the telescope (and everything attached to it) at the same orientation with respect to the celestial sphere. Figure 31.5A shows a picture of the constellation Orion made in this way. You can also attach a camera to a telescope with an alt–az mount, but here you need to point the camera close to the same direction that the telescope is pointing, and even so there will be some twisting of the image in long exposures because of the angle of the pivot to the celestial pole.

Photography is easier with a motor drive on the telescope, but it is important to keep in mind that the telescope must remain well balanced or you may strain the motor. The weight of the camera and any lenses should be counterbalanced by weights elsewhere on the telescope so there is no net force pulling the telescope to turn in one direction or another. The highest-magnification pictures can, of course, be taken with special adaptors that allow your camera to look through the telescope (Figure 31.5B). This is a challenge for the serious amateur astronomer. Be sure to try the simpler photographic techniques first—you will be amazed at some of the faint star clusters and nebulae that you can image. Many objects in the night sky are large enough that we could easily see them but for the limited sensitivity of our eyes.

As you continue your explorations of the night sky, you should look for local astronomy clubs and visit planetariums. There are also organizations like the American Association of Variable Star Observers (AAVSO), which will help you become involved in international efforts to monitor stars that have exhibited unusual behavior in the past. Through such monitoring projects, amateur astronomers make important contributions to the field of astronomy by alerting the astronomical community to unexpected events, such as outbursts from stars or the appearance of new comets.

Most of all, though, remember that amateur astronomy is about enjoying and appreciating the night sky.

31.5   SURFACE BRIGHTNESS

Always look well away from the Sun. It can damage your eyes permanently. It should never be looked at directly except with specialized equipment designed specifically for the purpose.

Look at a star with your naked eyes. Next, look at the same star through binoculars; you will notice it appears brighter. If, however, during the day you look at birds or trees through the same binoculars, they will appear larger, but not brighter. Now, with one eye, try looking at the blue sky through binoculars or a telescope, and look at the sky with the other eye unaided. The sky will appear the same or perhaps even brighter with the unaided eye. What these exercises demonstrate is that a telescope does not increase the surface brightness of an object.

We can express the **surface brightness** of the sky, a planet, or a distant galaxy as the number of photons received per square degree over the surface of the object. With a telescope you will collect more photons from each square degree than with your naked-eye observation; but the telescope also magnifies the image, spreading its photons out by a compensating amount. Hence, if we use a telescope with an aperture big enough to collect 100 times more photons than your naked eye can collect, the magnification of the image will spread the image's light over an area at least 100 times larger. As a result, the amount of light reaching one square millimeter at the back of your eye, or a CCD detector at the back of a telescope, will be no greater however big a telescope is used. This is why the sky does not appear any brighter through a telescope.

But why do stars appear brighter when magnified? Stars are so far away that they look like points of light to your eye. Even when they are magnified by a telescope to cover an area 100,000 times larger, so we may now be collecting 100,000 times more photons from the star, the magnified image of the star still remains imperceptibly small. The image size is smaller than the pixel size of a CCD or of a cell in your eye, either of which is then further limited by telescope diffraction (Unit 29) and by the blurring of the Earth's atmosphere (Unit 30). As a result, the increased numbers of photons appear still to be coming from one point, making the star look brighter, but not apparently bigger.

When you look through a telescope at a dim astronomical object, like a nebula or galaxy, it almost never matches its appearance in photographs. Photographs exhibit low surface brightness details you could never see with your eyes. If magnification cannot make an object appear brighter, how can photographs look so bright? The answer is that these photographs collect more photons in each square degree by making long exposures. Unlike our eyes, a CCD or photographic film can accumulate the photons over a long period. Even if very few photons arrive each second from a low surface brightness region, by training our CCD or open camera shutter on the region for hours, we will gather many photons. At the end of our collection period, their cumulative numbers will cause the area to appear brighter.

KEY TERMS

alt–az mount, 225

altitude, 225

averted vision, 223

azimuth, 225

cone, 222

cornea, 222

dark adaptation, 223

declination, 225

equatorial mount, 225

focal length, 224

fovea, 222

iris, 221

photopigment, 222

pupil, 221

retina, 222

right ascension, 225

rod, 222

surface brightness, 228

variable star, 223

QUESTIONS FOR REVIEW

1. What is meant by *dark adaption?*
2. What is averted vision?
3. What is focal length?

PROBLEMS

1. Name the most comparable term or piece of equipment from telescopes that corresponds to each of the following parts of the eye: iris, cornea, pupil, lens, retina, rods, fovea, brain. Explain your choices.
2. You are using an 8-inch f/10 Cassegrain telescope. What focal length eyepiece (in millimeters) will give you a magnification of 100?

TEST YOURSELF

1. In very dim light,
   a. your pupils are smallest.
   b. your rods and cones grow much more sensitive to light.
   c. your color vision is at its most sensitive.
   d. you can see things more clearly if you stare straight at them.
   e. All of the above
2. As a star rises and moves across the sky, which of the following change(s)?
   a. Its right ascension
   b. Its declination
   c. Its azimuth
   d. Both (a) and (b)
   e. None of the above
3. What is the primary advantage of a 40-cm reflector over a 20-cm refractor?
   a. The reflector will collect about 4 times more light.
   b. The reflector will let you see astronomical features that are twice as small.
   c. Reflectors transmit more light because they do not use any lenses.
   d. The reflector will provide twice the magnification.

Part Three

The content of some Units will be enhanced if you have previously studied some earlier Units in this textbook. These Background Pathways are listed below each Unit image.

Unit 38 Mercury

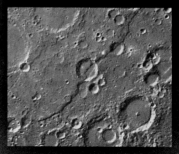

Background Pathways: Unit 37

Unit 32 The Structure of the Solar System

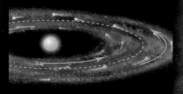

Unit 35 The Earth as a Terrestrial Planet

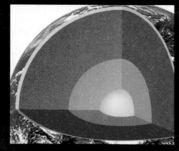

Background Pathways: Unit 32

Unit 39 Venus

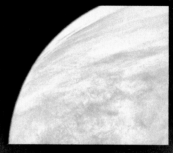

Background Pathways: Unit 36

Unit 33 The Origin of the Solar System

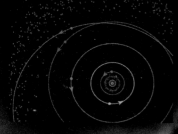

Background Pathways: Unit 32

Unit 36 Earth's Atmosphere and Hydrosphere

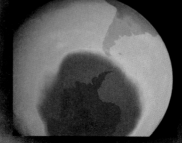

Background Pathways: Unit 35

Unit 40 Mars

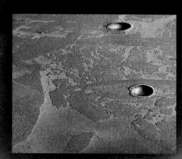

Background Pathways: Unit 36

Unit 34 Other Planetary Systems

Unit 37 Our Moon

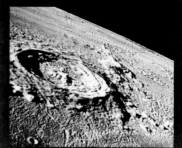

Background Pathways: Unit 35

Unit 41 Asteroids

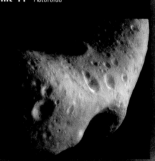

Stars and Stellar Evolution

Unit 59 Stellar Evolution

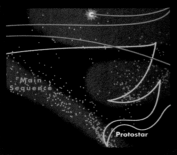

Background Pathways: Unit 58

Unit 60 Star Formation

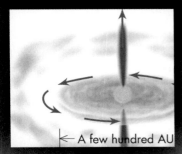

A few hundred AU

Background Pathways: Unit 59

Unit 61 Main-Sequence Stars

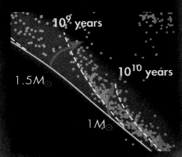

10^9 years

10^{10} years

$1.5 M_\odot$

$1 M_\odot$

Background Pathways: Unit 59

Unit 62 Giant Stars

Background Pathways: Unit 59

Unit 63 Variable Stars

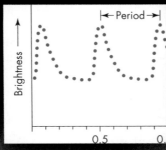

Period

Brightness

0.5   0.1

Background Pathways: Unit 57

Unit 64 Mass Loss and Death of Low-Mass Stars

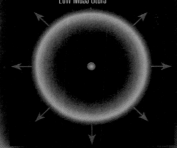

Background Pathways: Unit 62

Unit 65 Exploding White Dwarfs

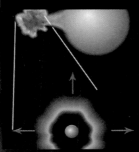

Background Pathways: Unit 64

Unit 66 Old Age and Death of Massive Stars

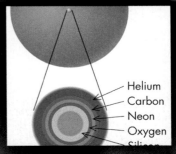

Helium
Carbon
Neon
Oxygen
Silicon

Background Pathways: Unit 59

Unit 67 Neutron Stars

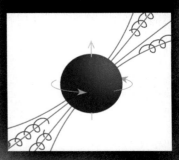

Background Pathways: Unit 66

Unit 68 Black Holes

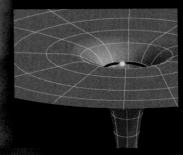

Background Pathways: Unit 67

Unit 69 Star Clusters

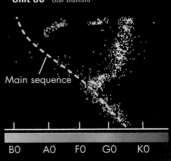

Main sequence

B0   A0   F0   G0   K0

Background Pathways: Unit 59

49

The Sun, Our Star

The Sun is a star—a dazzling, luminous ball of ionized gas. It is so large that a million Earths would fit inside its volume, and it is about 300,000 times more massive than the Earth. It is so bright that it will damage our eyes if we observe it directly. While the Sun exhibits some features in visible light (Figure 49.1), observations at other wavelengths reveal that the Sun's surface is violently agitated, with enormous storms blasting out fountains of incandescent gas (Figure 49.2). A single solar storm may release as much energy in a few minutes as the combined energy of all the earthquakes on Earth over the last 10 million years, and such storms are not rare. Yet despite their vast energy, these storms are dwarfed by the Sun's total flow of energy. Every second the Sun radiates a hundred times more energy than a single one of its solar storms.

Stars other than the Sun also generate power in enormous quantities, lighting up the universe despite their immense distance from one another. For centuries, the source of the Sun's and other stars' power was one of astronomy's greatest mysteries. Solving this mystery requires us to understand almost incomprehensible temperatures and pressures deep in a star's interior.

In its core the Sun releases the energy of 100 billion atomic bombs every second; yet it does not blow itself apart because although a star generates enormous power, this is counterbalanced by its enormous gravity. Stars pit titanic forces against one another. If these forces somehow became unbalanced, they might blast themselves apart or collapse in on themselves. Yet the Sun has remained stable for billions of years. Understanding this balance is one of the main themes of Part Four of this book.

In this unit we explore the structure of our star and the methods astronomers use to study parts of the Sun that cannot be seen directly. The subsequent two units examine the source of energy in the Sun and the activity that occurs on its surface. In later units we will discover that stars cannot maintain their precarious balance between gravity and pressure forever; and when the balance fails, cataclysmic events result. Along the way, we will learn of the likely fate of our own star.

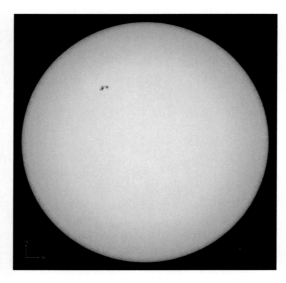

FIGURE 49.1
Visible light image of the Sun.

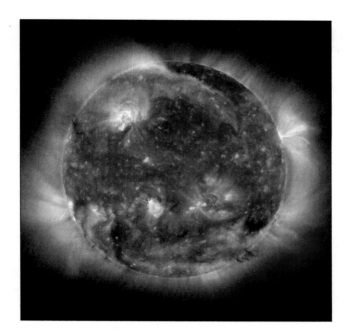

FIGURE 49.2
The complex stormy surface of the Sun is revealed in this false-color ultraviolet image made by the SOHO satellite at the same time as the visible light image in Figure 49.1. Blue colors in this image indicate the highest-energy photons, whereas red colors indicate the lower-energy photons.

49.1   SIZE AND STRUCTURE

Pressure is measured by the force exerted by a gas over an area. This can be expressed, for example, as newtons per square meter or pounds per square inch (see Unit 3).

The Sun dwarfs the Earth and even Jupiter, and its immense mass is what drives it to shine. The gravitational force (Unit 16) of this much concentrated matter crushes the gas in its interior to extremely high temperatures and densities. To offset that crushing force and prevent its own collapse, the Sun must generate a very high pressure within the gas. **Pressure** is the force produced by the particles in a gas as they collide and bounce off each other (or the walls of a container). A balloon is kept inflated by trillions upon trillions of molecules each bouncing off its inside surface, each giving a tiny "punch" outward.

A high pressure can produced if the particles are moving very fast. Thus, if the Sun can maintain a very high temperature, it can resist the gravitational pull inward. But hot objects always lose energy, and the Sun is no exception. We experience the Sun's lost energy as sunshine, and we welcome this as the source of our life. However, sunshine is a death warrant for the Sun because the energy it carries off must be replenished, or the Sun will collapse. The Sun must replace its lost energy, but only at the cost of slowly consuming itself—a necessity that is shared by all stars.

Before we discuss how the Sun replaces its lost energy, we need to understand better some of its overall properties. What forces are at work in its interior? How rapidly does it lose energy? What resources does it have available to supply its energy needs? Many of the Sun's properties, such as its internal temperature, cannot be observed directly and therefore must be deduced. Astronomers create computer models that use the laws of physics to extend our understanding. For example, depending on what temperature and composition are assumed for the Sun's core, the computer models predict slightly different results for the radius and surface temperature of the Sun.

Current models of the internal structure of the Sun are strongly constrained by both observation and theory, giving astronomers high confidence in the accuracy of the models. Table 49.1 shows how astronomers measure some of the Sun's physical properties. Figure 49.3 shows the Sun's portrait, illustrating its main features. It will be helpful many times in this unit to refer to this picture.

TABLE 49.1	Properties of the Sun	
Property	**Value**	**Method of Determination**
Distance	1 AU = 150 million kilometers or 93 million miles	Triangulation or radar (Unit 52)
Radius	7×10^5 km	From angular size and distance (Unit 10)
Mass	2×10^{30} kg	Modified form of Kepler's third law (Unit 17)
Average density	1.4 kg per liter	From radius and mass
Central density	160 kg per liter	Indirectly from need to balance internal pressure and gravity (Unit 49)
Surface temperature	About 6000 Kelvin	Color–temperature relation (Wien's law—Unit 23)
Central temperature	15 million Kelvin	Indirectly from need to balance internal pressure and gravity (Unit 49)
Composition	71% hydrogen, 27% helium, and 2% vaporized heavier elements such as carbon and iron	Spectra of gases in surface layers (Unit 55)
Luminosity	4×10^{26} watts	From amount of energy reaching Earth and inverse square law (Unit 54)

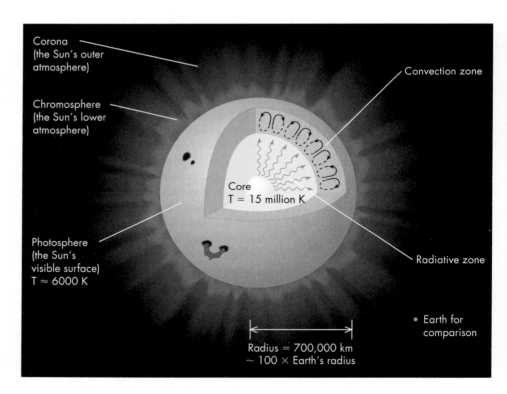

FIGURE 49.3
A cutaway sketch of the Sun.

49.2   THE SOLAR INTERIOR

When we look at the Sun, we see through the low-density, tenuous gases of its outer atmosphere. Our vision is ultimately blocked, however, as we peer deeper into the Sun because the material there is compressed to high densities by the weight of the gas above it. In this dense material, the atoms are sufficiently close

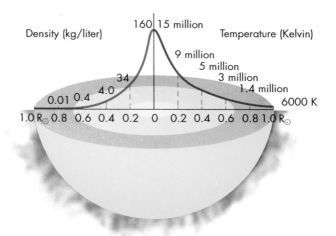

FIGURE 49.4

Plots of how density and temperature change through the Sun.

together that they strongly absorb the light from deeper layers, blocking our view of them much as a fog obscures our view of what lies beyond a certain distance.

The Sun's "surface" is about 700,000 km (430,000 miles) from its center. This is not a surface in the usual sense of the word. The glowing gas there is thin—only about one thousandth the density of the air we breathe. However, this region is where light can escape into space as the sunlight we see. This layer, where the Sun's gas changes from opaque to transparent—where the photons we see come from—is called the **photosphere.** The photosphere looks like a surface, but it is actually a region about 500 kilometers (300 miles) thick, with gas that grows denser below and thinner above.

From the upper part of the photosphere to its lowest part, the gas density increases by about a factor of 10. If we could see farther into the Sun, we would find that its density rises steadily toward its center, compressing the gas, much as a pile of laundry flattens clothes on the bottom most, while those on top remain puffed up. A similar compression occurs in the atmosphere of the Earth and other planets, but the greater mass of the Sun leads to a vastly greater compression of its gas. While the Sun has an average density of about 1.4 kilograms per liter—close to that of Jupiter—near its core, the Sun's density reaches 160 kilograms per liter—about 20 times the density of steel! Despite this great density, *the Sun is gaseous throughout* because its high temperature gives the atoms so much energy of motion that they are unable to bond with one another to form a liquid or solid substance. In fact the temperatures are so high that the atoms are ionized, and therefore the matter is properly termed a **plasma.**

Not only does the density of the Sun rise as we plunge deeper into its interior, the temperature rises too. The average temperature of the photosphere is about 5780 K; but at its top, the photosphere's temperature is about 4500 K, while near the bottom, the temperature is about 7500 K. Below the photosphere heat is retained more effectively by the denser gas, and the temperature climbs until at its core the temperature soars to about 15 million K. Figure 49.4, based on theoretical calculations, illustrates how the temperature and density change through the Sun. No probe has made (or is ever likely to make) the measurements directly, but we are confident that they are correct—because the Sun needs such high temperatures and densities to keep it from collapsing under its own gravity.

49.3   ENERGY TRANSPORT

Your own experience and experiments in the laboratory show that heat always flows from hot to cold. Applying this principle to the Sun, we can therefore infer that because its core is hotter than its surface, heat will flow outward from its center,

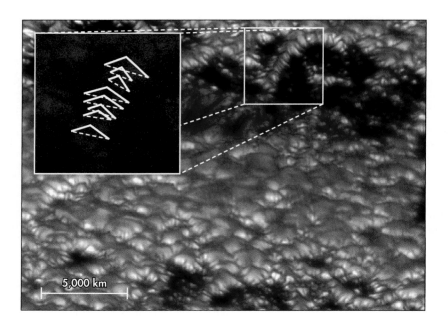

FIGURE 49.5
Image of a portion of the Sun's surface showing the granulation on the surface. This image was made using a telescope with adaptive optics to capture high-resolution details. The inset black-and-white image shows a region where analysis shows the height of the granulation features. The triangles represent the heights of these features, which are from 200 to 450 km tall.

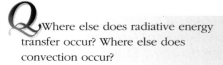

Where else does radiative energy transfer occur? Where else does convection occur?

as illustrated in Figure 49.3. Near the core, the energy is carried by photons through what is called the **radiative zone.** Because the gas there is so dense, a photon travels only about a centimeter before it is absorbed by an atom and stopped. The photon is quickly re-emitted, but it leaves the atom in a random direction—perhaps even back toward the Sun's center—and then is almost immediately reabsorbed by another atom. This constant absorption and reemission slows the rate at which photons escape from the Sun, like people lost in a dense forest walking randomly between trees until eventually they get close enough to the edge of the forest to see out. Even though photons travel at the speed of light between absorptions, it takes them about 100,000 years, on average, to move from the core to the surface. Today's sunshine was born in the Sun's core before the birth of human civilization!

The movement of photons outward from the Sun's core is slowed in a region that begins about two-thirds of the distance out from the Sun's center. In this region the gas is cooler and the atoms are less completely ionized than in the radiative zone; as a result the atoms are even more effective at blocking photons. The energy from the Sun's interior is therefore trapped, and it heats the gas there. Hotter regions in the gas expand, become buoyant, and rise like a hot air balloon. The gas rises and cools, then sinks back down in a region called the **convection zone** (Figure 49.3).

We can infer the gas's motion in the convection zone by observing the top layer of it—the photosphere. There we see numerous bright regions surrounded by narrow darker zones called **granulation,** as shown in Figure 49.5. The bright areas are bubbles of hot gas, up to about 1000 km (600 miles) across, which rise up through the convection zone. Upon reaching the surface, these hot bubbles radiate their heat to space, causing them to cool. That cooler matter then sinks back toward the hotter interior, where it is reheated and rises again to radiate away more heat. These bubbles of hot gas generally last only 5 to 10 minutes before they sink back into the roiling depths below the visible photosphere.

49.4   THE SOLAR ATMOSPHERE

Astronomers refer to the lower-density gases that lie above the photosphere as the Sun's *atmosphere.* This region marks a gradual change from the relatively dense gas of the photosphere to the extremely low-density gas of interplanetary space.

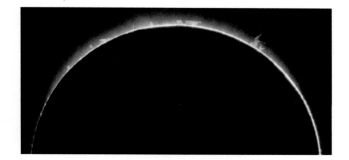

FIGURE 49.6
Photograph of a portion of the solar chromosphere during a total solar eclipse.

A similar transition occurs in our own atmosphere, where gas density decreases steadily with altitude, and eventually merges with the near-vacuum of space.

The Sun's atmosphere consists of two main regions. Immediately above the photosphere lies the **chromosphere,** the Sun's lower atmosphere. It is usually invisible against the glare of the photosphere, but can be seen during a total eclipse of the Sun as a thin red zone around the Sun (Figure 49.6) that is about 2000 kilometers (1200 miles) thick.

The chromosphere's red color is the source of its name, which literally means "colored sphere." The color comes from the strong red emission line of hydrogen, H-alpha (Unit 24). Telescopic views reveal that the chromosphere contains hundreds of thousands of thin columns or spikes called **spicules** (Figure 49.7). Each spicule is a jet of hot gas that grows during several minutes to be thousands of kilometers tall before it cools and sinks back to the surface.

Astronomers can determine a great deal about a gas by studying the emission lines it produces (Unit 24), and in the chromosphere the emission lines reveal a surprising reversal in the gas temperature. By the top of the photosphere, the temperature has declined to about 4500 K, but in the chromosphere the temperature begins climbing again. At the top of the chromosphere, only about 2000 kilometers above the photosphere, the temperature reaches 50,000 K. Above this, in a thin transition region at the top of the chromosphere, the density drops rapidly and the temperature shoots up to about 1 million degrees as we enter the **corona,** the Sun's outer atmosphere.

The corona's extremely hot gas has such low density that under most conditions we look right through it. But like the chromosphere, we can see it during a total solar eclipse when the Moon covers the Sun's brilliant disk. Then the pale glow of the corona can extend far beyond the Sun's edge to distances several times larger than the Sun's radius (Figure 49.8). Pictures of the Sun made at X-ray wavelengths show that the corona is not uniform; it has streamers that follow the direction of the Sun's magnetic field. Charged particles, such as those in the ionized gas of the Sun, experience a force as they move through a magnetic field that tends to make them follow the direction of the field. The importance of the magnetic field for surface features on the Sun will become even clearer when we examine activity on the Sun's surface (Unit 51).

Studies of the ionized gas in the chromosphere and corona show that it is stirred by the magnetic fields in complex ways (Figure 49.9), and this may cause the heating of these regions (Unit 51). Because the corona is so tenuous, it contains little energy despite its high temperature. It is like the sparks from a Fourth of July sparkler: Despite their high temperature, you hardly feel them if they land on your hand because they are so tiny and carry little total heat.

The corona contains large cooler regions where the magnetic field is weak, called **coronal holes.** The corona's high temperature generates pressures high enough to drive gases away from the Sun in the coronal holes because the magnetic field does not trap the gas. The flow of gas away from the Sun is known as the

Q What are the speeds of winds in the most severe storms on Earth? How do these compare with the speeds of the motion in spicules?

FIGURE 49.7
Photograph of spicules in the chromosphere made at the wavelength of hydrogen's H-alpha spectral line. The spicules are the thin, stringy features that look like tufts of grass.

FIGURE 49.8
Photograph of the corona during a total eclipse of the Sun in 1988.

FIGURE 49.9
Magnetic loops in the Sun's lower atmosphere. Gas trapped along the loops is heated by magnetic activity and in turn heats the Sun's corona. This image was made by the SOHO satellite at ultraviolet wavelengths.

solar wind, which results in a gradual loss of mass from the Sun. The expanding gas has a very low density (only a few hundred thousand atoms in a liter, compared to about 10^{22} molecules in a liter of the air we breathe). The amount of material lost from the Sun is "small": about 1.5 million tons each second. This number sounds large, but even after 10 billion years this would amount to only about 0.02% of the Sun's mass.

Near the Sun, the pressure within the corona may be only slightly larger than the Sun's gravitational attraction. The solar wind therefore begins its outward motion slowly. The gravitational pull of the Sun weakens with distance, but the pressure in the corona remains relatively high, so the solar wind gradually accelerates. At the distance of the Earth's orbit, the solar wind speed reaches about 500 kilometers per second (300 miles per second). The flow of the solar wind pushes comets' tails away from the Sun (Unit 47) and interacts with the Earth's magnetic field and ionosphere to create the aurora (Unit 36). Beyond the Earth, the wind coasts at a relatively steady speed that carries it beyond the orbit of Neptune. At some point it impinges on the interstellar gas surrounding the Solar System, where the gases collide and heat. In 2003 the *Voyager I* spacecraft, then nearly three times farther from the Sun than Neptune, detected indications of a changing environment, but scientists do not yet agree on whether the spacecraft had begun to enter the interstellar medium.

49.5 PRESSURE BALANCE

The outward pressure force balances the inward gravitational force everywhere inside the Sun.

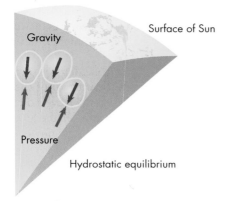

Center of Sun

FIGURE 49.10

A sketch illustrating the condition of hydrostatic equilibrium, the balance of pressure (blue arrows) and gravitational force (purple arrows) in the Sun.

The internal structure of the Sun depends on a balance between two forces. The force of gravity pulls every part of the Sun inward. The high-temperature gas inside the Sun produces pressure that pushes the matter outward. These forces must be in balance, or else the Sun will contract or expand. This balance between pressure and gravity is called **hydrostatic equilibrium.**

In the Sun, as in virtually all stars and planets, the balance of hydrostatic equilibrium requires that at all points the outward force created by pressure exactly balance the inward force of the Sun's gravity (Figure 49.10). Without such a balance, the Sun would change rapidly. For example, if its pressure were too weak, the Sun's own gravity would quickly crush it. Therefore, to understand the Sun, we need to discuss in more detail how its pressure arises.

Pressure in a gas comes from collisions among its atoms and molecules. If the gas is squeezed, atoms are pushed toward each other. As they collide, they rebound, resisting the compression. You experience this if you squeeze a balloon—the pressure you exert meets increasing resistance, or perhaps a section of the balloon you are not pressing may stretch outward in response to the increased pressure. The strength of the pressure depends on how often and how hard the collisions occur. Raising the density of a gas increases the rate of collisions by moving atoms closer together. Raising the temperature speeds atoms up, making them collide harder and more often (Figure 49.11). Thus, the

FIGURE 49.11

Sketch illustrating the ideal gas law. Gas atoms move faster at the higher temperature, so they collide both more forcefully and more often than atoms in a cooler gas. Therefore, other things being equal, a hotter gas exerts a greater pressure.

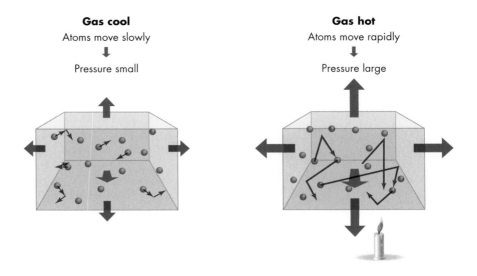

In chemistry, the ideal gas law is often written $PV = NkT$, where P stands for pressure, V stands for volume, N stands for the number of atoms, k is a constant, and T is the temperature. Because (N/V) is a measure of the density, the equation can be rearranged as we have shown it here.

How do you suppose the interior conditions in a very massive star would differ from the conditions in the Sun?

strength of the pressure is proportional to the density times the temperature of the gas:

$$\text{Pressure} = \text{Constant} \times \text{Temperature} \times \text{Density}.$$

This relationship is known as the **ideal gas law.** This law describes the behavior of gases: In response to a change of any of the three quantities involved, the other quantities must change so that the equation remains true. We will not need to carry out any specific calculations with this formula, but its principles are important. If pressure is increased, the density and/or the temperature must increase. Or if external conditions hold the pressure steady, and, for example, the temperature doubles, the gas must expand to half its former density. This is what happens when the air in a hot-air balloon is heated—as the air grows hotter, it expands (pushing some out of the balloon), and the remaining air is less dense, making the balloon buoyant.

Although the ideal gas law shows us that the pressure inside the Sun depends on temperature, it does not by itself show us how hot the temperature or how high the density needs to be. Because the Sun has a huge mass (and therefore a huge gravity crushing it), we know that it needs a huge temperature and density to offset that crushing force. The values noted earlier, of 15 million K for the temperature and about 160 kilograms per liter for the density, provide the necessary pressure. Based on the ideal gas law alone, though, the same pressure could be supplied if the temperature were 10 times lower and the density 10 times higher. However, if we assumed the Sun were 10 times more dense, we would derive a much higher mass than is measured for the Sun.

The computer models not only use the ideal gas law, but also consider how the Sun's internal gravity changes based on the density structure in the interior of the star, and how the temperature structure responds to the way the energy travels out from the center. The high temperature found is a necessary condition for the Sun to be stable. But maintaining that high temperature requires an extremely powerful energy source because energy is constantly being lost from the Sun as it radiates energy out into space.

49.6   SOLAR SEISMOLOGY

Just as scientists can study the Earth's interior by analyzing earthquake waves, or waves in a stream reveal the presence of submerged rocks, astronomers can learn about the Sun's interior by analyzing the waves that travel through the Sun. By analogy with the study of such waves in the Earth, astronomers call the study of such waves in the Sun **solar seismology.**

The study of waves in the Sun began in the 1960s, when astronomers noted that the photosphere of the Sun vibrates, oscillating up and down by several meters over periods of several minutes. These oscillations are similar to earthquake waves on the Earth (Unit 35). However, instead of arising from the shifting of rocky material in the Earth, waves in the Sun arise as huge convecting masses of material rise and fall in its outer regions. These motions in turn generate waves that travel through the body of the Sun, causing other portions of the Sun to shake in response.

The rising and falling surface gas makes a regular pattern (Figure 49.12), which can be detected as a Doppler shift (Unit 25) of the moving material. Astronomers next use computer models of the Sun to predict how the observed surface oscillations are affected by conditions in the Sun's deep interior. With this technique, astronomers can measure the density and temperature deep within the Sun. The results provide an independent confirmation of our theoretical models of the Sun's interior.

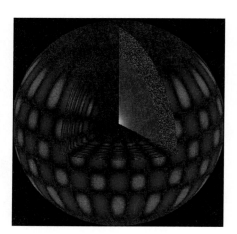

FIGURE 49.12
Computer diagram of solar surface waves.

KEY TERMS

QUESTIONS FOR REVIEW

1. How big is the Sun compared to the Earth?
2. What are the photosphere, chromosphere, and corona?
3. How do densities and temperatures vary throughout the Sun?
4. What visible evidence do we have that the Sun has a convection zone?

PROBLEMS

1. The Sun's angular size is ½ degree, and that its distance is 1.5×10^8 kilometers. Use this information to calculate the Sun's diameter.
2. How much mass would the Sun have to lose each second in order to lose 10% of its mass in 10 billion years?
3. At the rate the Sun is losing mass in the solar wind, how many years does it take the Sun to lose as much mass as the mass of the entire Earth?

4. Suppose you were an astronomy student on Jupiter. Use the orbital data for Jupiter (distance from Sun = 5.2 AU; period = 11.8 years) to measure the Sun's mass using the modified form of Kepler's third law.

TEST YOURSELF

1. The diameter of the Sun is about how large compared with the Earth's?
   a. Twice as big
   b. Half as big
   c. 10 times as big
   d. 100 times as big
   e. 10,000 times as big

2. According to the ideal gas law, if the temperature of a gas is made 4 times higher, which of the following is a possible result?
   a. Its pressure increases by 4 times and its density remains the same.
   b. Its density increases by 4 times and its pressure remains the same.
   c. Its pressure and density both double.
   d. Its pressure increases by 4 times while its density decreases by 4 times.
   e. Its pressure and density both decrease by 2 times.

3. The solar wind
   a. is caused by the sun's gravity.
   b. is caused by the corona's pressure.
   c. causes comets' tails to swing toward the sun.
   d. will deplete the Sun in 10 billion years.
   e. (a), (c), and (d)
   f. None of the above

50

The Sun's Source of Power

Background Pathways: Unit 4

Energy that leaves the Sun's hot core eventually escapes into space as sunshine. That energy loss in the core must be replaced, or the Sun's internal pressure would drop and the Sun would begin to shrink under the force of its own gravity. The Sun is therefore like an inflatable chair with tiny leaks through which the air escapes. If you sit in the chair, it will gradually collapse under your weight unless you pump air in to replace that which is escaping. What acts as the energy pump for the Sun?

Although the immense bulk of the Sun hides its core from view, we have a number of clues to the possible sources of energy. From a combination of theoretical models and direct observations, we can deduce what the core's temperature and density must be (Unit 49). From the spectra of its atmospheric gases, we know the Sun's composition. From the amount of sunlight we receive, we know how much energy the core must be generating. Moreover, in recent years a new area of astronomy has been developed based on measuring the numbers of a subatomic particle called a **neutrino** (Unit 4) that the Sun emits. Observations of such neutrinos allow us to "view" reactions occurring in the Sun's interior more directly than ever before.

These astronomical discoveries, combined with the discovery of nuclear energy in the 1900s, lead us to a picture of almost unbelievable violence in the core of the Sun. Every second over 4 million tons of matter are annihilated—not just turned into vapor or ionized or ejected into space, but turned into light—so that the Sun's mass gradually declines. Over 40 million years the Sun obliterates a mass equivalent to the entire Earth! Such mass loss may seem large, but the Sun has such a tremendously large mass that it can afford this extravagant destruction of its mass in its fight to keep from collapsing under its own weight.

50.1 THE MYSTERY BEHIND SUNSHINE

Early astronomers believed that the Sun might burn ordinary fuel such as coal. But even if the Sun were pure coal, it could burn only a few thousand years given its prodigious energy output. In the late 1800s another proposal was that the Sun is not in hydrostatic equilibrium, but that gravity slowly compresses it, making it shrink. In this hypothesis, compression heats the gas and makes the Sun shine, much as the giant planets generate heat in their interiors (Unit 43). However, gravity could power the Sun by this mechanism for only about 10 million years, and it would have to be shrinking by about 10 km each year, which modern observations rule out. Therefore, something else must supply its energy.

We know the Sun has been shining with approximately the same luminosity for billions of years because we have geological and fossil evidence of water on Earth dating back that far. For Earth to have maintained a climate with liquid water on

m = mass

E = energy

c = speed of light = 3×10^8 m/sec

Q How do we know what the Sun is made of?

An explosion of one kiloton (1000 metric tons) of TNT releases about 4.2×10^{12} joules of energy.

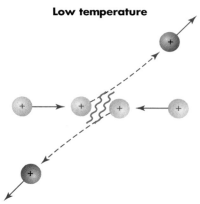

Low temperature

Nuclei move slowly and electric force repels them and pushes them apart. **No** fusion.

High temperature

Nuclei move faster and electrical force repelling them is overwhelmed. They collide and **fuse.**

FIGURE 50.1

How a high temperature acts to overcome the electrical repulsion of the nuclei and brings them close enough for the nuclear force to fuse them.

its surface, the Sun's power output must have remained fairly steady over that whole time. In 1899 T. C. Chamberlin suggested that subatomic energy—energy from the reactions of atomic nuclei—might power stars, but he could offer no explanation of how the energy was liberated. In 1905 Einstein proposed that energy might come from a body's mass. His famous formula,

$$E = m \times c^2$$

states that a mass, m, can become an amount of energy, E, equal to the mass multiplied by the square of the speed of light, c. It is important to understand that this formula simply states a basic "exchange rate" between energy and mass, similar to the way we might convert between dollars and euros. Einstein's formula indicates how many kg $\cdot$ m^2/sec^2 or joules of energy (Unit 3) correspond to one kilogram of mass. The term c^2 in the equation is the factor we use in calculating the exchange.

For example, if you could convert 1 gram (10^{-3} kg) of mass—about the amount of matter in a paperclip—into energy, it would release an energy of

$$E = 0.001 \text{ kg} \times (3 \times 10^8 \text{ m/sec})^2$$
$$= 9 \times 10^{13} \text{ kg} \cdot \text{m}^2/\text{sec}^2$$

or 9×10^{13} joules. This is the equivalent of approximately 20 kilotons of TNT, about as powerful as the atomic bomb that destroyed Hiroshima. You could grind the same paperclip into fine metal powder and burn it—a chemical process. This would release less than a ten-billionth as much energy—about as much as burning a match. With the understanding that mass may be converted into energy, scientists realized that if the Sun can convert even a tiny fraction of its mass into energy, it would have an enormous source of power. Einstein's equation does not say what process might release so much energy, but it provides a clue where to look: reactions that result in a significant change in mass.

In 1919 the English astrophysicist A. S. Eddington, a pioneer in the study of stars, developed a hypothesis for a reaction that might provide a significant source of energy: the conversion of hydrogen into helium. Hydrogen contains one nuclear particle, while helium contains four nuclear particles; however, the mass of helium is significantly less than four times the mass of hydrogen. Eddington recognized that if hydrogen nuclei could be combined to make helium nuclei, it would involve the loss of enough mass to provide the energy necessary to power the Sun. It was not until the late 1930s that the German physicists Hans A. Bethe and Carl F. von Weizsäcker worked out the details of how such a conversion might take place. They showed that the Sun generates its energy by converting hydrogen into helium through a process called **nuclear fusion,** a process that bonds two or more atomic nuclei into a single heavier one.

Under normal conditions, hydrogen nuclei repel one another, pushed apart by the electrical charge of their **protons.** Protons are positively charged nuclear particles, and like charges repel each other. This force of repulsion grows greater the closer together the protons get, so they must be moving toward each other at extremely high speeds for the nuclei to come into close contact, about 10^{-15} m apart. When this happens, the nuclear force of attraction—the **strong force** (Unit 4)—overcomes the electrical repulsion between the protons. The strong force can also bind protons together with another kind of nuclear particle, a **neutron,** which has no electrical charge but is otherwise similar to a proton. The strong force is about 100 times stronger than the electrical repulsion between neighboring protons, but it remains strong over only a very short distance, then dies away rapidly at larger distances.

Fusion is possible in the Sun because its interior is so hot. At very high temperatures—above about 5 million Kelvin—atomic nuclei move so fast that they collide at speeds that bring them close enough together for the strong force to take hold (Figure 50.1). The two separate nuclei are then able to collide and merge, or fuse, into a single new nucleus, and the nuclear energy binding the particles together is released in the form of a gamma-ray photon. This is a little like the potential energy released when a rock is dropped and gravity pulls it down to crash into the ground. However, the strong force is 10^{40} (ten thousand trillion, trillion, trillion) times stronger than the gravitational force, so the energy released is correspondingly immensely larger. But because this nuclear fusion process requires such a high temperature, the only place in the Sun hot enough for fusion to occur is its core.

50.2   THE CONVERSION OF HYDROGEN INTO HELIUM

Key

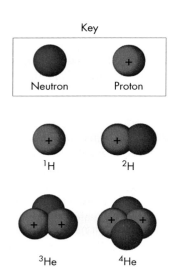

FIGURE 50.2
Schematic diagrams of the nuclei of hydrogen, its isotope deuterium, and two isotopes of helium.

Before we can fully understand how fusion creates energy, we need to look at the structure of hydrogen and helium as they are found in the Sun's core. (See also Unit 4.2.) The common form of hydrogen consists of one proton and an orbiting electron, and that of helium consists of two protons, two neutrons, and two orbiting electrons (Figure 50.2). In the Sun's hot interior, however, atoms collide so violently that the electrons are generally stripped off; that is, the atoms are completely ionized.

Hydrogen and helium always have one and two protons, respectively, but they have other forms, **isotopes,** with different numbers of neutrons. To identify the isotopes, we write their chemical symbol with a superscript that shows the total number of protons and neutrons. The usual form of hydrogen with one proton and no neutrons is written ^{1}H, whereas the form of hydrogen containing one proton and one neutron is ^{2}H. Most isotopes are not given separate names, but ^{2}H is so common that it has been given its own name: **deuterium.** This isotope of hydrogen is more common than all but about a half dozen elements in the universe.

The most common form of helium, with two protons and two neutrons, is written ^{4}He, whereas helium with two protons but only one neutron is ^{3}He. When speaking, we refer to these as "helium four" and "helium three," respectively. These isotopes of hydrogen and helium play a critical role in the energy supply of the Sun.

Hydrogen fusion in the Sun occurs in three steps called the **proton–proton chain.** In the first step, two ^{1}H nuclei (protons) collide and fuse to form the isotope of hydrogen, ^{2}H or deuterium. During the collision one proton turns into a neutron through the weak force (Unit 4), and two other particles are ejected: a **positron** (the "antiparticle" of the electron, denoted as e^{+}) and a neutrino (denoted by ν, the Greek letter "nu"). This step is depicted in Figure 50.3A and can be written symbolically like this:

$$^{1}\text{H} + {}^{1}\text{H} \rightarrow {}^{2}\text{H} + e^{+} + \nu + \text{Energy.}$$

The terms to the left of the arrow are the normal hydrogen nuclei that start the process. The terms to the right are the deuterium, positron, neutrino, and the potential energy of the nuclear bond that is released.

An indication that energy is released by this reaction is that the mass of ^{2}H (and the by-products of the reaction) is less than the mass of the two ^{1}H. The amount of energy released can be found from Einstein's formula $E = m \times c^{2}$. That energy gives the particles large kinetic energies that sustain the high temperature in the core and ultimately end up as thousands of visible-wavelength photons in the

Each positron produced in the first step of the proton–proton chain almost immediately annihilates one of the many electrons contained in the dense gas in the Sun's core. This creates two energetic photons (gamma rays). This side reaction contributes even more energy than the fusion of protons into deuterium.

ANIMATIONS

Proton–proton chain

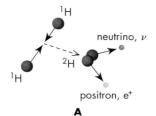

A

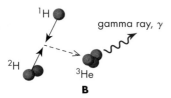

B

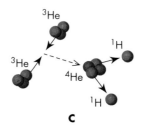

C

FIGURE 50.3
Diagram of the proton–proton chain.
(A) Hydrogen (H) nuclei first combine to
make deuterium (^{2}H). (B) Deuterium and
hydrogen combine to make ^{3}He. (C) ^{3}He
nuclei combine to form ^{4}He.

Q The term *solar energy* is also used
to refer to the generation of power on
Earth from sunlight. This is done
using collectors that convert the
energy of photons from the Sun into
electrical current. What other kinds of
power generation ultimately rely on
energy that came from the Sun?

sunlight we see. The neutrinos produced in this reaction also carry away a small part of the nuclear energy, but they play no further role in the Sun's energy generation. However, we will encounter them again: They can be detected when they leave the Sun, helping astronomers to learn more about conditions in the Sun's interior.

Once deuterium is created, the second step of the proton–proton chain proceeds very rapidly. The ^{2}H nucleus collides with one of the numerous hydrogen nuclei, ^{1}H, to make the isotope of helium containing a single neutron, ^{3}He. This process releases a high-energy gamma ray, denoted by the Greek letter gamma, γ. Figure 50.3B shows this step, which can be written as follows:

$$^1H + {}^2H \rightarrow {}^3He + \gamma + \text{Energy}.$$

Here again the resulting particle, ^{3}He, has a smaller mass than the particles from which it was made, and several times more energy is released than from the first reaction.

The third and final step in the proton–proton chain is the collision and fusion of two ^{3}He nuclei. Here the fusion results not in a single particle, but rather in one ^{4}He and two ^{1}H nuclei. You can think about this reaction as the attempt to form a nucleus with four protons and two neutrons, except that two protons are ejected by their electric repulsion, as shown in Figure 50.3C. This reaction, which releases about twice as much energy as the second reaction, is written as

$$^3He + {}^3He \rightarrow {}^4He + {}^1H + {}^1H + \text{Energy}$$

where again the final mass is less than the initial mass.

Scientists who want to understand the details of fusion in the Sun study each of the steps in the proton–proton chain individually; but for understanding the larger picture of the Sun's energy generation, we are concerned primarily with the overall energy production. We can find the quantity of energy released by comparing the initial and final masses of the reactions and using the mass–energy relationship $E = m \times c^2$. Steps 1 and 2 use three ^{1}H, but the first two steps must occur twice to make the two ^{3}He nuclei needed for the last step. Therefore, six ^{1}H nuclei are used, but two are returned in step 3, and so a total of four ^{1}H nuclei are needed to make each ^{4}He nucleus.

The mass of a hydrogen nucleus is 1.673×10^{-27} kilograms, whereas the mass of a helium nucleus is 6.645×10^{-27} kilograms. Comparing their masses, we find the following:

$$
\begin{aligned}
4 \text{ hydrogen} &= 6.693 \times 10^{-27} \text{ kg} \\
-1 \text{ helium} &= -6.645 \times 10^{-27} \text{ kg} \\
\hline
\text{Mass lost} &= 0.048 \times 10^{-27} \text{ kg}
\end{aligned}
$$

Multiplying the mass lost by c^2 gives the energy yield per helium atom made. That is,

$$
\begin{aligned}
E &= 0.048 \times 10^{-27} \text{ kg} \times (3 \times 10^8 \text{ m/sec})^2 \\
&= 0.048 \times 10^{-27} \times 9 \times 10^{16} \text{ kg} \cdot \text{m}^2/\text{sec}^2 \\
&= 4.3 \times 10^{-12} \text{ joules per helium nucleus created.}
\end{aligned}
$$

To power the Sun at its current rate, it needs to generate 4×10^{26} joules of energy per second. This means that each second, about 4×10^{38} hydrogen nuclei are converted into 10^{38} helium nuclei, and about 4×10^9 kg—*four million metric tons*—of mass are converted into energy each second. The total energy released is equivalent to exploding 100 billion-megaton H-bombs per second!

Our sunshine has a violent birth.

50.3   SOLAR NEUTRINOS

We saw in the previous section that the Sun makes neutrinos as it converts hydrogen into helium. Neutrinos are unusual subatomic particles. They have no electric charge and only a tiny mass, and they travel at nearly the speed of light. This gives them phenomenal penetrating power. They escape from the Sun's core through its outer 700,000 kilometers, and into space like bullets through a wet tissue. They pass straight through the Earth and anything on the Earth, such as *you*, and keep going. In fact, several trillion neutrinos from the Sun passed through your body in the time it took you to read this sentence.

The number of neutrinos coming from the Sun is a direct indication of how rapidly hydrogen is being converted into helium. This is a more immediate measure of the rate of current nuclear reactions than the light we see because photons take so long to work their way to space from the Sun's core as they interact with any intervening matter. Thus, if we can measure how many neutrinos come from the Sun, we can deduce the conditions in the Sun's core.

Counting neutrinos is extremely difficult. The elusiveness that allows neutrinos to slip so easily through the Sun makes them slip with equal ease through detectors on Earth. However, although they interact weakly, the rate of interaction is not zero. And because so many neutrinos are produced, we need detect only a tiny fraction of them to get useful information. Still, this requires very large detectors.

The detection of solar neutrinos was pioneered by the American physicist Raymond Davis Jr., who advanced the idea of placing tanks of material deep underground that could react with neutrinos from the Sun. Neutrino detectors are buried to shield them from the many other kinds of particles besides neutrinos that constantly bombard the Earth. For example, protons, electrons, and a variety of subatomic particles constantly shower our planet. These particles, traveling at nearly the speed of light, are called **cosmic rays** and are thought to be particles blasted across space by cataclysmic events, such as when a massive star explodes.

Cosmic rays can penetrate only a short distance into the Earth; so if a detector is located deep underground, nearly all the cosmic rays are filtered out. Neutrinos, on the other hand, are unfazed by a mere mile of solid ground. They have a small but predictable chance of interacting with the matter through which they pass, but that chance is so small that even if they encountered a wall of lead a light-year thick, most of the neutrinos would pass through!

Davis's experiment showed that the count of neutrinos was substantially lower than what physicists had predicted. However, his experiment relied on neutrinos produced in a side process of the proton–proton chain. Newer detectors are capable of detecting the neutrinos produced by the primary proton–proton interaction. Currently, the largest neutrino "telescopes" are the Super-Kamiokande detector, located deep in a zinc mine west of Tokyo, and the Sudbury Neutrino Observatory (Figure 50.4) more than a mile underground in a nickel mine in northern Ontario, Canada. The Sudbury detector contains about 1000 tons of *heavy water,* water consisting of molecules in which one of the hydrogen atoms in the molecule is the isotope ^{2}H (deuterium). This heavy water gets its name because the extra neutron in the deuterium nucleus gives the water a slightly higher density.

In a detector like the Sudbury Neutrino Observatory, occasionally a neutrino collides with a neutron in the heavy hydrogen, breaking the neutron into a proton and an electron. As the electron shoots off, it emits a tiny flash of light, which is recorded by photodetectors. These experiments detect solar neutrinos, but like

The penetrating power of neutrinos is the theme of a poem called "Cosmic Gall" by novelist and poet John Updike:

Neutrinos, they are very small.
They have no charge and have no mass
And do not interact at all.
The earth is just a silly ball
To them, through which they simply pass,
Like dustmaids down a drafty hall
Or photons through a pane of glass.
They snub the most exquisite gas,
Ignore the most substantial wall,
Cold-shoulder steel and sounding brass,
Insult the stallion in his stall,
And scorning barriers of class,
Infiltrate you and me! Like tall
And painless guillotines, they fall
Down through our heads into the grass.
At night they enter at Nepal
And pierce the lover and his lass
From underneath the bed—you call
It wonderful: I call it crass.

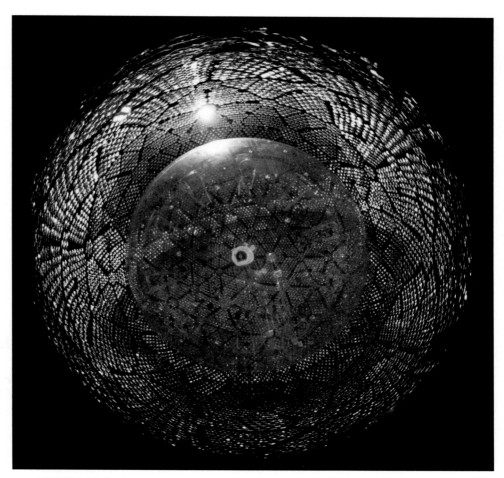

FIGURE 50.4

An inside view of the neutrino detector located in Sudbury, Canada. The sphere is 12 meters in diameter (about 40 feet). It is filled with heavy water, a form of water in which hydrogen atoms contain a neutron as well as the usual proton. When a neutrino strikes one of the neutrons, the neutron may break down into a proton and electron. As the electron streaks off, it produces a tiny flash of light. Ten thousand detectors (which form the grid visible around the sphere) record the emitted light, thereby allowing scientists to detect the neutrino's passage.

Davis's earlier experiment, they find only one-third the expected number. What does this low number mean? Is the Sun not fusing hydrogen into helium as predicted? Do neutrinos somehow escape detection? The solution to the puzzle offers us another chance to see the scientific process at work.

As a first step toward resolving the solar neutrino discrepancy, astronomers checked that their calculations for the predicted number of neutrinos were correct. All such checks led to roughly the same result, implying that there was no obvious flaw in their understanding of how the Sun works. Was it possible, then, that neutrinos had undiscovered properties that affected their detectability on Earth or their production in the Sun?

As scientists looked more closely at the properties of neutrinos, they realized that two other kinds of neutrinos were associated with different kinds of subatomic reactions (Unit 4). The nuclear fusion reactions in the Sun produce only one type of neutrino (called an *electron neutrino*), the type of neutrino that the early experiments were designed to detect. The standard theory held that the three kinds of neutrinos could not be converted into one another, but a new hypothesis suggested that they could. This hypothesis additionally predicted that the neutrinos have a mass, whereas the established model held that they had no mass. The new hypothesis suggested that the first experiments detected only one-third the expected number because the rest had been converted into the other types of neutrinos.

To test this idea, scientists built a second generation of neutrino detectors—the Sudbury detector is one of these—that could also detect the two other varieties of neutrino. By 2003 results from Sudbury clearly showed that the total number of neutrinos (of all three types) coming from the Sun agreed with the predictions of nuclear fusion rates. This gives astronomers a new and greater confidence in their understanding of the Sun. But in confirming the nuclear fusion model for energy generation in the Sun, they have helped to confirm a new hypothesis involving neutrinos, which modifies the previous standard model of subatomic particles. This new work gives us evidence that neutrinos must have a small mass, and the different varieties can be converted into each other.

50.4 THE FUSION BOTTLENECK

The fusion process in stars depends sensitively on all four of the fundamental forces. We have seen that gravity forces matter together and drives up its temperature to the point where protons can overcome the electromagnetic forces holding them apart. They then come so close to each other that the strong force can take hold. However, there is a problem with this process. Two protons cannot directly combine with each other because there is no stable isotope of helium with just two protons—no "^{2}He."

The first step of the proton–proton chain is actually a highly improbable interaction because in the extremely short moment of time while two protons are colliding and very close to each other, one of them must be transformed into a neutron. This is a process that requires the **weak force** (Unit 4), and such interactions are rare. On average, a particular proton in the Sun is likely to undergo this transformation while colliding with another proton only once in several billion years!

The conversion of a proton into a neutron *requires* energy:

$$^1H + Energy \rightarrow n + e^+ + \nu$$

where the "n" denotes a neutron. The amount of energy required for this process is large—it requires more than half as much energy as is released when a proton and neutron combine to make deuterium. This presents a barrier that dramatically slows down the rate of fusion, although it does not prevent the process from occurring because the overall process still has a net production of energy.

By contrast, the second and third steps of the proton–proton chain do not involve the weak force, and they typically occur just seconds after the first reaction. The weak force introduces a "bottleneck" in the rate of reaction. If the weak force were a bit stronger or weaker, stars would be very different than they are—they would burn at very different rates, or perhaps would not be able to undergo fusion at all. Thus, the weak force—as well as each of the other fundamental forces—is critical for making the Sun release its nuclear energy gradually over billions of years. This long time has allowed life to form and evolve on our planet.

KEY TERMS

cosmic ray, 408

deuterium, 406

isotope, 406

neutrino, 404

neutron, 405

nuclear fusion, 405

positron, 406

proton, 405

proton–proton chain, 406

strong force, 405

weak force, 410

QUESTIONS FOR REVIEW

1. What holds the Sun together?
2. Why must the interior of the Sun be so hot?
3. How is solar energy generated?
4. Why can the Sun not be powered by a chemical process such as the combustion of hydrogen and oxygen to form water?

PROBLEMS

1. Calculate the Sun's average density in kilograms per liter. The Sun's mass is approximately 2×10^{33} grams. Its radius is approximately 7×10^{10} cm. How does the density you find compare with the density of Jupiter?
2. Using $E = m \times c^2$, calculate how much energy is represented by the mass of the Earth, Jupiter, and the Sun. How many times more energy does Jupiter have than the Earth? How many times more energy does the Sun have than the Earth?

3. In chemical reactions, when energy is released, mass is lost according to Einstein's formula just as it is for nuclear energy. However, the amounts lost are so small that for centuries scientists thought that mass was conserved. When a kilogram of coal is burned, it releases about 40 million joules of energy.
    a. What amount of mass is lost?
    b. How many carbon atoms (mass $= 2.0 \times 10^{-26}$ kg) would it take to match the amount of mass that is lost?

TEST YOURSELF

1. The Sun produces its energy from
    a. fusion of neutrinos into helium.
    b. fusion of positrons into hydrogen.
    c. disintegration of helium into hydrogen.
    d. fusion of hydrogen into helium.
    e. electric currents generated in its core.

2. The Sun is supported against the crushing force of its own gravity by
    a. magnetic forces.
    b. its rapid rotation.
    c. the force exerted by escaping neutrinos.
    d. gas pressure.
    e. the antigravity of its positrons.

3. The number of neutrinos measured coming from the Sun permits us to measure
    a. how much deuterium is left in the Sun.
    b. the strength of the strong force within atomic nuclei.
    c. the current age of the Solar System.
    d. the risk of radiation exposure on the Earth.
    e. the rate of hydrogen fusion in the Sun's core.

51

Solar Activity

The Sun is a stormy place. The vast amount of energy pouring out of its interior heaves enormous masses of gas upward, which then plunge back down with forces far greater than any storm on Earth. Small hot spots send jets of white-hot ionized gas streaming up into the Sun's atmosphere, and occasional eruptions blast billions of tons of matter into space. We call these various disturbances **solar activity**.

The Sun's storminess exhibits dramatic kinds of activity that can take place in as little as a few minutes, or sometimes persist for months. As we look over even longer time scales, we find that the Sun goes through a regular pattern of changes every 11 years, and perhaps changes over longer periods still. From our safe distance on Earth, these phenomena are dramatic and lovely. Solar storms often take on forms quite unlike anything we might imagine from our experience on Earth because the Sun's hot ionized gas interacts with its strong, complex magnetic field. Moreover, solar activity can affect us directly: It can damage spacecraft, interfere with radio communications, trigger auroral displays, and alter Earth's climate.

Background Pathways: Unit 49

51.1 SUNSPOTS, PROMINENCES, AND FLARES

Sunspots are the most common type of solar magnetic activity. They are large, dark regions (Figure 51.1) ranging in size from a few hundred to many thousands of kilometers across. Sunspots last from a few days to over a month. They are darker than the surrounding gas because they are cooler (about 4500 K as opposed to the 5800 K of the normal photosphere), but they actually shine very brightly—they appear dark only by contrast. Sprinkling a few drops of water onto a hot electric stove burner will have a similar effect. Each drop momentarily cools the burner, making a dark spot. The exact causes of sunspots are still a puzzle four centuries after Galileo's first telescopic observations of them, although solar physicists are beginning to piece together the complex interactions that drive them.

One clue to the puzzle of sunspots was the discovery in the early 1900s that they contain intense magnetic fields. Astronomers can detect magnetic fields in sunspots and other astronomical bodies by the **Zeeman effect,** a physical process in which the magnetic field causes some of the spectral lines produced by atoms to split into two, three, or more components. The splitting occurs because the magnetic field alters the electron orbitals within an atom, which in turn alter the wavelength of its emitted light.

Figure 51.2A shows the Zeeman effect splitting spectral lines in a sunspot. The spectrum was taken from a region that cuts across a sunspot. The line is single outside the spot but triple within. By mapping the strength of such splitting across the Sun's face, astronomers can map the Sun's magnetic field, creating a **magnetogram,** as seen in Figure 51.2B. The colors in a magnetogram show the strength and polarity of the magnetic field. The *polarity* indicates whether the magnetic field is "north" or "south"; in other words, it indicates whether a compass needle would point toward or away from the region. The field is strong around spots and weak elsewhere.

Strong magnetic fields can affect the way hot gas circulates in their vicinity. The magnetic fields of sunspots are sometimes a hundred times stronger than the

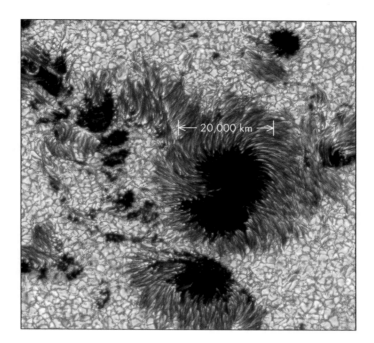

FIGURE 51.1

Image of a large group of sunspots. The darker areas are cooler gas, but they are still bright—they are dark in the image just by contrast to the surrounding hotter regions.

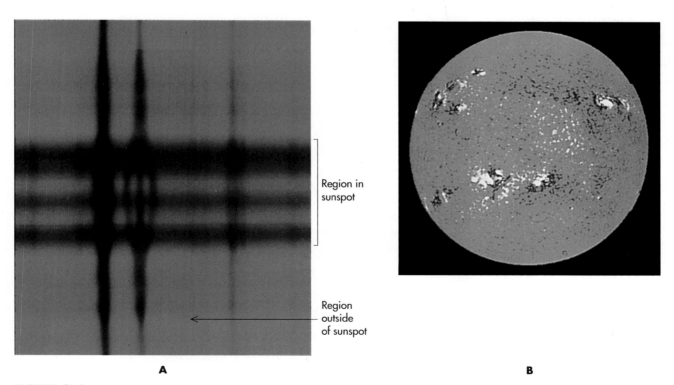

A   **B**

FIGURE 51.2

(A) Photograph of the Zeeman effect in a sunspot. The vertical marks are spectral lines of gases on the Sun's photosphere along a strip crossing through a sunspot. Notice that the line is split over the spot where the magnetic field is strong but that the line is unsplit outside the spot where the field is weak or absent. (B) Magnetogram of the Sun. Yellow indicates regions with north polarity, and dark blue indicates regions with south polarity. Notice that the polarity pattern of spot pairs is reversed between the top and bottom hemispheres of the Sun. That is, in the upper hemisphere, blue tends to be on the left and yellow on the right. In the lower hemisphere, blue tends to be on the right and yellow on the left.

Magnetic field

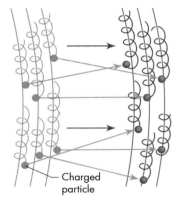

Charged
particle

Particles spiral
around the field lines.

FIGURE 51.3

Charged particles are tied to magnetic field lines, spiraling back and forth along them. If forces move the gas, it can drag the field with it—and vice versa.

normal field of the Sun, which is already a hundred times stronger than the Earth's magnetic field. In such intense fields, electrons and other charged particles spiral around the field, tightly tied to it, as shown in Figure 51.3. They are thereby forced to follow the magnetic field direction as they spiral along it. For example, the Earth's magnetic field deflects particles from striking the Earth in most locations. Instead the particles move along the magnetic fields until they reach low altitudes near the north and south magnetic poles, where they interact with the atmosphere and create the **aurora** (Unit 36.1). Likewise, in the Sun, if the magnetic field does not align with the direction in which the ionized gas is moving, it deflects the gas. Regions of strong magnetic fields may prevent hot gas in the interior of the Sun from rising to the surface, so that the surface cools and becomes comparatively darker, making a sunspot.

The magnetic field is not the whole explanation for sunspots, though. A region where the magnetic field is strong will normally push itself apart—just as magnets oriented in the same direction repel each other. Why, then, do sunspots persist for days or weeks, instead of rapidly weakening? This was one of the puzzles explored by a NASA spacecraft called the *Solar and Heliospheric Observatory (SOHO)*. Using detailed Doppler measurements of the flow and wave patterns of the photospheric gas, scientists were able to construct a three-dimensional image of the motions of the gas below the surface. An image of this structure is shown in Figure 51.4.

The cooling that occurs in the middle of the sunspot causes the gas there to sink, drawing in gas from the surrounding regions, much as a hurricane or tornado on

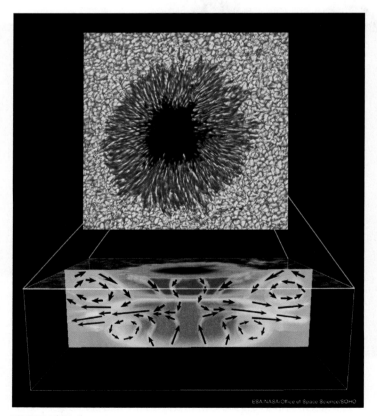

ESA/NASA/Office of Space Science/SOHO

FIGURE 51.4

Flows of gas underneath a sunspot measured by the spacecraft *SOHO* in 2001. The colored diagram at the bottom of the figure shows a slice through the region below the sunspot, showing where the gas is hottest (red) and coolest (blue). The arrows show the direction of gas flow.

Earth draws in air from the surrounding regions. *SOHO* measured flow speeds in and down toward the sunspot of about 5000 km/hour (3000 mph). The rapidly flowing gas drags along the magnetic field because, just as the magnetic field exerts a force on the charged particles, so do the charged particles exert a force on the magnetic field—another example of Newton's law of action and reaction (Unit 15.3). As a result, the flow of gas inward helps to trap and intensify the magnetic field, which keeps the hot gas from rising into the region of the sunspot from the layers below. This feedback between the magnetic field and the flow of gas is what makes sunspots persist on the Sun's surface.

The Sun's strong magnetic fields also shape the motions of ionized gas as it flows above the photosphere. **Prominences** are huge plumes of glowing gas that jut from the lower chromosphere into the corona (Figure 51.5A). You can get some sense of their immensity from the white dot, which shows the size of the Earth for comparison. Time-lapse movies show that gas streams through prominences, sometimes rising into the corona, sometimes raining down onto the photosphere. The flow is channeled by, and supported by, strong magnetic fields that arc between sunspots of opposite polarity. The pressure of the surrounding coronal gas

ANIMATIONS

Solar prominences

Q Prominences are measured to be less hot than the surrounding coronal gas. How then can they appear brighter than the corona in visible-wavelength images?

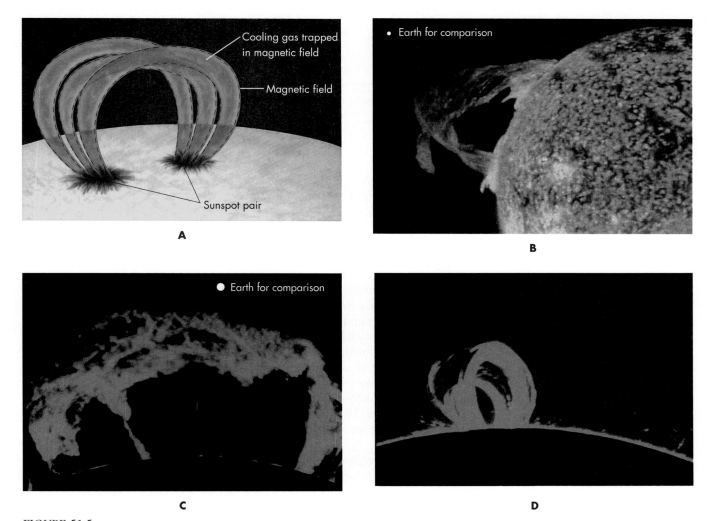

FIGURE 51.5
(A) Sketch illustrating how magnetic fields support a prominence. (B–D) Photographs of prominences. Image (B) was made at ultraviolet wavelengths, whereas (C) and (D) were made at the 656 nm wavelength spectral line of hydrogen (see Unit 24).

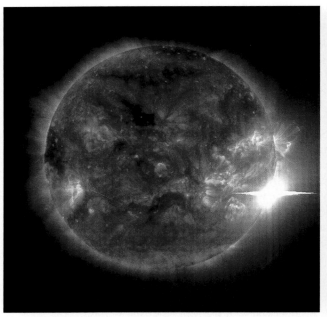

A

B

FIGURE 51.6
(A) A solar flare (lower right side of image) erupts from the Sun in this ultraviolet image.
(B) A coronal mass ejection is recorded at visible wavelengths. A disk blocks light from the Sun, and an ultraviolet image of the Sun made at the same time is superimposed.

helps to confine and support the gas in the prominence (Figure 51.5B–D). Under favorable conditions, the gas in a prominence may remain trapped in its magnetic prison and glow for weeks.

Sunspots also give birth to solar **flares,** which are brief but bright eruptions of hot gas in the chromosphere. Over a few minutes or hours, gas near a sunspot may brighten, emitting the energy equivalent of millions of atomic bombs. Such eruptions, though violent, are so localized they hardly affect the visible light output of the Sun at all. Generally you need a specialized telescope to detect the visible light of flares, though they can increase the Sun's radio, ultraviolet, and X-ray emission by factors of a thousand in a few seconds (Figure 51.6A).

Flares are poorly understood, but magnetic fields appear to play an important role in them too. One theory suggests that the field near a sunspot gets twisted by gas motions, a bit like winding up a rubber band–powered toy. Such twisting can go only so far before the rubber band breaks. So, too, the magnetic field can be twisted only so far before it suddenly readjusts, whipping the gas in its vicinity into a new configuration. The sudden motion heats the gas, and it expands explosively. Some gas escapes from the Sun and shoots across the inner Solar System to stream down on the Earth.

Closely related to flares, but on a much larger scale, are **coronal mass ejections.** These are enormous bubbles of hot gas, sometimes containing billions of tons of ionized gas, that are blasted from the corona out into space (Figure 51.6B). The mechanism for ejecting this gas is so powerful that the ionized gas drags the embedded magnetic field along with it. The cause of these ejections is still not well understood, but they seem to be triggered by flares. The most spectacular auroral events on Earth occur when one of these ejections strikes our planet (Figure 51.7).

FIGURE 51.7
Photograph of the great aurora of
March 1989.

The aurora usually begins about four days after the ejected outburst because it takes that long for the gas to traverse the 150 million kilometer distance between the Earth and the Sun, even though the gas is moving at a speed of 400 kilometers per second.

51.2   HEATING OF THE CHROMOSPHERE AND CORONA

Although the Sun's magnetic field cools sunspots and prominences, it heats the chromosphere and corona. To understand why, we need to recall that the temperature of a gas is a measure of how fast its particles are moving. Anything that speeds atoms up increases their temperature.

An analogy may help you understand how magnetic waves can heat a gas. When you crack a whip, a motion of its handle travels as a wave along the whip. As the whip tapers, the wave's energy of motion is transferred to an ever smaller piece of material. Having the same amount of energy but with less mass to move, the tip accelerates and eventually breaks the sound barrier. The whip's "crack" is a tiny sonic boom as the tip moves faster than the speed of sound—about 1200 kilometers per hour (about 700 miles per hour).

A similar speedup occurs in the Sun's atmosphere when magnetic waves, formed in the turbulent photosphere, move into the corona along the Sun's magnetic field lines. As the atmospheric gas thins, the wave energy is imparted to an ever smaller number of atoms, making them move faster, as illustrated in Figure 51.8. But "faster" in this case means hotter, so the upper atmosphere heats up as the waves travel into it, whipping the gas ions there back and forth.

Solar physicists estimate that heating by wave action alone does not offer enough energy to explain the high temperature of the corona. An additional source they propose is a steady energy input by a weak version of the magnetic twisting that produces solar flares. As with the flares, the heating would occur when the magnetic field rapidly realigns after being twisted by the motions of gas in the convection zone. Like heating by wave action, some of the energy of turbulent motion

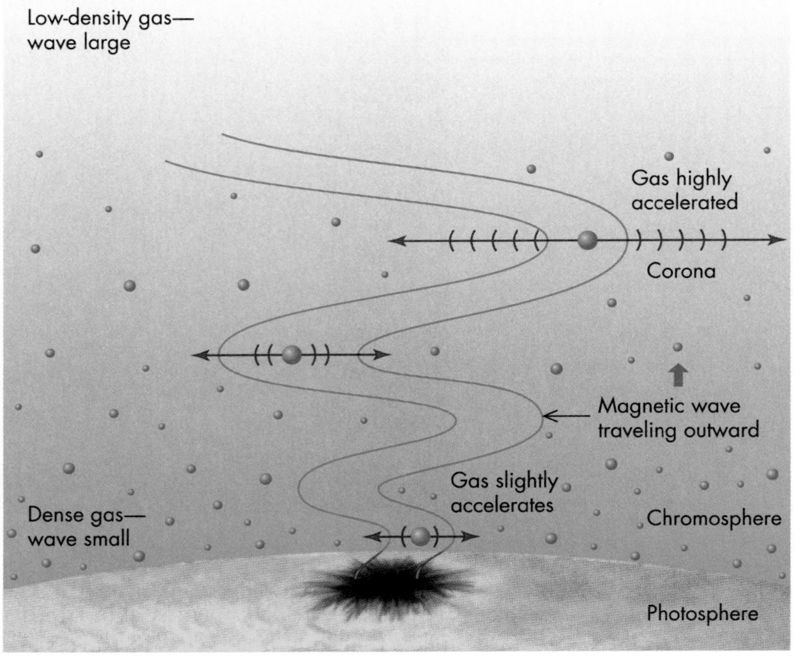

Low-density gas—
wave large

Gas highly
accelerated

Corona

Magnetic wave
traveling outward

Gas slightly
accelerates

Chromosphere

Dense gas—
wave small

Photosphere

FIGURE 51.8
Diagram illustrating how magnetic waves (blue) heat the Sun's upper atmosphere. As the waves move outward through the Sun's atmosphere, they grow larger, imparting ever more energy to the gas ions (green dots) through which they move. This wave motion accelerates the ions, which collide with other gas atoms, generating heat.

in the photosphere is transferred to the corona through the Sun's magnetic field. Thus, the high temperatures of the chromosphere and corona are further examples of the influence of the Sun's magnetic field and its convection zone. Astronomers infer that other stars with convection zones near their surface also have activity in their chromospheres and coronas because they too exhibit phenomena such as flarelike brightenings.

51.3   THE SOLAR CYCLE

Sunspot and flare activity change from year to year in what is called the **solar cycle.** This variability can be seen in Figure 51.9, which shows the number of sunspots detected over the last 140 years. The numbers clearly rise and fall approximately every 11 years. For example, the cycle had peaks in 1969, 1980, 1990, and 2001. The interval between peaks is not always 11 years: It may be as short as 7 or as long as 16 years.

Flares and prominences also follow the solar cycle, and climate patterns on Earth may too. For this reason, astronomers have sought to understand not only the cause of the solar cycle but also how it influences terrestrial climate.

As the Sun rotates, gas near its equator circles the Sun faster than gas near its poles; that is, the Sun spins differentially, a property common in spinning gaseous bodies. The Sun's differential rotation is such that its equator rotates in about 25 days and its poles in 30. So a set of points arranged from pole to pole in a straight line would move, over time, into a curve, as shown in Figure 51.10.

Differential rotation should similarly distort the Sun's magnetic field, "winding up" the field below the Sun's surface. Astronomers think such winding of the Sun's magnetic field may cause the solar cycle, though the exact mechanism is still not well understood. According to one hypothesis, the Sun's rotation wraps the solar

ANIMATIONS

Twisting of Sun's magnetic field

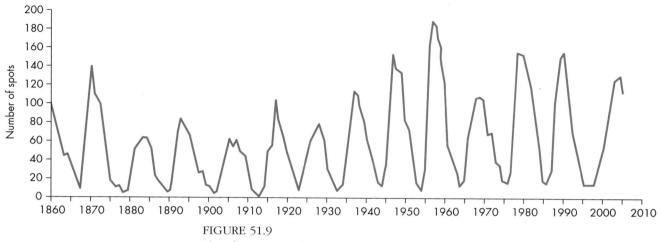

FIGURE 51.9

Plot of sunspot numbers showing solar cycle.

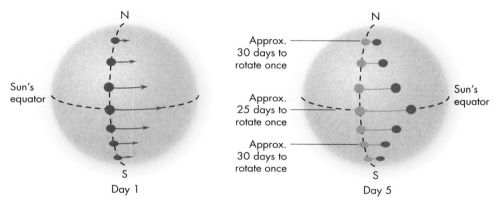

FIGURE 51.10

Sketch showing solar differential rotation. Points near the Sun's equator rotate faster than points near the poles.

magnetic field into coils below the surface (Figure 51.11), making the field stronger and increasing solar activity: spots, prominences, and flares. The wrapping occurs because the Sun's magnetic field is dragged by the circulating ionized gas, as discussed previously for the flows around sunspots. When the gas moves, so does the field, and vice versa. Sunspots form where the twisted magnetic field rises to the Sun's surface and breaks through the photosphere (last panel of Figure 51.11).

Over 11 years, the Sun's magnetic field grows increasingly strained, like the rubber band in a toy twisted tighter and tighter, as described earlier. The tightly twisted field breaks in places, producing flares and coronal mass ejections. After several years of these violent eruptions the magnetic field relaxes into an untwisted pattern again. Interestingly, the overall polarity of the Sun's magnetic field reverses after each of these peaks of activity, so that the "north magnetic pole" is in the opposite direction. Somehow the twisting of the magnetic fields reverses the internal circulation of the electric currents that produces the Sun's magnetic field.

The mechanism causing these reversals of the magnetic field is not yet understood, but it is probably similar to the reason that the Earth's magnetic field also flips direction. On Earth this occurs every 250,000 years, on average, as indicated by geologic tracers (Unit 35.5). Perhaps through understanding the Sun's magnetic field, we will come to understand the Earth's magnetic field better. This is not just a question of idle curiosity—when the Earth's magnetic field reverses, there is a period of

Q From time to time claims are made that something like, for example, the stock market correlates with the sunspot cycle. Do you think there *could* be a way for sunspots to cause changes in such things? How could you test your ideas?

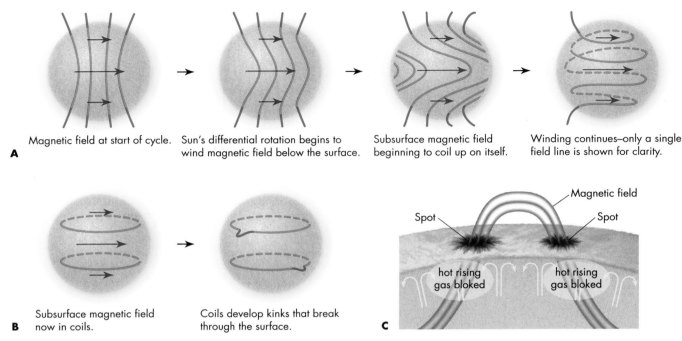

A

Magnetic field at start of cycle.   Sun's differential rotation begins to wind magnetic field below the surface.   Subsurface magnetic field beginning to coil up on itself.   Winding continues—only a single field line is shown for clarity.

B   Subsurface magnetic field now in coils.   Coils develop kinks that break through the surface.

C

FIGURE 51.11
Sketch showing (A) the possible winding up of the subsurface magnetic field; (B) fields penetrating the Sun's surface; and (C) formation of a spot pair.

up to several thousand years when the surface is poorly shielded from the solar wind and cosmic rays. Earth's magnetic field reversals are far more sporadic than the Sun's; but the last one occurred about 780,000 years ago, so we are long "overdue" for one. Being able to predict when the next one might occur, and how long and how dangerous the reversal period might be, is of vital interest to human life.

51.4   THE SOLAR CYCLE AND TERRESTRIAL CLIMATE

Climatologists have long debated whether the Sun's magnetic cycle affects the Earth's climate. One reason for the debate is that it is difficult to understand how changes in the Sun's magnetic field might alter our atmosphere. One possibility is as follows.

The Sun's magnetic field heats the corona. The corona drives the solar wind. The solar wind alters the Earth's upper atmosphere; in particular, it changes the way the temperature varies with altitude. This in turn alters the atmosphere's circulation, which may shift the jet stream to a new location. The jet stream steers storms and hence rainfall.

Although this hypothesis cannot be verified yet, many scientists think that solar activity affects our climate. The evidence to support this hypothesis is based in part on the work of E. W. Maunder, a British astronomer who studied sunspots. Maunder noted in 1893 that, according to old solar records, very few sunspots were seen between 1645 and 1715 (Figure 51.12). He concluded that the solar cycle turned off during that period. The period is now called the **Maunder minimum** in honor of his discovery.

The Maunder minimum coincides with an approximately 70-year spell of abnormally cold winters in Europe and North America. Glaciers in the Alps advanced; rivers froze early in the fall and remained frozen late into the spring. The

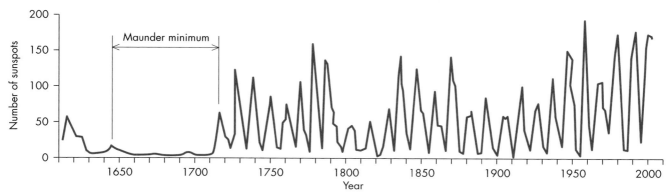

FIGURE 51.12

Plot illustrating that the number of sunspots changes with time, showing the Maunder minimum and the solar cycle.

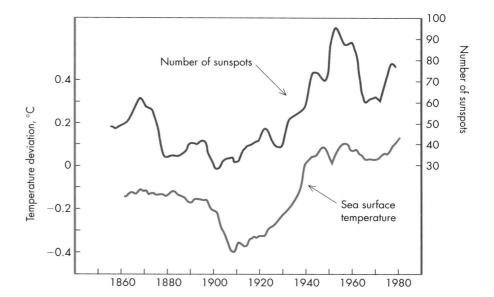

FIGURE 51.13

Curves showing the change in ocean surface temperature on Earth and the change in sunspot numbers over several decades. Notice that the curves change approximately in step. Astronomers deduce from these curves that solar magnetic activity affects our climate. (The spot numbers are averaged over 11-year intervals.)

Q Graphs like those in Figure 51.13 can be quite controversial. What kinds of changes in measurement techniques have occurred over the last 150 years? How might these changes have affected the values plotted?

Maunder minimum is in the middle of a period of several centuries when the Earth appears to have been cooler—meteorologists call this time the "little ice age." If only one such episode were known, we might dismiss the sunspot–climate connection as a coincidence; but three other cold periods have also occurred during times of low solar activity. This strengthens our belief that somehow the Sun's magnetic activity affects our climate. A recent hypothesis is that energetic particles from the Sun stimulate cloud formation on Earth, thereby slowing the escape of heat from our planet. Another hypothesis is that the Sun's total luminosity drops slightly during these periods of quiescence.

Although scientists are still unsure about what creates the link between solar magnetic activity and our climate, few now doubt that such a link exists—but there is substantial debate about whether its impact on climate is as large as other sources. Figure 51.13 shows how the ocean surface temperature (expressed as deviations from the normal average) changed from 1860 to 1980. The figure also shows how the number of sunspots, averaged over the 11-year cycles, changed over the same time span. Notice that when the Sun goes through cycles with a high number of spots, the ocean is warmer than average, and when the number of spots is low, the ocean is colder than average.

KEY TERMS

aurora, 414

coronal mass ejection, 416

flare, 416

magnetogram, 412

Maunder minimum, 420

prominence, 415

solar activity, 412

solar cycle, 418

sunspot, 412

Zeeman effect, 412

QUESTIONS FOR REVIEW

1. What is meant by *solar activity?*
2. Why do sunspots appear dark?
3. What role does magnetic activity play in solar activity?
4. How do we know there are magnetic fields in the Sun?
5. What is the solar cycle?
6. What is the period between maximum sunspot numbers? How does this differ from the period of the full solar cycle?
7. What is the Maunder minimum? Why is it of interest?

PROBLEMS

1. What is the speed of material ejected by a solar flare if it takes four days to reach the Earth?
2. The amount of light emitted by a hot region is proportional to the temperature to the fourth power ($L \propto T^4$). If a sunspot is at a temperature of 4500 K, how much less luminosity does it generate compared to if it were at 6000 K?
3. Describe three different impacts of solar magnetic activity on Earth. Discuss which of these is the most important in the short term and in the long term.

TEST YOURSELF

1. About how many years elapse between times of maximum solar activity?
   a. 3
   b. 5
   c. 11
   d. 33
   e. 105
2. Sunspots are dark because
   a. they are cool relative to the gas around them.
   b. they contain 10 times as much iron as surrounding regions.
   c. nuclear reactions occur in them more slowly than in the surrounding gas.
   d. clouds in the cool corona block our view of the hot photosphere.
   e. the gas within them is too hot to emit any light.
3. Differential rotation results in
   a. the solar wind.
   b. a wound-up magnetic field.
   c. the Maunder minimum.
   d. the Sun's generation of energy.
   e. All of the above

52

Surveying the Stars

Determining the distance of a remote object is one of the fundamental problems that astronomers must tackle. Knowing stars' distances is necessary for determining many of their most basic properties such as their diameters, masses, and the amount of light they emit.

In the last few decades, space exploration has allowed us to measure many distances within the Solar System directly. Spacecraft have traveled to most of the planets and their moons. As a result, we can time how long signals take to travel (at the speed of light) back and forth between the object being visited and the Earth. Ground-based technology has also advanced, so we do not have to rely on space probes to measure distances accurately. Instead, we can bounce radar signals off an asteroid, for example, and time how long the signals take to travel there and back. With this method we can determine a distance accurate to within a few meters (Figure 52.1).

Beyond the Solar System, however, distances are so vast that such direct techniques are impossible and will remain so for centuries to come. The situation is not hopeless, though. We can take a cue from the ancient Greeks, who were able to estimate the distances to the Moon and Sun (Unit 10) with only the crudest of astronomical instruments by applying geometry. The same geometric methods that a surveyor might use to measure distances on the Earth allow us to measure the distances of nearby stars, and such methods provide a fundamental stepping-stone to learning about the objects that lie far beyond the Solar System.

52.1 TRIANGULATION

Long before radar provided us with high-accuracy distance measurements, and long before space probes ventured out into the Solar System, some planetary distances were very accurately determined by a geometric method, used by surveyors, called **triangulation.** We construct a triangle in which one side is the distance we

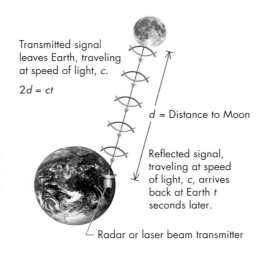

Transmitted signal leaves Earth, traveling at speed of light, c.

$2d = ct$

d = Distance to Moon

Reflected signal, traveling at speed of light, c, arrives back at Earth t seconds later.

Radar or laser beam transmitter

FIGURE 52.1

The distance to the Moon can be measured very precisely by bouncing radar or laser signals off it and timing how long they take.

423

Using trigonometry, the distance, d, across the gorge equals the baseline length, b, times the "tangent" of the angle A: $d = b \times \tan A$.

Scale drawing of measured triangle

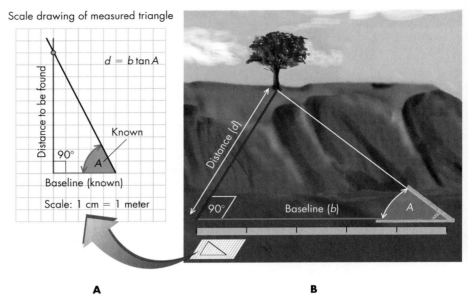

FIGURE 52.2
Sketch illustrating the principle of triangulation.

To make triangulation measurements more accurate, surveyors usually make the baseline as big as possible. Why would this improve the results?

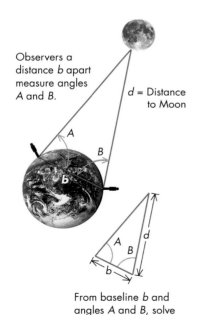

Observers a distance b apart measure angles A and B.

d = Distance to Moon

From baseline b and angles A and B, solve for d by trigonometry.

FIGURE 52.3
Finding the distance from the Earth to the Moon by triangulation.

seek, but cannot measure directly, and another side is a distance we can measure, as shown in Figure 52.2A.

Suppose we wanted to measure the distance across a deep gorge. We can construct an imaginary triangle to a point on the far side of the gorge. Two sides of this triangle span the gorge, while we can choose the length and direction of the third side—the **baseline**—along the edge we are on, as shown in Figure 52.2B. It simplifies our calculations if we choose the baseline so it is perpendicular to the line straight across the gorge. Then, by measuring the length of the baseline and the angle A between the baseline and the remaining side (the *hypotenuse*), we can determine the distance across the gorge from a scale drawing of the triangle.

Solving this type of problem is the motivation behind the area of mathematics called *trigonometry* (literally meaning "the measuring of triangles"). Trigonometric formulas allow us to calculate the length of the sides without requiring us to make precision scale drawings of the triangle. In the figure, the baseline is 5 meters, and the angle A is 63.4°. Constructing a right triangle with one of the angles at this value, we find that the unknown distance across the gorge must be two times the baseline, or 10 meters.

The Greek astronomer **Hipparchus** used triangulation to estimate the distance to the Moon as early as about 140 B.C.E. He compared the angle of the Moon with respect to the Sun, as seen from two locations during a solar eclipse. A similar experiment can be conducted today by two observers at different locations, simultaneously measuring the position of the Moon in the sky, who then compare their observations. From distant points on the globe, this difference can be as large as about 2°, an angle that can be measured with fairly simple instruments (Figure 52.3). The principle is no different from the idea of measuring the distance across a gorge. However, determining the length and angle of the actual baseline when the Earth's surface is curved makes this a more difficult calculation.

Triangulating distances to objects more distant than the Moon requires greater accuracy. Because of the much larger distances involved, the angles may be only a small fraction of an **arc minute.** An arc minute is 1/60th of a degree, or about 1/30th of the angular size of the Moon. Nonetheless, the ancient Greeks attempted to find the Sun's distance from Earth. Their result was far too small because instruments at the time were simply too crude to make a sufficiently accurate measurement of the angles. Because the Earth–Sun distance—known as the **astronomical unit** or AU—is so fundamental to our knowledge of the scale of the universe, we will discuss it in some detail here.

It turns out that we do not have to measure the distance to the Sun to determine the size of the AU. Johannes Kepler's discoveries in the early 1600s established the *relative* distances of the Sun and planets (Unit 12). For example, Kepler's law relating orbital size and the period of the orbit established that Mars's orbit is, on average, 1.52 times larger than the Earth's—1.52 AU—while Venus's is 0.72 AU. Therefore Mars is about 0.52 AU away from Earth at its closest approach and Venus is about 0.28 AU. So if we could triangulate the distance to Mars during one of its close approaches, we could determine the size of the AU.

Such a measurement of the distance to Mars was made during one of Mars's close approaches to Earth in 1672 by astronomers in Paris and South America, at sites separated by about 7000 km (about 4000 miles). We know today that the value they obtained was correct to about 10%, but it implied a distance to Mars—and therefore a size for the astronomical unit—that was almost 10 times greater than had been believed previously. This led many to question the accuracy of the value obtained. If the new distance was correct, all of the distances to the Sun that had been used for centuries were far too small. Recall that sizes of objects are measured by the angular size versus distance formula (Unit 10.4). If the distances were actually 10 times larger, then the sizes would have to be 10 times larger as well. Thus, not only was the Sun even bigger than previously thought, it also implied that Jupiter and Saturn were far larger than the Earth—all challenging ideas to accept in an age when even the concept of the Earth orbiting the Sun was new. Because of the controversial nature of these results, astronomers sought independent measurements to confirm these findings. It took almost a century, however, before the distance of Venus could be determined.

Even though Venus comes closer to Earth than Mars, attempts to triangulate its distance are difficult because when it is closest to Earth, it is lost in the glare of the Sun. However, the British astronomer Edmund Halley (of comet Halley fame) developed a method for determining both Venus's and the Sun's distance. He did this by timing how long Venus takes to move across the face of the Sun, from different locations on Earth, during one of Venus's rare **transits.** Because its orbit is tilted relative to our own, Venus usually lies slightly "north" or "south" of the Sun's disk as it passes between the Earth and Sun. But twice every 120 years or so it crosses directly in front of the Sun. This last happened in 2004 (Figure 52.4) and will next happen June 6, 2012. After that, it will not occur again until 2117.

For the transit of 1761, British astronomers Charles Mason and Jeremiah Dixon sailed to the southern tip of Africa, braving cannon fire (Britain was at war with

The exact values of the distances to Venus and Mars at closest approach vary because the orbits are elliptical, but Kepler also determined these shapes.

Q Why do you suppose scientists are usually reluctant to accept new results that are very different from previously accepted results?

Halley's technique is based on measuring the time it takes Venus to pass across the face of the Sun from points at different locations north and south on Earth. By comparing the duration, we can determine how much closer Venus was to crossing near the Sun's equator or pole from each location, giving the precise angular shift north or south relative to the Sun.

FIGURE 52.4
This image shows Venus (the round, dark circle) passing between us and the Sun on June 8, 2004. By precisely timing how long Venus takes to transit the Sun's disk from different locations on Earth, we can find Venus's distance.

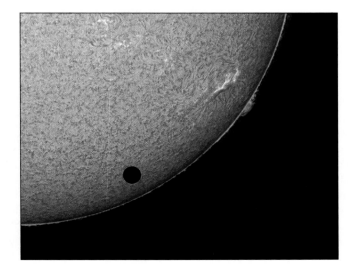

Mason and Dixon were summoned to the American colonies in 1763 to settle a dispute about the boundary between Pennsylvania and Maryland. This became known as the "Mason–Dixon Line," which figured prominently in the history of the United States in disputes between slave and free states.

France) to get to their destination. The observations of Mason and Dixon provided the long baseline needed for this astronomical surveying project. (Other astronomers were not so lucky, spending years traveling to remote parts of the globe, only to have clouds prevent them from making the critical measurements on the day of the transit.) The results of the Venus transit measurements confirmed the Mars measurements, yielding a size for the astronomical unit to within about 1% of the modern value.

In the end, however, the triangulation method of surveying is limited by the size of the baseline that we can obtain on the Earth. Even the most accurate measurements are nowhere near precise enough to tell us about the vast distances to stars if our baseline is at most the 12,750 km diameter of the Earth. A much bigger baseline is needed.

INTERACTIVE
Stellar parallax

ANIMATIONS
Parallax

52.2   PARALLAX

There is a baseline available to us that is more than 20,000 times larger than the size of the Earth—the Earth's orbit. As we move around the Sun, we shift our location in space by about 300 million kilometers (185 million miles) from one side of our orbit to the other. Even the ancient Greeks recognized that if the Earth moved around the Sun, stars should show shifts in their positions called **parallax** (see Unit 10).

An easy way to demonstrate the principle of parallax is to hold your thumb pointed up, motionless, at arm's length, and shift your head from side to side. Your thumb seems to move against the background even though in reality it is your head that has changed position. This simple demonstration also illustrates how parallax gives a clue to an object's distance. If you hold your thumb at different distances from your face, you will notice that the apparent shift in your thumb's position—its parallax—is also different. Its parallax is larger if your thumb is close to your face, and smaller if it is at arm's length (Figure 52.5). That is, for a given motion of the observer,

The smaller the parallax, the more distant the object.

A

FIGURE 52.5
When you hold your thumb at arm's length and shift your head from side to side or look through one eye and then the other, you see your thumb shift by a small amount relative to the background. This shift is *parallax*. If you hold your thumb close to your face and do exactly the same thing, your thumb shifts more against the background. This demonstrates that parallax is larger for nearer objects.

Viewed   Viewed
from right   from left

B

We might phrase this more technically as follows:

Distance is inversely proportional to parallax.

This law is just as true for stars as it is for your thumb.

Using parallax to find distance may seem unfamiliar, but parallax creates our stereovision—the ability to see things three-dimensionally. When we look around with two eyes, each eye sends a slightly different image to the brain. Your brain processes the two images and determines the distances to various objects in your field of view. You can demonstrate the importance to your brain of comparing the two images simultaneously by trying some experiments that involve covering one eye. Cover either eye, then drop a coin on a table in front of you. Hold your arm straight out in front of you, and with your first finger pointing down, try to lower it so that the tip of your finger comes down directly on the center of the coin. If you try this again with both eyes open, you'll see this is much easier using two eyes than one! With one eye closed, you may find yourself moving your head side to side (another way to estimate parallax) or looking for other visual cues to compare and adjust your hand's distance from you as you move it downward.

Try some other experiments with one eye closed. First, stand directly in front of a hanging cord (from a light bulb or window shade, for example). Hold one arm out to your side, with your finger pointed. Now bring your pointed finger in from the side, and try to touch the cord with the tip of your finger. Another experiment is to extend your arm in front of you, with your thumb pointed up. Look first through one eye and then the other, while keeping your head stationary. Notice that your thumb seems to shift to the left when looking with the right eye and then to the right when looking with the other eye.

Even after the Copernican revolution established that the Earth moves around the Sun, stellar parallax was not detected until centuries later. The failure to detect stellar parallax was one of the reasons that some astronomers continued until the 1700s to believe that the Earth did not move through space.

The stellar parallax technique is a little different from the triangulation method. Instead of measuring the angle between two views made along different lines of sight at the same time, we need to compare measurements made six months apart. For even the nearest star, the change in position, or **angular shift,** relative to distant stars is tiny.

To achieve such precision for observations made at different times, astronomers observe a star and carefully measure its position against background stars, as shown in Figure 52.6A. They then wait six months, until the Earth has moved to the other side of its orbit, and make a second measurement. As Figure 52.6B shows, the star will have a different position when compared with the background of stars, as seen from the two different points six months apart. The amount by which the star's apparent position changes depends on its distance from the Earth, just as the angular shift of your thumb depends on how far you hold it away from your eyes.

In the late 1500s, from his observatory at Uraniborg, Tycho Brahe carried out measurements of star positions that could have detected an angular shift of about 1 arc minute. When he failed to detect a shift, he concluded that the Earth must be stationary. Astronomers recognized that this failure to detect parallax might be because the stars are very far away, but this implied distances to the stars that seemed absurdly large to many of the time. In fact, it required the invention of the telescope and more than two centuries of improvements in optics and measurement techniques before the parallax shift was finally detected.

The distances of stars are so large that the parallax effect is extremely small—so small that it is measured in the fractions of a degree called **arc seconds.** One arc second is 1/3600th of a degree or 1/60th of an arc minute. It may help you visualize how tiny an arc second is if you keep in mind that one arc second is equivalent to the angular size of a U.S. penny at 4 km (2.5 miles) distance.

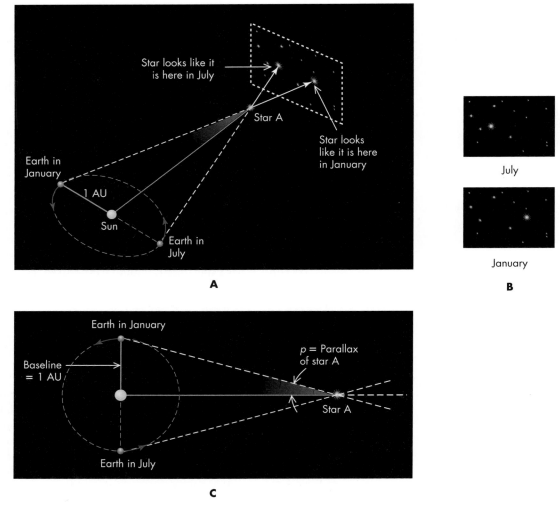

FIGURE 52.6
(A) Triangulation to measure a star's distance. The radius of the Earth's orbit is the baseline. (B) As the Earth moves around the Sun, the star's position changes as seen against background stars. (C) Parallax is defined as one-half the angle by which the star's position shifts. Sizes of bodies and their separation are exaggerated for clarity.

52.3 CALCULATING PARALLAXES

Q If you lived near one of these stars and could look back and observe the Earth orbiting the Sun, how would the motion of the Earth compare to the parallactic motion of your star as seen from Earth?

As the Earth orbits the Sun, stars appear to move in the sky in the opposite sense of the Earth's orbit. Each of them moves along a small circle, ellipse, or line, depending on their position relative to the Earth's orbital plane. For all but a few of them, the distances are so large that their positional shifts are virtually undetectable. For some nearer stars, however, we can detect a shift relative to the background stars.

Mathematically, astronomers define a star's parallax, p, as the angle by which a star shifts to each side of its average position (see Figure 52.6C). Because the distance d is inversely proportional to parallax, as discussed earlier, we can write

$$d = \frac{\text{constant}}{p}$$

where the constant depends on the size of the baseline (2 AU) and the units in which we measure angles and distances.

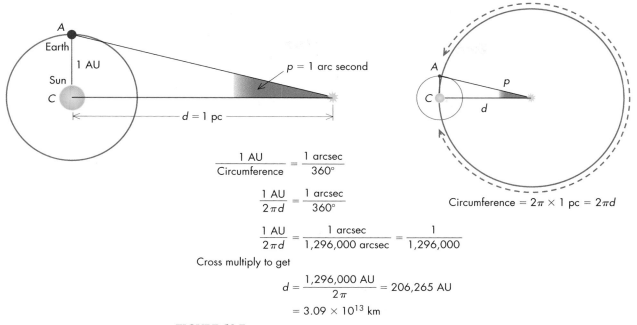

$$\frac{1 \text{ AU}}{\text{Circumference}} = \frac{1 \text{ arcsec}}{360°}$$

$$\frac{1 \text{ AU}}{2\pi d} = \frac{1 \text{ arcsec}}{360°}$$

$$\frac{1 \text{ AU}}{2\pi d} = \frac{1 \text{ arcsec}}{1{,}296{,}000 \text{ arcsec}} = \frac{1}{1{,}296{,}000}$$

Cross multiply to get

$$d = \frac{1{,}296{,}000 \text{ AU}}{2\pi} = 206{,}265 \text{ AU}$$

$$= 3.09 \times 10^{13} \text{ km}$$

FIGURE 52.7

How to determine the relation between a star's distance and its parallax.

To make this equation easier to use, astronomers invented a new distance unit, called the **parsec,** by setting the constant in this equation equal to 1. If we measure p in arc seconds and set the constant to 1, the value of d from this formula comes out as parsecs. That is,

$$d_{\text{pc}} = \frac{1}{p_{\text{arcsec}}}.$$

We have added the subscripts "pc" (for parsecs) and "arcsec" (for arc seconds) as reminders that for this formula to work, the parallax must be measured in arc seconds, which will result in a distance in units of parsecs.

Figure 52.7 shows how we can determine the size of a parsec from the geometry of a right triangle with a baseline of 1 AU and a parallax of 1 arc second. From the way we have defined a parsec, if we were one parsec from the Sun, then we would see the Earth move 1 arc second to either side of the Sun as it orbited. Therefore, 1 astronomical unit is to the circumference of a 1 parsec radius circle as 1 arc second is to 360 degrees, as shown in the figure. Using the known size of an AU then allows us to determine that the size of a parsec is

1 parsec = 3.09×10^{13} kilometers.

It turns out that the parsec is just a few times larger than the size of the light-year:

1 parsec = 3.26 light-years.

The word *parsec* comes from a combination of *parallax* and *arc second*. Most astronomers use parsecs when discussing distances—they are easy to calculate from parallax measurements. However, light-years are also frequently used, particularly when we discuss the time light takes to travel from an object to Earth—the conversion between travel time and distance is easy to calculate. It is a little like the difference in describing distances in meters versus feet.

To see how we can use this relation, suppose, for example, that we find from the shift in position of a nearby star that its parallax is 0.25 arc seconds. Its distance is then $d = 1/0.25 = 4$ parsecs. Similarly, a star whose parallax is 0.1 arc second is at a distance $d = 1/0.1 = 10$ parsecs from the Sun. Parallax measurements of nearby stars reveal that stars are typically separated by a few parsecs.

The star **Proxima Centauri** has the largest known parallax: 0.772 arc seconds. This indicates that it is at a distance $d = 1/0.772 = 1.30$ parsecs (or 4.22 light-years) from the Sun. Proxima Centauri is a dim companion of two slightly more distant stars, in orbit about each other, that have a parallax of 0.742 arc seconds, indicating a distance of 1.35 parsecs (or 4.40 light-years) from us. The larger of these two companions, Alpha Centauri A, is the third brightest star in the sky and is very similar to the Sun. This system of stars rises above the horizon only from latitudes south of 30° N and is too far south to be seen from most of the Northern Hemisphere.

Recognizing that these three stars are, in fact, close to one another in space is already telling us something interesting. Not all stars are as isolated from other stars as the Sun is. Surveying other nearby stars, we see that they are sometimes isolated, but more frequently they are members of small systems of stars.

Although the parallax–distance relation is mathematically a simple formula, obtaining a star's parallax to use in the formula is difficult because the angle by which the star shifts is extremely small. It was not until the 1830s that the first parallax was measured by the German astronomer Friedrich Bessel. Even now, the method fails for most stars farther away than about 100 parsecs. This is because the Earth's atmosphere blurs the tiny angle of their shift, making it very difficult to measure. Astronomers can avoid such blurring effects by observing from above the atmosphere. An orbiting satellite, Hipparcos—a clever acronym standing for HIgh-Precision PARallax Collecting Satellite in honor of the ancient Greek astronomer Hipparchus—has measured the parallax of almost 120,000 stars from space. With its data, astronomers can accurately measure distances to stars as far away as about 500 parsecs.

The tiny parallaxes of stars tell us that they are so far away that it is hard to even imagine the distances involved. Astronomers throughout history have had as difficult a time understanding the immensity of these distances as have *any* of us, when first hearing of such numbers. Modern measurements tell us that the Sun is about 150,000,000 kilometers (about 93 million miles) away, and that the next nearest star is about 42,000,000,000,000 km (about 25 trillion miles) away. One way of thinking of these huge distances is in terms of the time light takes to travel that far. Light moves so fast that it could circle the Earth almost eight times in one second. Light takes about 1.3 seconds to reach us at this speed from the Moon, and about 8½ minutes to reach us from the Sun. But even from the very nearest star, Proxima Centauri, light takes 4.22 years to arrive here at Earth, and for most of the stars we can see at night, their light takes hundreds or thousands of years to reach us.

52.4   MOVING STARS

Although it took more than two centuries after Tycho Brahe before telescopes and techniques were accurate enough to detect parallax, astronomers discovered other motions of the stars over a century earlier. Observed over many human lifetimes, the positions of stars appear to remain fixed on the celestial sphere. This was one of the reasons it was easy for early astronomers to suggest that what we see at night are glowing dots on a celestial sphere that surrounds us. If we could watch the stars for millions of years, however, we would see them moving about like bees in a swarm.

In the short history of human records of the sky, the motion of stars is almost imperceptible. Nonetheless, in 1718 Edmund Halley discovered that stars were shifting positions by comparing Tycho Brahe's star positions to those listed in ancient catalogs. This effect is called **proper motion**—the term *proper* here indicates

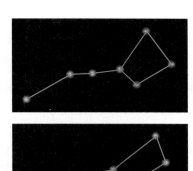

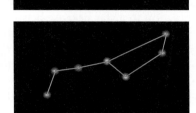

FIGURE 52.8
Proper motion of stars gradually alters their positions on the sky. The three figures show the pattern of these stars 50,000 years ago (top), today (middle), and 50,000 years in the future (bottom).

Q If a star has proper motion, is orbiting another star, and has a measurable parallax, what will its motion through the sky look like to us? How can we distinguish the different motions?

a property of the star, as opposed to an *apparent* effect caused by the Earth's own motion.

As a result of proper motion, the pattern of stars we see gradually changes. The configuration of constellations 50,000 years from now will be significantly different from their appearance today. The changing appearance of the Big Dipper, for example, is illustrated in Figure 52.8.

Because they are close to us, some stars move relatively rapidly over time on the celestial sphere. For example, Barnard's Star moves more than 10 arc seconds each year, and other nearby stars, such as Proxima and Alpha Centauri, also move by several arc seconds each year. Most stars, however, shift position much more slowly—not necessarily because their speed through space is slow, but because they are so far away that they must move a large distance for us see a significant angular shift. This is the same effect we see when a nearby car moves past us rapidly, while an airplane overhead seems to move much more gradually, even though it is traveling at a much greater speed than the car.

By the early 1800s careful examination of stars' positions also showed that some stars orbited each other (see Unit 56); and by the late 1800s astronomers were also able to detect stars' motions from their spectra, using the Doppler shift (Unit 25). This is the change in wavelength of light from a source as the source moves toward or away from us. If a source moves toward us, its wavelengths are shortened, whereas if it moves away, they are lengthened. The amount of wavelength shift depends on the source's speed along our line of sight—its **radial velocity.** Note that a star will have a radial velocity only if its motion has some component toward or away from us. Stars moving across our line of sight, maintaining a constant distance from Earth, have no radial velocity, although they will show proper motion.

What we learn of stars' motions by measuring Doppler shift and radial velocity can give us clues about an individual star's past. For example, we have found some stars that are moving at many hundreds of kilometers per second relative to all other stars. There is evidence that some of these "runaway" stars were flung out of orbit after a second star they were orbiting exploded!

52.5 THE ABERRATION OF STARLIGHT

In 1729, over a century before parallax was detected, the English astronomer James Bradley discovered a shift of the stars that he at first believed was the long-sought parallax. It proved to be a different phenomenon, but it nonetheless proved that the Earth orbited the Sun.

What Bradley observed was that all stars in the sky exhibit a shift in different directions depending on the time of year. The effect reaches a maximum of about 20 arc seconds for all stars—about 25 times larger than the largest parallaxes. Bradley soon found that the effect does not depend on the distance to the star, but instead happens for all stars by the same amount. This makes it a difficult effect to observe directly because we do not see it as a shift of one star relative to another star that appears near it in the sky.

What Bradley saw is easiest to visualize for a star that passes straight overhead. Imagine a telescope that is cemented into the ground and points straight up to the zenith. We would expect a star that passes straight overhead to pass through the middle of the field of view of the telescope once every rotation of the Earth. An example of the effect we might see instead is that the star was shifted 20 arc seconds

A star will pass through the center of the field of view of a stationary telescope once every **sidereal day** (Unit 7) of 23 hours 56 minutes.

to the north when it passed through the field of view in September. Then in December the star crosses through the field of view a little late; in March the star passes 20 arc seconds to the south of the zenith position, and in June it crosses through a little early.

If we marked all stars' positions throughout the year, we would see that they make little ellipses or circles on the sky. This is similar to the kind of motion that was expected from parallax, but it is the same for all stars lying in the same direction. Moreover, Bradley realized that the stars were always shifted in a direction *ahead* of the Earth's motion through space as the Earth orbits the Sun. To center a star in a telescope, you always have to point the telescope a little ahead of the actual direction toward the star.

Today astronomers call this shift in the positions of stars the **aberration of starlight,** and they recognize that it occurs because of the finite speed of light. We must tilt a telescope so it points ahead of the direction to a star because between the time light enters the front end of the telescope and the time it reaches the back end of the telescope, the Earth has shifted position in space (Figure 52.9). This is similar to the reason that you should hold an umbrella tilted slightly ahead of you if you are running through vertically falling rain. To keep the lower part of your body dry, you need to block the raindrops that are falling where your body *will* be.

If you ran through the rain in a large circle, you would tilt your umbrella ahead of you in whatever direction you ran, but this would cycle through north, then west, then south, then east, then north again as you completed one circuit. In addition, you would have to tilt the umbrella only slightly if you were running slowly,

Q Would the aberration of starlight be larger or smaller on Mars than on Earth? Why?

The shift of 20 arc seconds represents the angle of a right triangle with one side represented by $c = 300{,}000$ km/sec and another by the speed of the Earth $V_\oplus = 30$ km/sec. This angle is $360° \times (V_\oplus/c)/(2\pi) = 0.0057° = 20.6$ arc seconds.

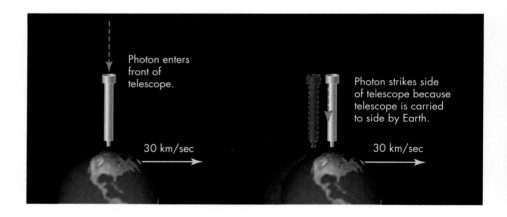

Photon enters front of telescope.

30 km/sec

Photon strikes side of telescope because telescope is carried to side by Earth.

30 km/sec

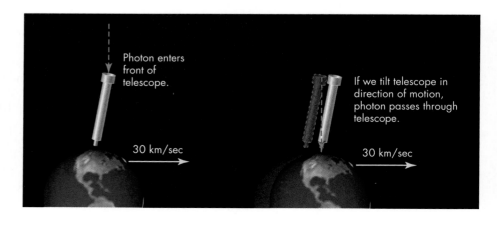

Photon enters front of telescope.

30 km/sec

If we tilt telescope in direction of motion, photon passes through telescope.

30 km/sec

FIGURE 52.9

The "aberration of starlight" is an effect of the Earth's motion through space. We must tilt telescopes toward the direction of the Earth's motion. Six months later, when the Earth is traveling in the opposite direction, we must tilt the telescope in the opposite direction.

but by a larger amount if you were running faster. Likewise, we tilt the telescope forward by an amount determined by the ratio of the Earth's speed through space and the speed of light. Even though it took another century to detect parallax, Bradley's discovery settled the question that the Earth is moving around the Sun.

KEY TERMS

aberration of starlight, 432	parallax, 426
angular shift, 427	parsec, 429
arc minute, 424	proper motion, 430
arc second, 427	Proxima Centauri, 430
astronomical unit, 425	radial velocity, 431
baseline, 424	transit, 425
Hipparchus, 424	triangulation, 423

QUESTIONS FOR REVIEW

1. Describe the method of distance measurement by triangulation.
2. How is a parsec defined? How big is a parsec compared with a light-year?
3. What is proper motion?
4. Explain the principle of aberration of starlight.

PROBLEMS

1. Make an approximate measurement of the distance from where you live to a neighboring building by triangulation. Make a scale drawing, perhaps letting 1 cm represent 1 meter (or you might let 1 inch represent 10 feet), and choose a baseline. Construct a scale triangle on your drawing and find how far apart the buildings are.
2. If a star's parallax is 0.1 arc seconds on Earth, what would it be if measured from an observatory on Mars?
3. Sirius has a parallax of 0.377 arc seconds. How far away is it?
4. A dim star is believed to be 5,000 parsecs away. What should its parallax be?

TEST YOURSELF

1. A star has a parallax of 0.04 arc seconds. What is its distance?
   a. 4 light-years
   b. 4 parsecs
   c. 40 parsecs
   d. 25 parsecs
   e. 250 parsecs
2. Star A is 10 times closer than star B. Star A's parallax is therefore _____ star B's.
   a. 10 times larger than
   b. 100 times larger than
   c. 10 times smaller than
   d. 100 times smaller than
   e. the same as
3. The most favorable conditions for observing the angular shift of a star occur when
   a. a large star moves across a field of smaller stars.
   b. a small star moves against a field of larger stars.
   c. Venus transits the Sun.
   d. a nearby star moves against a field of farther stars.
   e. nearby stars move relative to one another.

53 Special Relativity

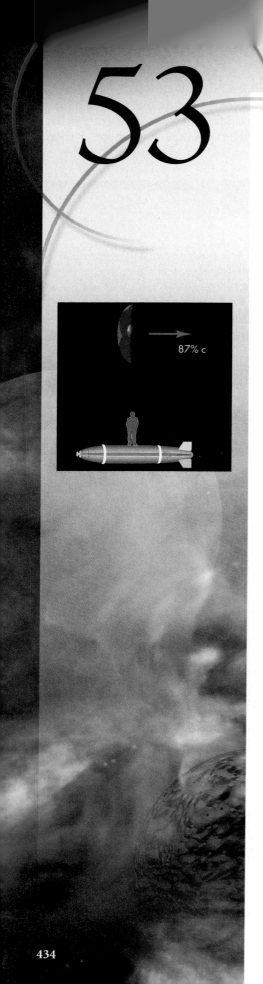

The universe is full of motion at every scale. Our predecessors gradually came to this understanding once they realized that the Earth moves through space. As the age-old notion of the Earth being fixed at the center of things crumbled, it soon became clear that the Sun is not stationary either. In fact, we now know that all stars move at tens or even hundreds of kilometers per second relative to one another, and that the large systems of stars known as galaxies move at even larger speeds.

All these motions create a problem of understanding: Is there some way to know what is moving and what is not? The answer to this initially appeared to be yes. Based on the way light moves through space, it appeared there was some medium that filled space that defined motion in an absolute way. Light, it was thought, moved through this medium at a constant speed, c. However, experiments to confirm this in the late 1800s seemed to require that light was doing impossible things—going faster and slower in different situations but always tricking the experimental apparatus into measuring the same speed.

In the early 1900s a young physicist named Albert Einstein suggested a revolutionary explanation for these strange results. His explanation, called **special relativity**, has changed some of our most basic ideas about the nature of space and time. One of the central ideas of special relativity theory is that our motion fundamentally alters both space and time for us. It has implications for the possibilities of ever traveling to distant stars, and fundamentally alters such concepts as how time progresses. To understand the revolutionary aspects of special relativity, we need to look at the motion of ordinary objects in a little more detail.

53.1 LIGHT FROM MOVING BODIES

Astronomers began detecting the motions of stars in the early 1800s (Unit 52). When they measured the motion of a star, they measured it relative to the Sun. They used our vantage point in the Solar System to define a local **rest frame**—a coordinate system tied to our position and motion. With respect to this frame we can measure the positions of other objects in space and the speeds at which they are moving relative to us. In our rest frame we observe, for example, that some stars are moving through space at high speed. However, it is important to recognize that an alien orbiting a "high-speed" star could define its own rest frame. To this alien, it would appear that the Sun was moving by *it* at high speed.

The motions of stars are also detected from their Doppler shifts (Units 25 and 52). The Doppler shift is a change in the wavelength of light we detect when stars move toward or away from us. This raises an important question: Does light from a star that is moving toward us reach us more quickly than light from a star that is moving away from us?

We might expect a boost or reduction in the speed of photons emitted by a source depending on whether it is moving toward or away from us, based on our experience at low speeds. For example, if you throw a javelin you will find that you can throw it faster and therefore farther if you are running toward your target. On

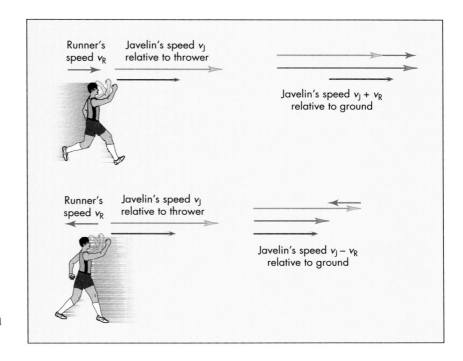

FIGURE 53.1
The speed of a javelin throw may be the same relative to the thrower no matter what direction he or she runs, but the speed of the javelin relative to the ground depends on the direction of motion.

Q What are other examples of situations when different speeds add and subtract from each other?

If light's speed depended on the radial velocity of a star, we would observe an increase in the amount of aberration of starlight (Unit 52) for stars moving away from us. The resulting speed of light would be slow, so the Earth's motion would be more significant in shifting its position as the light passed down through the telescope. This would be easily detectable with modern instruments, but no such effect is seen.

the other hand, if you are running *away* from your target, your throw will be weak and short even if you throw the javelin just as hard. This is because by running toward the target, you add your speed of running to the speed of the javelin. However, when you are running away from the target, your speed subtracts from the speed of the javelin. Thus, with respect to the ground, the javelin is traveling more slowly (Figure 53.1).

Using the term we introduced at the beginning of this section, the javelin is traveling faster or slower in the target's rest frame depending on whether you move toward the target or away from it. In your own rest frame, the javelin moves just as fast (you throw it just as hard), but the difference in your rest frames combines with the speed of the javelin to produce the overall speed.

The addition of relative speeds (you plus javelin) in this fashion is an example of what is sometimes called **Galilean relativity.** Galilean relativity gets its name because Galileo was one of the first scientists to recognize how motions add and subtract from one another, and how the measurements of these motions depend on the rest frame of the observer.

We might expect similar behavior for photons emitted by a star that is moving toward or away from us. If Galilean relativity applied to photons from a star moving toward us, we would expect to observe an increase in the speed of its photons toward us. Likewise, if it was moving away from us, we would expect a decrease in the speed of its photons. This change in speed might be difficult to measure directly, but it would have easily detectable effects on the appearance of stars that orbit each other.

Astronomers first detected stars that orbit each other in the early 1800s (Unit 56). Because of that orbital motion, during one part of its orbit a star will move toward us, while half an orbit later it will move away from us. Such motion is easily detected from the stars' changing Doppler shift. That motion should also change the speed of the star's photons that are moving toward us and create some very strange phenomena. If the photons coming from the star when it was moving toward us were traveling faster, they could overtake photons that left the star earlier

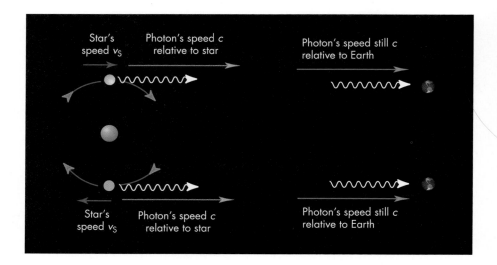

FIGURE 53.2

Light from a star in orbit reaches Earth at the same speed whether the star is moving toward or away from us.

when it was moving away from us. The result would be that we might see the orbiting star in different positions in a scrambled order, or even in two different parts of its orbit simultaneously!

The light we see does nothing of the sort; we see no changes in the apparent position of the star that would be expected for a changing speed of the light. The light apparently moves through space at the same speed no matter what the speed of the star (Figure 53.2). This is not as odd as it might first seem. For example, when a boat makes waves on a lake, the boat may move toward or away from us, but the waves travel through the water at the same speed. In the 1800s scientists reasoned that space must be something like the water in this example. They believed it must contain some substance that light waves travel through always at the same speed. They called this hypothesized substance the **æther.**

53.2   THE MICHELSON–MORLEY EXPERIMENT

By the late 1800s physicists were able to measure the speed of light with good precision, and the intriguing possibility of discovering the absolute motion of the Earth through the hypothesized æther became a possibility. As the Earth moved through the æther, from the perspective of the moving Earth the aether should be "flowing" by us. This flowing aether should affect the motions of photons moving through it much as flowing water affects the motion of boats traveling through it. In other words, the photons should travel at different speeds depending on the direction of their motion relative to the flow.

American scientists Albert Michelson and Edward Morley designed an apparatus in 1887 that could compare the speed of light moving in two perpendicular directions simultaneously (Figure 53.3). This was an extremely sensitive apparatus built on a massive granite slab that floated in a pool of mercury to isolate it from any vibrations, and they could rotate the slab. They reasoned that if it was rotated in the right direction, along one path through the apparatus, the light would travel parallel to the flow of the æther. Along the other path the light would travel perpendicular to the æther. Therefore, the speed of the light ought to be different along the different paths.

However, when Michelson and Morley conducted measurements using their apparatus, they detected no difference in the speed of light along the perpendicular

FIGURE 53.3
Photograph of Albert Michelson's and Edward Morley's experimental apparatus.

paths no matter how they oriented the apparatus. They realized that this could just be bad luck during the first experiment—the combination of the Earth's motion around the Sun and the Sun's motion through space might just happen to make the Earth stationary relative to the æther at that point in its orbit. So they repeated their experiment many times over a year. At some point they should have easily detected the motion of Earth relative to the æther, but they found no sign of any relative motion at all.

The Michelson–Morley experiment has been called the most famous "failed" experiment in history because it led to a revolution in physics. The results implied that there was no æther regulating the speed of light. However, if light was not moving relative to the æther, then physicists could not explain the constancy of the speed of light reaching us from sources moving at different speeds toward or away from us.

All experiments made then and now show that no matter how fast the source generating the light is moving, and no matter how fast the observer measuring the light is moving, the speed of light is always measured to be $c = 299,792,458$ meters per second. How can this be?

One explanation offered in the late 1800s was that motion through space somehow caused matter to contract in the direction of motion. If matter contracts when it is moving, this would change our perception of length so that we might be tricked into thinking the speed of light had not changed. For example, Michelson and Morley were searching for a difference in the speed of light in two perpendicular directions. If the apparatus were compressed in the direction it was moving through space but not in the perpendicular direction, the path the light traveled would be shorter. Such a contraction could potentially cancel out the effect that Michelson and Morley were searching for. The factor by which the apparatus would need to contract is known today as the **Lorentz factor.** The Lorentz factor, usually denoted by the Greek letter gamma (γ), hypothesizes that an object's length shortens in the direction that it is moving by a factor equal to

$$\gamma = \frac{1}{\sqrt{1 - V^2/c^2}}$$

where V is the speed of the object and c is the speed of light. The value of γ for different speeds V is plotted in Figure 53.4.

The Lorentz factor is close to 1 at small speeds. For example, at the 30 km/sec speed at which the Earth is orbiting the Sun, the Earth would contract by only a few centimeters in its direction of motion. However, at high speeds the contraction

The idea of a contraction caused by motion was first proposed by Irish physicist George Fitzgerald. The Dutch physicist Hendrik Lorentz developed a model for explaining how the contraction might arise.

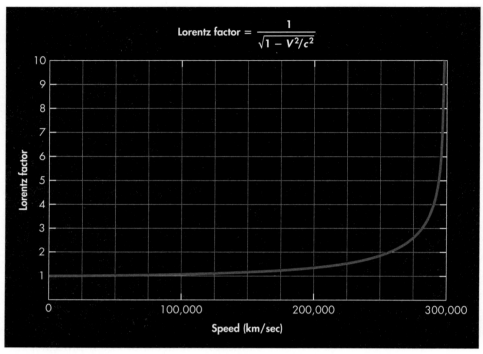

FIGURE 53.4

The Lorentz factor expresses the amount by which objects appear to compress in the direction of their motion. It is also the factor by which a moving clock appears to slow down, and the factor by which a moving object's mass appears to increase.

factor becomes very large, growing to infinity if the speed V were to reach the speed of light. The Lorentz contraction factor can explain the Michelson–Morley experiment, but it cannot explain a variety of other conflicting results that were found. It contains, however, an important idea that grew into a whole new concept about the nature of motion.

53.3   EINSTEIN'S THEORY OF SPECIAL RELATIVITY

In 1905 a 26-year-old graduate student named Albert Einstein (Figure 53.5) took on the problem of the seemingly inexplicable measurements of the speed of light. He was completing his physics degree while working in the Swiss patent office and supporting a new family. Yet in that one year alone, he completed his doctorate degree and wrote four papers in several areas of physics. Physicists widely agree that three of these papers were each worthy of a Nobel prize! He was little known at the time and had few colleagues with whom to discuss his ideas; but nonetheless in one of these papers he came up with a brilliant new approach to the question of the motion of light.

Einstein began by concentrating on the findings that light travels at the same speed no matter what the speed of its source or of the observer measuring the light. Even though many experiments had come to this conclusion, most physicists had assumed it was an impossibility and so were seeking other explanations—such as errors in the experiments or the Lorentz contraction. Einstein, instead of thinking of a constant speed of light for all observers as an impossibility, accepted this

FIGURE 53.5
Albert Einstein (1879–1955).

notion as correct, and proceeded to work out its consequences. He found that this led inevitably not only to a Lorentz-like contraction of space, *but also to a stretching of time by the same factor.* Even more important, he found that this contraction was not relative to some imagined æther filling space, but that these effects depended on just the relative motions of any two objects.

These alterations of both space and time affect everything we see that has any speed relative to us. They tell us that if a star or a rocket moves by us at high speed, we will see it squashed in its direction of motion by the Lorentz factor (Figure 53.6). If we could watch a clock tick or measure the speed of an astronaut's heart in the spaceship, we would discover that all these processes occur more slowly by the Lorentz factor.

What is remarkable is that the mathematics Einstein worked out showed that the situation is exactly symmetrical for the astronaut moving by us at high speed. That astronaut will sense herself being the one who is stationary and see us as moving by her at high speed. She will see us and the Earth contracted in the direction of "our" motion, and she will see our clocks and our hearts running slowly (Figure 53.6). So what two observers in relative motion see is parallel—

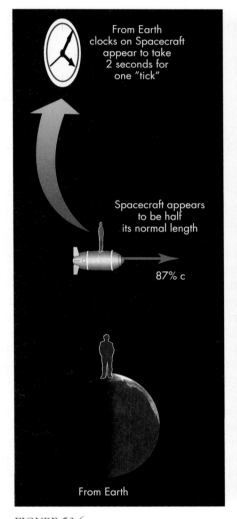

FIGURE 53.6

When a spacecraft travels by us, its length is contracted by the Lorentz factor, and clocks on board run more slowly by the same factor. From on board the ship it appears that the Earth is compressed and time is running slowly on Earth.

each would find that the other was the one undergoing the distortions of space and time.

An especially important feature of Einstein's work is the behavior of light. Suppose we or the passing astronaut shine a light beam at each other. We will each measure that the light is moving past us at the same speed, c. However, as we watch each other making these measurements, we will each think that the other is measuring a shorter distance and using a clock that runs too slow.

This theory of special relativity is far-reaching in its implications. The theory is the basis for Einstein's discovery of the relationship between energy and mass: $E = m \times c^2$ (Unit 50). Other fundamental quantities also change for moving objects. For instance, a moving object grows more massive, also by the same Lorentz factor that describes how lengths grow shorter and time slows.

These effects mean that nothing can reach the speed of light in our rest frame, let alone exceed it. The rocket ship traveling by us can fire its rockets and accelerate forever—but every time it goes a little faster, its time slows down more, and the rocket exhaust comes out more slowly and does not travel as far because of the contraction in length of the speeding ship. Meanwhile the ship's mass grows heavier and heavier, so it picks up less and less speed. The ship may approach the speed of light, but because the Lorentz factor goes to infinity, it would require infinite energy to reach the speed of light.

How do science fiction stories you have seen or read treat motion at speeds approaching c? Do any of these stories correctly depict the effects of special relativity?

53.4   SPECIAL RELATIVITY AND SPACE TRAVEL

The theory of special relativity may seem strange, but it has been tested by a century of high-precision experiments. There is not a single verified contradiction to it, and its predictions about such things as the slowing of time have been verified directly.

Special relativity is also about more than just perceived differences in space, time, and mass. For example, when atomic clocks (the most accurate clocks, used for establishing time worldwide) are flown between locations, it is found that their travel in airplanes leaves them a little bit slow relative to the network of fixed clocks maintained around the world. The moving clocks must be readjusted after each trip. The Lorentz factor for traveling at airplane speeds of about 1000 km/hour (600 mph) is just $\gamma = 1.0000000000004$, so every tick of the clock on an airplane takes about 4 ten-trillionths of a second longer than a tick of the clock in the ground-based network; but after several hours of flying the effect is measurable.

More intriguing is what happens at such high speed that the Lorentz factor is large. Experiments have demonstrated, for example, that a subatomic particle called a *muon* (Unit 4) normally has a lifetime of only about 2 millionths of a second before it decays. However, when muons are traveling at 99% of the speed of light (and therefore have a Lorentz factor of about 7), they live about 14 millionths of a second. This much longer lifetime allows them to travel distances that would be impossible for them within their normal lifetimes.

When speaking of microseconds, this change seems minor; but the same factor would apply for human space travel at 99% of the speed of light. If a spaceship could be built that traveled that fast, time would effectively run seven times more slowly on board compared with time here on Earth. If astronauts had food and air supplies for a 10-year trip, the Lorentz factor of $\gamma \approx 7$ would mean they could travel for $\gamma \times 10$ years ≈ 70 years in Earth time. At their speed

For $V = 0.99\,c$, the Lorentz factor is

$$\gamma = \frac{1}{\sqrt{1 - (0.99c)^2/c^2}}$$

$$= \frac{1}{\sqrt{1 - 0.9801}}$$

$$= \frac{1}{\sqrt{0.0199}} = 7.089.$$

At a speed of $V = 0.99\,c$, in 70 years a spaceship would travel $0.99 \times 70 = 69.3$ light-years.

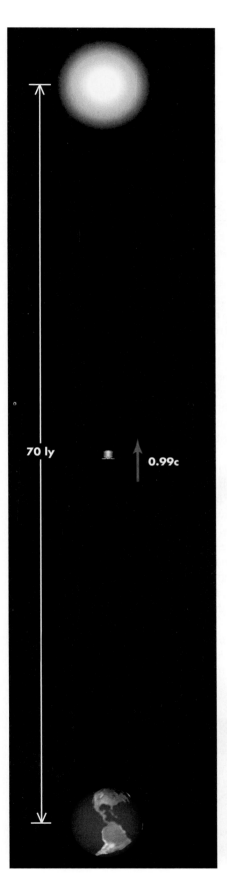

relative to Earth of 0.99 *c* they would be able to reach a distance of almost 70 light-years (Figure 53.7A) before they ran out of supplies. At speeds even closer to *c* the Lorentz factor becomes even larger (see Table 53.1) and the potential distances greater.

From the perspective of astronauts on a craft traveling at 0.99 *c*, it would not seem that time was passing any more slowly than normal. Nor would they feel that they or their ship was foreshortened or more massive. From their perspective the Earth and the star they are visiting and the distance between the two are *contracted* by the Lorentz factor (Figure 53.7B). The distance would look like 70 light-years when they were stationary (in the rest frame of Earth); but at their high speed, the distance would look only one-seventh as large!

These marvelous "tricks" of relativity open up possibilities for traveling distances far greater than we might once have imagined. Theoretically, it is not impossible to travel a million light-years away within your lifetime—although it is far beyond current technologies. It is also important to realize that such travel would have major challenges beyond simply reaching such high speeds. From the perspective of the spaceship traveling among the stars at near the speed of light, every atom and every dust particle in space along the ship's path has a mass increased by the Lorentz factor, and it is heading toward the ship at near the speed of light! This can impart to a pebble the impacting force of a ship-destroying asteroid.

Also, as intriguing as these possibilities are, they offer no time savings for the rest of us back on Earth. Consider again the astronauts traveling at 99% of the speed of light for 10 years to visit a distant star. If they then turned around and came home in another 10 years (by their reckoning), they would find that 140 years had passed on Earth. Everyone they knew would have grown old and died. They might even be younger than their great-great-grandchildren!

TABLE 53.1	The Lorentz Factor at High Speeds
Speed	Lorentz Factor
0.87 *c*	2.0
0.97 *c*	4.1
0.99 *c*	7.1
0.999 *c*	22.4
0.9999 *c*	70.7
0.99999 *c*	223.6

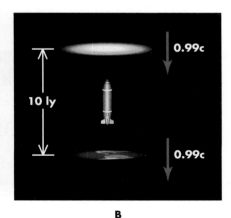

FIGURE 53.7
(A) From the rest frame of Earth and a star 70 light-years away, it appears that a spaceship traveling at 99% of *c* is shortened to one-seventh of its original length. (B) On the spaceship, once it is up to speed, it appears that the Earth and star are both moving at 99% of *c*. They and the distance between them are shortened to one-seventh of their original length.

A   B

53.5 THE TWIN PARADOX

Something may seem wrong about the description of space travel in the previous section because when we first introduced relativity we noted that the time stretching appears symmetrical. While from the rest frame of the Earth it appears as though the astronauts' time is running slowly, from the rest frame of the astronauts it appears that time on Earth is running slowly. This is sometimes presented as the **twin paradox:** If one of the astronauts left a twin back on Earth, how can we say that one ages more than the other?

The explanation of this seeming paradox is that the situations are not actually symmetrical. The astronauts experience accelerations as they first speed up their ship and then slow down at the star they visit. They turn their ship around and experience more accelerations as they again speed up and later slow down when they reach Earth again. The people remaining back on Earth experience none of these accelerations because they always remain in the same rest frame, and so the progression of time remains constant. By contrast, the astronauts' rest frame keeps changing, and in the end they return to the rest frame of the Earth.

Imagine, for example, that one of the astronaut twins sends messages once each day to her twin back on Earth, and meanwhile the twin sends messages once each day to his astronaut sister on the ship. As they part, they each receive the other's messages much more slowly—both because of the Lorentz factor *and* because the separation between the ship and Earth is growing larger and the messages take longer to reach each other. When the astronaut twin reaches the distant star, she sends her 10th-year message. Until this time, the situations are symmetrical. Both have received only a small fraction of each other's messages because of the Lorentz factor and growing separation.

The astronaut twin turns around and starts back, receiving the messages from her brother that were sent years earlier and have been on their way to her across the 70 light-year gap. They come more quickly now because she is approaching Earth, cutting the distance each message has to travel. She reads about her brother getting older and older, now seemingly very rapidly. In the meantime, her brother back on Earth is still reading her messages from the outgoing trip. He would not even receive the 10th-year message announcing that the ship had reached the other star before he dies. At the speed of light in Earth's rest frame, that 10th-year message would take 70 years to reach Earth. In fact, the spaceship, traveling at 99% of the speed of light, returns to Earth just shortly after the message arrives at Earth saying that the astronauts reached the star.

In the movies, space travel is fast and everyone ages at the same rate. In reality, traveling at high speeds means that the travelers must leave behind not only their homes but their own times and the people in them.

Q With everyone's time running at different speeds, can you imagine a story line that would make a good movie if it portrayed space travel accurately?

KEY TERMS

æther, 436

Galilean relativity, 435

Lorentz factor, 437

rest frame, 434

special relativity, 434

twin paradox, 442

QUESTIONS FOR REVIEW

1. What is Galilean relativity? Give an example of how it is used.

2. Describe what the Michelson–Morley experiment was trying to detect.

3. What is the Lorentz factor?
4. How are length, time, and mass altered according to special relativity?

PROBLEMS

1. What is the Lorentz factor for a spaceship traveling at 0.5 c?
2. What speed would give a Lorentz factor of about 1 million (10^6)?
3. The Earth is traveling at about 30 km/sec as it orbits the Sun.
    a. How much more slowly does time run on the Earth than the Sun?
    b. Why will the time be measured as slower on the Earth than the Sun? (Why is it *not* "just relative"?)
4. Suppose a one-gram paperclip struck a spaceship at 99% of the speed of light. Special relativity indicates that the paperclip's energy goes from $\gamma \times mc^2$ and drops down to its "rest mass energy" of mc^2 once it stops moving.
    a. How much energy is released in the collision?
    b. How much energy is this in kilotons of TNT?

TEST YOURSELF

1. When a spaceship is traveling at 99% of the speed of light (Lorentz factor = 7), an astronaut on board the ship will find that
    a. everything in the ship weighs seven times more.
    b. the ship has become very small—only one-seventh its original length.
    c. everyone on the ship is talking seven times more slowly than normal.
    d. All of the above
    e. None of the above. Everything seems normal to the astronaut on board.

2. Suppose Tom and Molly are both flying in spaceships toward each other at half the speed of light (0.5 c). If Tom shines a light toward Molly, what speed will Molly measure for the light coming toward her?
    a. 0.25 c
    b. 0.5 c
    c. 1.0 c
    d. 1.5 c
    e. 2.0 c

3. If Bob travels at close to the speed of light to another star and then returns, he will find that his twin sister Alice who remained on Earth is
    a. younger than him.
    b. older than him.
    c. the same age as him.
    d. He cannot return to Earth because it would violate special relativity.

54 Light and Distance

It is hard to believe that the stars we see in the night sky as tiny glints of light are in reality huge, dazzling balls of gas similar to our Sun. In fact, many of those gleaming specks are vastly larger and brighter than the Sun. Stars look dim to us only because they are so far away—several **light-years** (tens of trillions of kilometers or miles) to even the nearest. Such remoteness creates tremendous difficulties for astronomers trying to understand the nature of stars (Figure 54.1). Direct probing of such remote objects is impossible; but by studying the light they emit, we can learn about many properties of stars despite their vast distance.

Astronomers have discovered that most stars are like the Sun in many ways. They are composed mostly of hydrogen and helium, and they have similar masses. Most fuse hydrogen in their cores and emit visible-wavelength photons into space. A small percentage of stars have more than 30 times the Sun's mass ($30\ M_\odot$). These stars are substantially hotter than the Sun and are bluish. Other stars are much less massive than the Sun—only one-tenth its mass—and are cool, red, and dim. Moreover, even stars similar to the Sun in composition and mass may differ enormously from it in their diameter. For example, some giant stars have diameters hundreds of times larger than the Sun's—so big that were the Sun their size, it would extend beyond the Earth's orbit and swallow the inner planets. On the other hand, some dwarf stars are only about the size of the Earth.

Astronomers have learned about these properties of stars by using physical laws and theories to interpret measurements made from the Earth. This area of astronomy, known as **astrophysics**, has the remarkable ability to give us detailed "pictures" of stars even though nearly all of them look like points of light under even the highest magnifications of the largest telescopes.

Background Pathways: Unit 21

FIGURE 54.1

The brightest star in the image, slightly below center, is Alpha Centauri, one of the nearest stars to us, at a distance of a little over 4 ly. The star nearly as bright as Alpha Centauri, just up and to the right from it, is over 500 ly away. Comet Halley, on the left side of this photograph taken in April 1986, looks almost as bright as the stars but was at a distance of under 1 AU.

54.1 LUMINOSITY

One critical property for understanding a star is the total amount of energy it radiates out into space each second: its **luminosity,** usually abbreviated as *L*. A star's luminosity tells us how much energy is being generated within it, which is one of the most important differences between star types.

An everyday example of luminosity is the wattage of various lightbulbs: A typical table lamp bulb has a luminosity of 100 watts, whereas a bulb for an outdoor parking lot light may have a luminosity of 1500 watts. By comparison, stars are almost unimaginably more luminous. The Sun, for example, has a luminosity of about 4×10^{26} watts. This "wattage" is fairly typical of stars, although they may range many factors of 10 larger or smaller.

These numbers are so enormous that it is more convenient to use the Sun's luminosity as a standard unit, where $1\, L_\odot = 4 \times 10^{26}$ watts. When we speak, then, of a star with a luminosity of $2\, L_\odot$ or $100\, L_\odot$, we mean it has a luminosity of 8×10^{26} or 4×10^{28} watts, respectively.

Stars generate their luminosity by "burning," or more accurately fusing, the nuclei of atoms together in their cores. A star's luminosity indicates not only the rate at which it is emitting energy into space, but also the rate at which it is fusing atoms in its core. Over thousands of years of observation, nearly all of the stars we see in the night sky have remained steady in their brightness. From this constancy, we conclude that the energy generation process deep inside stars must be very stable, steadily replacing the energy radiated to space.

The Sun must convert 4 million tons of matter into energy every second (Unit 50). One of the most luminous stars known, discovered by the Hubble Space Telescope in 1997, has 10 million times the Sun's luminosity. At that phenomenal rate of energy release, every few months the star is annihilating as much mass as is contained in the entire Earth.

54.2 THE INVERSE-SQUARE LAW

When we look at stars at night, some look brighter than others, but this is not necessarily because they are more luminous. We all know that a light looks brighter when we are close to it than when we are far from it. As light travels outward, its energy spreads uniformly in all directions, as shown in Figure 54.2A. If you are standing near the source, the light will have spread out only a little, and so more light enters your eye; if you are farther away, less light enters your eye, which you will perceive as the light source being dimmer.

This dimming with increasing distance may be easier to understand if you think in terms of photons. Photons leaving a star or other light source spread out along straight lines in all directions. If you imagine a series of progressively larger spheres drawn around the light source, the same number of photons pass through each sphere in one second. However, because more distant spheres are larger, the number of photons passing *through a fixed area* grows smaller, as shown in Figure 54.2B. As you can see, fewer photons pass through one square meter on a sphere at each successive distance from the source. As a result, an observer collecting light with a telescope will find that if the telescope is far from a star, it will collect fewer photons than when it is near the star.

The amount of light reaching us from a star is called its **brightness** and is a function of distance. This is different from a star's luminosity, which is the total amount of light emitted. The luminosity is expressed by the number of watts of light energy

Some astronomers use the term *flux* or *apparent luminosity* to describe the brightness.

$$\frac{36}{1^2} = 36 \text{ photons/m}^2 \qquad \frac{36}{2^2} = 9 \text{ photons/m}^2 \qquad \frac{36}{3^2} = 4 \text{ photons/m}^2$$

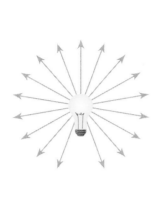

A   **B**   **C**

FIGURE 54.2

The inverse-square law. (A) Light spreads out from a point source in all directions. (B) As photons move out from a source, they are spread over a progressively larger area as the distance from the source increases. Therefore an area of one square meter, intercepts fewer photons the farther it is from the source. (C) At a distance d from a source of light, the area over which the light is spread equals $4\pi d^2$.

ANIMATIONS

Inverse-square law

B = Brightness of received light (watts/meter^2)

L = Luminosity of star (watts)

d = Distance to star (meters)

 A standard 40-watt lightbulb produces about 1 watt of luminous power at visible wavelengths. The difference between the electrical power consumption (40 watts) and the light delivered (1 watt) occurs because most of the power is radiated as heat at infrared wavelengths we cannot see. Stars also generate some of their light at infrared and ultraviolet wavelengths, but a star like the Sun emits most of its power at visible wavelengths.

generated by the star, but we describe brightness as the wattage received *per square meter*. That is, if we had a telescope with a collecting area of one square meter, and we collected 10 watts of light, we would conclude that the brightness was 10 watts per square meter. If we measured the same light source with a smaller telescope, with a collecting area of just 0.1 square meters, we would collect only one watt of light.

To picture how the brightness of a star depends on its distance from us, imagine all the photons emitted by a star in a single moment. At some later time, those photons will have traveled a distance d, forming a sphere surrounding the star. The star's luminosity L will be spread over the surface of a sphere of radius d, as shown in Figure 54.2C. We use d to stand for radius here to emphasize that we are speaking of a distance. The surface area of the sphere is given by the geometric formula $4\pi d^2$, so the brightness we observe, B, is calculated as follows:

$$B = \frac{L}{4\pi d^2}.$$

This relationship is called the **inverse-square law** because distance appears in the denominator as a square. The inverse-square law relates an object's luminosity (the total energy it emits) to its distance and its brightness (how bright it looks to us). Thus, the inverse-square law puts the everyday experience of how distant objects look dimmer into a mathematical form.

To get an idea of the brightness levels of stars, you could perform the following experiment. Take a light source that generates 1 watt of luminosity and place it at various distances until it looks similar in brightness to a bright star. It turns out that you will have to place it about 2 km away to look about as bright as the brightest stars in the sky. By applying the brightness formula to this known source, you would find that its brightness is

$$B = \frac{1 \text{ watt}}{4\pi(2000 \text{ m})^2} = 2 \times 10^{-8} \text{ watt/m}^2.$$

This tells us that the brightest stars in the sky deliver only 2 hundred-millionths of a watt per square meter. Because the pupil of your eye is at most about 8 mm in diameter, your eye is sensing only about a *trillionth* of a watt when you look at a bright star.

The inverse-square law is one of the most powerful tools of astrophysics. We can measure B for a star by using a **photometer,** a device similar to the electronic exposure meter in a camera. A photometer measures the amount of electromagnetic energy that strikes it each second. When calibrated for the size of the telescope, this can be converted into watts per square meter. If we can also determine the star's distance, d, using a technique such as parallax (Unit 52), for example, then we can calculate the star's luminosity, L, by rearranging the inverse-square law:

$$L = B \times 4\pi d^2.$$

Such measurements indicate that although many stars have luminosities comparable to the Sun's, the entire range of measured luminosities is enormous. Some stars are thousands of times less luminous than the Sun, while others are millions of times more luminous. Determining what causes this wide range of luminosities is one of our challenges in understanding stars.

54.3   DISTANCE BY THE STANDARD-CANDLES METHOD

We use brightness to estimate distance in many situations. For example, suppose you look at two street lights that have the same wattage bulb, but one is close and one is far away. From how bright they appear, you can estimate how much farther away the dim one is than the bright one (Figure 54.3). In fact, if you drive at night, your life depends on making such distance estimates when you see traffic lights or oncoming cars. Astronomers use a more refined version of this idea to find the distances to stars and galaxies.

The inverse-square law provides a way to determine a star's distance if we know the star's luminosity and can measure how bright it appears. Finding the distances of stars based on their luminosities works only if we know the luminosities with some assurance. For example, if we determine that the properties of some star are almost identical to the Sun's, we might reasonably infer that its luminosity would be the same as the Sun's too. Then, knowing the star's luminosity, L, we can measure its brightness, B, and solve for its distance from the inverse-square law. Rearranging that equation, we get

$$d = \sqrt{\frac{L}{4\pi B}}.$$

$$B \times \left(\frac{d^2}{B}\right) = \frac{L}{4\pi d^2} \times \left(\frac{d^2}{B}\right)$$

$$\sqrt{d^2} = \sqrt{\frac{L}{4\pi B}}$$

$$d = \sqrt{\frac{L}{4\pi B}}$$

For example, Alpha Centauri A is a star whose color and spectrum have been found to be very similar to the Sun's. Using a photometer, we find that its brightness is about 3×10^{-8} watts per square meter. If we assume it has the same luminosity as the Sun, we can plug the numbers into the above equation:

$$d = \sqrt{\frac{4 \times 10^{26} \text{ watts}}{4\pi \times 3 \times 10^{-8} \text{ watts/m}^2}} = \sqrt{1 \times 10^{33} \text{ m}^2} = 3 \times 10^{16} \text{ m}.$$

FIGURE 54.3
Street lights in this photograph illustrate that the brightness of light we see from sources with the same luminosity diminishes with distance.

How could you determine the light output of a candle compared to that of a lightbulb?

Converting this to parsecs, we find a distance of about 1 pc, which is slightly less than Alpha Centauri's measured distance of 1.35 pc. The match is not perfect because Alpha Centauri is slightly more luminous than the Sun; however, the distance is fairly close to the correct value.

Before parallaxes were measured, astronomers applied very early photometric estimates made by eye (discussed in detail in the next section) and reached similar conclusions about the distances of stars. Of course, most stars are *not* as similar to the Sun as is Alpha Centauri A, so the particular distances derived for them were often much less accurate than the distance we just found for Alpha Centauri. Nevertheless, astronomers began to appreciate just how far away the stars must be for us to be receiving so little energy from them compared with the Sun.

In this early history of photometry, scientists used as their standard of light generation a particular kind of candle that burned wax at a specified rate. These were well-calibrated light sources and were known as "standard candles." Today, if we can find any particular type of star whose luminosity is well determined and can therefore serve as a reference for other stars, we call them **standard candles** too. The Sun is *not* considered a good standard candle because many stars look quite similar to it, yet have substantially different luminosities than the Sun. On the other hand, some stars have peculiar characteristics—such as unusual aspects of their spectra, or curious ways in which their brightness pulsates—that allow them to be identified as unique. Tests have shown that some of these unique classes of stars have consistent luminosities, making them good standard candles.

We can use standard candles to find distances just as we did for Alpha Centauri. This distance-finding scheme is referred to as the *method of standard candles*. This is a powerful method for the following reason: If we identify a set of stars that all have the same luminosity, *we can determine the distances to all of them if we can find the distance to any one of them.*

Care must be taken, though! The accuracy of the standard-candle method relies on the degree to which the stars actually have the same luminosity. If we mistakenly group stars of dissimilar luminosities, we will get incorrect distances. This kind of mistake has happened several times in the history of astronomy, forcing us to revise distance estimates, and even forcing us to revise our notions of the size of the entire universe!

54.4   THE MAGNITUDE SYSTEM

The magnitude system is convenient in some ways—it avoids the need to express brightness and luminosity with scientific notation—but it is complicated to use in many other ways. It is not necessary to use it, although it is handy to know a little bit about the values used because star brightness levels are often reported in magnitudes.

About 140 B.C.E., the Greek astronomer Hipparchus measured the apparent brightness of stars using units he called *magnitudes*. He designated the stars that looked brightest "magnitude 1" and the dimmest ones he could just barely see "magnitude 6." For example, Betelgeuse, a bright red star in the constellation Orion, is magnitude 1, and there are only about two dozen first-magnitude stars over the whole celestial sphere. The somewhat dimmer stars in the Big Dipper's handle are approximately magnitude 2, whereas the stars in the bowl of the Little Dipper range from magnitude 2 to 5 (Figure 54.4). From a dark clear sky, you may be able to see about 4500 stars, about two-thirds of which are sixth magnitude.

Astronomers still use Hipparchus's scheme to measure the brightness of astronomical objects, but they now use the term **apparent magnitude** to emphasize that they are measuring only how bright a star *appears* to an observer. A star's apparent magnitude is a way of indicating its brightness. Astronomers use the magnitude system for many purposes (for example, to indicate the brightness of stars on star charts, like the fold-out chart in the back of the book), but it has several confusing properties. First, the scale is "backward" in the sense that bright stars have "smaller" magnitudes while dim stars have "larger" magnitudes. Moreover,

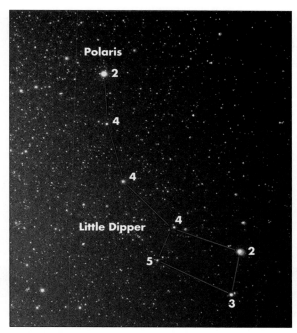

FIGURE 54.4

In Ursa Minor, the bowl of the "Little Dipper" is made of stars of magnitude 2, 3, 4, and 5 at each of its corners. Polaris is also a second-magnitude star.

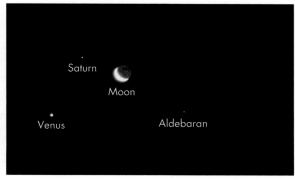

FIGURE 54.5

Venus, Saturn, and the Moon in close alignment with the star Aldebaran.

Q Where else do we use systems like the magnitude system, where a step is used to indicate a change by some factor?

modern measurements show that Hipparchus underestimated the magnitudes of the brightest stars, and so the magnitudes now assigned to the brightest few stars are negative numbers!

Magnitudes are not easy to work with because *differences* in magnitudes correspond to *ratios* in brightness. In particular, a difference of five magnitudes corresponds to a ratio of 100 in brightness. So when we say a star is five magnitudes brighter than another, we mean it is a factor of 100 brighter. That is, if we measure the brightness of a first-magnitude star and a sixth-magnitude star, the first-magnitude star is 100 times brighter than the sixth.

Each magnitude difference corresponds to a factor of about 2.512 (the fifth root of 100) in brightness. A first-magnitude star is 2.512 times brighter than a second-magnitude star and is 2.512 × 2.512, or 6.310, times brighter than a third-magnitude star. Table 54.1 lists the ratios that correspond to various differences in magnitude. For example, let us compare the apparent brightness of the planet Venus with the star Aldebaran (Figure 54.5). At its brightest, Venus has an apparent magnitude of −4.2; Aldebaran's apparent magnitude is 0.8, and so the difference in their magnitudes is 0.8 − (−4.2) = 5.0. We see from Table 54.1 that Venus at its brightest appears 100 times brighter to our eye than Aldebaran.

When using magnitudes it is important to understand the distinction between what astronomers call **absolute magnitude** and the apparent magnitude we observe. Absolute magnitude is defined as the apparent magnitude a star would have if it were situated at a distance of 10 parsecs from us. Although we cannot, of course, move stars to whatever distance we choose, absolute magnitude serves as a way of indicating a star's luminosity in units that are mathematically compatible with the apparent magnitude system. Table 54.2 illustrates how absolute magnitude is related to a star's luminosity.

TABLE 54.1	Relating Magnitudes to Brightness Ratios
Magnitude Difference	**Ratio of Brightness**
0	$2.512^0 = 1{:}1$
0.5	$2.512^{0.5} = 1.58{:}1$
1	$2.512^1 = 2.512{:}1$
2	$2.512^2 = 6.31{:}1$
3	$2.512^3 = 15.85{:}1$
4	$2.512^4 = 39.81{:}1$
5	$2.512^5 = 100{:}1$
10	$2.512^{10} = 10^4{:}1$
20	$2.512^{20} = 10^8{:}1$

TABLE 54.2	Relating Absolute Magnitude to Luminosity
Absolute Magnitude	**Approximate Luminosity in Solar Units**
−5	10,000
0	100
5	1
10	0.01

KEY TERMS

absolute magnitude, 449

apparent magnitude, 448

astrophysics, 444

brightness, 445

inverse-square law, 446

light-years, 444

luminosity, 445

photometer, 447

standard candle, 448

QUESTIONS FOR REVIEW

1. Describe the difference between luminosity and brightness.

2. Why is the Sun not a good standard candle?

3. What is the difference between apparent magnitude and absolute magnitude?

PROBLEMS

1. The faintest visible stars deliver only about 10^{-14} watts to your eye. A typical visible-wavelength photon has an energy of about 3×10^{-19} joules. How many photons per second are you seeing?

2. In the text we estimated the distance to Alpha Centauri assuming it had the same luminosity as the Sun. Use the measured brightness (3×10^{-8} watts per square meter) and its distance (4.4 light-years) to calculate its actual luminosity. How many times bigger than the Sun's luminosity is this?

3. Suppose star A is 36 times brighter than star B, but we have reason to believe that they both have the same luminosity.
   a. Which is more distant?
   b. How many times more distant is it than the other star?

4. Star C and star D are at the same distance from us, but star D is 10,000 times more luminous than star C. How do their brightness levels compare?

5. How do the magnitudes of star C and D in the previous problem compare?

TEST YOURSELF

1. Brightness is used by astronomers as a measure of
    a. watts received per square meter.
    b. a star's total luminosity in watts.
    c. a star's distance.
    d. absolute magnitude.
    e. a star's surface temperature.

2. The luminosity of a star depends on
    a. the star's distance from us.
    b. the inverse-square law.
    c. the energy the star generates each second.
    d. the star's rate of hydrogen to helium fusion.
    e. (c) and (d)

3. If a star is 5 *magnitudes* brighter than another, it is a *factor* of _____ brighter.
    a. 2.512
    b. 5
    c. 10
    d. 100
    e. None of the above

PLANETARIUM EXERCISES

All the planetarium exercises use the Starry Night planetarium software provided with your text. You should install the program and familiarize yourself with the various parameters and controls before trying these exercises. When the program asks you to enter a password, click "OK" to run Starry Night.

Set the location from which you are viewing to the nearest large city or your latitude and longitude. Accomplish this by clicking on the Location box on the toolbar. Verify that the local time is correct.

Go outside, and with the help of the star chart in the back of the book, locate a constellation. Sketch that constellation along with your estimate of relative star brightness. Rank the stars in order of decreasing brightness and use the magnitude scale to label your estimates for their brightness. Load the planetarium program and center on the constellation you chose. Place the cursor on each of the stars and compare your estimates of magnitude and order of brightness to the real numbers in the information boxes. Are there any discrepancies in your estimates? Any surprises?

55

The Temperatures and Compositions of Stars

Background Pathways:
Units 23 and 24

If we were studying flowers or butterflies, we would want to know something about their appearance, size, shape, colors, and structure. So, too, astronomers want to know the sizes, colors, and structures of stars. Such knowledge not only helps us better understand the nature of stars, but also is vital in unraveling their stories. Obtaining such knowledge is not straightforward, though, because stars are so far away that we cannot examine them directly.

A star's **spectrum** is perhaps the single most revealing thing we can examine to learn about the star. The spectrum depicts the amount of light emitted at each wavelength (Units 23 and 24) and tells us directly about a star's temperature and composition, as we shall explore in this unit. Beyond this, it is possible through detailed analyses of the spectrum to determine a star's luminosity, its velocity in space, and information about its mass and radius. This wealth of information arises because the physical conditions at the surface of a star determine the characteristics of the light emitted and absorbed by the atoms there.

55.1 INTERACTIONS OF PHOTONS AND MATTER IN STARS

Photons are generated deep in a star's extremely hot interior and then move outward through successive layers of ionized gas which have lower temperatures and densities. As the photons make their way outward, they typically travel just a centimeter or so before interacting with one of the numerous tightly packed electrons and ions. At each layer, the photons exchange energy with the matter there, and all reach a shared temperature.

The last stop photons make in their journey outward from the star's interior is in the **photosphere** (Unit 49). The photosphere is the layer where photons can escape to space—essentially because the gas above this region is thin enough for photons to pass through with little likelihood of additional interactions (Figure 55.1). Because photons that we see come from the photosphere, it is what we see as the "surface" of the star, even though there is gas of progressively lower densities continuing out beyond this radius. When we examine starlight, its properties reflect the temperature and composition in the photosphere.

Not all photons have an equal chance of escaping to space from the photosphere. The photosphere is relatively cool (compared to the interior of the star), so

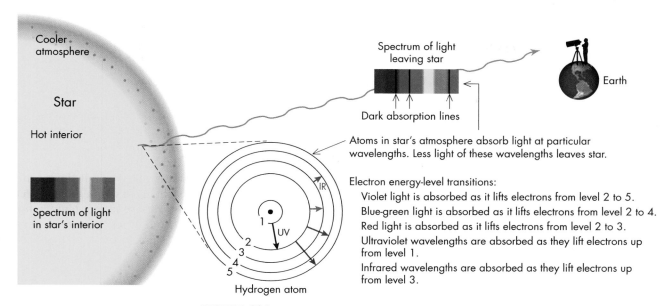

FIGURE 55.1

Formation of stellar absorption lines. Atoms in the cooler atmospheric gas absorb radiation at wavelengths corresponding to jumps between electron orbits. Absorption lines of hydrogen are illustrated.

Q We detect *some* photons from a star that have wavelengths matching the wavelengths of absorption lines in the star's atmosphere. At what elevation would you expect these photons to come from compared to photons that are far from any absorption line?

some electrons recombine with atoms there. Some stars' photospheres are so cool that atoms even combine to form molecules there. These atoms and molecules will absorb photons that have energies that match the energies needed to lift electrons into higher energy levels. Because a photon's energy is determined by its wavelength or color, when we observe the spectrum of light from a star, we find that certain colors are more strongly blocked. This creates dark **absorption lines** in the star's spectrum, as shown in Figure 55.1.

Each type of atom—hydrogen, helium, calcium, and so on—absorbs photons with a unique set of wavelengths. For example, hydrogen absorbs photons with wavelengths of 656.3, 486.1, and 434.1 nanometers, in the red, blue-green, and violet parts of the spectrum, respectively. Gaseous calcium, on the other hand, absorbs strongly at 393.3 and 396.8 nanometers, producing a strong double line in the violet portion of the spectrum. Because each type of atom absorbs a unique combination of wavelengths of light—its "fingerprint"—the dark lines in the spectrum depend on the composition of the gas in the photosphere.

In principle then it is possible to measure a star's composition by comparing the absorption lines in its spectrum with tables that list the wavelengths of lines made by each kind of atom. When we find matching absorption lines, we infer that the element exists in the star. To find the quantity of each atom in the star—each element's abundance—we use the darkness of the absorption line. A darker line generally implies a greater amount of that particular element, although some spectral lines are naturally stronger than others, so we refer to laboratory and theoretical work to interpret the line strengths under conditions such as those in the atmosphere of a star.

The strength of the absorption lines is heavily influenced by the temperature of the star's atmosphere. The temperature of a gas can change the strength of the spectral lines because the hotter the gas, the more frequent and higher the speed of collisions between the atoms. This alters the energy levels of the electrons and therefore the spectral lines we observe. We therefore need to know a star's temperature to interpret its spectrum.

55.2   STELLAR SURFACE TEMPERATURE

Stars are extremely hot by earthly standards. The surface temperature of even cool stars is far above the temperatures at which most substances vaporize, so using a physical probe to take a star's temperature would not succeed even if we had the means to send the probe to the star. If we want to know how hot a star is, we must once more rely on indirect methods. Yet the method used is familiar. You use it yourself in judging the temperature of an electric stove burner or a piece of glowing charcoal. When it glows bright orange, it is hotter than when it glows a dull red.

The typical temperatures in a star's photosphere are so hot that an electric stove burner would be vaporized. Even the coolest stars have surface temperatures as hot as the tungsten filament in a bright lightbulb, and other stars are so hot that they glow an intense blue or violet. Thus, an object's temperature can often be deduced from the color of its emitted light.

Hotter objects emit more blue light and cooler ones emit more red light.

You can see such color differences if you look carefully at stars in the night sky. For example, Rigel and Betelgeuse, the two brightest stars in the constellation Orion, have distinctly different colors (Figure 55.2). Rigel has a blue tint whereas Betelgeuse is reddish, so even our eyes can tell us that stars differ in temperature.

We can use color in a more precise way to measure a star's temperature with Wien's law (Unit 23). Wien's law relates an object's temperature to the wavelength,

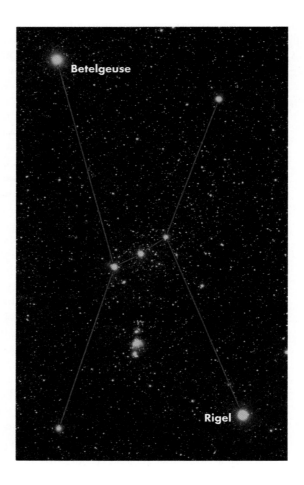

FIGURE 55.2

The constellation Orion contains both a bright red star, Betelgeuse, and a bright blue star, Rigel.

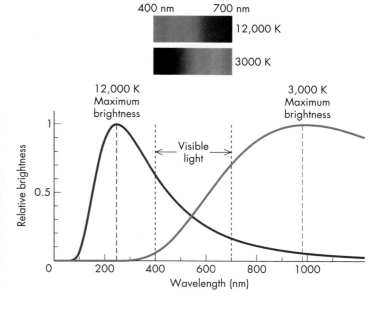

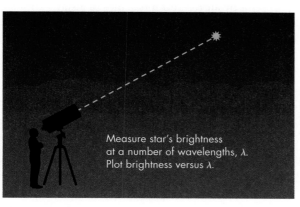

Measure star's brightness at a number of wavelengths, λ. Plot brightness versus λ.

FIGURE 55.3

Measuring a star's surface temperature using Wien's law. Spectra for radiation produced by a 12,000 K source and a 3,000 K source are shown. The wavelength at which a spectrum peaks allows us to estimate the source's temperature. The curves shown extend from ultraviolet to infrared wavelengths. Over the range of visible wavelengths, the brightness of the emission climbs from one end to the other, giving stars with these temperatures a blue-white or red-white color overall. (Note: brightness of hot and cool sources not to same scale.)

T = Temperature of continuum source (Kelvin)

λ_{max} = Wavelength of maximum brightness (nanometers)

λ_{max}, at which it radiates most strongly, as shown in Figure 55.3. Wien's law lets us calculate the temperature T from λ_{max} using this formula:

$$T = \frac{2.9 \times 10^6 \text{ K} \cdot \text{nm}}{\lambda_{max}}.$$

This law applies specifically to sources of thermal radiation, and the photospheres of stars are fairly good examples of this.

As a demonstration of Wien's law, consider Rigel again. Detailed measurements of its spectrum show that it radiates most strongly (has a wavelength of maximum brightness, λ_{max}) at about 240 nanometers, in the ultraviolet. Its temperature is therefore about

$$T(\text{Rigel}) = \frac{2.9 \times 10^6 \text{ K} \cdot \text{nm}}{240 \text{ nm}} = \frac{2,900,00}{240} \text{ K} = 12,000 \text{ K}.$$

The spectrum of light for 12,000 K thermal radiation is illustrated in Figure 55.3. Notice that even though the spectrum reaches a maximum outside the range of visible light, visible light is produced. The visible light is stronger toward the blue end of the spectrum, so Rigel looks blue to us.

The star Betelgeuse in Orion looks red to our eyes, but its wavelength of maximum brightness is actually in the infrared, at about 970 nm (Figure 55.3). So if we calculate its surface temperature, we find

$$T(\text{Betelguese}) = \frac{2,900,000}{970} \text{ K} = 3,000 \text{ K}.$$

This is about the temperature of the filament inside a standard 100 watt lightbulb. Sometime when you observe stars, compare the color of a distant incandescent bulb with Betelgeuse or another of the reddish stars you see in the sky. That reddish color also helps to explain why photographs taken using indoor lighting often look yellow or orange compared with those taken in sunlight. Like Betelgeuse, lightbulbs emit radiation mostly in the infrared. We see their visible light because they also emit electromagnetic radiation at shorter wavelengths than λ_{max}.

Wien's law is important for interpreting stellar colors; but as we have seen here, many stars have their maximum brightness outside the visible-wavelength range. In addition, the absorption of light by atoms and molecules in a star's atmosphere may make it difficult to measure the peak wavelength of the thermal radiation. Fortunately, studies of the spectra of stars have revealed another way to determine stars' temperatures.

55.3 THE DEVELOPMENT OF SPECTRAL CLASSIFICATION

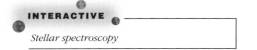

INTERACTIVE

Stellar spectroscopy

The classification of stellar spectra is important to astronomers today because it offers a quick way to determine a star's temperature. But this discovery took decades of study. **Stellar spectroscopy**—the study and classification of spectra—originated early in the 19th century when the German scientist Joseph Fraunhofer discovered absorption lines in the spectrum of the Sun and later in other stars.

In the late 1800s Henry Draper, a physician and amateur astronomer, began recording spectra with the then-new technology of photography. By observing a field of stars through a thin prism, he spread each star's light into a spectrum (Figure 55.4). On his death, Draper's widow endowed a project at Harvard to create a compilation of stellar spectra. In this work, stars were assigned to alphabetic types running from A to Q, collecting together similar-looking spectra in each type. These classifications were based mostly on the strength of the spectral lines—how dark they were—particularly the lines of hydrogen that are seen in the visible part of the spectrum. A-type stars have the strongest lines of hydrogen, while O-type stars, for example, have very weak lines.

About 1901, Annie Jump Cannon (Figure 55.5), the astronomer doing most of the classification for the Draper Catalog, determined that the types fell in a more

FIGURE 55.4
This is a photograph of the Hyades cluster made through a thin prism. The prism spreads the light of each star into a spectrum.

FIGURE 55.5
Annie Jump Cannon (1862–1941) founded the modern system of classifying stars.

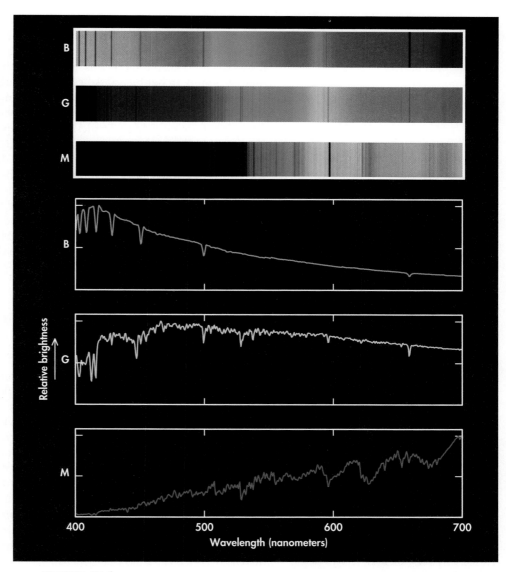

FIGURE 55.6

Spectra of stars of type B, G, and M. The B star is the hottest of the three, while the M star is the coolest. Cooler stars tend to have more absorption lines, but some lines are darker in hotter stars.

Q If the Sun had a temperature like Rigel or Betelgeuse, do you think our eyes would have evolved differently? Why or why not?

reasonable sequence if she rearranged them by temperature. She eliminated a number of types that had similar temperatures and reordered the rest into the sequence, from hottest to coolest, of O-B-A-F-G-K-M. Her work is the basis for the stellar **spectral types** we use today.

An example of the differences between spectra is given in Figure 55.6, which shows the spectra of stars with three different temperatures. The B-type spectrum is for a hot star like Rigel that produces mostly violet and blue light. The G-type spectrum is similar to the Sun's, with a temperature of about 6000 K and radiation peaks near the middle of the visible-wavelength range. The spectra are shown in Figure 55.6 both as they appear through a prism and as line tracings showing the relative intensity at each wavelength. The M-type spectrum is for a cool star like Betelgeuse. The B-type spectrum has only a few lines, but they are strong, and their pattern indicates they come from hydrogen. The M-type spectrum, however, shows a welter of lines with no apparent regularity.

The complicated history of the development of the spectral types resulted in the odd nonalphabetical progression for the spectral types—O-B-A-F-G-K-M. So much effort, however, had been invested in classifying stars using this system that it was easier to keep the types as assigned, with their odd order, than to reclassify them. Cannon, in her life, classified nearly a quarter million stars!

In the late 1990s astronomers started using sensitive infrared detectors that allow detection of stars even cooler than the M-type stars. Some of these stars are so cool that not only do molecules form in their atmospheres, but solid dust particles condense there as well. To continue the progression of spectral types astronomers have designated two new types—L and T—to describe these cool objects. The full spectral type sequence is therefore now **O-B-A-F-G-K-M-L-T,** from hottest to coolest.

As a help to remember the peculiar order, astronomers and students used a mnemonic. Before the discovery of the L and T types, the most widely known mnemonic was "Oh, Be A Fine Girl/Guy. Kiss Me," which was printed in textbooks throughout the 20th century. Perhaps you can develop a better mnemonic for the 21st century.

55.4   HOW TEMPERATURE AFFECTS A STAR'S SPECTRUM

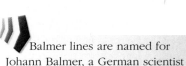

Balmer lines are named for Johann Balmer, a German scientist who first studied their pattern.

To see why temperature affects the lines in a star's spectrum, recall that a photon can be absorbed by an atom only if the photon's energy exactly matches the energy difference between two electron orbitals. For an atom to absorb the photon, it must already have an electron in the orbital with the lower energy level. An atom may be abundant in a star's atmosphere but create only weak lines at a particular wavelength simply because the gas is so hot or so cold that its electrons are in the "wrong" level to absorb light at that wavelength.

This was first worked out in the 1920s by the U.S. astronomer Cecilia Payne (later Payne-Gaposhkin; Figure 55.7). She pointed out that the strength of the hydrogen lines in a star's spectrum depend strongly on the star's temperature. She recognized that it would be a mistake to assume that a star had little hydrogen if these lines were weak unless the appropriate corrections for temperature had been made.

The absorption lines of hydrogen that are observable at visible wavelengths are made by electrons orbiting in the hydrogen atom's second level (see Figure 55.1). These lines are sometimes called the **Balmer lines** to distinguish them from other hydrogen lines with ultraviolet and infrared wavelengths. The Balmer lines occur at wavelengths where light has exactly the amount of energy needed to lift an electron from a hydrogen atom's second energy level up to the third or higher levels. These lines are no more important than other transitions in hydrogen that occur at infrared or ultraviolet wavelengths, but because their wavelengths are in the visible spectrum, they are much more easily observable.

If the hydrogen atoms in a star have very few electrons orbiting in level 2, the Balmer absorption lines will be weak, even if hydrogen is the most abundant element in the star. This may happen if a star is either very cold or very hot. In a cool star, most hydrogen atoms have their electrons in level 1, the lowest energy level. In a hot star the atoms move faster, and when they collide, electrons get excited ("knocked") into higher energy orbits: the hotter the gas, the higher the orbit. In fact, in very hot stars, electrons may be knocked out of the atom entirely, in which case the atom is said to be *ionized*. As a result of this excitation, proportionally more of the electrons in a very hot star will be in level 3 or higher. Hydrogen has the maximum number of its electrons in level 2 when its temperature is about

FIGURE 55.7
Cecilia Payne (1900–1979) discovered that hydrogen is the most abundant element in stars. She went on to become the first woman to be made a full professor at Harvard.

10,000 K. Hence it is in this temperature range that the Balmer lines are strongest. If we are therefore to deduce correctly the abundance of elements in a star, we must correct for such temperature effects.

After accounting for these effects, Payne discovered that virtually all stars are composed mainly of hydrogen. This idea was controversial at the time because most astronomers thought that stars were made primarily of heavy elements, and she was discouraged from publishing her conclusions by her thesis adviser. However, several years after her thesis was published, her ideas gained wide acceptance. Payne's idea led to our current understanding that about 71% of a star's mass is hydrogen, 27% is helium, and the remaining 2% is a mixture of all the other elements known to exist.

55.5   SPECTRAL CLASSIFICATION CRITERIA

Astronomers have put the stellar spectral types in order of temperature by using the pattern of lines in a star's spectrum. The similarity in most stars' compositions means that the main cause of differences in the spectra is the temperature of the photosphere. Figure 55.8 shows spectra of the seven main types—O through M—and illustrates that the differences in the line patterns are easy to see.

For example, A-type stars show strong hydrogen Balmer lines because they are near the ideal temperature for hydrogen to be in energy level 2. O stars have weak absorption lines of hydrogen but detectable absorption lines of helium, the second most abundant element, because they are so hot. At their high temperature, O stars' hydrogen atoms collide so violently and are excited so much by the star's intense radiation that the electrons are stripped from most of the hydrogen, ionizing it. With its electron missing, a hydrogen atom cannot absorb light. Because most of the O stars' hydrogen is ionized, such stars have extremely weak hydrogen absorption lines. Helium atoms, on the other hand, are more tightly bound and typically retain at least one of their electrons, allowing them to produce absorption lines.

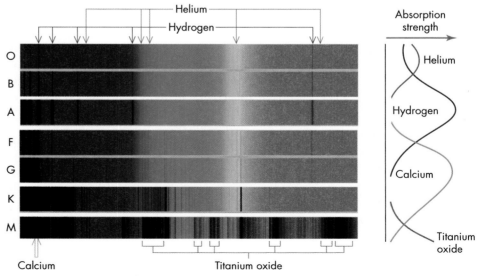

FIGURE 55.8

The stellar spectral types. The types are shown in order from hottest (O) to coolest (M). Several absorption lines are indicated, and their changing strength is illustrated at right.

TABLE 55.1	Summary of Spectral Types	
Spectral Type	Temperature Range (K)	Features
O	Hotter than 25,000	Ionized helium, weak hydrogen
B	11,000–25,000	Neutral helium, hydrogen stronger
A	7500–11,000	Hydrogen very strong
F	6000–7500	Hydrogen weaker, metals—especially ionized Ca—moderate
G	5000–6000	Ionized Ca strong, hydrogen weak
K	3500–5000	Metals strong, CH and CN molecules appearing
M	2200–3500	Molecules strong, especially TiO and water
L	1300–2200	TiO disappears. Strong lines of metal hydrides, water, and reactive metals such as potassium and cesium
T	900?–1300?	Strong lines of water and methane

Lines of some elements like ionized calcium become more prominent at lower temperatures, but in cooler stars the hydrogen Balmer lines grow much weaker. In this case the lines are weak because the hydrogen's electrons are mostly in level 1 and therefore require much more energetic ultraviolet photons to raise them to a higher energy level. The K and M stars have such cool atmospheres that some molecules, like titanium oxide, are able to form. These molecules often have rich and complex patterns of absorption lines.

The even cooler L stars show strong molecular lines of iron hydride and chromium hydride, while T stars show strong absorption lines of methane. These features, and those found in the other spectral types, are summarized in Table 55.1. Comparing the strength of spectral lines to determine a spectral type is now often done automatically by computers that scan a star's spectrum and match it against standard spectra stored in memory.

We have seen that O stars are hot and M stars are cool; but what, in fact, are their temperatures? Application of Wien's law and theoretical calculations show that temperatures range from more than 25,000 K for O stars to less than 3500 K for M stars, with A stars being about 10,000 K and G stars, such as our Sun, being about 6000 K. Because a star's spectral type is set by its temperature, its type also indicates its color. We know that hotter objects are blue and cooler objects are red, and indeed we find that O and B stars (hot types) are blue, while K and M stars (cool types) are red.

To distinguish finer gradations in temperature, astronomers subdivide each type by adding a numerical suffix—for example, B1, B2, . . . , B9—with the smaller numbers indicating progressively higher temperatures. With this system, the temperature of a B1 star is about 20,000 K, whereas that of a B5 star is about 13,500 K. Similarly, our Sun, rather than being just a G star, is a G2 star.

Finally, we note that a small number of stars have spectra and temperatures that do not fit into the normal classification sequence. They have compositions or temperatures that are quite different from the stars in the standard spectral sequence, and they are sometimes assigned their own classification letters. Some of these stars are now understood to be standard stars that have undergone dramatic changes late in their lifetimes; others remain mysteries that are still being explored.

Q: What might a cool star's spectrum look like if it contained no elements other than hydrogen and helium? What might a hot star's spectrum look like if it contained a large fraction of heavy elements?

KEY TERMS

absorption line, 453

Balmer lines, 458

O-B-A-F-G-K-M-L-T, 458

photosphere, 452

spectral type, 457

spectrum, 452

stellar spectroscopy, 456

QUESTIONS FOR REVIEW

1. What different ways are there to measure a star's temperature?

2. Why do stars have dark lines in their spectra?

3. What are the stellar spectral types? Which are hot and which are cool?

4. What distinguishes the spectral types of stars?

PROBLEMS

1. The star Rigel radiates most strongly at about 200 nanometers. How hot is it?

2. The bright southern star Alpha Centauri radiates most strongly at about 500 nanometers. What is its temperature? How does this compare to the Sun's?

3. If a T star has a surface temperature of 1000 K, at what wavelength should it be brightest?

4. A star has very faint Balmer lines. What are the next steps involved in determining its spectral type?

TEST YOURSELF

1. A star radiates most strongly at 400 nanometers. What is its surface temperature?
   a. 400 K
   b. 4000 K
   c. 40,000 K
   d. 75,000 K
   e. 7500 K

2. Which of the following stars is hottest?
   a. An M star
   b. An F star
   c. A G star
   d. A B star
   e. An O star

3. A star has lines of ionized helium. Ionizing helium requires a very high temperature. It is therefore most likely to belong to spectral type
   a. O.
   b. A.
   c. G.
   d. M.
   e. Z.

56

The Masses of Orbiting Stars

Gravity holds stars
orbit around each

Background Pathways: Unit 17

Many stars are not alone in space; rather, they have one or more stellar companions held in orbit about each other by their mutual gravitational attraction (Figure 56.1). Astronomers call such stellar pairs **binary stars**. Binary stars are extremely useful because, from their orbital motion, astronomers can determine their masses. To understand how, recall that the gravitational force between two bodies depends on their masses. The gravitational force, in turn, determines the stars' orbital motion. Therefore, if we can measure that motion, we can work backward to find the mass.

A star's mass is not just another piece of data in our catalog of stellar properties; it is the most critical property for determining the structure and fate of a star. A star's mass tells us how much weight squeezes the interior of the star, thereby driving the nuclear reactions that generate the heat and pressure that hold the star up (Unit 49). And the mass tells us how much hydrogen "fuel" is potentially available for the nuclear reactions needed to keep the star from collapsing under its own weight (Units 50 and 61).

Binary star systems are quite common—at least 40% of all stars known have orbiting companions, and the fraction may be much larger. Typically the stars in a binary system are a few AU apart. In many cases more than two stars are involved. Some stars are found in triplets, others quadruplets, and in at least one case a six-member system is known. The masses we determine for these stars give us mass estimates for stars of similar type that do not have companions.

56.1 TYPES OF BINARY STARS

A few binary stars are easy to see with a small telescope. Mizar, the middle star in the handle of the Big Dipper, is a good example and is illustrated in Looking Up #2. When you look at this star with just your eyes, you will also see a dim star Alcor near it, but this is not the binary companion. Alcor and Mizar form an **apparent double star:** two stars lying in the same direction but not actually in orbit around each other. With a telescope, however, we can resolve Mizar to reveal that it consists of two much closer stars in orbit around each other, Mizar A and Mizar B: a **true binary.**

We can directly observe the orbital motion of Mizar A and Mizar B around each other by comparing images made years apart. Such binary stars are called **visual binaries** because we can see two separate stars and their individual motions. In visual binary systems we can usually observe the two stars completing an orbit around each other over several years. Some widely separated binary stars, like Mizar A and B, have completed only a fraction of their orbit since their discovery in 1617. Astronomers can trace a partial arc of the orbit, and they estimate that the total length of the orbital period is a few thousand years.

Some binary stars are so close together that their light blends into a single blob that defies separation with even the most powerful telescopes. In such cases their orbital motion cannot be seen directly, but it may nevertheless be inferred from their combined spectra. As each star moves along its orbit, it alternately moves toward and then away from the Earth. This motion creates a Doppler shift, and so the spectrum of the star pair shows two sets of spectral lines that shift relative to each other. While one star's spectral lines shift to longer wavelengths (because the star is moving away from us), the other's spectral lines shift to shorter wavelengths (because it is approaching us). Then, in a cyclic fashion, half an orbit later, the pattern reverses (Figure 56.2). Astronomers call such star pairs **spectroscopic binaries,** and by observing a full cycle of their spectral shifts, we can determine the orbital motion of the stars. These stars are typically close together, and the orbital periods are short. For example, Mizar A is itself a spectroscopic binary with a period of just 20 days.

Eclipsing binaries are pairs of stars whose orbits are oriented exactly edge-on to our line of sight; so the stars eclipse each other sequentially. From the duration of the eclipses, it is possible to determine the sizes of the stars (Unit 57). When an eclipsing binary is also detected as a spectroscopic binary, we can determine the parameters of the system better than for other spectroscopic binaries. This is because we need to know the orientation of the orbits to know whether the Doppler shift reflects the full speed of orbital motion or just the part of the motion that happens to be along our line of sight toward the star. In an eclipsing system we know that the orbit is edge on to us.

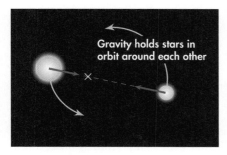

FIGURE 56.1

Two stars orbiting in a binary system, held together as a pair by their mutual gravitational attraction.

56.2 MEASURING STELLAR MASSES WITH BINARY STARS

Q When we observe spectroscopic binaries, we often do not know how the plane of the orbit is oriented to our line of sight. How will the Doppler shift change if the same orbit is oriented edge-on or face-on or somewhere in between?

From their knowledge of orbital motion as just described, astronomers can find the mass of a stellar pair using a modified form of Kepler's third law (Unit 17). Kepler demonstrated that the time required for a planet to orbit the Sun is related to its distance from the Sun. If P is the orbital period and a is the semimajor axis (half the long dimension) of the planet's orbit, then $P^2 = a^3$, a relation called Kepler's third law.

Newton discovered that Kepler's third law could be generalized to apply to *any* two bodies in orbit around each other. If their masses are M_A and M_B and they follow an elliptical path, then

$$(M_A + M_B) \times P^2 = a^3$$

where a is expressed in astronomical units, P in years, and the masses M_A and M_B in solar masses. This relationship is our basic tool for measuring stellar masses.

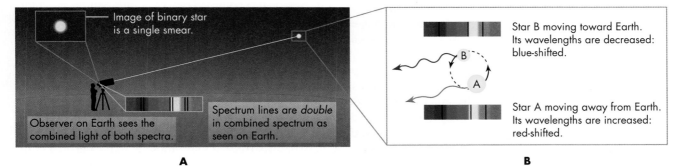

A   **B**

FIGURE 56.2

Spectroscopic binary star. (A) The two stars are generally too close to be separated by even powerful telescopes. (B) Their orbital motion creates a different Doppler shift for the light from each star. Thus, the spectrum of the stellar pair contains two sets of lines, one from each star, shifted relative to each other by different amounts depending on the stars' movements.

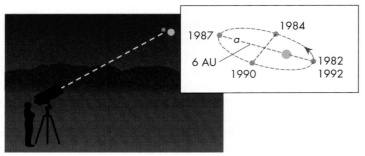

Plot of star positions ⟶ Period of 10 years

Measure semi-major axis = a = 6 AU

Use modified form of Kepler's third law

$$m + M = \frac{a^3}{P^2}$$
$$= \frac{6^3}{10^2}$$
$$= 2.16 \, M_\odot$$

FIGURE 56.3
Measuring the combined mass of two stars in a binary system using the modified form of Kepler's third law.

When Kepler devised his third law, he did not include a term describing mass. Why was he able to do this for the Solar System?

M_A = Mass of star A (solar masses)

M_B = Mass of star B (solar masses)

P = Period of orbit (years)

a = Semimajor axis of orbit (AU)

To find the mass of the stars in a visual binary, astronomers first plot their orbital motion, as depicted in Figure 56.3. It may take many years to observe the entire orbit, but eventually we can determine the time required for the stars to complete an orbit, P. From the plot of the orbit, and with knowledge of the stars' distance from the Sun, astronomers next measure the semimajor axis, a, of the orbit of one star about the other.

Consider, for example, the orbit of the two stars that comprise the visual binary star Alpha Centauri (shown in Looking Up #8: Centaurus and the Southern Cross). The two components have an orbital period of 68 years and a semimajor axis of 20.6 AU. We can then find their combined mass, M_A and M_B, by solving the modified form of Kepler's law, and we get (after dividing both sides by P^2)

$$M_A + M_B = \frac{a^3}{P^2}.$$

Inserting the measured values for P and a, we see that

$$M_A + M_B = \frac{20.6^3}{68^2} = 1.9 \, M_\odot.$$

That is, the combined mass of the stars is 1.9 times the Sun's mass.

56.3   THE CENTER OF MASS

Kepler's law allows us to find the combined mass of two stars that orbit each other. Additional analysis of stars' orbits allows us to find their individual masses. We can tell their masses relative to each other by the amount each star moves. If one star is much more massive than the other, then the massive one will hardly move while the less massive one exhibits nearly all of the motion—like the planets orbiting the Sun. If the two stars are equal in mass, they will both move the same amount.

Even in the Solar System, where we usually speak of the planets "orbiting the Sun," we should properly say that the Sun and planets orbit their common center of mass. The Sun is so much more massive than the planets that this point turns out to be nearly inside the Sun. The Sun "wobbles" because of the planets, primarily Jupiter. This kind of wobble seen for other stars is what allows us to detect planets orbiting them, even though we cannot see the planets directly (Unit 34).

Pairs of stars do not have such disparities in mass, though, so they orbit a point more nearly equidistant between them called the **center of mass,** as depicted in Figure 56.4. This is located along a line joining the two stars at a position that depends on the stars' relative masses. It is like the point of balance on a children's playground "see-saw" or "teeter-totter." If one star is three times more massive than the other, the point of balance will be three times closer to the more massive star.

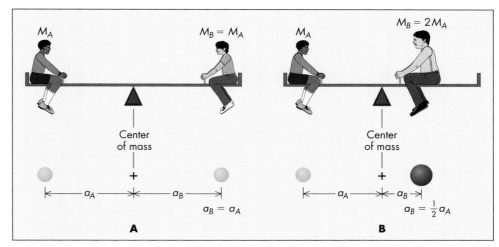

FIGURE 56.4

The location of the center of mass of two bodies depends on their relative mass. (A) If the masses are equal, the center of mass is halfway between. (B) If one body is more massive than the other, the center of mass is closer to the more massive body.

Note that $a_A + a_B$ add up to the semimajor axis a used in Kepler's formula.

Q The Greek scientist Archimedes is supposed to have said, "Give me a lever and a place to stand and I will move the Earth." Might this be possible? How?

We can generalize this idea into a mathematical formula. Suppose two stars have masses M_A and M_B (Figure 56.4). Their distances from the center of mass are a_A and a_B. The *larger* mass has the *smaller* distance from the center of mass, so we can write this as an inverse relationship:

$$\frac{a_A}{a_B} = \frac{M_B}{M_A}.$$

So, for example, if two stars are equal in mass ($M_A = M_B$), then $a_A = a_B$, and they will orbit a point exactly halfway between them. On the other hand, if star *B* is two times less massive than star *A* ($M_B = \frac{1}{2} \times M_A$), then star *B* will orbit two times farther from the center of mass than its orbital companion ($a_B = 2 \times a_A$). Similarly, an adult who weighs twice as much a child will overbalance the child on a see-saw unless he sits half as far from the pivot point as the child.

For spectroscopic binaries, we can compare the relative speed of each star. Each orbits the center of mass in the same period of time, but the more massive one has less distance to move in its orbit. So, for example, we might detect star *A* moving toward us at 2 km/sec while star *B* moves away from us at 8 km/sec (Figure 56.5). Half an orbit later we see star *A* moving away from us at 2 km/sec while star *B* moves toward us at 8 km/sec. This shows us that star *B* is moving 4 (= $\frac{8}{2}$) times more than star *A*. Therefore, star *B* is the less massive of the two and is four times less massive than star *A*.

For Alpha Centauri, measurements show that star *A* orbits about 0.7 times as far from the center of mass as star *B*. So $a_A = 0.7 \times a_B$. We therefore know that $M_B = 0.7 \times M_A$. From the previous section we also know that their combined mass equals 1.9 solar masses: $M_A + M_B = 1.9\ M_\odot$. Therefore,

$$M_A + 0.7 \times M_A = 1.7 \times M_A = 1.9\ M_\odot.$$

So Alpha Centauri A must be about 1.1 solar masses, while Alpha Centauri B is about 0.8 solar masses.

From analyzing many star pairs, astronomers have discovered that star masses range mostly from about 0.1 to 50 $M_\odot$. Our star, the Sun, is fairly average in mass, as it is in many other ways.

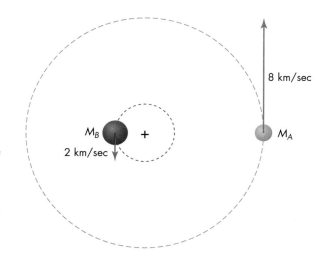

FIGURE 56.5

As two stars orbit their center of mass, the more massive star completes a smaller orbit about the center of mass and therefore travels at a slower speed than the less massive star. The Doppler shift of the more massive star is also smaller in proportion to the stars' relative masses.

 KEY TERMS

apparent double star, 462

binary star, 462

center of mass, 464

eclipsing binary, 463

spectroscopic binary, 463

true binary, 462

visual binary, 462

 QUESTIONS FOR REVIEW

1. What is a binary star?
2. How do visual and spectroscopic binaries differ?
3. Why are binary stars useful to astronomers?
4. What is the center of mass?

PROBLEMS

1. Two stars are in a binary system. Find their combined mass if they have
   a. An orbital period, P, of 5 years and an orbital separation, a, of 10 AU.
   b. An orbital period, P, of 2 years and an orbital separation, a, of 4 AU.
2. Mizar A and B are about 400 AU apart currently. Their orbital period is estimated to be between 2000 and 4000 years. What range of masses does this suggest for the pair?
3. Two stars are orbiting each other, and the sum of their masses is 6 times the Sun's mass. It is found that star A of the pair is orbiting 3 arc seconds from the center of mass, while star B is orbiting 6 arc seconds from the center of mass. What are the two stars' masses?

TEST YOURSELF

1. Calculating a star's mass from its binary orbital motion is important because
   a. such systems are rare, so we don't often get to collect this information.
   b. then we can predict the star's ultimate fate.
   c. then we can tell which star is orbiting the other.
   d. it's the only way we can tell if there's a second star in the system.
   e. it's the only method available for detecting apparent double stars.

2. Eclipsing binaries
   a. can be spectroscopic binaries.
   b. can be visual binaries.
   c. are edge-on to our line of sight.
   d. in some cases allow us to determine a pair's exact orbital parameters.
   e. All of the above

3. Which of the following is not a class of orbiting star systems?
   a. A sequential binary
   b. An eclipsing binary
   c. A spectroscopic binary
   d. A true binary
   e. A visual binary

57

The Sizes of Stars

A property of stars as basic as diameter seems like it should be easy to measure. However, the great distances of stars have made this one of the most challenging stellar properties to determine through direct observation. Astronomers have devised a variety of techniques for determining stars' sizes from observations, but measuring stellar sizes remains near the limits of our present technologies.

Astrophysics offers another solution. A great success of astrophysics in the 1800s and 1900s toward determining the nature of stars was the discovery of a physical law that describes the amount of light generated by hot objects. This law, discovered in the late 1800s, is called the *Stefan-Boltzmann law*. It tells us the radii of stars if we know their temperatures and luminosities.

We begin this unit by reviewing techniques for making direct measurements of stars' sizes. We conclude by showing how we can use the physical laws determined in Earth's laboratories to give us this important information about distant stars.

Background Pathways: Units 54 and 55

57.1 THE ANGULAR SIZES OF STARS

In principle, we can measure a star's radius from its angular size and distance (Unit 10). Unfortunately, the angular sizes of all stars except the Sun are extremely tiny because they are so far from the Earth. Even under the highest magnification in the most powerful telescopes, stars generally look like a smeary spot of light. The size of the smeared spot has nothing to do with the star's size, but is caused by the blurring effects of our atmosphere and by a physical limitation of telescopes called *diffraction* (Unit 29).

Astronomers have developed techniques for reducing the blurring effect of the Earth's atmosphere or avoiding it altogether by putting telescopes into space (Unit 30). However, diffraction presents a more fundamental limit. Bending of light at the edge of a telescope's aperture smears the light by an amount that depends inversely on the telescope's diameter. That is, diffraction effects are less severe in bigger telescopes. To measure the angular size of stars, however, we need telescopes with truly immense diameters. For example, measuring the angular size of a star like the Sun, if it were 50 light-years away, would require a telescope 300 meters in diameter, about three times the size of a football field—nearly impossible with current technologies, and unimaginably costly even if it were possible!

To avoid the need for enormous telescopes, astronomers have devised an alternative way to measure the angular size of stars: by using not one huge telescope, but two (or more) smaller ones separated by a large distance (for example, several hundred meters). This method, called **interferometry** (Unit 29), allows astronomers to measure angular sizes with a precision equal to those from a single telescope whose diameter is equal to the distance that separates the two smaller ones. Two smaller telescopes separated by 300 meters can measure details almost equivalent to those from a single 300-meter diameter telescope. A computer combines the information from the two telescopes to determine the angular size of the star. Such interferometric observations are still hampered by our atmosphere, and

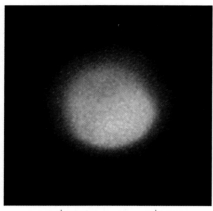

FIGURE 57.1

Image of the star Betelgeuse made by speckle interferometry. Betelgeuse is about 130 pc distant. The lack of detail in this image reflects the current difficulty in observing the disk of any star but our Sun.

0.05 arc seconds

they do not really provide a complete "picture" of the star; but they permit us to measure the diameters of many larger stars.

Another technique for determining stellar diameters is called **speckle interferometry.** In speckle interferometry, a single telescope is used to capture many short-exposure images of the object of interest. Each separate image is called a "speckle"—it is actually an image of the star formed by momentary focusing of light in the turbulent atmosphere of the Earth. By combining all of these images with a computer, astronomers can compensate for much of the blurring effects of the atmosphere. You can see a typical example of this process in the "deblurred" image of Betelgeuse, shown in Figure 57.1. Betelgeuse has the largest detected angular size of any star, but this is still only about 1/20th of an arc second, so the image has very little detail. Observations of other stars, which have smaller angular sizes, reveal even less detail about their appearance.

With interferometry, astronomers have measured the radii of over a hundred stars, but something is quite unusual about the stars that have been detected. We are detecting very few of the nearest stars—most are at large distances like Betelgeuse, which has a distance estimated from its parallax (Unit 52) to be about 130 pc. Using the distance versus angular size formula (Unit 10), this implies a diameter of over 6 AU—that's bigger than the orbit of Mars! If the Sun were that big, all of the inner planets would be swallowed up and destroyed. Almost all other stars with detectable diameters are also very large compared with the Sun. Interferometry is not allowing us to detect the diameters of stars the size of the Sun, so we need other techniques to analyze smaller stars in the stellar population.

57.2   USING ECLIPSING BINARIES TO MEASURE STELLAR DIAMETERS

Astronomers have found that many stars are in orbit around a companion star, much as the Earth orbits the Sun. Such gravitationally bound star pairs are called **binary stars** (Unit 56), and they allow astronomers to measure a number of stellar properties, including stellar diameters. On rare occasions, the orbit of a binary star pair will be oriented so that the plane of their orbit happens to lie almost exactly edge-on to our perspective from the Earth. When this occurs, as the stars orbit, one will eclipse the other as it passes between its companion and the Earth. Such systems are called **eclipsing binary stars,** and if we watch such a system, its light will periodically dim. The star Algol (Arabic for "Demon Star") in the constellation Perseus (see Looking Up #3) is the brightest example of this phenomenon. Normally, Algol is fairly bright—it is the second brightest star in Perseus—but once every 68 hours it dims to one-third its usual brightness for 10 hours, becoming the seventh brightest star in that constellation.

During most of the orbit of an eclipsing binary system, we see the combined light of both stars; but at the times of eclipse, the brightness of the system decreases as one star covers part or all of the other. This produces a cycle of variation in light intensity called a **light curve.** Figure 57.2A is a graph of such change in brightness over time.

The duration of the eclipses depends on the diameter of the stars. Figure 57.2B shows why. The eclipse begins as the edge of one star first lines up with the edge of the other, as shown in the figure. Dimming increases until the smaller star is completely in front of the larger star; the light level remains low until the smaller star reaches the other edge of the large star. Then the process reverses until we see all the light from both stars again. A dip in brightness is also seen when the smaller star passes behind the larger star. The relative size of these dips depends on

ANIMATIONS

Eclipsing binary

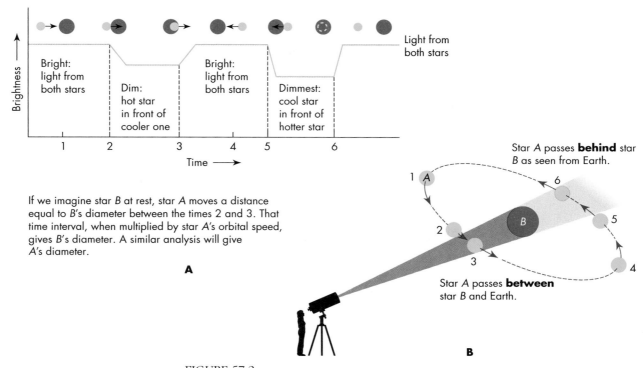

If we imagine star *B* at rest, star *A* moves a distance equal to *B*'s diameter between the times 2 and 3. That time interval, when multiplied by star *A*'s orbital speed, gives *B*'s diameter. A similar analysis will give *A*'s diameter.

A

Star *A* passes **behind** star *B* as seen from Earth.

Star *A* passes **between** star *B* and Earth.

B

FIGURE 57.2
An eclipsing binary and its light curve. From the durations of the eclipses, the diameter of each star may be found. In this illustration, the yellow star is hotter than the red star.

Orbital speeds can be measured using Doppler shifts (Unit 25) of the stars relative to each other.

which star has the hotter surface temperature—not necessarily on which star is bigger. For example, when a large cool star is eclipsed by a smaller hotter one, the reduction of light may be quite small. The pattern of the light curve also depends on whether the stars pass completely behind each other, as we have assumed here, or if they only partially eclipse each other.

If we know how fast the stars are orbiting each other, we can estimate the size of each star. Suppose the stars are moving at a speed of 100 km/sec around each other, and the light takes 10,000 seconds (about 3 hours) to dim from its full brightness at the start of the eclipse to its dim state at full eclipse. This is how long it takes for the smaller star's disk to move sideways by its own diameter across the edge of the larger star—either in front of or behind it. The star's diameter is therefore equal to the product of its speed and the time for the dimming. In this example the distance is 100 km/sec × 10,000 sec = 1,000,000 km. The total duration of the eclipse is the time it takes for the disk of the smaller star to move from being entirely uncovered on one side of the larger star, across the larger star, to being completely uncovered on the other side—which is the sum of the diameters of both stars. Suppose this takes 100,000 sec (about 28 hours); then the sum of the diameters is = 100 km/sec × 100,000 sec = 10,000,000 km. Combining the two results, we find that the larger star has a diameter of 9,000,000 km.

This method of using eclipses lets us estimate the sizes of stars that are much smaller than are detectable with interferometric methods. It has drawbacks, though. For example, it applies only to the small subset of stars that happen to be in binary systems with orbits having just the right characteristics. Moreover, the method has uncertainties because the orbit may not be exactly edge-on, in which case the duration of the eclipse does not reflect the total diameter of the star.

Q Would you expect the kinds of stars that occur in eclipsing binary systems to be typical of all stars?

57.3   THE STEFAN-BOLTZMANN LAW

σ = Constant = 5.67×10^{-8} watts per meter^2 per Kelvin^4

T = Temperature (Kelvin)

In the late 1800s scientists discovered how the luminosity of a hot object depends on its temperature. They found that bodies radiate an amount of energy that depends on their temperature—each square meter of the body radiating an amount equal to

$$\text{Luminosity (per square meter)} = \sigma T^4$$

as shown in Figure 57.3A. This relationship, named for its discoverers, is called the **Stefan-Boltzmann law** (Unit 23). It expresses the fact that most matter—as long as it is not a thin gas—emits electromagnetic radiation that increases rapidly as the temperature climbs.

The Stefan-Boltzmann law does not apply to all materials. For example, it does not accurately describe the radiation from hot, low-density gas, such as that found in fluorescent lightbulbs or interstellar clouds. This is because the emission from a thin gas comes primarily from spectral lines of individual atoms. The Stefan-Boltzmann law applies only to the broad thermal radiation produced by atoms constantly colliding with one another and hence vibrating at a wide range of frequencies. Thus, although the Stefan-Boltzmann law applies to the gas in a star's photosphere, it does not apply to the radiation emitted in the low-density layers above it (such as in the Sun's chromosphere). In these less dense regions, individual electron jumps are the way electromagnetic energy is emitted and absorbed.

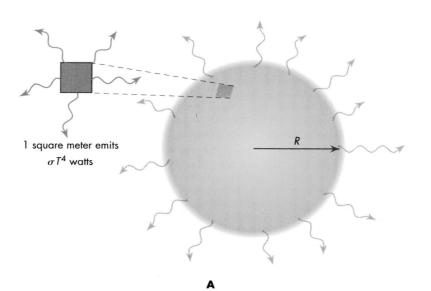

Total energy radiated per second by the star is its

Luminosity = L

L = Energy emitted by 1 square meter × Number of square meters of its surface

= σT^4 × Star's surface area

For a spherical star of radius R, the surface area is $4\pi R^2$

Thus, $L = \sigma T^4 \times 4\pi R^2$

or

$L = 4\pi R^2 \sigma T^4$

A   **B**

FIGURE 57.3

The Stefan-Boltzmann law can be used to find a star's radius. (A) Each part of the star's surface radiates σT^4. (B) Multiplying σT^4 by the star's surface area ($4\pi R^2$) gives its total power output—its luminosity, $L = 4\pi R^2 \sigma T^4$. To find the star's radius, we solve this equation to get $R = \sqrt{L/(4\pi \sigma T^4)}$. Finally, we can use this equation and measured values of the star's L and T to determine R.

L = Luminosity of star (watts)

σ = Constant = 5.67×10^{-8} watts per meter^2 per Kelvin^4

T = Surface temperature of star (Kelvin)

R = Radius of star (meters)

Q What other equations have you used that have three variables? Can you always solve for one variable if you know the other two?

$$L \div (4\pi\sigma T^4) = 4\pi R^2 \times \sigma T^4 \div (4\pi\sigma T^4)$$

$$R^2 = \frac{L}{4\pi\sigma T^4}$$

Taking square root of both sides gives

$$R = \frac{\sqrt{L}}{\sqrt{4\pi\sigma} \times T^2}$$

A convenient form of the Stefan-Boltzmann equation is to solve for the radius in terms of the Sun's measured properties:

$$\frac{R}{R_\odot} = \sqrt{\frac{L}{L_\odot}} \times \left(\frac{T_\odot}{T}\right)^2$$

We can apply the Stefan-Boltzmann law to a star's luminosity as follows. According to the law, if a star has a temperature T, each square meter of its surface radiates an amount of energy per second given by σT^4. We can find the total energy the star radiates per second—its luminosity, L—by multiplying the energy radiated from one square meter (σT^4) by the number of square meters of its surface area (Figure 57.3B). If we assume that the star is a sphere, its surface area is $4\pi R^2$, where R is its radius. Its luminosity, L, is therefore

$$L = 4\pi R^2 \times \sigma T^4.$$

That is, a star's luminosity equals its surface area times σT^4. This relation between L, R, and T may at first appear complex, but the equation's meaning is as follows: Increasing either the temperature or the radius of a star makes it more luminous. Making T larger makes each square meter of the star brighter. Making R larger increases the number of square meters.

The Stefan-Boltzmann equation has three "variables"—the quantities luminosity, radius, and temperature—that may vary from star to star or even from one time to another in a particular star. Whenever we have an equation with three variables, if we can measure two of them, we can find the third one from the equation. This means that we can find a star's radius from the Stefan-Boltzmann equation because we have ways of determining a star's luminosity (from its brightness and distance) and temperature (from its color or spectrum). Mathematically, this permits us to solve the equation for R if we know T and L.

We can rearrange the Stefan-Boltzmann equation algebraically and solve for a star's radius:

$$R = \frac{\sqrt{L}}{\sqrt{4\pi\sigma} \times T^2}.$$

We can test this formula on the Sun. Its luminosity is measured to be $L_\odot = 4 \times 10^{26}$ watts, while its surface temperature is measured to be $T_\odot = 5800$ K (Unit 49). We calculate then that

$$R_\odot = \frac{\sqrt{L_\odot}}{\sqrt{4\pi\sigma} \times T_\odot^2} = \frac{\sqrt{4 \times 10^{26}}}{\sqrt{7.13 \times 10^{-7}} \times 5800^2} = \frac{2 \times 10^{13}}{2.8 \times 10^4} = 7.1 \times 10^8 \text{ m}$$

or 710,000 km. This is quite close to the Sun's measured radius of 696,000 km, demonstrating the power of this equation.

To gain a feel for this equation, think about two different cases: (1) when the temperature remains the same or (2) when the luminosity remains the same. If two stars have the same temperature, the more luminous one has the larger radius. On the other hand, for two stars that have the same luminosity, the hotter one must be smaller (because T is in the denominator). We can use this form of the Stefan-Boltzmann equation to solve for the radius of any star whose distance, brightness, and temperature are known—which is many hundreds of thousands of stars.

The Stefan-Boltzmann law is very useful, but it relies on some assumptions about how stars generate light. Eclipsing binary and interferometry measurements provide an important check on the technique. The observations by all three methods show that stars differ enormously in radius. Although many stars have radii similar to the Sun's, some are hundreds of times larger; astronomers call them *giants*. Some are hundreds of times smaller and are called *dwarfs*. We will discover in subsequent units that a star like the Sun will in future stages turn into both!

KEY TERMS

binary star, 468

eclipsing binary star, 468

interferometry, 467

light curve, 468

speckle interferometry, 468

Stefan-Boltzmann law, 470

QUESTIONS FOR REVIEW

1. How does interferometry work to measure stars' diameters? What sort of stars does this method work for?

2. What is an eclipsing binary? What can be learned from eclipsing binaries?

3. The Stefan-Boltzmann law is based on what assumptions?

4. How does the Stefan-Boltzmann law allow us to measure stars' sizes?

PROBLEMS

1. Alpha Centauri is actually two stars that orbit each other. Recent measurements show that Alpha Centauri A has an angular size of 0.0085 arc seconds, while Alpha Centauri B has an angular size of 0.0060 arc seconds. The pair are at a distance of 4.4 light-years. What are their diameters compared to the Sun's?

2. A star with the same color as the Sun is found to produce a luminosity 81 times larger. What is its radius compared to the Sun's?

3. A star with the same radius as the Sun is found to produce a luminosity 81 times larger. What is its surface temperature compared to the Sun's?

4. The surface temperature of Arcturus is about half as hot as the Sun, but it is about 100 times more luminous. What is its radius compared to the Sun's?

5. If a star's surface temperature is 30,000 K, how much power does a square meter of its surface radiate?

TEST YOURSELF

1. If the surface temperature of a star is doubled, but its radius remains the same, its new luminosity is _____ its old luminosity.
    a. 16 times smaller than
    b. 4 times smaller than
    c. the same as.
    d. 4 times larger than
    e. 16 times larger than

2. The Stefan-Boltzmann Law applies to
    a. individual atomic spectral lines.
    b. a star's chromosphere.
    c. a star's photosphere.
    d. clouds of interstellar gas.
    e. All of the above

3. If a star's luminosity increases, we can conclude that
    a. its radius has increased.
    b. its temperature has increased.
    c. its temperature or its radius has increased.
    d. its temperature and its radius both must have increased.
    e. None of the above

58 The H-R Diagram

By the early 1900s astronomers had learned how to measure stellar temperatures, masses, radii, and compositions, but understood little of how stars worked. We have now reached a similar point in our study of stars. We know that stars exhibit a wide range of properties. But why does such variety occur?

An important method in science for clarifying a wide array of data is to look for patterns and relationships between the different variables. A basic tool for doing this is to plot one variable against another in an *X–Y* diagram. Astronomers early in the 20th century discovered that a plot of temperature versus luminosity gave new insights into the nature of stars. That plot is now known as the **H-R diagram**—the letters H and R standing for the initials of the Danish astronomer Ejnar Hertzsprung and the U.S. astronomer Henry Norris Russell, who independently discovered its usefulness.

To construct an H-R diagram of a group of stars, astronomers plot each star according to its spectral type (or temperature) on the *X* axis and its luminosity on the *Y* axis, as shown in Figure 58.1. By tradition, they put bright stars at the top and dim stars at the bottom—O stars (hot, blue) on the left and M stars (cool, red) on the right. Notice that temperature therefore increases to the left, which is the opposite of the usual convention in graphs.

The H-R diagram shows that stars do not have just any combination of temperature and luminosity. Instead, there are patterns in this diagram, suggesting relationships between these quantities for certain classes of stars. The H-R diagram is like a crime detective's bulletin board. We pin up the various pieces of information we have about stars and see what they suggest about the stars. For example, it takes many billions of years for a star like the Sun to be born and to die. So we cannot follow a single star in its lifetime. But if we look at all the stars with masses similar to the Sun's, we can begin to piece together clues about how a star with the Sun's mass might appear at different stages of its life.

**Background Pathways:
Units 54–57**

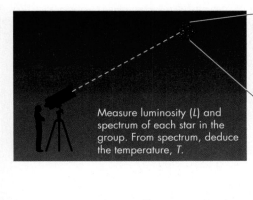

FIGURE 58.1

Constructing an H-R diagram from the spectra and luminosities of a set of stars.

473

58.1   ANALYZING THE H-R DIAGRAM

Figure 58.2 shows an H-R diagram based on modern measurements of parallax, brightness, and color from the Hipparcos satellite. These data have been converted into luminosities and temperatures, and one point is plotted for each star. The majority of the stars lie along a diagonal line in the diagram that runs from upper left (hot and luminous) to lower right (cool and dim). This line is called the **main sequence,** and stars within this region of the diagram follow a trend that you might expect: The hotter stars are brighter than the cooler stars. It is important to keep in mind that the H-R diagram does *not* depict the position of stars at some location

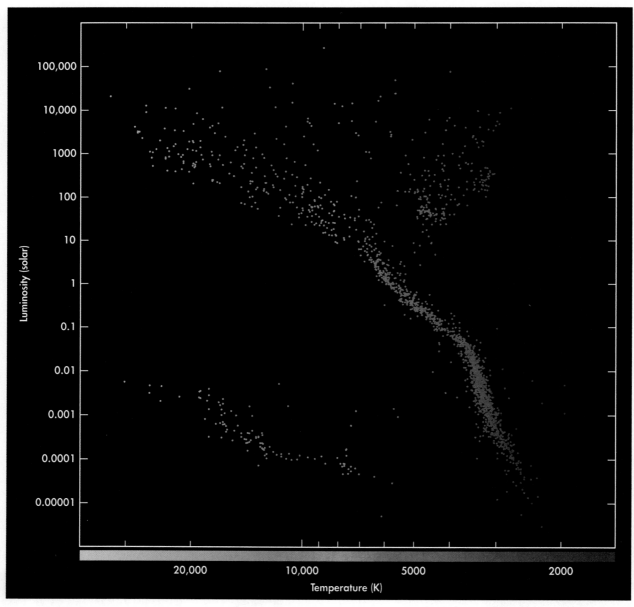

FIGURE 58.2

Hertzsprung-Russell diagram of bright and nearby stars based on data from the Hipparcos satellite.

The H-R diagrams have been constructed using both bright and nearby stars to illustrate the full range of stellar properties. The brightest stars in the sky are about the Sun's luminosity or greater, while the nearest stars are almost all less luminous than the Sun (see Unit 69).

Q Think of two properties that describe humans. If you made a plot of these two properties, what would the diagram look like? What could you learn about people from it?

INTERACTIVE

H-R diagram

Q If we made a plot of stars' distances versus their temperatures, what sort of patterns would we find?

in space. It shows merely a correlation between stellar properties, much as a height–weight table does for people, where we can see the trend that in general taller people are heavier than shorter people.

Although about 90% of the stars in most random samples will lie along the main sequence, some others are in the upper right part of the diagram, where stars are cool but very luminous, and some are below and to the left of the main sequence, where stars are hot but dim. To understand what makes stars in these two regions different, we turn to the Stefan-Boltzmann Law (Unit 57). This equation shows that if two stars have the same surface temperature but differ in their size, the larger one emits more energy—that is, it is more luminous—than the smaller one. For stars of the same temperature, more luminous implies a larger radius.

We can apply this reasoning to the H-R diagram. In Figure 58.3 we have labeled a number of the better-known stars. Consider first the stars on the right (cool) side of the diagram. For example, Proxima Centauri and Antares lie along an approximately vertical line. These stars lie at the same horizontal position on the H-R diagram because they have the same surface temperature. However, Antares, being nearer the top of the diagram, is more luminous than Proxima; that is why it is plotted in the upper part of the diagram. Because these two stars have the same temperature, one star must have more surface area than the other to be more luminous. In other words, the more luminous stars above the main sequence have a larger radius. Because Antares is large and red, it is called a **red giant.**

We can go one step further, however. By comparing how much more luminous one of these stars is, we can calculate how much larger its radius is. For example, we see that Antares has a luminosity about 100 million times larger than Proxima. This implies that Antares's surface area is 100 million times larger. If we treat the stars as spheres, then their surface area depends on the square of their radii. The area of a sphere is $4\pi R^2$, where R is its radius. A little math shows that if these two stars differ by 100 million (10^8) in area, they must differ by $\sqrt{100,000,000} = 10,000$ in radius. If the Sun were as big as Antares, it would swallow the Earth.

Not all giant stars are red. They span a range of temperatures and colors. For example, the North Star, Polaris, is a giant. However, it has about the same temperature and color as the Sun—yellow—but it is about 50 times larger. To distinguish between different classes of giants, we sometimes refer to them by their actual color—a blue or yellow giant, for example.

A similar analysis shows that the stars lying below the main sequence must have small radii if they are both hot and dim. Because these stars are so hot compared to main-sequence stars of the same luminosity, they are much bluer. Many of the first of these stars discovered were very hot, glowing with a white heat, and were therefore called **white dwarfs.** With modern observations we can detect some that have surface temperatures cooler than the Sun; but nonetheless, stars falling in this region below the main sequence are all called white dwarfs (Figure 58.3). For example, Sirius B, a dim companion of the bright star Sirius, is a white dwarf. Its radius is about 0.008 the Sun's—a little smaller than the Earth's radius—and its surface temperature is more than 20,000 Kelvin.

We can illustrate the range of star diameters by drawing lines in the H-R diagram that represent a star's radius. That is, we use the Stefan-Boltzmann equation ($L = 4\pi R^2 \times \sigma T^4$) and set R constant. As we make T larger, L must get larger. The result is a line for each value of R that slopes from upper left to lower right as shown in Figure 58.4. Any stars lying along such a line have the same radius. If we move up or to the right in the H-R diagram, we find stars of progressively larger radius. Thus the largest stars lie in the upper right, while the smallest stars lie in the lower left of the diagram. We can see that the giants span a wide range of radii. By contrast, the white dwarfs are all about the same size, about 1/100th of the Sun—about the size of the Earth. The main-sequence stars increase gradually from about under 1/10th the Sun's radius at the cool end to nearly 10 times larger at the hot end.

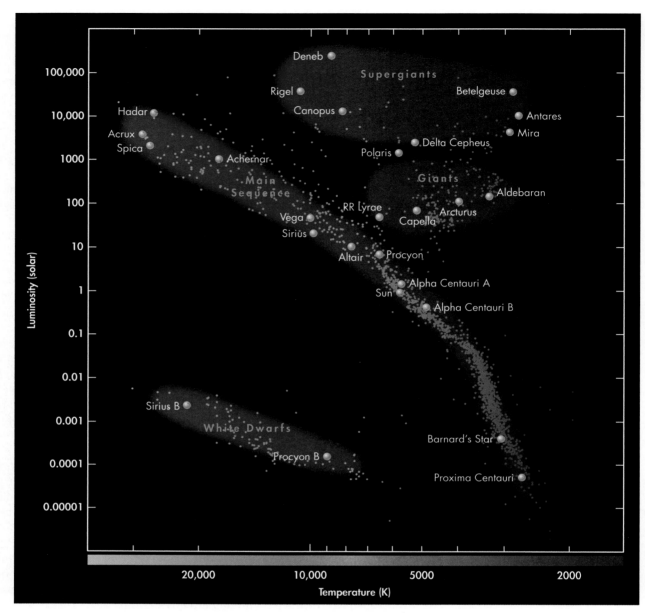

FIGURE 58.3

The H-R diagram. The long diagonal line from upper left to lower right is the main sequence. Giant stars lie above the main sequence. White dwarf stars lie below it. Notice that bright stars are at the top of the diagram and dims stars at the bottom. Also notice that hot (blue) stars are on the left and cool (red) stars are on the right.

58.2   THE MASS–LUMINOSITY RELATION

Some of the stars we have plotted in the H-R diagram are in binary systems, allowing us to determine their masses (Unit 56). When we label the stars in the H-R diagram by their masses, we notice that the masses are not randomly scattered around the diagram. As on a detective's bulletin board, we see certain patterns emerging.

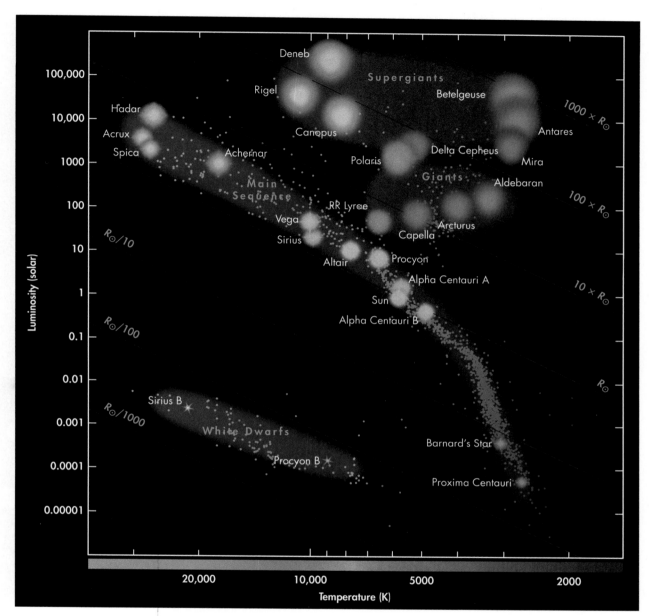

FIGURE 58.4
Lines showing where in the H-R diagram stars of a given radius will lie.

Stars near the same position on the main sequence all have approximately the same mass. Thus, main-sequence stars of the same spectral type as the Sun all have approximately the same mass.

Farther up and left along the main sequence we find stars of steadily larger masses, while farther down and right on the main sequence we find that the stars have smaller masses. From top to bottom along the main sequence, the star masses decrease from about 20 to 0.2 $M_\odot$ as shown in Figure 58.5A. If we expand our sample of stars, we find that we can extrapolate this trend to even more luminous stars that lie along the line of the main sequence to masses of more than 50 $M_\odot$, while there are extremely dim stars extending down from the main sequence that have masses under 0.1 $M_\odot$.

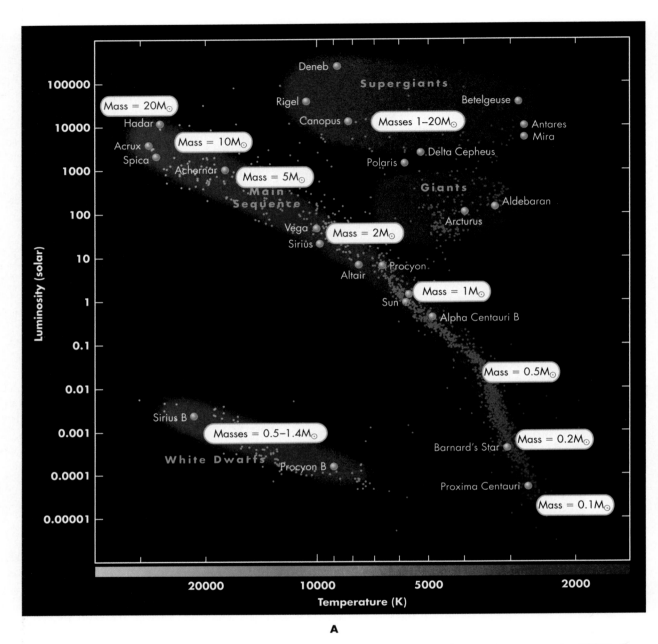

A

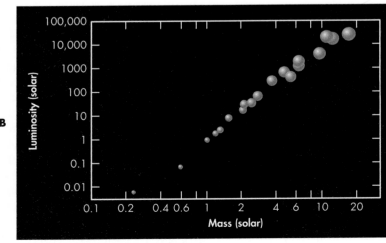

B

FIGURE 58.5

(A) On the H-R diagram, high-mass stars lie higher on the main sequence than low-mass stars. (B) The mass–luminosity relation shows that along the main sequence more massive stars are more luminous. The law does not work for red giants or white dwarfs.

This trend does not hold for stars in the rest of the H-R diagram. The giants have a wide range of masses, similar to that of the stars on the main sequence but in no particular order. White dwarfs, on the other hand, tend to all be around one solar mass or less, but again in no organized sequence. We will note these facts about the giants and white dwarfs, but set these types aside for the moment and concentrate on the main-sequence stars.

The trend of masses among main-sequence stars is illustrated in Figure 58.5B, where the stars' luminosities are plotted against their masses. This pattern was first recognized in 1924 by the English astrophysicist A.S. Eddington. The relationship between mass and luminosity can be expressed by a formula. If M and L are given in solar units, then for main-sequence stars

$$L \approx M^{3.5}.$$

This **mass–luminosity relation** is approximate and varies between $L \approx M^3$ and $L \approx M^4$ over different portions of the main sequence, but it gives a fairly accurate description of how the luminosity of a main-sequence star is related to its mass. If we apply this relation to the Sun, whose mass and luminosity are each 1 in these units, the relation is obeyed because 1 raised to any power still equals 1.

If we consider a more massive star, for example one 10 times as massive as the Sun, then the mass–luminosity relationship tells us that $L \approx 10^{3.5} \approx 3 \times 10^3$. That is, its luminosity is about 3000 $L_\odot$. Alternatively, a star with a mass of just 1/10th the Sun's mass would have a luminosity $L \approx 0.1^{3.5} \approx 3 \times 10^{-4}$. That is, its luminosity is about 0.0003 $L_\odot$.

We can understand in general terms how such a mass–luminosity relationship arises. For increasingly larger stellar masses, the gravitational pull among all the parts of a star squeezes the interior more strongly so that the gas is driven to a higher density and a higher temperature. A higher temperature generates a higher luminosity (recall the Stefan-Boltzmann law), so more massive stars are more luminous.

L = Luminosity in solar luminosities

M = Mass in solar masses

Note that taking M to the power of 3.5 is the same as multiplying $M \times M \times M \times \sqrt{M}$. On many calculators you can carry out this calculation using a key labeled x^y. To find $10^{3.5}$ you would type this into the calculator:

$\boxed{1}\,\boxed{0}\,\boxed{x^y}\,\boxed{3}\,\boxed{.}\,\boxed{5}\,\boxed{=}$

58.3   LUMINOSITY CLASSES

Because the luminosity of a star is so important in characterizing it, astronomers have sought other ways to estimate it. We cannot always measure stars' luminosities because their distance may be unavailable. Fortunately, in the late 1800s, Antonia Maury, an astronomer at Harvard, discovered that the more luminous the star, the narrower were the absorption lines in its spectrum, as illustrated in Figure 58.6.

The width of spectral lines is affected by the density of the gas producing them. This occurs because in a dense gas the atoms collide more frequently, "jostling" the electrons' orbitals. This slightly alters each atom's energy levels, making the photon energies it can absorb slightly larger. The net effect in the entire star's atmosphere is to smear the range of photon energies that are absorbed in the photosphere and to broaden the absorption lines. By contrast, when the atoms are spread farther apart, they collide less frequently and tend to absorb only the exact energies for that type of atom, leading to narrow absorption lines.

Maury's discovery implies that more luminous stars are less dense. We can show why this must be true for giant stars. Recall that the average density of a body is its mass divided by its volume. Giant stars have masses similar to those of main-sequence stars. So with similar masses but much larger volumes, the giant stars are much less dense. For example, the average density of main-sequence stars is roughly 1 kilogram per liter, whereas the density of a typical giant star is about 10^{-6} kilogram per liter—about a million times less dense. The atmospheres of giant stars are much less dense than the atmospheres of main-sequence stars.

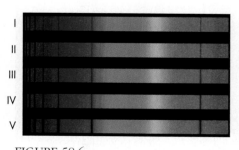

FIGURE 58.6

Spectral lines are narrow in the most luminous stars (I) and wide in main-sequence (dim) stars (V).

Using the relationship between spectral line width and luminosity just described, astronomers divide stars into five luminosity classes, denoted by the Roman numerals I through V. Class V stars are the most dense and dimmest (for a given surface temperature), and class I stars are the largest and brightest. Class I stars—**supergiants**—are split into two classes, Ia and Ib. Figure 58.7 and Table 58.1 show the correspondence between class and luminosity, and although the scheme is not very precise, it allows astronomers to get an indication of a star's luminosity from its spectrum. A star's luminosity class is often added to its spectral type to give a more complete description of its light. For example, our Sun is a G2V star, whereas the blue supergiant Rigel is a B8Ia star.

The high luminosity of the giants is a surprise given their low densities. After all, in discussing the mass–luminosity relation, we argued that the more luminous main-sequence stars became that way because of the greater compression of their cores. The giants seem at first to defy this relationship—growing more and more

Q If the density of two stars' atmospheres is the same, but the temperature of one is higher, how would this affect the width of the absorption lines of the hotter star? Why?

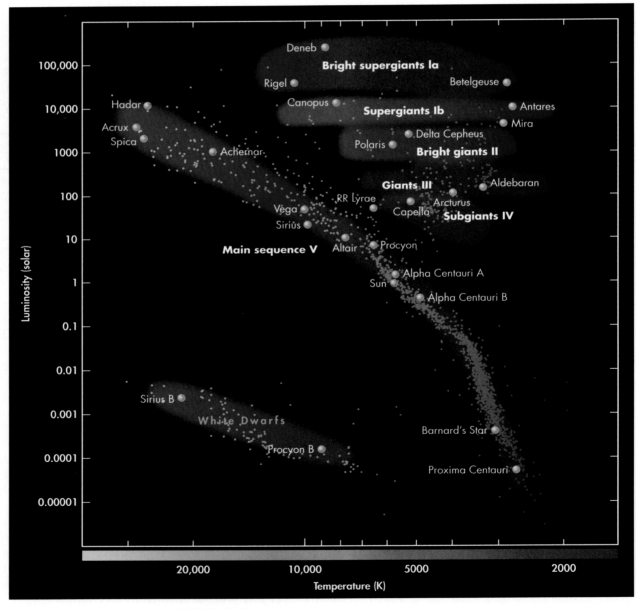

FIGURE 58.7
Stellar luminosity classes.

TABLE 58.1	Stellar Luminosity Classes	
Class	**Description**	**Example**
Ia	Bright Supergiants	Betelgeuse, Rigel (brightest stars in Orion)
Ib	Supergiants	Antares (brightest star in Scorpius)
II	Bright giants	Polaris (the North Star)
III	Ordinary giants	Arcturus (brightest star in northern constellation Boötes)
IV	Subgiants	Acrux (brightest star in the southern constellation Crux)
V	Main sequence	The Sun, Sirius (brightest star in sky, in Canis Majoris)

luminous at lower and lower densities. The key to this puzzle is that the density of a star's outer atmosphere does not necessarily tell us about the density deep in a star's core. For some reason the giant stars have a very different structure than the main-sequence stars—another clue to be noted in our detective work as we begin to unravel the stories of stars.

KEY TERMS

H-R diagram, 473   red giant, 475

main sequence, 474   supergiant, 480

mass–luminosity relation, 479   white dwarf, 475

QUESTIONS FOR REVIEW

1. What is the H-R diagram? What are its axes?
2. What is the main sequence?
3. How do we know that giant stars are big and dwarfs small?
4. How does mass vary along the main sequence?
5. What is the mass–luminosity relation?

PROBLEMS

1. If a star has a luminosity of 3000 $L_\odot$, what is its mass in relation to the Sun's?

2. Calculate the luminosity of a star that has a mass twice the Sun's. By what factor is such a star brighter than the Sun?

3. Use the data in Table 58.2 to plot an H-R diagram. Which star is a red giant? Which is a white dwarf? *Note:* Plotting will be much easier if you plot the logarithm of the luminosity; that is, express it in powers of 10 and use the power. For example, if the luminosity is 100, plot it as 2 for 10^2. Alternatively, use a pocket calculator as follows: Enter the luminosity in solar units and hit the "log" key. If the luminosity is 300,000, the answer you get should be 5.477.

TABLE 58.2	Data for Plotting an H-R Diagram		
Star Name	**Temperature**	**Spectrum**	**Luminosity (Solar Units)**
Sun	6000	G2	1
Sirius	10,000	A1	25
Rigel	12,000	B8	50,000
Betelgeuse	3000	M2	50,000
40 Eridani B	15,000	DA	0.01
Barnard's Star B	3000	M3	0.001
Spica	25,000	B1	2000

These data have been rounded off to make plotting easier.

TEST YOURSELF

1. A star that is cool and very luminous must have
   a. a very large radius.
   b. a very small radius.
   c. a very small mass.
   d. a very great distance.
   e. a very low velocity.

2. In what part of the H-R diagram might you find a white dwarf?
   a. Upper left
   b. Lower left
   c. Upper right
   d. Lower right
   e. Just above the Sun on the main sequence

3. In a sample of nearby stars, about what percentage will lie on the main sequence?
   a. 3.5%
   b. 9.0%
   c. 35%
   d. 50%
   e. 90%

59

Stellar Evolution

To us, the stars appear permanent and unchanging; but they, like us, are born, grow old, and die. Over their lifetimes they slowly change, transforming themselves in many basic ways but taking millions or billions of years to do so. Over the time span of one human life, it is rare to see change in any one star. Even during recorded human history, few stars have been seen to change significantly. We would need to go back to times before humans evolved to notice major changes in the properties of the stars we see today. Astronomers refer to the changes that occur while stars age as **stellar evolution**.

The English astronomer Sir William Herschel offered an analogy to the effort of humans seeking to understand the evolution of stars, and in particular our own Sun. He suggested it is like being allowed to spend a brief time in a forest, and then trying to deduce the life story of trees from the information you've collected. One minute in the forest compared to the lifetime of a tree would be about the equivalent of the entire human record of astronomical observations compared with the lifetime of the Sun.

Figuring out a tree's life story from a minute's observation might sound like an impossible task, but think of all we could see in a forest. We could match the leaves and bark to identify and track the life stages of a particular type of tree. We might see some trees of that type as saplings, others that were fully grown, and yet others as rotting stumps. We might find seeds on the tree and on the ground, and perhaps even some seedlings. We might see a leaf fall. If we piece together our snippets of information correctly, we could develop a description of a tree's life without ever having witnessed an individual tree change at all.

This then is the challenge of stellar evolution—to piece together the story of stars from the vast array of data we have gathered. In this unit we summarize the story of stellar evolution; the individual pieces of the story are explored in more detail in the units that follow.

Background Pathways: Unit 58

59.1 STELLAR EVOLUTION: MODELS AND OBSERVATIONS

Astronomers can deduce some features of how a star evolves from their observations. For example, the existence of main-sequence stars, red giants, and white dwarfs suggests to astronomers a picture of how stars age, much as a snapshot of a baby with its parents and grandparents suggests to a viewer a picture of human aging. However, we need not wait vast spans of time to watch a star age. We can "see" a star's aging with computer calculations that solve the equations that govern the star's physics. **Stellar models,** as such calculations are called, allow us to trace a star's life from birth to death.

The stellar models suggest that the entire life story of a star is largely determined at its birth. How a star evolves and then dies depends primarily on how much material the star contains—its mass. Mass is the most critical factor in a star's evolution because it determines the force with which the matter in a star is squeezed, and it also indicates the total amount of fuel that is available for the star to consume in its lifetime.

A star evolves in response to the gravity that is constantly squeezing it. At each moment the star must have a balance between the gravitational forces that compress it and the internal pressure forces that resist compression. The gravitational forces and pressures inside an object as massive as a star are intense. For the Sun, the stellar models indicate the matter at its core is squeezed to a density 20 times greater than steel (Unit 49), and there are stars that reach much higher densities.

In the competition between gravity and pressure, the force of gravity never runs down or gets used up. It crushes the matter together inexorably. Meanwhile, a star's heat energy steadily escapes into space (mostly as starlight), so the gas pressure will drop unless the star can replenish its lost energy.

Technically, when we refer to an object as a "star," we are speaking of an object like the Sun that uses nuclear fusion to generate the heat that maintains the pressure resisting gravity (Unit 61). Nuclear energy is a finite resource, and over its lifetime a star will exhaust all possible sources of nuclear fuel (Units 62 and 66). As a star is forming, when it is a *protostar,* the heat is generated from the process of contraction itself (Unit 60). At the end of its lifetime, a star becomes a **stellar remnant** that is supported by forces between the atoms and particles that it is composed of (Units 64 and 67)— more like the forces that keep the Earth from collapsing under its own weight.

The stellar models show that any change in the source of energy or the way a star produces its internal pressure will result in a different internal structure of the star. Observations show that sometimes these changes are gradual, but sometimes they occur violently.

Note that the term *evolution* is used differently by astronomers than biologists. In astronomy *stellar evolution* refers to the gradual changes that occur during a single star's lifetime, whereas in biology evolution refers to the gradual changes that occur between generations as a result of natural selection.

59.2   THE EVOLUTION OF A STAR

Figure 59.1 depicts what we might see if we could watch the changes that occur for a star like the Sun over many billions of years. This figure is based on the findings of observations and stellar models over the last century. We summarize these findings here.

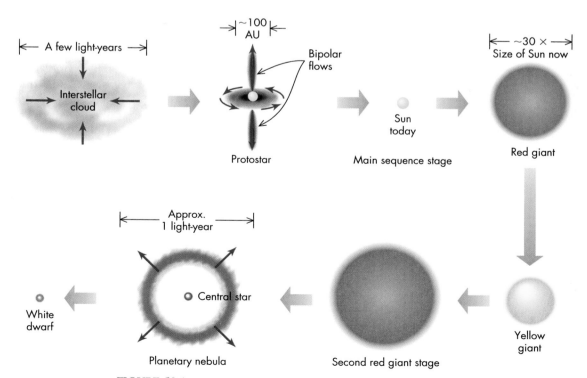

FIGURE 59.1

Evolution of a 1 solar mass star such as the Sun, based on computer calculations.

Evolution of a low-mass star

The formation of a star begins when a region of interstellar gas—an **interstellar cloud**—starts to contract. Initially this contraction is very slow, and other events may disrupt it. Eventually, though, if the knot of material grows dense enough, the gravitational force increases to a point where the material falls together, collapsing in on itself as the gravitational force grows steadily stronger. The gas heats up as it falls together and collides. The hot gas generates enough pressure to slow down the collapse to a more gradual contraction. The cloud at this point has become a **protostar.** A protostar gradually contracts under its own weight. The contraction converts gravitational potential energy (Unit 20) into thermal energy or heat. This is the same energy that is released when you drop a stone to the ground, but on a vastly larger scale. During the protostar stage, inflowing matter becomes redistributed, sometimes in surprising ways. For example, observations show that some of the gas is heated and driven back outward in **bipolar flows,** driven off in opposite directions from the protostar. That heat provides the pressure that counters gravity's inward pull, but it is a losing battle. Because the star must keep shrinking to generate this heat, gravity's hold grows steadily stronger. The formation process is described further in Unit 60.

A protostar becomes a **main-sequence star** when its core grows hot enough to start fusing hydrogen into helium (Unit 61). The amount of energy available from fusion reactions is enormous, so the star becomes stable for a long time. However, eventually it will consume all the hydrogen in its core and so must resume the process of compressing its core. Compressing a gas causes it to heat up, and this heating can raise the temperature in regions surrounding the core so that they now become hot enough to fuse hydrogen. The core may grow hot enough for helium fusion to begin there as well. This yields even more energy than before, which makes the star brighter yet. As the energy floods through the star's outer layers, they heat and swell, and the surface becomes cooler. The star becomes a **red giant** (Unit 62).

Many details of stars' evolution differ slightly depending on their exact mass, but the differences grow greater late in their lifetimes. In particular, the ultimate fate of stars is much different for stars whose mass is more than eight times greater than the Sun's. We will therefore divide stars into two groups—**low-mass stars,** such as the Sun, and **high-mass stars** whose mass is greater than about eight times the Sun's mass.

High-mass stars are blue during their main-sequence stage because their surfaces are hotter than those of low-mass stars, but the more important differences lie deep inside. Whereas stars like the Sun consume their cores' hydrogen in about 10 billion years, high-mass stars may use up their hydrogen in only a few million years (Unit 61). High-mass stars fuse their fuel quickly to supply the energy they need to support their great weight. The sequence of stages that high-mass stars experience as they evolve beyond the main-sequence stage also differs from that of low-mass stars, as illustrated in Figure 59.2. High-mass stars evolve through these stages very rapidly.

In both high- and low-mass stars the core eventually compresses and heats to the point that the helium can begin fusing, although this happens in somewhat different sequences in each. In low-mass stars the core takes a long time to reach helium fusion temperatures. During this time the star grows into a red giant. But when helium fusion begins, the energy released expands the core, and the outer layers shrink, making the star's surface hotter and hence more yellow than red. The cores of massive stars are able to reach helium fusion temperatures fairly rapidly, allowing them to expand directly into *yellow giants*. The surfaces of these yellow giants are often seen to pulsate in and out, and their total light output varies, making them **variable stars** (Unit 63).

When a star exhausts its helium, its core again shrinks, and the star grows into a red giant—for the second time in low-mass stars. These red giants are so luminous now that their radiation drives off the outer layers of gas in a **stellar wind.** Among low-mass stars, the stellar wind carries away so much material that the intensely hot core of the star is laid bare. Its energetic photons may ionize the inner parts of this wind of gas flowing away from the star. This faintly luminous cloud of gas surrounding the dying star is called a **planetary nebula.** The core remains as a small,

The term *planetary nebula* can be easily misinterpreted. The name arose because to astronomers with early telescopes these objects looked somewhat like the planets, but they have no relation otherwise to planets.

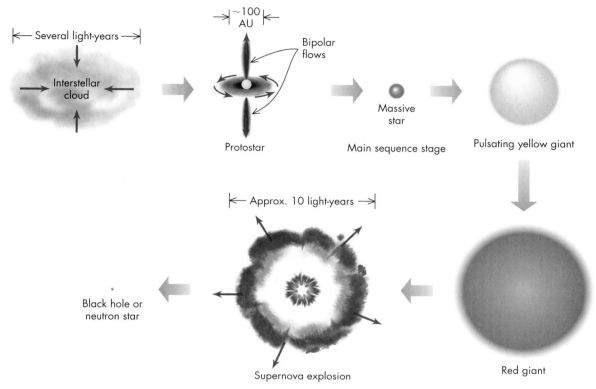

FIGURE 59.2
Evolution of a high-mass star, based on computer calculations.

dense stellar remnant: a **white dwarf** (Unit 64), which gradually cools and dims. However, a stellar remnant may be brought back to life in explosive fashion if a companion star deposits material on its surface. If enough material is deposited, the white dwarf may collapse and blast itself apart in a **supernova** explosion (Unit 65).

High-mass stars fuse and generate a sequence of heavier and heavier elements in their cores—elements such as carbon, oxygen, and iron. They fuse these fuels rapidly to supply the energy they need to support their great weight. However, they exhaust their fuel and die quickly (Unit 66). Thus, the stars that begin their lives on the blue end of the main sequence burn out quickly and therefore die "young" by the standards of other stars. They also die more violently because of the tremendous forces at work inside them. At the end of their lives the cores of high-mass stars undergo dramatic implosions that lead to supernova explosions. The stellar remnants left behind, **neutron stars** and **black holes,** are some of the strangest objects known (Units 67 and 68).

59.3   TRACKING CHANGES WITH THE H-R DIAGRAM

A good way of tracking the changes a star undergoes as it evolves is to plot its "path" in the H-R diagram. A star's position in the diagram changes whenever its surface temperature or luminosity changes, which it usually does in response to any changes in its interior.

It may be helpful to consider a similar sort of diagram for a population of people. Suppose we plotted a "height–strength diagram" for a random sample of people. On the X axis we plot height from tall to short (in the backward tradition of the H-R diagram), while on the Y axis we plot the amount of weight the person is able to lift.

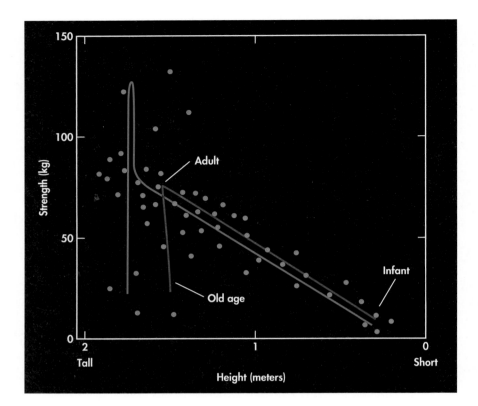

FIGURE 59.3
A hypothetical diagram of the heights of people versus their strength as measured by the amount of mass they can lift. Each person in a sample is marked by a green dot. Paths of two individuals throughout their lifetimes are also shown.

Q If stars move through almost all parts of the H-R diagram during their lifetimes, why do we see most of them concentrated in isolated regions of the diagram when we plot the temperatures and luminosities of a sample of stars?

Q Try drawing what you would expect a height–strength diagram to look like for different populations (such as preschoolers, college students, retirees). What characteristics would allow you to identify the age of the people plotted in such a diagram if you were not told their ages in advance?

The analogy to our traditional H-R diagram would be to find a large sample of people, measure their heights and the amount of weight they can lift, and plot one point for each person. A hypothetical diagram like this is shown in Figure 59.3.

If we took a random sample of people, everyone from infants to the elderly, we would generally find a trend that taller people can lift more weight, something like the main sequence in the H-R diagram. Some especially muscular people can lift a much larger weight than we would predict from their heights, while others are particularly weak for their height—analogous to the giant and dwarf regions of the H-R diagram.

If you traced various individuals' heights and strengths throughout their lifetimes, you would see that they "move" through this diagram. Perhaps they follow a relatively normal pattern like that illustrated by the red line in Figure 59.3, growing up and gaining strength as they grow taller, but losing strength (and maybe even a little height) in old age. Some individuals might have jobs that require a lot of lifting, and therefore they grow stronger after reaching adulthood, moving above the "main sequence" into the upper part of the diagram. Perhaps they suffer debilitating injuries, which reduce their strength drastically so that they drop below the main sequence. Every person's track through this diagram might be a little different, although the common features of their "evolution" will create certain patterns in the diagram.

Similarly, an individual star's path through the H-R diagram can be quite complex. The paths of a low-mass and a high-mass star through the H-R diagram are illustrated in Figure 59.4. Compare these paths to Figures 59.1 and 59.2. Each is a way of illustrating a star's lifetime.

Finally, we can also find clues to the changes that result with aging if we examine groups of stars or people who are all of a similar age. If we draw samples of humans from a nursery school, a college, and a retirement community, we find very different distributions of height and strength in each group. Similarly, we can test our ideas about stellar evolution if we can identify groups of stars that are all of similar ages, where physical traits will also differ within the group. We will return to the idea of testing sets of stars of the same age in Unit 69.

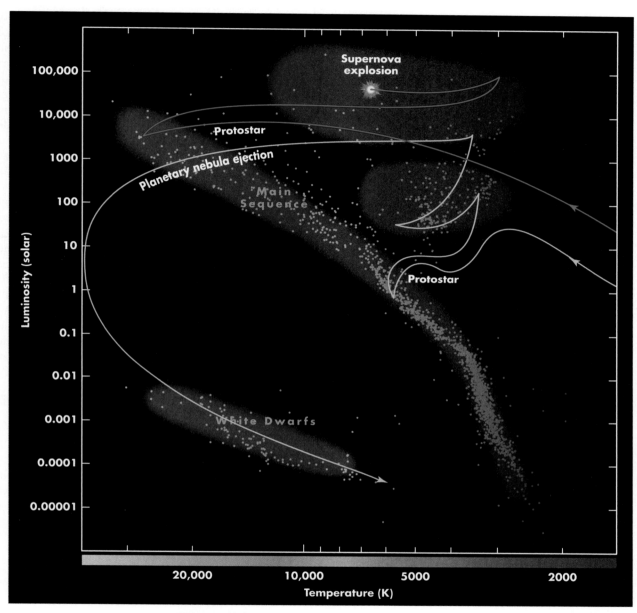

FIGURE 59.4

Schematic diagram of the evolutionary tracks of a low-mass star (yellow) and high-mass star (blue) through the H-R diagram.

59.4   THE STELLAR EVOLUTION CYCLE

A star forms from gas in an interstellar cloud that is drawn together by gravity. This process heats the gas to the point where it can fuse hydrogen into helium. Low-mass stars fuse the helium into carbon and sometimes oxygen before dying, but high-mass stars can fuse a series of heavier elements.

Stellar winds from low-mass stars during their red giant phase (Unit 64) return some of these newly created elements to become part of the interstellar gas. The internal structure of stars frequently does not allow much mixing between layers of stars, so most of the material returned to interstellar space by the winds is the outer atmospheric gases—which are changed little from when the star formed. However, the more massive of the low-mass stars eject more than 80% of their mass before they shut down, and this may include a significant amount of carbon and oxygen. Others blast all the heavy elements they formed into space when a companion star dumps material onto their surface, inducing them to undergo a supernova explosion.

High-mass stars build in their cores a series of layers of heavy elements: carbon, neon, silicon, and eventually iron. Once an **iron core** forms in a high-mass star, it is not long before the star must die, because iron cannot release energy through fusion to continue supporting the star. When the iron core collapses under its own weight, it triggers a supernova explosion. The heavy elements produced during the high-mass star's life are blasted into space. Some heavy elements, such as gold, are formed during the explosion itself. Supernovae therefore enrich the gas in interstellar clouds with heavy elements.

In the end, some of the mass that initially formed stars is locked away in stellar remnants—the white dwarfs, neutron stars, and black holes—that remain after the stars die. However, most of the material is returned to interstellar space, enriching the gas there with heavy elements formed through nuclear fusion. The material flowing outward from these dying stars may also run into interstellar clouds, compressing them and triggering star formation. In this way dying stars may initiate the birth of new generations of stars as well as providing elements vital to human life such as carbon and iron. Each new generation of stars incorporates a greater amount of these heavy elements, so that stars in later generations, such as the Sun, are able to have planets made of heavy elements, such as the Earth, and carbon-based life forms, such as us. We therefore owe our existence to stellar evolution.

KEY TERMS

bipolar flow, 484

black hole, 485

high-mass star, 484

interstellar cloud, 484

iron core, 488

low-mass star, 484

main-sequence star, 484

neutron star, 485

planetary nebula, 484

protostar, 484

red giant, 484

stellar evolution, 482

stellar model, 482

stellar remnant, 483

stellar wind, 484

supernova, 485

variable star, 484

white dwarf, 485

QUESTIONS FOR REVIEW

1. What determines when a star becomes a main-sequence star?
2. What is a protostar? What is a planetary nebula?
3. How can we track stellar evolution in an H-R diagram?
4. What kinds of stars end up as neutron stars or black holes?

PROBLEMS

1. Sketch an H-R diagram of the evolutionary path of the Sun.
2. Sketch an H-R diagram of the evolutionary path of a 15 $M_\odot$ star.

3. Visit a spot where trees are growing. List the different things you find there and the evidence you can put together about their relative place in the life cycle of trees.

TEST YOURSELF

1. Suppose we could measure one property of a protostar. Which would tell us most about its future evolution?
   a. Its temperature
   b. Its radius
   c. Its color
   d. Its luminosity
   e. Its mass

2. Blue main-sequence stars
   a. may end their lives as supernovae.
   b. have longer lives than most other types of stars.
   c. have cores that are hotter than a red giant's.
   d. are blue because they are fusing helium.
   e. All of the above

3. A star with half the mass of the Sun has
   a. too little mass to make it onto the main sequence.
   b. enough mass to end as a supernova.
   c. too little mass to fuse carbon in its core.
   d. enough mass to become a black hole.
   e. All of the above

60

Star Formation

|← A few hundred AU

Background Pathways: Unit 59

We described in Unit 33 the idea that the Solar System formed from a large cloud of gas, contracting because of its gravity. We saw how it spun faster as it contracted, much like water going down a drain, but with the bits of material that did not fall into the star becoming the planetary system.

This hypothesis that the Solar System formed from a gas cloud was proposed in the 1700s, long before astronomers had any evidence of gas between the stars. But the idea made sense because it successfully explained so many features of the Solar System. Today we can study interstellar gas clouds directly and hunt for clouds that may be forming stars. The idea of a collapsing cloud of gas producing a star is convincingly borne out by our current observations, and with these data we find a rich variety of unexpected details in the ways stars form.

60.1 INTERSTELLAR GAS CLOUDS

Stars—whether they are of high or low mass—form from **molecular clouds.** These clouds are the densest and coldest kinds of interstellar clouds, in which atoms combine to form molecules and heavier elements combine to form small solid particles called **interstellar dust.** Molecular clouds are often many light-years across and can contain anywhere from a few to tens of thousands of solar masses of cold gas and dust. These clouds orbit along with the stars inside our galaxy. The gas is mostly hydrogen (71%) and helium (27%); the dust is composed of solid, microscopic particles of silicates, carbon, and iron compounds with a coating of ices. The presence of dust in these clouds blocks most light from entering the clouds, while the molecules can radiate energy even at low temperatures. The combination of these effects allows the gas to cool to low temperatures, which is important for star formation.

Molecular clouds are generally very cold, often only about 10 Kelvin—not far above absolute zero. At such low temperatures, atoms and molecules in the gas cloud move too slowly to generate much pressure. As a result, the pressure may be barely enough to support the gas cloud against its own gravity, so if the gas cloud grows even slightly more dense, it can collapse. Such a collapse might be triggered by a collision with a neighboring cloud, strong stellar winds from a nearby star, or even the explosion of such a star. Whatever disturbance triggers the collapse, the cloud is unlikely to collapse as a whole. Instead, denser regions in the cloud will begin to contract on their own, and these may fragment even further because the rate of collapse is faster wherever the density is higher.

Astronomers picture the transformation from molecular cloud to star proceeding in several distinct stages. In the first stage, a dense clump within the much larger molecular cloud begins to collapse, its gas drawn inward by gravity. In the second stage, any rotation of the gas clump makes it flatten into a disk, as described in Unit 33's discussion of the origin of the Solar System. After about a million years, the disk forms a small, hot, dense core at its center called a **protostar,** which marks the third stage of stellar formation. These stages are illustrated in Figure 60.1.

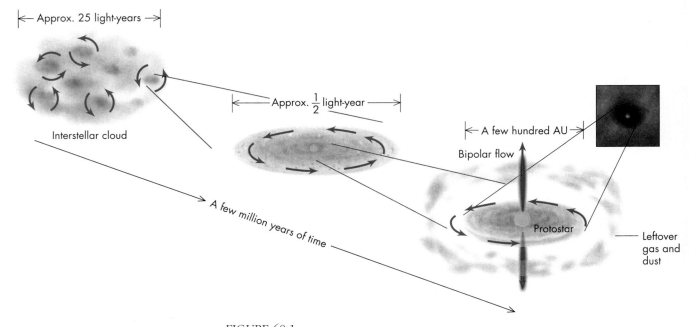

FIGURE 60.1

An artist's sketches depicting the birth of a star. The disk around a star forming in Orion is shown at far right in a Hubble Space Telescope image.

FIGURE 60.2

False-color radio map of a star-forming molecular cloud. The red regions show the densest concentrations of molecules, where stars are beginning to form.

At any given time, many clumps exist within a molecular cloud. Astronomers map the structure of these clumps at radio wavelengths, detecting emissions from molecules in the cold gas (Figure 60.2). The radiation coming from these molecules allows us to trace where the gas is densest. And the emission of this radiation is one of the important ways that the clouds can lose energy and continue to collapse.

Gravitational energy is released whenever something falls. For example, if a cinder block falls onto a box of tennis balls, the impact scatters the balls in all directions, giving them kinetic energy—energy of motion. As material collapses inward toward a forming protostar, clumps of material collide; the rapid motions of the atoms and molecules in the gas appear as heat, and the temperature of the gas rises. If a cloud had no way to radiate the heat produced by this gravitational collapse, the cloud would grow hotter until its pressure was high enough to stop it from contracting any further. The emission of radiation allows the cloud to continue contracting. Because it radiates away energy, this is sometimes called "cooling" radiation. However, in allowing the cloud to contract, the gravitational force grows stronger so the collisions grow more violent, and the cloud grows steadily hotter as it contracts.

A large, cold molecular cloud breaks up into separate clumps, each clump collapsing on its own. The time that it takes each clump to collapse to a protostar is approximately tens to hundreds of thousands of years—short by astronomical standards. The result is that protostars generally form in groups, not in isolation, and all the stars within a group form at approximately the same time. These groups of stars often persist for millions of years or more after the stars form, and they can tell us a great deal about star formation (Unit 69).

60.2 PROTOSTARS

The protostar stage begins once a dense core forms at the center of an interstellar cloud. The protostar is no longer collapsing rapidly, but it is trapping internal heat and only gradually shrinking. If a protostar were removed from its environment, it would not look much different from an ordinary star. As it contracts, its surface temperature grows to exceed 2000 K, but its heat still comes from its gravitational contraction rather than fusion. This initially low surface temperature does not allow a protostar to produce much visible light. Even as it heats up, though, the dust in the dense molecular cloud that usually surrounds a protostar blocks most of the visible light that is produced. For example, the Hubble Space Telescope photograph of the Eagle Nebula (Figure 60.3) shows a star-forming region. A few young stars are visible, but many more forming stars are hidden in the dark pillars of gas and dust in the image.

Protostars inside these dark clouds are not completely hidden, however. Astronomers can detect infrared emission from them, as seen in Figure 60.4. This image was made with one of the 8 meter diameter telescopes of the VLT. Many of the stars in this false-color image lie behind the Eagle Nebula. But comparison of the visual and infrared images reveals a number of protostars (bright orange here) that are hidden by the dark veil of gas and dusk pillars in the visible-wavelength image.

Protostars contract relatively slowly at this stage. They may be more than 10 times larger than their eventual size as stars, but at this stage not much cloud material falls directly onto their surface layers. Instead, the surrounding material drawn inward by the protostar's gravity joins the growing disk of matter circling the protostar. Several star systems at this stage of evolution have been found with

Q If parallaxes are not available, how might it be possible to decide whether stars seen at infrared wavelengths are in a molecular cloud instead of behind it?

FIGURE 60.3
The Eagle Nebula, a region where stars are forming from molecular clouds. Dust hides many protostars in dense portions of the cloud in this visible-wavelength image made with the Hubble Space Telescope.

FIGURE 60.4
A false-color infrared image of the Eagle nebula made with the VLT (Very Large Telescope) 8 m telescope in Chile.

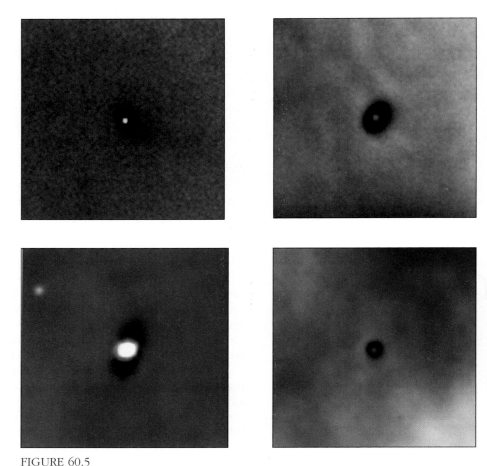

FIGURE 60.5

Dark dusty disks surrounding protostars in the Orion Nebula, imaged with the Hubble Space Telescope. The disks show up in silhouette against the background glow of the nebula.

the Hubble Space Telescope (Figure 60.5) in the Orion Nebula. This is a large star-forming cloud, which to your unaided eye looks like the middle star of the sword in the constellation Orion (see Looking Up #6). In Figure 60.5 you can see several glowing young stars surrounded by dark disks that appear in silhouette against the background glow of the Orion Nebula.

Gravity forces the protostar to continue contracting, and its core grows hotter. This type of compression heating is something you can observe for yourself with a bicycle pump—the pump cylinder grows warmer as you compress the gas to high pressure to fill a tire. The heat does not come from friction between parts of the pump rubbing together. You can verify this if you use the pump without attaching it to a tire. Gravity squeezes the star, much as you use your weight and muscles to squeeze the gas inside the pump cylinder, and the gas responds by heating up. The unusual aspect of gravitational contraction is that the more the star contracts, the stronger the gravity becomes, so the core continues to grow hotter.

When the temperature in the protostar's core reaches about 1 million Kelvin, some nuclear fusion begins. This temperature is not sufficient to begin the proton–proton chain (Unit 50), but fusion of naturally occurring deuterium (the isotope of hydrogen with one proton and one neutron) begins. Because naturally occurring deuterium is less than one ten-thousandth ($<10^{-4}$) as abundant as hydrogen, it is

Q For the type of fusion currently taking place in the Sun, the first step is to **create** deuterium by colliding hydrogen atoms. This is a very slow process because it involves the weak force. Once deuterium is made, however, the subsequent steps in the fusion process to build helium are quite rapid.

not a long-lived energy source. Nevertheless, this fusion is a new, powerful energy source that sustains the temperature and pressure in the core, slowing further contraction for a while.

The protostar phase is not quiet. Protostars with masses similar to the Sun are observed to eject gas in opposite directions called a **bipolar flow,** as shown in Figure 60.6. This gas is driven out more than a light-year from the protostar. In some cases narrow jets of gas (Figure 60.7) are seen shooting outward at speeds of

FIGURE 60.6
A bipolar outflow from a forming star is seen in this false-color infrared image produced by the Spitzer Space Telescope. The inset image is a visible-wavelength image of this region, showing the dark cloud in which the protostar is hidden.

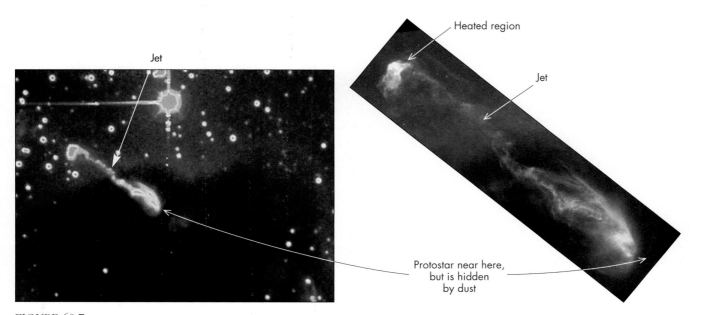

FIGURE 60.7
False-color image of a protostar and jet. Where the jet strikes a nearby interstellar cloud, the gas is heated and glows.

several hundred kilometers per second. Where the jets of material run into surrounding interstellar clouds, they heat and ionize the gas, producing bright glowing spots. The cause of these outflows and jets is not yet fully understood. They are probably powered by energy that is released as matter falls from the inner parts of the disk surrounding the star, crashing into the protostar's surface layers and becoming very hot. Their narrow shape may be caused by the star's magnetic field, which confines ionized gas to flow out along the directions of the magnetic poles, much as such fields confine prominences on the Sun (Unit 51). The disk and infalling matter surrounding a protostar may also funnel the hot gas into the direction of least resistance, perpendicular to the disk.

Another phenomenon observed among some protostars is very strong winds of gas. These are similar to the solar wind but are far more intense, and they seem to occur only episodically during a protostar's life. In addition, protostars are also often seen to vary erratically in brightness. Such variable stars are called **T Tauri stars** after a star in the constellation Taurus, which is one of a group of young stars that exhibit this behavior. Some of this variability is probably caused by magnetic activity, much like that observed on our Sun. The inflow of material from the disk onto the star may not be uniform—the gas in the disk may be "lumpy"—and this could contribute to the variability as well.

If T Tauri stars are plotted on an H-R diagram, they lie a little above the lower main sequence because they are still shrinking and are therefore larger and brighter than main-sequence stars. Figure 60.8 shows the H-R diagram for a very young star

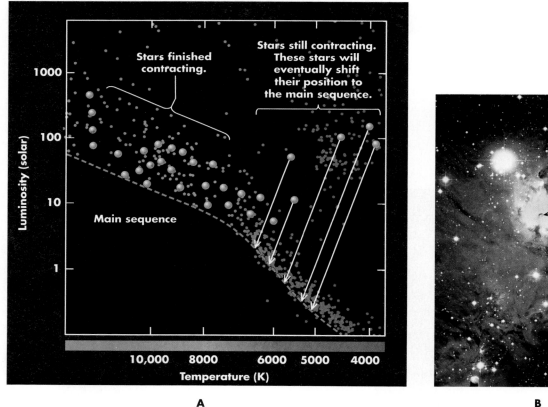

A   **B**

FIGURE 60.8

(A) The H-R diagram for young stars and protostars in a region of star formation shown in (B). (B) This glowing region is named NGC 2264. The gas is heated mostly by recently formed massive stars. The low-mass stars are still in their protostar phase, some still in small dark clumps called Bok globules.

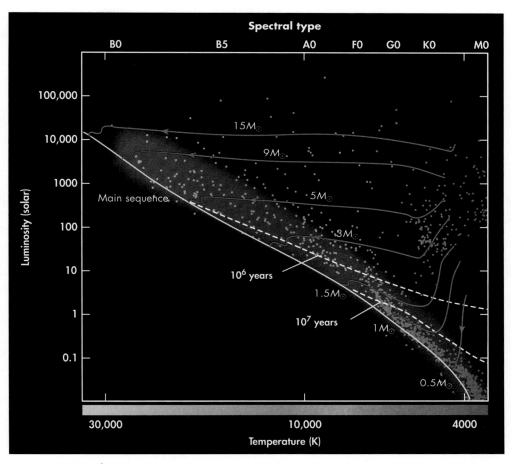

FIGURE 60.9

Evolutionary tracks for protostars in the H-R diagram. The solid red lines show the changing state of the surface temperature and luminosity for stars of various masses. The white dashed lines show where a star can be found along these tracks 1 million and 10 million years after the cloud begins to collapse.

Q If protostars lie above the main sequence, why is it that they are not seen filling this region in most H-R diagrams (such as those shown in Unit 58)?

group and a photograph of the cluster itself. (Notice that the cluster is still surrounded by gas left over from its formation.) Such H-R diagrams help verify our ideas about star formation because older clusters never contain T Tauri stars.

The duration of the protostar stage is difficult to predict precisely; it depends on the physical circumstances within the cloud. Computer models indicate that the mass of the forming star is the most important variable in determining how long the protostar stage lasts. The evolutionary tracks different stars follow through the H-R diagram are illustrated in Figure 60.9. For a star like the Sun, the protostar phase lasts approximately 10 million years. For high-mass stars it may be less than a tenth as long, whereas for very low-mass stars it may be more than 10 times longer. The difference in the duration of the protostar stage explains why the higher-mass stars have already reached the main sequence in Figure 60.8A, while lower-mass stars in that star-forming region are still in their protostar stage.

The varying timescales for formation imply that the more massive stars may also influence the formation of lower-mass stars. For example, in the image of the star cluster in Figure 60.8B, even though many stars have already formed, you can see small, dark blobs that appear to be lower-mass protostars that have not yet become hot enough to emit visible light. These tiny, dark blobs are called **Bok globules** in honor of Bart Bok, the Dutch-born U.S. astronomer who first studied them in detail. Radiation and stellar winds from more massive stars that form first may limit the growth of the slower-forming low-mass stars.

60.3   STELLAR MASS LIMITS

Observationally, we find few stars outside the range of about 0.2 to 20 $M_\odot$. Lower-mass stars are dim and hard to detect, while higher-mass stars are rare. However, it is not just observational limitations that determine this range. Theoretical studies of the physical conditions necessary to allow a star to form indicate that stars must have masses between 0.08 and 150 $M_\odot$.

Q ▸ How would young brown dwarfs differ in appearance from protostars with masses greater than 0.08 $M_\odot$?

Stars below 0.08 $M_\odot$ do not occur because their mass is too small to compress the gas in their cores enough to reach hydrogen fusion temperatures. Although lower-mass objects form, they never establish a stable balance between gravity and fusion's internal heat production. Some unstable fusion of hydrogen may occur in slightly less massive objects, and fusion of deuterium may occur in objects down to about 0.016 $M_\odot$ (about 17 times Jupiter's mass). However, naturally occurring deuterium is too scarce for this to be a stable state. Objects with masses between 0.016 and 0.08 $M_\odot$ fall between our definitions of planet and star and have been dubbed **brown dwarfs.** These are dim, cool objects that infrared technologies have begun detecting only since the mid-1990s. Most of the objects in the L and T spectral types (Unit 55) appear to fall in this mass range. These objects are probably very numerous, but astronomers are still in the early stages of characterizing this population of "failed stars."

By contrast, the upper limit to star masses might be described as too much of a good thing! A star more massive than 150 $M_\odot$ never stabilizes for the following reasons: If a giant cloud of gas begins to contract, the intense gravitational compression drives up its temperature and luminosity very rapidly. The center of a contracting clump of gas more massive than 150 $M_\odot$ emits such intense radiation that it heats the surrounding gas to an extremely high temperature. This raises its pressure so much that no additional material can fall in, setting a limit on the amount of material the largest stars can accumulate. Furthermore, once formed, high-mass stars become so luminous (recall the mass–luminosity relation) that their radiation rapidly drives gas from their outer layers into space in a strong stellar wind. The most massive stars—or would-be stars—essentially "shine" themselves apart.

In reality, conditions governing the protostar stage are rarely conducive to the formation of stars much larger than 30 solar masses. Moreover, such massive stars burn themselves out extremely rapidly (Unit 61). Nevertheless, some isolated examples have been found, like the star Eta Carina (see Looking Up #8: Centaurus and the Southern Cross), which is estimated to have a mass between 100 and 150 times the Sun's mass.

KEY TERMS

bipolar flow, 493

Bok globules, 495

brown dwarf, 496

interstellar dust, 489

molecular cloud, 489

protostar, 489

T Tauri star, 494

QUESTIONS FOR REVIEW

1. Describe how a protostar is thought to form.

2. At what wavelengths is it best to observe star formation? Why?

3. What heats a protostar?

4. What determines the upper and lower mass limits of stars?

PROBLEMS

1. The jets from a T Tauri star are measured to have speeds of about 300 km/sec. At this speed, how long would it take the ejected material to travel one light-year away from the protostar?

2. Suppose a molecular cloud begins with a mass of 1 $M_\odot$ and has a radius of one light-year. Imagine a dust grain at the outer edge of the cloud pulled straight inward, accelerating the entire way. When it reaches a radius of 1 million kilometers from the center of the forming star, how fast will it be moving? (*Hint:* This problem is much easier to answer through the conservation of energy.)

3. Compare the average density of a newly formed star of mass 20 $M_\odot$ (radius $\approx$ 10 $R_\odot$) and 0.1 $M_\odot$ (radius $\approx$ 0.1 $R_\odot$) to the Sun's density. You should find that the higher-mass star has a lower density. Relate these relative densities to the formation processes that determine the upper and lower mass limits for stars.

4. Inflate a balloon and carefully measure its size. Put it in a freezer for a few hours. Does it look the same when it is cold? How does this relate to how stars form in cold regions of space?

5. Take a plastic bottle and put a little soapy water in it. Run your finger across the mouth to make a soap film. Now, without breaking the film, run hot water over the bottle. What happens to the soap film? How does this relate to what happens to a star when it is heated?

TEST YOURSELF

1. Which of the following is *not* true about molecular clouds?
   a. They contain helium.
   b. They contain iron.
   c. Their temperature is near absolute zero.
   d. Some of their particles are coated in ice.
   e. The entire cloud contracts as a whole.

2. T-Tauri stars
   a. lie on the main sequence.
   b. vary in brightness.
   c. are dimmer than main-sequence stars.
   d. are typically found in older clusters.
   e. All of the above except (a)

3. Why are there no stars less massive than 0.08 $M_\odot$?
   a. With such a low mass they take longer than the age of the universe to contract.
   b. Their mass is so small that deuterium fusion blasts them apart.
   c. They exist, but they are so dim that we cannot detect any.
   d. They cannot compress their cores to hydrogen fusion temperatures.
   e. All of the above

PLANETARIUM EXERCISES

All the planetarium exercises use the Starry Night planetarium software provided with your text. You should install the program and familiarize yourself with the various parameters and controls before trying these exercises. When the program asks you to enter a password, click "OK" to run Starry Night.

Set the location from which you are viewing to the nearest large city or your latitude and longitude. Accomplish this by clicking on the Location box on the toolbar. Verify that the local time is correct.

Locate at least two star-forming regions using the program. Center on objects like M42 (Orion), M17 (Swan), or M16 (Eagle) and mark their positions on the star chart at the back of your book. Using this information, observe the objects with a telescope and make a sketch of both nebulosity and star position and brightness. Compare what you saw to the images and discussion in the text. (These nebulae can also be found in Looking Up #6 Orion and Looking Up #7 Sagittarius.)

61 Main-Sequence Stars

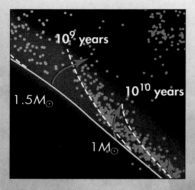

Background Pathways: Unit 59

A star stops being a protostar and becomes a **main-sequence star** when it begins fusing hydrogen into helium in its core. Fusion generates the heat needed to maintain pressure in the core, which stops the protostar's contraction. At this stage of its life, the star's interior structure is much like the Sun's: It contains a core, where hydrogen is fusing into helium, and an envelope of gas around the core, where energy is transported to the star's surface. The properties of the core and envelope, however, depend greatly on the star's mass.

61.1 MASS AND CORE TEMPERATURE

In examining the physical properties of stars along the main sequence, we find that the higher the mass of a star, the hotter and more luminous it is. This relation between mass, temperature, and luminosity results from the fact that the gravity and pressure inside a star must be in balance, a condition called **hydrostatic equilibrium.** The principles governing the structure of a star are illustrated in Figure 61.1. With

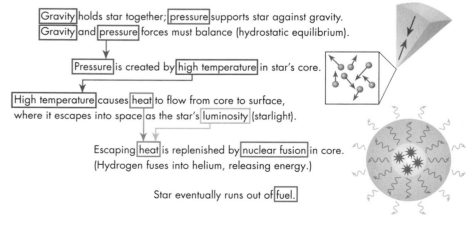

Principles Governing the Structure of a Main Sequence Star

Gravity holds star together; pressure supports star against gravity.
Gravity and pressure forces must balance (hydrostatic equilibrium).

Pressure is created by high temperature in star's core.

High temperature causes heat to flow from core to surface, where it escapes into space as the star's luminosity (starlight).

Escaping heat is replenished by nuclear fusion in core. (Hydrogen fuses into helium, releasing energy.)

Star eventually runs out of fuel.

Note: High-mass stars require more pressure to support their greater mass.
Greater pressure is produced by higher temperature.
Higher temperature produces higher luminosity.
Higher luminosity leads to faster fuel usage.
Faster fuel usage means a high-mass star burns out sooner than a low-mass star.

FIGURE 61.1
Outline of the processes that govern the structure of stars.

computer modeling of stars, we can determine how these processes balance with one another and learn many details about a star's internal structure—its density, temperature, pressure, and energy generation.

The lowest-mass stars take the longest time to complete the protostar phase and enter the main-sequence phase. Their weaker gravity causes them to contract so slowly that their core temperature rises very gradually. As a consequence, a 0.1 $M_\odot$ star must compress the gas in the core to a very high density—about 500 kilograms per liter—just to reach a temperature where hydrogen fusion can begin. By contrast, massive protostars contract rapidly and reach high temperatures in their cores without compressing the gas to nearly as high a density. A 10 $M_\odot$ star will need to reach a density of only a few kilograms per liter before fusion begins.

The higher density of low-mass stars may seem a little surprising at first: It would seem that high-mass stars would compress matter more. Later in their lifetimes that does eventually prove true; but during the main-sequence phase, the highest-mass stars are lowest in density. They achieve hydrostatic equilibrium despite their lower density because they reach a higher temperature in their cores, and that produces the balancing pressure. At the bottom of the main sequence, the cores may be only 5 million Kelvin. At the upper end, the core temperature is about 10 times hotter. Thus, high-mass stars have hotter, lower-density cores than do low-mass stars.

The rate of nuclear fusion rises rapidly with temperature, and additional fusion mechanisms become possible at high temperatures as well. The high temperatures in high-mass stars drive the nuclear reactions to generate the *very* large luminosities observed in massive stars.

61.2 STRUCTURE OF HIGH-MASS AND LOW-MASS STARS

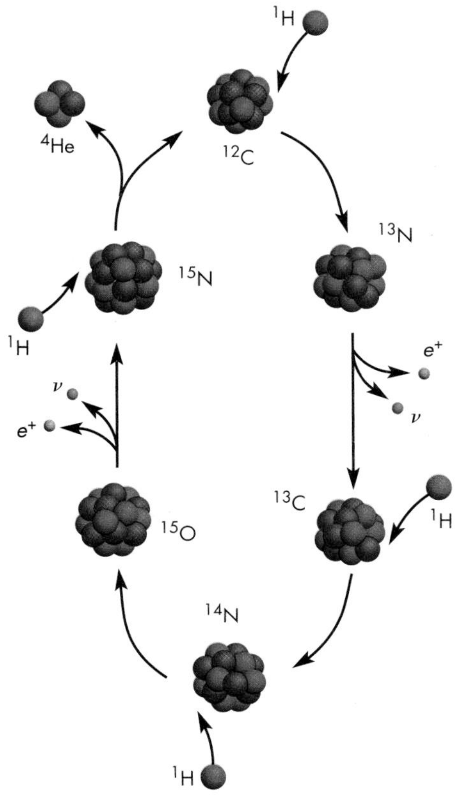

In the Sun and other low-mass stars, hydrogen is converted to helium through the **proton–proton chain** (Unit 50). In the proton–proton chain, two hydrogen nuclei, or protons, fuse to form the isotope of hydrogen ^{2}H (deuterium). A third proton combines with the deuterium to make ^{3}He, and the ^{3}He fuses to form ^{4}He. At the higher temperatures in the cores of stars more massive than about 2 $M_\odot$, a different set of fusion reactions converts hydrogen to helium.

In these larger stars, where core temperatures rise above about 20 million Kelvin, hydrogen fusion takes place primarily by means of the **CNO cycle.** In this cycle, illustrated in Figure 61.2, carbon atoms already present in the star's core act as catalysts to aid the reaction through a series of steps that build nitrogen and oxygen nuclei. No carbon, nitrogen, or oxygen is created or destroyed in the end by the CNO cycle, and the net energy release is essentially the same per created helium atom as in the proton–proton chain. However, at high temperatures the CNO cycle is much faster than the proton–proton chain reaction.

The high rate of energy production at these high temperatures makes the internal structure of high-mass stars different from that of low-mass stars. The photons (gamma rays) generated by the fusion reactions cannot carry away the enormous energy output rapidly enough from the core. In the

FIGURE 61.2
The nuclear reactions of the CNO cycle. Hydrogen nuclei (H) combine with carbon, nitrogen, and oxygen (C, N, and O) in a cycle whose net result is to produce helium (^{4}He).

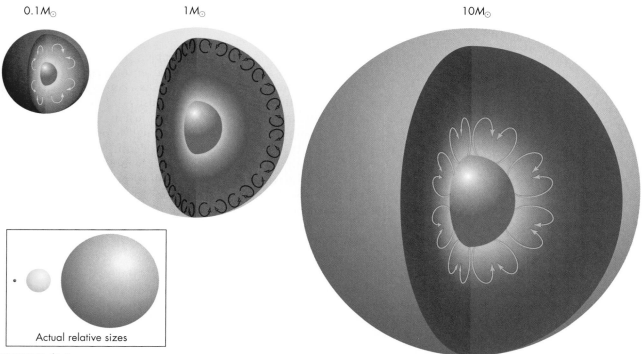

FIGURE 61.3

The comparative structure of main-sequence stars of different mass. Note that the relative sizes are not accurate for clarity in showing their internal structure. The inset box shows the stars' true relative sizes.

cores of stars, the photons travel in short hops, only a centimeter or so at a time, before being absorbed by another atom. The failure of the photons to carry the energy away quickly enough causes clumps of gas to become superheated. These rise toward the star's surface, traveling a sizable fraction of the star's radius, and then sinking back down after releasing their heat and cooling—the process of **convection.**

Convection occurs wherever the rate of energy flow by photons becomes too slow. Similar convection motions occur just below the surface in stars like the Sun, where the cooler surface layers strongly absorb the photons and impede the flow of energy by radiation (Unit 49). In stars less massive than the Sun, the surfaces are even cooler, and the convection zone reaches proportionately deeper into the star. These structural differences are illustrated in Figure 61.3.

A star's core is the region where temperatures are high enough for hydrogen to fuse into helium. The core typically contains only about one-tenth of the star's total mass. For a star like the Sun, the convection zone does not carry gas all the way from the surface into the core. So as the hydrogen fuel is gradually consumed in the core, the Sun's convection is unable to reach deep enough to mix unfused hydrogen from the outer parts into the core. Therefore the core can run out of hydrogen fuel, even though there is a large amount in surrounding regions.

Computer models indicate that the highest- and lowest-mass stars can replenish their cores with hydrogen to varying degrees. Convection within the inner regions of massive stars can mix some of the unburned hydrogen from the surrounding layers into the core and thereby increase the amount of hydrogen that is available for fusion up to several tenths of the star's mass. In stars less massive than about 0.4 $M_\odot$, the gas in the outer layers and core mixes completely. Such stars are said to be **fully convective.** These stars can fuse most of their hydrogen. By contrast, more massive stars, whose surface gas does not mix down into the core, can be starved for hydrogen in their cores even though their outer layers remain hydrogen-rich.

Q If all stars were fully convective, how would we be able to determine this from studying stars' spectra?

61.3   MAIN-SEQUENCE LIFETIME OF A STAR

While a star is fusing hydrogen to helium in its core, its structure remains nearly constant. Therefore, it remains in almost the same position in the main sequence on the H-R diagram. The length of time a star spends fusing hydrogen to helium in its core is called its **main-sequence lifetime.** This also explains why the main sequence stands out so clearly when we plot the luminosities and temperatures of a random set of stars. Because all stars spend most of their lives in the main-sequence phase, and because the external properties of these stars remain nearly constant, we find most stars in the main-sequence region of the H-R diagram.

Any individual star's main-sequence lifetime depends on its mass and luminosity. To see why, we can use a simple analogy. Suppose we want to know how long a camping lantern will run on a tank of propane fuel. That time depends on how much fuel it has and how rapidly the fuel is consumed. For example, if the tank contains 2 liters of propane, and at maximum brightness the fuel burns at rate of 0.2 liters per hour, the lamp will last for a time

$$t = \frac{\text{Amount of fuel}}{\text{Rate of burning}} = \frac{2 \text{ liters}}{0.2 \text{ liters/hour}} = \frac{2}{0.2} \text{ hours} = 10 \text{ hours}$$

before running out of propane. Clearly we can increase the amount of time the lamp will run if we increase the amount of fuel or dim the light so it burns the fuel more slowly. We can apply the same formula to a star to determine its main-sequence lifetime if we know how much fuel it has and how rapidly the fuel is being consumed.

The amount of fuel a star has is set by its mass, or to be a little more precise, the amount of a star's mass that is hydrogen—about 71% of the mass of most stars. This represents the total amount of potential fuel, but it is only within the core—approximately the central tenth of the star's overall mass for most stars—that temperatures are high enough for fusion to occur. This means only 1/10th of 71%, or about 7%, of a star's mass is available as fuel. Some stars increase this amount by mixing gas from the outer parts of the star into the core, but this results in significant adjustments only for very high- and very low-mass stars.

The rate at which a star consumes its fuel is directly indicated by its luminosity: More luminous stars fuse their fuel faster. This makes sense because the energy that a star puts out is supplied by its fuel. Again, it is like a lantern. If it is brighter, it is burning the fuel faster; if it is dimmer, it is burning the fuel more slowly.

Thus, a star's lifetime, t, is proportional to its mass divided by its luminosity, M/L. To work out the actual times involved, we have to determine what mass of hydrogen is needed to produce a given quantity of light. This is determined by the amount of nuclear energy released in the fusion process. For the Sun, we have determined that 360 million tons of hydrogen are being fused each second to generate its observed luminosity (Unit 50). This creates 356 million tons of helium and "4 million tons of light"—that is, according to Einstein's law, $E = m \times c^2$, 4 million tons of mass are turned into energy each second.

Taking the amount of hydrogen in the Sun's core (7% of its total mass of 2×10^{30} kg) and dividing it by the amount of fuel consumed each second (360 million tons per second), we find the total number of seconds in the Sun's main-sequence lifetime is

$$t_\odot \approx \frac{0.07 \times 2 \times 10^{30} \text{ kg}}{360 \times 10^9 \text{ kg/sec}} = 3.9 \times 10^{17} \text{ sec} = 1.2 \times 10^{10} \text{ yr.}$$

For the values given, we obtain a lifetime for the Sun of about 12 billion years on the main sequence. Other, more detailed stellar evolution models indicate that slightly less hydrogen will be fused in the core before the Sun leaves the main-sequence phase,

t = Main-sequence lifetime

M = Star's mass (in solar masses)

L = Star's luminosity (in solar luminosities)

The mass–luminosity relation (Unit 58) states that $L \approx M^{3.5}$ for a star whose mass and luminosity are measured in solar units. Using this relationship to substitute for the luminosity, we find that a star's lifetime is approximately

$$t \approx \frac{M}{M^{3.5}} \times 10^{10} \text{ years}$$

$$\approx \frac{1}{M^{2.5}} \times 10^{10} \text{ years.}$$

Q Suppose in the remote future we wanted to extend the Sun's lifetime. A government official proposes collecting interstellar hydrogen and dumping it on the Sun. If this were possible, would it be a good idea? Why or why not?

Q If a star is red, what can you say about its age?

so the Sun's main-sequence lifetime is rounded off to about 10 billion (10^{10}) years. Given its present age, the Sun is roughly halfway through its main-sequence lifetime.

We can estimate the main-sequence lifetimes of other stars by comparing them with the Sun. If we measure M and L in solar units, a star's main-sequence lifetime is

$$t = \frac{M}{L} \times 10^{10} \text{ years.}$$

This assumes that all stars fuse the same fraction of their mass as the Sun does, so it will underestimate the lifetimes of the highest- and lowest-mass stars somewhat; but it gives a good lifetime estimate for most stars.

We might expect lifetimes to be longer for massive stars because they have more fuel to fuse. However, the luminosities of massive stars go up by even more than their masses. For example, the star Sirius has an estimated mass of about 2 $M_\odot$, but its luminosity is about 20 $L_\odot$. Therefore it has 2 times more fuel, but it is fusing it 20 times faster, so its lifetime will be 2/20ths as long as the Sun's. In other words, Sirius will have a main sequence lifetime, t, of

$$t(\text{Sirius}) = \frac{2}{20} \times 10^{10} \text{ years} = 10^9 \text{ years.}$$

So Sirius is expected to live on the main sequence only about a tenth as long as the Sun. For a star whose mass is 10 $M_\odot$ and whose luminosity is 10^4 $L_\odot$, the main-sequence lifetime is a mere 10^7 years.

These results demonstrate that massive stars have much shorter lifetimes than the Sun. Despite having more fuel, they fuse it much faster to supply their greater luminosity. In other words, massive stars are like gas-guzzling cars that, despite having large fuel tanks, run out of fuel sooner than fuel-efficient cars with smaller tanks. This brevity of massive stars' lives implies that those we see must be relatively "young" by the standards of the Sun. Because massive stars are blue, we can conclude that, in general,

Blue stars have formed recently.

Because of their recent formation, we can understand why this type of star is often seen associated with clouds of interstellar gas. They die so quickly that there has been little time for the interstellar gas to disperse in response to stellar winds, or for the star to drift away from the cloud if it has any relative motion. Therefore blue stars are frequently embedded in matter left over from their formation, as you can see in Figure 61.4. In these recently formed groups of stars, the hot blue stars produce ultraviolet light that ionizes the interstellar hydrogen, producing the pink glow of interstellar hydrogen seen in the images.

61.4   CHANGES DURING THE MAIN-SEQUENCE PHASE

Stars change relatively little as they age during their main-sequence phase, but they do change. Stellar structure models tell us that as a star depletes the hydrogen in its core, the hydrogen nuclei become more spread out and less likely to collide. The decline in energy production causes the star's core to contract slightly, compressing and heating the core. This drives up the temperature until the overall rate of reactions once again generates enough heat to support the star. Over the long lifetime of a star in its main-sequence phase, it will gradually increase in luminosity by a factor of about 3. The stellar models show that its surface temperature will drop slightly over this time as the atmosphere expands in response to the hotter core. Tripling the luminosity is not a minor change, but on the H-R diagram, which displays a factor of over a million in luminosity, such a change is hardly noticeable (Figure 61.5).

FIGURE 61.4
Photograph of two young star clusters still surrounded by the interstellar gas from which they formed. NGC 6559/IC 1274–75 is on the left, and NGC 2264 is on the right.

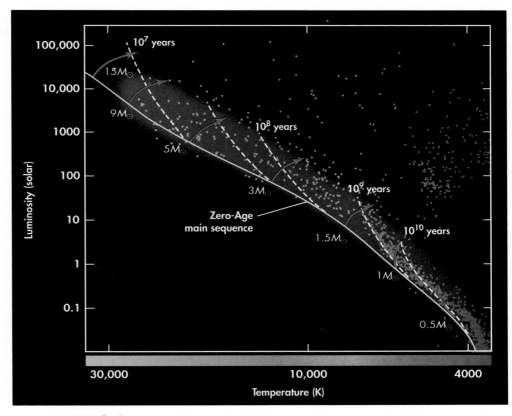

FIGURE 61.5
Stellar evolution while stars are on the main sequence. Luminosities typically increase by about a factor of 3 during this phase. Tracks of stars of different masses are shown in red. The white dashed lines indicate where the stars will be along the tracks at different times after they begin main-sequence fusion.

Q If the Sun's power output has changed so much over the last several billion years, how could there have been liquid water on the Earth during its early history—as indicated by the geological record?

It is possible to estimate how much a star has aged by its shift from the expected initial characteristics when it began fusing hydrogen—also know as its **zero-age main sequence** (or ZAMS) position in the H-R diagram. This requires such precise measurements of a star's mass, luminosity, temperature, and chemical composition that relatively few stars have such age estimates.

Astronomers estimate that the Sun is about 1.4 times more luminous today than when it began on the ZAMS. Its luminosity will continue to increase, approximately doubling from its current value over the remaining 5 or 6 billion years of its main-sequence lifetime. The increase will grow fairly steep toward the end of this time. As the Sun's core temperature climbs, the CNO cycle will become the dominant source of its energy production.

Although these changes are small compared to the range of other stars' luminosities, they are very significant for the environment of the Earth and other planets. Just a few percentage points of increase in the Sun's luminosity would drive the Earth's temperature up by about 2°C (about 4°F), enough to cause polar ice cap melting to change ocean levels dramatically. Unless compensating effects occur on Earth, its climate and habitability will change in the distant future.

KEY TERMS

CNO cycle, 499

convection, 500

fully convective, 500

hydrostatic equilibrium, 498

main-sequence lifetime, 501

main-sequence star, 498

proton–proton chain, 499

zero-age main sequence (ZAMS), 504

QUESTIONS FOR REVIEW

1. What determines when a star becomes a main-sequence star?
2. What determines how long a star stays on the main sequence?
3. How does convection differ in very low-mass stars, low-mass stars, and high-mass stars?

PROBLEMS

1. Calculate the approximate main-sequence lifetime of a 0.5 solar mass star, using the mass–luminosity relationship (Unit 58) to estimate its luminosity.
2. The star Alpha Centaurus A has the same spectral type as the Sun, but it is about 20% more luminous. What might explain this difference?
3. Calculate the predicted luminosity of the 2.0 $M_\odot$ star Sirius from the mass–luminosity equation. Its observed luminosity is about 20 $L_\odot$. How might you explain the difference?

TEST YOURSELF

1. A star whose mass is 2 times larger than the Sun's has a main-sequence lifetime
   a. many times longer than the Sun's.
   b. a few percent longer than the Sun's.
   c. a few percent shorter than the Sun's.
   d. many times shorter than the Sun's.

2. Blue main-sequence stars have
   a. a large mass and a short lifetime.
   b. a large mass and a long lifetime.
   c. a low mass and a short lifetime.
   d. a low mass and a long lifetime.
   e. You can't say, because colors of main-sequence stars are unrelated to their masses or lifetimes.

3. *Hydrostatic equilibrium* means that
   a. a star is not convecting in its interior.
   b. a star has achieved a state where water is a fusion by-product.
   c. electrons are in their highest possible energy levels before ionization.
   d. gravity and pressure are in balance.
   e. fusion is occurring at a steady rate.

62 Giant Stars

Although the main-sequence stage of a star's life is one of relative stability, the next stage of its life is one of constant change. When hydrogen runs out in the core of a main-sequence star, the force of gravity remains, and the star's core resumes the gravitational contraction that was taking place during the protostar phase. But an unexpected thing happens. Even though the core of the star begins contracting and heating again, the surface of the star expands and cools, and the star grows into a **giant**. This peculiar behavior needs to be explained so we can understand many of the unusual features of the giant phase of a star's life.

The giant phase never establishes the stability of the main-sequence phase because its new sources of fuel run out rapidly and the star is constantly readjusting its structure. Stellar models indicate that a star spends about 10 to 20 percent as long as a giant as it does as a main-sequence star.

Background Pathways: Unit 59

62.1 RESTRUCTURING FOLLOWING THE MAIN SEQUENCE

During a star's main-sequence phase, its structure is stable. This stability occurs because the nuclear fusion in the star's core acts as a thermostat. For example, if a star contracts slightly, the compression of the gas in its core heats it, driving up the rate of nuclear reactions. Consequently, the core produces more heat and pressure, reversing the contraction. A slight expansion of a star reduces nuclear reactions, leading to less pressure, which reverses the expansion. A main-sequence star is therefore quite stable. However, once nuclear reactions cease in the core, the thermostat is broken, and this leads to unusual effects, which are the hallmarks of the next phase of a star's life.

When a main-sequence star has consumed the hydrogen in its core, the energy production there halts. With no energy generation in the core, the weight of the star's outer layers pressing downward overwhelm the pressure supporting its core. The core is squeezed smaller. This compresses the gas and therefore heats it (as discussed also in Units 49 and 60). The result for the star is that its core temperature rises. However, because the core contains little hydrogen, the star has no way to generate energy to halt this compression process. As a result, the core of the star resumes the contraction that stopped when it began to fuse hydrogen as a new main-sequence star.

Although the core itself has no hydrogen to fuse, regions outside the core are still rich in hydrogen because they never reached temperatures high enough for hydrogen to fuse. Now, however, as the interior of the star shrinks and is compressed, the regions immediately surrounding the core are also compressed and heated to fusion temperatures. Hydrogen fusion begins in a layer outside the core in a process that is commonly called hydrogen **shell fusion** (Figure 62.1). Although this shell fusion generates heat and pressure, it is outside the core, and therefore it cannot provide the internal pressure needed to stop the core's contraction.

This situation is unlike the thermostat effect that makes a main-sequence star so stable. The core continues contracting and growing hotter, causing more and more

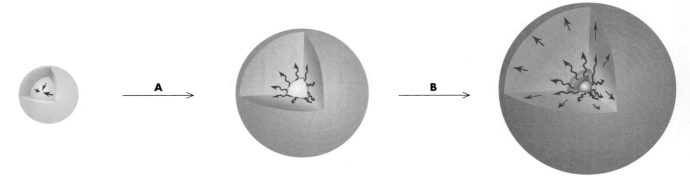

As fuel is exhausted, outward pressure in core drops, and gravity compresses core.

Contracting core rises in temperature. Surrounding shell begins fusing hydrogen. Outer layers are pushed outward by higher pressure.

Core shrinks and heats further. Shell fusion grows stronger. Atmosphere expands and cools further.

FIGURE 62.1

(A) The core of a star begins to shrink as a star uses up the hydrogen in its core. This compresses and heats the core. (B) The heated core ignites the surrounding gas and the outer layers of the star expand, turning it into a red giant.

Some energy is also released from gravitational potential energy as the star's core contracts.

hydrogen to fuse in the regions surrounding the core. The star grows steadily more luminous, but the rise in energy production now does nothing to reverse the contraction. For low-mass stars like the Sun, the luminosity may increase by a factor of a thousand. In other words, it will be fusing 1000 times more hydrogen each second in its shell region during its giant phase than it fused each second in its core during its main-sequence phase.

High-mass stars entering the giant phase generally show relatively little change in luminosity. They have already consumed more of their hydrogen outside the core because of greater convection activity during their main-sequence lifetimes, and the shell fusion during this phase does not represent as substantial an increase over their vigorous main-sequence rate of fusion.

Shell fusion pushes the surface layers outward for both low- and high-mass stars, and this is reflected by a significant drop in surface temperature. This drop in surface temperature is perhaps the most counterintuitive aspect of the giant phase. The core shrinks and heats up, but the outer layers of the star expand and cool.

We can understand this unusual behavior as follows. The shell fusion region raises the temperature and pressure of the gas in layers of the star that are above it. That stronger pressure pushes the surrounding gas outward, causing the star's surface to expand. The star may grow in radius by a factor of anywhere from five to several hundred, depending on the star's mass. This expansion leaves the outer layers much farther from the source of the star's luminosity, the fusion in and around its core.

At this larger distance from the core, even though the star has grown more luminous, the intensity of radiation is reduced. This is a consequence of the inverse square law (Unit 54.2): $B = L/4\pi d^2$, in which the brightness of light is inversely proportional to the square of the distance from the source of light. The luminosity coming from the interior of the star is spread out over a much larger area because the distance d from the core has grown much larger. The decreased intensity of radiation results in surface layers that are cooler, and because cooler bodies radiate more strongly at longer wavelengths, the star becomes redder. Thus, the main-sequence star has evolved into a red giant.

In the H-R diagram, the star moves up and to the right of the main-sequence region along a part of its evolutionary track called the **red giant branch.** A star like the Sun will grow steadily larger and more luminous over nearly a billion years as it searches for a new equilibrium, its core gradually shrinking and heating

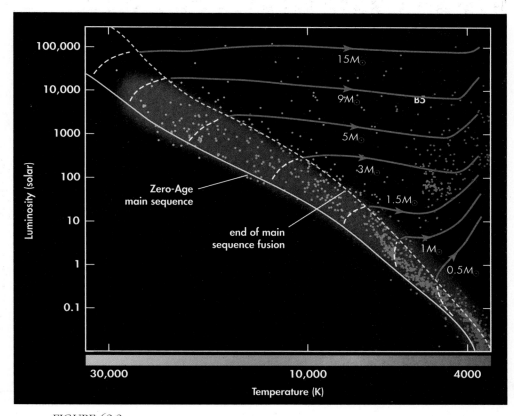

FIGURE 62.2

Evolutionary tracks followed by stars with masses from 0.5 to 15 times the Sun's mass after main sequence fusion ends. The stars move into the red giant region of the H-R diagram following the tracks shown in red.

throughout this time. This leads to substantial changes in the star's position on the H-R diagram, as illustrated in Figure 62.2.

62.2   HELIUM FUSION

A red giant star has a very different structure from that of a main-sequence star. As its name implies, its surface is cool and its radius is large: in some cases more than a thousand times larger than the Sun's. But this enormous size is deceptive. Most of the star's volume is taken up by its very low-density, tenuous atmosphere, and most of its mass lies in a tiny, hot, compressed core within the central 1% of its radius. Hydrogen shell fusion supplies most of the energy during the beginning of the giant phase, but if the star's core becomes hot enough it will begin to generate energy by the fusion of helium.

Helium nuclei can fuse to form heavier elements, but the process requires a much higher temperature than hydrogen fusion. Two nuclei fuse when they are brought close enough together for the strong force (Unit 4) to bind them. That force, which is what holds the protons and neutrons together in a nucleus, operates only over very short distances. If nuclei are more than a few diameters apart, the nuclear strong force is too weak to bond them. In fact, they will be repelled by the electrical force between the similarly charged protons, as illustrated in Figure 62.3. The electric repulsion grows as the number of protons increases. The repulsion is smaller for hydrogen with its single proton, but larger for heavier elements with more protons. It is more difficult, therefore, for helium nuclei, with their two protons, to fuse than for hydrogen atoms to do so.

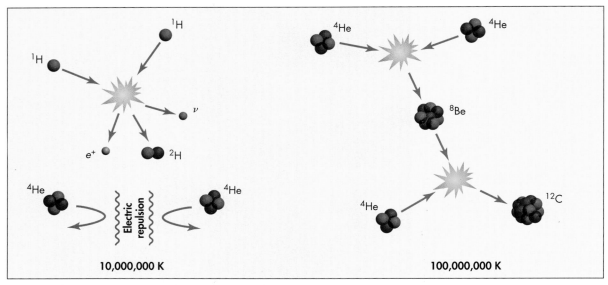

FIGURE 62.3

At 10 million Kelvin hydrogen nuclei can fuse, but the larger electric repulsion between helium nuclei keeps them from fusing. At 100 million Kelvin helium nuclei can fuse, making first beryllium and then carbon.

The triple alpha process requires two steps in rapid succession. First, two helium nuclei fuse to make a beryllium nucleus, ^{8}Be. However, this nucleus is unstable and breaks back apart into the helium nuclei in only about 2 millionths of a second, so another helium nucleus must collide with it in that short interval. If ^{8}Be were stable, how might this change stars' giant phase?

Helium fusion occurs when three ^{4}He nuclei combine to make a carbon nucleus, ^{12}C. This is called the **triple alpha process.** Helium nuclei are known as *alpha particles* because they were detected in some of the first nuclear radioactivity experiments, but their identity was unknown for several years, so they were given this alternative name. For the triple alpha process to become a steady source of energy, a temperature in excess of 100 million Kelvin is required. However, despite operating at such a high temperature, the triple alpha process releases much less energy than hydrogen fusion—only about one-tenth the amount of energy per kilogram of fuel as hydrogen fusion yields. Thus, although the initiation of helium fusion stabilizes the core against further contraction, the triple-alpha process does not produce the majority of a giant star's luminosity.

Calculations indicate that the compression that occurs in stars more massive than about 0.5 $M_\odot$ is sufficient to reach temperatures that allow helium fusion, but the process begins very differently in low- and high-mass stars. For example, a 10 $M_\odot$ star needs to compress its core to only about one hundred times its main-sequence density before helium begins to fuse, because its core is already extremely hot. This makes it fairly easy for a high-mass star to compress and heat its core to the point where it can fuse helium. A low-mass star like the Sun, however, must compress its core by about a factor of *ten thousand* before it becomes hot enough for helium fusion. This requires the Sun's core to reach a density of about 1 million kilograms per liter. A teaspoonful of such matter would weigh as much as a truck! As you might suppose, such densities strongly affect the properties of the gas. To see how, we need to look more closely at the nature of extremely dense gases.

62.3 ELECTRON DEGENERACY AND HELIUM FUSION IN LOW-MASS STARS

To see why gases behave differently when they are densely packed, imagine a few dozen tennis balls inside a large box. If you shake the box, the tennis balls bounce about inside the box. This is similar to the behavior of the particles in a normal gas.

You can make the box smaller, and the tennis balls will still bounce—until you make the box so small that the balls are packed tightly against one another. At that point the balls no longer bounce, and they prevent you from making the box any smaller.

Gases display a similar behavior—they can be compressed until the particles are so closely packed that they fill the volume. At that point the gas becomes almost rigid and extremely difficult to compress.

The particles that resist being packed so closely together are the electrons in a hot ionized gas. The behavior of subatomic particles like electrons is different from how solid balls interact. The electrons obey what is called an **exclusion principle:** No two electrons can occupy the same space if they have the same energy. Note that this rule *does* allow electrons to occupy the same volume of space if they have *different* energies.

If a gas is compressed to the point where electrons of the same energy are trying to occupy the same space, they behave like the tennis balls, resisting coming any closer together. A gas in which its electrons are packed together like this is called a **degenerate gas.** This packing limit makes the gas as "stiff" as a solid. As a result, a degenerate gas does not contract when it cools any more than a cooled brick shrinks. This stiffness has important consequences for nuclear fusion in a degenerate gas.

For stars of less than about 0.5 $M_\odot$, the compression makes the core degenerate before it grows hot enough to start helium fusion. Hydrogen fusion in the shell continues briefly after degeneracy is reached, but without the rising temperature coming from the continued contraction of the core, there is nothing to keep the surrounding gas hot enough for fusion to continue in the shell. As the core cools, hydrogen fusion slows in the shell and then halts. The star then eventually cools and dims.

The cores of stars with masses from about 0.5 to 2 $M_\odot$ get compressed to the degeneracy limit when they are close to the helium ignition temperature, and their fate is quite different. Normally, the thermostatic feedback behavior of nuclear fusion in a gas prevents the nuclear reactions from becoming too violent. If the gas becomes hot, the increase in nuclear fusion heats the gas, raising its pressure so the gas expands. The expansion cools the gas and reduces the rate of nuclear fusion. This is not true when electron degeneracy pressure supports the core: The pressure resisting contraction remains almost independent of the temperature.

When the core becomes degenerate, it no longer compresses easily, but compression is not completely eliminated. Normally the electrons in a gas are all in the lowest energy level possible, so a degenerate gas cannot be compressed more unless some of its electrons are raised to higher energies. As a result, whenever the electrons in the core absorb energy, they can move to higher energy levels, and they can be packed more tightly according to the exclusion principle. As the core is heated by gravitational contraction and the radiation from the surrounding fusion shell, the electrons absorb some of this energy. The core shrinks, the gravity grows more intense, and the greater compression on the hydrogen-fusion shell makes it hotter so it generates more energy, which continues the process—although the compression proceeds more slowly than before the electrons became degenerate.

Under these circumstances, the initiation of helium fusion can occur explosively. When the helium fusion temperature is reached, the energy released by the fusion reactions is soaked up by the electrons. This relieves some of the degeneracy, so the core shrinks. But this drives the temperature up, causing more rapid helium fusion, which gives the electrons more energy, allowing more compression and heating. This runaway process accelerates the star's energy production explosively in what is called a **helium flash.**

During the helium flash, a star's energy production increases by many thousand times in just a few minutes. The outburst is totally hidden from our view by the

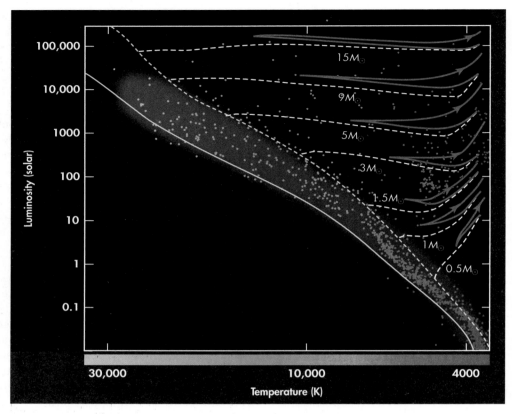

FIGURE 62.4

Evolutionary tracks (in red) of stars in H-R diagram after fusion of helium in the core first halts core contraction then later as a helium fusion shell forms.

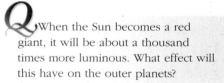

Q When the Sun becomes a red giant, it will be about a thousand times more luminous. What effect will this have on the outer planets?

star's outer layers, just as a firecracker set off under a mattress creates little visible disturbance. When the released energy heats the core enough that the gas is no longer degenerate, it stabilizes. With continued energy generation from helium fusion, the gas can expand and adjust to maintain a more stable pressure balance with the gravitational weight of the layers pressing down on it.

The star's outer layers adjust in turn—shrinking and rising in temperature even as the core expands and cools. The star's surface grows hotter and therefore changes color from red to yellow, and the star drops slightly in luminosity. The tracks of such stars in the H-R diagram are shown in Figure 62.4. The stars shift rapidly down and to the left as they adjust to helium fusion in the core. As the helium is consumed in the core, their luminosities climb again, and they travel back up the diagram, growing to be even more luminous red giants than before.

62.4 HELIUM FUSION IN HIGH-MASS STARS

Stars more massive than about two solar masses do not have a helium flash. When they exhaust the hydrogen in their core, they leave the main sequence and shell fusion begins. Meanwhile, their surfaces expand and cool, just like the low-mass stars. However, the cores of high-mass stars are already so hot as main-sequence stars that they ignite helium with relatively little compression, and so they do not become degenerate at this stage of their lives. The large compressive forces in these stars may even raise the temperature high enough, about 200 million Kelvin, for more *alpha capture* fusion reactions to take place, providing an additional source of energy. For example, an additional ^{4}He nucleus can fuse with ^{12}C to form oxygen, ^{16}O. This tends to occur where both carbon and helium are present—at the inner boundary of the hydrogen fusion shell where newly formed helium is deposited onto the core, most of which has already been converted into carbon.

The time it takes massive stars to accomplish these internal changes is only about 1% of their main-sequence lifetime. As a result, their positions in the H-R diagram shift rapidly from the main-sequence region over to the red giant region. After their cores are compressed sufficiently to initiate helium fusion, their luminosity generally stabilizes at a slightly higher value than when the star was on the main sequence. Luminosity is generated by both core helium fusion and hydrogen shell fusion. The surface layers of these stars all stabilize around a temperature of about 4000 K. The radii, however, may grow to a few hundred times the original radius—the larger the mass, the larger the radius. This relatively stable helium fusion phase lasts 10 to 20% as long as the star's main-sequence lifetime before the helium in the core is consumed.

KEY TERMS

degenerate gas, 509

exclusion principle, 509

giant, 505

helium flash, 509

red giant branch, 506

shell fusion, 505

triple alpha process, 508

QUESTIONS FOR REVIEW

1. What makes a star move off the main sequence?

2. Where do main-sequence stars end up as they evolve?

3. Why do high- and low-mass stars evolve differently as they become giants?

4. What happens to a solar mass star when it starts to fuse helium in its core? What does it turn into?

PROBLEMS

1. When the Sun expands to be a red giant, its radius may be 100 $R_\odot$. What will the average density of the Sun be at that time?

2. When the Sun becomes a red giant, its core region (containing about 20% of its mass) may contract to 0.01 $R_\odot$. What will the density of the core be at that time?

3. By how many times its current radius would the Sun have to expand before its outer atmosphere reached the Earth?

4. If the Sun becomes a red giant that has a radius 100 times larger than its current radius and a luminosity 1000 times its current luminosity, what will its surface temperature be? (Use the Stefan-Boltzmann equation from Unit 57 and solve for T.)

TEST YOURSELF

1. As a star like the Sun evolves into a red giant, its core
    a. expands and cools.
    b. contracts and heats.
    c. expands and heats.
    d. turns into iron.
    e. turns into uranium.

2. High-mass stars entering the giant phase
    a. intensify greatly in luminosity.
    b. contract significantly.
    c. have most of the hydrogen left in their cores.
    d. experience a drop in their surface temperatures.
    e. All of the above

3. Helium fusion
    a. begins as soon as hydrogen is depleted in a star's core.
    b. can never happen once a core is degenerate.
    c. happens only with high-mass stars.
    d. eventually happens with all stars.
    e. None of the above

63 Variable Stars

Many stars pass through a stage sometime during their lives in which their luminosity varies. For many stars this stage occurs after they enter the red giant phase. Astronomers call stars whose luminosities change **variable stars**. All stars vary slightly in brightness—even the Sun varies by a few percentage points because of its magnetic activity cycle (Unit 51)—but this kind of variation is detectable only with careful photometric measurements. All stars also experience gradual changes in their luminosity and temperature while they are on the main sequence, and faster changes while they are giants.

Although at some level virtually all stars vary, traditionally the stars cataloged as variables change in brightness over a decade or less. Furthermore, these changes can be seen just by looking through a telescope and comparing the variable star with neighboring stars. An example of a variable star is shown in Figure 63.1—the star Mira (*MY-rah*) in the constellation Cetus. For hundreds of years it has been known to regularly brighten and fade in a little less than a year.

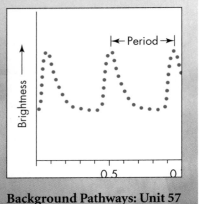

Background Pathways: Unit 57

63.1 CLASSES OF VARIABLE STAR

The first variable stars identified were stars that brightened dramatically, often from previous invisibility, before fading again. A "new star" suddenly appearing this way in 1572 was called a *nova stella* by Tycho Brahe (Unit 12). Today this is shortened to **nova.** These are associated with explosive events sometimes marking the final stages of a star's lifetime, and they may remain visible for weeks or

FIGURE 63.1

These two photographs show the star Mira in the constellation Cetus. Mira varies over a period of 332 days from being visible to the unaided eye to being about 100 times fainter.

months before fading away. Actually the nova seen in 1572 was an especially luminous type that today is called a **supernova.** Tycho Brahe's measurements of the brightening star's position relative to other stars showed that it had no parallax—demonstrating that it must belong to the celestial sphere. Tycho's discovery was revolutionary at the time because it proved that even the highest celestial realm was changeable, running counter to beliefs of the day.

To characterize a star's variability, astronomers measure its brightness at frequent intervals and plot these against time. This results in a graph called a **light curve.** Over a hundred types of variable star patterns have been identified, but these can be divided into a few classes. Stars like novae belong to the class of **irregular variables,** which undergo unpredictable outbursts of brightness that do not follow a repeating pattern. Another example of an irregular variable is a T Tauri star (Unit 60), which is a stage some protostars pass through. These objects probably brighten as material falls irregularly onto the forming star. Dozens of patterns have been identified in the irregular class, and these are often associated with very young or very old stars.

Some stars also vary because of periodic eclipses by other stars in orbit about them (Unit 56), but the majority of cataloged variable stars fall in the class of **pulsating variable.** These stars change in luminosity in a rhythmic pattern. Based on the shapes of their light curves, astronomers have identified more than a dozen types of regularly pulsating variable stars. For example, some stars pulsate with a period of about half a day, while others take more than a year to complete a cycle of pulsation. The interval of time it takes the pattern of brightness to repeat is called the **period,** as shown in Figure 63.2.

The Latin heritage of *nova* is remembered by most astronomers in the plural form *novae,* although *novas* is also acceptable. The same applies to other Latin-derived words, such as *supernovae* versus *supernovas.*

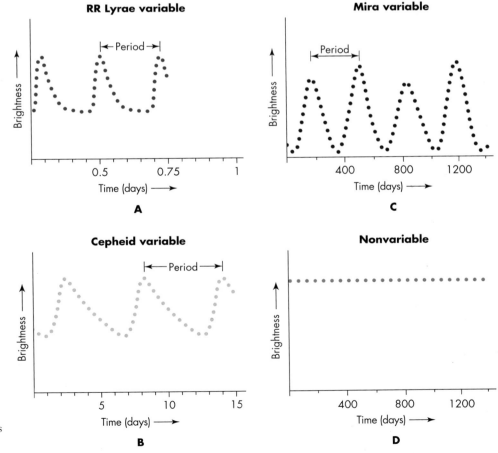

FIGURE 63.2
Schematic light curves of (A) an RR Lyrae variable; (B) a Cepheid variable; (C) a Mira variable; and (D) a nonvariable star. The timing and unique pattern of variability allow astronomers to identify these stars.

When we analyze the temperatures and luminosities of pulsating variable stars, we find that they may also vary in size, rhythmically swelling and shrinking. These changes in diameter are caused by the pressure rising below the stars' surfaces, which makes the surfaces expand outward. Some classes of these stars may more than double in size over their period, but the size variations are more commonly less than 20%. Such variations in size also cause a star's luminosity to change. If the star expands, its surface area increases so that it has more area to radiate. This would make the star more luminous. But as the gas expands, its temperature usually drops, sometimes to half its former value. This would make the star dimmer. The result of both changes is that, as the star pulsates, its luminosity changes according to the Stefan-Boltzmann law (Unit 57) in a more complex way. Mira variables' luminosities can vary by more than a factor of 100, although the majority of variables exhibit a factor of two or less.

63.2   YELLOW GIANTS AND PULSATING STARS

Pulsating variable stars are important to astronomers because they can serve as "standard candles" (Unit 54) and thereby offer a way to measure distances within the Milky Way and even to other galaxies (Unit 74). When such stars are plotted on the H-R diagram according to their average luminosities and temperatures, many of them lie within a narrow strip in the giant region (Figure 63.3) called the **instability strip.** Because many of these stars have temperatures of about 5000–6000 K, they are yellow. Most pulsating variables are giants (Unit 62) and are sometimes called **yellow giants** or **yellow supergiants,** according to their luminosity.

One type of pulsating star useful for finding distances is the **RR Lyrae** (pronounced *LIE-ree*) **variable.** RR Lyrae variables have a mass comparable to the Sun's and are yellow-white giants with about 40 times the Sun's luminosity. The period of their pulsation cycle—from bright to dim and back to bright, as illustrated in Figure 63.2—is about half a day. They are named for RR Lyrae, a star in the constellation Lyra, the harp, which was the first star of this type to be identified. From their known luminosity, we can apply the standard-candles method (Unit 54) to RR Lyrae variables. Knowing that their luminosity is $40 \times L_\odot = 1.6 \times 10^{28}$ watts, we can plug this value into the equation for distance ($d = \sqrt{L/4\pi B}$), and if we measure the brightness B we can find the distance.

Other important pulsating stars for finding distances are the **Cepheid** (pronounced *SEF-ee-id*) **variables.** Cepheid variables are yellow supergiants that are more massive than the Sun, typically with about 10,000 times its luminosity. They are named for the star Delta Cepheus (see Looking Up #1), and their periods range from about 1 to 70 days. Their luminosities vary widely, but they obey a relationship (discussed in the next section) that allows us to determine individual luminosities.

Giant stars pulsate because their atmospheres trap some of their radiated energy. This heats their outer layers, raising the pressure and making the layers expand. The expanded gas cools and the pressure drops, so gravity pulls the layers downward and recompresses them. The recompressed gas begins once more to absorb energy, leading to a new expansion. These stars continue alternately to trap and release the energy, and so they continue to swell and shrink, as shown in Figure 63.4.

A covered pan of boiling water behaves similarly. The lid will trap the steam so that pressure inside rises. Eventually the pressure becomes strong enough to tip the lid, and steam escapes. The pressure decreases, and the lid falls back. It again traps the steam, the pressure builds up, and the cycle is repeated. In pulsating stars, the role of steam in this process is played by the star's radiation, and the role of the lid is played by partially ionized helium gas in the star's atmosphere. This is sometimes called the **valve mechanism** because it is like a valve that opens to relieve excessive pressure.

If an RR Lyrae variable is measured to have a brightness of 10^{-11} watts/m^2, we would calculate a distance of
$$d = \sqrt{L/4\pi B} = \sqrt{1.6 \times 10^{28}/10^{-11}}$$
$$= \sqrt{16 \times 10^{38}} = 4 \times 10^{19} \text{ m, or}$$
about 1300 parsecs.

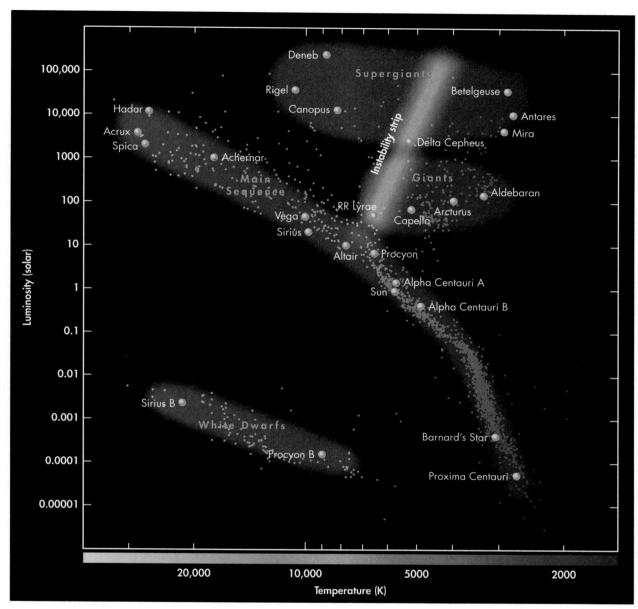

FIGURE 63.3
The instability strip in the H-R diagram. Stars that fall in this band generally pulsate.

Q If you put a heavy lid on a pan of boiling water instead of a light lid, how would this change the frequency of the steam escaping? What does this suggest about the variability of a star with a greater depth of atmosphere that must be adjusted?

Stars in the instability strip have surface temperatures of about 5000 K. The reason these stars are susceptible to pulsation is that not far below their surface, stars of this temperature have an unstable region where partially ionized helium is able to absorb and then release the energy flowing out from the center of the star. This unstable region can alternately push the outer layers of the star upward and release the pressure so they fall back. The instability strip shifts to lower temperatures for higher-luminosity, larger-radius giants because their surface gravity is weaker, so the partially ionized helium layer can be deeper in the star and still provide enough pressure to push the surface layers upward.

When stars enter their giant phase, they often move along complex evolutionary paths through the giant region of the H-R diagram (Unit 62). When a star crosses through the instability strip, it begins to pulsate, and it continues to pulsate until its surface temperature changes enough to remove it from the instability strip.

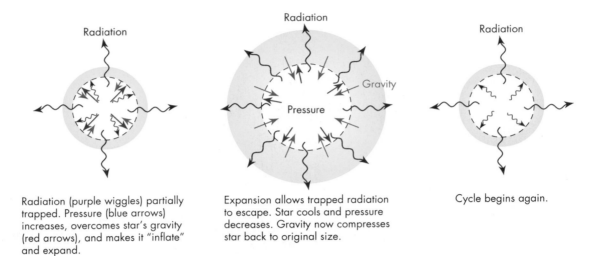

Radiation (purple wiggles) partially trapped. Pressure (blue arrows) increases, overcomes star's gravity (red arrows), and makes it "inflate" and expand.

Expansion allows trapped radiation to escape. Star cools and pressure decreases. Gravity now compresses star back to original size.

Cycle begins again.

FIGURE 63.4
Schematic view of a pulsating star.

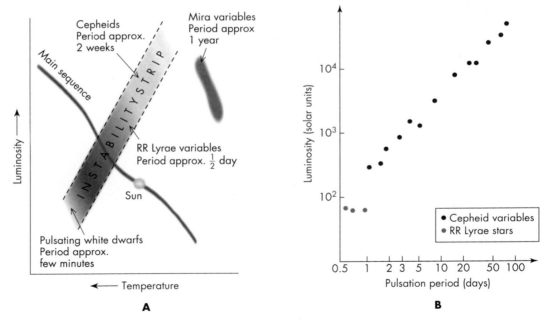

FIGURE 63.5
Properties of variable stars. (A) The instability strip in the H-R diagram. Stars in this narrow region pulsate. The locations of a few other types of variable stars are also indicated. (B) The period–luminosity law. More luminous stars tend to pulsate more slowly.

When a low-mass star crosses the instability strip, it becomes an RR Lyrae variable. When a high-mass star crosses, being more luminous, it lies above the RR Lyrae variables in the instability strip and instead becomes a Cepheid variable.

The amount of time a given star spends in the instability strip depends on its mass. Cepheids evolve across the strip in less than 1 million years. RR Lyrae variables spend more time in the strip, perhaps a few million years. In either case, stars pulsate for only a brief portion of their lives.

Astronomers have identified several other types of pulsating variables besides Cepheids and RR Lyrae stars. For example, Mira variables, which have pulsation periods of about a year, lie in the upper right portion of the H-R diagram. Another kind of variable star, the ZZ Ceti stars, are pulsating white dwarfs with periods as short as a few minutes. These lie along an extrapolation of the instability strip to the other side of the main sequence in the lower part of the H-R diagram (Figure 63.5A). The location suggests that they, too, may be driven by the valve mechanism, although under different physical circumstances than occur in the Cepheids and RR Lyrae variables.

63.3   THE PERIOD–LUMINOSITY LAW

Cepheids and some other classes of pulsating variable stars obey a law that relates their luminosity to their period—the time it takes them to complete a pulsation. Observations (and theoretical calculations) indicate that the more slowly a Cepheid pulsates, the more luminous it is (Figure 63.5B). This **period–luminosity relation** arises because, other things being equal, more luminous stars have a larger radius than less luminous ones. This follows simply from the fact that a bigger star has more surface area, so it can emit more light.

Why, though, does its larger radius make the larger star pulsate more slowly? Again, the answer is simple: because an object's gravity weakens with increasing distance. Thus, a large-radius star has weaker surface gravity than a small-radius star. So if the layers of a star with a large radius are pushed outward, the feeble gravity pulls them inward more slowly than in a small-radius star. Therefore, the pulsations in a big star take longer than in small, less luminous stars.

The period–luminosity law is one of the astronomer's most powerful tools for determining a Cepheid's luminosity. Those with a period of a few days have a luminosity as small as a few hundred solar luminosities. On the other hand, those with a period of a few months have a luminosity of tens of thousands of solar luminosities. By observing a Cepheid's light curve, we can therefore determine its luminosity.

Some other classes of variables, such as Miras, also obey a period–luminosity relationship, although a different law than the one Cepheids obey. On the other hand, RR Lyrae stars have a nearly fixed luminosity, independent of their pulsation period. In short, if we can identify the class of a variable star from its light curve, we can measure the star's luminosity from the period–luminosity law appropriate to it. This knowledge of the star's luminosity makes it a standard candle (Unit 54.3), so variable stars are one of the astronomers' critical tools for finding distances to objects too far away to measure by parallax.

Q If you extrapolated the Cepheid period–luminosity relationship to a luminosity like the Sun's, how fast would you expect it to vary?

Q At optical wavelengths, Miras sometimes vary in brightness by a factor of 10,000. At infrared wavelengths the variation is much smaller. What might explain this difference?

KEY TERMS

QUESTIONS FOR REVIEW

1. What is a variable star?
2. What is the period of a variable star?
3. Where in the H-R diagram are variable stars found?
4. What is a pulsating star?

Phase	Luminosity	Temperature
0.00	10,000 $L_\odot$	6700 K
0.25	7,800 $L_\odot$	5900 K
0.50	6,100 $L_\odot$	5400 K
0.75	5,000 $L_\odot$	5500 K
1.00	10,000 $L_\odot$	6700 K

TABLE 63.1   Temperature and Luminosity of a Variable Star

PROBLEMS

1. An interferometer measured the change in radius of one Cepheid as about 10% during its 10-day period. The star was measured to have an average radius of 40 million kilometers. How fast was the surface of the star moving outward, on average, if it took 5 days to expand by 10%?

2. A variable star varies in luminosity and temperature throughout its cycle according to the values listed in Table 63.1. The *phase* indicates the fraction of time between one maximum and the next. Graph the luminosity and temperature change, then plot the positions of the star in the H-R diagram. Connect the sequential points with a line to show the pattern of variation.

3. Suppose a Cepheid has a period of 5 days. Estimate its luminosity from Figure 63.5B, then determine its distance if its measured brightness is 4×10^{-12} watts/m^2.

4. Calculate the radius of the star in the previous problem at each phase using the Stefan-Boltzmann law. When is the star largest? When is it smallest?

TEST YOURSELF

1. Cepheid variables are important to astronomers because
   a. they are one of the only types of stars whose radii are known.
   b. they demonstrate the presence of spots on stars other than the Sun.
   c. the precise rate of their pulsation allows them to be used as time standards.
   d. we can use them to estimate accurate distances.
   e. All of the above

2. The period–luminosity relation indicates that
   a. a dimmer star pulsates more slowly.
   b. a brighter star rotates more quickly
   c. a larger star rotates more quickly.
   d. a dimmer star rotates more slowly.
   e. a brighter star pulsates more slowly.

3. Variable stars change in
   a. luminosity.
   b. temperature.
   c. size.
   d. density.
   e. All of the above

PLANETARIUM EXERCISES

All the planetarium exercises use the Starry Night planetarium software provided with your text. You should install the program and familiarize yourself with the various parameters and controls before trying these exercises. When the program asks you to enter a password, click "OK" to run Starry Night.

Set the location from which you are viewing to the nearest large city or your latitude and longitude. Accomplish this by clicking on the Location box on the toolbar. Verify that the local time is correct.

A number of variable stars, such as Algol, Delta Cepheus, Beta Lyra, or Eta Aquila, are bright enough that you can see their variations compared to neighboring stars if you can observe them every night for at least a week. Use the planetarium program to locate one of these variable stars that will be visible in your night sky by using the Find option under Settings and typing in the name. Identify other stars in the vicinity that have a range of magnitudes slightly brighter and slightly fainter than the star you have chosen.

Observe your chosen star each night and estimate its brightness in comparison to these stars. The listed stars are all visible with the unaided eye, but binoculars or a small telescope may make this easier. Make a light curve of your star's brightness by plotting your brightness estimates versus the date.

64 Mass Loss and Death of Low-Mass Stars

Background Pathways: Unit 62

A low-mass star (less massive than about 8 $M_{\odot}$) reaches the end of its life when its core can no longer contract. This prevents its core temperature from rising further, and as a result the star's nuclear fusion ends. The end does not come quietly, however.

In its terminal stages such a star ejects great quantities of matter, forming dramatically shaped and brightly colored objects (see Figure 64.1) known as **planetary nebulae**. How a star ejects its outer layers is still not fully understood, but in this unit we discuss some of the ideas proposed by astronomers to explain what happens when a low-mass star approaches the end of its life. The planetary nebula signals the death of a star. The remnant it leaves behind, a **white dwarf**, is initially hot but does not generate power, so it gradually cools and fades from view.

FIGURE 64.1
Matter ejected from a dying star can produce complexly shaped planetary nebulae.

519

64.1 THE FATE OF STARS LIKE THE SUN

When a star like the Sun begins to fuse helium in its core, it is nearing the end of its life. As death approaches for the Sun, its evolution will speed up. This is shown in Figure 64.2 by the schematic evolutionary track through the H-R diagram. We can summarize the Sun's life as follows.

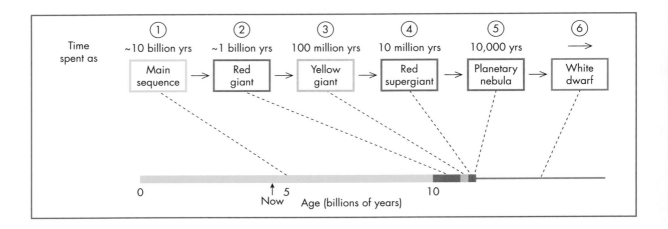

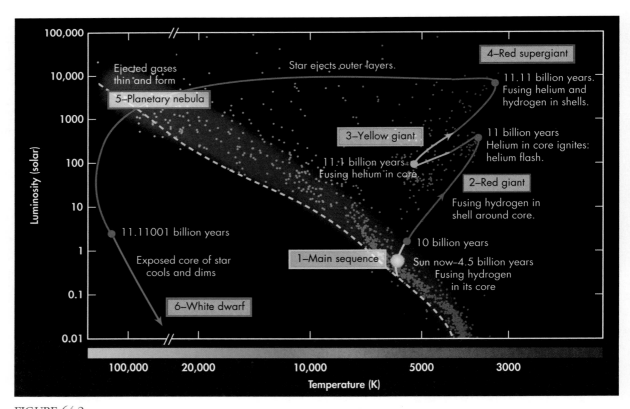

FIGURE 64.2

The evolution of a low-mass star such as the Sun plotted in an H-R diagram. The time line along the top illustrates the shorter and shorter times spent in each stage.

The Sun will spend approximately 10 billion years in its main-sequence phase, consuming the hydrogen in its core. It will then spend about one-tenth as long—about a billion years—as a red giant. Because the Sun's core will become degenerate (Unit 62.3) before helium fusion begins, it will spend most of its red giant phase just fusing hydrogen into helium in a shell surrounding the core. However, its core will finally grow hot enough to fuse helium. The onset of helium fusion is marked by the *helium flash* (Unit 62). After that event, the Sun will rapidly readjust its structure from that of a red giant to that of a smaller yellow giant. The evolution during that phase will be even faster because helium yields less energy than hydrogen when it is fused, so the helium must be fused faster to produce enough energy to support the star. Helium fusion will occur in the core for about the last 100 million years of the Sun's giant phase.

When the helium in the Sun's core is consumed, the core will again contract and heat, similar to the onset of the red giant phase. At the same time, the Sun's atmosphere will reexpand to an even greater diameter than it had when it first entered the red giant stage. The Sun may expand so much by the end of this phase that it will be classified as a **supergiant** (Unit 58) with a luminosity nearly 10,000 times greater than it has now. However, the core's contraction will not raise temperatures enough to begin fusing carbon. During this contraction, the core's density will climb to more than 2 million kilograms per liter—approximately 30 tons per cubic inch! This phase has an even shorter duration of about 10 million years. By the end of this phase, electron degeneracy pressure will halt the compression, and the Sun's core will begin to cool.

Astronomers have discovered that when a star is a red giant or supergiant, it possesses a strong stellar wind. The winds during the red giant phase are far stronger than the solar wind, and they may strip away most of a red giant's mass (discussed in the next section). In fact, so much material is removed from the star over time that the gravitational force squeezing the core drops substantially, greatly altering its evolution. This too will happen to our Sun as it nears the end of its life. All life on Earth will almost certainly have been extinguished by this time: After millions of years of such intense luminosity, the Earth's surface will be nearly molten. However, distant astronomers will observe more and more of the Sun's outer layers flowing into space and thinning out, exposing deeper levels of what remains of the Sun. Ultraviolet photons from the very hot interior of the Sun will now be able to travel out large distances, ionizing some of the Sun's former atmosphere—now expanded far beyond Pluto's orbit. This gas will glow briefly as a planetary nebula (Section 64.3). Finally the remainder of the core will be exposed and will gradually cool to become a white dwarf (Section 64.4). These final stages are illustrated in Figure 64.3.

When the Sun is driving off its atmosphere at the end of the red supergiant stage, it will shift rapidly all the way across the H-R diagram horizontally from right to left before dropping into the white dwarf region (Figure 64.2). Stars maintain their luminosity during the brief transition to the planetary nebula phase, but the observed color and surface temperature of the central star will rapidly climb as deeper layers are exposed. Once exposed, though, the core—now a white dwarf—steadily cools, dropping in luminosity and surface temperature.

A similar sequence is followed by other low-mass stars, although the time taken to pass through the phases depends on mass, like all the other stages of stellar evolution—taking several times longer for lower-mass stars than for higher-mass stars. Low-mass stars larger than about 4 $M_\odot$ have further evolutionary possibilities, depending on how much mass they lose during their red giant stage. When helium fusion has ended, their cores may contract enough and grow hot enough to reach the fusion temperature of carbon (about 600 million Kelvin). Some astronomers hypothesize that the onset of carbon fusion may be far more catastrophic than the helium flash. In fact, they describe this as a **carbon detonation**

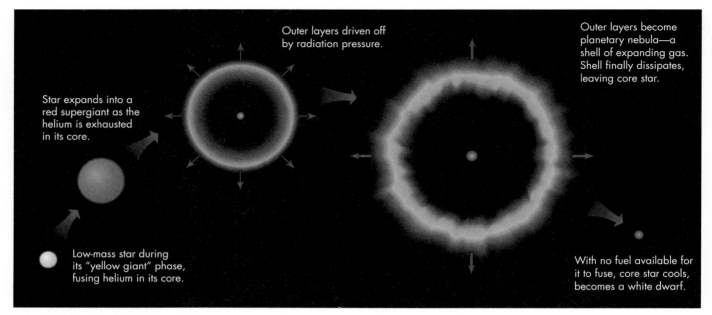

FIGURE 64.3

Final stages of the evolution of a low-mass star. After helium is used up in its core, the star expands to a red supergiant, gradually driving off its outer atmosphere. The remaining core of the former star becomes a white dwarf.

because most of the accumulated carbon reaches fusion temperature almost simultaneously. According to some computer models, this sudden ignition may release so much energy that it blasts the star apart. Other computer models suggest that the onset of carbon fusion may be more gradual, or that it might not occur at all because stars of this size lose too much mass in stellar winds. This remains an area of active research in both observational and theoretical astronomy.

64.2   EJECTION OF A STAR'S OUTER LAYERS

After helium is exhausted at the core of a yellow giant, the star's core contracts again. It heats to temperatures well over 100 million Kelvin, allowing helium and hydrogen shell fusion to occur in layers surrounding the core. As a result, the star's atmosphere may grow to be more than a hundred times larger than its main-sequence diameter.

At this huge size, the surface gravity of the star is quite weak, so heated gas may reach escape velocity, helping to drive a **stellar wind** similar to the solar wind (Unit 49). However, another mechanism for producing a stellar wind may be even more important for red giants. As the star swells, its outer layers cool to about 2500 K—so cool that carbon and silicon atoms condense and form **grains,** much as water in our atmosphere forms snowflakes when it cools. These are not like compact grains of sand, however, but rather are expected to be like flakes of carbon and "rock," very loosely assembled as their atoms and molecules stick together.

The grains do not fall into the star. As illustrated in Figure 64.4, they are pushed outward by the flood of photons pouring from the star's luminous core, just as the dust in a comet is blown by the Sun's radiation into a tail that streams away from the Sun (Unit 47). The rising grains in a red giant drag gas with them, driving it into space. This forms a strong stellar wind, in which as much as about 10^{-4} solar masses of material flow each year into space, typically at tens of kilometers per

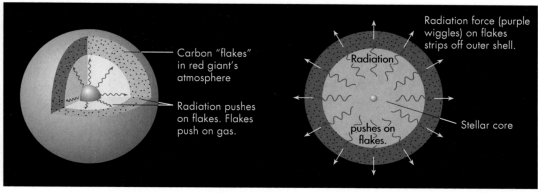

FIGURE 64.4

Artist's conception of how radiation pressure in a luminous red giant pushes on carbon flakes in its atmosphere and drives off the gas. Magnetic forces probably play a role too. The ejected gas forms a shell around the star—a planetary nebula.

second. If the flow remained constant at this rapid rate, a star like the Sun would lose its entire mass in just 10,000 years. Observations, however, indicate that the flow strengthens and weakens during the star's supergiant phase. The strong stellar winds leave the star (or what remains of its dense inner regions) in the middle of a growing expanse of gas—gas that used to be the star's atmosphere (Figure 64.4).

64.3   PLANETARY NEBULAE

As the atmosphere of a star flows away into space, the pressure compressing the core of the star decreases. The core of the star is still furiously hot, and fusion of hydrogen and helium is taking place in shells around the core. However, with the declining compression, these inner regions begin to expand and cool enough that the rate of fusion declines. At the same time, the expanding atmosphere thins and becomes progressively more transparent, which allows high-energy photons from the hot core to travel farther within the remaining atmosphere. These energetic photons drive what is called a *fast wind*, pushing much of the remaining atmosphere of the star outward at high speed. The fast wind runs into the slower-moving gas of the star's earlier stellar wind, clearing out the region surrounding the star and exposing the star's core.

Because the core is so hot, its radiation is rich in ultraviolet light. Photons at these energetic wavelengths heat and ionize the inner portion of the expanding cloud of gas around it, causing it to glow. Astronomers have observed many such glowing shells around dying stars (Figure 64.5) and call them *planetary nebulae*. This term is unfortunate because planetary nebulae have nothing to do with planets. The usage survives from times when astronomers had only poor telescopes, through which planetary nebulae looked like small disks, similar to planets.

The unusual colors of planetary nebulae come from a mix of emission lines (Unit 24) of a number of elements, but particularly oxygen (green) and nitrogen (red), which have been enriched by fusion reactions during the star's late stages. These spectral lines occur in addition to the more common emission lines of hydrogen that give rise to the pink-colored glow from many ionized gas clouds (the Balmer lines, Unit 24). Initially astronomers had difficulty identifying many of the spectral lines in planetary nebulae. The oxygen and nitrogen lines are sometimes called *forbidden lines* because they do not normally occur under laboratory conditions. Even low-density gases studied in a lab are much denser than in a planetary

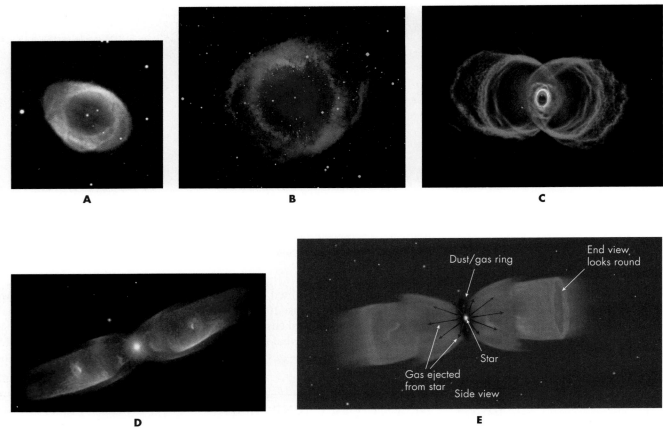

FIGURE 64.5
Pictures of several planetary nebulae. (A) The Ring Nebula. (B) The Helix Nebula. (C) The Hourglass Nebula. (D) The Butterfly Nebula. Notice the central star in each. Other stars that look as if they are inside the shell are foreground or background stars. (E) Sketch of a planetary nebula.

Q Take a clear, cylindrical glass and shine a light through it so it casts a shadow. What kinds of shapes can you make with the shadow?

nebula. When the gas is very diffuse, uncommon electron energy transitions can occur in the long time between atoms colliding with each other.

For many years, astronomers thought that the typical planetary nebula was ejected uniformly in all directions, so that it formed a huge "bubble" around the star's core, as you might deduce from Figure 64.5A. Astronomers knew of oddly shaped planetary nebulae but thought them to be the exception. This view has changed markedly over the last decade, especially now that more detailed images (such as Figure 64.5C and D, made with the Hubble Space Telescope) are available. Most of the planetary nebulae revealed by these photos show that the shell is not spherical, but instead that the gas has been ejected primarily in two directions. Such ejection creates not a bubble but two oppositely directed cones, as sketched in Figure 64.5E. Why this happens is not entirely clear, although a ring of dust and gas in orbit around the star—perhaps part of a planetary system—may physically block the ejection in those directions. Magnetic fields in the stars probably also play a role, much as they are thought to affect the bipolar flow seen ejected from some protostars (Unit 60). Some of the more complex geometries may be caused as the stellar wind interacts with a companion star. Spherically shaped planetary nebula, perhaps the result of isolation from other bodies, may be the exception because we suspect that few stars are truly isolated.

Planetary nebulae eventually grow so big and diffuse that light from the central star no longer ionizes the gas much, and it fades to invisibility. This occurs on a very short time scale by the standards of stellar evolution. In a mere 15,000 years, gas traveling at 10 km/sec in the expanding envelope around the star will have expanded to a light-year in diameter—about the largest size seen among planetary nebulae. We can directly detect this expansion in less than a human lifetime by careful comparison of pictures made several decades apart. The expanding gas mingles with the general interstellar gas, and the cycle of a star's life is complete: gas to star and back to gas. Only the hot core of the star remains: a tiny, glowing ball that gradually cools.

64.4   WHITE DWARFS

The hot core of the former star at the center of a planetary nebula becomes a white dwarf. A white dwarf's dim light comes not from fusion, for it has exhausted all its fuel supplies. Rather, the light is produced by the residual heat left from when the star was fusing hydrogen and helium. This light slowly drains away the remaining heat from the star's interior.

White dwarfs are hot but tiny objects that lie in the lower part of the H-R diagram, as illustrated in Figure 64.2. The composition of a white dwarf is typically the same as that of the core of the star from which it evolved. As we have discussed earlier, such cores are composed mainly of carbon and oxygen, the end products of the parent star's nuclear fusion. The white dwarf's surface is a very hot, thin layer of hydrogen and helium, but its carbon–oxygen core has too little mass, and therefore too little gravity, to contract and heat itself to the ignition temperature of carbon. So, unable to fuse any further, the star's "corpse" gradually cools and grows dim.

The stars that reach this final stage may begin with masses up to 8 $M_\odot$, but with the mass lost from stellar winds during the red giant phase, the remaining mass is found to be less than 1.4 $M_\odot$. The diameter of a white dwarf is about a hundredth of the Sun's—roughly the same size as the Earth. Their tiny size gives white dwarfs so small a surface area that they are very dim despite having surface temperatures as hot as about 100,000 K. Even the brightest white dwarfs are about 100 times fainter than the Sun.

Very low-mass stars (less than about 0.5 $M_\odot$) may become white dwarfs following a somewhat different sequence of events than other low-mass stars. These stars are relatively cool throughout, so their gas, like the relatively cool matter just below the Sun's surface, must carry energy by convection. The convection currents in these small stars extend throughout their spheres, mixing their gas and preventing a core from forming. Moreover, their weak gravity cannot compress their spent fuel much, a necessary step for becoming a red giant. They may simply cool off and progress directly to the white dwarf phase without ejecting a planetary nebula. Because low-mass stars evolve so slowly, however, they require longer than the current age of the universe to reach this point in their evolution.

What happens to a white dwarf as time passes? Although it is initially very hot (it was, after all, the core of a star), it has no fuel to fuse to stay hot. So like a dying ember, it simply cools. Astronomers calculate that it takes about 10 million years for the temperature of a white dwarf to drop to 20,000 K. This may sound long to us, but for a low-mass star, it is the equivalent of about one month in a human lifetime. Although a white dwarf cools steadily, it still emits tiny amounts of visible light even after billions of years. Some astronomers are hunting for the dimmest of these stars as a way to test theories about the age of the universe.

White dwarfs gradually evolve toward what astronomers call **black dwarfs**—the dead, dark corpses of former stars. Cool white dwarfs are probably very abundant in our galaxy. Some estimates of the number of stars that die each year suggest that as much as half our galaxy's mass might be composed of dead white dwarfs (Unit 78).

Q: It is estimated that the Sun will lose about half of its mass by the time it becomes a white dwarf. How will this mass loss affect the orbits of the planets that survive its red giant phases?

KEY TERMS

black dwarf, 525

carbon detonation, 521

grains, 522

planetary nebula, 519

stellar wind, 522

supergiant, 521

white dwarf, 519

QUESTIONS FOR REVIEW

1. What is a planetary nebula?
2. What is one explanation for how a low-mass star expels its outer layers to make a planetary nebula?
3. What is left when a planetary nebula dissipates?

PROBLEMS

1. As a dying star moves across the H-R diagram during its planetary nebula ejection phase, it maintains an almost constant luminosity as successively deeper and hotter layers inside the star are exposed. If it begins in the upper right part of the H-R diagram as a 3000 K, 100 $R_\odot$ red giant, calculate the central star's radius when its surface temperature is
   a. 6000 K.
   b. 30000 K.
   c. 100,000 K.
2. How long will it take a planetary nebula shell moving at 20 kilometers per second to expand to a radius of one-fourth of a light-year?

TEST YOURSELF

1. Which of the following sequences correctly describes the evolution of the Sun from birth to death?
   a. White dwarf, red giant, main-sequence, protostar
   b. Red giant, main-sequence, white dwarf, protostar
   c. Protostar, red giant, main-sequence, white dwarf
   d. Protostar, main-sequence, white dwarf, red giant
   e. Protostar, main-sequence, red giant, white dwarf
2. A planetary nebula is
   a. another term for the disk of gas around a young star that forms planets.
   b. the cloud from which protostars form.
   c. a shell of gas ejected from a star late in its life.
   d. what is left when a white dwarf star explodes as a supernova.
   e. the remnants of the explosion created by the collapse of the iron core in a massive star.

3. How does a 4 solar mass star end up as a 1 solar mass white dwarf?
   a. A stellar wind carries away most of the star's mass when it is a red giant.
   b. Helium fusion turns most of the star's mass into energy.
   c. When an object contracts, its mass turns into density instead.
   d. Neighboring stars and planets pull the star apart when it is a giant.
   e. It spins so much faster as it contracts that much of the gas is thrown off.

PLANETARIUM EXERCISES

All the planetarium exercises use the Starry Night planetarium software provided with your text. You should install the program and familiarize yourself with the various parameters and controls before trying these exercises. When the program asks you to enter a password, click "OK" to run Starry Night.

Set the location from which you are viewing to the nearest large city or your latitude and longitude. Accomplish this by clicking on the Location box on the toolbar. Verify that the local time is correct.

Planetary nebulae are found throughout the sky. Locate two planetary nebulae, such as M57 (Ring), M27 (Dumbbell), or M97 (Owl), by centering on them and locating their positions on the star chart at the back of the book. You can do this by entering their names using the Find option in the Select menu.

You can also find many other planetary nebulae in the planetarium program. In the Settings menu, select "NGC-IC . . ."; on the panel that appears, remove check marks from "Which objects to draw" for all except planetary nebulae. Next click on the "Labels . . ." button and make sure that "NGC-IC" is selected. Click "Apply" and then "OK." Click "Apply" and "OK" on the first panel as well. When you examine the chart now, the planetary nebulae will be marked. Find a part of the sky that is visible to you and locate planetary nebulae there.

Observe and sketch two planetary nebulae with a moderate-size telescope and compare what you saw to the images and information in the text.

65

Exploding White Dwarfs

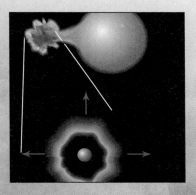

Background Pathways: Unit 64

After a low-mass star has run out of nuclear fuel, the stellar remnant left behind is a **white dwarf.** This fate awaits the Sun about 5 billion years from now (Unit 64). Gravity will crush its remains into an object about 100 times smaller than the present Sun and roughly the size of the Earth. Initially hot, the white dwarf will gradually cool and fade.

Because most stars eventually reach the white dwarf stage, the galaxy is littered with these tiny dead stars. They shine only with heat left over from their previous phase. Their matter, too, is unusual. In crushing a white dwarf to its small dimensions, gravity squeezes the matter to densities far higher than anything we can create in Earth laboratories. For example, a piece of white dwarf material the size of an ice cube would weigh about 20 tons.

Most white dwarfs are extremely dim. None are visible to the naked eye, and most are challenging to detect even telescopically. Over time, isolated white dwarfs cool off and disappear from sight to become *black dwarfs*. However, some white dwarfs have an opportunity for a new life—and a spectacular death—if they accrete fresh matter.

65.1 NOVAE

A white dwarf can accrete matter when it has a nearby companion star. This is a possibility for a fairly large number of white dwarfs because many stars begin life in multiple-star systems (Unit 56). For example, the nearest white dwarf, Sirius B, orbits the brightest star in the sky, Sirius A. Stellar winds from Sirius A may deposit material on the white dwarf in a process astronomers call **mass transfer.**

Mass transfer can grow strong if the white dwarf's companion enters the red giant phase as illustrated in Figure 65.1. As the atmosphere expands, it can eventually

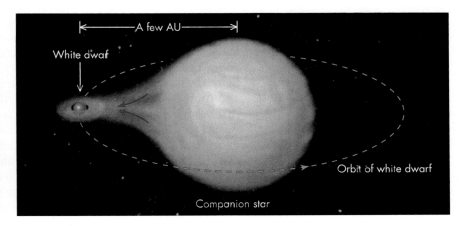

FIGURE 65.1

A white dwarf accreting (pulling in) mass from a binary star companion, which in this case is shown as a red giant. The typical separation of the stars is a few AU.

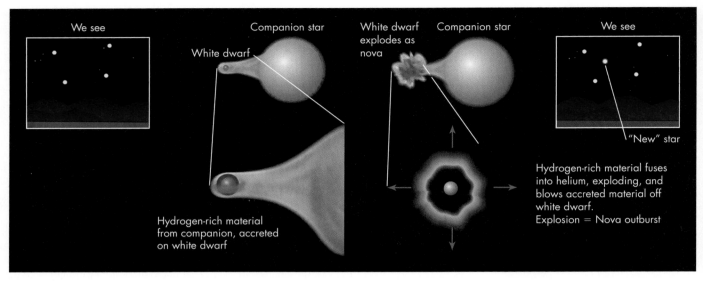

FIGURE 65.2
A white dwarf exploding as a nova. The hydrogen that falls onto the surface of a white dwarf from its companion may suddenly fuse into helium, creating an explosion that makes the star brighten.

ANIMATIONS

Mass transfer

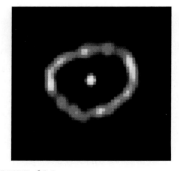

FIGURE 65.3
These two images (made with the Hubble Space Telescope) were taken $7\frac{1}{2}$ months apart. They show the shell of gas expanding away from a white dwarf after a nova.

expand out to a distance where the gravitational pull from the white dwarf exceeds the pull of the red giant itself. The red giant's atmosphere can grow to fill a volume called the **Roche** (pronounced *rohsh*) **lobe** before it starts "spilling" onto the companion star. This teardrop-shaped volume is named after the French astronomer Edouard Roche, who calculated the gravitational effects in a rotating binary system. The process of sweeping up additional matter tends to add a drag on the white dwarf's orbit, gradually drawing the two stars closer together and further increasing the rate of transfer.

Coming from the companion's outer layers, the gas deposited on the white dwarf is generally rich in hydrogen. This gas forms a layer on the white dwarf's surface, where gravity compresses and thereby heats it. Additional heating occurs as more matter falls onto the star's surface. Such matter gains a great deal of kinetic energy as it falls and releases that energy as heat when it strikes the star's surface (Unit 20).

Most of the gas falling toward a white dwarf from a companion star spirals around the white dwarf before it reaches the white dwarf's surface. Astronomers call this spiraling gas an **accretion disk.** Gas in the accretion disk orbits the white dwarf rapidly and collisions occur violently, strongly heating the gas as it spirals in toward the star's surface.

Observations suggest that as the gas accumulates on the white dwarf's surface, the gas may eventually grow hot and dense enough to begin fusing the recently added hydrogen into helium. Even if the fusion begins only in a small region initially, the heat released will trigger fusion in surrounding areas. As a result, more and more of the hydrogen ignites in a **thermonuclear runaway,** a detonation that blasts some of the gas into space. This forms an expanding shell of hot gas, as Figure 65.2 illustrates. This shell of gas radiates far more energy than the white dwarf itself. In fact, the luminosity may grow to 100,000 times its previous level, so that the explosions may be visible to the naked eye.

When early astronomers saw such an event, they called it a *nova stella,* from the Latin for "new star," because the explosion would make a bright point of light appear in the sky where no star was previously visible (see also Unit 63 on variable stars). Today we shorten *nova stella* to **nova.** Modern observations allow us to observe the expanding shell of material blasted outward as a nova by the white dwarf (Figure 65.3).

Q: Other astronomical objects, such as forming stars, also have disks of material around them. What circumstances in these two situations cause them to give rise to similar phenomena?

Novae may occur repeatedly for the same white dwarf, depending on the rate at which material is transferred from the companion to the white dwarf. **Recurrent novae** generally occur at intervals of a decade to a century, and it is possible that many novae that have been recorded only once will repeat in the future. The timing depends on how rapidly material accumulates on the surface, whether it has time to cool, and how much the rate of mass transfer fluctuates—in other words, many factors that are difficult to predict.

65.2   THE CHANDRASEKHAR LIMIT

Despite the ability of a nova to blast away some of the material accumulated from a companion star, a white dwarf gradually increases its mass through mass transfer. This gradual accumulation of mass may have dire consequences for the white dwarf.

Because white dwarfs are very compact and have no fuel supply, their structure differs significantly from that of ordinary stars. Although they are in hydrostatic equilibrium, with pressure balancing gravity, their pressure arises from **electron degeneracy,** the peculiar interaction that prevents electrons of the same energy from sharing the same volume (Unit 62.3). This gives white dwarfs an odd property: Increasing their mass makes them shrink. Even more crucial, however, white dwarfs must have a mass below a critical limit, or they collapse. To see why, we must consider conditions in a white dwarf's interior.

White dwarfs are very dense, having formed from the dense cores of their parent stars. If we divide the mass of a white dwarf by its volume, we find that its density is about 10^6 kilograms per liter—one ton per cubic centimeter (about 16 tons per cubic inch). This high density packs the star's particles so closely that they are separated by less than the normal radius of an electron orbit. Even as the white dwarf cools, then, the electrons cannot drop into orbits around their atomic nuclei, as we would normally expect when matter cools.

Q: Some astronomers have suggested that cooled white dwarfs are made of diamond. If we could travel to such a white dwarf, why might it be impractical to mine it?

If mass is added to a white dwarf, the extra mass increases the star's gravity. This increases the white dwarf's gravitational potential energy and drives some electrons into higher energy levels. The electrons that shift into higher energy levels can now overlap with neighboring electrons, allowing the white dwarf to compress into a slightly smaller volume. When the white dwarf's mass grows large enough, however, any further additional mass will cause the white dwarf's gravity to increase so much that it will contract more and more in a runaway process. The mass at which this occurs was determined by theoretical calculations made in 1931 by the young Indian astrophysicist Subrahmanyan Chandrasekhar (pronounced *chahn-drah-say-car*). Chandrasekhar won the Nobel Prize for physics in 1983 for these calculations and related work. The limiting mass of a white dwarf is now called the **Chandrasekhar limit** in his honor.

For stars made of a helium–carbon mix—the expected composition of many white dwarfs—the Chandrasekhar limiting mass is about 1.4 $M_\odot$. If a white dwarf's mass exceeds this limit, it begins to contract uncontrollably—until other forces come into play. Chandrasekhar's work has been confirmed observationally by the fact that all white dwarfs for which we can measure the mass (those in binary systems) always have masses below the limit he determined theoretically.

65.3   SUPERNOVAE OF TYPE IA

As a white dwarf accumulates mass from a companion star, it may eventually approach the Chandrasekhar limit. When the white dwarf exceeds the limit, the white dwarf collapses in on itself. The resulting compression of the matter drives the temperature up, rapidly climbing to billions of Kelvin.

A star becomes a white dwarf because it did not originally have enough mass to compress and heat its core to temperatures that would allow carbon or heavier elements to fuse. It is therefore full of fuel. Carbon and oxygen in the collapsing white dwarf rapidly reach their ignition temperature and begin nuclear fusion. Carbon and oxygen can fuse to form silicon either directly (^{12}C + ^{16}O → ^{28}Si) or through a chain of reactions (as in the core of a high-mass star; see Unit 66.1). Much of the silicon in turn fuses into nickel (2 ^{28}Si → ^{56}Ni). The energy released in these fusion reactions is enough to blow the entire white dwarf apart, resulting in a **Type Ia supernova** (Figure 65.4). The isotope of nickel, sprayed into space by the explosion, is highly radioactive and rapidly decays into cobalt (^{56}Co), which then decays to iron (^{56}Fe), each decay adding additional energy to the outburst.

A Type Ia supernova explosion can produce nearly 10^{10} times the Sun's luminosity at its brightest. This makes the supernova nearly as bright as an entire galaxy of stars. Figure 65.5 is a photograph of a distant galaxy with a Type Ia supernova: The exploding star gleams like a beacon. Supernovae can also occur when the core of a massive star is no longer generating heat and pressure from fusion in its core. When the core becomes massive enough, it too will collapse (Unit 66.3). This also produces supernovae, although these have a variety of names. They are called Type Ib, Type Ic, and Type II supernovae (Unit 66).

At their brightest, Type Ia supernovae are about two times brighter than Type II supernovae at visible wavelengths; however, the total power output is greater in Type II, in which much more of the power goes into the energy of expansion and neutrinos.

FIGURE 65.4

A white dwarf exploding as a Type I supernova. If too much hydrogen from a companion accumulates on the white dwarf, it may raise its mass to a value above the Chandrasekhar limit (about 1.4 $M_\odot$). A white dwarf more massive than this limit collapses, heats from the resulting compression, and explodes.

FIGURE 65.5

A Type Ia supernova in a distant galaxy. The bright white supernova was seen in 2002 superimposed on a galaxy of approximately a hundred billion stars.

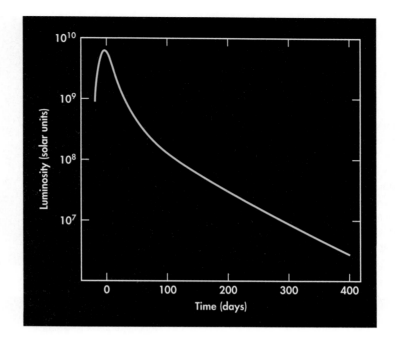

FIGURE 65.6

The light curve of a Type Ia supernova. The exploding white dwarf shines many billion times brighter than the Sun initially, then dims steadily following a characteristic curve.

Q Could the Sun become a nova or supernova under any circumstances?

Type Ia supernovae indicate their distinctive origin from a collapsing white dwarf. First, the spectra of Type Ia supernovae show no signs of hydrogen because they are the remnants of stars that have lost their outer atmospheres of hydrogen in the stellar wind and planetary nebula phases. Second, the spectra contain strong lines of silicon, the element we expect to form from the rapid fusion of carbon and oxygen during the fatal last moments. Finally, the characteristics of the Type Ia **light curve**—the way the luminosity changes over time (Unit 63)—closely match theoretical predictions for the thermonuclear runaway of a white dwarf of this mass. The light curve of a Type Ia supernova is illustrated in Figure 65.6.

These characteristics not only confirm the white dwarf origin of the Type Ia supernovae, but allow astronomers to distinguish them when they occur. Studies of other galaxies indicate that Type Ia supernovae are the most common. No supernova has been seen in our own galaxy for over 400 years, although the last two appear to be Type Ia—one in 1572 known as Tycho's supernova and another in 1604 known as Kepler's supernova. Both occurred before the invention of the telescope.

Type Ia supernovae appear to be excellent "standard candles" (Unit 54.3), meaning that their luminosities are all similar. This is because the pattern of their collapse and explosion occurs the same way each time. The mass gradually increases to the Chandrasekhar limit, and then the collapse occurs. In other words, the Chandrasekhar limit is a leveling factor, which causes each Type Ia supernova to ignite with roughly the same raw materials at its disposal.

Type Ia supernovae leave no star behind except the former companion of the white dwarf. The matter from the white dwarf becomes an expanding cloud of gas, known as a **supernova remnant.** The supernova remnant from the explosion of 1572 is shown in Figure 65.7. Astronomers have also detected a star moving away from the point of the explosion at over 100 kilometers per second—apparently the former companion of the white dwarf, flung out of orbit when its partner exploded. All that remains of the white dwarf is a cloud of rapidly expanding gas rich in leftover carbon and oxygen, plus the silicon, iron, and other elements made during the final moments of nuclear fusion during the explosion. The silicon in a chunk of rock and the iron in your blood were probably made when a white dwarf met its doom.

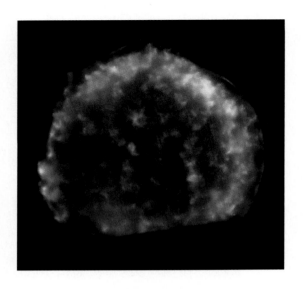

FIGURE 65.7
X-ray false-color image of the remnant of the Type Ia supernova of 1572 seen in Cassiopeia. X-ray spectra of the 10–20 million K gas reveal silicon, iron, nickel, and other heavy elements throughout the expanding cloud of gas, which is approximately 20 light-years across. (The lower part of the image of the remnant was cut off at the edge of the detector.)

KEY TERMS

accretion disk, 528

recurrent nova, 529

Chandrasekhar limit, 529

Roche lobe, 528

electron degeneracy, 529

supernova remnant, 531

light curve, 531

thermonuclear runaway, 528

mass transfer, 527

Type Ia supernova, 530

nova, 528

white dwarf, 527

QUESTIONS FOR REVIEW

1. What are the approximate mass and radius of a white dwarf compared with those of the Sun?

2. How does a white dwarf form?

3. Can a white dwarf have a mass of 10 solar masses? Why?

4. What is meant by *degeneracy pressure*?

PROBLEMS

1. Calculate the density of a white dwarf of 1 solar mass that has a radius of 10^4 kilometers.

2. If the maximum luminosity of a Type Ia supernova is about $10^{10} \, L_\odot$, how far away would one be if it was as bright as the brightest stars in the sky (about 10^{-7} watts/meter^2)? Express your answer in light-years.

3. How does the acceleration of gravity (Unit 16) on the surface of a white dwarf of 1 solar mass, which has a radius of 10^4 kilometers, compare to g, the acceleration due to gravity on the surface of the Earth?

4. What is the escape velocity (Unit 18) from the surface of a white dwarf of 1 solar mass that has a radius of 10^4 kilometers?

TEST YOURSELF

1. If mass is added to a white dwarf, which of the following does *not* occur?
   a. Its radius increases.
   b. Its mass increases.
   c. Its density increases.
   d. It may exceed the Chandrasekhar limit and collapse.
   e. Some of its electrons gain higher energies.

2. Why do Type Ia supernovae have no hydrogen in their spectra?
   a. Type Ia supernovae have converted all of their hydrogen into iron.
   b. Type Ia supernovae have lost their hydrogen atmospheres during their planetary nebula phase.
   c. Type Ia supernovae do not have enough power to excite the hydrogen lines.
   d. Type Ia supernovae are so powerful that they fuse all of the hydrogen in the explosion.
   e. Type Ia supernovae become black holes, which swallow up the hydrogen.

3. Which of the following is *not* part of the reason for a white dwarf exploding?
   a. Mass transfer
   b. Fusion of oxygen and carbon
   c. Decreasing gravitational pressure
   d. Exceeding the Chandrasekhar limit
   e. Exceeding the pressure of electron degeneracy

Background Pathways: Unit 59

66 Old Age and Death of Massive Stars

High-mass stars (mass greater than about 8 $M_\odot$) have violent ends. Like low-mass stars, they shed mass as they age. And like low-mass stars, as a high-mass star approaches the end of its life, the temperature in its core grows hotter. It grows so hot, in fact, that it can fuse not only helium but also carbon and heavier elements to provide the energy it needs—for a while. Less energy is available, and the demands of the star grow steadily larger.

Once a massive star forms iron in its core, the end is near because elements heavier than iron do not release energy when they fuse. The inability of an iron core to provide energy triggers the collapse and explosion of these stars. When massive stars explode, they spew into space the heavy elements they have formed inside, adding heavy elements to the interstellar medium. They also leave behind some of the densest objects in the universe—neutron stars and black holes—which are explored in Units 67 and 68.

66.1 FORMATION OF HEAVY ELEMENTS

Because the late stages in the life of a massive star are strongly linked to its production of heavy elements, we need to look in more detail at how such elements form. Heavy elements form when two or more lighter nuclei combine and fuse to create a heavier one. Because of the electrical repulsion between the positively charged protons, these reactions require successively higher particle speeds for fusion to occur among increasingly larger nuclei. To see why, consider the electrical repulsion between two carbon nuclei: With 6 protons each, the repulsion is $6 \times 6 = 36$ times stronger than the repulsion between two hydrogen nuclei, which have only 1 proton each. The collision speed needed to overcome this repulsion can be achieved if the gas is hot enough because the hotter a gas, the faster its particles move. To fuse progressively heavier nuclei, the star's core must grow progressively hotter. For example, fusing carbon requires about 600 million Kelvin, whereas fusing heavier elements may require temperatures greater than a billion Kelvin.

In nuclear reactions of this sort, the sum of the number of protons and neutrons remains the same. However, the weak force (Unit 4) allows neutrons and protons to be converted into each other. Under most conditions, neutrons are converted to protons because the reverse reaction, converting a proton to a neutron, absorbs energy rather than releases it. As a result, the new nucleus (or nuclei) must have a number of protons and neutrons equal to the sum of the numbers found in the parent nuclei. For example, suppose two ^{12}C nuclei fuse. Carbon has 6 protons and 6 neutrons, so the fusion reaction begins with 12 of each nuclear particle. Their fusion could produce magnesium, ^{24}Mg, which has 12 protons and 12 neutrons in its nucleus. Studies show that the more likely reaction is to produce a neon nucleus and a helium nucleus, $^{20}Ne + ^{4}He$ (also containing the same total number of protons and neutrons), thus creating a new, heavier nucleus as well as a helium nucleus that may go on to have fusion reactions with other particles.

	TABLE 66.I	Major Fusion Reactions in Stars			
Reaction	Primary Fusion Products	Minimum Temperature	Energy Released (% of Mass)	Duration of Fusion in 25 $M_\odot$ Star	
Hydrogen fusion	$4\,^{1}\text{H} \rightarrow\,^{4}\text{He}$	5 MK	0.71%	7,000,000 yr	
Helium fusion	$3\,^{4}\text{He} \rightarrow\,^{12}\text{C}$	100 MK	0.065%	} 700,000 yr	
Alpha capture	$^{12}\text{C} +\,^{4}\text{He} \rightarrow\,^{16}\text{O}$	200 MK	0.016%		
Carbon fusion	$^{12}\text{C} +\,^{12}\text{C} \rightarrow\,^{20}\text{Ne} +\,^{4}\text{He}$	600 MK	0.021%	300 yr	
Neon fusion	$^{20}\text{Ne} +\,^{20}\text{Ne} \rightarrow\,^{24}\text{Mg} +\,^{16}\text{O}$	1500 MK	0.012%	8 months	
Oxygen fusion	$^{16}\text{O} +\,^{16}\text{O} \rightarrow\,^{28}\text{Si} +\,^{4}\text{He}$	2000 MK	0.032%	3 months	
Silicon fusion	$^{28}\text{Si} +\,^{28}\text{Si} \rightarrow\,^{56}\text{Fe}$	2500 MK	0.034%	1 day	

Key: H = hydrogen; He = helium; C = carbon; O = oxygen; Ne = neon; Mg = magnesium; Si = silicon; Fe = iron.

The most common reactions occurring in stars, including those for earlier phases of a star's lifetime, are listed in Table 66.1, along with the approximate temperatures required for these fusion processes to take place. Besides the reactions listed in the table, a variety of less common interactions that create other combinations of nuclei can also occur. For example, as you can see in the table, silicon fusion produces the element iron, but it may also produce nickel, although normally in smaller quantities. Similarly, oxygen fusion produces sulfur in addition to silicon. Note that helium nuclei (alpha particles) are created in a number of these fusion reactions, and these are usually quickly "captured" in further fusion interactions with the heavier elements.

Each of these reactions converts some of the nuclei's mass into energy, in accord with Einstein's law $E = m \times c^2$. The percentage of mass turned into energy is listed in the second to last column of Table 66.1. Nearly 0.7% of the fusing nuclei's mass is converted into energy during hydrogen fusion. However, the sum of all of the other reactions converts only about 0.2% of the matter's mass into energy.

With this introduction to nuclear reactions in massive stars, we can now turn to how such reactions affect and are affected by the evolution of massive stars.

The typical life of a massive star is plotted in the H-R diagram shown in Figure 66.1. The massive star begins its life on the upper main sequence as a blue star. As its core hydrogen is consumed, it leaves the main sequence, swelling and cooling to become a yellow supergiant while it fuses helium in its core. When the helium there is consumed, the star's core contracts and heats again. This drives the atmosphere to swell up, and the star becomes a red supergiant. When the core reaches carbon-fusing temperatures, the star stabilizes again. After carbon is exhausted, the core begins to fuse other fuels, all the while contracting and growing hotter.

Despite all these changes, the star's luminosity remains approximately constant as it ages, but it changes in both size and surface temperature. These changes in radius and temperature cause the star's position to shift back and forth across the H-R diagram as it adjusts to its core's changing temperature and energy output. Computer models yield a variety of results for these stages for several reasons: The amount of mass loss during the supergiant stage is uncertain, and it appears that factors like the rotation rate of the star and its initial composition may have major effects. Many models suggest that the massive star may loop back to the blue side of the H-R diagram, and there is observational evidence that this occurs as well, discussed in section 66.3. As a result, a massive star follows a zigzag path in the H-R diagram, much like a low-mass star but at a higher luminosity.

As the star progressively consumes one nuclear fuel after another, the "ashes" of one set of nuclear reactions become the fuel for the next set. Oxygen, neon, magnesium, and eventually silicon are formed; but because progressively higher

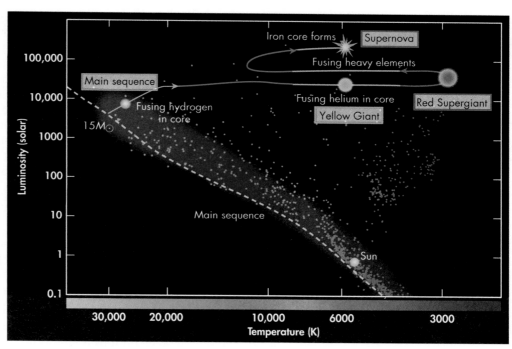

FIGURE 66.1

H-R diagram of a massive star's evolution.

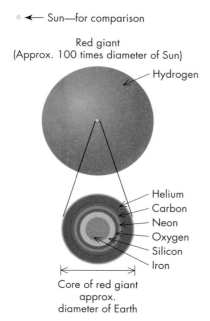

FIGURE 66.2

The layered structure of a massive star as it burns progressively heavier fuels in its core. (These inner zones are magnified enormously for clarity.)

temperatures are needed for each new fusion process, each new fuel is confined to a smaller and hotter region around the star's core. The star develops a layered structure, as illustrated in Figure 66.2. The star's surface, where the temperature is too low for nuclear reactions to occur, remains hydrogen. But beneath the surface lies a series of nested shells, each made of an element heavier than the one surrounding it. At the interior base of the hydrogen atmosphere, the temperature is high enough for hydrogen fusion, and the hydrogen fusion shell "eats" its way out into the surrounding atmosphere, leaving behind the helium it produces. At the base of the helium layer, helium fusion produces carbon and some oxygen (through the alpha capture process illustrated in Table 66.1). At the base of that layer, carbon fusion produces a neon layer, and so on.

By the time the star is fusing silicon into iron, its core has shrunk to a diameter smaller than the Earth's, and the core's central temperature is about 2.5 billion K. At such temperatures, the silicon is being consumed at a fantastic pace. To appreciate how rapidly the star consumes its silicon, consider the following: A massive star of 25 $M_\odot$ takes about 8 million years from its birth to form an iron core. The main-sequence phase lasts about 7 million years; the helium fusion phase lasts only about 700,000 years; the carbon fusion phase lasts just a few hundred years, and the silicon fusion phase lasts only about one day! These times, listed in Table 66.1, are based on averages of several different recent computer models of stellar evolution.

The decreasing time involved in each phase occurs for several reasons. First, the star must produce ever larger amounts of heat and energy to support the core against the greater intensity of gravity as the core compresses. Second, less energy is released per kilogram of material fused in most of the heavy-element fusion reactions than in hydrogen or helium fusion. Third, much of the energy released in these heavy-element reactions is carried away by neutrinos, which travel out of the star, reacting very little with its gases—and therefore offers little help in heating the star and providing the pressure needed to support it.

Q The final stages of fusion may generate very large amounts of energy in a short time, yet the star does not brighten significantly. What factors might explain this?

Q The even-numbered elements are more common throughout the universe than the odd-numbered elements. How might this be understood from the fusion reactions we have examined?

Astronomers call the formation of heavy elements by such nuclear fusion processes **nucleosynthesis.** The details of these reactions were first worked out in the 1950s by a team of astronomers—E. Margaret Burbidge, Geoffrey R. Burbidge, William A. Fowler, and Fred Hoyle. They proposed that all of the chemical elements in the universe heavier than helium were made by nuclear fusion in stars. The observed abundances of the elements in the universe agrees extremely well with this hypothesis. The fact that carbon and oxygen are the third and fourth most common elements in the universe, and that the Earth is made largely of elements such as iron and silicon, are direct consequences of nucleosynthesis in stars. However, we have yet to explain how that matter gets out of the star and into interstellar space, and we will see that what is fortunate for us is fatal for the star.

66.2   CORE COLLAPSE OF MASSIVE STARS

The formation of an iron core signals the end of a massive star's life. Iron cannot fuse and release energy; the iron nucleus turns out to be the most tightly structured of all nuclei that can be formed through nucleosynthesis. In larger nuclei, the electrical repulsion between protons begins to weaken the binding of the nucleus because the strong force, which binds protons and neutrons together, weakens much more rapidly with distance than does the electrical force of repulsion. As a result, attempting to fuse additional nuclei onto an iron nucleus weakens the bonds and absorbs energy, rather than releasing it. Thus, nuclear fusion stops with iron, and a star with an iron core is out of fuel.

At this point, the reaction of the star at first looks the same as when previous fuels were exhausted. The star's core shrinks and heats, and just as for low-mass stars, the shrinkage presses the matter tighter and tighter until electron degeneracy (Unit 62.3) finally halts the contraction. However, the pressure is so enormous and the most energetic electrons have so much energy that a new reaction can occur. Protons and electrons may themselves combine to become neutrons by this reaction:

$$^1H + e^- + Energy \rightarrow n + \nu.$$

Besides converting protons and electrons into neutrons and neutrinos, this reaction *absorbs* energy, and that absorption triggers a catastrophe for the star.

To understand why this seemingly ordinary result has such dire consequences, consider the following: Each layer of the star is supported by the one below it, with all of them finally supported by electron degeneracy pressure at the very center, where the density reaches over 1 million kilograms per liter. When the electrons are swallowed up by protons to become neutrons, the pressure provided by electron degeneracy disappears. The star begins to implode.

Neutrons also have a degeneracy limit, but the volume each degenerate neutron occupies is about a billion times smaller than the volume the electrons filled. This effectively creates a hole in the middle of the star. In a fraction of a second, the core is thus transformed from a sphere of iron, with a radius of 10,000 km, into a sphere of neutrons with a radius of just 10 km held up by neutron degeneracy pressure. This is like pulling out the bottom brick from an enormous stack. Nothing remains to support the star, and so it begins falling inward.

The plunging outer layers of the star strike the neutron core, reaching speeds of over 100,000 km/sec—a substantial fraction of the speed of light. Smashing into the extremely dense core at these speeds, some of the material "bounces" back outward, colliding with other infalling matter. The impact heats the infalling gas to billions of degrees. Several solar masses of matter are almost instantaneously raised to

Q Hold a small rubber ball on top of a basketball, and drop them together toward the floor. What happens to the small ball? Does that help you understand what happens to the outer layers of a supernova as they collapse on the core?

fusion temperature and detonate all at once. Meanwhile an enormous quantity of neutrinos is pouring out from the core—one for each of the protons that turned into a neutron. Normally neutrinos interact so weakly that they escape freely from a star; but at the enormous densities of the collapsing neutron core, many of the neutrinos flooding out from the core collide with the infalling matter and give up their energy to it. This also heats the infalling matter and begins to drive it outward.

All of these processes generate more energy in a few seconds than the Sun generates in its entire lifetime. This creates a powerful blast wave that travels outward from the core into the atmosphere of the star. The blast wave tears through the surface of the supergiant in an eruption that sends most of the star's atmosphere flying outward into space. This catastrophe marks the death of a high-mass star. Astronomers call such an event a **supernova.**

66.3 MASSIVE STAR SUPERNOVA EXPLOSIONS

The explosion of a massive star mixes the elements synthesized by nuclear fusion during the star's evolution with the star's outer layers and blasts them into space. This incandescent spray expands away from the star's collapsed core at more than 10,000 kilometers per second. Depending on the star's mass, many solar masses of matter may be flung outward, ultimately mixing with surrounding material and, in time, forming new generations of stars. Interstellar gas is thereby enriched in heavy elements: the atoms needed to build the rock of planets and the bones of living creatures here on Earth.

The supernova outburst itself generates even heavier elements than were originally created in the star's core. Elements such as gold, platinum, and uranium are created in a supernova explosion from the abundant energy of its blast. This occurs because the explosion creates free neutrons, which rapidly combine with atoms from the star to build up elements heavier than iron.

White dwarfs can also produce supernova explosions if a companion star deposits enough matter. Because there are so many more low-mass stars, these are the most common type of supernova, called Type Ia (Unit 65). Astronomers refer to supernovae that result from the core collapse of a massive star as **Type II supernovae,** as well as the less common classes of Type Ib and Ic supernovae. Astronomers originally distinguished between these Types I and II because of differences in the spectra of the light emitted by the two kinds of supernovae. Type II supernovae show spectral lines of hydrogen, produced by the superheated gas of the star's former atmosphere, whereas Type I do not exhibit hydrogen lines.

Not all massive stars exhibit hydrogen lines in their spectra when they explode. These supernovae, which are classified as Type Ib and Type Ic, are suspected to be massive stars undergoing core collapse (like Type II supernovae) that have lost their hydrogen atmospheres to a companion star or stellar winds. Astronomers can distinguish these from Type Ia supernovae because they do not exhibit some of the secondary characteristics of Type Ia, such as silicon or helium lines.

The supernovae of massive stars come from stars that have a wide range of masses and histories. A star of 50 $M_\odot$ will build up the critical mass for the iron core to collapse much more quickly than a star of 15 $M_\odot$. When the iron core in each of these stars is about to collapse, their atmospheres are structured in significantly different ways in response to their many layers of shell fusion and the amount of atmosphere lost in stellar winds. A massive star supernova blast may therefore look quite different for stars of different masses. These differences are

Some of the light seen coming from supernovae in the weeks and months following the explosion has been shown to originate from heavy radioactive elements. These must have been produced in the explosion because they decay rapidly.

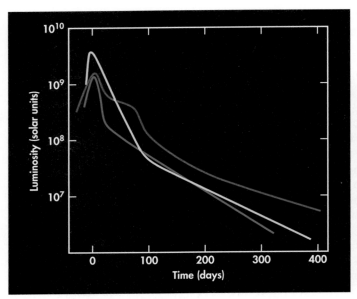

FIGURE 66.3
Some examples of the light curves of supernovae of massive stars.

FIGURE 66.4
Photograph of a portion of the Large Magellanic Cloud (a small, nearby galaxy) before Supernova 1987A exploded, and a second photograph taken during the event.

illustrated by the wide variety of light curves—the amount of light emitted as a function of time—by massive star supernovae, as shown in Figure 66.3.

For the supergiant star, the supernova explosion is a quick and glorious death. In a few minutes it releases more energy than it has generated by nuclear fusion during its entire existence. These supernovae are generally slightly less luminous at visible wavelengths than the Type Ia supernovae, but they are more energetic. Most of their energy is carried by the burst of neutrinos. By some estimates, almost 99% of the energy of a Type II supernova is carried away by the neutrinos produced when the protons and electrons combine. Several neutrino detectors around the world detected just such a burst in February 1987 when a supernova—SN 1987A—blew up. Several hours later, light from the exploding star became visible as the blast wave emerged through the surface of the supergiant star.

Images of this star had been obtained before it exploded, making it the first supernova for which we had "before" and "after" photographs (Figure 66.4). A surprise for astronomers was that the star that exploded was a blue supergiant. Before that time it had been widely expected that supernova would occur in red supergiants. It is now understood that stars may be almost anywhere in their evolutionary "journey" through the H-R diagram (Figure 66.1) when the iron core develops.

Supernova 1987A did not even occur in our own galaxy, but in the Large Magellanic Cloud, a small galaxy neighboring our own (Unit 76). Nonetheless, even though the blast took place about 160,000 light-years away—hence 160,000 years ago—trillions of the neutrinos from that explosion passed through every person on Earth!

66.4   SUPERNOVA REMNANTS

Gas ejected by a supernova blast plows into the surrounding interstellar space, sweeping up and compressing other gas that it encounters. Because massive stars have such short lives, they are often surrounded by large amounts of the gas and dust in the interstellar cloud from which they formed (Unit 60). The **supernova remnant**—a huge, glowing cloud of stellar debris and interstellar gas—may expand to a diameter of several light-years within a century, but its expansion slows as it runs into more surrounding gas. Figure 66.5 shows photographs of three Type II supernova remnants of different ages. Notice how ragged the remnants in Figure 66.5 look compared with the smoothly ejected bubbles of planetary nebulae (Unit 64)—evidence that massive stars die more violently than low-mass stars.

One such violent outburst was seen almost a thousand years ago in 1054 by astronomers in China and elsewhere in the Far East. This is perhaps the most famous of all supernova remnants, now known as the Crab Nebula (Figure 66.5B). With even a small telescope, you can still see the glowing gases ejected from that dying star (see Looking Up #5 Taurus). Another intriguing supernova shown in Figure 66.5A, Cassiopeia A, was first seen as a radio source—the brightest radio source in the sky outside the Solar System. Astronomers have determined from its expansion rate that it exploded about 300 years ago, making it the most recent known supernova in our galaxy. However, there are no clear records that it was seen. The explosion appears to have been hidden from view by dense dust surrounding the young massive star.

A supernova explosion marks the death of a massive star, but a part of the star may survive—the collapsed core of neutrons that initiated the explosion (Figure 66.6). Such *neutron stars* (Unit 67) have been detected, for example, in the middle of the

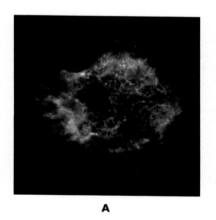

The long and intriguing history of the observations of the Crab Nebula supernova is described in Unit 26.4.

A

B

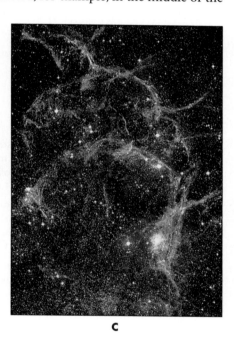

C

FIGURE 66.5

A sequence of Type II supernova remnants of increasing age. (A) Cassiopeia A is about 300 years old; (B) the Crab Nebula is about 1000 years old; (C) the Vela remnant is about 10,000 years old. The image of Cassiopeia A is a false-color X-ray image—the red on the left is emission from iron, and bright green is emission from silicon and sulfur.

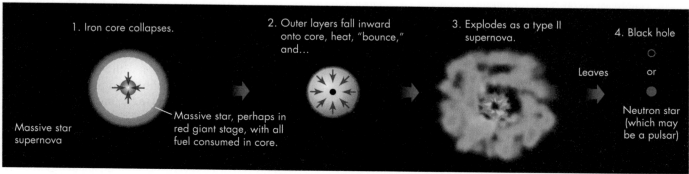

FIGURE 66.6

Formation of a neutron star or black hole by the collapse of the iron core of a massive star.

expanding debris of the three supernova remnants in Figure 66.5. In some other supernova remnants, no remaining core is detected. It is possible that these stars destroyed themselves entirely, like the Type Ia supernovae; or perhaps they remain but are hidden from sight in the form of an even more compressed body—a strange object known as a black hole (Unit 68).

KEY TERMS

nucleosynthesis, 536

supernova, 537

supernova remnant, 539

Type II supernova, 537

QUESTIONS FOR REVIEW

1. What makes a high-mass star's core collapse?

2. What is a supernova explosion?

3. What is the observational difference between Type I and Type II supernovae?

PROBLEMS

1. If every proton in the core of a massive star turns into a neutron and releases one neutrino, how many neutrinos are produced? Assume the core contains 2 solar masses, and half the mass is made of protons ($M_{proton} = 1.67 \times 10^{-27}$ kg).

2. How long does it take the blast wave from the center of a star undergoing a supernova to reach the star's surface if the blast wave is traveling at 10,000 km/sec and the star is about 100 times the size of the Sun?

3. The Crab Nebula is today measured to be about 10 light-years across. If it exploded in 1054, what has its average speed been in kilometers per second?

4. It is estimated that in the final stages of fusion, a giant star produces a core of iron more massive than the Sun. From the energy release given in Table 66.1,
   a. How much energy is this (in joules)?
   b. How long would it take the Sun to emit this much energy at its rate of 4×10^{26} joules per second?

TEST YOURSELF

1. Stars like the Sun probably do not form iron cores during their evolution because
   a. all the iron is ejected when they become planetary nebulae.
   b. their cores never get hot enough for them to make iron by nucleosynthesis.

   c. the iron they make by nucleosynthesis is all fused into uranium.
   d. their strong magnetic fields keep their iron in their atmospheres.
   e. None of the above

2. What is the heaviest element that fusion can produce in the core of a massive star?
   a. Carbon
   b. Gold
   c. Iron
   d. Nickel
   e. Silicon

3. Protons combine with electrons to form neutrons
   a. just before the formation of an iron core.
   b. when neutrinos are replaced by neutrons.
   c. just before a supernova explosion.
   d. just after a supernova explosion.
   e. (b) and (d)

PLANETARIUM EXERCISES

All the planetarium exercises use the Starry Night planetarium software provided with your text. You should install the program and familiarize yourself with the various parameters and controls before trying these exercises. When the program asks you to enter a password, click "OK" to run Starry Night.

Set the location from which you are viewing to the nearest large city or your latitude and longitude. Accomplish this by clicking on the Location box on the toolbar. Verify that the local time is correct.

Locate a supernova remnant, such as M1 (the Crab), the Veil Nebula, or IC 433, by using the Find option in the Selection menu and typing in the name. Determine a time when it will be visible for you during the night and identify its location on a star chart so that you can find it with a telescope. Observe and sketch the supernova remnant with a moderate-size telescope and compare what you saw to the images and information in the text.

67 Neutron Stars

When a massive star forms an iron core late in its lifetime, the pressures at the center of the star can force electrons to merge with protons to make neutrons, and the whole star collapses to an extraordinarily small size. The resulting star resembles a giant atomic nucleus, with a density a billion times greater than a white dwarf—about a trillion *tons* per liter!

In 1934 Walter Baade (pronounced *BAH-deh*) and Fritz Zwicky, astrophysicists at Mount Wilson Observatory and the neighboring California Institute of Technology, respectively, suggested that when a massive star reaches the end of its life, its gravity will crush its core and make the star collapse. They predicted that the collapse of the core would trigger a titanic explosion, which they named a *supernova* (Unit 66). Almost as an afterthought, they speculated that the collapsed core might be so dense that the star's protons and electrons would be driven together and merge into neutrons, forming a **neutron star.**

Astronomers readily accepted Baade and Zwicky's idea that a massive star dies as a supernova because such violent explosions had been observed (Unit 66); but they paid little attention to their suggestion that a neutron star might be born in the blast. Such stars would have radii of only about 10 kilometers (about 6 miles), incredibly tiny compared with even white dwarfs, and were thought to be unobservable if they existed at all. The idea of these tiny, collapsed stars lay dormant for over 30 years.

Background Pathways: Unit 66

67.1 PULSARS AND THE DISCOVERY OF NEUTRON STARS

In 1967 a group of English astronomers led by Anthony Hewish was observing fluctuating radio signals from distant, peculiar galaxies. Jocelyn Bell, a graduate student working with the group, noticed an odd radio signal with a rapid and astonishingly precise pulse rate of one burst every 1.33 seconds. The precision of the pulses led some astronomers to wonder whimsically if they had perhaps stumbled onto signals from another civilization, and informally the signal became known as LGM-1, for "little green men #1." Over the next few months Hewish's group found several more pulsating radio sources, which came to be called **pulsars** for their rapid and precisely spaced bursts of radiation (Figure 67.1).

The ensuing discovery of many more pulsars, and the unchanging repetitiveness of their signals, convinced astronomers that they were observing a natural phenomenon. But it was difficult to come up with a plausible hypothesis of an astronomical phenomenon that could be so regular and rapid.

The solution to this puzzle came from the work of the Italian astronomer Franco Pacini and the Austrian-born British astronomer Thomas Gold. Their work led to the idea that a pulsar is a rapidly spinning neutron star. This idea soon gained support from the discovery of a pulsar at the center of the Crab Nebula, a supernova remnant.

The pulses from a spinning neutron star are produced by its magnetic field. Collapsed to such a small radius, the star's magnetic field is squeezed into a far smaller volume, amplifying the field's strength to about 1 trillion times that of the Earth's

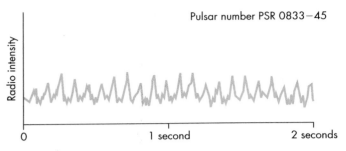

FIGURE 67.1
Pulsar signals recorded from a radio telescope.

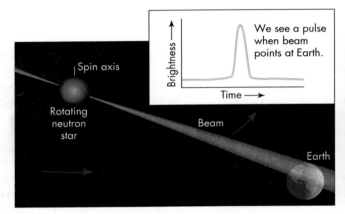

FIGURE 67.2
A pulsar's pulses are like the flashes of a lighthouse as its lamp rotates.

ANIMATIONS

Pulsar model

To convert days to seconds, multiply by the number of hours per day, the number of minutes per hour, and the number of seconds per minute: 25 days = 25 × 24 × 60 × 60 seconds = 2,160,000 seconds.

magnetic field. A neutron star is thus an extremely powerful magnet. Its intense magnetic field causes it to emit radiation in two narrow beams along its north and south magnetic poles (section 67.2). Like the Earth's and most other astronomical bodies' magnetic fields, a pulsar's magnetic field is not aligned with the star's rotation axis; so as the star spins, its beams sweep across space. We see a burst of radiation at radio wavelengths when a beam points at the Earth. A pulsar shines like a cosmic lighthouse whose beam swings around as its lamp rotates, as illustrated in Figure 67.2.

But what could make a neutron star spin so fast? For example, the pulsar at the center of the Crab Nebula rotates 30 times per second. By comparison, the Sun and most other stars take weeks to make a single rotation. However, when a massive star collapses to a tiny radius, the **conservation of angular momentum** (Unit 20) requires it to spin faster. This is the same effect that allows ice skaters to spin rapidly by pulling their arms and legs close to their axis of rotation.

An object's angular momentum is approximately given by the body's mass, M, times its equatorial rotation velocity, V, times its radius, R:

$$\text{Angular momentum} \approx M \times V \times R.$$

The principle of the conservation of angular momentum states that the product of these three quantities must remain constant unless some force brakes or accelerates the spin. Therefore, in the absence of external forces, if a fixed mass contracts, it must increase its rotational speed by the same factor that its radius decreases—if R shrinks, V must increase to keep the angular momentum constant. Therefore the contracting core of a star must spin faster and faster.

Imagine the Sun contracted from its current radius of about 700,000 km to just 7 km, similar to the radius of a neutron star. Because its radius would be 100,000 times smaller, conservation of angular momentum requires that the speed of material rotating at its equator would be 100,000 times greater. The period—the time it takes a point on the equator to come back around again—would therefore be $(100,000)^2 = 10,000,000,000 = 10^{10}$ times faster. So instead of rotating once every 25 days, the collapsed Sun would rotate once every 0.0002 seconds—about 5000 times per second!

Such a rapid spin is what is expected when the iron core of a massive star collapses. Pulsars probably begin spinning at such a rate, although friction gradually slows them down. The Crab Nebula pulsar period, for example, is currently observed to be slowing down by about 10^{-5} seconds each year, so when it formed nearly a thousand years ago, it may have been spinning as fast as our calculation here suggests.

Q Astronomers suspect the slowdown rate of pulsars is probably greater when they are young. Why might this be?

INTERACTIVE

Neutron stars

67.2   EMISSION FROM NEUTRON STARS

When a magnetic field moves, it creates an electric field—a principle we use here on Earth by spinning magnets in dynamos to generate electricity. Similarly, the rapid spin of a neutron star and its intense magnetic field generates powerful electric fields. These fields rip positively and negatively charged particles off the star's surface and accelerate them to nearly the speed of light. The electrically charged particles are channeled by the pulsar's intense magnetic field to travel along the magnetic field lines. This is similar to how the Earth's magnetic field directs solar wind particles *inward,* creating the aurora near the magnetic poles. The magnetic field of a spinning pulsar, in a like manner, generates two narrow beams of charged particles at the star's magnetic poles.

As the charged particles move along the pulsar's magnetic field, they generate radio waves along their direction of motion. This is somewhat like a radio broadcast antenna on Earth, where electric currents are pulsed through the antenna, accelerating electrons in it; the accelerated electrons in turn produce the radio waves we detect. In a pulsar, too, the charges radiate as they accelerate, in this case along the magnetic field, spiraling as they go (Figure 67.3A). The pulses of radiation we see are the collective emission from myriad charges pouring off the neutron star's surface. This radiation is beamed because the charges are traveling along the field lines that emerge from the star's surface at each magnetic pole. Because of this beaming, we see only neutron stars that have a magnetic pole that points toward Earth at some time as they rotate. Pulsars radiating in other directions are invisible to us, so many more neutron stars must exist than the ~1500 that we have detected to date.

The emission created by the accelerating charges is called **synchrotron radiation,** named for a kind of particle accelerator used in physics experiments that accelerates charged particles using a magnetic field. This kind of radiation produces electromagnetic waves across a broad and continuous range of wavelengths. Synchrotron radiation is different from thermal radiation, which also produces radiation over a broad range of wavelengths (Unit 23), because its properties depend on the charged particles' acceleration and on the strength of the magnetic field, rather than on the temperature of a heated gas.

Most pulsars generate synchrotron radiation that is detectable only at radio wavelengths. However, some young pulsars, like the Crab Nebula, generate synchrotron radiation across the entire electromagnetic spectrum, including visible light and gamma rays. A high-speed camera has captured the flashes of visible light from the Crab Nebula's pulsar as it spins (Figure 67.3B).

In everyday life, spinning objects slow down. Pulsars are no exception. As a pulsar spins, it drags its magnetic field through the particles that boil off its surface

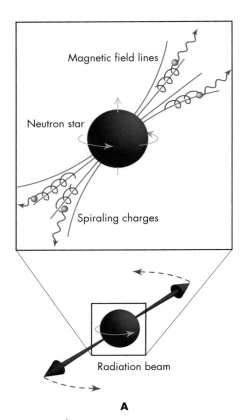

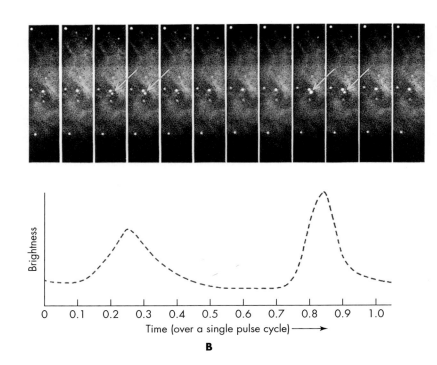

FIGURE 67.3

The strong magnetic field of a spinning pulsar generates a powerful electric field, which strips charges from its surface. (A) The charges spiral in the star's magnetic field and emit radiation along their direction of motion. Because the charges stream in a narrow beam confined by the magnetic field, their radiation is also in a narrow beam at each magnetic pole of the star. (B) The rapid rotation of the pulsar in the Crab Nebula makes the star appear to turn on and off 30 times per second as its beams of radiation sweep across the Earth.

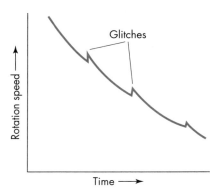

FIGURE 67.4

A pulsar's rotation speed gradually slows, but occasionally it suddenly speeds up during a glitch. This probably occurs as the crust readjusts its shape.

into the surrounding space. That drag slows down the pulsar. Astronomers measure this **spindown** of a pulsar by precisely timing the interval between pulses. Such measurements indicate that the time interval—the period of a spinning pulsar—gradually lengthens. The slowing rotation also reduces the energy of the radiation that the pulsar emits. Thus young, rapidly spinning pulsars emit electromagnetic waves from visible light to radio waves, whereas old, slowly spinning ones generate only radio waves.

Although pulsars gradually slow down overall, occasionally they suddenly speed up by a small amount, as illustrated by the graph in Figure 67.4. Such jumps in rotation speed are called **glitches,** and they can tell us about the interior structure of the neutron star. The sudden speed-up occurs because the neutron star adjusts its shape during its overall slowdown. When it is spinning rapidly, the pulsar is larger at its equator because of the centripetal force outward. The Earth, Sun, and other spinning planets all bulge outward at their equators because of this same effect (Unit 42), but at the high spin rate of a pulsar, the distortion is much larger. As the spin slows, the neutron star settles toward a more spherical shape, and therefore its equatorial radius shrinks slightly. Because the matter moves closer to the axis of rotation, conservation of angular momentum causes the star to speed up a small amount before continuing its gradual slowdown.

Q If the Earth's polar caps melt, will the Earth's rotation speed change? How does this relate to the changes of rotation speed of a pulsar?

These glitches reveal that the crust of the neutron star is rigid because it does not deform gradually. Computer models suggest it is probably made of iron, a few hundred meters thick, as illustrated in Figure 67.5. The material inside appears to be a "fluid" of neutrons. The models suggest the neutron star also has a gaseous atmosphere—about 1 millimeter thick.

As the pulsar's rotation slows, its radiation weakens, and ultimately the star becomes undetectable. Thus, a pulsar "dies" by becoming invisible. But just as a white dwarf can be resurrected with infalling gases from a nearby companion star, so too can a pulsar in a binary system be reawakened. As material from the companion spirals into a pulsar, it adds angular momentum to the pulsar, causing it to spin faster and faster. The fastest pulsars known, called **millisecond pulsars** because their periods are just a few milliseconds long, all appear to have been "revived" by this process.

Millisecond pulsars tend to be the exception, however, because pulsars are not frequently found in binary star systems. The supernova explosion that creates a pulsar usually expels so much mass that the gravitational attraction of the pulsar is no longer strong enough to hold the pair of stars together. The sudden loss of mass can be compared to a ball spun around at the end of a string: If the string breaks, the ball flies off with the speed at which it was orbiting. After the mass loss of a supernova explosion, the pulsar may fly off at its orbital speed—perhaps several hundred kilometers per second. It rapidly escapes from the debris of the explosion that spawned it. Such speeding pulsars have been detected moving at hundreds of kilometers per second through our galaxy.

The nearest known neutron star to the Solar System probably escaped from a binary star system after it became a supernova. This object has the catalog name RX J1856.5–3754, and its fast motion is reflected by its shifting position seen in Hubble Space Telescope measurements made several years apart. Its distance is measured to be about 60 parsecs (about 200 light-years). RX J1856.5–3754 is not a pulsar, but the extremely faint thermal emission from its hot surface has been detected at visible and X-ray wavelengths. Its surface temperature is estimated to be about 700,000 K. Even though it is so hot, it is thought to be 1 million years old based on its trajectory and distance from the nearest star-forming region. Observations with the VLT 8 m telescope in Chile show that it is creating a "bow wave" like a boat as it plows through interstellar gas (Figure 67.6). One intriguing aspect

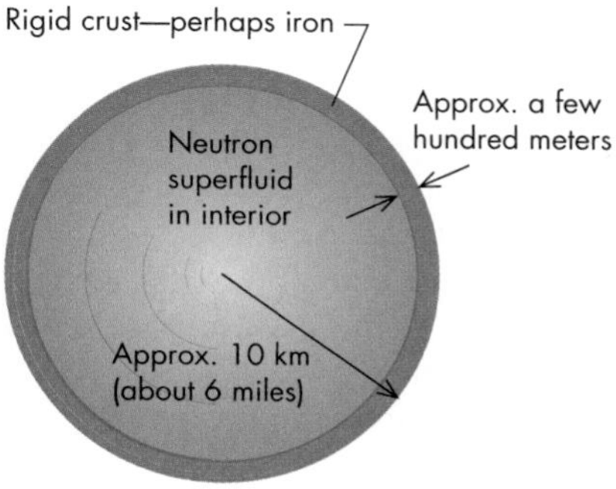

FIGURE 67.5
Schematic structure of the interior of a neutron star.

Rigid crust—perhaps iron
Approx. a few hundred meters
Neutron superfluid in interior
Approx. 10 km (about 6 miles)

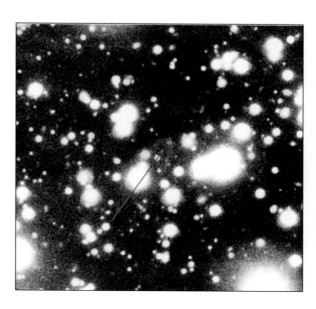

FIGURE 67.6
The nearest known neutron star to the Solar System, RX J1856.5–3754, is an extremely faint object shown in this image made with the VLT. An arrow points to the neutron star and shows its direction of motion through space. As it plows through interstellar gas at high speed, it creates a "bow wave," seen here in pink from hydrogen emission.

Q What effects might a supernova or magnetar burst have on the Earth if it occurred only a few parsecs away? Is such an event likely to have occurred over the Earth's history?

of this neutron star is that its diameter appears to be smaller and its temperature higher than most models have predicted. According to some computer models, these observations suggest that the neutrons in the star have split up into their constituent quarks. If true, this would be unlike any form of matter previously studied.

67.3   HIGH-ENERGY PULSARS

Astronomers have identified some additional classes of neutron stars that produce very high-energy pulsed radiation at X-ray and gamma-ray wavelengths. These appear to be powered by different sources than normal pulsars—from either accreting matter or extremely strong magnetic fields.

X-ray pulsars generate pulses of X-ray radiation as their name implies. However, besides the higher energy of the radiation, their pulsations are more erratic. The strength of the pulses is sometimes fairly steady and sometimes arrives in strong bursts, and the rate of pulsation is seen to speed up sometimes and to slow down at others. These appear to be neutron stars that remain in a binary star system after the supernova explosion.

An X-ray pulsar generates its emission from small "hot spots" near its magnetic poles. Gas from a companion star flows in along magnetic field lines. As the gas falls toward the neutron star, the star's intense gravity accelerates the gas. As it crashes onto the neutron star at the magnetic poles (Figure 67.7), the gas is compressed and heated intensely, causing it to emit X-rays. As the neutron star spins, the hot regions near each pole move into and then out of our field of view, creating the X-ray pulses that we observe. These X-ray pulsars are often measured to be speeding up—and perhaps someday will be millisecond pulsars.

Another type of pulsar is known as a **magnetar.** These appear to be neutron stars with extremely intense magnetic fields. The magnetic field of a magnetar may be a thousand times stronger than the already intense magnetic fields of a normal pulsar, and more than 10 billion times stronger than the Sun's magnetic field. Magnetars have been identified from extremely intense bursts of X-ray and gamma-ray radiation. These bursts last for minutes, pulsing during the burst at the rate at which the neutron star is known to be spinning. During a burst, the magnetar can generate as much energy as it takes the Sun hundreds of thousands of years to produce. An outburst in December 2004 was so powerful that it briefly shut down a number of satellites and ionized gas in the Earth's upper atmosphere. These effects occurred despite a huge distance to the magnetar—about 50,000 light-years.

Only a few magnetars have been identified to date, perhaps because their strong magnetic fields cause them to spin down much faster than normal pulsars. The leading hypothesis for explaining the enormous outbursts is that these neutron stars undergo massive glitches as they slow down, and these create huge shifts and rearrangements of the magnetic field. These might be analogous in some ways to the flare activity seen on the Sun, but vastly stronger because of the enormous strength of the magnetic field.

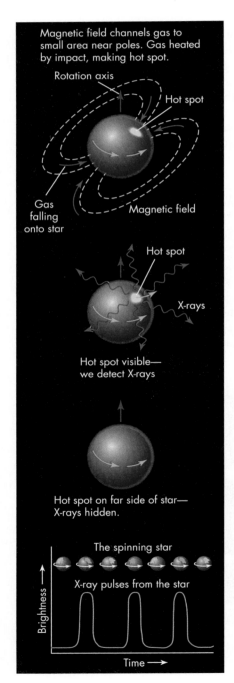

FIGURE 67.7

Gas falling onto a neutron star follows the magnetic field lines and makes a hot spot on the star's surface, creating X-rays. As the star rotates, we observe X-ray pulses.

KEY TERMS

conservation of angular momentum, 542

glitch, 544

magnetar, 546

millisecond pulsar, 545

neutron star, 541

pulsar, 541

spindown, 544

synchrotron radiation, 543

X-ray pulsar, 546

QUESTIONS FOR REVIEW

1. What are the mass and radius of a typical neutron star compared with those of the Sun?

2. How does a neutron star form?

3. How do we observe neutron stars?

4. What is a pulsar? Does it pulsate?

5. Are all neutron stars pulsars? Are all pulsars neutron stars?

PROBLEMS

1. Calculate the escape velocity (Unit 18) from a white dwarf and a neutron star. Assume that each has 1.4 solar masses. Let the white dwarf's radius be 10^4 kilometers and the neutron star's radius be 10 kilometers.

2. The mass of a neutron is 1.67×10^{-27} kg. How many neutrons are in a neutron star that has 1.4 solar masses?

3. The volume of a neutron is about 10^{-45} cubic meters. Suppose you packed the number of neutrons you found in the previous problem into a sphere of radius R. The volume of a sphere is $4/3 \pi R^3$. What is R in kilometers?

4. Suppose a neutron star is spinning with a period of 1.000 seconds. If its radius shrinks by 1%, what will its new period be?

TEST YOURSELF

1. Which of the following has a radius (linear size) closest to that of a neutron star?
   a. The Sun
   b. The Earth
   c. A basketball
   d. A small city
   e. A gymnasium

2. What causes the radio pulses of a pulsar?
   a. The star vibrates like the quartz crystal in a watch, which sends electromagnetic waves through space.
   b. As the star spins, beams of radio radiation from it sweep through space. If one of these beams points toward the Earth, we observe a pulse.
   c. The star undergoes nuclear explosions that generate radio emission.
   d. The star's dark orbiting companion periodically eclipses the radio waves emitted by the main star.
   e. Convection inside the pulsar produces twisted magnetic fields much like what occurs during the Sun's sunspot cycle, but at a much faster rate.

3. If a spinning star suddenly collapses to a radius one-tenth of its former value, then
   a. its gravitational pull also decreases by a factor of 10.
   b. its mass also decreases by a factor of 10.
   c. the speed of rotation at the equator increases by a factor of 10.
   d. the period of rotation increases by a factor of 10.
   e. All of the above

68

Black Holes

A white dwarf is supported by pressure between the electrons in its interior. If enough mass is added to it, the gravitational force grows so large that the star implodes into a neutron star. What happens if we gradually add mass to a neutron star? Will it collapse to a body with a yet smaller size? The answer appears to be that something fundamentally stranger happens. The nature of space and time around the object changes in such a way that the collapsing object disappears from view. It becomes an object that astronomers call a **black hole** from which nothing can escape, not even light.

This strange fate is not limited to massive stars, but occurs for any concentration of enough mass in one place. When gravity grows large enough, it changes the nature of space in a way that prevents any other possible force from supporting the star. The effect that mass has on space was another of the extraordinary discoveries made by Albert Einstein in the early 1900s. He developed the **general theory of relativity** to describe these effects. General relativity is a theory of gravity. It explains not only what happens to turn an object into a black hole, but how any amount of mass or energy alters the nature of space and time in its vicinity.

Black holes represent the most extreme example of how space can be altered by matter. This is the fate that awaits the most massive stars. Some astronomers think that stars that begin life more massive than about 25 solar masses can collapse directly into black holes, with no accompanying supernova explosion. Likewise, a neutron star that grows too massive may simply disappear into the black hole that forms. Collisions of matter outside the black hole may be extremely violent, so the vicinity of black holes is not peaceful. However, general relativity makes it clear that no explosion is possible for matter that falls in too far. Such matter is essentially lost from the universe as we know it.

Background Pathways: Unit 67

68.1 THE ESCAPE VELOCITY LIMIT

To understand black holes, we need to review the concept of escape velocity (Unit 18). **Escape velocity** is the speed a mass must acquire to avoid being drawn back by another object's gravity. For a body of mass M and radius R, the escape velocity, V_{esc}, for an object moving away from that body is

$$V_{esc} = \sqrt{\frac{2GM}{R}}.$$

In this equation G is Newton's gravitational constant, which has a value of 6.67×10^{-11} if V_{esc} is in meters per second, R is in meters, and M is in kilograms. For the Earth, the escape velocity is about 11 kilometers per second; for the Sun, it is about 600 kilometers per second.

You can see from the formula that because R is in the denominator, the escape velocity for an object of a particular mass will be larger as its radius shrinks. This occurs because the same mass packed into a smaller radius creates a larger force of gravity at its surface, making it more difficult to escape from the object.

A white dwarf whose mass is the same as the Sun's has an escape velocity much larger than the Sun's 600 km/sec because the white dwarf's radius is so small. In

Just as for Einstein's theory of special relativity (Unit 53), the term *theory* is used here in its technical sense—a thoroughly tested and verified hypothesis. The theory of general relativity is not just speculation, but has been verified by a wide range of experiments. Some modern technologies like Global Positioning System (GPS) units would not work if general relativity was not taken into account.

If we set the escape velocity to the speed of light, $c = \sqrt{2GM/R_S}$, then we can solve for the Schwarzschild radius, first by squaring both sides of the equation:

$$c^2 = \frac{2GM}{R_S}$$

$$c^2 \times \frac{R_S}{c^2} = \frac{2GM}{R_S} \times \frac{R_S}{c^2}$$

$$R_S = \frac{2GM}{c^2}$$

Q Thinking about the gravitational force at different distances, why might passing close to a small black hole be more destructive to a spaceship than passing close to a large one?

fact, because its radius is about 100 times less than the Sun's, and because the escape velocity depends on the square root of the radius, its escape velocity is larger by a factor of $\sqrt{100} = 10$, making it about 6000 kilometers per second. An object leaving the surface of a white dwarf at a velocity of 6000 kilometers per second would therefore just overcome the white dwarf's gravity and be able to escape into space. Let's now make the same calculation for a neutron star. If the neutron star has a radius 10^5 times smaller than the Sun's (or about 10 km) and it has a mass about twice the Sun's, its escape velocity is about $\sqrt{2 \times 10^5}$ larger than the Sun's escape velocity, or about 270,000 kilometers per second. This is approximately 90% of the speed of light! Hence if a neutron star is compressed just slightly, its escape velocity could exceed the speed of light. Such an object becomes a black hole.

As long ago as 1783, the English cleric John Michell discussed the possibility that objects with escape velocities exceeding the speed of light might exist. A little more than a decade later, the French mathematician and physicist Pierre Simon Laplace entertained the same idea. Following their logic, we can calculate the radius, R_S, at which a body of a given mass would become a black hole by equating its escape velocity to the speed of light. We find then that a mass M becomes a black hole if it is compressed to a radius

$$R_S = \frac{2 \times G \times M}{c^2}.$$

We label this radius R_S and call it the **Schwarzschild radius** in honor of the German astrophysicist Karl Schwarzschild, who was studying how to modify Newton's theory of gravity to account for effects near the speed of light and when gravity is extremely strong. His results, published in 1915, showed that the simple approach we have adopted here provides the correct result.

In principle, a body of any mass can be turned into a black hole if it is compressed to a small enough radius. For example, suppose we wanted to find out how small we would need to squeeze the Sun to make it a black hole. The Sun has a mass of 1.99×10^{30} kg, so if we plug in this mass and that of the speed of light (3.00×10^8 m/sec), we find that the Schwarzschild radius for the Sun is

$$R_S(\text{Sun}) = \frac{2 \times G \times M_\odot}{c^2}$$

$$= \frac{2 \times 6.67 \times 10^{-11} \times 1.99 \times 10^{30}}{(3.00 \times 10^8)^2}$$

$$= 2950 \text{ m} = 2.95 \text{ km}.$$

So if the Sun could be compressed to a radius slightly smaller than 3 kilometers (1.9 miles), it would become a black hole. Similarly, a 2 solar mass object would become a black hole if it were compressed to a radius of about 6 km; a 3 solar mass object would become a black hole if it were compressed to a radius of 9 km; and so on.

Stars can be turned into black holes if they are compressed to sizes on the order of kilometers. However, black holes can be created, in principle, for any size mass. We could even turn the Earth ($M_\oplus = 5.97 \times 10^{24}$ kg) into a black hole if it could be compressed to a radius

$$R_S(\text{Earth}) = \frac{2 \times 6.67 \times 10^{-11} \times 5.97 \times 10^{24}}{(3.00 \times 10^8)^2}$$

$$= 0.0088 \text{ m} = 8.8 \text{ mm}.$$

Thus, an Earth-mass black hole would have a radius of just under 1 cm—about the size of a marble! However, forces and circumstances far beyond anything we think exists today would be required to turn the Earth into a black hole. Nonetheless, some astronomers hypothesize that early in the history of the universe such forces might have existed and might have formed Earth-mass black holes that survive to the present. At the other extreme, there is strong evidence inside galaxies for black holes with masses millions of times greater than the Sun, although the mechanism for forming these massive objects is not yet clear.

The pathway for a high-mass star to become a black hole is much clearer. Suppose we consider a neutron star whose mass is 2 $M_\odot$ that formed after a supernova explosion. If it had a companion star that remained in orbit with it after the explosion, mass transfer could begin adding mass to the neutron star, just as happens to some white dwarfs. Material falling onto the neutron star would be crushed to neutron star densities by the intense gravity, and gradually the neutron star's mass would increase. The escape velocity from the surface of the neutron star would grow larger and larger and then some last bit of matter would fall on the surface, pushing the mass above about $3M_\odot$ and the escape velocity over the speed of light.

At that point, an odd thing happens. When we think of particles like neutrons pushing against each other, the "push" is a force between the neutrons. These forces must themselves travel from one point to another, and they can travel no faster than the speed of light. When the escape velocity exceeds the speed of light, none of the neutrons can communicate its "push" to the particle above it. No force can be communicated outward, so the star collapses inward, disappearing from sight like a boat sinking below the surface of water. What happens next is unobservable, but it is thought that the collapse continues until the matter collapses to a single point—what astronomers call a **singularity.**

Interestingly, the most massive stars may have the quietest endings. A massive star forms an inert iron core as the final step in the fusion reactions in its core (Unit 66). In the final few days, the iron core may grow to several solar masses in a star that began its life with more than about 25 $M_\odot$. When this large iron core cools, its internal pressure drops, and the core collapses catastrophically as for other high-mass stars. However, instead of halting when the core reaches the density of nuclear matter, the greater mass compresses the core to within its Schwarzschild radius almost immediately, and collapses to a singularity. Computer models do not agree on the details of what happens to surrounding material in such a collapse. Some models suggest that there may be no supernova explosion at all—the energy and neutrinos are simply swallowed inside the black hole before they can drive the star's surviving atmosphere outward. Other models predict that the rotation of the atmosphere may cause it to collapse violently to a flat disk surrounding the newly formed black hole, generating a brief but powerful burst of energy beamed along the rotation axis in opposite directions. This hypothesized collapse, called a **hypernova,** may explain powerful bursts of gamma rays that have been detected from distant galaxies (Unit 30.4).

Black holes are remarkable objects, and they reveal that our understanding of gravity based on Newton's law (Unit 16) is incomplete. To understand better how matter behaves in gravitational fields, we need to look in a little more detail at the relation between mass, gravity, and space.

68.2   CURVED SPACE

In 1916 Albert Einstein proposed his theory of general relativity, which provided a new mathematical and physical understanding of gravity. General relativity describes how mass alters the nature of space, creating what astronomers call **curvature of space.** In particular, according to general relativity, a black hole is a place where the curvature of space has become so extreme that a hole forms.

An analogy may help illustrate how gravity and curvature of space are related. Imagine a water bed on which you have placed a baseball (Figure 68.1). The baseball makes a small depression in the otherwise flat surface of the bed. If a marble is now placed near the baseball, it will roll along the curved surface into the depression. The bending of its environment made by the baseball therefore creates an "attraction" between the baseball and the marble. Now suppose we replace the baseball with a bowling ball. It will make a bigger depression, and the marble will roll in farther and be moving faster as it hits the bottom. We therefore infer from the analogy that the strength of the attraction between the bodies depends on the amount by which the surface is curved. Gravity also behaves this way, according to the general theory of relativity. According to that theory, mass creates a curvature of space, and gravitational motion occurs as bodies move along the curvature.

We can extend our analogy to black holes by supposing that we remove the bowling ball and put on the water bed a large boulder. The boulder presses down so hard on the vinyl that it tears a hole in it. A marble placed on the water bed will roll into the depression made by the boulder and disappear into the hole it made. So too, a black hole creates a "rip" in space where the curvature has become so strong that the structure of space is disrupted.

General relativity is a major departure from Newton's ideas about gravity. For example, in the old idea of a black hole, like that envisioned by Michell and Laplace, photons trying to leave the black hole should behave like rocks thrown upward at less than the escape velocity. That is, they would reach some maximum height and then fall back down again. If this were the correct description of how a black hole works, it would be surrounded by photons that had traveled out and were momentarily stopped before they fell back in again. This contradicts observations that photons must always travel at the speed of light.

Problems such as this are resolved by general relativity. Another way of describing the curvature of space is that space can have a motion of its own. To see how, imagine a huge lake in which the water steadily drains out of a hole deep in its middle. Suppose you were in a motorboat on the lake: You would find the boat being pulled in toward the middle of the lake. At the edges of the lake this pull would be gentle, but as you drifted closer to the position of the drain, the current would grow faster. If the flow was powerful enough, you might even find that at a certain distance from the drain your boat was being pulled in by a flow of water so rapid that, even when operating at its top speed, the boat could make no progress away from the drain.

At the Schwarzschild radius of a black hole, space is flowing inward at the speed of light, just as the water in the imaginary lake is flowing toward the drain. A photon inside the Schwarzschild radius finds itself, like the boat, fighting a current moving faster than it can travel. Even though the photon is moving at the speed of light, the space through which it is traveling is "falling" into the hole at the speed of light, and thus the photon makes no progress outward and cannot escape (Figure 68.2A).

Black holes are not the only astronomical objects that curve space. The inward motion of space occurs around all masses. If you were in a motorboat traveling across the imaginary lake in our example, you would find that even a small flow of water toward the drain would bend the path of your boat into a curve. Photons behave similarly as they move by any massive object. As a photon travels through space, its path is bent by the inward flow of space toward the mass, deflecting the direction of the light as illustrated in Figure 68.2B.

A total solar eclipse in 1919 gave astronomers their first opportunity to test general relativity's prediction that light would be deflected. Astronomers had accurate positions of the stars that would be seen near the Sun during the eclipse. They discovered that the stars were all deflected by exactly the amount predicted by Einstein's theory. Today astronomers do not need to wait for an eclipse to see objects whose light passes close to the Sun where the deflections can be measured. Distant

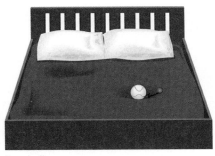

Baseball:
Marble rolls into depression

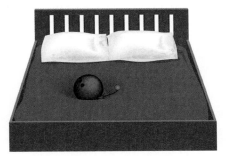

Bowling ball:
Marble rolls in faster

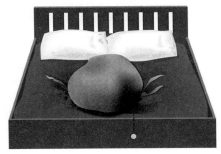

Big rock:
Marble disappears into hole

FIGURE 68.1

Objects on a water bed make depressions analogous to the curvature of space created by a mass. According to the general theory of relativity, that curvature produces the effect of gravity. We can see the similarity by placing a marble at the edge of the depression and watching it roll inward, as if attracted to the body that makes the depression. Bigger bodies make bigger depressions, so a marble rolls in faster. However, a very big body may tear the water bed, creating an analog of a black hole.

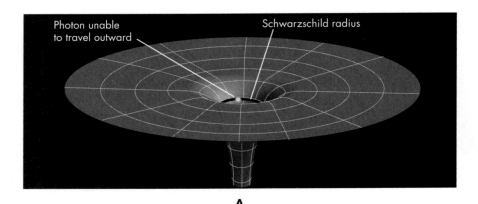

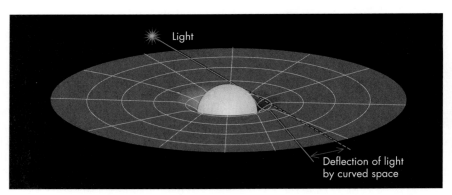

B

FIGURE 68.2
(A) Curved space around a black hole produces such a strong inward motion of space at the Schwarzschild radius that a photon moving outward at this radius cannot make any progress away from the black hole, and all photons and matter inside this radius must flow inward.
(B) Curved space around the Sun bends a ray of light passing near the Sun.

The motion of space into a black hole does not "remove space" from the universe like water flowing down a drain. Another analogy for this motion is an escalator that carries matter downward unless it is traveling upward at a greater speed than the escalator is moving down.

Q: Suppose you were traveling in a spaceship and calculated that your trajectory might carry you within a million kilometers of the center of a 5 $M_\odot$ black hole, or the same distance from the center of a 5 $M_\odot$ main-sequence star. Which would present the bigger danger of crashing?

objects producing radio waves, for example, show these same deflections. Astronomers have also detected the deflection of light around many other bodies. The deflections can sometimes even focus and brighten the light from a background object, as a lens does. This allows astronomers to detect objects that might not otherwise be visible (Unit 78).

The extreme curvature of space in a black hole, which prevents light from escaping from it, creates a kind of boundary around it that astronomers call the **event horizon.** Just as the curvature of the Earth's surface blocks our view of what lies beyond the horizon, so too the curvature of space at the Schwarzschild radius prevents our seeing beyond the event horizon into the interior of the black hole. All that happens within the black hole is hidden forever from our view. No radiation of any sort, nor any material body—rocket, spacecraft, or other object—can break free of its gravity. Because we cannot observe the interior of a black hole even in principle, there are only a limited number of physical properties we can ascribe to it. For example, it is meaningless to ask what a black hole is made of. It could be made of neutrons or cornflakes—only the amount of mass is important, not what it is composed of.

We can measure the mass of a black hole because its mass generates a gravitational field. In fact, at large distances the gravitational field generated by a black hole is no different from that generated by any other body of the same mass. For example, if the Sun were suddenly to become a black hole with the same mass it has now, the Earth would continue to orbit it without any change: It would *not* be pulled in. What makes black holes special is that they are so small that it is possible for another object to get to within an extremely small radius around them. And when the separation between objects becomes small, gravity's force grows strong because the gravitational force increases as separations become smaller.

68.3   OBSERVING BLACK HOLES

Besides having a gravitational field detectable beyond the Schwarzschild radius, black holes may also have an electric charge if, for example, an excess of positive charges falls into them. They may also have a spin—which will change the shape of the event horizon to be fatter at the equator than at the poles, like other spinning masses.

ANIMATIONS

X-ray binary system

An object that emits no light or other electromagnetic radiation is not easy to observe. But just as you can "see" the wind by its effect on leaves and dust, so too astronomers can see black holes by their effects on their surroundings.

Suppose a massive star in a binary system undergoes core collapse, directly forming a 10 $M_\odot$ black hole. Gas from the companion star may be drawn toward the hole by its gravity, just as we know happens for neutron stars and white dwarfs. The infalling matter swirls around the black hole and forms an **accretion disk** whose inner edge lies just outside the Schwarzschild radius, as depicted in Figure 68.3. Here, where the disk orbits at nearly the speed of light, turbulence and friction heat the swirling gas to a furious 10 million K, causing it emit to X-rays and gamma rays.

As the black hole orbits its neighbor, the X-ray–emitting gas may disappear from our view as it is eclipsed by the companion star. An X-ray telescope trained on such a star system will show a steady X-ray signal that disappears at each eclipse, as shown in Figure 68.4. Such a signal might be the sign of a black hole, and at least three cosmic X-ray sources bear that signature. But how do we know an X-ray source is not just a neutron star?

Because these X-ray sources are binary stars, we can measure their masses. If X-ray–emitting gas surrounds a body we cannot see, but whose mass exceeds 3 $M_\odot$, we can be reasonably confident that the invisible object is a black hole. This is because if more than 3 $M_\odot$ of matter is packed together at neutron star densities, its radius will be smaller than the Schwarzschild radius and it therefore must become a black hole. For example, Cygnus X-1, the first X-ray source detected in the constellation Cygnus (see Looking Up #4: The Summer Triangle), consists of a B supergiant star and an invisible companion whose mass—based on an application of the modified form of Kepler's third law—is at least 6 $M_\odot$. Nothing we know of but a black hole can be so massive and yet invisible. An even better candidate is A0620-00, an X-ray–emitting object in the constellation Monoceros, the Unicorn. For A0620-00, the invisible star's mass is estimated to be 16 $M_\odot$.

Although no radiation can leave the interior of a black hole, the English physicist Stephen Hawking discovered that black holes emit radiation through a different process. Hawking noted that at the subatomic level, empty space undergoes constant fluctuations. For example, a particle and its antiparticle (such as an electron and a positron—see Unit 4) will spontaneously form. Such particle–antiparticle pairs are called **virtual particles** because they exist as a tiny energy fluctuation,

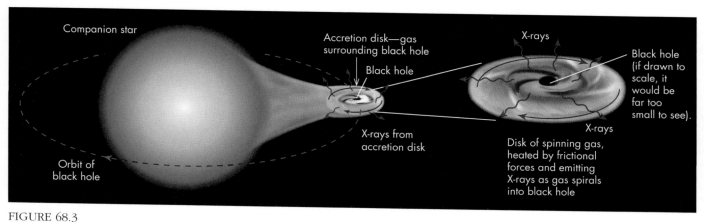

FIGURE 68.3

Black holes may reveal themselves by the X-rays emitted by gas orbiting them in an accretion disk. (Sizes and separations are not to scale.)

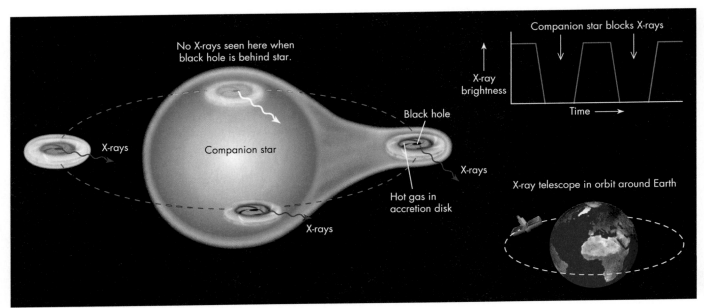

FIGURE 68.4

X-ray emission from gas around a black hole. The signal drops as the hole is eclipsed by its companion star. (Sizes and separations are not to scale.)

although for only a minuscule fraction of a second. Conservation of energy requires them to rapidly recombine, annihilating each other and returning the energy they have momentarily "borrowed."

If such a virtual particle–antiparticle pair forms outside the event horizon of a black hole, one of the pair may fall into the black hole while the other escapes into space. This prevents them from recombining. But because energy must be conserved, some energy must be lost from the black hole, with the consequence that the black hole drops in mass according to $E = m \times c^2$. The particles created by this process do not actually come out of the black hole, but from a distance the black hole would look this way. According to Hawking's model, the predicted radiation is extremely weak, and it has never been detected. However, such energy loss from a black hole means that a black hole will not last forever.

68.4 STRETCHING OF SPACE AND TIME BY GRAVITATION

Einstein's theory of general relativity has become one of the cornerstones of modern physics. It predicts a number of surprising phenomena that have now been measured and precisely agree with Einstein's predictions, such as the gravitational bending of light discussed in the previous section.

When Einstein introduced the theory, he showed that it solved a long-standing problem in explaining a small deviation in the orbit of Mercury. Mercury's orbit does not repeat itself precisely as predicted by Kepler's laws (Unit 12), but instead shifts direction slightly with each orbit. Astronomers tried to explain this shift using Newton's laws to adjust for the influence of the other planets, but this still left an unexplained deviation.

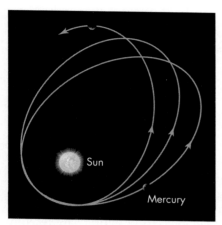

FIGURE 68.5

The orientation of Mercury's orbit shifts slightly each orbit by an amount predicted precisely by general relativity.

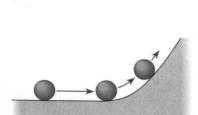

Ball rolling up a hill loses energy and slows down.

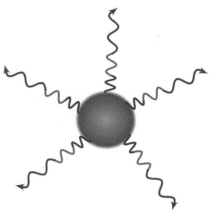

Light moving away from a star loses energy.

FIGURE 68.6

Time runs slower where the gravitational field is stronger, so an electromagnetic wave emitted at a particular frequency from the surface of a star will have a lower frequency (longer wavelength) at larger distances. The waves lose energy as they travel away from the star similar to how a ball loses energy rolling up a hill.

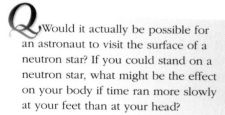Would it actually be possible for an astronaut to visit the surface of a neutron star? If you could stand on a neutron star, what might be the effect on your body if time ran more slowly at your feet than at your head?

There is a gravitational Lorentz factor γ_G that is mathematically similar to the Lorentz factor of special relativity (Unit 53):

$$\gamma_G = \frac{1}{\sqrt{1 - R_S/R}}$$

where R is the distance from the center of a massive object, and R_S is the object's Schwarzschild radius. This factor describes the amount by which both time and space are stretched by gravity.

General relativity predicts that the space near the Sun—deeper in the Sun's gravitational field—is stretched more than the space far away. In addition, general relativity predicts that time runs slower where gravity is stronger. Because Mercury's orbit is elliptical, when it is closest to the Sun at its perihelion (Unit 38), it is traveling through stretched-out space at a rate that is slightly too slow. As a result it does not travel as far as expected, making the orientation of the orbit shift slightly as illustrated in Figure 68.5.

The distortion of time by gravity is also visible in the way it affects the wavelength of light emitted in a region where gravity is strong. Physical processes, including the frequency of vibrations in atoms, all occur more slowly where gravity is strong. When light is emitted from the surface of a white dwarf, for example, the spectral lines are all found to be at lower frequencies (longer wavelengths) than we would normally expect. This effect is called a **gravitational redshift** because it shifts visible light toward the red end of the spectrum. For example, hydrogen on the surface of a white dwarf may emit a spectral line of a known frequency from its surface; but when that light reaches a distant observer whose clock is running faster, the waves have a lower frequency, as illustrated in Figure 68.6. Gravitational redshift has been directly measured in physics experiments even between the top and bottom of a building. It also results in measurable differences between the time measured on a satellite and on the ground.

This stretching of time means that if one astronaut stayed in a spaceship and watched her twin brother travel down to the surface of a neutron star, the twin would appear to be moving, speaking, and breathing more slowly than normal when he was in the strong gravitational field. On the other hand, to the twin visiting the surface of the neutron star, it would appear that his sister on the spaceship was moving, speaking, and breathing faster than normal. When he returned to the spacecraft, he would discover that less time had passed according to his clocks than had passed on the spaceship. He would be slightly younger than his sister.

The degree by which time slows depends on how close you get to the Schwarzschild radius. To an outside observer, time appears to stop at the Schwarzschild radius; while far from any masses, time runs at its normal rate. In principle then, if you could spend some time just outside the Schwarzschild radius of a black hole, when you came back out (*if* you could get back out) hundreds of years might have passed in the rest of the universe!

68.5   GRAVITATIONAL WAVES

Another prediction of Einstein's theory of general relativity is that when one object orbits another, their motion generates **gravitational waves.** Just as ripples spread away from a stone tossed into a pond, so too gravitational waves spread across space, stretching and distorting the space through which the waves move (Figure 68.7). These waves have not yet been directly detected, but U.S. scientists have recently built a large gravitational wave detector called LIGO, the Laser Interferometer Gravitational Wave Observatory, to search for them. LIGO consists of 4 km-long vacuum tunnels in which light is reflected back and forth between mirrors. If a gravitational wave of sufficient strength passes through the detector, it will shift the distance between the mirrors in a predictable way, slightly altering the light patterns between the mirrors. The detector has begun operating, and its sensitivity is being steadily increased. It is expected to eventually have the sensitivity to measure these passing distortions of space and time.

Although gravitational waves have not yet been directly detected, indirect evidence of such waves has been observed from rapidly orbiting compact stars. Such a detection is possible because as the waves carry gravitational energy away from the orbiting stars, the stars spiral inward toward each other. Exactly such behavior was detected in 1974 by the U.S. astronomers Joseph Taylor and Russell Hulse, who discovered a pulsar in a binary system with another neutron star. Their observations showed that as the neutron stars orbit each other, they are gradually losing energy and the two stars are spiraling toward each other. Taylor and Hulse were able to measure this inward motion extremely precisely because the pulsar member of the pair acts as an extraordinarily precise clock.

Taylor and Hulse showed that the rate of orbital decay exactly matches the rate of predicted energy loss in the form of gravitational waves. This has become one of the strong tests supporting the theory of general relativity. For this discovery, Taylor and Hulse were awarded the 1993 Nobel Prize in physics.

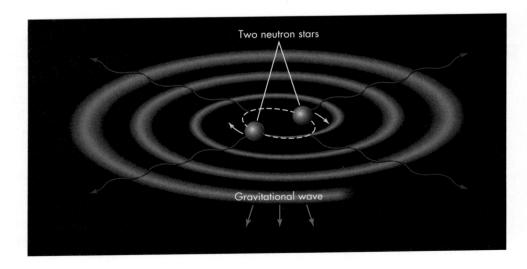

FIGURE 68.7
Gravitational waves generated by a rapidly orbiting pair of neutron stars. Such waves are "ripples" in space—and like ripples in water, they make objects move slightly as they pass.

KEY TERMS

accretion disk, 553

black hole, 548

curvature of space, 550

escape velocity, 548

event horizon, 552

general theory of relativity, 548

gravitational redshift, 555

gravitational wave, 556

hypernova, 550

Schwarzschild radius, 549

singularity, 550

virtual particle, 553

QUESTIONS FOR REVIEW

1. What is a black hole?
2. What is the Schwarzschild radius?
3. What is the event horizon?
4. What is an accretion disk?

PROBLEMS

1. Calculate the Schwarzschild radius of a 3 solar mass object.
2. Calculate *your* Schwarzschild radius. How does that compare to the size of an atom? How does it compare to the size of a proton?
3. Some galaxies show evidence of very massive black holes at their center. Calculate the radius of a billion solar mass black hole. Calculate its density in kilograms per liter. (*Reminder:* There are 1000 liters in a cubic meter.)
4. Calculate the density of a black hole with a billion (10^9) times the mass of the Sun. (Use $4/3\pi R_S^3$ as the volume of the black hole.) Compare your result to the density of water, and explain whether larger black holes have higher or lower densities.
5. If a neutron star is 10% bigger than its Schwarzschild radius, how slowly would a clock run on its surface? (Use the formula $\gamma_G = 1/\sqrt{1 - {}^{R_S}/_R}$ to calculate the gravitational time-stretching factor.)

TEST YOURSELF

1. What evidence leads astronomers to believe that they have detected black holes?
   a. They have seen tiny dark spots drift across the faces of some distant stars.
   b. They have detected pulses of ultraviolet radiation coming from dark regions of space.
   c. They have seen X-rays from accretion disks orbiting dark massive objects.
   d. They have seen a star suddenly disappear as it was swallowed by a black hole.
   e. They have looked into a black hole with X-ray radar telescopes.

2. The Schwarzschild radius of a body is
   a. the distance from its center at which nuclear fusion ceases.
   b. the distance from its surface at which an orbiting companion will be broken apart.
   c. the maximum radius a white dwarf can have before it collapses.
   d. the maximum radius a neutron star can have before it collapses.
   e. the radius of a body at which its escape velocity equals the speed of light.

3. If by some unknown process the Sun suddenly collapsed in on itself and became a black hole tomorrow, the planets would
   a. move off into space in the direction they were moving when the Sun collapsed.
   b. be dragged inward and sucked into the black hole.
   c. be blasted into pieces by the neutrinos emitted by the black hole.
   d. continue orbiting the black hole just as they orbit the Sun now.
   e. also collapse into black holes because of the gravitational waves generated.

Star Clusters

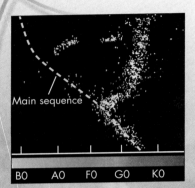

Main sequence

BO AO FO GO KO

Background Pathways: Unit 59

Stars usually form in large groups out of huge clouds of interstellar gas (Unit 60). In principle, a small cloud of gas might collapse to form a single star, but this appears to be rare. More commonly a cloud containing thousands or even millions of solar masses of gas and dust begins to contract, and as it contracts under its own gravity, smaller clumps begin to form. As stars form, the most massive ones collapse fastest, and their initial high luminosity and stellar winds may shred the rest of the cloud into smaller pieces.

The resulting groups of stars are called **star clusters**. Within a star cluster, each star moves along its own orbit about the center of mass of the cluster, held to the cluster by the group's gravity. Many clusters are loosely bound, and after the remnants of the original gas cloud are driven away—by stellar winds, radiation pressure and exploding stars—the stars may drift apart.

Star clusters provide an important place to test ideas about stellar evolution. Because the stars within a cluster form at approximately the same time, they are all nearly the same age. If, for example, we expect one type of star to die more quickly than another, we can test that hypothesis by observing many star clusters to see which kinds of stars have survived and which have died at various cluster ages. Clusters tell us about the relative numbers of stars of each type that normally form, and they provide an important link for finding distances and establishing new standard candles.

69.1 TYPES OF STAR CLUSTERS

Star clusters range from loose associations of tens of stars to dense concentrations of hundreds of thousands of stars. The spacing between the stars in a cluster varies enormously. In some clusters the stars are loosely packed, so the cluster is only slightly denser than its surroundings. In other clusters, however, the stars are so closely spaced that their separation may be as little as a tenth of a light-year. Yet even in such dense star clusters, the spacing between stars is still large compared with the sizes of stars. Thus, although the stars in a cluster sometimes look quite crowded in a photograph, the likelihood of a physical collision between them is near zero.

The star cluster nearest to us is about 75 light-years away and is about 30 light-years across. It is so close that it covers a large portion of the sky. Called the **Ursa Major group,** it includes most of the stars of the Big Dipper (see Looking Up #2: Ursa Major) and a number of stars from neighboring constellations. About a hundred stars have been identified as members of this group by their common location and motion through space, including the brightest star in the constellation Corona Borealis and the second brightest in Auriga.

You can see several clusters that are richer than the Ursa Major group with your unaided eyes. One of the most obvious is **the Pleiades** (Figure 69.1A), or the "Seven Sisters," named for the daughters of the giant Atlas who, in Greek mythology, carried the world on his shoulders. The Pleiades (pronounced *PLEE-ah-deez*) are visible as

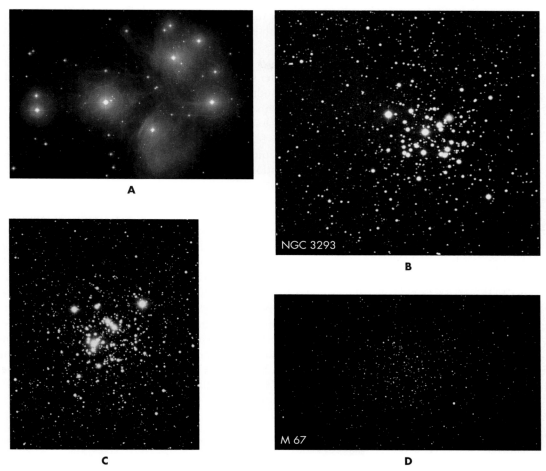

A

NGC 3293

B

C

D

FIGURE 69.1

Open clusters. (A) The Pleiades in the constellation Taurus; (B) NGC 3293 in Carina; (C) NGC 4755, the "Jewel Box," in Crux; (D) M67 in Cancer.

Note that the brightest stars in the Pleiades form almost a miniature "dipper" shape. They are sometimes confused with the "Little Dipper" of the constellation Ursa Minor.

a tiny group of stars north of the V in the constellation of Taurus (see Looking Up #5: Taurus). The V in Taurus is another cluster of stars called the Hyades (pronounced *HI-ah-deez*), who in mythology were half sisters of the Pleiades. The two clusters both happen to lie in the same constellation, but they are not otherwise associated. The Hyades are much closer—about 150 light-years away from us—while the Pleiades are 440 light-years distant. These clusters are highest in the evening sky in December and January. With binoculars, you can see that the Pleiades and Hyades each contain hundreds of stars—far more than the half dozen or so stars visible to the unaided eye.

Over a thousand other **open clusters** are cataloged in our galaxy; they contain up to a few thousand stars in a volume with a radius of typically 7 to 20 light-years. They are called *open* because their stars are scattered loosely, as you can see in the sampling of other open clusters shown in Figure 69.1. Astronomers think that open clusters form when giant, cold interstellar gas clouds are compressed and collapse, breaking up into hundreds of stars whose mutual gravity binds them into the cluster. Once formed, the stars of an open cluster continue to move through space together; but over hundreds of millions of years the stars gradually drift off on their own, so the cluster eventually dissolves. The numerous blue stars in the Pleiades suggests that this cluster is fairly young, perhaps only some tens of millions of years old. The Hyades and the Ursa Major group, by contrast, are both estimated to be many hundreds of millions of years old, and they appear to be less tightly bound than the Pleiades. Our own Sun was probably a member of such a star group, but its companion stars have long since scattered across space.

Other stars sometimes occur in loose groups called **stellar associations** that are a few hundred light-years across. Associations typically spread out from a single large open cluster near their center and may contain other, smaller star groupings. Moreover, the stars in associations are usually still mingled with the massive clouds of dust and gas from which they formed. The stars in associations probably form at about the same time, perhaps from the same triggering event. They have at most a very weak gravitational link to one another and often have different motions, so they disperse rapidly.

A far denser type of cluster, called a **globular cluster,** is much more strongly bound together, and is larger in size and stellar content than an open cluster. These clusters contain from a few hundred thousand to several million stars and have radii of about 40 to 160 light-years. The stronger gravity in globular clusters pulls their stars into a dense ball, as you can see in Figure 69.2. About 150 globular clusters have been cataloged. Although they are part of our galaxy, most of them are so distant that despite the many stars they contain, they are not easily seen without a telescope. A notable exception to this is the globular cluster Omega Centauri in the southern constellation Centaurus (Figure 69.2A; also see Looking Up #8: Centaurus and the Southern Cross). Cataloged as the 24th brightest "star" in the constellation, it is clearly fuzzy even to the naked eye. In northern skies in the constellation Hercules, a globular cluster cataloged as M13 can sometimes be seen by eye as a faint fuzzy patch (Figure 69.2B). A small telescope shows both of these clusters to be a large swarm of stars.

Q What would the sky look like if you lived on a planet orbiting a star within a globular cluster?

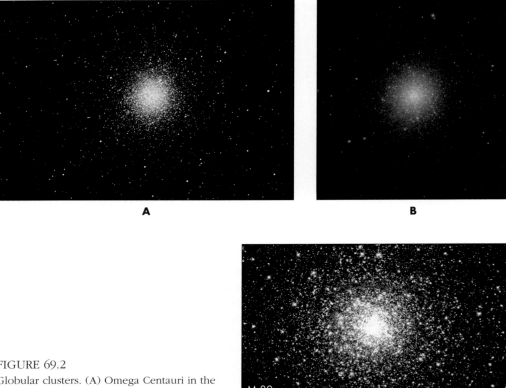

A   **B**

C

FIGURE 69.2

Globular clusters. (A) Omega Centauri in the constellation Centaurus; (B) M13 in Hercules; (C) M80 in Scorpius.

TABLE 69.1	Properties of Clusters and Associations	
Type	Number of Stars	Radius*
Open cluster	Tens to a few thousand	7–20 ly (2–6 pc)
Globular cluster	10^5–10^6	40–160 ly (12–50 pc)
Associations†	5–70 O or B stars	65–325 ly (20–100 pc)

*Because star clusters do not have sharp edges but instead gradually thin out from their center, quoted dimensions differ substantially.
†Astronomers identify several other types of associations as well. For example, T associations are regions with above-average numbers of T Tauri stars.

Table 69.1 summarizes some properties of open clusters, globular clusters, and stellar associations.

69.2   TESTING STELLAR EVOLUTION THEORY

Because the stars in a given cluster are approximately the same age, a cluster's H-R diagram shows a snapshot of the state of evolution of its stars. For example, in a young cluster, we expect that more massive stars will have completed their contraction stage from protostars and will lie on the main sequence, and no stars will have had time to become red giants and shift off the main sequence. If we look at an older cluster, however, some massive stars will have consumed their core hydrogen and evolved off the main sequence. We can use such differences between cluster H-R diagrams to check our theory of stellar evolution. We can do this by calculating evolutionary tracks (such as those in Figures 64.2 and 66.1) for every star on the main sequence to show where each star will be at 10 million years, 100 million years, and so on. The resulting curves show us what the H-R diagram of the entire cluster should look like at 10 million years, 100 million years, and so on after its birth.

Astronomers can deduce the age of a star cluster from the pattern of its H-R diagram. To understand why a cluster's H-R diagram reveals its age, recall that the main sequence is determined by the location of stars fusing hydrogen in their cores. All the stars in a newly formed cluster lie on or near the main sequence. But not all the cluster's stars use up their fuel at the same rate. Massive stars use up their fuel more rapidly (to maintain their high luminosity) than low-mass stars do. With their hydrogen used up, high-mass stars leave the main sequence and turn into red giants. The low-mass stars, on the other hand, still have hydrogen to fuse, so they remain on the main sequence longer.

Because the high-mass stars in an older cluster turn into red giants (and therefore lie off the main sequence), a line in the H-R diagram connecting the position of the cluster's stars bends away to the right, as shown in Figure 69.3. The point where that line bends away from the main sequence is called the **turnoff point.** A star just below the turnoff point is not yet old enough to have used up the hydrogen in its core, but it is about to run out. That is, its age is just a tiny bit less than its total main-sequence lifetime. But all stars in the cluster are the same age—namely, the age of the cluster. To get the cluster's age, we therefore determine how long a star at the turnoff point can live by calculating its main-sequence lifetime from its mass and luminosity, as described in Unit 61. The answer we get is the age of the star cluster.

Figure 69.4 shows a series of H-R diagrams for clusters ranging from very young to very old (a few million years for the youngest to more than 10 billion years for

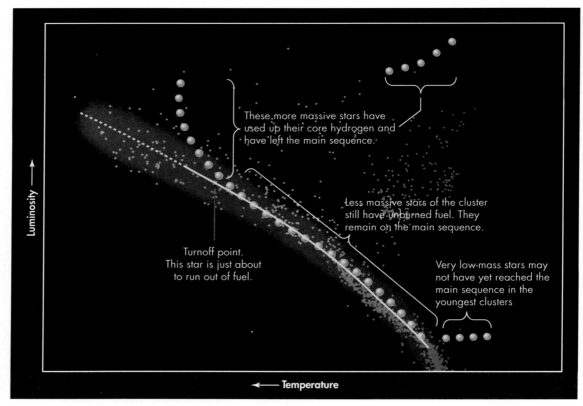

FIGURE 69.3
The pattern of stars in the H-R diagram of a star cluster indicates its age.

the oldest). Notice that old clusters have few, if any, stars on the upper part of the main sequence. On the other hand, short-lived stars are still present on the upper main sequence of young clusters.

When we compare such curves with H-R diagrams of actual star clusters, as in Figure 69.5, the match is excellent. The examples in the figure are very similar to the first and last panels of Figure 69.4, and their estimated ages are 10 million and 7 billion years for the open cluster and globular cluster, respectively. If our theory of stellar evolution was wrong, the shapes would be unlikely to agree so well. In addition, the models help astronomers understand why so many kinds of stars exist. For example, the models not only offer a natural explanation of features such as main-sequence, red giant, and white dwarf stars; they also help us see how totally different objects such as pulsating variables, planetary nebulae, and supernovae fit into the overall scheme of stellar evolution. This success in interpreting such stellar diversity is evidence that our theory of stellar evolution is essentially correct. Moreover, this success lets astronomers measure the age of a star cluster.

69.3   THE INITIAL MASS FUNCTION

The stars we see at night turn out to be very unrepresentative of the general stellar population. When we look at the night sky, we see the stars whose brightness is great enough for us to detect their light. We primarily see the stars that are unusually luminous and that, therefore, stand out relative to the others. As a result, when we look at the night sky, the stars we see tend to be giants. This is illustrated in Figure 69.6,

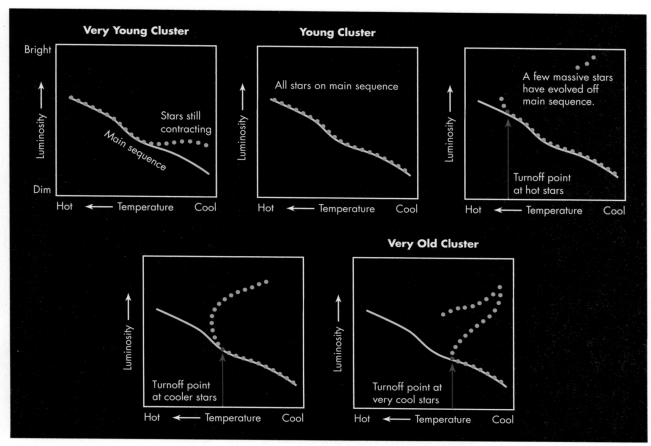

FIGURE 69.4

Schematic H-R diagrams of clusters of different ages illustrating the turnoff point and how it shifts down and to the right for older clusters.

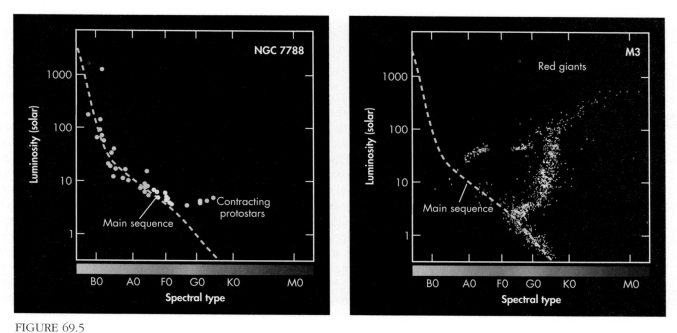

FIGURE 69.5

The H-R diagrams of two clusters. NGC 7788 is an open cluster in Cassiopeia with an estimated age of about 10 million years. M3 is a globular cluster in Canes Venatici with an estimated age of 7 billion years.

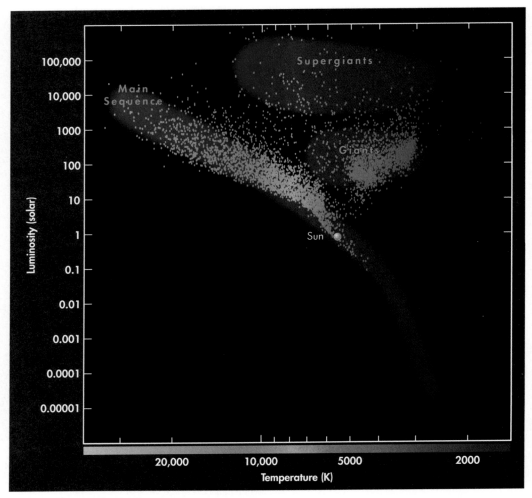

FIGURE 69.6

H-R diagram of the stars that are visible to the unaided eye in the night sky. Nearly all are more luminous than the Sun.

If star A is a million times more luminous, then $L_A = 10^6 \times L_B$. If star B is a thousand times closer than star A, then $d_A = 10^3 \times d_B$. Comparing these, we find the following:

$$B_A = \frac{L_A}{4\pi d_A^2}$$

$$= \frac{10^6 \times L_B}{4\pi(10^3 \times d_B)^2}$$

$$= \frac{10^6 \times L_B}{10^6 \times 4\pi d_B^2}$$

$$= \frac{L_B}{4\pi d_B^2} = B_B$$

which shows an H-R diagram of all the stars in the sky that are visible to the unaided eye. Notice that the Sun is one of the least luminous among all of these stars.

The class of stars we see at night are rare; but because we can see them out to very large distances, we see greater numbers of them than the far more numerous stars that are too dim to be visible. How bright a star looks to us is determined by its luminosity and distance, as we saw in discussing the inverse-square law (Unit 54.2): $B = L/4\pi d^2$. From this law we can see that if star A is a million times more luminous than star B, but star B is a thousand times closer than star A, they will appear to have the same brightness because $1000^2 = 1,000,000$. So according to the inverse-square law, star A will appear just as bright to us even if it is a thousand times farther away than star B. That is, great luminosity can compensate for distance.

Many giant stars are *more* than a million times more luminous than the low-luminosity stars. As a result, when we see a set of stars that looks bright to us, we are looking at the most luminous stars at distances more than a thousand times greater than those of the low-luminosity stars. In other words, the stars we see are heavily biased in favor of luminous stars. Astronomers call this problem of sample biasing a **selection effect**.

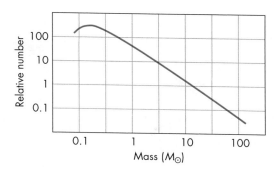

FIGURE 69.7

The initial mass function of stars indicates the relative number of stars of each mass. Studies of star clusters indicate that there are *many* more low-mass stars than high-mass stars, as indicated by the curve.

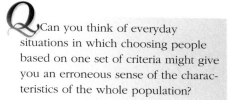

Can you think of everyday situations in which choosing people based on one set of criteria might give you an erroneous sense of the characteristics of the whole population?

What does the number of white dwarfs in Figure 69.8 versus the number of red giants suggest about the relative duration of these stages of stellar evolution?

When astronomers observe a star cluster, they avoid this problem because all of the stars are at (nearly) the same distance. If we can count all the stars within the cluster, we have an unbiased sample. The relative numbers of stars of different luminosities in the sample will then indicate the true proportion of stars of each luminosity. If we use the mass–luminosity relation (Unit 58) to deduce the masses of the main-sequence stars from their luminosities, we can estimate how many stars of each mass were born when the cluster formed. Astronomers call such a census of a cluster's stars its **initial mass function.**

Studies of star clusters show that stars similar to or smaller than the Sun vastly outnumber more massive stars, as shown by the initial mass function plotted in Figure 69.7. The most numerous stars turn out to be dim, cool, red dwarfs—stars that lie on the main sequence but whose mass is below 0.5 $M_\odot$. Astronomers do not yet understand what determines the distribution of stellar masses, but it appears to be similar in most clusters, so it must be a property of how a collapsing interstellar cloud breaks up into stars.

The initial mass function found from these studies indicates that the number of stars having a certain mass increases sharply as mass decreases. Typically there are about 20 times more stars with masses between 1 and 2 solar masses than between 10 and 20 solar masses. The relative numbers of stars of different masses are also suggested by Figure 69.8, which shows an H-R diagram for all the stars currently known within 10 parsecs of the Sun.

Despite their great numbers, low-mass stars are hard to see because they are so intrinsically dim. Not a single main-sequence M star is visible to the unaided eye, even though these are by far the most common type of star. Even the star nearest to the Solar System, the M-type star Proxima Centauri (see Looking Up #8: Centaurus and the Southern Cross), is more than 10,000 times too dim to be seen with the unaided eye. The initial mass function suggests that M-type stars (masses ranging from 0.08 to 0.48 $M_\odot$) outnumber all other types of stars combined by a factor of 10. In fact, the combined mass of all stars of this dim spectral type is approximately half the combined mass of all types of stars.

Objects less massive than about 0.08 $M_\odot$ were once also believed to be rare, but astronomers now think they may be common and were simply missed in previous stellar surveys. How could astronomer miss these objects if they are so abundant? Recall that low-mass stars tend to be cool. In fact, if a contracting mass of gas is less massive than about 0.08 $M_\odot$, it is so cool that it is unable to fuse hydrogen into helium and is termed a *brown dwarf* (Unit 60). These are in some ways more like huge planets than tiny stars, and their extreme dimness and small size make them difficult to detect. However, searching the sky at the wavelengths at which brown dwarfs are brightest greatly increases the chance of finding them. What wavelengths are best? Because they are so cool, brown dwarfs are brightest in the infrared, so it is with infrared surveys such as 2MASS (2 Micron All Sky Survey) that

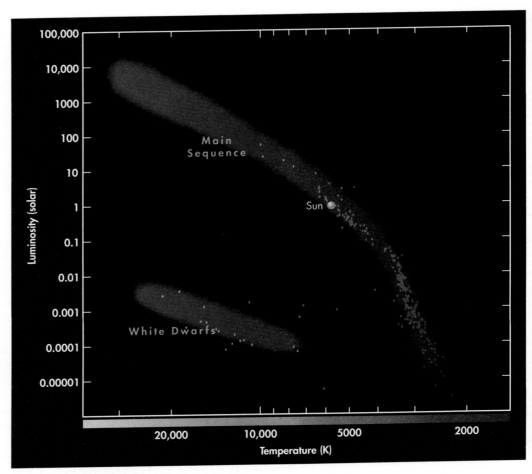

FIGURE 69.8

H-R diagram of the known stars within 10 parsecs of the Sun. Few of the nearest stars are more luminous than the Sun.

astronomers are now finding large numbers of these "failed" stars. Studies of the Pleiades and other young open clusters at infrared wavelengths have revealed many probable brown dwarfs. Astronomers now think brown dwarfs may outnumber ordinary stars, making them one of the most common astronomical objects in our galaxy, although it appears that they do not contain as much mass in total as is contained in ordinary stars.

In almost any unbiased sample of stars, we find that most of the mass is contained in the lowest-mass objects, yet most of the light comes from the highest-mass objects. In this wide range of properties, the Sun is sometimes called a "typical" star. It is true that the Sun sits somewhere in the middle of the range of stellar properties, yet it is typical neither in mass nor in luminosity. The Sun is more massive than about 97% of all stars, yet more than 99% of the total luminosity in a typical collection of stars is produced by stars more massive than the Sun. However, when we add up the total mass and the total luminosity of stars in the neighborhood of the Sun, it happens that the ratio of mass to luminosity is similar to that of the Sun. In other words, if a thousand solar masses are contained in the collection of stars, then this star collection will generate about a thousand solar luminosities of light. Hence, the mass and luminosity of the stars in our neighborhood balance in a way that makes the Sun seem fairly average.

KEY TERMS

globular cluster, 560

initial mass function, 565

open cluster, 559

the Pleiades, 558

selection effect, 564

star cluster, 558

stellar association, 560

turnoff point, 561

Ursa Major group, 558

QUESTIONS FOR REVIEW

1. How do star clusters form?

2. What are the different types of star clusters?

3. How are star clusters useful for testing stellar evolutionary theory?

PROBLEMS

1. Sketch an H-R diagram for three clusters of stars. Make your sketches for one that is 10^8 years old, one that is 10^9 years old, and one that is 10^{10} years old. Explain what parts are similar and different in each.

2. If a globular cluster contains 1 million stars in a volume with a radius of 100 light-years, what is the average separation between stars? Calculate this by assuming that each star fills one millionth of the total volume of the cluster. The diameter of the sphere it occupies is then the average separation. (The volume of a sphere is $4/3\pi R^3$.)

3. Suppose we were observing the globular cluster in the previous problem through a telescope. At the distance of the globular cluster, its radius appears to be 100 arc seconds, so it covers an area on the night sky of $\pi\,(100\text{ arcsec})^2$.
   a. Supposing each star has the radius of the Sun, what is its radius in arc seconds?
   b. What fraction of the area of the globular cluster is covered by the summed area of the million stars?

TEST YOURSELF

1. The most numerous type of star is
   a. about half the Sun's mass.
   b. about the Sun's mass.
   c. about twice the Sun's mass.
   d. about 20 times the Sun's mass.
   e. None of the above. The stars are fairly evenly distributed as far as mass goes.

2. When looking at the night sky with our unaided eyes, most of the stars we see are
   a. brown dwarfs.
   b. lower-mass than the Sun.
   c. about the Sun's mass.
   d. more luminous than the Sun.
   e. young, hot O stars.

3. Most of the mass in stellar populations comes from the
   a. most luminous stars.
   b. lowest-luminosity stars.
   c. most massive stars.
   d. hottest stars.
   e. red giants.

PLANETARIUM EXERCISES

All the planetarium exercises use the Starry Night planetarium software provided with your text. You should install the program and familiarize yourself with the various parameters and controls before trying these exercises. When the program asks you to enter a password, click "OK" to run Starry Night.

Center the program on a nearby cluster such as the Hyades, the Pleiades, or the Big Dipper. (You can do this by using the Find option in the Settings menu and typing in the name. You may need to turn off "Daylight" and "Horizon" in the Sky menu.) Zoom in until the cluster fills a large portion of your field of view. On the toolbar click the "H-R" button. This generates an H-R diagram for the stars in your display that have the necessary data.

Point to several stars in the cluster to find their distances, then set the distance cutoff (click "Options") for the H-R diagram to exclude stars that are not part of the cluster. Examine the H-R diagram of the cluster stars and compare them. Use these data to estimate the age of one cluster compared to another based on where the stars turn off the main sequence as described in the text.

Choose several different regions of the sky and generate H-R diagrams for the stars. Do the diagrams show any differences? Try choosing a field with many stars and then adjust the distance limits to different ranges. How does the H-R diagram change? How can you explain the differences you find?

Part Five

The content of some Units will be enhanced if you have previously studied some earlier Units in this textbook. These Background Pathways are listed below each Unit image.

Unit 70 Discovering the Milky Way

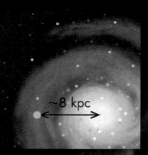

~8 kpc

Unit 73 Mass and Motions in the Milky Way

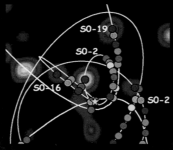

S0-19

S0-2

S0-16

S0-2

Background Pathways: Unit 70

Unit 76 Galaxy Clustering

Background Pathways: Unit 75

Unit 71 Stars of the Milky Way

Background Pathways: Unit 70

Unit 74 A Universe of Galaxies

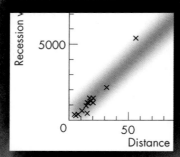

Recession

5000

0   50
Distance

Background Pathways: Unit 70

Unit 77 Active Galactic Nuclei

Background Pathways: Unit 74

Unit 72 Gas and Dust in the Milky Way

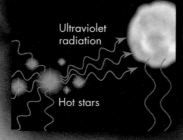

Ultraviolet radiation

Hot stars

Background Pathways: Units 24, 70

Unit 75 Types of Galaxies

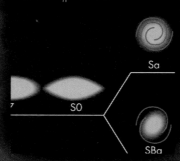

Sa

S0

SBa

Background Pathways: Units 71, 74

Galaxies and the Universe

Unit 78 Dark Matter

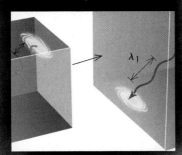

Background Pathways: Unit 77

Unit 79 Cosmology

Unit 80 The Edges of the Universe

Background Pathways: Unit 79

Unit 81 The Beginnings of the Universe

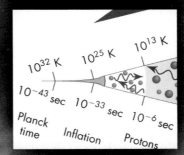

Background Pathways: Unit 79

Unit 82 The Fate of the Universe

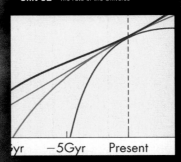

Background Pathways: Unit 78

Unit 83 Astrobiology

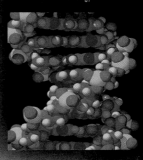

Unit 84 The Search for Life Elsewhere

Background Pathways: Unit 83

Unit 79 Cosmology

Background Pathways: Unit 74

70

Discovering the Milky Way

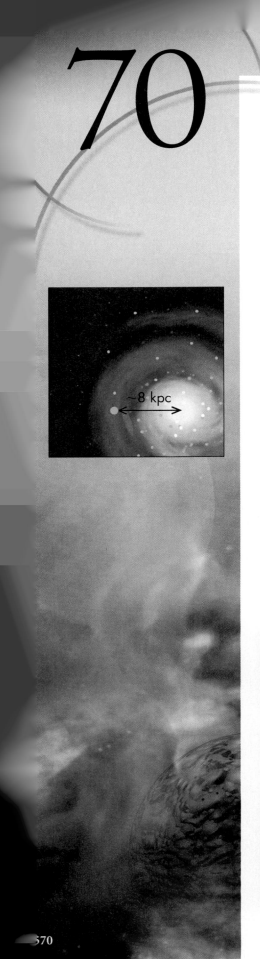

~8 kpc

On a clear, moonless night, far from city lights, you can see a pale band of light spangled with stars stretching across the sky (Figure 70.1). The ancient Hindus thought this shimmering river of light in the heavens was the source of the sacred river Ganges. To the ancient Greeks, this dim celestial glow looked like milk spilled across the night sky, so they called it the **Milky Way**. Astronomers also call this our **galaxy** from the Greek word for "milk" (*galactos*).

A view of the Milky Way on a clear, dark night is one of nature's finest spectacles. The band stretches in a full circle around us on the celestial sphere, but it is at a different angle from either the Earth's equator (the celestial equator) or the Solar System (the ecliptic). The orbits of the planets as well as the Earth's equator are tilted by about 60° with respect to the circle of the Milky Way. As a result, when you observe the Milky Way it crosses the sky at different angles, depending on when or where you see it.

Superimposed on the dim background glow are most of the bright stars and star clusters that we can see, which all belong to our galaxy. Here and there dark blotches interrupt the glowing backdrop of stars, as you can see in Figure 70.1. The Incas of ancient Peru, who observed the Milky Way from their temple observatories in the Andes, gave these dark areas names, just as peoples of the classical world named the star groups. Today we know the dark regions are clouds of dust and gas that give birth to new stars.

Our galaxy is an enormous system of stars—home to the Sun and hundreds of billions of other stars. It contains huge interstellar clouds, stars that are forming, and stars that are

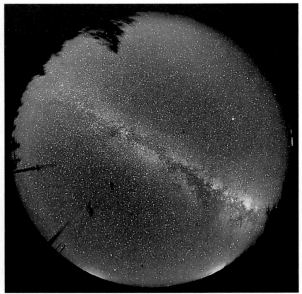

FIGURE 70.1
Wide-angle photo of the Milky Way taken from Mount Graham in Arizona. The dark ring around the edge of the picture is the horizon. The white regions along the horizon at the bottom of the figure are distant city lights.

dying. Yet the galaxy is more than just a collection of all these other things. It has its own structure on a scale far larger than stars or solar systems. Understanding that structure provides clues to why and where stars form and the Sun's relationship to other stars.

70.1   THE SHAPE OF THE MILKY WAY

The pattern of the Milky Way can be seen on the foldout chart at the back of the book. It is shown as an irregular S-shaped blue region crossing north and south of the celestial equator.

ANIMATIONS

The Milky Way

When Galileo Galilei pointed a telescope at the Milky Way in 1609, he discovered that it contains millions of stars too dim to be seen as individual points of light with the unaided eye (Figure 70.2). This observation confirmed the hypothesis proposed 2000 years earlier by the Greek philosopher Democritus (who also proposed the idea that matter was made of atoms). In the 21st century we know that these stars, along with our Sun, form a huge, slowly revolving disk—our galaxy.

Our understanding of the Milky Way as a star system developed further in the mid-1700s when Thomas Wright, an English astronomer, and Immanuel Kant, a German philosopher, independently suggested that the Milky Way is a flattened swarm of stars. They argued that if the Solar System were near the center of a spherical cloud of stars, we would see roughly the same number of stars in all directions. However, from inside a disk-shaped system, we would see vastly more stars in directions toward the outer edge of the disk than in directions perpendicular to it, as illustrated in Figure 70.3. This explains why the stars of the Milky Way appear to stretch around us in a great circle.

A simple analogy may help you better understand this argument. Imagine making a lemon gelatin fruit salad in a big, round flat dish. Imagine that you have spread blueberries throughout the dish. If an ant fell into the salad and was clinging to a blueberry in it, the ant would see few berries if it looked up through the top of the salad or down through the bottom. However, it would see lots of blueberries if it looked in directions that lie in the plane of the dish. So too it is with us and stars in the disk of the Milky Way.

FIGURE 70.2

Through a telescope the Milky Way is resolved into millions of stars. Several clusters of stars, some pink from the emission lines of hydrogen, are seen scattered across the image.

FIGURE 70.3

If the Milky Way had a spherical shape, we would see about the same number of stars in every direction. However, because it is a disk, we see stars concentrated into a band around the sky.

70.2   STAR COUNTS AND THE SIZE OF THE GALAXY

The first quantitative attempts to determine the Milky Way's size and shape were made in the 1780s. William and Caroline Herschel seem an unlikely pair to have revolutionized astronomy, but this brother and sister team of musicians made a hobby of studying the skies. William (1738–1822) and Caroline (1750–1848) spent their free nights scanning the stars with high-quality telescopes built by William, discovering comets and the planet Uranus (Unit 44). Their growing fame won them support from the king of England, and they became full-time astronomers.

William decided to attempt to measure the shape of the Milky Way by observing hundreds of areas over the sky and counting the stars. On the assumption that other stars were like the Sun, he concluded that the number of stars that he could see in each direction would tell him the extent of the star system in that direction. When Caroline cataloged and indexed the results of William's counts, they found very similar star counts throughout the band of the Milky Way, but about a fifth as many stars when looking at directions 90° away from it. Based on the number of stars along many directions, he produced the cross-sectional diagram shown in Figure 70.4. From his understanding about the characteristics of stars, William suggested that the Milky Way was a disk about 2500 parsecs in diameter, with the Sun near the center, and about one-fifth as thick as it was wide.

The left side of this diagram illustrates one of the problems with this method. The stars extend out a large distance, except right in the middle, where their distribution looks like a pair of open alligator jaws. This odd shape was caused by a dusty interstellar cloud that blocked light in that direction—however, astronomers did not learn about interstellar clouds until 150 years later. William recognized another problem after he had built larger telescopes. He could see even more stars, and he realized his earlier observations had not seen to the edge of the Milky Way. Nevertheless, this remained the best picture of the star system in which we live for over 100 years.

It was not until the early 1900s that much more extensive studies gave us a clearer idea of the size of the Milky Way. By this time it was clear that stars did not all shine with the same luminosity as the Sun, and that stars had to be studied to much fainter limits to detect the "edge" of the galaxy. The Dutch astronomer Jacobus C. Kapteyn (pronounced *CAP-tine*) carried out an extensive study along the lines of the Herschels', but with modern instruments, photography, and knowledge of the different types of stars. Because most of the stars were too distant to make direct parallax estimates, Kapteyn made his distance estimates for various types of stars by determining how much they appeared to shift their position on the sky (their *proper motion*— Unit 52.4). If two stars are moving through space at the same speed relative to the Sun, the one that is farther away will appear to have a smaller motion across the sky. This approach avoids the problem of assuming stars all have the same luminosity.

Kapteyn's resulting model (Figure 70.5) revised the size of the Milky Way greatly, suggesting that it was 18,000 parsecs in diameter, again with the Sun fairly near the center. This model became known as **Kapteyn's Universe** because at the

Q If the Earth's axis were tilted 90° with respect to the circle of the Milky Way, how would it appear in the sky? What if the Earth's axis pointed toward the Milky Way?

Q Why are proper motions smaller, on average, for more distant stars?

FIGURE 70.4

A copy of Herschel's sketch of the Milky Way made in 1784. This cross-sectional view was based on the number of stars he could see in different directions. From this picture, he correctly deduced that the Milky Way was wider than it was thick. However, Herschel *incorrectly* concluded that the Sun (the orange dot) is near the Milky Way's center. He was led astray by his lack of knowledge that dust clouds blot out distant stars and prevent us from seeing our galaxy's true extent.

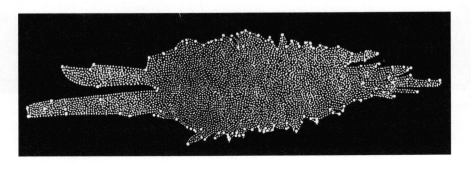

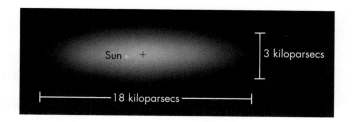

FIGURE 70.5

"Kapteyn's Universe" was a model that proposed the Milky Way was a roughly disk-shaped system about 3 kpc thick with the Sun near its center.

time it was not known that there were other galaxies beyond the Milky Way. Galaxy dimensions are so huge that even parsecs are inconvenient for measuring their size, so astronomers often use **kiloparsecs** (kpc) for that purpose. One kiloparsec = 1000 parsecs ≈ 3300 light-years. Thus, the Herschel model of the Milky Way was just 2.5 kpc in diameter, while the Kapteyn Universe was 18 kpc in diameter.

In both the Kapteyn and the Herschel models, our galaxy is depicted essentially as a disk of stars. Astronomers usually refer to this flattened disk as the **galactic plane.** Within the disk, different kinds of stars are concentrated more or less tightly in the galactic plane. Young stars, gas, and dust are found close to the plane on average, but older stars span a much larger range of distances from the plane, extending up to several kiloparsecs.

70.3   GLOBULAR CLUSTERS AND THE SIZE OF THE GALAXY

At about the same time Kapteyn was working, the U.S. astronomer Harlow Shapley argued that the Milky Way was even larger—about 100 kpc across—and that the Sun was not near the center but rather was about two-thirds of the way out in the disk.

Shapley used an entirely different method from the Herschels or Kapteyn to determine the Milky Way's size. He studied the locations of **globular clusters** (Unit 69)—these are dense groupings of up to a million stars (an example of a globular cluster is shown in Figure 70.6). Because these clusters contain so many stars, they are very luminous and can be seen at large distances—across the galaxy and beyond. Moreover, many of them have orbits that carry them far outside the galactic plane, so we have a clear view of them above or below the dense disk of the Milky Way. Shapley argued that these massive star clusters must orbit the center of the Milky Way.

Shapley noticed that the globular clusters are *not* scattered uniformly across the whole sky but are concentrated in the direction where the Milky Way looks brightest to us, toward the constellation Sagittarius (see Looking Up #7: Sagittarius). He hypothesized, therefore, that the system of globular clusters was centered on the middle of the galaxy. By mapping where the clusters lay, he could deduce the true distribution of the stars and therefore the size of the Milky Way.

To map the positions of the clusters, Shapley needed to estimate their distances from the Sun. He did this by observing the variable stars in them (Unit 63). Certain kinds of variable stars have predictable luminosities, and globular clusters often contain a class known as RR Lyrae variables. From the luminosity and apparent brightness of these stars, he could calculate their distance using the inverse-square law. The distance to the stars in a cluster gave the cluster's distance. With good estimates of the clusters' distances, Shapley plotted where they lay in the Milky Way. Figure 70.6 shows his results. The clusters fill a roughly elliptically shaped region, and the Sun lies *not* at the middle of the region, but about two-thirds of the way from its center. Shapley therefore concluded that our galaxy is roughly 100 kpc in diameter and that the Sun is nearer its edge than its center.

The great difference between Shapley's and Kapteyn's findings about the Milky Way created a major controversy among astronomers. As in so many other

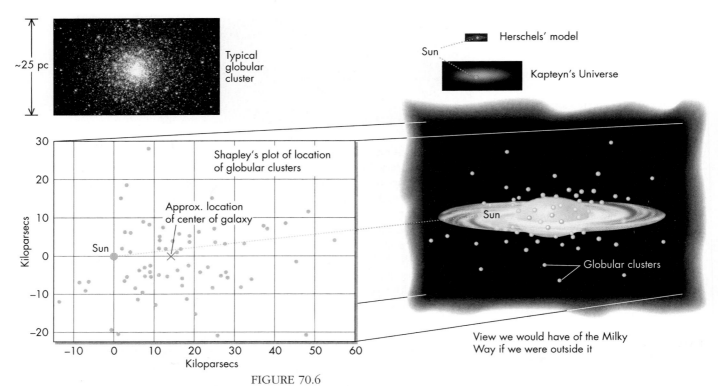

FIGURE 70.6

Schematic version of Shapley's plot of globular clusters, from which he inferred the size of the Milky Way and the Sun's location in it. Notice that the clusters fill a roughly elliptical region and that the Sun in *not* at the center of their distribution.

controversies, both sides were partially right and partially wrong. However, Shapley's results were closer to the truth. The main reason both results differed from what we know today is that neither recognized the effects of **interstellar dust**—small particles of solid matter that are found in deep space (Unit 72). Close to the galactic plane, the dust blocks light from distant stars, so Kapteyn's Universe is limited to the portion of the Milky Way's disk that surrounds the Sun, making it appear that the Sun is near the center. Along sight lines to the globular clusters, which are not so deeply embedded in the galactic plane, dust still dims their light slightly, making them look a little more distant than they really are. As a result, Shapley overestimated the distance to the globular clusters, which led him to deduce that our galaxy is bigger than it really is. Several decades later, astronomers also discovered that the class of variable stars he used to estimate the cluster distances was less luminous than he had assumed, and this also caused him to overestimate the distances.

Modern estimates place the center of the galaxy about 8 kpc in the direction of the constellation Sagittarius. Observations indicate that the outer part of the galaxy has no distinct edge—the density of stars steadily declines, becoming so sparse that it is difficult to detect stars at about twice the Sun's distance from the center. Overall, though, the Milky Way's diameter is about 30–40 kpc, or roughly 100,000 light-years.

70.4   GALACTIC STRUCTURE AND CONTENTS

Today our understanding of the structure of the Milky Way is aided greatly by studies of other galaxies. In the early 1900s it was not widely accepted that there actually *were* other galaxies; this is explored further in Unit 74. Because we live inside

Approx. 50,000 light-years

FIGURE 70.7
Photograph of the spiral galaxy NGC 5236 (also known as M83), illustrating its spiral arms.

the Milky Way, we cannot observe our galaxy in its entirety—it is difficult "to see the forest for the trees." Having other examples of galaxies gives us a better idea of the kinds of structures we might find in the Milky Way.

Many external galaxies are flat disks with conspicuous **spiral arms** (Unit 75), as shown in Figure 70.7. Such arms are hard to detect in our own galaxy because our location in the Milky Way is such that we cannot get an overview of its disk. Nevertheless, given that the Milky Way is a flat disk and that other disk galaxies have spiral arms, it is reasonable to deduce that ours does too, and observations of the locations of interstellar gas, dust, and bright stars are consistent with this picture. By combining observations of our own galaxy with the general features known to occur in other galaxies, astronomers can assemble a more detailed model for the Milky Way, as described next.

The Milky Way consists of three main parts, illustrated in Figure 70.8: a **disk** about 30–40 kpc in diameter, a more spherically distributed component called the **halo,** and a flattened, somewhat elongated **bulge** of stars at its center. Within the disk, numerous bright young stars collect into spiral arms that wind outward from near the center. Our Solar System lies between two arms about 8 kpc from the center. You may find the following analogy helpful in visualizing the scale of our galaxy: If the Milky Way were the size of the Earth, the Solar System would be the size of a large cookie.

Mingled with the stars of the disk are huge clouds of gas and dust that amount to about 15% of the disk's mass. We can see some of these clouds by the visible light they emit, whereas others reveal themselves because their dust blocks the light of background stars, as can be seen in Figure 70.2. In fact, dust scattered throughout the galaxy prevents us from seeing farther than about 3 kpc away from the Sun in almost any direction close to the galactic plane. For example, we cannot see our galaxy's **nucleus**—its core—at visible wavelengths. However, radio, infrared, and X-ray telescopes can "see" through the dust and reveal that the core of our galaxy contains a dense swarm of stars and gas as well as a supermassive black hole (Unit 73).

Q What would the Milky Way look like in the night sky to an observer on a planet at the very edge of the galaxy? What would it look like to an observer at the galactic nucleus?

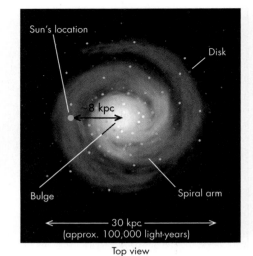
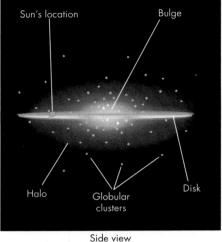

FIGURE 70.8
Artist's sketch of the Milky Way showing top and side views, illustrating the disk, bulge, and halo. Notice how thin the disk is and how the halo surrounds the disk much as a bun surrounds a hamburger. Because the Milky Way's stars gradually thin out and do not just stop at some distance, its size is labeled only approximately.

KEY TERMS

bulge, 575

disk, 575

galactic plane, 573

galaxy, 570

globular cluster, 573

halo, 575

interstellar dust, 574

Kapteyn's Universe, 572

kiloparsec, 573

Milky Way, 570

nucleus, 575

spiral arm, 575

QUESTIONS FOR REVIEW

1. What does the Milky Way look like in the night sky?
2. How do we know our galaxy is a flat disk?
3. How did Shapley deduce the Milky Way's size and the Sun's position in it?
4. What effects did dust have on the determination of the size of the Milky Way?
5. What are the major components of the Milky Way?

PROBLEMS

1. How long does it take light from the center of the Milky Way, 8 kpc away, to reach the Solar System?
2. Suppose a galaxy like the Milky Way was about 2 million light-years away (the distance to the next nearest galaxy similar in size to the Milky Way). What would its angular size be, given that its disk is about 100,000 light-years across?

3. If we made a model of the Milky Way that had the diameter of the Earth, how big would the Earth itself be in this shrunken model? (Assume that the Milky Way's diameter is 50 kpc for this problem.)

TEST YOURSELF

1. One way astronomers deduce that the Milky Way is a disk is that they
   a. see stars arranged in a circular region around the north celestial pole.
   b. see far more stars along the band of the Milky Way than in other directions.
   c. see a large dark circle silhouetted against the Milky Way in the Southern Hemisphere.
   d. see the same number of stars in all directions in the sky.
   e. None of the above

2. Our Solar System is approximately _____ from the center of the Milky Way galaxy.
   a. 4.3 pc
   b. 8 kpc
   c. 30–40 kpc
   d. 1 Mpc
   e. 13.7 Mpc

3. One of the reasons that Kapteyn underestimated the size of the Milky Way and Shapley overestimated it is that they did not recognize the _____.
   a. effects of the motion of the Sun from the center.
   b. dimming effect of interstellar dust.
   c. age of the globular clusters.
   d. existence of dark matter.

71

Stars of the Milky Way

The Milky Way contains many types of stars: giants and dwarfs, hot and cold, young and old, stable and exploding. All of these star types combine to define the overall population of stars in our galaxy. Astronomers have also undertaken a "stellar census," which counts the relative numbers of each type of star in different regions of the Milky Way.

By analyzing the types of stars and their relative numbers, astronomers can deduce some of the history of our galaxy, much as archaeologists can learn about life in an ancient city from the kinds of buildings it contained. Such studies of stellar populations reveal that despite the wide range of star types and the enormous range of luminosities, the typical star in the Milky Way is rather small, dim, and cool.

Background Pathways: Unit 70

71.1 TWO STELLAR POPULATIONS: POP I AND POP II

Hidden in the great diversity of star types is an underlying simplicity that astronomers first noted in the 1940s. At that time, the 100-inch reflector at Mount Wilson Observatory, located outside Los Angeles, California, was the largest telescope in the world and the best for observing galaxies. The glow from city lights, however, made it hard to see faint galaxies. But blackouts during World War II darkened the night sky, and Walter Baade (*BAH-deh*), an astronomer at Mount Wilson, took advantage of the darkness to make a series of photographs of neighboring galaxies. He noticed that stars in these nearby galaxies were segregated by color. Red stars were concentrated in the bulges and halos of the galaxies, whereas blue stars were concentrated in their disks and especially in their spiral arms. To distinguish these groups, Baade called the blue stars in the disk **Population I,** and the red stars of the bulge and halo **Population II** (Pop I and Pop II for short). While pursuing his studies of the stellar populations of other galaxies, he realized that Milky Way stars showed the same division. Furthermore, when he examined the properties of Pop I and II stars in our galaxy, he found many striking differences between the two populations in addition to their color and location.

Population I and Population II stars differ in many respects: color, age, location in the galaxy, motion, and composition. Population I stars are young, typically less than a few billion years old, and many of them are blue. They lie in the plane of the galaxy's disk and follow approximately circular orbits, as shown by the blue orbits in Figure 71.1. Their atmospheric composition is like the Sun's, mostly hydrogen and helium; but about 3% of their mass consists of heavy elements such as carbon and iron.

Population II stars are generally red and old—more than about 10 billion years old. They lie in the bulge and halo of the galaxy, moving along highly elliptical orbits that are often tilted strongly with respect to the galactic disk, as shown by the red orbits in

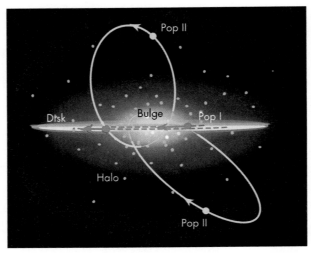

Side view

A

Top view

B

FIGURE 71.1
Stellar orbits in the Milky Way. Pop I stars orbit in the disk (blue lines). Pop II stars orbit in the halo (yellow lines). Notice the elongation and inclination of their orbits.

Pop I and II

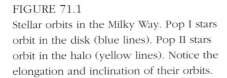

Suppose the Sun were a Pop II star in the halo of the Milky Way. In what ways would that make studying the Milky Way easier? Is it likely there are Earth-like planets orbiting halo stars?

Figure 71.1. Population II stars are almost entirely hydrogen and helium, with only a few hundredths percent of their mass composed of heavy elements—roughly a hundred times less than stars like the Sun. Table 71.1 summarizes some of these properties.

Division of all stars into two broad categories is an oversimplification. For example, the Sun's age does not fit precisely into either category, but it is considered a Population I star because of its heavy element content and approximately circular orbit in the disk. To avoid forcing a star into a category in which it does not fit, astronomers sometimes subdivide Pop I and Pop II into "extreme" and "intermediate" Pop I and Pop II—the extremes being the youngest and oldest of Pop I and II, respectively, while the intermediate classes fall in between. Also, some astronomers use the term "old-disk population" to describe stars like the Sun that are located in

TABLE 71.1	Properties of Population I and II Stars	
Property	**Pop I**	**Pop II**
Age	Young (< few billion years)	Old (> 10 billion years)
Color	Blue (overall)	Red
Location	Disk and concentrated in arms	Halo and bulge
Orbit	Approximately circular in disk	Plunging through disk on approximately elliptical orbit
Heavy element content	High (similar to Sun)	Low (10^{-2} to 10^{-3} Sun's)

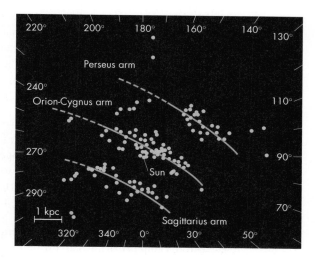

FIGURE 71.2

Young star clusters and HII regions show parts of the spiral arms in the Sun's neighborhood. The arms are given names based on constellations in which parts of them appear bright to us.

the disk but are not so young. In short, some stars exhibit properties intermediate between Pop I and Pop II stars. Nevertheless, the fairly sharp division in properties between the two population categories seems to relate to major changes in the structure of the Milky Way during its history. Many astronomers suspect that the two different populations indicate that star formation has not occurred continuously in the Milky Way. Population II stars probably formed in a major burst at the galaxy's birth during its initial collapse, whereas Population I stars formed much later and continue forming today. This hypothesis explains the differences between the two populations, as we will discuss in greater detail.

One of the first uses astronomers made of stellar population differences was to map the Milky Way's spiral arms. As Baade noted, the spiral arms of other galaxies gleam with the blue light of their Pop I stars. The blue color of Pop I is produced primarily by the luminous stars with spectral types O and B (Unit 55) that use up their fuel rapidly, "burning out" in just tens of millions of years. Therefore, by measuring the location of O and B stars near the Sun, astronomers can make a picture of the spiral arms in the Milky Way. This method is limited, however, because dust in space prevents us from seeing even the most brilliant O and B stars if they are farther from the Sun than about 3 kpc. Nevertheless, maps such as that shown in Figure 71.2 provided some of the first direct evidence that we live in a spiral galaxy.

The stars in open clusters and globular clusters (Unit 69) also exhibit the differences between Pop I and II. Open clusters are almost all Pop I, located within the Milky Way's disk, whereas globular clusters are always Pop II and orbit in the Milky

Way's halo. This difference in their stellar populations makes clusters especially useful to astronomers for studying the structure of the Milky Way: Globular clusters outline the halo and bulge, while young, open clusters trace the galaxy's arms.

71.2   FORMATION OF OUR GALAXY

The structure of the Solar System also provides clues to its origin (Unit 33). Its disklike shape, the common age of the planets, and the existence of two different families of planets allowed astronomers to develop the solar nebula theory.

One of the major unsolved problems in astronomy is how galaxies form. The process is presumably a large-scale version of star formation (Unit 60). That is, a gas cloud collapses under the influence of gravity and breaks up into stars, but this does not explain why the galaxy contains two main groups of stars, Population I and II, that differ so greatly in their properties. The British astronomer Donald Lynden-Bell and the U.S. astronomers Olin J. Eggen and Allan R. Sandage proposed in 1961 a two-stage collapse model to explain the birth of the Milky Way.

In the two-stage collapse model, our galaxy began as a vast, slowly rotating gas cloud, perhaps a few million light-years (a million parsecs) in diameter, containing several hundred billion solar masses of gas. The cloud was composed of almost pure hydrogen and helium. As gravity began shrinking this immense cloud, clumps of gas within it grew in density and gradually turned into stars, as illustrated in Figure 71.3A.

These first stars would have had a composition unlike stars today. The heavy elements are produced by nuclear fusion inside stars, so the first stars would have been composed of essentially pure hydrogen and helium. According to some computer simulations, they may have been exceptionally massive compared to stars forming today. Because of their greater mass, they evolved quickly and blew up as supernovae (Units 65 and 66), which added the first heavy elements to the gas cloud even before it had finished collapsing.

It may seem odd that stars could form and die before the cloud's collapse was complete. But the collapse of such an immense cloud takes hundreds of millions of years, whereas massive stars can evolve and die in less than a few tens of millions of years.

As the cloud's collapse continued, the forming stars and globular clusters now contained at least a few heavy elements. Formed from gas falling in from all directions, the resulting inward trajectories of these bodies would have given them

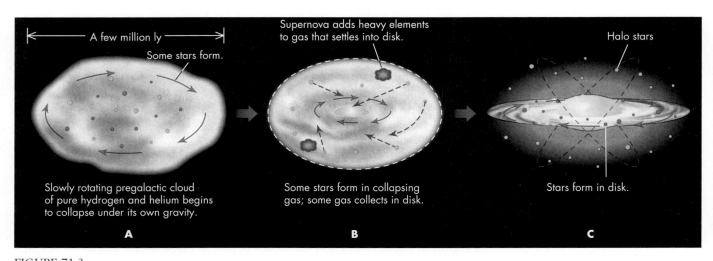

FIGURE 71.3

(A) Birth of the Milky Way as a gas cloud collapses. A first generation of stars (Population II) forms. (B) Collapse continues with more Population II stars forming and some exploding as supernovae. Heavy elements created by the stars exploding as supernovae are mixed into the gas. Gas not used in making stars settles into a rotating disk. (C) Population I stars form in the disk.

Q If you randomly selected two Population I stars, one red and one blue, which would you expect to have higher metallicity?

highly elliptical orbits. Hence their orbits continue to carry them in and out of the inner part of the galaxy, distributed all around the galaxy in the halo. These are the Population II stars that we see today.

Star formation during this initial stage did not turn all of the collapsing gas into stars. The remaining clouds of gas continued to collapse inward. These clouds collided with each other, losing orbital energy in the process and settling into a smaller rotating disk of gas, as sketched in Figure 71.3B. That disk gas, now rich in heavy elements from the death of Pop II stars, then began to form the first Pop I stars. These stars were born with the motion of their parent gas and continued circling the center of the young galaxy. Pop I stars, therefore, contain more heavy elements than Pop II stars and move along approximately circular orbits lying in the disk, as illustrated in Figure 71.3C.

Computer simulations show that the collapse model can explain many of the features of the galaxy. For example, Figure 71.4 shows that if an unstructured region of gas of the appropriate size and rotation speed collapses, it will form a thin disk with spiral arms, surrounded by a halo.

A few features of the Milky Way call for some refinements to the two-stage collapse model we have just described. First, according to the model, all Population II stars should be about the same age, having formed during the relatively brief period of the galaxy's initial collapse. But observations of Pop II stars show that they formed over a significantly longer time span than the model predicts. Second,

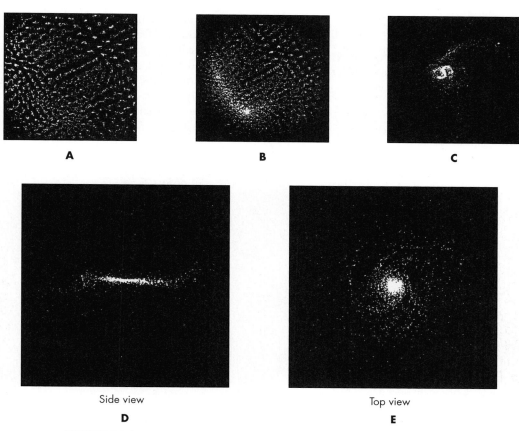

| A | B | C |

Side view   Top view
D   **E**

FIGURE 71.4

Computer simulation of the birth of a galaxy similar to the Milky Way. The top three frames (A), (B), and (C) show the initial state of the gas and the development of a clump that becomes the galaxy, illustrated in the bottom two frames (D) and (E). Notice the thin disk and spiral arms.

studies of the ages and distribution of stars in the disk and halo show some complex patterns that do not obviously result from the collapse model. For example, there appear to have been episodes of intense star formation, and some streams of stars follow unusual orbits within the galaxy. Third, the model predicts that some of the very first generation of stars—those with virtually no heavy elements at all—would have had such low masses that they would have lived for many billions of years. In fact, some should have survived to the present, but no such stars have ever been observed.

A hypothesis for explaining some of these anomalies is that the Milky Way has collided with other galaxies throughout its history. Galaxy collisions are unlike the collision of two solid bodies. Because of the large spaces between stars, if two galaxies collide, individual stars will almost never strike each other, although the gas and dust between the stars *will*. The compression of the interstellar medium in both galaxies may lead to a huge burst of star formation. Computer modeling of such collisions suggests that the two galaxies often merge into a single larger system, a process that astronomers call **galactic cannibalism.**

We see evidence that some small galaxies are merging with the Milky Way even now. For example, astronomers recently have detected a small "dwarf" galaxy on the far side of the Milky Way. The dwarf galaxy is elongated, probably because the Milky Way's gravity has stretched it and is pulling it apart. Indeed, a long stream of stars within the Milky Way's halo appears to have been pulled from the smaller galaxy (Figure 71.5). Other streams of stars seen in the Milky Way may be all that remains of other smaller galaxies that have been pulled apart and merged with our galaxy. Astronomers conclude that large galaxies like the Milky Way may have begun life with the collapse of a single gas cloud, but they continue to grow by swallowing smaller galaxies.

This description still leaves unanswered the question of why we see no stars formed from pure hydrogen and helium. We should expect such a composition if our galaxy formed from a cloud in which no stars had yet made heavier elements. All stars so far found in the Milky Way contain at least a small percentage of heavy elements (typically at least one thousandth the amount seen in the Sun), although two stars have recently been observed with a mere 3×10^{-6} times the heavy element abundance of the Sun.

Q How might our galaxy be different if stars formed at a significantly faster rate? At a significantly slower rate?

FIGURE 71.5
Astronomers have mapped out a stream of stars that were pulled out of a small galaxy in its encounter with the Milky Way. Based on the estimated distances of the stars in the stream, the figure shows what the stream of stars (colored red) might look like from outside the Milky Way (shown as the blue spiral). The small galaxy is in the process of being swallowed up by the Milky Way.

Astronomers have devised several explanations for why they have so far seen no stars composed purely of hydrogen and helium—which are termed **Population III** stars. For example, the gas may have been too hot or turbulent to make small stars that could have lived long enough for us to see them today. Alternatively, Pop III stars may have had their surface layers contaminated by gas ejected at the death of their more massive brethren. For example, the two stars just noted are estimated to be about 13 billion years old. Only stars less massive than the Sun can survive so long (Unit 61), and such low-mass stars take much longer to contract and form than more massive stars forming out of the same collapsing cloud (Unit 60). Thus, Pop III stars may exist but are masquerading as Pop II stars because they contain heavy elements that were not part of their original makeup. Finally, such pure hydrogen and helium stars may simply be very rare and therefore as yet undiscovered.

71.3   THE AGE AND FUTURE OF THE MILKY WAY

We can use the oldest stars in the Milky Way to estimate its age. Using the techniques to measure stellar ages in star clusters (Unit 69), astronomers calculate that our galaxy's most ancient stars are about 13 billion years old, a number we will take as the approximate age of our galaxy.

As stars formed, the amount of gas remaining was depleted, so the rate of star formation is unlikely to have remained fixed. Dying stars return a large fraction of their material to space as planetary nebula shells, supernova remnants, and stellar winds; but with each generation of stars, more and more of the galaxy's mass is locked away from further use in stellar remnants such as white dwarfs, neutron stars, and black holes (Unit 59). A larger fraction remains tied up in the very low-mass stars that can live even longer than the current age of the galaxy.

Throughout this process, the interstellar gas becomes steadily more enriched with heavy elements so that later generations of stars have a greater percentage of **metals,** which for an astronomer means any element heavier than helium. At the time of our galaxy's birth, the birthrate was probably very high and has since declined steadily. The enrichment of metals in the Sun suggests that it is about a "fifth generation" star, in the sense that its enrichment of metals would require that the gas had cycled through about five stars before forming the Sun and Solar System.

By dividing the number of stars in the Milky Way by its age, astronomers estimate that, on average, about 10 stars have been born each year. A current star formation rate of a few stars per year is consistent with studies of star-forming clouds today. This seemingly small number of new star births will be sufficient to keep our galaxy shining for many billions of years into the future. At current rates the Milky Way may run out of material for making new stars about 10 billion years from now. The lowest-mass stars will continue to shine for several hundred billion years, and the galaxy will grow dimmer and redder. In about a trillion years even these low-mass stars will have died, and all that will remain are the dark remnants of former stars. The Milky Way will fade, slowly spinning in space, a dark disk of stellar cinders.

The future will not be entirely quiet. Large merger events can be anticipated. Astronomers predict that in a few billion years two fairly large satellite galaxies—the Large and Small Magellanic Clouds—will spiral into our galaxy and will eventually merge with it. When they do, they will bring fresh gas and will probably initiate a period of strong star formation. In about 10 billion years, the Milky Way will probably collide with a nearby galaxy larger than our own, the Andromeda Galaxy (Unit 76), and eventually merge with it. And so this cannibalism will continue, the galaxies consuming one another even as their stars fade and die.

Q When the Milky Way is very old, the interstellar gas may contain little hydrogen. In what ways might this affect the evolution of stars?

KEY TERMS

galactic cannibalism, 582

metals, 583

Population I (Pop I), 577

Population II (Pop II), 577

Population III (Pop III), 583

QUESTIONS FOR REVIEW

1. What are some differences between Pop I and Pop II stars?

2. Describe one model for the origin of the Milky Way. How does this model explain the difference between Pop I and Pop II stars?

3. What are Pop III stars?

4. What is the eventual fate of the Milky Way?

PROBLEMS

1. Conservation of angular momentum (Unit 20) requires that the product of the speed at which an object is orbiting times the distance from the point around which it is orbiting remain constant. If the Sun is currently orbiting at 220 km/sec, 8 kpc from the center of the Milky Way, how fast did the material that formed the Sun move if the gas began 400 kpc distant—about halfway to the next nearest large galaxy?

2. Suppose that after X billion years, the Milky Way converted half of its gas into stars. In the next X billion years, it converted half of the remaining gas into stars—and so on, each X billion years. The Milky Way is estimated to be about 13 billion years old, and only about 15% of its disk remains today in the form of gas.
   a. What value of X would give the Milky Way's current gas fraction? (*Hint:* You can calculate this mathematically, or you can draw an approximate graph of the declining gas fraction and estimate X.)
   b. In how many billion years will the gas fraction be just 1%?

3. Suppose we found a nearby Pop II star that is on a highly elliptical orbit, but it is currently near its most distant point from the center of the Milky Way in its orbit. To stars like the Sun in orbit in the disk, what would be the approximate relative speed of this Pop II star? In what direction?

TEST YOURSELF

1. A young blue star moving along a circular orbit in the disk is
   a. a Pop I star.
   b. a Pop II star.
   c. a Pop III star.
   d. not observable from Earth.
   e. None of the above

2. Which of the following may explain the failure to observe Population III stars?
   a. We do not expect to observe Population III stars because pure hydrogen and helium stars never formed.
   b. Population III stars had their surfaces contaminated by remnants of massive stars.
   c. The first stars to be formed in the young Milky Way were very massive and therefore short-lived.
   d. All of the above
   e. Only (b) and (c) are acceptable explanations.

3. The greater number of heavy elements seen in the spectra of Pop I stars relative to Pop II stars is explained by the fact that Pop I stars
   a. are older, and have therefore fused more hydrogen into heavy elements.
   b. formed more recently, and have therefore been made from enriched interstellar gas.
   c. have planetary systems, and debris from these fall onto the star's surface.
   d. are colder and therefore exhibit strong "metal" lines.

72

Gas and Dust in the Milky Way

Ultraviolet
radiation

Hot stars

Background Pathways:
Units 24, 70

The space between stars is not empty but contains gas and dust particles that comprise what astronomers call the **interstellar medium** or **ISM**. By terrestrial standards, this space is almost a perfect vacuum. On the average, each liter contains only a few thousand atoms of gas. For comparison, the air we breathe contains about 10^{22} atoms per liter. The density of gas in interstellar space compared to the Earth's air is like having one marble in a box 8 kilometers (5 miles) on a side compared to the same box being filled completely with marbles.

Despite the interstellar medium's very low density, it plays several important roles in the Milky Way. First, it is the material from which new stars form. With no interstellar gas, star birth would stop. Second, we wouldn't exist if there had been no interstellar dust, because the atoms that comprise the Earth and our bodies came primarily from such material.

One property of interstellar matter is that it limits what we can observe within the Milky Way. Interstellar dust blocks our view of distant parts of our galaxy and impedes our ability to accurately measure the size of the Milky Way and distances to the stars within it. Before we can understand how astronomers overcome these observational difficulties, it will help to look more closely at the nature of interstellar matter.

72.1 PROPERTIES OF THE INTERSTELLAR MEDIUM

Interstellar matter is not spread smoothly through the Milky Way. Gravity pulls most of it into a thin layer in the disk, as can be seen by the dark band of dust in other galaxies (Figure 72.1). Within this layer, gravity and gas pressure further

FIGURE 72.1
Interstellar matter, outlined by dark dust clouds, is concentrated in the plane of a galaxy, forming a dark band across it. This galaxy is sometimes called the "Sombrero galaxy." Technically, it is called either NGC 4595 or M104.

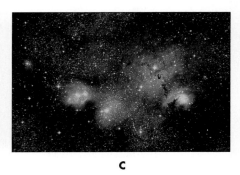

A   B   C

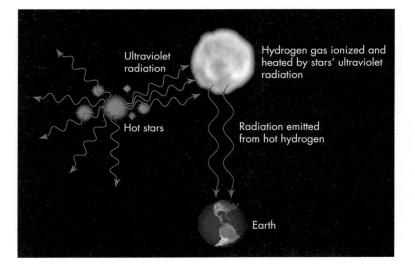

Ultraviolet radiation

Hydrogen gas ionized and heated by stars' ultraviolet radiation

Hot stars

Radiation emitted from hot hydrogen

Earth

D

FIGURE 72.2

(A), (B), (C) Emission nebulae. Notice the bright (and hot) stars near the glowing gas. (D) These stars emit ultraviolet radiation, which heats and ionizes the nearby hydrogen gas. Electrons dropping from hydrogen's third to second energy level generate the pink light.

Stars' orbits may in fact be scattered into new directions when they make close gravitational encounters with other massive objects in the galaxy, so their orbits tend to spread out rather than settle into the plane.

Pink dots in the photographs of other galaxies are usually emission nebulae as well.

Q Under what circumstances could an interstellar cloud include all three types—emission, reflection, and dark nebulae?

clump the gas and dust into clouds, which are embedded in a very low-density background gas.

Interstellar matter behaves quite differently from stars as it orbits within the galaxy because the gas and dust particles fill space and collide with each other—losing energy and settling down into the common orbital plane. Stars, by contrast, are so dense relative to the gas that they can pass through the galactic plane, which offers too little "friction" to slow them down significantly.

Sometimes the gas in an interstellar cloud is hot and emits visible light, as you can see in Figure 72.2. Such a glowing cloud is called an **emission nebula** (*nebula* is the Latin word for "cloud"). The power source causing the gas to light up is usually young, luminous O or B stars. They produce ultraviolet radiation that can ionize the hydrogen and other atoms in these clouds. An example of such a gas cloud is the Orion Nebula, which is easy to see with a small telescope or even a pair of binoculars (see Looking Up #6: Orion). If you look at the middle star in Orion's sword you will see a pale, fuzzy glow created by luminous interstellar gas surrounding a dense star cluster.

Sometimes the gas in an interstellar cloud is not hot enough to emit visible light of its own, but dust particles in it may reflect light from a nearby star, in which case it is called a **reflection nebula** (Figure 72.3). The reflection effect is something like the beams of light you see coming from the headlights of a car as it drives through fog.

If a cold interstellar cloud is sufficiently dense, the dust in it may block the light of background stars and appear as a **dark nebula** against the starry background (Figure 72.4). Clouds like these are too cold to emit visible light, but they emit lower-energy radiation at infrared or radio wavelengths. Several dark nebulae are also visible against the emission nebulae seen in Figure 72.2. Dark nebulae are the starting point for the formation of the next generation of stars.

FIGURE 72.3

Reflection nebulae. (A) Light from a star near a dust cloud is reflected from the dust, making it visible. (B) The Pleiades star cluster. Note the bluish reflection nebula around many of the brighter stars.

FIGURE 72.4

A dark nebula in the Milky Way. Dust blocks our view of the background stars. The cluster in NGC 6520. The dark nebula is Barnard 86. A few stars lie between us and the nebula.

Radiation is both emitted and absorbed by gas in an interstellar cloud, giving astronomers a way to measure many of the cloud's properties. An atom or molecule will absorb the specific wavelengths of light that correspond to energy-level changes of its electrons (Unit 24). Therefore, when astronomers observe the spectrum of a star behind an interstellar cloud, they see the dark lines of the interstellar gas's absorption superimposed on the star's own spectral signature (Figure 72.5). On the other hand, if the gas has been heated, it produces emission lines. By studying the absorption and emission lines of interstellar gas, astronomers can deduce the composition, temperature, density, and motion of the cloud. Such spectra show that interstellar gas has a composition similar to that of the Sun and other stars—about 71% hydrogen and 27% helium, with the remainder made up of heavier elements.

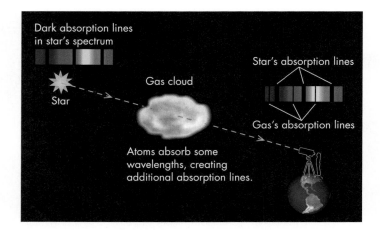

FIGURE 72.5

Gas atoms in an interstellar cloud between us and a distant star absorb some of the star's light, adding their own absorption lines to those in the star's spectrum.

Dust particles also absorb and emit light. But because dust particles are solids, they do not generate clear spectral signatures as a gas does. Still, astronomers can determine that dust particles are probably made of silicates and carbon compounds, coated with a thin layer of ices made of molecules such as water, carbon monoxide, and methyl alcohol. Based on the way they interact with light, these particles must be extremely tiny, ranging from about a micrometer to a nanometer or so in diameter (not much larger than some large molecules).

Although interstellar clouds are alike in composition, they differ greatly in size, ranging from a fraction of a light-year to a few hundred light-years in diameter. They also range widely in mass from about one to hundreds of thousands of solar masses. Finally, they also have a wide range of density and temperature. For example, some regions have temperatures of millions Kelvin while others are only a few degrees above absolute zero.

The complexity of the interstellar medium is a little like "weather." If we could watch a high-speed movie of interstellar clouds, we would see that they constantly form and dissolve, driven by powerful physical phenomena such as stellar winds and supernova explosions. Although matter in interstellar space undergoes huge changes in temperature and density, this occurs on timescales of millions of years. Our observations of interstellar clouds reveal a momentary snapshot of a dynamic and turbulent medium.

72.2   HEATING AND COOLING IN THE ISM

The temperature of any object is set by a balance between the energy it receives (heating) and the energy it loses (cooling), and interstellar clouds are no exception. Gas can be heated in several ways. Two especially important heat sources are radiation and winds from hot stars, and blasts and radiation from supernovae. These processes add energy to the cloud's gases. Atoms and molecules are physically set in motion, and the high-energy photons from hot stars or explosions give atoms so much energy that they are ionized and eject electrons at high speed. If we could watch microscopically, we would see this energy translated into the random motions of thermal energy as electrons, atoms, and molecules collide with one another and rebound in different directions.

Heating of interstellar gas is strongly affected by the distance from the source and by shielding. Greater distances spread out the impact of photons, winds, and blast waves—the energy being spread over a larger and larger surface area with distance. **Shielding** occurs wherever material near a star "casts a shadow" that blocks energy from reaching interstellar clouds that are farther away. Dust is a particularly important source of shielding. Just as dust blocks the light of some stars from us, it can shield material behind it.

Cooling occurs when the gas emits radiation of its own. Radiation often results from collisions between gas atoms or molecules. To see how the thermal energy generated by a collision is turned into radiation energy, we look again at the interactions microscopically. If two molecules collide, part of the kinetic energy of their collision may be absorbed by electrons being knocked into higher energy levels. Energy is conserved because the two molecules rebound from each other with slightly lower speeds than they collided, reducing the thermal energy. The electrons later drop down to lower energy levels, emitting one or more photons.

The net effect of collisions between molecules is to transfer thermal energy into radiation energy. Indeed, the radiation that allows us to see a cloud is also the means by which the cloud cools. Because collisions are more frequent where the atoms or molecules are closer together—that is, where interstellar matter is denser—high-density regions tend to cool more rapidly. This is especially interesting because cool high-density regions are also where gravitational forces are largest, so these regions tend to cool even more as they grow denser.

Dust grains also heat up when they absorb radiation or undergo collisions. The grains cool by emitting thermal radiation, generally at infrared wavelengths for the typical temperatures inside interstellar clouds (Unit 23). In dense clouds containing lots of dust, the combination of high density and shielding allows the gas and dust deep in the cloud to cool down to only a few degrees Kelvin (about $-270°C$ or $-450°F$).

Applying these ideas, we find that near hot, blue stars (spectral type O or B, Unit 55), gas and dust will itself be hot, typically about 10,000 Kelvin. Such temperatures may even be hot enough to vaporize the dust grains or boil off some of their surface molecules. O and B stars are especially effective at heating nearby material because they emit large amounts of ultraviolet radiation, which has short wavelengths that make it much more energetic than visible light. Ultraviolet light with a wavelength less than 91.2 nm is so energetic that when it is absorbed by hydrogen, it tears the electron free of the nucleus, ionizing the gas. The free electrons collide with atoms of oxygen, nitrogen, and other elements and excite them, making them emit light of their own. Eventually the electrons recombine with the ionized hydrogen atoms and emit more radiation as they drop down from higher-energy orbitals down to lower-energy orbitals. Visible-wavelength photons are produced when the electrons drop down to the second energy level from higher levels, producing the pink light of hydrogen Balmer lines (Unit 55.4). This gives these nebulae their characteristic pink color, as seen in Figure 72.2. Because hydrogen in these hot gas clouds is ionized, they are sometimes called **Hɪɪ regions** (pronounced H-two)—the H denoting the hydrogen and the ɪɪ indicating that it is ionized.

The free electrons in Hɪɪ regions are themselves a powerful source of radio emission. Astronomers can detect that emission and thus see, in radio wavelengths, Hɪɪ regions that are otherwise hidden from us by dust in the interstellar medium. Because these Hɪɪ regions are generally located in spiral arms, radio maps of their location help to reveal the spiral structure of our galaxy. In fact, maps made by combining radio and optical observations of Hɪɪ regions provide some of the best maps we have of the Milky Way's spiral structure (Figure 72.6).

72.3   INTERSTELLAR DUST: DIMMING AND REDDENING

Although only about 1% of interstellar matter is in the form of dust grains, the effects of dust are strong. Not only does it dim the light of distant stars, it also alters the light's color, much as haze in our atmosphere dims and reddens the setting Sun. These effects occur because visible light interacts strongly with dust.

When a photon strikes a dust particle, it might be absorbed or reflected in a random direction in a process called **scattering.** Even if there is no absorption,

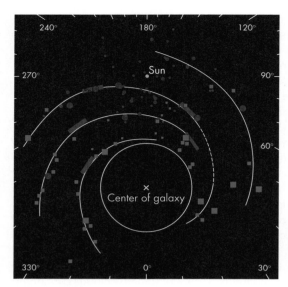

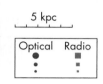

FIGURE 72.6

Map of the Milky Way made by combining radio observations (radio can penetrate dust and thereby allow us to see all the way across the Milky Way's disk) and optical observations of the position of HII regions. Such regions outline the spiral arms of other galaxies. In our galaxy, too, they form a spiral pattern.

Q Suppose we observe a main-sequence star behind a thin dust cloud, so its light is slightly reddened and dimmed. If we used its color to predict its luminosity, would we tend to overestimate it or underestimate it? What if we tried to use its brightness to estimate its distance?

scattering dims a star's light because some of the radiation heading toward us is sent off in random directions and is lost to our sight. A thin cloud of dust will scatter just a fraction of the light headed in our direction, which we observe as a reflection nebula (Figure 72.3). However, if a dust cloud is extremely thick or dense, the photons are scattered repeatedly and their energy is eventually absorbed by the dust grains. This may dim the light of background stars to the point where they become invisible in a dark nebula (Figure 72.4).

A dust particle has the strongest scattering effect on electromagnetic waves whose wavelength is close to or slightly smaller than the size of the particle. If the dust particle is much smaller than the wavelength of radiation, it is much less effective at scattering the radiation. An analogy to this is that a boat will block short-wavelength ocean waves, which "slap" against the side of the boat, while it does little to affect long-wavelength waves, which raise and lower the boat and continue past it.

A sunny day shows the effect that small particles in our atmosphere have on sunlight. Here oxygen and nitrogen molecules scatter short-wavelength blue photons of sunlight much more strongly than longer-wavelength red photons. The sky is not blue because the molecules in the atmosphere are glowing with this color. Instead the blue photons we see are part of the mix of all wavelengths coming from the Sun. Because of the size of Earth's atmospheric molecules, the Sun's blue wavelengths are scattered in different directions while other colors (wavelengths) pass through to the surface. This process is similar to the effect of small dust grains in interstellar space, and this is why reflection nebulae usually have a bluish color (Figure 72.3).

Because shorter-wavelength photons are scattered in random directions, the balance of the remaining light contains a relatively larger number of longer-wavelength photons. This effect is called **reddening** because the color balance shifts toward the red end of the visible spectrum. It might more accurately be called "de-bluing" because no red color is added to the light, but instead the bluer colors are subtracted by scattering. The effect of reddening by dust in the Earth's atmosphere can make the Sun look yellow, orange, or even red as more of the short-wavelength photons are scattered. The effect is strongest at sunset when the sunlight reaching us has traveled a long path through our atmosphere, so that more of the short-wavelength photons are scattered away from their path toward us.

Small interstellar dust grains are also very effective at scattering short-wavelength blue and ultraviolet light. As a result, when we look at a star through a cloud containing dust, the light appears reddened. Light is scattered and absorbed at all wavelengths to some degree, so there is an overall dimming, or **extinction,** of the light as well. If the cloud is thick enough, the extinction may be so strong that little or no light of any wavelength will pass through it, and the cloud will be seen as a dark nebula. Both reddening and extinction can be seen around the edges of the dark nebula shown in Figure 72.7.

The extinction effects of dust on visible wavelengths of light make distant parts of our galaxy hard to study. However, the longer wavelengths of infrared and radio waves are relatively unaffected because they are much larger than the dust particles. As a result, infrared pictures reveal the structure of the Milky Way far better than visible-light pictures can, as shown in Figure 72.8. The disk and central bulge of our galaxy are easy to see in this wide-angle computer image of stars mapped by infrared cameras.

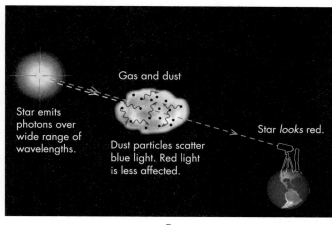

A   **B**

FIGURE 72.7

(A) A string of dark clouds called the "Snake Nebula" blots out the light from background stars. Short-wavelength light is blocked more strongly than long wavelengths, so near the edge of the clouds, the background stars are reddened. (B) Dust in an interstellar cloud scatters the blue light from a distant star, removing it from the radiation reaching Earth and making the star look redder than it really is.

FIGURE 72.8

An all sky view of our galaxy, as we see it from Earth. The image was made by plotting on a map of the whole sky the millions of stars observed in the infrared by the 2-Micron All Sky Survey (2MASS). It is therefore like a map of the Earth, flattened out to show all continents. The map clearly reveals the flatness of our galaxy and the bulge of its central regions.

72.4   RADIO WAVES FROM COLD INTERSTELLAR GAS

Unless an interstellar cloud is within a few parsecs of a hot star, it may cool down to just a few degrees Kelvin. Such cold material emits no visible light because it has too little energy to generate visible photons. However, even at such low temperatures it may still emit the low-energy, long-wavelength radiation that we can detect with radio telescopes.

One of the most important sources of radio emission is cold hydrogen atoms, which radiate at a wavelength of 21 centimeters. This radiation is called **HI** (pronounced H-one) **emission** because cold hydrogen atoms with their electrons attached are called HI to distinguish them from ionized hydrogen atoms, HII.

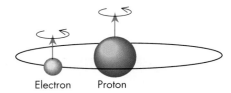

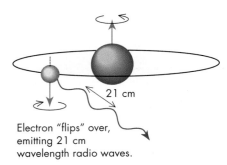

Electron "flips" over,
emitting 21 cm
wavelength radio waves.

FIGURE 72.9
Radio radiation at 21 centimeters is
emitted by cold hydrogen as an electron
"flips" and spins in the opposite direction.

Q Why are H, C, N, and O the most
common constituents of interstellar
molecules?

H I emission arises because subatomic particles such as protons and electrons possess a property, known as *spin*, that makes them behave like tiny magnets. The energy of the magnetic interaction between the proton and electron is a tiny bit higher in the hydrogen atom when the proton and electron spin in the same direction. If the electron flips its spin direction, as shown in Figure 72.9, the energy is slightly lower. The atom emits the energy difference as a radio wave with a wavelength of 21 centimeters.

This **21-centimeter radiation** has proved extremely valuable for studying the Milky Way and other galaxies. First, this type of radiation is not absorbed by interstellar dust. Second, hydrogen is abundant in space, so the signal of 21-centimeter radiation is strong. From the strength of this signal, astronomers can deduce the amount of hydrogen in a region. If the gas is moving, the wavelength will be Doppler-shifted (Unit 25), from which the gas's motion toward or away from us can be determined. Radio observations thereby allow astronomers to map not only where the gas is concentrated in the Milky Way but also how it is moving. Astronomers still face complications in interpreting these measurements because the 21-cm radiation coming from a particular direction in the Milky Way is a blend of the signals from nearby and more distant gas along our line of sight. However, by modeling known motions in the galaxy, astronomers can estimate the distances and produce maps of the entire Milky Way at the 21-cm wavelength. Such maps show that H I gas is confined to a thin disk, and its distribution is suggestive of a spiral pattern, confirming the picture of our galaxy deduced by other means.

In addition to 21-centimeter radiation, many other wavelengths of radio radiation are emitted by interstellar gas. In the densest and coldest clouds, molecules form. These regions are known as **molecular clouds,** and molecules within them emit primarily at radio wavelengths. More than 100 interstellar molecules have been identified so far, including common compounds such as molecular hydrogen (H_2), carbon monoxide (CO), formaldehyde (HCHO), and ethyl alcohol (CH_3CH_2OH), along with even more complex molecules. Table 72.1 lists a number of the more common molecules detected.

Collisions between molecules may make them spin or vibrate, which are forms of energy that they can emit at a variety of radio wavelengths. For example, carbon monoxide is a common interstellar molecule, and it can emit at wavelengths of 2.6 mm, 1.3 mm, and a series of shorter wavelengths as its spin slows (Figure 72.10). Astronomers can analyze these other waves, just as they do 21-cm radiation, to learn about gas in cold dense clouds. In fact these clouds do not usually emit much 21-cm radiation because their atomic hydrogen has combined to form molecular hydrogen. The observations of molecules therefore provide information about the densest interstellar environments, which complements the 21-cm observations.

TABLE 72.1	Some Interstellar Molecules
Molecule	**Chemical Formula**
Molecular hydrogen	H_2
Hydroxyl radical	OH
Cyanogen radical	CN
Carbon monoxide	CO
Water	H_2O
Hydrogen cyanide	HCN
Ammonia	NH_3
Formaldehyde	HCOH
Acetylene	HCCH
Formic acid	HCOOH
Methyl alcohol (methanol)	CH_3OH
Ethyl alcohol (ethanol)	CH_3CH_2OH

FIGURE 72.10
Emission of radio waves by spinning molecules in cold interstellar clouds. Collisions set the molecules spinning. When their rotation speed changes, they emit radio waves. The wavelength emitted depends on the kind of molecule and by how much its rotation speed changes.

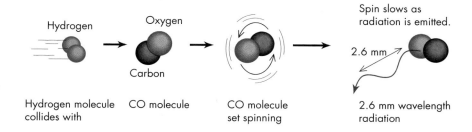

Hydrogen

Oxygen

Spin slows as radiation is emitted.

Carbon

2.6 mm

Hydrogen molecule collides with

CO molecule

CO molecule set spinning

2.6 mm wavelength radiation

KEY TERMS

dark nebula, 586

emission nebula, 586

extinction, 590

H I emission, 591

H II region, 589

interstellar medium, 585

ISM, 585

molecular cloud, 592

reddening, 590

reflection nebula, 586

scattering, 589

shielding, 588

21-centimeter radiation, 592

QUESTIONS FOR REVIEW

1. How do we know interstellar matter exists?

2. How do we know that some interstellar matter is dust?

3. How does interstellar dust affect our observations of stars and the Milky Way?

4. What evidence makes astronomers believe the Milky Way has spiral arms?

PROBLEMS

1. Convert the following radio wavelengths to frequencies (see Unit 22):
   a. 21-cm line of atomic hydrogen.
   b. 2.6-mm line of carbon monoxide.

2. Find the ratio of the energy of a 21-cm photon to the energy of an electron shifting from the second energy level of hydrogen to the ground state, which produces a photon of 122-nm wavelength.

3. The dust in an interstellar cloud blocks blue light in the following way: For every 1 parsec light travels through the cloud, only 90% of the light continues. Thus after 2 pc, 90% of the remaining 90%, or 81% ($0.9 \times 0.9 = 0.81$), remains.
   a. After 10 pc, you might initially have expected ten 10% reductions to have removed all of the blue light. How much blue light actually remains?
   b. The same cloud removes about 7% of the red light every parsec, so after 1 pc, the ratio of blue to red has dropped to $90/93 \approx 97\%$ of its unreddened value. What is the ratio of blue to red after 5 pc? After 10 pc?

TEST YOURSELF

1. Astronomers know that interstellar matter exists because
   a. they can see it in dark clouds and clouds that absorb light.
   b. the matter creates narrow absorption lines in the spectra of some stars.
   c. they can detect radio waves coming from atoms and molecules in the cold gas.
   d. spacecraft have sampled clouds near Orion.
   e. All of the above except (d)

2. What is the source of energy that makes emission nebulae glow?
   a. Heat from hot white dwarfs within the nebula
   b. Nearby hot O and B type stars
   c. Nearby black holes
   d. Intense radio waves
   e. Gravitational compression

3. The typical size of the interstellar dust particles is _____ and they consist of _____.
   a. 1 cm; silicates and carbon compounds
   b. 1 mm; hydrogen and helium
   c. about a micrometer or less; silicates and carbon compounds
   d. About a nanometer or less; hydrogen and helium

73

Mass and Motions in the Milky Way

Background Pathways: Unit 70

Astronomers have measured that our Sun and neighboring stars move around the center of the Milky Way at a speed of about 220 kilometers per second. The Sun's large orbital speed (more than seven times the speed of the Earth around the Sun) may make you think the Milky Way spins rapidly; but because our galaxy is so huge, the Sun takes approximately 220 million years to complete one trip around it. Since the extinction of the dinosaurs about 65 million years ago, our galaxy has made less than one-third of a revolution.

Were it not for the collective gravity of all the components of the Milky Way, the motions of its stars and gas would cause the galaxy to fly apart and disperse within a few billion years. At the same time, were it not for the motion, gravity would draw all the stars and gas inward, causing the galaxy to collapse. The motion and gravity are in balance, and astronomers can use this balance to determine our galaxy's mass by studying the motions of the stars within it.

73.1 MASS OF THE MILKY WAY AND THE NUMBER OF ITS STARS

The rotation of matter in a flattened disk around the galaxy's center is somewhat similar to the orbital motion of the planets around the Sun. Unlike a spinning solid disk, such as a wheel or a Frisbee, different parts of the Milky Way complete their orbits in different amounts of time. This is similar to how planets near the Sun complete their orbits faster than planets farther out—so too do stars near the center of the galaxy complete their orbits faster than stars at the edge of the galaxy. The differing times taken by the inner and outer parts of the galaxy to complete an orbit are a phenomenon called **differential rotation.**

We can describe the unique properties of the Milky Way's rotation by observing stars at a variety of distances from the galactic center to obtain a relation between orbital velocity and radius. Astronomers call this relation a **rotation curve.** Rotation curves for a solid disk and the Solar System are illustrated in Figure 73.1A and B. When we measure the speeds of stars rotating around the center of the Milky Way, though, we see neither a steadily rising rotation speed nor a steadily declining one. The stars orbit at about the same speed at most radii (Figure 73.1C).

The Milky Way's rotation curve reflects the distribution of mass in our galaxy. It is very different from the distribution of mass in the Solar System, where more than 99% of the mass is concentrated in the central object (the Sun). A planet orbiting the Sun is subject to the gravitational pull essentially of just that one object. In the Milky Way, on the other hand, an orbiting star is held in its orbit by the collective gravitational pull of all the stars and other matter lying inside its orbit.

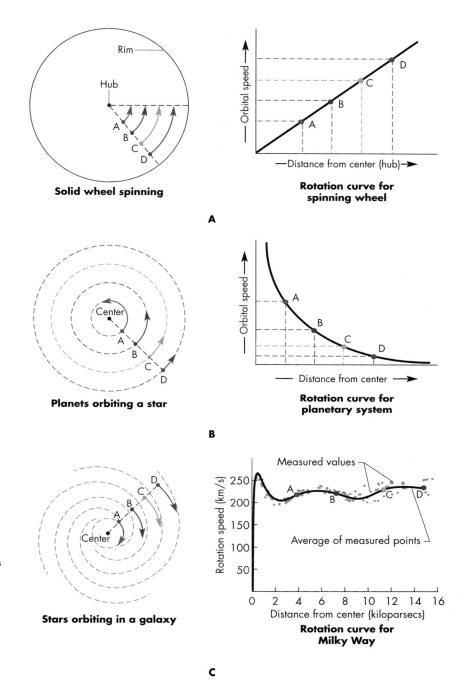

FIGURE 73.1

(A) When a wheel rotates, the outer parts move faster, creating a rising rotation curve. (B) Planets orbiting a star have slower speeds the farther they are from the star. (C) The Milky Way's rotation speed remains nearly constant at all radii as illustrated by actual data measured within our galaxy.

For example, there might be a total of 1 billion solar masses out to a radius of 1 kpc, but 2 billion solar masses within a radius of 2 kpc of the center, so a star orbiting at 2 kpc will be feeling the pull of much more mass. As a reminder that the amount of mass depends on the radius out to which we measure it, we will write $M_{<R}$. The subscript "$<R$" indicates "within a radius smaller than R." This peculiar notation will be helpful when we discuss the Milky Way and other galaxies because modern observations suggest that galaxies have no clear edge, and their masses increase as we measure them out to larger radii.

Astronomers can calculate the Milky Way's mass from the gravitational attraction needed to hold the Sun and other stars in orbit (Figure 73.2). This is yet another

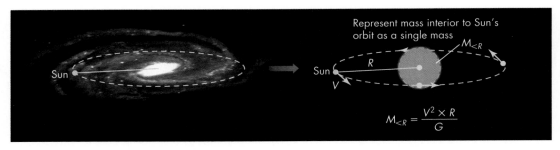

FIGURE 73.2

The Sun orbits all of the mass interior to its orbit $M_{<R}$. We can represent all of this mass by a single mass (the orange circle in the figure) and apply Newton's law to find the mass within the Sun's orbit.

$M_{<R}$ = Mass within radius R (kilograms)

R = Radius (meters)

V = Velocity of rotation (meters per second)

$G = 6.67 \times 10^{-11}$ m³/kg · sec²

application of Newton's law of gravitation (Unit 17). From the speed V of a star orbiting at radius R from the center of the galaxy, we can calculate the mass it is orbiting as follows:

$$M_{<R} = \frac{V^2 \times R}{G}$$

where G is Newton's gravitational constant. This formula specifies the mass out to the radius R. By looking at the rotation speeds of other stars at different distances from the center, we can build up a picture of the mass structure of the galaxy.

If we put in the rotation speed of the Sun (220 km/sec = 2.2×10^5 m/sec) and distance from the galaxy's center (8 kpc = 2.5×10^{20} m), we find that the Milky Way's mass out to 8 kpc is

$$M_{<8\text{ kpc}} = \frac{(2.2 \times 10^5 \text{ m/sec})^2 \times 2.5 \times 10^{20} \text{ m}}{6.67 \times 10^{-11} \text{ m}^3/\text{kg} \cdot \text{sec}^2}$$

$$= 1.8 \times 10^{41} \text{ kg}.$$

This is 9×10^{10} times the Sun's mass ($M_{\odot} = 1.99 \times 10^{30}$ kg). Thus, there are 90 billion solar masses of material in the Milky Way out to the Sun's orbit—the total of stars, gas, and any other matter.

This enormous total of material estimated to be within the Sun's orbit is fairly consistent with estimates of the number of stars and other known material in the Milky Way. What is peculiar, though, is that outside the Sun's orbit, the rotation speed remains high even though the number of stars declines rapidly toward our galaxy's outer reaches. At twice the Sun's distance from the center (16 kpc), the measured rotation speed remains at about 220 km/sec. So in the calculation everything remains the same except we multiply by a radius that is two times larger. The mass within 16 kpc is therefore twice the mass within 8 kpc even though there appears to be only a fraction as many stars in this outer region.

We can even estimate the Milky Way's mass out to about 100 kpc from its center—beyond where we detect any stars—by examining the speeds of small satellite galaxies orbiting the Milky Way. We do not have very complete information about their orbits, but the Milky Way's gravitational effect on them implies that the Milky Way contains approximately 2 trillion solar masses ($2 \times 10^{12} M_{\odot}$). This is about 20 times more mass than we just found within the Sun's orbit, which is difficult to understand because we see almost nothing at these large distances that could account for so much mass. This discrepancy leads astronomers to conclude that the Milky Way's halo contains much more matter than the disk, and most of

Because 16 kpc = 5.0×10^{20} meters, we find that

$M_{<16\text{ kpc}}$

$= \dfrac{(2.2 \times 10^5 \text{ m/sec})^2 \times 5.0 \times 10^{20}\text{m}}{6.67 \times 10^{-11} \text{ m}^3/\text{kg} \cdot \text{sec}^2}$

$= 3.6 \times 10^{41}$ kg

$= 2 \times M_{<8\text{ kpc}}$

Q If there was no mass beyond the Sun's orbit around the center of the Milky Way, what would the rotation curve look like?

this material emits no as-yet detectable light. They therefore describe this unseen mass as **dark matter.**

Our measurements suggest that dark matter fills the Milky Way. Its density is highest near the center of the Milky Way, but in those regions the density of normal matter is much greater than the dark matter density. Based on rotation curve measurements, it appears that the overall fraction of the mass that consists of dark matter is greater in the outer parts of the galaxy, eventually representing the majority of mass. The dark matter apparently continues well beyond where we detect the outermost stars and gas.

Astronomers have few clues about the nature of dark matter. Some observations suggest that about half of the Milky Way's dark matter might consist of dim, cold stars (perhaps old white dwarfs); but these observations are disputed, and the nature of the rest of the dark matter remains unexplained. Dark matter thus remains a mystery, showing that even today we have much to learn about our galaxy. We return to the question of dark matter in Unit 78.

Despite the uncertainty about the Milky Way's total mass, in the inner portions of the disk it appears that stars dominate the overall mass, so we can still estimate the number of stars from the rotation speed. From the rapid decline in stars seen at radii larger than the Sun's, we can deduce that nearly all of the Milky Way's stars are contained within about twice the Sun's distance from our galaxy's center. In total, the disk contains about 100 billion stars, along with about an equal amount of mass consisting of interstellar matter and stellar corpses. Each object moves along its own orbit around the center of the galaxy, held in its path by the collective gravity of all the other objects.

Q Would it be difficult to detect planets elsewhere in the galaxy? Are they likely to make up much of the dark matter? Why or why not?

73.2   THE GALACTIC CENTER AND EDGE

Within this spinning cloud of stars, the Sun is nearly lost, like a speck of sand on a beach. But unlike grains of sand that touch others, stars are widely separated. For example, in the vicinity of the Sun stars are typically more than a parsec apart. The Sun's nearest neighbor is 1.3 pc, a separation akin to sand grains spaced 10 km (6 miles) apart. Near the core of the Milky Way, stars are packed far more densely, with a separation roughly 100 times smaller than that near the Sun. Here the typical separation can be represented by sand grains placed at opposite ends of a football field.

The density of stars at the Sun's orbital radius can be measured in terms of the number of stars within a cube one parsec on a side (stars/pc^3). This is found to be about 0.1 stars/pc^3 or equivalently one star every 10 cubic parsecs. At twice the Sun's orbital radius (about 16 kpc) from the center, stars are spread even more thinly—less than 0.01 stars/pc^3—much as our atmosphere thins out as it merges into space. Thus, the Milky Way has no sharply defined outer edge.

Toward the center of the galaxy, the density in the disk rises to about 1 star/pc^3 just outside the bulge—about 2 kpc from the center. Within the bulge the density is about 10 times higher still. At the center of the galaxy the density rises sharply, exceeding 100,000 stars/pc^3 in the central few parsecs, and the stars there orbit around the center quite rapidly.

What lies at the very center of the Milky Way galaxy? The core is difficult to observe because interstellar dust clouds almost completely block the visible light emitted by objects there (see Looking Up #7: Sagittarius). But radio, infrared, X-ray, and even gamma-ray telescopes allow astronomers to see through the dust, and they reveal a dense swarm of stars and gas swirling around the center of our galaxy (Figure 73.3). Moreover, these telescopes reveal that at the core is an intense source of radio waves known as **Sagittarius A***, which is abbreviated "Sgr A*" (but usually pronounced *sadj ay star* by astronomers). The position of this mysterious object is shown in Figure 73.4, an infrared image of the stars within the central 2 light-years of the galaxy.

Sgr A* appears to mark the exact center of the Milky Way. Gas and stars orbit it, reaching speeds of thousands of kilometers per second. Many of those orbits have been tracked with infrared cameras over the last decade (Figure 73.5). We can use these stars' orbital motions to estimate the mass of Sgr A* using Newton's law of

FIGURE 73.3

The central region of the Milky Way as observed at X-ray wavelengths by the orbiting Chandra X-ray telescope. The supermassive black hole at the Milky Way's core lies inside the bright white patch at the center. The image shows an area 900 light-years across. In this false-color image, the colors represent X-rays of different energies. The small bright spots are white dwarfs, neutron stars, and probably gas surrounding a few black holes. The diffuse glow is from extremely hot gas surrounding our galaxy's core.

FIGURE 73.4

This infrared image shows the inner 2 light-years of the core of the Milky Way. The position of Sagittarius A*—the center of the galaxy—is shown by two small arrows.

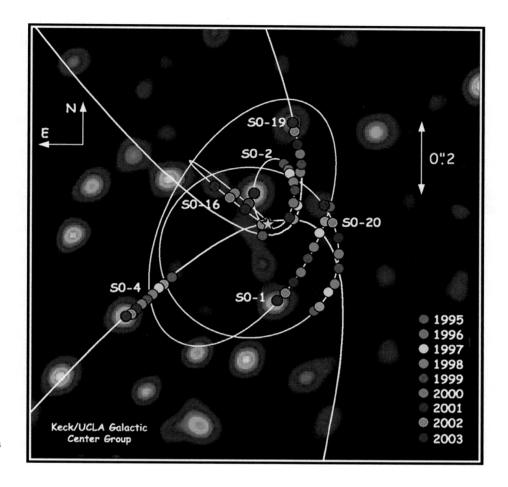

FIGURE 73.5

A plot of the position of several stars as they orbit the massive black hole at the core of the Milky Way. The scale bar corresponds to about 1600 AU. The stars are colored to show their positions in successive years. (The yellow "star" marks the center of the Milky Way.)

gravitation, just as we did to find the mass of the galaxy. For example, the object SO-2 in Figure 73.5 was measured to be traveling at 6000 km/sec (6×10^6 m/sec) when it was about 20 billion kilometers (2×10^{13} m) from Sgr A*. The mass that SO-2 is orbiting, lying within this distance, should thus be

$$M_{\text{Sag A}^*} = \frac{(6 \times 10^6 \text{ m/sec})^2 \times 2 \times 10^{13} \text{ m}}{6.67 \times 10^{-11} \text{ m}^3/\text{kg} \cdot \text{sec}^2}$$

$$= 1 \times 10^{37} \text{ kg}.$$

This is about 5 million solar masses. Yet no object is visible there, only a strong source of radio emission. That emission comes from a very small object, less than 10 AU in diameter (about the size of Jupiter's orbit). The almost inescapable conclusion is that a huge black hole lies at the center of the Milky Way and holds SO-2 in its grip. In fact, these observations are perhaps the best evidence yet that black holes exist.

How might such a supermassive black hole form? When a massive star reaches the end of its life, it may collapse, leaving a black hole as a remnant. Such black holes contain only 10 or so solar masses; but in the central regions of a galaxy, where there is a greater density of objects, ordinary black holes may grow vastly more massive by drawing in nearby gas with their gravitational attraction. In normal star clusters, stars virtually never collide. But stars are packed more tightly in the center of the Milky Way than within clusters, and they may have been even more tightly packed at the Milky Way's center when it formed. In such a crowded environment, stars may have collided, tearing away one another's gas, which then fell into the growing black hole. Once a black hole starts growing, it will not stop until it runs out of available matter, and some computer simulations suggest a 10 solar mass black hole could grow to $10^6 \ M_\odot$ in less than 1 billion years.

/73.3   DENSITY WAVES AND SPIRAL ARMS

The motions of stars and gas give important clues to the structure of the Milky Way. Although the stars in the Milky Way's disk orbit its center approximately in circles, we also know that the Milky Way displays concentrations of stars in its spiral arms. Spiral arms appear to be a common feature of galaxies, so it appears that some force must drive stars to collect together in this pattern.

The presence of a feature like spiral arms is puzzling because we found that the stars in the Milky Way exhibit differential rotation—stars take different periods of time to orbit the center of the Milky Way. The Sun takes about 220 million years, but a star at a radius of 4 kpc takes about 110 million years, and a star at 12 kpc takes about 330 million years. Yet a spiral arm may stretch from closer than 4 kpc to beyond 12 kpc—so within a few hundred million years it should be pulled apart completely because the inner part of the arm has moved around the galaxy several times while the outer part has not orbited even once.

We know that gravity holds a galaxy together, so we might guess that the gravity of the stars in a spiral arm might hold the arm together. Gravity does in fact help hold a spiral arm together, but not as a fixed rotating mass of material. Instead the stars and matter in different regions become alternately more or less tightly packed together due to gravity's effects, creating a phenomenon known as a **density wave.** A sound wave passing through the air is another kind of density wave—molecules in the air are alternately pressed closer together and then spread farther apart. Density waves in a rotating galaxy behave in a somewhat different fashion.

As stars circle the galaxy, a group of closely spaced stars at a given location will make the local gravity slightly stronger than elsewhere in the disk. That excess gravity draws stars toward the region so that the clump grows. However, stars do not remain in the clump because they continue along their orbits around the galaxy. Likewise, the same material does not remain in a spiral arm. Rather, all material just lingers a little longer in that portion of its orbit because of the extra concentration of mass in the arm. Thus, the arms maintain themselves because their gravity slightly modifies the orbits of stars and gas as they continue to circle the center of the galaxy.

In the density-wave model, waves of stars and gas sweep around the galactic disk. The waves are not an up-and-down motion like an ocean wave; instead, they are places where the density of stars and gas is large compared to the surroundings. The dense region can be thought of as the wave crest, with the concentration of stars in the wave crest resulting in what we observe as a spiral arm. Spiral waves are not unusual in rotating matter; for example, liquid in a food blender often forms a spiral wave with the liquid moving through it.

To better understand how arms form in the Milky Way, consider the following analogy. As cars move along a freeway, most move at nearly the same speed and keep approximately the same separation. If one car moves slightly more slowly than the others, however, traffic will begin to bunch up behind it. Cars can pass the slower vehicle but must change lanes to do so. The result is a clump of cars behind the slow one. But that clump is not composed permanently of the same cars. Cars join the clump from behind, pass the slower car, and then leave the clump at its front. Of course, in the spiral arms no stars move more slowly than all the rest; rather the *concentration* of stars in the arm is moving more slowly than the stars themselves.

Stars are not the only objects that are subject to the effects of density waves. When gas clouds drift through a spiral arm, they are initially accelerated into the arm by its excess gravity, and then slightly slowed as they move out the other side. The result is that the gas becomes compressed, and there is also a higher probability that clouds will collide with one another. This may in turn trigger a cloud's collapse so that it begins to form stars. The brilliant short-lived O and B stars that result illuminate the arms and make them stand out against the more ordinary stars of the disk. They are concentrated in the arms because they die before they have a chance

The winding dilemma

Q Where else in nature do we see spiral patterns? What causes them? Are there any connections to the spiral pattern in the Milky Way?

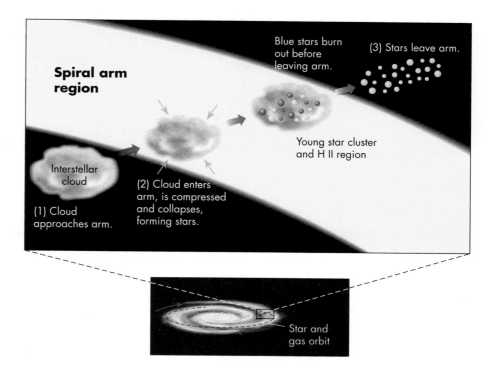

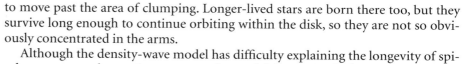

FIGURE 73.6

Sketch of a density wave in a galaxy, showing the progression from dust and gas to stars across the arm.

FIGURE 73.7

Ragged arms in the spiral galaxy M33, perhaps the result of a self-propagating star formation.

to move past the area of clumping. Longer-lived stars are born there too, but they survive long enough to continue orbiting within the disk, so they are not so obviously concentrated in the arms.

Although the density-wave model has difficulty explaining the longevity of spiral arms, many observations support the model. For example, by combining radio and visual pictures of a galaxy, astronomers can see a progression through the spiral arm as gas enters on one side, is compressed, and forms stars, which then take a few million years to sweep through the arm. On one side of the arm we can see gas and young stars; on the other side we see older, more evolved stars, as is sketched in Figure 73.6. Doppler-shift studies of the gas motion also clearly show the gas moving into the arm on one side and leaving on the other.

Although most spiral galaxies probably make their arms this way, another hypothesis has been proposed to explain the very ragged arms in some galaxies, such as the one illustrated in Figure 73.7. This second hypothesis is called the **self-propagating star formation** model.

In the self-propagating model, star formation starts at some random point in the disk of a galaxy when a gas cloud collapses and turns into stars. As the stars heat the gas around them and explode as supernovae, the blasts they generate make the surrounding gas clouds collapse and turn into stars. The original stars die, but the new stars—formed as the old ones evolve and explode—trigger more gas clouds to collapse and form additional stars. In this fashion, the region of star formation spreads across the galaxy's disk much as a forest fire burns outward in a circle through a forest. But unlike a forest where the trees stand still, stars orbit in a galaxy, and those near the center orbit in less time than those farther out. This draws out the zone in which star formation occurs into a spiral shape by the difference in rotation rate between the inner and outer parts of the disk. The result is spiral arms, as illustrated in Figure 73.8.

Astronomers suspect that both density waves and self-propagating star formation operate in many galaxies, but in some galaxies one of the processes may have a stronger effect. As we expand our view to examine more galaxies, it appears that there are some in which star formation has proceeded in different ways, leading to galaxies that do not much resemble our own. Our understanding of how galaxies form and evolve is an area of active research and current debate.

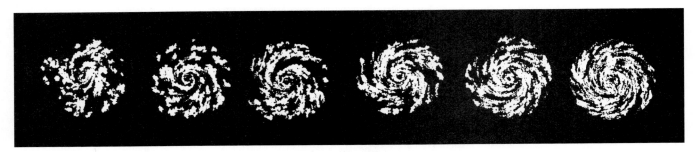

FIGURE 73.8
Winding of arms in a computer model of self-propagating star formation.

KEY TERMS

dark matter, 597   rotation curve, 594

density wave, 600   Sagittarius A* (Sgr A*), 597

differential rotation, 594   self-propagating star formation, 601

QUESTIONS FOR REVIEW

1. How can we determine the Milky Way's mass?

2. What is dark matter? Why do astronomers conclude the Milky Way may contain such unobserved material?

3. What is the evidence for a black hole at the center of the Milky Way?

4. What are possible causes of the Milky Way's spiral arms?

PROBLEMS

1. Given that the Sun moves in a circular orbit of radius 8 kiloparsecs around the center of the Milky Way, and its orbital speed is 220 kilometers per second, work out how long it takes the Sun to complete one orbit of the galaxy.

2. Near the center of the Milky Way, the rotation speed climbs in proportion to the radius; so at 400 pc from the center, the speed is twice as fast as at 200 pc from the center.
   a. What does this imply about the mass within 400 pc versus the mass within 200 pc?
   b. Is there any differential rotation in this region? Explain.

3. Suppose there was no additional mass in the Milky Way beyond the distance of the Sun's orbit. What would the rotation speed be at a radius of 16 kpc (twice as far from the center as the Sun)?

4. What is the Schwarzschild radius of a black hole with the mass of the black hole thought to be at the center of the Milky Way? (See Unit 68.)

5. Cut three circles of paper or cardboard whose diameters are 8, 10, and 12 cm. Punch a hole in the middle of each and fasten them together through their centers. Fill in a circle near the edge of the 10-cm disk that extends across the visible portion of this disk onto the two other disks. This dot represents a large region of star formation. To model the differential rotation of our galaxy, assume the disks rotate with a "flat rotation curve." If the 8-cm disk makes half a rotation, estimate how much the other two disks will have rotated in the same time. Draw how the star-forming region has become distorted. How does this relate to ideas for spiral structure in galaxies?

TEST YOURSELF

1. Astronomers think the Milky Way has spiral arms because
   a. they can see them unwinding along the celestial equator.
   b. radio maps show that gas clouds are distributed in the disk with a spiral pattern.
   c. young star clusters, H II regions, and associations outline spiral arms.
   d. globular clusters outline spiral arms.
   e. Both (b) and (c) are correct.

2. The modified form of Kepler's third law allows astronomers to determine the Milky Way's
   a. mass.
   b. age.
   c. composition.
   d. shape.
   e. number of spiral arms.

3. Astronomers think that spiral arms form because
   a. of shock waves from the black hole at the center of the galaxy.
   b. younger stars travel more slowly than older stars.
   c. dust and gas do not orbit at the same speed as the stars.
   d. density waves create a stellar pile-up.
   e. stars must travel up and down instead of in circles along a flat plane.

74

A Universe of Galaxies

Background Pathways: Unit 70

Beyond the edge of the Milky Way and filling the depths of space are billions of other star systems similar to our own. These remote, immense star clouds are called "external galaxies" or simply galaxies. When they were first discovered, it wasn't clear what to call them because until early in the 1900s the Milky Way was thought to be the entire universe, the only galaxy. These other objects, some of which were called "spiral nebulae," were argued to be "island universes," which was an apt description because each is an island of stars in a vast dark space.

Today we have the technology to detect more galaxies than the total number of stars in the Milky Way, and every one of these galaxies contains a similar vast number of stars as does the Milky Way. It is interesting to realize that when Einstein made his great discoveries about space and time in the early 1900s (Units 53 and 68), he thought that the universe consisted of a flattened collection of just a billion stars surrounded by a vast emptiness. It was not until the 1920s that the picture of the universe as we understand it today began to take shape. In this picture, each **galaxy** is an immense cloud of hundreds of millions to hundreds of billions of stars, with each star moving along its own orbit, held within its galaxy by the combined gravitational force of all the other stars and matter. Thus each galaxy is an independent, isolated star system.

74.1 EARLY OBSERVATIONS OF GALAXIES

All galaxies are extremely distant from our own, with even the nearest being more than 150,000 light-years away. Their great distance makes galaxies look dim to us, and only a few can be seen with the unaided eye. For example, on clear, dark nights in autumn and winter in the Northern Hemisphere, you can see the Andromeda Galaxy (Figure 74.1) with just your eyes (see Looking Up #3: M31 and Perseus). It looks like a pale elliptical smudge on the sky. Early observers such as the tenth-century Persian astronomer Al-Sufi noted this object, and it appeared on early star charts as the "Little Cloud." From the Southern Hemisphere, you can easily see by eye the Large and Small Magellanic Clouds, two small satellite galaxies of the Milky Way (Figure 74.2).

One thing you can learn when looking at these galaxies by eye or through a telescope is that galaxies, even though magnified, still shine with a pale glow—similar to the pale light from the Milky Way itself. Even though they contain billions of stars, they do not appear bright. The reason for this is that the large distances between stars in any galaxy mean that the light is always spread out. They look pale through a large telescope because, although the starlight is magnified, so are the vast spaces between the stars. Astronomers refer to this property as the **surface brightness** of galaxies (also see Unit 31.5). The surface brightness does not change with distance or magnification because the tiny fraction of a galaxy's total area that is covered by the area of stars remains the same.

FIGURE 74.1
Photograph of M31, the Andromeda galaxy, the nearest spiral galaxy to us.

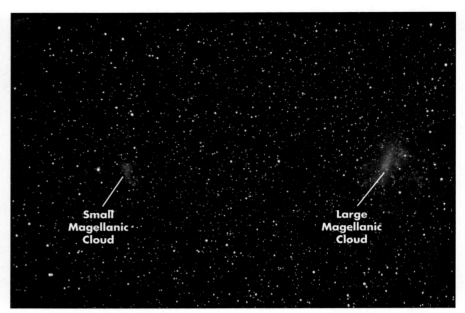

FIGURE 74.2
Photograph of the Large and Small Magellanic Clouds.

The faintness of galaxies' surface brightness makes them challenging to observe, even with a good telescope. The study of external galaxies began in the eighteenth century when the French astronomer Charles Messier (pronounced *mess-yay*) accidentally discovered many galaxies during his searches for new comets. He realized that some of the faint, diffuse patches of light never moved; so to avoid confusing them with comets, he assigned them numbers and made a catalog of their positions. Although many of Messier's objects have since been identified as star clusters or glowing gas clouds in the Milky Way, several dozen are galaxies. These and the other objects in the Messier catalog are still known by their Messier or M number, such as M31, the Andromeda Galaxy mentioned earlier.

In the early part of the nineteenth century other astronomers, such as William and Caroline Herschel, began to map faint objects in the heavens systematically, finding and cataloging numerous astronomical objects, including galaxies. The Herschels' work was continued by William's son John, and in the late 1800s was revised and added to by John Dreyer to include many thousands of galaxies and additional nebulae. This compilation is now known as the New General Catalog (or NGC for short), and it gives many galaxies their name, such as NGC 1275. Many galaxies appear in more than one catalog and so bear several names. For example, Messier's M82 is the same galaxy as NGC 3034.

For these early observers, the nature of the galaxies was a mystery. They looked different from other types of cataloged objects—planetary nebulae, globular clusters, emission nebulae, and so on—and because individual stars could not be resolved, many believed they were clouds of glowing gas rather than the merged light of billions of stars. One notable aspect of galaxies was that they were found only in portions of the sky away from the plane of the Milky Way. Even a modern catalog of deep-sky galaxy observations contains very few galaxies near the galactic plane, as shown in Figure 74.3A. Many astronomers interpreted this **zone of avoidance** around the Milky Way as evidence that the galaxies must somehow be interacting with (and therefore be near) the Milky Way. They asked, If these objects were far away, how could they "know" to avoid this direction? What astronomers did not realize at the time was that the zone of avoidance is simply an effect of the dust in our own galaxy blocking the light from distant galaxies (Figure 74.3B). This is the same effect that blocks our view of the center of the Milky Way.

Some astronomers capitalize *Galaxy* when referring to the Milky Way to distinguish it from other galaxies.

N.A. Sharp *P.A.S.P.* 98 740 (1986)
34729 galaxies (limiting size ~ 0.3′, limiting mag. ~ 15.5)

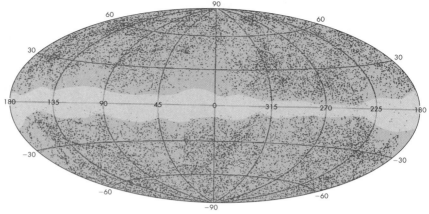

Galactic coordinates

A

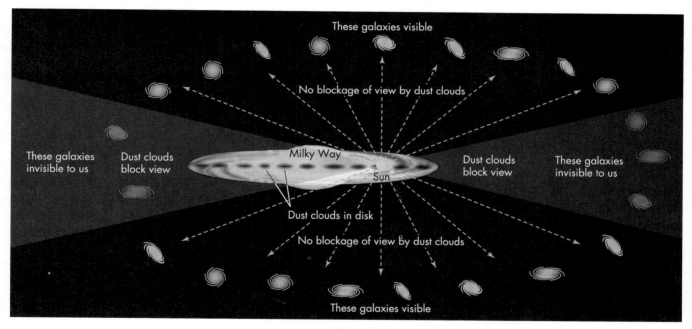

B

FIGURE 74.3

Dust limits our view in the disk of our galaxy and creates the zone of avoidance. We see no galaxies in the plane of our own galaxy because dust in the disk blocks our view. (A) A plot showing how galaxies are distributed on the sky. Notice how few lie along the central line (the galactic equator). This blank region is the zone of avoidance. (B) A sketch illustrating why we see so few galaxies along the plane of the Milky Way. The gray blobs represent dust clouds that block our view toward directions that lie in the Milky Way's plane. The galaxies are in reality about the same size as the Milky Way and are, of course, much farther away than can properly be shown in the figure.

As early as 1755, the German philosopher Immanuel Kant, whose views on the origin of the Solar System were far ahead of their time, suggested that galaxies might be remote star systems—island universes—similar to the Milky Way. However, until the 1920s, there was little observational evidence to clearly show whether galaxies were huge distant systems of stars like the Milky Way or were small nearby objects associated with our own galaxy, and astronomical opinion was strongly divided.

This controversy was aired in detail during a series of debates organized by the National Academy of Sciences in which two U.S. astronomers, Harlow Shapley (Unit 70) and Heber Curtis, took opposing sides. Shapley argued that the Milky Way was huge and that other galaxies (then called spiral nebulae) were merely small, nearby companions. Curtis, on the other hand, argued that the Milky Way was smaller than Shapley claimed, that the nebulae were star systems like our own, and that they were immensely distant from it.

The debate reflects the challenges of an observational science like astronomy. For example, Curtis argued that objects such as M31 must be far away based on a supernova that had been observed in 1885. Assuming this was as luminous as the supernovae that had been observed in our own galaxy hundreds of years earlier (Unit 65), he found that M31 must be very distant; and if it was distant, then M31 must be huge. Shapley countered that the object seen in 1885 was not a supernova but a nova—a much less luminous explosion—and therefore M31 was nearby, perhaps at a distance more like he'd found for the globular clusters orbiting the center of the Milky Way. Both arguments were reasonable for the time and fit into the model each side had developed.

Shapley based part of his argument on observations of motions inside galaxies—observations that were later found to be incorrect. The motions that were claimed to have been detected would have ruled out the possibility that the galaxies were far away. Curtis correctly doubted these reported motions, even though the report was made by a well-respected astronomer of the time. As in any mystery story, not all clues are useful or correct.

Historians differ on who "won" the debate. Shapley was more nearly correct about the Milky Way's size and structure (also see Unit 70). Curtis was right about the nebulae being distant systems similar in structure to the Milky Way.

One of the biggest problems for both sides in the debate was the lack of understanding about the dimming caused by dust in the Milky Way. Curtis adopted *Kapteyn's Universe,* a model of a small Milky Way (Unit 70), whereas Shapley interpreted the zone of avoidance as evidence that the galaxies were nearby. The other major problem was the lack of a definitive method for measuring the galaxies' distances. Breakthroughs in the 1920s solved both of these problems. Astronomers discovered how starlight was dimmed by dust in the Milky Way, and they were also able to find *standard candles* (Unit 54.3) in external galaxies. With these discoveries, the true nature of galaxies as enormous systems containing billions of stars was finally confirmed.

Q If the intergalactic space between the Milky Way and M31 contained dust, how would that have affected Curtis's measurement of M31's distance?

74.2   THE DISTANCES OF GALAXIES

The enormous distances to galaxies do not let us use the methods that we use to find distances to stars. Parallax cannot be used to measure such vast distances because the angle by which the galaxy's position changes as we move around the Sun is too tiny to detect. Individual stars within an external galaxy are extremely difficult to detect because the angular separations are so small that atmospheric blurring (Unit 30.2) blends their light together. To apply the method of standard candles, then, astronomers must find stars of known luminosity that are extremely bright, so that their light can be picked out from the sea of surrounding stars.

In using the standard-candle method to find a galaxy's distance, astronomers use the principle that the farther from us an object is, the dimmer it looks. More precisely, if we know the luminosity of a light source—our standard candle—and how bright it looks to us, we can find its distance by using the inverse-square law.

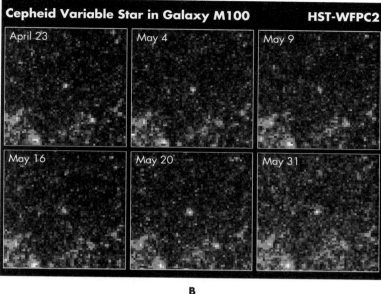

A

B

FIGURE 74.4

(A) Cepheids in the Milky Way have known distances, d_*, so we can determine their luminosities. Knowing those luminosities, we can find the distance to a galaxy, d_g, if we can measure the brightness of a Cepheid in it. (B) Hubble Space Telescope observations of a Cepheid variable in the galaxy M100.

d_g = Distance to Cepheid and galaxy (meters)

L_c = Luminosity of Cepheid from period–luminosity relationship (watts)

B_c = Measured brightness of Cepheid (watts/m²)

Why do multiple measurements give a more accurate value?

Astronomers use many different astronomical objects as standard candles, but one of the most reliable is the class of luminous, pulsating variable stars known as Cepheids (Unit 63). Cepheids have luminosities more than ten thousand times the luminosity of the Sun, and because of that brilliance, it is possible to detect them in nearby galaxies. Furthermore, because their brightness changes as they pulsate, their light can be distinguished from other stars. The period–luminosity relation (Unit 63.3) then allows us to use the time these stars take to go from bright to dim to bright again (their period) to determine their luminosity.

Once astronomers know the variable star's luminosity, they can find its distance from a simple measurement of how bright it appears. For example, suppose we observe a Cepheid in a distant galaxy, as illustrated in Figure 74.4A. From observations of Cepheids in our own galaxy, we know what luminosity a Cepheid has based on its period: the period–luminosity relationship. We can determine the luminosity L from the period of variation and measure the brightness B directly. With these values we can determine the distance to the Cepheid using the standard-candle method:

$$d_g = \sqrt{\frac{L_C}{4\pi B_C}}.$$

Astronomers generally measure the distances of many Cepheids within the same galaxy to get an accurate average for the galaxy's distance.

Cepheids are very luminous, but only a few galaxies are close enough that we can make successful ground-based measurements of their light. The first such measurement was achieved by the U.S. astronomer Edwin Hubble in 1923 for M31, finally proving that it was millions of light-years away and even larger than the Milky Way. Over the subsequent half century, Cepheids were observed in only a handful more galaxies. It required the Hubble Space Telescope to carry out similar observations for dozens of more distant galaxies, such as that shown in Figure 74.4B. The Cepheid in the figure is in the galaxy M100 and was measured to have a period of 51 days. The period–luminosity relationship indicates that this Cepheid has a luminosity of $3 \times 10^4 \, L_\odot$ (or 1.2×10^{31} watts). The apparent brightness measured with the Hubble Space Telescope was 3.4×10^{-18} watts per square meter. Therefore, the distance to the Cepheid (and its galaxy) is

$$d_g = \sqrt{\frac{1.2 \times 10^{31} \text{ watts}}{4\pi \times 3.4 \times 10^{-18} \text{ watts/m}^2}} = 5.3 \times 10^{23} \text{ m} = 1.7 \times 10^7 \text{ parsecs.}$$

Thus, the galaxy M100 is about 17 million parsecs (56 million light-years) distant. Even for the Hubble Space Telescope, though, Cepheids are undetectable in all but the nearest few hundred galaxies.

To measure greater distances, astronomers must use even more luminous standard candles, such as supernovae. However, we cannot predict when a supernova of known luminosity might occur in a galaxy of interest. This might have been an insurmountable problem but for a very strange discovery.

Q When Hubble originally measured the distance to M31, the values of the luminosity in the period–luminosity relationship he was using were four times lower than they are known to be today. How would this have affected his distance measurement?

74.3 THE REDSHIFT AND HUBBLE'S LAW

In the early 1900s the U.S. astronomer Vesto Slipher discovered that the spectral lines of nearly all of the galaxies he observed were strongly shifted toward longer wavelengths—to the red end of the visible spectrum. This so-called *redshift* is seen in all galaxies, except for a few nearby ones whose motion is influenced by the gravitational attraction of the Milky Way and its neighbors.

Redshift is defined as the fraction by which the light's wavelength changes—that is, the change in wavelength divided by the original wavelength. For example, suppose we observed a galaxy's spectrum and saw that the pattern of spectral lines was shifted to longer wavelengths (Figure 74.5). Among these we might identify an absorption line that normally occurs at 500 nm, but is shifted to 505 nm in this galaxy. The line has thus been shifted by 5 nm, 1% of its normal wavelength of 500 nm. This galaxy's redshift is 1% or 0.01.

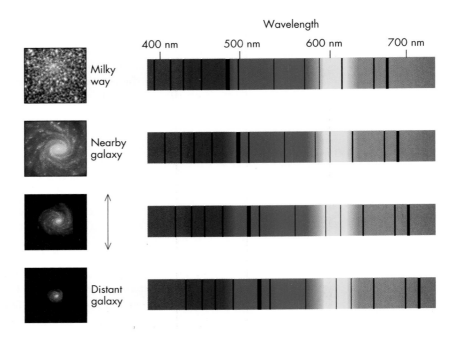

FIGURE 74.5
Galaxies show a larger redshift of their spectral lines the more distant they are. The spectrum of the integrated light from the Milky Way (which has zero redshift) is shown for comparison.

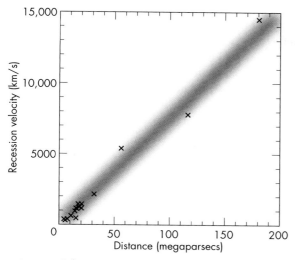

FIGURE 74.6

The Hubble law. Galaxies farther away recede faster.

Astronomers use the letter z to stand for redshift, and mathematically we can define the redshift as follows:

$$z = \frac{\Delta\lambda}{\lambda}.$$

In this formula, $\Delta\lambda$ stands for the change in wavelength, and λ is the original, unshifted wavelength. If we had a second galaxy where the wavelength of the 500 nm line had shifted to 515 nm, we would say $\Delta\lambda = (515 \text{ nm} - 500 \text{ nm}) = 15 \text{ nm}$, and $\lambda = 500$ nm, so

$$z = \frac{15 \text{ nm}}{500 \text{ nm}} = 0.03.$$

This galaxy's redshift is 0.03 or 3%. Incidentally, we chose a 500 nm line for this example for numerical simplicity. If we looked at any other spectral line, we would find that they have all been shifted by 3% relative to their normal value, whether they were generated at radio wavelengths or X-rays.

The discovery of galaxy redshifts seemed even stranger when Edwin Hubble and others found that the redshift was larger for dimmer (and therefore presumably more distant) galaxies. They interpreted the redshift as a Doppler shift, and the increasing redshifts then implied that the speed with which a galaxy moves away from us—its **recession velocity**—increases with distance, as Figure 74.6 shows. That is,

The faster a galaxy recedes, the farther away it is.

Assuming the Doppler effect, Hubble calculated the recession velocity $V = c \times z$. A 1% redshift implies a recession velocity of 1% of the speed of light or about 3000 kilometers per second. A 3% redshift implies a recession velocity of about 9000 km/sec.

Galaxy redshifts look like Doppler shifts, but today we understand from Einstein's theory of general relativity (Unit 68) that the redshifts are actually caused by a different effect: the expansion of the space between the galaxies as the light travels through it. Does this sound like nothing more than a semantic difference? It turns out to be fundamentally important for understanding the universe as a whole. We will use the notation and language of velocities here that were adopted by Hubble and are still used by astronomers today, but we must keep in mind that astronomers today interpret them quite differently. In particular, astronomers now view redshift as arising from the expansion of space. We will return to the implications of the redshift for understanding the universe in Unit 79.

In 1920 Edwin Hubble discovered that a simple formula related the recession velocity, V, and the distance, d. When he graphed the recession velocities and distances of a number of galaxies (Figure 74.6), he found that they lay along a straight line. The line can be described by the equation

$$V = H \times d$$

where H is a constant. Because of his discovery, this relation is called **Hubble's law,** and H is called **Hubble's constant.**

The value of the Hubble constant depends on the units used to measure V and d. Astronomers generally measure V in kilometers per second and d in **megaparsecs** (abbreviated Mpc), where 1 Mpc is a million parsecs (= 3.26 million light-years). In these units, H is about 70 km/sec per Mpc.

If $V = c \times z$, then for $z = 3\%$, we find

$$V = 3.00 \times 10^8 \text{ m/sec} \times 0.03$$
$$= 9 \times 10^6 \text{ m/sec} = 9000 \text{ km/sec}.$$

V = Recession velocity of galaxy (km/sec)

H = Hubble's constant (km/sec per Mpc)

d = distance to galaxy (Mpc)

As recently as the 1990s, astronomers could not agree on the value of H to within a factor of two. The reason for this controversy was that to determine H, we must know both the distance and velocity of at least a few galaxies to calibrate the law. H is simply the average value of each calibration galaxy's velocity divided by its distance. However, as we have just seen, accurately measuring the distance to even a nearby galaxy is difficult. New techniques for determining H have resolved most of this uncertainty, although some groups still argue for significantly different values. The current best value for Hubble's constant, based on galaxies for which Cepheid variable distances have been calculated, is $H = 70$ km/sec per Mpc, with an uncertainty possibly as large as 10 km/sec per Mpc.

Once a value for H is known, we can turn the method around and use the Hubble law to find the distance to a galaxy. The method is as follows: Take a spectrum of the galaxy whose distance you want. From the spectrum, measure the shift of the spectral lines. From the Doppler-shift formula, calculate the recession velocity. Finally, use that velocity in the Hubble law to find the distance. Mathematically, we divide both sides of the Hubble law by H, obtaining

$$d = V/H.$$

We insert the measured value of V and our choice of H and solve for the distance. The power of this formula is also why getting H right is so important—if H is incorrect, our whole scale of the universe will be off.

As an example, suppose we find from a galaxy's spectrum that its recession velocity is 49,000 kilometers per second. We find its distance by inserting this value for V in the previous equation and then divide by H:

$$d = \frac{49,000 \text{ km/sec}}{H}.$$

For $H = 70$ kilometers per second per Mpc, the value of d is

$$d = \frac{49,000 \text{ km/sec}}{(70 \text{ km/sec})/\text{Mpc}} = 700 \text{ Mpc}.$$

The galaxy's recession velocity implies a distance of about 700 Mpc.

In applying this method, we must be careful about units. When we divide V in kilometers per second by H in km/sec per Mpc, the answer we get will be in Mpc. The 700 we just found is not in parsecs or light-years but in Mpc. To get the distance in light-years, we multiply by 3.26 light-years per parsec to get about 2300 million light-years or 2.3 *billion* light-years. Detecting a galaxy this far away was time-consuming until the 1990s, but newer technologies can measure thousands of galaxies at these distances in a single night of observation.

The Hubble law does not work for nearby objects. For example, M31, the spiral galaxy nearest to us, is moving *toward* the Milky Way at about 90 km/sec, so it has a "negative redshift." We cannot apply Hubble's law to it, or we would get the absurd result that M31 has a negative distance. Similarly, we cannot apply Hubble's law to a nearby star that is moving away from the Sun, even though its spectrum is redshifted. Its redshift is caused by orbital motion inside our galaxy, and therefore, again, Hubble's law does not apply.

Hubble's law applies only to redshifts caused by the general motion of galaxies away from each other—what astronomers term the *overall expansion of the universe.* When we are observing distant galaxies, the redshift we measure is caused in part by the expansion of the universe, and in part by the galaxy's orbital motion around another galaxy or system of galaxies. To make the distinction clear, astronomers refer to the first as the **expansion redshift,** and to the second simply as the *Doppler shift.* If we assume the redshift we measure is entirely due to expansion, the distance we calculate from Hubble's law might be too high or too low by an amount depending on the size of the Doppler shift. For an isolated galaxy, the

Doppler shift motion is usually under 100 km/sec. So if we calculate a velocity of 700 km/sec for such a galaxy from its redshift, the true recession velocity might be anywhere in the range of 600 km/sec to 800 km/sec. By using 700 km/sec in the Hubble law, we will find this distance:

$$d = \frac{V}{H} = \frac{700 \pm 100 \text{ km/sec}}{70 \text{ km/sec per Mpc}} = 10 \pm 1.4 \text{ Mpc}.$$

The uncertainty in the expansion redshift produces an uncertainty in the distance of the orbital speed divided by H.

This uncertainty may cause significant errors in the distances of a galaxy orbiting within a large system of galaxies where orbital speeds in excess of 1000 km/sec have been observed. If we assumed these galaxies' redshifts were caused entirely by expansion, we could make an error of up to $(1000/70) \approx 14$ Mpc in calculating their distances. This is one reason it was so difficult to determine the Hubble constant accurately—the nearby galaxies with good distance measurements all have uncertainties in their expansion redshifts. On the other hand, with an accurate value of Hubble's constant, we can use the redshift to find the distance to a galaxy several hundred megaparsecs away to an accuracy of a few megaparsecs if the galaxy is isolated. No other method for finding the distances of remote galaxies is as accurate or as easy to use.

KEY TERMS

expansion redshift, 610

galaxy, 603

Hubble's constant, 609

Hubble's law, 609

megaparsec (Mpc), 609

recession velocity, 609

redshift, 608

surface brightness, 603

zone of avoidance, 604

QUESTIONS FOR REVIEW

1. How do astronomers measure the distance to nearby galaxies?
2. What is the Hubble law?
3. Why are astronomers uncertain about the value of the Hubble constant?

PROBLEMS

1. Suppose that a galaxy's 21-cm line of atomic hydrogen is shifted to 22 cm.
    a. What is the redshift of the galaxy?
    b. What is its recession velocity?
2. A galaxy has a recession velocity of 28,000 kilometers per second. What is its distance in megaparsecs?

3. What is the approximate recession velocity of a galaxy that we are seeing as it looked 1 billion years ago?

TEST YOURSELF

1. A galaxy's spectrum has a redshift of 7,000 kilometers per second. If the Hubble constant is 70 kilometers per second per megaparsec, how far away from Earth is the galaxy?
    a. 10^6 megaparsecs
    b. 100 megaparsecs
    c. 0.01 megaparsecs
    d. 10^8 megaparsecs
    e. 10^{-6} megaparsecs

2. The nearest galaxy to ours is approximately how far away?
    a. 1,000,000,000 light-years
    b. 1,000,000 light-years
    c. 150,000 light-years
    d. 100 light-years
    e. 4.3 light-years

3. Which of the following would an astronomer be likely to refer to as a standard candle?
    a. The Sun
    b. An S0 galaxy
    c. A Cepheid variable
    d. The Andromeda Galaxy
    e. A wax cylinder approximately 1000 light-years tall

75

Types of Galaxies

Many external galaxies are flattened systems with spiral arms, similar to the Milky Way in size and shape. Other external galaxies are distinctively different from our own. For example, some are not disk-shaped at all. Rather, they are shaped more like a rugby ball or a distorted sphere, with their stars distributed in a vast, smooth elliptical cloud surrounding a dense central core of stars. Others are neither disks nor smooth elliptical shapes but are completely irregular in appearance.

Galaxies differ in more than shape; they also differ in their content. Some contain mostly old stars, but others contain predominantly young stars. Some are rich in interstellar matter, whereas others have hardly any at all. Some galaxies—despite having as much total mass as the Milky Way—have formed far fewer stars, leaving them dim and scarcely visible to us. Others have a small but tremendously powerful energy source at their core that emits as much energy as the entire Milky Way but from a region about 0.001% its size, less than a few light-years in diameter. Many of these *active galactic nuclei* eject gas in narrow jets at nearly the speed of light (Unit 77).

Astronomers can see all this diversity in galaxies, but we are only beginning to understand how it arises. Unlike stars, whose structure and evolution are well understood, many aspects of galaxies remain a mystery. For example, astronomers do not all agree on what causes some galaxies to be disk-shaped while others are more nearly egg-shaped. Although this lack of certainty can be confusing, it offers us a chance to see how scientists work. In particular, we will learn how astronomers begin with basic observations and construct hypotheses to explain what they observe. For example, many galaxies appear to be undergoing collisions with their neighbors. Could past interactions of this kind account for the different galaxy shapes?

Background Pathways: Units 71, 74

75.1 GALAXY CLASSIFICATION

Early nineteenth-century observers noticed that not all galaxies look the same. However, it was Edwin Hubble, working first as a graduate student at the University of Chicago and later at Mount Wilson in California in the 1920s, who demonstrated that galaxies could be divided conveniently on the basis of their shape into three main types: spirals, ellipticals, and irregulars.

The first type has two or more arms winding out from the center. Astronomers call these **spiral galaxies,** a term often abbreviated to simply **S.** The Milky Way is probably of this type, and Figure 75.1 shows photographs of several others.

The second type shows no signs of spiral structure. These galaxies have a smooth, featureless appearance and a generally elliptical shape, as can be seen in Figure 75.2. Accordingly, astronomers call them **elliptical galaxies,** abbreviated as **E.**

Galaxies of the third major type show neither arms nor a smooth uniform appearance. In fact, they generally have stars and gas clouds scattered in random

FIGURE 75.1

Photographs of typical spiral galaxies. (A) NGC 4565; (B) NGC 4414; and (C) NGC 1288. These examples show spiral galaxies with their disks oriented from nearly edge-on (A) to nearly face-on (C).

FIGURE 75.2

Photographs of a few elliptical galaxies: NGC 205, M49, and M87.

patches. For this reason, they are called **irregular galaxies** (Figure 75.3), abbreviated as **Irr.**

In addition to these three main types, Hubble recognized two additional galaxy types closely related to spiral systems. The first of these is called a **barred spiral galaxy,** abbreviated as **SB.** These have arms that emerge from the ends of an elongated central region, or bar, rather than from the core of the galaxy (Figure 75.4). It is this bar that gives them their name, and they are denoted SB galaxies to distinguish them from normal spirals. Although Hubble treated SB galaxies as a separate class, astronomers today do not think they are so distinct. For example, computer models of galaxies show that most disk-shaped systems will form a bar if they undergo a gravitational disturbance, perhaps as the result of a close encounter with a neighboring galaxy. Even our own Milky Way appears to have a weak bar. Computer models also suggest that over hundreds of millions of years, the bars may appear and disappear; so an ordinary spiral galaxy may change into a barred spiral and then back again.

FIGURE 75.3
Photographs of two irregular galaxies (NGC 6822 and the Small Magellanic Cloud).

FIGURE 75.4
Barred spirals. NGC 1365 has a prominent bar, while NGC 5236 (M83) has a
weaker bar—more like the one thought to be at the center of the Milky Way.

FIGURE 75.5
An S0 galaxy, NGC 1201.

Hubble also identified another kind of galaxy similar to spirals. These are disk
systems with no evidence of arms (Figure 75.5), and they are called **S0** ("*S-zero*")
galaxies. Hubble thought they were intermediate in type between S and E type
galaxies. Today, however, astronomers think they are spiral galaxies whose gas has
been stripped by their motion through the hot, tenuous gas in intergalactic space.
The impact of such gas from outside "sweeps" gas and dust out of the moving spiral, much as a leaf-blower cleans debris from a sidewalk. Without gas, the disk
galaxy can make no new stars and hence shows no obvious spiral structure.

Hubble took his classification system one step further, seeking a pattern that might
lead to an understanding of why galaxies exhibit such diversity in their appearance.
He noticed that both spiral and elliptical galaxies can be subdivided into additional
classes that exhibit a gradation of properties within these classes. The subdivisions
among spirals are indicated by lowercase letters, ranging from Sa with a large bulge

and tightly wound arms, to Sd with a small bulge and loosely wrapped arms. The ellipticals were subdivided according to how circular or elongated they appeared. Hubble hypothesized that a sequence of galaxy types that showed a smooth transition of appearances might represent an evolutionary sequence of galaxies. He developed such a sequence, starting with spherical E type galaxies to flatter E galaxies, to S0 galaxies, then the sequence of spirals, and finishing with irregulars. Hubble proposed that as a galaxy aged, its type might evolve through this sequence. This hypothesis seems plausible if you look at the "tuning fork" diagram that Hubble proposed (Figure 75.6), but astronomers now think it is incorrect. Nevertheless, the diagram is still a useful way to organize the galaxy types and subtypes that Hubble described.

Although Hubble's classification system covers the most conspicuous galaxy types quite well, it omits two forms that astronomers today think are important: **low surface brightness galaxies** (dim, often large-diameter systems with far less star formation than ordinary galaxies) and **dwarf galaxies** (perhaps the building blocks of ordinary galaxies). Examples are shown in Figure 75.7. These types of

Surface brightness is discussed in Unit 31. Unlike the apparent brightness of a galaxy, why does the surface brightness not change with distance?

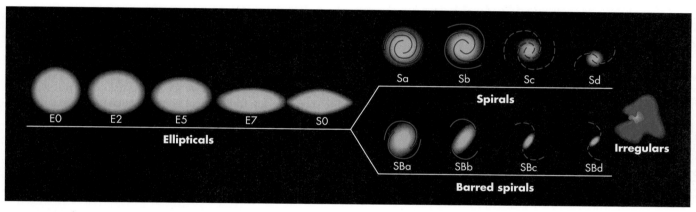

FIGURE 75.6

The Hubble tuning-fork diagram. Elliptical galaxies are subdivided according to how "flattened" they look, while both regular and barred spirals are subdivided into subclasses according to how large their bulges are and how tightly wound their arms are.

FIGURE 75.7

(A) Dwarf galaxy, Leo I, a nearby galaxy found in 1977. (B) Comparison between two distant galaxies of matching size. One is a low surface brightness spiral of the type often missed in optical surveys.

galaxies are very numerous, but their properties make them harder to detect, so few were known before modern observing technologies developed.

75.2   DIFFERENCES IN STAR AND GAS CONTENT

Since Hubble's time, astronomers have discovered that spiral, elliptical, and irregular galaxies differ not only in their shape but also in the type of stars they contain. For example, spiral galaxies contain a mix of young and old stars (Pop I and II—see Unit 71.1); but elliptical galaxies contain mostly old (Pop II) stars, and irregulars contain mostly young (Pop I) stars.

This difference in the kind of stars in the different galaxy types is understandable in terms of their different gas content. To make young (Pop I) stars, a galaxy must contain dense clouds of gas and dust. Spiral systems typically have at least 10% of the mass of their disk in the form of such interstellar clouds, and so they can easily make the young stars of their spiral arms. On the other hand, ellipticals contain much less interstellar matter. In fact, astronomers used to think that the E galaxies contained virtually no interstellar gas or dust. However, more recent observations made with X-ray telescopes show that E galaxies often contain very low-density but very hot (10^7 K) gas. Such gas contains few or no high-density cold clumps that might collapse gravitationally to form stars, so it is no surprise that elliptical galaxies rarely contain young, blue stars like the ones we see in the disks of spiral systems.

Although young, blue stars are rare in elliptical systems, such stars are common in irregular galaxies. These Irr systems also often contain large amounts of interstellar matter, amounting sometimes to more than 50% of their mass. Accordingly, astronomers believe that these galaxies are "young" in the sense that they have not yet used up much of their gas in making stars.

Apart from their different star and interstellar matter content, S, E, and Irr galaxies have few other differences to generalize about. For example, if we look at the mass and radius of galaxies, we do not find that all ellipticals are huge and all spirals are small. Instead, we discover that elliptical galaxies range enormously in size. Some are not much larger than globular clusters and contain only 10^7 solar masses of material; others are monster galaxies, 10 to 100 times the mass of the Milky Way. This spread in sizes can be seen in Figure 75.8, which shows the

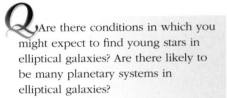

Q Are there conditions in which you might expect to find young stars in elliptical galaxies? Are there likely to be many planetary systems in elliptical galaxies?

FIGURE 75.8

Photograph of the central 1 Mpc region of the Virgo galaxy cluster, illustrating the range in sizes of the galaxies. All of the galaxies are at about the same distance, so their relative apparent sizes do not differ much due to distance. Some of the smallest dots are foreground stars in the Milky Way, but many are dwarf galaxies within the system of galaxies.

center of a nearby cluster of galaxies in which most of the members are elliptical systems.

What fraction of galaxies falls into each type classification? Even such a basic question has no clear answer. If we take a census of the galaxies near the Milky Way, we find that most galaxies are dim dwarf E and Irr systems, sparsely populated with stars. These dim galaxies are undetectable at greater distances, where we can see only the more luminous galaxies. If we take a census of those luminous objects, we find that the majority of galaxies are spirals, but the percentage depends on whether we look at galaxies that occur in large groups or in more sparsely populated regions. In the largest groups of galaxies, fewer than 50% of the members are spirals or S0 galaxies, whereas in the sparsely populated regions, the proportion of spirals is about 80%. The diversity of galaxy form and content requires an explanation, and a major question in extragalactic research today is why galaxies come in such distinct and different forms.

75.3   WHAT CAUSES THE DIFFERENT GALAXY TYPES?

For many decades astronomers thought that rotation played the main role in determining how galaxies evolved to their present-day type. When they measure the rotation speed of galaxies, astronomers find that the disks of spirals rotate relatively rapidly compared with ellipticals of the same size. On the other hand, the halos and bulges of the spirals do *not* rotate especially fast. In the model where a spiral galaxy's halo forms first during the initial stages of collapse, it is not obvious that there is much difference between a spiral and an elliptical in their early stages. Some factors other than rotation must also play a role in determining a galaxy's type.

The Hubble Space Telescope provides a way of studying galaxies when they were very young, allowing us to see how galaxies have changed throughout the history of the universe. Because light takes a finite time to reach us, when we observe a very distant galaxy, we are seeing it as it was when the light left it. This was the idea behind the Hubble Deep Field project—a very long observation of a small "blank" patch of sky. By collecting light for over 100 hours, it was possible to detect extremely distant and therefore very young galaxies. A section of the Hubble Deep Field is shown in Figure 75.9.

The time it takes the light to reach us is the distance divided by the speed of light: $t = d/c$. Solving for the time is particularly easy if we write the distance in light-years because the distance in light-years also tells us how long ago the light left the galaxy. For example, suppose we observe a galaxy that is about 3000 Mpc (3 billion parsecs) distant. Multiplying by 3.26 light-years per parsec, we find that the galaxy is about 10 billion light-years away from us. Thus, the light we see today left the galaxy 10 billion years ago, and we are seeing the galaxy as it looked at that time. At such a distance a galaxy has a very small angular size; but with the Hubble Space Telescope, which is unencumbered by the blurring effects of our atmosphere, astronomers can see details in such a very distant—and thus very young—galaxy.

Although present observations do not yet reveal to us the very first stages of the collapse of primordial gas clouds into galaxies, the Hubble Space Telescope has allowed astronomers to observe galaxies just past the collapse stage, less than a billion years after the galaxies first formed. Interestingly, these very young galaxies are much smaller than systems such as the Milky Way. Moreover, their shapes often do not fit into the Hubble scheme of three main classes. Finally, counts of these objects indicate that in the distant past there were far more galaxies than we find today. These earliest galaxies must have somehow disappeared or changed their form.

The position of the Hubble Deep Field is shown in Looking Up #2: Ursa Major. If you hold a grain of sand at arm's length, you could completely cover the area observed.

FIGURE 75.9
The Hubble Deep Field is a small region observed for more than 100 hours to detect extremely distant galaxies. The galaxies in this image are not close together in space, but are spread out over more than 10 billion light-years along our line of sight.

How can galaxies disappear? Perhaps they grow dim as they use up their gas and can no longer make new stars, or they may collide and merge into bigger systems. Making deductions from what is currently known, astronomers think that the **merger** explanation is more likely. It would account both for the formation of large systems such as the Milky Way and for the smaller number of galaxies we see later. But a major question still remains: What causes one of these merging systems to eventually become a spiral galaxy, whereas another becomes an elliptical galaxy?

75.4   GALAXY COLLISIONS AND MERGERS

Our own galaxy shows signs of past mergers. For example, astronomers can see streams of stars in our galaxy's halo, which are suspected to be debris from small galaxies that have fallen into the Milky Way and have been torn apart by its gravity. Could mergers also give rise to elliptical galaxies? The answer seems to be yes.

For example, Figure 75.10 shows two spiral galaxies whose gravitational force has drawn them together and merged them into a single new galaxy that looks much like an elliptical system. Observational evidence such as this and computer simulations have encouraged astronomers to propose new hypotheses for the origin of spiral and elliptical galaxies. The basic idea is that most galaxies are born as disklike or irregular systems with essentially no bulge. Subsequently, collisions and mergers between these disk systems create elliptical galaxies, as shown schematically in Figure 75.11.

This hypothesis explaining the origin of spiral and elliptical systems has an added attraction for astronomers. It explains several puzzling features of the large spiral galaxies such as our own Milky Way. For example, astronomers at one time thought that the stars in our galaxy's bulge formed in one single, massive event at the galaxy's birth. However, stars in the bulge have a wide range of ages, suggesting

Q Hubble proposed that his tuning-fork diagram might represent evolutionary changes, with E galaxies gradually changing to S galaxies. What evidence based on the amount of interstellar matter and kind of stars in E and S galaxies indicates that this is unlikely?

FIGURE 75.10
A merger in which two spirals are colliding and forming an elliptical galaxy.

that the bulge grew gradually over time. Such gradual growth is easy to explain by mergers. Each time a large galaxy (such as the Milky Way) swallowed a smaller companion, some of the gas that was in the companion sank to the center of the larger system, where it formed new stars. Successive mergers led to successive generations of stars so that the bulge stars have the wide age range that is observed.

A final feature of the merger model is that it allows some ellipticals to turn back into spirals. This can happen if an elliptical captures a gas-rich galaxy. The gas and young stars of the captured spiral then become the disk of a "new" spiral while the original elliptical becomes the bulge and halo.

Even when a merger does not occur, computer simulations indicate that when two galaxies pass near each other, their gravitational force alters the orbits of their stars and thereby changes the shape of each galaxy. Figure 75.12 shows a simulation of

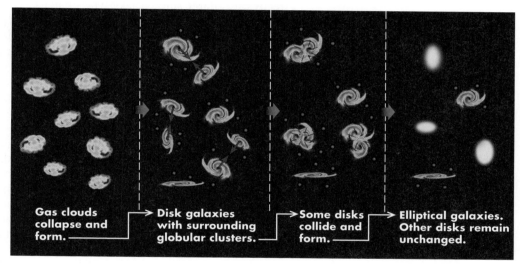

FIGURE 75.11
Elliptical galaxies are created by the collision and merger of disk systems.

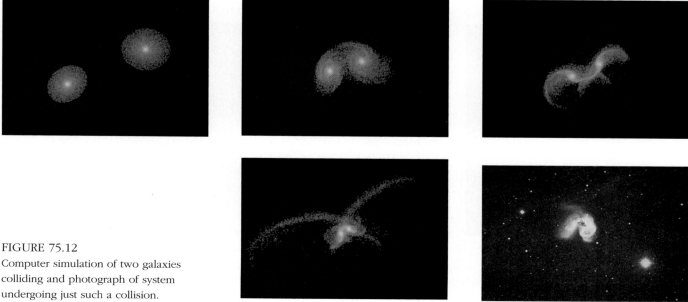

FIGURE 75.12
Computer simulation of two galaxies colliding and photograph of system undergoing just such a collision.

ANIMATIONS

Galaxies with tails

such a close passage, and you can see the stars torn from each galaxy by their mutual gravitational force and flung outward in long, luminous arcs. The photograph (the fifth picture in the sequence) shows two real galaxies with just such plumes of stars.

Although interactions of this type may radically alter the appearance of galaxies, they do not generally disrupt the individual stars. Because stars are so far apart within a galaxy, most stars move harmlessly past one another. Galaxy interactions are much like tossing two handfuls of sand together: Most of the sand particles simply pass by one another without colliding. Dust and gas clouds in the galaxies are not so lucky. Filling each galaxy's space much more completely than stars, clouds do collide, and their impact may compress them enough to trigger a burst of star formation. Such **starburst galaxies,** as these collisionally stimulated galaxies are called, are among the most luminous galaxies known.

Galaxy interactions can create other bizarre forms, as Figure 75.13 illustrates. Figure 75.13A shows a picture of what can happen when a small galaxy plows directly into a large spiral galaxy. The small galaxy has punched a hole through the larger one, creating what astronomers call a **ring galaxy.** The hole results not because the smaller galaxy destroyed the stars in the other galaxy's disk. Rather, it made the hole by shifting the orbits of stars that were near the center of the larger galaxy into wider orbits farther from the center—more like the motion of waves that travel out from the spot where a rock strikes a pond. Figure 75.13B shows another bizarre result of galactic interaction: a pair of galaxies—"The Mice"—distorted by each other's gravity.

Mergers probably affect just about every galaxy sometime during its life. If a large galaxy collides with a small galaxy, the small one is captured and absorbed by the larger one in a process called **galactic cannibalism.** Repeated cannibalism by a galaxy may turn it into a giant, which probably explains why some galaxy groups have an abnormally large elliptical galaxy at their center. In addition to evidence of past mergers, astronomers have detected evidence that our own galaxy is currently devouring its smaller neighbors. For example, radio observations show that the Milky Way is in the process of tearing apart the Magellanic Clouds (Figure 75.14). Our neighboring galaxy, M31, also appears to have recently cannibalized a small galaxy. With the Hubble Space Telescope, astronomers have detected a small

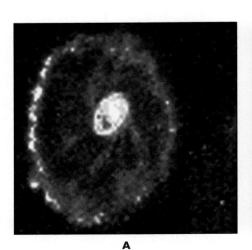

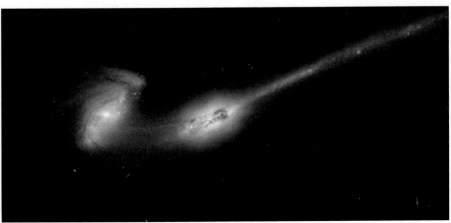

A   B

FIGURE 75.13
(A) A ring galaxy. (B) The Mice—a tidally distorted system.

FIGURE 75.14
This is a false-color radio map of hydrogen gas being stripped out of the Large and Small Magellanic Clouds by the gravitational (tidal) pull of the Milky Way. The tail of gas is about 100 kpc long.

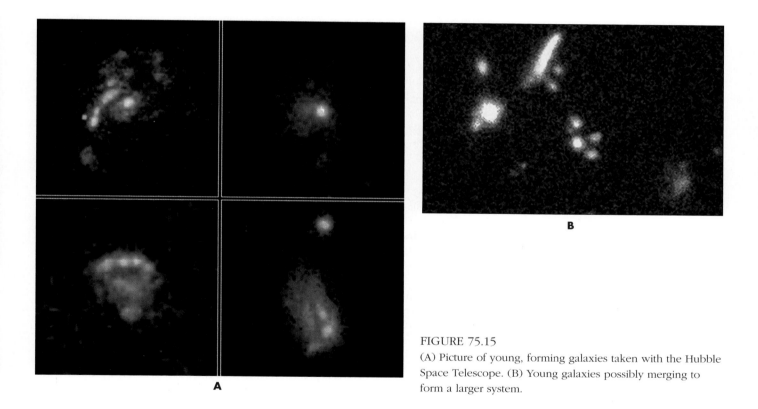

A

B

FIGURE 75.15
(A) Picture of young, forming galaxies taken with the Hubble Space Telescope. (B) Young galaxies possibly merging to form a larger system.

Q. A number of galaxies can be found in Looking Up #1: Northern Circumpolar Constellations; #2: Ursa Major; and #8: Centaurus and the Southern Cross. What type of galaxy does each appear to be? Do you see any evidence of merging?

bright clump of stars near the center of M31 that may be the remains of a galaxy it has absorbed.

Striking pictures of more distant systems made with the Hubble Space Telescope (Figure 75.15) suggest that interactions were even more frequent in the early history of galaxies. From these images, it appears that irregular galaxies represented a higher proportion of the galaxy population in the past than today, implying that with time galaxies change their form. Counts of the number of these galaxies also indicate that there were many galaxies within a given volume, which would have made interactions more likely in the past than now. These were not necessarily direct collisions, but included more distant interactions, called harassment, in which the gravitational force of passing galaxies altered each other's shape. Signs of such collisions in these distant galaxies are seen as bent arms, twisted disks, and other irregular shapes.

KEY TERMS

barred spiral (SB) galaxy, 613

dwarf galaxy, 615

elliptical (E) galaxy, 612

galactic cannibalism, 620

irregular (Irr) galaxy, 613

low surface brightness galaxy, 615

merger, 618

ring galaxy, 620

S0 galaxy, 614

spiral (S) galaxy, 612

starburst galaxy, 620

QUESTIONS FOR REVIEW

1. What are the three main types of galaxies?
2. How do the basic galaxy types differ in shape, stellar content, and interstellar matter?
3. What happens to the stars and gas when two galaxies collide?

PROBLEMS

1. If a galaxy the size of the Milky Way (about 30 kpc in diameter) was placed at various distances, what would its angular size (Unit 10.4) appear to be? Express your answers in arc minutes.
   a. At 10 Mpc?
   b. At 1000 Mpc?
2. For the majority of galaxies astronomers observe, the light travel time is less than a billion years. This is only a fraction of the Sun's age, and therefore we do not expect many differences in the galaxies' properties due to their age. What redshift would a galaxy have that we are seeing as it was 1 billion years ago?
3. The Hubble Deep Field covers an area of about 5 square arc minutes and contains about 3000 galaxies. If we observed the entire sky to this depth, about how many galaxies would we find?
4. The radial velocity of M31 toward the Milky Way is about 100 km/sec. If it maintains that speed, how long will it take to travel the 700 kpc distance between the two galaxies?
5. Assume the disk of the Milky Way has a diameter of 40 kpc and has a surface area found by πR^2. Likewise, each star covers an area determined by $\pi R_\odot{}^2$.
   a. If the Milky Way contains 100 billion stars each the size of the Sun, what fraction of the galaxy's disk is actually "covered" by a star?
   b. The fraction found in (a) expresses the approximate chance that when a single star passing through the Milky Way's disk, it will strike a star. If another galaxy with the same

number of stars passed through the Milky Way, how many stars do you estimate would undergo direct collisions?

TEST YOURSELF

1. A large galaxy contains mostly old (Pop II) stars spread smoothly throughout its volume, but it has little dust or gas. What type of galaxy is it most likely to be?
   a. Irr
   b. S
   c. SB
   d. E
   e. All of the above are possible.
2. In starburst galaxies we do not observe
   a. high luminosity.
   b. stars colliding and exploding.
   c. two galaxies colliding.
   d. a greater proportion than usual of stars forming.
3. Which of the following would be a probable result of a collision between two galaxies?
   a. Complete destruction of all the stars
   b. Formation of a giant spiral from two elliptical ones
   c. Formation of a single open star cluster
   d. Formation of a ring galaxy
   e. All of the above

PLANETARIUM EXERCISES

All the planetarium exercises use the Starry Night planetarium software provided with your text. You should install the program and familiarize yourself with the various parameters and controls before trying these exercises. When the program asks you to enter a password, click "OK" to run Starry Night.

Set the location from which you are viewing to the nearest large city or your latitude and longitude. Accomplish this by clicking on the Location box on the toolbar. Verify that the local time is correct.

Locate a galaxy, such as M31, M81, M104, or M33, by using the Find option in the Selection menu and typing in the galaxy's name. Observe one that is visible for you as indicated by the planetarium program. Correlate the information about the position of your galaxy with the star chart, and locate this galaxy with a small telescope or binoculars, preferably on a clear, moonless night. Observe the galaxy and sketch what you see; then compare it to the images in the text.

76 Galaxy Clustering

Galaxies are not spread uniformly across the sky. Just as stars often lie in clusters, so too galaxies often lie in **galaxy clusters**, and galaxy clusters themselves lie in clusters of galaxy clusters called **superclusters**. Galaxy clusters and superclusters are held together by the mutual gravity of the galaxies and other matter within them. Each galaxy within a cluster moves along its own orbit, just as within a galaxy each star moves on its own orbit.

The clustering of galaxies can be seen when we plot the brightest galaxies on the sky. Each dot in Figure 76.1 represents one galaxy. The clusters seen in this figure are typically 10 million light-years across and may contain anywhere from a handful to a few thousand member galaxies. These immense collections of galaxies are not themselves stationary. They are separating from one another, moving apart in the overall expansion of the universe discovered by Edwin Hubble in 1920.

In this unit we take the next steps in the revolution of our thinking that began when Copernicus recognized that the Earth is just one of the planets. In the early 1900s we did not know about the universe beyond the Milky Way. We find now that our galaxy is not the largest galaxy in it its own rather small galaxy cluster.

Background Pathways: Unit 75

76.1 THE LOCAL GROUP

On the small end of galaxy clustering, astronomers often refer to clusters as **galaxy groups.** Our own galaxy, the Milky Way, belongs to the (unimaginatively named) **Local Group.** The Local Group contains over 40 known members, as sketched in

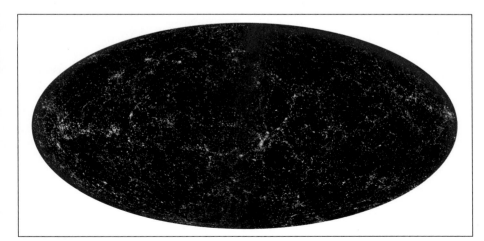

FIGURE 76.1
An all-sky plot of the positions of over a million galaxies detected by 2MASS, an infrared survey of the sky. Each white dot represents a galaxy, or in clusters and superclusters of galaxies, they merge to form brighter areas. The blue shaded portion of the diagram shows where the plane of the Milky Way crosses through this map.

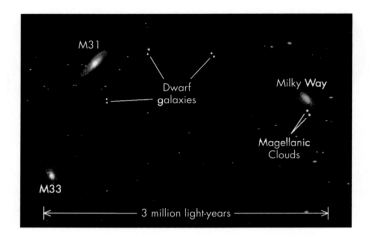

FIGURE 76.2
Artist's view of the Local Group.

FIGURE 76.3
The Local Group's largest member is M31. The Milky Way is intermediate in size between it and M33. Two small satellites of M31 are visible in its picture. Images of three other dwarf galaxies are also shown with approximately the correct relative size.

Figure 76.2. Its three largest members are spiral galaxies: M31, the Milky Way, and M33. The smaller members include the satellite galaxies orbiting M31 and the Milky Way, and dozens of small, low-luminosity **dwarf galaxies** scattered about the group (Figure 76.3).

The most famous of the dwarf galaxies are the **Magellanic Clouds,** which are close to the south celestial pole. They are about one-tenth the size of the Milky Way. They appear to be irregular galaxies, but infrared photographs of the large cloud show that it has a faint barlike structure. Both galaxies are approaching the Milky Way in a very elliptical orbit and are being pulled into a distorted shape by the Milky Way's gravity (Unit 75).

The nearest galaxy to the Milky Way was discovered in 2003 (Figure 76.4). Known as the Canis Major dwarf galaxy, it orbits only about 13 kiloparsecs (42,000 light-years) from the center of the Milky Way. It was not detected earlier because it is merging with the disk of the Milky Way and therefore is largely hidden by dust in our galaxy.

FIGURE 76.4

The nearest galaxy to the Milky Way is called the Canis Major dwarf galaxy. It is hidden from view at visual wavelengths because it is so close to the galactic plane that it is obscured by dust. The image here is based on infrared data from the 2MASS survey. This galaxy appears to be merging with the Milky Way.

 How could we determine whether most of the mass of the Local Group is in the three spirals or in the several dozen smaller galaxies?

The Local Group is particularly interesting to astronomers not only because it is our home galaxy cluster, but also because it allows us to study the demographics of galaxies. That is, we can see the true population of galaxies in this region of space, close to home. For example, most of the galaxies in the Local Group are faint dwarf galaxies such as Leo I (Figure 76.3). Such galaxies are too dim to be visible except when very near. Without the Local Group to let us observe the abundance of these small systems, we might grossly underestimate their true numbers in other galaxy clusters. When we examine much larger clusters, which may contain hundreds or in some cases thousands of member galaxies like the few largest in the Local Group, there are likely to be many times more uncounted dwarf galaxies.

76.2   RICH AND POOR GALAXY CLUSTERS

The largest groups of galaxies are called **rich clusters** because they contain hundreds to thousands of member galaxies (Figure 76.5). The great mass of such a galaxy cluster draws its members into an approximately spherical cloud, with the most massive galaxies near the center.

In contrast, galaxy clusters that contain relatively few members, such as the Local Group, are called **poor clusters.** Poor clusters have so little mass that their gravitational attraction is relatively weak, and their member galaxies are not held in so tight a grouping. They generally have a ragged, irregular appearance, with no central concentration.

Not only do rich and poor clusters differ in their numbers of galaxies, but they also tend to contain different types of galaxies. Rich clusters contain mainly elliptical and S0 galaxies. Moreover, the few spiral galaxies that they contain tend to be found in the outer parts of the cluster. On the other hand, poor clusters tend to have a high proportion of spiral and irregular galaxies. Spirals are rare in the inner regions of rich clusters because of galaxy collisions. In the core of a rich cluster galaxies are close together, so collisions between them are frequent. As we discussed in Unit 75, collisions may change spiral galaxies into ellipticals. Thus,

FIGURE 76.5
The Hercules Cluster, a rich cluster of galaxies.

FIGURE 76.6
M87, at the center of the Virgo cluster, is a giant elliptical.

although rich clusters may have once contained many spirals, today they contain few, and those are the "lucky" ones that escaped running into a neighbor. The many collisions in the crowded inner parts of these clusters also create giant elliptical galaxies by cannibalism as small galaxies merge with the larger galaxies they are orbiting. A giant elliptical galaxy is shown in Figure 76.6. It is at the center of the rich cluster nearest to us, the Virgo cluster at a distance of about 15 megaparsecs (50 million light-years).

Rich and poor clusters differ in yet another way. Observations with X-ray telescopes indicate that rich clusters often contain large amounts (10^{12} to 10^{14} $M_\odot$) of extremely hot intergalactic gas that emits X-rays (Figure 76.7A). In fact, some clusters contain 10 times more matter in hot gas than they do in the stars of all the galaxies within the cluster. Astronomers are not certain where that gas originates. Some of it probably comes from within the cluster's galaxies. For example, when a star explodes as a supernova, some of the gas it ejects may leave the star's galaxy and end up in the intergalactic space within the cluster. On the other hand, some of the gas may be material left over from when the galaxies formed, or it may be interstellar gas stripped from galaxies when they collide.

Astronomers are still uncertain how galaxy clusters form. But evidence is accumulating that clusters form by galaxies first attracting one another into small groups. Over time, these small groups pull in more galaxies and merge with other groups to make larger clusters, eventually growing into the rich clusters we see today. Much of this evidence comes from our ability to "see back in time" by looking out in space. In particular, galaxies in nearby clusters are generally fairly evenly spread within the clusters. Galaxies in remote (younger) clusters, however, sometimes have a clumpy distribution in the cluster, as if they have recently joined it and have not had time to spread themselves smoothly. Moreover, X-ray observations of these clusters (Figure 76.7B) show that the gas is sometimes "clumpy," and the clumps are colliding with one another.

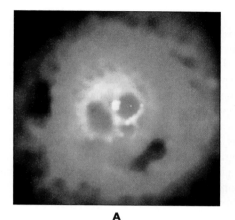

A

B

FIGURE 76.7

(A) A false-color X-ray image of the Perseus galaxy cluster. The orange glow is hot gas in the cluster. The picture shows a region about 600,000 light-years across. (B) A combined false-color image of a distant cluster in the constellation Hydra. X-ray emission is shown in violet, while the optical emission from galaxies is shown in other colors. This cluster is more than 8 billion light-years away, so it shows the appearance of the cluster when it was less than half its current age.

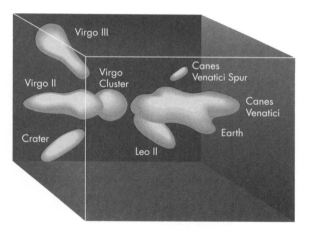

FIGURE 76.8

A depiction of the structure of the Local Supercluster. A number of groups and moderately rich clusters surround the rich Virgo cluster, which contains most of the supercluster's mass.

76.3   SUPERCLUSTERS

The great mass of galaxy clusters creates gravitational interactions between them and holds them together into clusters of clusters, or *superclusters*. These large structures contain a half dozen to many dozen galaxy clusters spread throughout a region of space, ranging from tens to hundreds of millions of light-years across. For example, the Local Group is part of the **Local Supercluster,** which contains the Virgo galaxy cluster and more than a dozen other clusters spanning nearly 30 Mpc. It is sketched in Figure 76.8.

The presence of large concentrations of galaxies can also be deduced from the motions of galaxies. In general, galaxies are moving away from one another as part of the overall expansion of the universe, and their distances follow the Hubble law (Unit 74.3). However, in regions where there is a large mass concentration, we see deviations from the velocities predicted by Hubble's law. Such a deviation has been found for the motions of the Local Group itself, along with the Virgo cluster and other galaxies and clusters out to a distance of about 50 Mpc from the Milky Way.

These peculiar motions suggest the presence of a huge concentration of mass in the direction of the constellation Centaurus. Our whole local supercluster has a deviation of over 500 km/sec in that direction. To cause an entire supercluster to move at such a speed would require an enormous gravitational pull. The source of this enormous pull has been named the **Great Attractor** by astronomers. The direction of our motion is approximately toward a position near the center of the all-sky map shown in Figure 76.1. There is a fairly large concentration of galaxies in this direction, although many galaxies are hidden from our view in this direction because of obscuration by the plane of the Milky Way.

The Great Attractor is probably not a single large mass, but an assemblage of many superclusters covering a large region centered about 50 Mpc (about 160 million light-years) away from us. It may contain an excess of about $3 \times 10^{16}\ M_\odot$—the equivalent of 10,000 or more galaxies like the Milky Way—compared to regions of comparable size in other directions. The Great Attractor may hint at an even larger kind of structure than a supercluster—a cluster of superclusters!

76.4 LARGE-SCALE STRUCTURE

On the largest scales, astronomers have traced a variety of immense structures out to the limits of where we have surveyed galaxies. Figure 76.9 shows a map of the distribution of galaxies out to a distance of several billion light-years. On this map each galaxy is represented by a single dot. Its distance from us is based on its recession

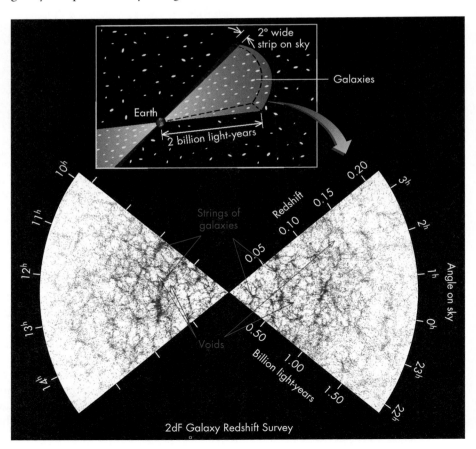

FIGURE 76.9

A map of how galaxies are distributed in space. The shape of the region shown in the figure is like two slices of pizza, tip-to-tip, as illustrated in the upper sketch. Each dot in the lower figure represents one of some 100,000 galaxies that lie in the wedge-shaped regions that extend out to about 2 billion light-years from Earth. Note the nearly empty regions (voids), the stringy structures, and the overall cobwebby appearance. The distances to the galaxies were measured from their redshifts. (The angles refer to Right Ascension coordinates on the sky.)

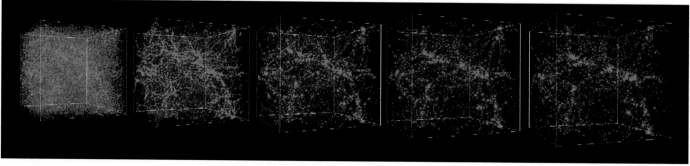

FIGURE 76.10

A computer simulation of the growth of structure as the universe expands. You can see how an initially smooth distribution of matter becomes stringy, much like the features in Figure 76.8. Each box spans about 140 million light-years. The first box on the left represents a time when the universe was roughly one-tenth its present age. The last box on the right shows that same piece of the model of the universe as it would be today.

Q What is the difference between the information shown in a map like that in Figure 76.9 and that in Figure 76.1?

velocity (Unit 74). This map represents a thin slice of the sky in which the redshifts of about 100,000 galaxies have been observed by a large team of astronomers. It gives us a cross-sectional look at the distribution of galaxies out to distances of about 700 Mpc (about 2 billion light-years).

On this scale individual galaxies and clusters are no longer obvious, but we *can* see that they often appear to be arranged into long filaments or shells surrounding regions nearly empty of galaxies. These latter spaces are called **voids,** and although they are not totally empty, the few galaxies they contain are generally small and dim.

Large-scale structures such as superclusters, voids, and filaments are at the frontier of our knowledge of the distribution of galaxies in space. The patterns of large-scale structure we see at this scale are probably the consequence of the initial conditions in the universe, and astronomers have proposed a variety of hypotheses to explain the features we see. For example, when the huge voids were first found, some astronomers suggested that they might be evidence that even before galaxies began forming, huge explosions pushed the gas out of these regions. Other astronomers have hypothesized that these features were probably produced by gravitational interactions of dark matter (Unit 78).

To attempt to understand large-scale structure, astronomers have turned to computer simulations. In such simulations, astronomers write out the equations that govern the motion of matter (basically Newton's laws) and then solve the equations, assuming the material filling space is expanding so as to mimic the general expansion of the universe. They then watch what happens as the gravitational force of the material acts within the simulated volume. Figure 76.10 shows a series of images from such a simulation. You can see that gravity draws matter that initially had a relatively smooth distribution into a network of strings. Astronomers identify the places where strings meet as locations where galaxies form. They adjust the equations by varying the amount of ordinary and dark matter and initial lumpiness, thus experimenting to discover what mix best creates structures similar to what is actually observed. The outcome of many such simulations by many astronomers confirms that no explosions are needed; but without dark matter, the distribution of galaxies fails to match what we observe.

76.5   PROBING INTERGALACTIC SPACE

The large-scale structure of galaxies indicates where the most luminous galaxies are concentrated, but astronomers suspect that a large amount of matter must lie between the galaxies in intergalactic space. Observations using a variety of techniques are beginning to uncover large reservoirs of matter that never formed into the collections of stars we call galaxies.

We can see the hot intergalactic gas inside rich clusters of galaxies from the X-rays it emits. Outside these regions, intergalactic gas is much cooler and more difficult to detect. On the other hand, cool gas can be detected spectroscopically by its absorption against a bright background source. **Quasars** are bright sources of light that are extremely distant and therefore useful as probes of intergalactic space along our line of sight toward them. Quasars are intriguing objects in their own right (Unit 77), but for absorption studies we are interested only in the fact that they are extremely distant light sources.

Because quasars are so distant, their light passes through many other galaxies and intergalactic gas clouds as it travels to Earth. Even if an intervening galaxy or cloud is too dim to see, its matter will imprint a signature on the quasar's spectrum, allowing us to detect these otherwise invisible objects. Because these absorbing objects all have different redshifts, each produces a set of absorption lines shifted to different wavelengths. This results in a whole "forest" of absorption lines, each one corresponding to a cloud of gas along the line of sight, as illustrated in Figure 76.11.

Most of these absorption lines come from clouds that have not yet formed into galaxies themselves but are condensing and cooling. Studies of these clouds suggest that they follow a large-scale structure similar to that of galaxies. Intergalactic gas is located in huge sheets, filaments, and voids out to the very largest distances we can see—and therefore the earliest times we can see in the universe's history. The amount of this gas appears to decline with time—in other words, there is less of it in the regions nearer to us. Simulations suggest that over time intergalactic gas accretes onto galaxies, providing a source of additional gas for galaxies and perhaps affecting their evolution. These simulations also suggest that such accretion may continue even to the present, and astronomers are searching for evidence of this very tenuous gas and its possible effects on the Milky Way and other nearby galaxies.

Q How might the absorption lines produced by a spiral galaxy in front of a quasar look different from those produced by a small cloud of gas?

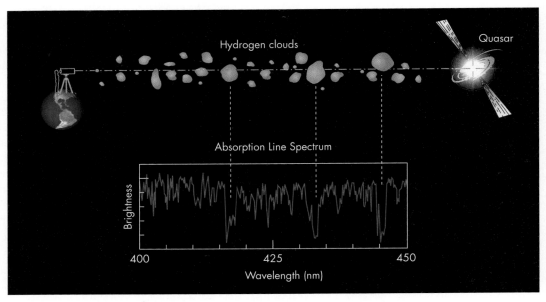

FIGURE 76.11
The light from a distant quasar produces a uniform light source against which we can see hundreds of absorption lines produced by hydrogen clouds along the line of sight to the quasar.

KEY TERMS

dwarf galaxy, 624

galaxy cluster, 623

galaxy group, 623

Great Attractor, 628

large-scale structure, 629

Local Group, 623

Local Supercluster, 627

Magellanic Clouds, 624

poor cluster, 625

quasar, 630

rich cluster, 625

supercluster, 623

void, 629

QUESTIONS FOR REVIEW

1. What is the Local Group?

2. What is the Local Supercluster?

3. What are the differences between rich and poor clusters?

4. What is large-scale structure?

5. What produces quasar absorption lines?

PROBLEMS

1. Some galaxies in the Virgo cluster are found to be moving at 1000 km/sec through the center of the cluster. How long would it take to move 1 Mpc at that speed?

2. The Great Attractor, located some 50 Mpc away, is suspected to have caused the Local Group to accelerate to a speed of 500 km/sec over the last 13 billion years.
   a. What is the average acceleration over that time, expressed in units of meters per second^2?
   b. Using Newton's law of gravitation, how big a mass would be needed at that distance to produce the observed acceleration?

3. The spectral line of hydrogen producing the quasar absorption lines in most studies has a laboratory wavelength of 122 nm. If a hydrogen cloud is at a distance of 1 billion light-years,
   a. use Hubble's Law to find the recession velocity of this cloud.
   b. What is the redshift of the cloud?
   c. At what wavelength will the absorption occur?

TEST YOURSELF

1. The Local Group
   a. contains about 30 member galaxies.
   b. is a poor cluster.
   c. is the galaxy cluster to which the Milky Way belongs.
   d. mostly contains galaxies much smaller than the Milky Way.
   e. All of the above

2. Why are quasars used to observe the absorption lines produced by intergalactic gas clouds?
   a. Because this is the gas left over from when quasars exploded.
   b. Because quasars and the gas are of similar age and hence are near each other.
   c. Because the hotter gas in galaxies bends the quasar's light away from us.
   d. Because quasars are attracting the clouds.
   e. Because quasars are very distant and bright.

3. The majority of galaxies in rich clusters are _____, whereas poor clusters contain high proportions of _____ galaxies.
   a. elliptical and S0 type; spiral and irregular
   b. elliptical and spiral; irregular
   c. spiral; irregular
   d. elliptical and irregular; spiral and S0 type
   e. spiral and irregular; elliptical and S0 type

77

Active Galactic Nuclei

Background Pathways: Unit 74

If our eyes could see radio wavelengths instead of visible light, the night sky would look completely different to us. For example, many of the brightest sources at radio wavelengths vary in brightness, so that if we worked out a pattern of constellations, we would find in a few years' time that some "stars" had become so much brighter, and some so much fainter, that the constellations would no longer look the same. And when we tried to measure the distances to these "stars," we would find that many of the brightest are among the most distant objects in the universe.

The strangeness of the radio sky became apparent during the 1950s and 1960s as the early technology for making radio observations steadily improved. Astronomers found that some radio sources were associated with objects already cataloged at optical wavelengths (such as galaxies, nebulae, and the galactic center). But others—including many of the brightest radio sources—appeared to have no optical counterpart. Later some of the strongest radio signals were identified with dim "stars" that proved to be extremely distant, implying that these objects were the most luminous objects in the universe.

Today we recognize these luminous sources as **active galactic nuclei** or **AGNs**. They are tiny regions in the centers (nuclei) of galaxies that emit abnormally large amounts of energy. Not only is the emitted radiation intense, but it usually fluctuates in intensity as well. At least 10% of all known galaxies have active cores, and in many instances they exhibit intense radio emission and other activity outside their cores in addition to their AGNs.

77.1 ACTIVE GALAXIES

Over the past half century different classes of galaxies with unusually vigorous energy output have been discovered. Their discovery occurred in a variety of different ways at different wavelengths; at first they were thought to be independent classes of objects. Although the amount of activity that these galaxies exhibit varies, the past two decades' evidence suggests that most of this activity probably shares a common explanation. Here we describe two major classes of activity found in nearby galaxies.

A **Seyfert galaxy** is a spiral galaxy whose nucleus is abnormally luminous (Figure 77.1). These unusual galaxies are named for the U.S. astronomer Carl Seyfert, who first drew attention to their peculiarities in the 1940s. The core luminosity of a typical Seyfert galaxy is immense, amounting to the entire radiation output of the Milky Way but coming from a region less than a parsec across. Moreover, the radiation is at many wavelengths: optical, infrared, ultraviolet, and X-ray. Despite its immense luminosity, the radiation from a Seyfert galaxy's core fluctuates rapidly in intensity, sometimes changing appreciably in a few minutes.

FIGURE 77.1

The Seyfert galaxy NGC 7742, imaged with the Hubble Space Telescope, is a spiral galaxy with an extremely luminous nuclear region.

In addition to the intense source of radiation in their nuclei, Seyfert galaxies also contain gas clouds moving at speeds of about 10,000 kilometers per second near their cores. In some instances, the core is seen to be ejecting gas in a pair of oppositely directed jets that collide with surrounding gas in the galaxy.

Another type of activity is seen in **radio galaxies.** As their name indicates, they emit large amounts of energy in the radio part of the spectrum. Some emit millions of times more energy at radio wavelengths than do ordinary galaxies. This energy is generated primarily in two types of regions: the galaxy's nucleus and enormous regions outside the galaxy and on opposite sides of it, as you can see in Figure 77.2. This figure shows false-color images in which the intensity of radio emission is shown in red while the visible emission (showing light from stars in the galaxy) is shown in blue. The **radio lobes,** as the regions of strong emission outside the galaxy are called, may span hundreds of thousands of parsecs (about a million light-years). On the other hand, the core source is typically less than one-tenth of a parsec across.

What causes the intense emission of radio energy from these galaxies? Astronomers can tell from its spectrum that the emission is **synchrotron radiation,** generated by electrons traveling at nearly the speed of light and spiraling around magnetic field lines. The electrons are part of a hot ionized gas that is somehow shot out of the core in narrow **jets** (such as the one shown in Figure 77.2C) that eventually collide with intergalactic gas in the vicinity. There the jets of ionized gas spread out to form the radio-emitting lobes that we observe.

Clearly there are some similarities between radio galaxies and Seyfert galaxies. They both involve rapidly moving gas and a small, bright nuclear region. Furthermore, some Seyfert galaxies have been found that have jets emanating from the nucleus, though smaller than those seen in radio galaxies. Some further clues about such strange behaviors of galactic nuclei come from some of the most distant objects we know of in the universe.

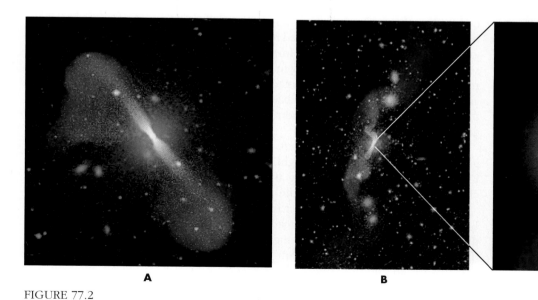

A   **B**   **C**

FIGURE 77.2

These images show two radio galaxies, NGC 5532 (A) and NGC 383 (B & C). The figures show the radio emission in red—the brightness indicating its intensity—superimposed on an optical image, shown in blue. The extended red regions in A and B are called radio lobes. Image (C) is a close-up of the central region of the radio emission in NGC 383, showing the bright nuclear source and jets of hot plasma being shot out of the center of the galaxy.

77.2   QUASARS

Quasars are extremely luminous, extremely distant, active galactic nuclei. They were originally identified because many of them are among the brightest radio sources found in early surveys of the "radio sky." Quasars get their name by contraction of the term *quasi-stellar radio source,* where *quasi-stellar* (meaning "almost starlike") refers to their appearance in the first photographs taken of them (Figure 77.3), in which they look like stars—that is, like points of light. Early photographs also revealed that some quasars have small wispy clouds near them or tiny jets of hot gas coming from their cores, like the one seen in Figure 77.3 coming from the quasar.

Pictures of quasars lack detail because these objects are so far away. We deduce their immense distance from the huge redshift of their spectra. (Recall from Unit 74.3 that a galaxy's distance can be found from its redshift using the Hubble law.) In fact, quasars as a class have the largest redshifts known for any kind of astronomical object. The redshifts of quasars are so large that their visible spectra are often shifted completely to infrared wavelengths, and what we observe at visible wavelengths was originally emitted by the quasar in the ultraviolet. The visible-light spectra astronomers observed were therefore unfamiliar to astronomers. In 1963 the Dutch-American astronomer Maarten Schmidt realized that they were highly redshifted and therefore extraordinarily distant objects.

On the basis of its redshift, the most distant quasar yet observed emitted the light we see today nearly 13 billion years ago, less than 1 billion years after astronomers estimate that the universe began (Unit 79). To be visible at such an immense distance, an object must be immensely luminous, and quasars turn out to be about 1000 to 100,000 times more luminous than the Milky Way. In fact, the brightest quasars have a power output equivalent to a supernova explosion occurring every hour!

During the first decades after quasars were discovered, their extraordinary properties caused some astronomers to question the large distances that had been deduced for them. Alternative hypotheses were offered, suggesting that quasars were nearby stars and that their observed redshifts had a different cause—such as a large gravitational redshift (Unit 68). However, the Hubble Space Telescope was able to settle the question in the 1990s by making high-resolution images of quasars' surroundings (Figure 77.4). These images revealed that quasars lie at the centers of galaxies, like Seyfert and radio galaxies. They resemble radio galaxies in other ways too, producing jets of material and radio lobes, for example, although quasars are generally much more luminous.

Despite their huge power output, quasars appear to be very small. Like Seyfert galaxies, quasars fluctuate in brightness (Figure 77.5). Slow changes occur over months, but short-term flickering also occurs in periods as brief as hours. Rapid changes of this sort give clues to the size of the emitting object.

For example, suppose an object is one "light-hour" in diameter (about 7 AU). Light from its near side therefore reaches us one hour earlier than the light from its distant side. Even if every part of the object lit up simultaneously, after light from the near side reached us, light from the distant portions of the object would not reach us for an additional hour (Figure 77.6). And if it turned off all at once, it would again take an hour for the object to drop back down to its original brightness. An object's light cannot fully appear or disappear in less time than it takes light to travel across it. We can write this as a formula:

$$\text{Diameter} < c \times \Delta t$$

where c is the speed of light and Δt (pronounced "delta t") is the time interval over

Quasars often have redshifts greater than 1. This does not mean that they are traveling at speeds greater than the speed of light; it indicates that the universe has more than doubled in size since light left the quasar (see Unit 79).

If a quasar is seen as it looked 13 billion years ago, was it 13 billion light-years distant when the light was emitted? Is it this far away now?

FIGURE 77.3

The quasar 3C273 from a ground-based telescope image. It is the brightest visually of any quasar, but is a dim "star." Even though it is one of the brightest radio sources in the sky, it is a thousand times fainter than the dimmest star visible to the unaided eye. "3C" stands for the Third Cambridge Catalog of Radio Sources.

A similar equation could be written for sound waves if we replaced c by the speed of sound. Imagine, for example, everyone in a football stadium snapping their fingers simultaneously. The sound you would hear outside the stadium would be spread out by the sound travel time across the stadium.

which the quasar's brightness changes substantially. We write that the diameter is less than $c \times \Delta t$ because the actual diameter can be smaller than this limit if the luminosity changes are not instantaneous.

The light variability intervals imply that many quasars are smaller than the size of the Solar System. If a quasar shows substantial variations in $\Delta t = 10$ hours (36,000 seconds), its diameter is smaller than

$$c \times 36{,}000 \text{ sec} = 3.0 \times 10^8 \text{ m/sec} \times 3.6 \times 10^4 \text{ sec} = 1.1 \times 10^{13} \text{ m.}$$

An object this diameter would fit within the orbit of Pluto. Astronomers have confirmed the remarkably small size of a number of quasars using radio interferometer telescopes. The incredible power output of a quasar comes from a very small object.

Today we recognize a near continuum of properties spanning the range from nearby radio galaxies and Seyfert galaxies to quasars. What, though, could possibly explain such extraordinary luminosity from such a small object? Many ideas have been tried. For example, at one time, astronomers thought that the luminosity was

FIGURE 77.4
The quasar PG 0052+251 imaged with the Hubble Space Telescope. The image reveals that the quasar lies in the nucleus of a spiral galaxy. ("PG" stands for Palomar Green survey.)

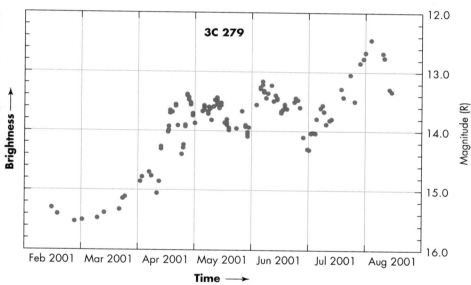

FIGURE 77.5
Plot of the light variation of a quasar.

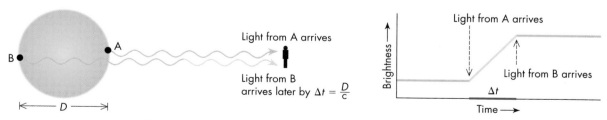

FIGURE 77.6
Sketch illustrating that when a light source turns on, we see the light from its near side before we see the light from its far side, and so we do not see its light turn on instantly. For the objects that we encounter in everyday life, the delay is so short we do not notice it. However, for astronomical bodies, the effect can create delays ranging from days to months, from which we can deduce that the emitting body is light-days or light-months across.

produced by a single star of 10^8 solar masses that had managed to form in the center of the galaxy. Astronomers have also considered more exotic explanations. For example, a few scientists have proposed a new class of objects called "white holes." These would be the opposite of black holes in the sense that they would spew material out into space rather than draw it in. This is an intriguing idea, but no other evidence for them exists, and we gain little understanding of an extraordinary phenomenon if we resort to inventing a fantastic new kind of object.

77.3   CAUSE OF ACTIVITY IN GALAXIES

Because radio galaxies, Seyfert galaxies, and quasars share many common features, astronomers have sought a unified model for their activity that could explain all three.

Any successful model for active galactic nuclei must explain how such a small central region can emit so much energy over such a broad range of wavelengths. For example, at least one quasar is a powerful source of both radio waves and gamma rays. Astronomers therefore hypothesize that the core must contain something very unusual. No ordinary single star could be so luminous. No ordinary group of stars could be packed into so small a region. One kind of object that is very small but that can also emit intense radiation is the accretion disk around a black hole. When matter falls toward a black hole, it generally does not fall straight in, but ends up orbiting just outside the black hole, where it collides with other matter and emits intense X-ray radiation.

The need for an extreme object in the cores of these objects is demonstrated by the very high speeds of jets measured in a number of quasars and at least one Seyfert galaxy. Clumps of gas shoot out from the AGN at speeds clocked to be greater than 90% of the speed of light! Whatever is in the core must be able to accelerate matter to almost the speed of light.

Based on the accumulated evidence—the huge energy output, the small size, the enormous speeds—astronomers have ruled out all "ordinary" objects. The best remaining hypothesis is that active galactic nuclei contain an immense black hole. According to this model, the black hole would contains a hundred million or more solar masses around which swirls a huge accretion disk of gas that produces the energy we see. Such a black hole would have a radius about the size of the Earth's orbit around the Sun, while the accretion disk would have a radius of tens to hundreds of AU. Gas within the disk spirals toward its center, drawn inward by the black hole's gravity, and is ultimately swallowed. Orbiting under the black hole's fatal attraction, the gas orbits faster and faster, growing hotter and hotter as it is pulled deeper into the gravitational field of the black hole—converting the gravitational energy into kinetic and thermal energy. The gas is heated to incandescence by frictional forces in the disk and emits the light and jets that we observe.

Figure 77.7, a picture made with the Hubble Space Telescope, supports this model. The picture shows just such a disk in the central regions of the radio galaxy NGC 4261. Gas falling into the disk releases gravitational energy, which heats the material to temperatures perhaps as hot as several million Kelvin. Although most of the mass ultimately falls into the black hole, some material from the accretion disk boils off into space. If a thick accretion disk surrounds the black hole, it may channel such escaping gas into narrow jets from the central regions of the top and bottom surfaces of the disk (Figure 77.8).

Although this model nicely explains the jets from active galactic nuclei, astronomers have recently proposed a revised theory. In this new model, the accretion disk has a strong magnetic field that is twisted as the disk spins. Magnetic fields trap charged particles that exist in a hot ionized gas, as also occurs in the

In jets that point nearly toward the Earth, the high-speed motion can sometimes make clumps of gas appear to be moving faster than the speed of light. This is an illusion of timing because the gas is moving toward us, so light emitted from it at later times has less distance to travel before reaching us.

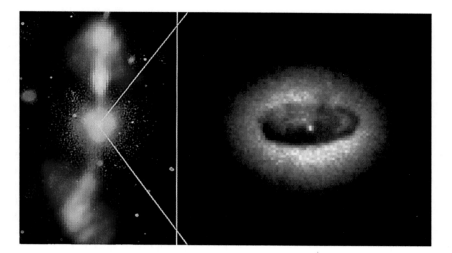

FIGURE 77.7

An image (at far right) made with the Hubble Space Telescope of an accretion disk in the center of the radio galaxy NGC 4261. The inset image on the left shows a wider view of this galaxy's central region and shows the jets of hot gas being ejected from the central regions of the disk.

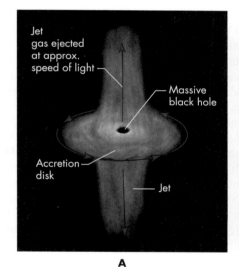

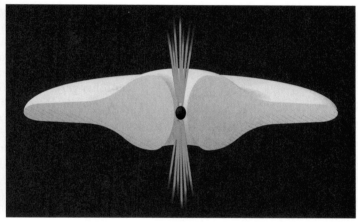

FIGURE 77.8

(A) Sketch of jets formed by an accretion disk. (B) The accretion disk may help to funnel the hot gas into the jet.

magnetic field of the Sun (Unit 51) and pulsars (Unit 67). The spinning and twisting of the magnetic field allows it to capture hot gas from the accretion disk and pump it outward along the field. This hypothesis may better explain the narrowness of the jets along with their kinks and turns. Moreover, it provides a powerful supply of energy for the jet—the rotational energy of the spinning black hole.

Whichever hypothesis is correct, left to itself, the accretion disk gradually drains into the black hole. Once this supply is exhausted, with no more infalling matter, activity ceases. The disk, however, may be replenished with matter from at least two sources. Sometimes stars within the galaxy may approach too closely to the black hole and be torn apart by its tidal forces. Their gaseous debris then forms a new accretion disk, providing a new source of energy. Alternatively, collisions with neighboring galaxies may occasionally add new material as the active galaxy cannibalizes its smaller neighbors. Each such event can give the central black hole a new source of matter for its accretion disk, causing the active galactic nucleus to increase in brightness.

Such a model for AGNs may sound extreme, but our own galaxy appears to have a black hole at its center, albeit a much smaller one than the one we are describing here (Unit 73.2). According to one hypothesis, our own galaxy's central black hole formed in several steps. First, a massive star in the central region of the Milky Way reached the end of its life, exploded as a supernova, and formed a black hole perhaps as small as 5 $M_\odot$. Normally a black hole like this has little opportunity to grow—it is only 30 km in diameter, so the odds of anything hitting it are small. However, at the center of the galaxy, matter is packed densely and was falling toward this region during the early stages of galaxy formation. As a result, an initially small black hole at our galaxy's center had a special opportunity to grow by swallowing infalling matter. Bit by bit, the mass of the black hole (and therefore its radius) increased, making it easier for the hole to attract and swallow yet more material. Eventually the hole became large enough to swallow entire stars, at which point it grew rapidly to a million or more solar masses. During its growth the hole was surrounded by an accretion disk. Our Milky Way probably had an active galactic nucleus. As the Milky Way aged, its black hole consumed the gas around it, and the activity subsided to the low level we see today.

Other galaxies may have had more mass at their cores to begin with, or they may have gained mass by many mergers with small neighboring galaxies. This inflow of extra mass thereby allowed their central black holes to grow far larger than the black hole in the Milky Way. In either case, we expect that the central black hole is active mainly when a galaxy is young and there is substantial matter accreting onto the black hole. After hundreds of millions of years, any material that has an orbit carrying it near the black hole will have been swallowed by the black hole or ejected by the process that produces the AGN jets. With no more matter feeding the black hole, it grows quiescent. However, each time the galaxy undergoes a collision with another galaxy, additional matter may end up on orbits that lead it close to the central black hole, reawakening the activity.

Even though only about 10% of galaxies show signs of activity, astronomers suspect that most (perhaps even all) large galaxies have a supermassive black hole at their cores. They come to this conclusion from the observation that stars and gas at the core of many galaxies are orbiting the core at extremely high speed, as in the Milky Way (Unit 73.2). To keep such rapidly moving material from flying out of the galaxy's core requires a large mass. Binding the stars and gas in such tight but high-speed orbits requires (in many cases) more than a million times the mass of the Sun. Yet this huge mass seems to emit no light or other radiation, and black holes are the only known objects that are both massive and dark enough.

We just described how such a huge black hole might have formed at the center of the Milky Way by the steady accumulation of mass into an initially small black hole. But another recent discovery has led astronomers to wonder if supermassive black holes might form in some other way. Observations made with the Hubble Space Telescope show a relationship between the mass of the black hole and the mass of its host galaxy's central stellar bulge. They find that the more massive the black hole, the more massive the bulge. Why does such a relation exist? Does the black hole form first and shape the bulge, or vice versa? Astronomers do not yet know, but one piece of evidence suggests that the central black hole may grow along with the bulge when one galaxy merges with another. This evidence comes from X-ray observations of a galaxy that appears to contain a pair of orbiting massive black holes in the process of merging.

If black holes are so common in galaxies, why are none of the active galactic nuclei close to us as luminous as quasars? The youth of quasars may be the key. On average quasars are more than 9 billion light-years away. Because quasars are so distant, we see such objects the way they were in the distant past, not the way they are now. Recall that when we look at an object that is 9 billion light-years away, we are seeing light that has taken 9 billion years to reach us. Thus, we are seeing the object as it was 9 billion years ago. The fact that most quasars are many billions of light-years away implies that the phenomenon causing them occurred billions of

Q Suppose you currently lived in one of the galaxies that we are currently observing as a quasar. What would your galactic center look like to you today? What would the Milky Way look like to you?

years ago, when the universe was younger. Objects that once were quasars may surround us, but their activity has long since died away.

So despite the reliance of our model of AGNs on an object as peculiar as a supermassive black hole, the model is consistent with observations of quasars and Seyfert and radio galaxies. Consistency is not proof, but the supermassive black hole model is the strongest hypothesis for explaining AGNs.

KEY TERMS

active galactic nucleus (AGN), 632

jet, 633

quasar, 634

radio galaxy, 633

radio lobe, 633

Seyfert galaxy, 632

synchrotron radiation, 633

QUESTIONS FOR REVIEW

1. What are the three main types of active galaxies?
2. What is a quasar?
3. Why do some astronomers think that quasars are very distant?
4. How is it known that active galaxies have small core regions?
5. What mechanism has been suggested to power active galaxies?

PROBLEMS

1. If a quasar is generating 10,000 times more luminosity than a galaxy, how many times farther away would it be if it appeared to be the same brightness?
2. The radius of a black hole is approximately 3 km times the mass of the black hole in solar masses (Unit 68).
   a. Find the diameter of a black hole with a mass of 1 billion solar masses in astronomical units.
   b. For an accretion disk 10 times the radius of the black hole, how many light-hours across would it be?
3. Hydrogen has a spectral line at 122 nm in the ultraviolet portion of the spectrum. What are the largest and smallest

redshifts a quasar could have for this line to be seen in the visible portion of the spectrum (400–700 nm)?

4. An AGN is observed to double in brightness in a week and then to grow dim again in the next week. What is an upper limit to the probable size of the AGN? Give your result in AU.

TEST YOURSELF

1. A spiral galaxy has a small bright central region, and its spectrum shows that it contains hot, rapidly moving gas. It is most likely a _____ galaxy.
   a. barred spiral
   b. Seyfert
   c. radio
   d. dwarf
   e. quasar
2. What produces the synchrotron radiation seen in radio galaxies?
   a. High-speed electrons spiraling around the magnetic field lines
   b. Supernova explosions
   c. Visible starlight that has been heavily redshifted
   d. Hydrogen molecules
   e. Radiation from a massive black hole at the center of the galaxy
3. Because the large redshifts of quasars arise from the expansion of the universe, we can conclude that
   a. quasars must be very small.
   b. quasars must be within the Local Group.
   c. quasars must be single stars with extremely large masses.
   d. quasars must be moving toward Earth with a large radial velocity.
   e. quasars must be very luminous.

78 Dark Matter

When we map the positions of galaxies throughout the universe, it is something like studying the lights seen from an airplane flying over a city at night. The pattern of light suggests the general shape of things, but there is a whole landscape that can only be guessed at. Astronomers have begun to realize that galaxies hint at a very strange landscape, in which most of the matter in the universe is of a type unlike anything we know.

We can deduce the existence of this other kind of matter because it exerts a gravitational force on the things we see, such as stars and galaxies, even though it is not seen in visible light. Initially astronomers labeled this as **dark matter** to indicate that they knew mass was present but that it emitted no light. But a growing body of evidence says that it is a kind of matter that in fact *cannot* be seen. This unusual matter has mass that produces gravity, but it cannot coalesce into stars. It may be made of particles that do not interact with photons, and therefore it may be all around us without being detected.

It is interesting to consider how far we have moved from our Earth-centered view of the universe in our exploration of galaxies. We have learned that the Sun lies in the outskirts of a galaxy, which is not a particularly significant galaxy, which is in a minor cluster of galaxies. And now we are realizing that the kind of matter that makes up everything we know is just a minor kind of matter in the universe. This is the Copernican revolution taken to extremes! The idea that dark matter dominates the universe is revolutionary, and revolutionary ideas require exceptionally strong evidence. But the evidence that has accumulated in favor of this idea is so strong that it is now widely accepted among astronomers. In the following sections we will see how astronomers have come to this conclusion. But before we describe the search for dark matter, it will be helpful to review how astronomers measure a galaxy's mass.

Background Pathways: Unit 77

78.1 MEASURING THE MASS OF A GALAXY

We can measure the mass of a galaxy the same way we measure the mass of a star or planet: by applying the modified form of Kepler's third law (Unit 17). This method uses the principle that the orbital motion of one object around another is set by their mutual gravity. Their gravity is in turn determined by their combined mass. Thus, from the orbital motion of an object we can find its mass. We saw in Unit 73 that when astronomers use orbital motion to measure the mass of the Milky Way, they find a puzzling discrepancy between the mass they calculate and the mass they can account for.

Studies of other galaxies reveal essentially the same discrepancy. Almost without exception, the mass calculated from the modified form of Kepler's law is larger than the mass of stars and interstellar matter detectable by optical, infrared, and radio telescopes. Galaxies seem to contain a significant amount of matter that cannot be seen. This discrepancy is very large, typically amounting to 10 times a galaxy's visible mass. This means that when we look at the starlight from a galaxy, we may in some cases be observing only one-tenth of its matter.

The most direct evidence for dark matter comes from **rotation curves** of spiral galaxies. A galaxy's rotation curve is a plot of the orbital velocity of the stars and gas moving around it at each distance from its center (curve A in Figure 78.1). A rotation curve reveals how a galaxy's mass is distributed because the speed at each radius from the galaxy's center reveals how much mass must lie interior to the orbit at that distance (Unit 73). We can understand how that dependence arises by thinking about what keeps a star in orbit.

According to Newton's first law, if an object does not move in a straight line (for example, a star moving in a circular orbit around a galaxy), a force must be acting on it. We know that for a star orbiting in a galaxy, that force is supplied by gravity. We also know that the strength of the gravitational force depends on the mass of the attracting body (in this case, the galaxy). It turns out, however, that to a good approximation, only the material between the star's orbit and the galaxy's center contributes to the gravitational force. If the speed of rotation is V at radius R, we can write this as an equation for the mass interior to that radius as in Unit 73:

$$M_{<R} = \frac{V^2 \times R}{G}$$

where the subscript "$<R$" is used as a reminder that the measurement represents the mass interior to this radius.

Stars near the center of a galaxy will therefore have only a small force acting on them because there is little mass between them and the center. Because the force acting on them is small, such stars cannot be orbiting very rapidly, or they would have flown out of the galaxy. On the other hand, stars orbiting far from the center should also feel a small gravitational force acting on them from the galaxy. They feel the effect of the galaxy's full mass, but the gravitational force on them is nevertheless weak because they are so far from the main mass (recall that the force of gravity grows rapidly weaker at greater distances from a mass). These outermost stars must also be orbiting slowly if they are not to fly out of the galaxy. Detailed mathematical models bear out this analysis and indicate that a galaxy's rotation curve should start at small velocities near the core. The curve should then rise to higher velocities at middle distances and finally drop to lower velocities again far from the center, as indicated in curve B of Figure 78.1.

However, when astronomers measure the rotation curves of galaxies, they do not find this behavior. In work pioneered by the U.S. astronomer Vera Rubin, it was discovered that the rotation curves do rise near the center, but instead they then flatten out and almost never drop off to low velocities again. These **flat rotation curves** are found in nearly all spiral galaxies studied, as shown by a number of examples in Figure 78.2. Rotation curves have been studied for many galaxies, to distances far beyond those we have studied in the Milky Way, and they maintain flat rotation curves out to radii exceeding 50 kpc (about 160 thousand light-years).

A constant rotation speed in the mass formula implies that the interior mass climbs steadily as the radius increases. This is because with a constant value of V the mass formula indicates the interior mass rises in proportion to the radius R. For

$M_{<R}$ = Mass within radius R (kilograms)

R = Radius (meters)

V = Velocity of rotation (meters per second)

$G = 6.67 \times 10^{-11}$ m³/kg-sec²

INTERACTIVE

Dark matter

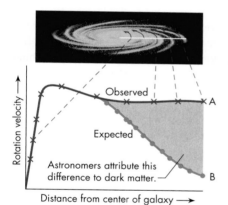

FIGURE 78.1

A schematic galaxy rotation curve. The line with dots represents the curve expected if the galaxy's mass comes only from its luminous stars. The line with crosses represents the observed curve, implying that the galaxy contains dark matter.

FIGURE 78.2

The rotation curves of an assortment of spiral galaxies. The curves shown are offset vertically for clarity. Each rises to a rotation velocity of about 200 km/sec and remains there to the limits of the observations.

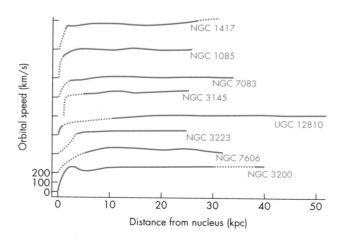

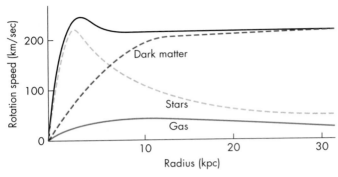

FIGURE 78.3
The rotation curve of a typical spiral galaxy is shown in black. The amounts of rotation produced by the detected stars and gas are shown. Dark matter becomes dominant beyond several kiloparsecs from the center of the galaxy.

Q If there is an equal amount of dark matter between radii of 10 and 20 kpc as between 40 and 50 kpc, what does that imply about the *density* of the dark matter at different radii?

example, the amount of matter measured out to 10 kpc from a galaxy's center might be found to be $10^{11}\ M_\odot$. Then out to 20 kpc the constant velocity implies there is $2 \times 10^{11}\ M_\odot$; out to 30 kpc there is $3 \times 10^{11}\ M_\odot$, and so forth. The amount of mass measured rises steadily the farther out we look.

This can be contrasted with a galaxy's light curve. In a typical spiral galaxy, about 90% of the light comes from the inner 10 kpc, while 99% is contained within about 20 kpc. The number of stars in the outer regions declines very rapidly, so they cannot contribute enough additional mass at larger radii to explain the flat rotation curve. Astronomers have hunted for other components, such as interstellar gas, which often extends out farther than the stars, but this also provides too little mass. The predicted rotation speeds based on stars and gas alone are illustrated in Figure 78.3.

Analyzing elliptical and irregular galaxies is a little more complex because the stars and gas do not rotate in circles. However, the same mass formula with some modifications for such random orientations of the velocities can be applied. Again, the results suggest that stars alone cannot account for the orbital speeds observed.

What can explain such a major discrepancy between theory and observation? If our laws of physics are correct, only one thing can explain such behavior: Galaxies must contain large amounts of unseen mass in their outer parts. That is, they contain dark matter, which exerts a gravitational force on stars in the outer parts of the galaxy and holds the stars in orbit despite their large velocity. The large gravitational force needed to bind the outermost stars in a galaxy at their high orbital speeds implies a huge mass of dark matter (many times the galaxy's luminous mass). We deduce that each galaxy is embedded in its own **dark matter halo,** vastly more massive and larger than the visible galaxy.

The term *halo* can be a little confusing because it sometimes makes people think of a ring. The dark matter is actually distributed throughout a galaxy. The rotation curve studies imply that the dark matter density is highest at the center of the galaxy. Nonetheless, it represents only a small fraction of the mass in the inner parts of a galaxy because the stars and gas there are packed tightly. At larger radii, however, the amount of stars and gas drops off rapidly, while the dark matter content declines only gradually. In the outermost parts of the galaxy, dark matter is left as the primary component.

78.2   DARK MATTER IN CLUSTERS OF GALAXIES

The mass formula was based on assuming circular orbits; but even if we simply require that the mass be great enough to prevent a star or galaxy from escaping, the required mass is only a factor of 2 smaller (see Unit 18 on orbital and escape velocities).

Galaxies in clusters often have speeds of more than a thousand kilometers per second with respect to one another. These great speeds reflect the enormous mass present in the cluster, pulling whole galaxies into high-speed orbits. If we add up the masses of all the galaxies in a cluster, we come up far short of what is needed to hold the cluster together. Just as stars within a galaxy orbit too rapidly for the gravitational force that can be attributed to its luminous mass, so too galaxies in a cluster have speeds too large to be kept from flying away by the observed mass.

This stark discrepancy was discovered in the 1930s by the Swiss astronomer Fritz Zwicky. Once again, applying the mass formula, we can estimate the mass interior to any radius in the cluster by measuring the speeds of the galaxies at that radius. The discrepancy is immense—a typical cluster must have a mass more than 100 times larger than would be predicted from its combined starlight. Galaxy clusters, like galaxies themselves, must contain huge amounts of dark matter.

For decades astronomers searched for normal kinds of matter to explain the discrepancy by developing techniques to study clusters at other wavelengths. Hot intergalactic gas was one possibility, and X-ray telescopes do reveal the presence of a lot of such gas in clusters (Unit 76). The amount of mass estimated to be within this hot gas is actually larger than that contained within the galaxies themselves, but it is still only about a tenth of the amount of mass predicted from the galaxies' orbital speeds.

Hot gas actually provides additional evidence for the dark matter in galaxy clusters. To hold the hot gas within the cluster and prevent its expansion, the cluster must exert an inward gravitational pull on the gas that matches the pressure within the gas. (This is the same condition of *hydrostatic equilibrium* that pertains to our Sun and other stars—Unit 49.) The gravity needed to confine the gas depends on the cluster's total mass; and once again, the calculated mass greatly exceeds the mass directly observable: the signature of dark matter.

Combining the evidence for dark matter in galaxies and galaxy clusters, it appears that these systems contain about 10 times more dark matter than any kind of normal matter that is detectable via electromagnetic emission. The dark matter appears to be distributed relatively smoothly within these systems; it is most dense toward the center of each galaxy as well as at the center of a cluster, but it extends out much farther than the galaxy's stars and other matter. The picture that emerges is that if we could see dark matter, it would look something like the artist's depiction in Figure 78.4. The stars we see would be concentrated in only the regions where the dark matter is most dense.

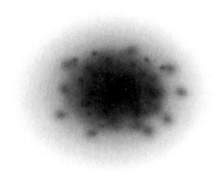

FIGURE 78.4
Artist's depiction of what the dark matter distribution might look like in a galaxy cluster and surrounding galaxies.

78.3   GRAVITATIONAL LENSES

Dark matter may not be observable by electromagnetic radiation, but it can be detected in a different way. It not only affects star and galaxy orbits, but it can change the path of light through space. This gravitational bending of light is one of the predictions of *general relativity*, Albert Einstein's theory of gravity (Unit 68).

The bending of light was detected soon after Einstein developed his theory of gravity, and in 1937 Zwicky suggested the possibility of a **gravitational lens.** A gravitational lens would focus the light from an object behind it, making background objects look brighter. A gravitational lens does not require dark matter in principle, but generally the masses of galaxies and galaxy clusters without it are found to be too small to produce detectable lensing effects. The idea was mostly forgotten for several decades.

Then, in the late 1970s, a quasar was discovered that seemed to have a nearby companion quasar with an essentially identical spectrum but with a slightly different brightness and shape. The existence of two so nearly identical quasars so close together is extremely improbable. Today several dozen such quasar pairs and even "quintuplets" are known (Figure 78.5A). They are not actually companions. Instead, they are images of a single quasar created by a gravitational lens.

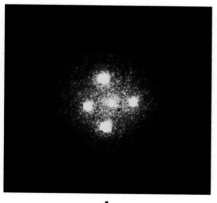

A

B

FIGURE 78.5
(A) An Einstein cross, a complex image of a quasar created by a gravitational lens.
(B) Sketch of how a gravitational lens forms an image.

FIGURE 78.6
The arcs in this picture of the galaxy cluster Abell 2218 are the images of distant galaxies distorted by the gravitational lens effect created by the galaxy cluster. (One of the arcs is quite red, visible to the left of the center of the image.)

INTERACTIVE

Gravitational lensing

An ordinary lens forms an image because light bends as it passes through the lens's curved glass. A gravitational lens forms an image because light bends as it passes through the curved space around a massive object such as a galaxy. The galaxy's gravitational force bends the space around it so that light rays that would otherwise travel off in other directions and never reach the Earth are bent so that they do reach us, as depicted in Figure 78.5B.

Gravitational lenses are more like a wavy piece of glass than a magnifying glass. Because the matter is distributed in a complex way, it usually produces highly distorted images of a background object. For example, Figure 78.6 shows the arcs created when light from a distant galaxy was bent into elongated shapes as it passed through a galaxy cluster. The shape of these arcs depends on the mass of material within the cluster, allowing astronomers to measure the cluster's mass. The large amount of bending observed provides additional proof that galaxy clusters contain the large amounts of dark matter implied by the high speeds of the galaxies.

The detection of gravitational lenses produced by galaxy clusters confirmed the dark matter hypothesis. Astronomers had developed various alternative hypotheses to explain their observations. For example, one hypothesis argued that the motions in clusters did not indicate the amount of mass present because the clusters were not held together by their gravity, but were instead expanding. Another hypothesis suggested that the laws of motion developed by Newton behaved differently at the large distances in galaxy clusters, so the formula we used for finding the mass would not apply. Neither of these hypotheses predicted the large bending of light seen. The agreement of the light-bending measurements with other results shows that there is in fact an immense amount of unseen mass within the clusters.

78.4   WHAT IS DARK MATTER?

What could this dark matter be? Astronomers can rule out ordinary dim stars because such objects would emit at least some detectable infrared radiation. Even the very dim brown dwarfs (Unit 69) have been ruled out in the Milky Way's halo by deep infrared observations with the Hubble Space Telescope. Astronomers also can rule out cold gas because that would be detectable by radio telescopes. And hot gas would be detectable by optical, radio, or X-ray telescopes.

It is possible to imagine other objects that would be difficult to detect, such as planet-sized bodies, very old and cold dead white dwarfs or neutron stars, or black holes. This idea is sometimes called the **MACHO** hypothesis—standing for **massive compact halo object.** It is difficult to think of plausible evolutionary scenarios for why such objects would be present in the outer parts of a galaxy in great numbers. This would require a star formation scenario in which the earliest stars all burned out quickly—otherwise we would expect to see many normal stars in the same proportions that they occur in the inner parts of the galaxy.

Astronomers are currently searching for MACHOs within the Milky Way's halo using the gravitational lensing technique. Although the most dramatic gravitational lenses are created by massive galaxy clusters, astronomers have recently developed techniques sensitive enough to detect lensing caused by a star or even a planet. If a dark MACHO in the halo passes in front of a background star as seen from Earth, the gravitational force of the MACHO can bend the light of the background star and focus it, brightening the light we see.

Astronomers have set up telescopes to monitor the light from millions of stars within the Magellanic Clouds, nearby companion galaxies of the Milky Way (Unit 76.1). If a MACHO passes in front of one of those stars, it will focus the star's light for several days until it moves out of alignment. The brightening caused by lensing has a precise pattern, and several dozen such brightening events have been detected.

Even though the foreground objects are themselves too dim to see, astronomers can estimate their mass from the brief lens effect they create. Moreover, from a statistical analysis of the number of such brightenings, astronomers can deduce the number of objects creating the lens effect. From the evidence so far available, the objects creating the lens effect are low-mass stars. This research is ongoing, but it does not appear that the mass responsible for these events can account for more than about 10% of the dark halo mass.

In the novels of Sherlock Holmes, the fictional detective often states that "when you have excluded the impossible, whatever remains, however improbable, must be the truth." The nature of dark matter appears to be one of these improbable truths.

The prevailing hypothesis today is that dark matter is entirely unlike the matter we are familiar with. Consider, for example, how difficult it is to detect neutrinos (Unit 4). Neutrinos can pass through the Sun and Earth, almost as if they were totally transparent. Evidence from studies of solar neutrinos (Unit 50) indicates that neutrinos have mass, but this mass is so tiny that the mass limits we can set on neutrinos would not allow them to account for the dark matter.

However, some physicists speculate that more massive particles may, like the neutrino, interact very weakly with normal matter. This has been dubbed the **WIMP** hypothesis—standing for **weakly interacting massive particles.** Such particles are proposed in various models of particle physics. They have not been detected, but physicists around the world have designed experiments to try to detect them.

If dark matter is made of WIMPs, then it would be more accurate to describe a galaxy as a dense region of these particles. After all, the rotation curve measurements indicate that they represent about 90% of a galaxy's mass. The normal matter—the stuff that stars, planets, and we are made of—was probably attracted by the gravity of this dark matter and fell into the region, collecting at the center. Normal matter passes through this sea of WIMPs, registering its presence only by the effect its mass has on the matter we can detect. Hence, WIMP particles would be present within the Solar System, but their density is so low as to be virtually undetectable.

The WIMP hypothesis and other similar hypotheses, based on currently undetected particles, are the leading ideas for explaining dark matter. However, the case is not yet settled, and the nature of dark matter remains one of the central mysteries in astronomy today.

Q Suppose the dark matter was made of old white dwarfs and neutron stars. What would this predict about the amount and kinds of stars that formed early in the Milky Way's history?

Q Could planets around other stars be detected with the lensing technique?

KEY TERMS

dark matter, 640

dark matter halo, 642

flat rotation curve, 641

gravitational lens, 643

massive compact halo object (MACHO), 645

rotation curve, 640

weakly interacting massive particle (WIMP), 645

QUESTIONS FOR REVIEW

1. What is meant by dark matter?
2. What is a gravitational lens?
3. What is a galaxy rotation curve?
4. What are MACHOs and WIMPs?

PROBLEMS

1. A spiral galaxy is observed in which the rotation speed remains 250 km/sec from 1 kpc out to the largest distances at which we can detect anything. How much mass exists out to 10 kpc? 11 kpc? 12 kpc? Can you generalize your results as to how much additional matter there is for each kiloparsec farther out you look?

2. In the Virgo cluster there are galaxies measured to be traveling at about 1000 km/sec at a distance of 500 kpc from the center of the cluster. If you apply the mass formula to this velocity and radius, what interior mass do you find?

3. Suppose the Milky Way contains 10 billion solar masses of dark matter at radii between 8 and 9 kpc from its center.
   a. What is the density of matter in the "shell" between the radii of 8 and 9 kpc? (*Hint:* Subtract the volume of a sphere of radius 8 kpc from the volume of a sphere of radius 9 kpc.)

   b. How much mass of dark matter would there be in a spherical volume the size of the Earth's radius? Compare your result to the mass of the Sun.

4. Most spiral galaxies show a decline in the amount of light they produce that drops by a factor of 2 every 2 kpc farther out. For example, 3×10^{10} $L_{\odot}$ are produced within 2 kpc of the center, 1.5×10^{10} $L_{\odot}$ are produced between 2 and 4 kpc of the center; half as much again between 4 and 6 kpc; and so on. Compare the amount of luminous matter found this way to the total amount of matter contained within these same regions if the galaxy has a flat rotation curve with a rotation speed of 250 km/sec. Find the ratio of dark matter to luminous matter within 2 kpc of the center; between 2 and 4 kpc from the center; between 10 and 12 kpc from the center.

TEST YOURSELF

1. Astronomers think that dark matter exists because
   a. they can detect it with radio telescopes.
   b. the outer parts of galaxies rotate faster than expected on the basis of the material visible in them.
   c. the galaxies in clusters move faster than expected on the basis of the material visible in them.
   d. it is the only way to explain the black holes in quasars.
   e. Both (b) and (c) are correct.

2. Most of the mass of a galaxy is
   a. contained in the massive O and B stars in the galaxy.
   b. contained in the cold interstellar gas.
   c. contained in the central black hole of the galaxy.
   d. contained in the dark matter of the galaxy.
   e. contained in the disk of the galaxy.

3. Gravitational bending of light does *not*
   a. show that space can be curved.
   b. give the illusion that quasars have companions.
   c. allow astronomers to detect MACHOs.
   d. provide a means to measure the mass of galaxy clusters.
   e. produce the appearance of constant rotation speeds in spiral galaxies.

79

Cosmology

Cosmology is the study of the structure and evolution of the universe as a whole. Cosmologists seek answers to questions such as these: Is the universe infinite? Does it have an edge? Has it existed forever, or does it have a definite age? How did it form? What will happen to it in the future? Given our insignificant size in the cosmos, such questions may seem futile or even arrogant, but most cultures have tried to answer them. Many of the attempted answers have become part of humanity's religious heritage. In cosmology we address some of these same questions based on astronomical evidence and the scientific method. It is important to set aside personal beliefs when examining what the weight of scientific evidence indicates.

We have made many surprising discoveries. Current evidence indicates that the universe was born about 13.7 billion years ago out of a hot, dense, violent state of matter and energy called the **Big Bang**. It has been expanding ever since, and it is filled with radiation from the early stages of that explosion. That radiation carries an imprint of information about the earliest stages of structure formation that we will describe in Unit 80. Within the last few decades, cosmologists have begun to extend our knowledge of the universe to the very beginning of time. They have discovered that the Big Bang may have been born out of even more turbulent events known as the *inflationary stage* (Unit 81), when the entire universe that we see today may have fit in a volume smaller than a proton! Our cosmological models even allow us to predict the ultimate fate of the universe (Unit 82). Understanding how the evidence leads us to such remarkable deductions is the goal of this and the next three units.

Background Pathways: Unit 74

79.1 EVOLVING CONCEPTS OF THE UNIVERSE

Over the centuries, our understanding of the universe has steadily changed. For millennia the idea of a central, stationary Earth seemed natural—a moving Earth seemed absurd, both to scientists and to others expressing their common sense. As concepts such as inertia (Unit 14) developed, the idea of a moving Earth no longer seemed physically impossible, and by the early 1600s the idea of planets orbiting the Sun became plausible. Such a model explained the planet's motions and other observed phenomena more simply, although it required a break with common beliefs of the time.

This idea was not yet supported by strong evidence—such as the predicted parallax effect—and therefore it was not accepted by all. Some astronomers continued to work with a geocentric model, which was still taught in many textbooks until the early 1700s. However, additional evidence accumulated, such as the aberration of starlight and parallax of stars (Unit 52), that so strongly contradicted the old Earth-centered model that it was finally abandoned.

Even as the debate over geocentric and heliocentric models continued, others began proposing even more complex cosmologies that recognized the Sun as one of many stars. One of the most commonly accepted cosmologies was that the universe consisted of the stars of the Milky Way with the Sun near its center, and this island of stars was surrounded by an empty void. An alternative idea was proposed by the German philosopher Immanuel Kant in the mid-1700s, anticipating some of our current ideas. He suggested that the Milky Way was merely one of a multitude of galaxies. This idea was not widely accepted until the 1920s, when astronomers could demonstrate that galaxies were millions of light-years from the Milky Way and comparable to it in size (Unit 74).

The pace of changes in our models of the universe has grown more rapid in the last century. Even as we were just learning to accept the idea that the universe contains billions of galaxies, the Milky Way being just one of them, observations revealed a strange phenomenon—that the other galaxies are moving away from us at high speeds (Unit 74.3). Edwin Hubble first demonstrated this in the 1920s, and it meshed with ideas about the nature of space and gravity developed by Albert Einstein around the same time. This new picture, of an expanding universe, had implications for how the universe must have begun.

The extraordinary conclusions we have reached about the universe today are rooted in observations that began as we asked simple questions about how the things we see came to be, and then followed a train of reasoning and pursued observations that led to the present model. Given the history of changing ideas about the universe, it needs to be stressed that we are speaking of models of the universe, rather than certain knowledge about a clearly defined "Universe," which we might write with an uppercase "U." This is *not* to say that the scientific evidence supporting modern cosmology is weak, but rather that our understanding of what the entirety of the universe comprises continues to evolve and be refined.

The scientific approach requires that we weigh the evidence and work with the best current model, *and* that we accept that the model may change as new evidence accumulates. In addressing the fundamental questions about the universe, it would be unscientific to work toward "proving" one model of the universe. This is a subtle point: Individual astronomers do attempt to argue the merits of particular models, but as scientists they accept that their models may be inaccurate or incomplete, and that a model should not be selected because of personal beliefs.

79.2 THE RECESSION OF GALAXIES

Nearly all galaxies have a redshift of their spectral lines. Astronomers in the early 1900s interpreted this as meaning that most of them are moving away from us (Unit 74). Hubble showed that the redshift of a galaxy increases with distance, d, in such a way that the recession velocity, V, of a galaxy obeys the Hubble law:

$$V = H \times d$$

where H is the Hubble constant.

The nature of this equation is startling. It seems to say that all the galaxies in the universe are flying away from us, making it appear as if the Milky Way is at the center of the universe—that a vast explosion (the Big Bang) has sent galaxies flying away from us in all directions. But this is incorrect.

A simple example illustrates why the Milky Way's position is not special. Imagine a string of galaxies distributed evenly along a line, as shown in Figure 79.1. Each galaxy is separated from its nearest neighbor by 10 megaparsecs. From the galaxy marked as the Milky Way in the figure, we would observe that each galaxy has a recession velocity given by Hubble's law: V = (70 km/sec per Mpc) × (distance). That is, A recedes from us at 700, B at 1400, and C at 2100 km/sec. Next, suppose we could communicate with an alien in galaxy C and ask what it sees. It would see us receding at 2100 km/sec, galaxy A receding at 2100 − 700 = 1400 km/sec, and galaxy B receding at 2100 − 1400 = 700 km/sec. In fact, the alien would see galaxies receding from it exactly the same way we see galaxies receding from us.

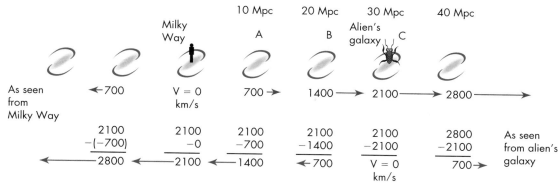

FIGURE 79.1

A line of galaxies illustrating that in a universe that obeys a Hubble law of expansion ($V = Hd$), an observer in any galaxy sees the same law.

Q If astronomers calculated the Hubble constant when the universe was half its current age, would the value they found for *H* be the same?

Furthermore, the galaxies seen by the alien will obey precisely the same form of Hubble's law.

A similar argument can be made for an observer in any of the galaxies along this line. Each will see all of the other galaxies flying away from it. This argument can be extended to two or three dimensions, and in each case for whatever galaxy we imagine being "home," we will observe all the other galaxies "flying away" from us with speeds that agree with Hubble's law. A universe obeying Hubble's law has no preferred "center" of expansion. Astronomers describe this lack of a preferred location as the **cosmological principle.** It is a statement of cosmic modesty—an extension of the Copernican revolution (Unit 11.4): There are no special positions in the universe. *We* are not at the center of the universe, nor is anyone else.

If there is no central location from which galaxies are moving away, then it does not make sense to think of the universe's expansion as being like the explosion of a bomb, sending fragments in all directions from a central blast point. But how then *do* we explain this motion? Einstein's **general theory of relativity** offers an answer.

Einstein developed general relativity in 1916, just a few years before the nature and motion of galaxies were discovered. He was trying to solve other puzzles having to do, for example, with gravity and the way it affects light (see Unit 68). The resolution of these problems led him to develop the idea that space is curved and can itself have motion.

The idea that space can have motion is alien to us, and it may even seem a bit silly. How can "nothing" be moving? But motion of space is needed to explain many observed phenomena. We can detect changes in the arrival of light signals that seem to imply that the light has changed speed as it moved along its path through different portions of space. Yet we can measure the speed of light extremely accurately, and every experiment shows that it moves through space at a constant speed. The explanation for the changes in arrival time is that the light traveled through a region where the space itself was moving relative to us. Its arrival time is advanced or delayed much as an airplane's arrival time may be affected if it is traveling through air currents that are moving toward or away from us—even though the airplane maintained a steady speed relative to the air throughout its trip.

Returning to the example of the galaxies located along a line (Figure 79.1), one way we might picture the idea of space having motion is to think of the galaxies as buttons strung along a rubber band, as shown in Figure 79.2. If the rubber band is steadily stretched, each button remains stationary relative to the rubber band, but the space *between* the buttons expands. Buttons that are farther apart on the rubber

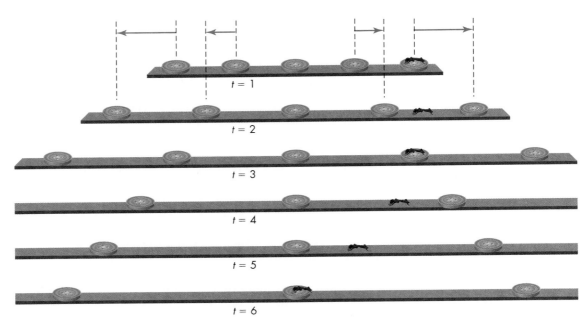

FIGURE 79.2

Galaxies in expanding space are like buttons on a stretching rubber band. An ant walking at a steady speed on the rubber band is like a photon traveling through expanding space: It moves at a speed relative to a distant button that is a combination of its own speed and the speed of the space it is traveling through at that moment.

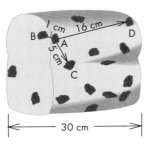

Raisin bread dough before rising

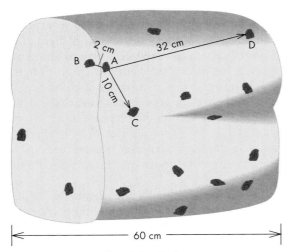

Raisin bread dough after rising

band move away from each other faster because there is more of the stretching rubber between them—following a Hubble-like law.

The effect of expanding space on an object moving through space can be illustrated by an ant walking along the rubber band in our analogy. The ant starts walking from one end of the rubber band toward the middle, but what was initially a small distance grows steadily larger. If the rate of expansion is fast enough, the ant may at first be carried farther from its goal by the stretching band. Eventually, though, as it keeps walking, the ant reaches parts of the rubber band that are not moving away from its target so quickly, and then it makes more rapid progress toward its goal.

The ant walking on the rubber band also illustrates what happens to the motion of a photon traveling through expanding space. Even though the photon is traveling steadily at a constant speed c through space, it makes slower progress than we would expect if space was motionless. In motionless space, a photon leaving a galaxy and traveling toward another would need to traverse only the original distance separating the two galaxies. This would be like the ant in Figure 79.2 hopping off the rubber band and walking on a tabletop to reach the "stationary" middle button—it would get there much more quickly.

The rubber-band analogy can be extended to two dimensions by imagining a stretching rubber sheet with the buttons glued to it. A three-dimensional analogy would be the relative motion of raisins in the dough of rising raisin bread (Figure 79.3). The separations between

FIGURE 79.3

As a loaf of raisin bread dough rises, the raisins move farther apart from each other at a rate proportional to their separation.

buttons on the rubber sheet or raisins in the dough grow because of the expansion of the medium in which they are set.

However, these analogies present a problem. The rubber band, rubber sheet, and the dough all have edges. Based on these analogies, then, we would expect there to be some galaxies close to the edge of the universe; if we lived in such a galaxy we would see galaxies in one direction but not the opposite. Because we do not see such an imbalance from the Milky Way, does this imply that our location is close to the center of the universe after all? We can avoid a conflict with the cosmological principle that this might imply by hypothesizing that the universe is infinite, or at least so large that none of the edges is anywhere close to being visible to us. (We will examine some alternative possibilities in Unit 80 by using the idea that space can be curved.)

In a similar fashion as these analogies, general relativity predicts that space itself is expanding, and galaxies are carried apart by that expansion, not by their own motion *through* space. However, although space expands, matter is not "glued" in place like the buttons or raisins in our analogies. Galaxies, stars, planets, and other bodies can move within the expanding space. In fact, the force of gravity that attracts them toward other nearby objects is often strong enough to overcome the expansion of space that would otherwise carry them apart. As a result, these objects can move toward each other despite the expansion of space, gathering into clusters, galaxies, and solar systems.

79.3   THE MEANING OF REDSHIFT

The expansion of space affects more than just motion. It also produces an observable effect on the wavelengths of light. As light waves travel through expanding space, they are stretched out by its expansion. The redshift we see from distant galaxies is caused by this stretching. Hubble interpreted this redshift as the recession velocity V used in Hubble's law (Unit 74); but as we discussed in the previous section, this interpretation does not explain the light travel time.

The stretching of wavelengths is the same effect you would find if you drew a wiggly line on the rubber sheet in the earlier analogy. As the rubber sheet is stretched, the crests and troughs of the wiggles become more widely spaced. The ant walking on the stretching rubber band would similarly feel its feet being spread apart as it walked (although it would pull them together after each step). The effect on a light wave traveling between galaxies is illustrated in Figure 79.4—the wavelength of the light is stretched by the same amount as the space itself.

Expansion and redshift

FIGURE 79.4

As space expands, it stretches radiation moving through it, making the wavelengths longer. Because longer wavelengths are associated with cooler objects, the stretching of the radiation has the effect of cooling it.

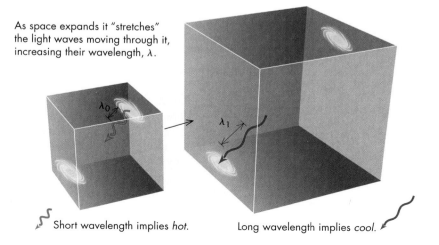

As space expands it "stretches" the light waves moving through it, increasing their wavelength, λ.

λ_0

λ_1

Short wavelength implies *hot*.

Long wavelength implies *cool*.

Understanding that the redshift we observe is caused by the stretching of wavelengths in expanding space gives us a better way to interpret redshift. The redshift indicates how much the universe has expanded since light left the object we are observing. Astronomers determine the redshift z from the fractional change in the wavelength λ of light emitted by a galaxy. We can calculate the redshift from the following formula:

$$z = \frac{\text{Change in wavelength}}{\text{Normal wavelength}} = \frac{\Delta\lambda}{\lambda}$$

$$= \frac{\text{Change in size of universe}}{\text{Size of universe when light left galaxy}}$$

For example, if we detect a spectral line that was emitted at 500 nm but is redshifted to 550 nm from a distant galaxy, the change in wavelength $\Delta\lambda$ is 50 nm. Therefore the redshift is

$$z = \frac{550 \text{ nm} - 500 \text{ nm}}{500 \text{ nm}} = \frac{50}{500} = 0.1,$$

and we can say that the universe has expanded by 10% since the light left that galaxy.

Astronomers find it helpful to use z when they describe the expansion of the universe because the redshift is related to the changing size of the universe. In particular, z compares the size of the universe when we receive the light from a galaxy to the size of the universe when the light was emitted by the galaxy. That relation is

$$\frac{\text{Size of universe today}}{\text{Size of universe when light left galaxy}} = z + 1.$$

Because the universe has no measurable edge, for *size* here we might substitute "the average distance between galaxies." In the preceding example, with $z = 0.1$, we can say that the universe is 1.1 times bigger today than when light left the galaxy.

We observe some extremely distant galaxies and quasars for which $z > 1$. This does *not* mean that they are traveling faster than the speed of light. Rather, the result is telling us that $z + 1 > 2$—that is, space has expanded by more than a factor of 2 since light left those objects. If a galaxy has a redshift $z = 1$, then the size of the universe has exactly doubled since light left the galaxy. For $z = 2$, space has expanded by $(z + 1) = 3$ times since light left the galaxy, or we might equivalently say that the universe was three times smaller when light left that galaxy.

The current record holder for redshift is a quasar with $z = 6.4$. For this redshift, the universe today is 7.4 times larger than when the light was emitted. This means that the matter in the universe was packed much more tightly together when the light we see left that quasar. The average separation between galaxies was just $1/7.4 = 0.135$ times the current separation. The smaller separation would have been in every direction, so the *volume* of space (length $\times$ width $\times$ height) would be just $0.135 \times 0.135 \times 0.135 = 0.0025$ times what it is today.

There are hints of galaxies at even higher redshifts, when galaxies were even closer together, but the signals are very weak, and the identification of spectral lines has not yet been confirmed. Galaxies at high redshifts look very different from nearby galaxies, in part because the matter was just in the beginning stages of gathering together to form a galaxy (Unit 75.4). If we look out farther and farther, to earlier and earlier times in the age of the universe, ultimately we should see a time when the whole universe was packed together at extremely high density.

Hubble interpreted the redshift as a Doppler shift and would have said $z = 0.1$ implies galaxy was traveling away from us at 10% of the speed of light, or about 30,000 km/sec.

If the size of the universe today is R_0 and the size of the universe when light was emitted from a galaxy was R, then

$$z = \frac{R_0 - R}{R} = \frac{R_0}{R} - 1.$$

Adding 1 to both sides of the equation, we find that

$$\frac{R_0}{R} = z + 1.$$

The Solar System formed about 4.5 billion years ago—about one-third the age of the universe. Was the separation between the Sun and Earth smaller then? What about the size of the Earth?

79.4 THE AGE OF THE UNIVERSE

$$d = Vt$$

Therefore, $t = \dfrac{d}{V}$

But according to the Hubble law,

$$V = dH$$

Therefore, $t = \dfrac{\cancel{d}}{\cancel{d}H} = \dfrac{1}{H}$

FIGURE 79.5

Estimating the age of the universe by the recession of one galaxy from another galaxy.

Because Speed = Distance/Time or $V = d/t$, multiply both sides of the equation by t/V to get $t = d/V$.

The model of the universe we have just discussed, as an expanding space peppered with galaxies, allows cosmologists to estimate its age and to calculate conditions near the time of its birth. Alexander Friedmann, a Russian scientist, made such calculations in the 1920s, but his work attracted little attention. Not until 1927, when Abbé Georges Lemaître, a Belgian cosmologist and priest, independently made similar calculations, did astronomers appreciate that conditions at the birth of the universe could be deduced from what is observed today.

Lemaître pointed out that because the separations of galaxies are growing today, the galaxies must in the past have been closer together, as we saw in the last section. In fact, if you simply follow the expansion implied by Hubble's law back in time, there must have been a period long ago when all of the galaxies in the universe were crowded together. Galaxies and stars as we know them today could not have existed in such a dense environment. With so much mass in such a relatively small volume, the matter that now composes the far-flung galaxies and their stars must have been packed into an extremely dense ball that Lemaître called the *Primeval Atom*. From this dense state, the universe must have expanded at a tremendous speed—the Big Bang.

It is straightforward to estimate the time since the Big Bang as follows. We know the rate at which space is expanding between any pair of galaxies—how fast they are moving away from each other. We solve for how long it would take them to get this far apart moving at their current speed, just as we might solve for how long a car takes to travel 200 km at a speed of 50 km per hour.

Consider two representative galaxies separated by a distance d and separating with a velocity V (Figure 79.5). We will assume that V has remained constant and call the time we calculate that it took the galaxies to reach this separation the **Hubble time,** t_H. We can calculate the Hubble time as follows: The time it takes to travel a particular distance, d, is the distance divided by the speed. (For example, 200 km divided by 50 km per hour gives 4 hours.) Therefore, the Hubble time is

$$t_H = d/V.$$

Hubble's law, however, tells us that a galaxy's velocity grows with distance according to the formula $V = H \times d$. Substituting this into our equation, we find

$$t_H = \frac{d}{H \times d} = \frac{1}{H}.$$

The distance d cancels out, so it doesn't matter which two galaxies we choose—they began moving apart a time equal to $1/H$ ago. In other words, the inverse of Hubble's constant indicates a length of time that approximately measures the age of the universe.

To get a value for t_H in years, we have to carry out some unit conversions because H is expressed in units of km/sec per Mpc. One megaparsec is 3.09×10^{19} km, so if the Hubble constant is 70 km/second/Mpc, this gives

$$t_H = \frac{1 \text{ Mpc}}{70 \text{ km/sec}} = \frac{3.09 \times 10^{19} \text{ km}}{70 \text{ km}} \text{ sec} = 4.41 \times 10^{17} \text{ sec}.$$

Because there are 3.16×10^7 seconds per year, $1/H = 1.40 \times 10^{10}$ years. Therefore we find that the age of the universe is about 14 billion years.

In this simple calculation we assumed that the expansion of space has remained constant. Gravity can slow the expansion, but other effects on space can cause acceleration (Unit 82.3). Detailed calculations that take into account these different

TABLE 79.1	Relationship Between Redshift and Time		
Redshift z	Size of Universe Compared to Now	Length of Time Ago	Time After Big Bang
0	1	Today	13.7 billion years
0.1	0.91	1.3 billion years	12.4 billion years
0.5	0.67	5.0 billion years	8.7 billion years
1.0	0.50	7.7 billion years	5.9 billion years
2.0	0.33	10.3 billion years	3.4 billion years
3.0	0.25	11.5 billion years	2.2 billion years
5.0	0.17	12.5 billion years	1.2 billion years
7.0	0.13	12.9 billion years	800 million years
10	0.09	13.2 billion years	500 million years
1000	0.001	13.7 billion years	400,000 years

Q What were your beliefs about the nature of the universe prior to reading about modern cosmology? Are you willing to change any of them if the scientific evidence disagrees with them?

effects give an answer that is very close to the age we have calculated: 13.7 billion years. Table 1 shows how much the universe has expanded for different values of the redshift according to the best current model.

KEY TERMS

Big Bang, 647
cosmological principle, 649
cosmology, 647

general theory of relativity, 649
Hubble time, 653

QUESTIONS FOR REVIEW

1. What does cosmology study?
2. Why do astronomers think the universe is expanding?
3. What causes redshift?
4. How old is the universe? How is its age found?

PROBLEMS

1. When Hubble first estimated the Hubble constant, galaxy distances were still very uncertain, and he got a value for H of about 600 km/sec per Mpc. What would this have implied about the age of the universe? What problems would this have presented for cosmologists?

2. Hubble's constant is still not precisely known. Calculate t_H for two other values of Hubble's constant currently in favor by some astronomers:
   a. $H = 71$ km/sec per Mpc.
   b. $H = 65$ km/sec per Mpc.

3. What wavelength would visible light (400–700 nm) have if it were seen from a galaxy formed just half a billion years after

the Big Bang? What if it came from matter when the universe was just 400,000 years old? (Refer to Table 79.1.)

TEST YOURSELF

1. From what evidence do astronomers deduce that the universe is expanding?
   a. They can see the disks of galaxies getting smaller over time.
   b. They see a redshift in the spectral lines of distant galaxies.
   c. They see the edge of the universe moving away from us.
   d. They can see distant galaxies dissolve, pulled apart by the expansion of space.
   e. All of the above

2. If we discovered tomorrow that the distances we measured to galaxies were incorrect, and Hubble's constant was 35 km/sec per Mpc instead of 70 km/sec per Mpc, we would conclude that the universe's age is_____ previously thought.
   a. 4 times older than
   b. 2 times older than
   c. the same as
   d. 2 times younger than
   e. 4 times younger than

3. If a galaxy has a redshift $z = 3$, what fraction of the universe's current size was the universe when the light was emitted from that galaxy?
   a. One-fourth
   b. One-third
   c. One-half
   d. Three-quarters
   e. One-ninth

80

The Edges of the Universe

Background Pathways: Unit 79

Some simple questions can have profound answers. One such question puzzled astronomers for centuries: Why is it dark at night? At first glance this question seems to have an obvious answer. However, answering it requires a surprising amount of cosmology, involving the expansion and curvature of space, the speed of light, the cosmic density of matter, and the age of the universe.

To understand why the sky is dark at night, we need to know what conditions were like when all of the matter in the universe was packed together just after the expansion of the universe began. Extrapolating to what the physical conditions must have been like at these times, cosmologists have been able to make several predictions about characteristics of the universe. The confirmation of these predictions has provided strong support for the Big Bang theory of the universe's origin.

In this unit we explore the light that has been traveling through the universe since shortly after the Big Bang. As we look out into space we see galaxies in every direction out to the limits of our technology. As we probe these great distances, we are also looking out into the past, and we can see back to a time just a few hundred thousand years after the Big Bang. Before that time all of the matter in the universe was merged into a great ocean of hot gas, and light could not travel freely through space. We can even detect small fluctuations in this hot gas that provide clues about how galaxies first formed.

Light comes to us from all directions from hundreds of billions of galaxies that have been emitting light for more than 13 billion years. We begin then by trying to understand the question we posed: Why is the sky dark at night?

80.1 OLBERS' PARADOX

Our discussion of why the sky is dark at night requires that we look first at the distribution of the objects that generate visible light. We show a map of galaxies on the largest scales in Figure 80.1A. This wide-angle view shows the distribution of over 1 million galaxies as seen looking up out of the disk of our own galaxy, the Milky Way. Around the edge of the picture, you can see only a few spots of light because dust in the Milky Way hides the galaxies in that part of the sky. But the unobscured portion of the picture shows an important clue about the structure of the universe.

No matter what direction you look (ignoring the dust layer of the Milky Way), you see approximately the same number of galaxies (Figure 80.1B). That is, averaged over a very large area, the universe is more or less the same in all directions, and galaxies are spread throughout the universe much as raisins are spread throughout raisin bread. They may clump a bit in one region or another, but overall, they are fairly evenly distributed. If we look over large enough distances, even the superclusters and voids of large-scale structure (Unit 76.4) average out, and the matter in the universe appears to be approximately smoothly distributed, or **homogeneous,** out to the largest distances we can measure.

FIGURE 80.1

(A) A computer-generated picture showing the location on the sky of nearly a million galaxies. This is the view looking up and out of the Milky Way. Note the relatively uniform distribution of the galaxies. (B) A sketch illustrating the uniformity of the universe. Roughly the same number of galaxies occur in all directions.

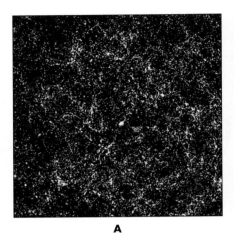

A

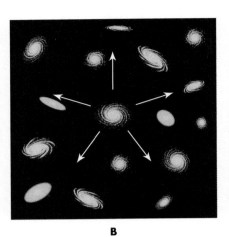

B

Although the paradox bears Olbers' name, it was discussed by various astronomers over the two previous centuries. Kepler pointed out that in an infinite universe, the night sky should be bright, and he therefore concluded the universe was finite.

Q Olbers' paradox was originally phrased in terms of individual stars, not galaxies. How would this have altered the thinking about the paradox?

The distance to the cosmic horizon can be measured in different ways. For example, the most distant regions we can see today emitted the light when they were only about 50 million light-years away. Those regions are today about 50 *billion* light-years away.

This homogeneity presents a problem, though. If there are galaxies in every direction, then our sky should be completely covered by them. In fact, if the universe has no edge, then in every direction we look, our line of sight should eventually encounter a star. So the sky should be bright everywhere. This is known as **Olbers' paradox** after the German astronomer Heinrich Olbers, who in 1823 popularized the problem. He calculated that if the universe extends forever and has existed forever, the night sky should be blazingly bright. The paradox arises from reasonable premises (or so it seemed at that time) that lead to a conclusion clearly at odds with the simple fact that the sky is dark at night.

The argument is easiest to understand from a simple analogy. Suppose you stand in a small grove of trees and look out between the tree trunks to the surrounding landscape. If the grove is small, your line of sight will be blocked by trees in a few directions, but it will pass between the trees in most directions (Figure 80.2). If the grove is larger, more distant trees in it will block your view in what previously were clear gaps. In that case, no matter where you look, your line of sight will be intercepted by a tree trunk.

Now suppose that rather than looking between trees, you look out into space through the galaxies and stars that comprise the universe. If space extends sufficiently far and is populated with galaxies homogeneously, then no matter in what direction you look, every line of sight will ultimately intercept a star—just as in a sufficiently large forest every line of sight ultimately hits a tree. Even though the stars in distant galaxies are faint to us, there are so many of them that their collective brightness should be large. Therefore, the sky should be covered with starlight, glowing brilliantly with no dark spaces in between. In other words, the night sky should not be dark.

This argument must contain a false assumption because the most obvious observation one can make about the sky is that it is dark at night. That is the paradox: The night sky should be bright, but it is not. Where, therefore, is the error in our reasoning?

The first way the paradox can be avoided is if there are no stars beyond some distance. That is, just as we can see out of the woods if there are only a few trees, so too we will see a dark night sky if we run out of stars and galaxies beyond some distance. However, we do not see any evidence of an edge to the universe.

There is an edge in a second sense—an edge in time. Even if the universe is infinite in extent, if it has a finite *age*, then we can see only the stars whose light has had time to reach us within the time limit set by the age of the universe. That is, if the universe is 13.7 billion years old, we can see light from a star that is 10 billion light-years away, but not one that is 20 billion light-years away. In other words, the universe has to have existed long enough for the light to arrive. The maximum distance that light can travel in the universe's age defines the **cosmic horizon.** Astronomers call the

In a small grove of trees, only a few block your view.
Lots of space to view between trees.

A

In a larger forest, more distant trees block your view.
No open space visible between trees.

B

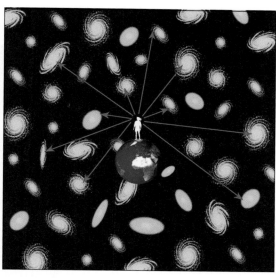

In a large enough universe, every line of sight eventually
encounters a star within a galaxy

C

FIGURE 80.2
(A) An observer in a small grove of trees can see out through gaps
between them to the surrounding countryside. (B) In a larger forest,
distant trees block the gaps so that no matter where you look your line
of sight ends on a tree. (C) So, too, in a universe that extends forever
and has existed forever, your line of sight will always end on a star.
Therefore, the night sky should be bright.

Q Is our cosmic horizon the same as that of alien astronomers living in another galaxy? Why or why not?

space within the cosmic horizon the **visible universe** (Figure 80.3). As the universe ages, the visible universe grows larger. At present, we don't know how large the entire universe might be, but it is suspected to be far, far larger than the visible universe. This would suggest that if we could wait long enough, in the future the night sky *might* become bright.

However, realizing that we are looking back in time presents another problem. In every direction we look, we should eventually see back to early times when the universe was filled with hot gas from the Big Bang, and this should be just as bright as the surface of a star. Olbers' paradox predated the discovery that the universe expanded from a fiery hot state, but it appears that we need to understand the effects of that expansion to complete our explanation of why the sky is dark at night.

80.2 THE COSMIC MICROWAVE BACKGROUND

The idea that we should see light from the Big Bang was first suggested by George Gamow (pronounced *GAM-off*), a Russian astrophysicist who fled to the United States before World War II. Gamow built upon Lemaître's idea that the young universe was extremely dense (Unit 79.4), and he went on to conclude that it must also have been extremely hot. Gamow's argument was based on the observation that compression heats a gas, and so the enormous compression of the early universe must have heated the matter to very high temperatures at the earliest times.

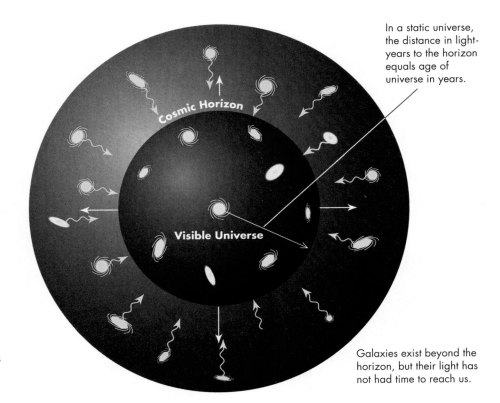

In a static universe, the distance in light-years to the horizon equals age of universe in years.

Cosmic Horizon

Visible Universe

Galaxies exist beyond the horizon, but their light has not had time to reach us.

FIGURE 80.3
We cannot, even in theory, see beyond the cosmic horizon. Light from there takes more time than the age of the universe to reach us.

The last scattering epoch is also sometimes called the *recombination epoch* in reference to the electrons and ions combining to form atoms. This name can be misleading, though, because this was actually the first time these particles ever combined.

When we observe out to large enough distances, then, because we are seeing the universe as it was when it was young, we should see evidence of initially high temperatures. However, the radiation from this hot gas has been diluted and redshifted to long wavelengths by the expansion of the universe. When we look out at night, every line of sight *does* encounter a hot glowing surface, and the sky *is* bright, but at radio wavelengths. Thus, the darkness of the night sky results from the limitations of our senses. If we look at radio wavelengths, we should see the afterglow of the Big Bang.

This idea was developed in 1948 by two of Gamow's collaborators, the U.S. physicists Ralph Alpher and Robert Hermann. They showed that this radiation would come not directly from the first moments of the Big Bang but rather from a later time, after the universe had been expanding for several hundred thousand years.

Just minutes after the Big Bang, the temperature of the universe would have been billions Kelvin. After a few hours it cooled to 100 million K; after about 10 years, about 100,000 K; and after 400,000 years, about 3000 K. This cooling is straightforward to predict from laboratory observations of expanding gases. During this whole time, the high temperature of the gas kept it ionized. In an ionized gas, photons interact strongly with the free electrons and, as a result, can travel only a short distance (Figure 80.4A). Because light could not travel far through this hot gas, it was essentially opaque.

Throughout the first few hundred thousand years of the universe's existence, the temperatures and densities of the original gas were similar to those inside a star. As time progressed, the gas temperature and density dropped, making the conditions more similar to those farther from the core of a star. The conditions at 400,000 years finally reached temperatures like those at the surface of a star—low enough that the ionized hydrogen and helium combined with the free electrons, and the universe became transparent (Figure 80.4B). This period is called the **last scattering epoch** because photons had their final interaction with the expanding gas of the Big Bang.

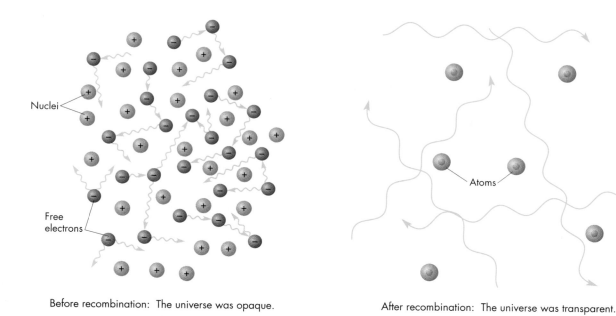

Before recombination: The universe was opaque.

A

After recombination: The universe was transparent.

B

FIGURE 80.4

(A) In the young universe, free electrons in the hot ionized gas constantly interacted with photons, making the universe opaque. (B) When the universe became cool enough that electrons combined with the nuclei to make neutral atoms, the universe became transparent.

Even though we cannot see to times before 400,000 years, a variety of clues have survived, and these help take us back to less than one second after the Big Bang occurred. These clues are explored in Unit 81.

Thus, after 400,000 years, the photons from the Big Bang were able to move freely through space—unless they hit something, like a telescope on Earth! These photons are the oldest radiation we can see. Just as we cannot see photons directly from the core of the Sun, we cannot see the photons from the Big Bang itself. In both cases the photons travel through a hot ionized gas that constantly interacts with the photons for hundreds of thousands of years before the photons can move freely through space. Through these interactions the photons exchange energy with the gas particles and reach the same temperature. When the photons reach a place or a time where the gas they encounter is thin, the photons undergo a final scattering and then travel freely through space. The photons travel away, carrying with them energies set by the temperature of the last atoms they interacted with.

So what should this radiation look like? According to Wien's law, the thermal radiation from the photosphere of a star or the last scattering epoch should have a spectrum that is most intense at a wavelength determined by the temperature (Unit 23). In particular, hot matter radiates most strongly at short wavelengths. Just after the last scattering epoch, the radiation would have been at a temperature of about 3000 K and therefore looked much like the surface of an M-type star like Betelgeuse (Unit 55). If we lived then, we would have experienced Olbers' picture of a sky everywhere as bright as the surface of a star. The radiation at that time would have peaked at a wavelength of about 1000 nm.

The expansion of space modifies this radiation. The redshift of the wavelengths makes the radiation appear cooler. The universe has expanded by a factor of about 1000 since the last scattering epoch, so the wavelength of the radiation has also stretched by a factor of 1000 (Unit 79). According to Wien's law, the temperature of thermal radiation drops in proportion as the wavelength grows longer, so the apparent temperature of this radiation would drop by a factor of 1000—from 3000 K down to about 3 K today.

If the universe began with a Big Bang, it should today be filled with long-wavelength radiation created when the universe was very young. This prediction

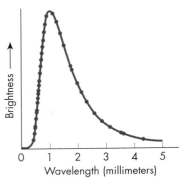

FIGURE 80.5
Spectrum of the cosmic microwave background. The shape of the spectrum matches perfectly a thermal spectrum for a temperature of 2.7 K. Notice the wavelength at which it peaks. Red dots show data points schematically.

was confirmed in 1965, when Arno Penzias and Robert Wilson of the Bell Telephone Laboratories accidentally discovered cosmic microwave radiation having just that property.

Penzias and Wilson were trying to identify sources of background noise on telephone satellite links. They detected a radio signal with the unusual property that its strength was constant no matter in which direction of the sky they pointed their detector. They soon demonstrated that the signal came not from isolated objects such as stars or galaxies but rather from all of space. Initially mystified by the signal, they discovered that scientists at Princeton University had repeated Alpher, Hermann, and Gamow's calculations and were searching for the predicted radiation. Once aware of that work, Penzias and Wilson realized that the background interference they had found was radiation created in the young universe. For their discovery of the **cosmic microwave background,** or **CMB** for short, they won the 1978 Nobel Prize in physics.

Figure 80.5 shows measurements of the intensity of the CMB and demonstrates that it has exactly the thermal spectrum that theory predicts. Moreover, the radiation is most intense at about 1 millimeter (10^6 nanometers), in the microwave part of the radio spectrum. Knowing this wavelength and using Wien's law, we find that the temperature describing the radiation is about 3 K, within a few degrees of the value predicted nearly 20 years earlier. A more precise measurement shows that the temperature that describes the CMB is 2.726 K, only a little warmer than absolute zero, the lowest temperature anything can have.

Today the cosmic microwave background results provide one of the cornerstones that support the Big Bang theory. In earlier years, though, Gamow's predictions were considered fantastic. The high temperature and rapid expansion from a hot, dense state led the English astronomer Sir Fred Hoyle (a proponent of a competing theory called *steady state cosmology*) to jokingly refer to this theory of the birth of the universe as the *Big Bang,* a name that has stuck despite his sarcasm.

80.3   THE ERA OF GALAXY FORMATION

After the cosmic microwave background scattered off the rapidly thinning and cooling gas, matter condensed to form the stars and galaxies we see today. How that process occurred is one of the major areas of research in modern astronomy.

On the basis of the ages of stars within galaxies, astronomers deduce that the Milky Way and other galaxies are at least 10 to 12 billion years old. We can observe galaxies when they were very young in deep images such as the Hubble Deep Field (Unit 75) and Ultra Deep Field (Figure 80.6), where we find some extremely faint galaxies at redshifts that imply they were formed less than 1 billion years after the Big Bang. Galaxies must therefore have formed rapidly once the universe had cooled enough. To understand how the universe evolved from a sea of hot gas so quickly into galaxies, astronomers are carrying out simulations using the most advanced supercomputers available.

Presumably gravity pulled gas clouds into protogalaxies, much as gravity pulls interstellar clouds into protostars; but there appear to be important differences. In particular, the amount of matter observed in the universe exerts too feeble a gravitational attraction for this process to have been able to form galaxies within the age of the universe, let alone within the first billion years. Most astronomers therefore conclude that additional matter must be present that provided a strong enough gravitational pull to speed up galaxy formation. This need for unseen matter to aid galaxy formation strengthens astronomers' suspicions that dark matter (Unit 78) plays a major role in the structure of the universe.

FIGURE 80.6

The Hubble Ultra Deep Field. This image shows a portion of the sky in the constellation Fornax that you could easily cover with this dot (•) held at arm's length. The image shows about 10,000 galaxies. Some are so far away that we are seeing them the way they appeared nearly 13 billion years ago, when the universe was only about 1/20th of its present age.

The dark matter's gravity could draw itself into clumps, which then might pull in ordinary matter to form galaxies. Computer simulations show that the dark matter could have begun clumping even before the last scattering epoch, precisely because it does not interact with electromagnetic radiation as ordinary matter does. As a result, the large-scale organization of matter into denser regions was able to begin sooner than 400,000 years after the Big Bang. These higher-density regions began pulling in normal matter as soon as it was no longer being buffeted by high-energy photons. The clumping of the dark matter itself responded to variations in density that came out of even earlier moments immediately after the Big Bang (see Unit 81).

Computer simulations show that the clumping of dark matter would make some regions slightly more dense than others. The greater compression of the gas in these regions would make them slightly hotter and brighter—small fluctuations in the temperature of the CMB. Astronomers have found exactly such clumps, as shown in Figure 80.7. This figure displays the CMB over one region of the sky, the bright regions being the dense clumps where galaxies and clusters of galaxies are going to form.

The size of these clumps is determined by the size of the cosmic horizon during the last scattering epoch. The cosmic horizon would have had a radius of about 400,000 light-years at this time because that was as long as the universe had lived. This means that regions of gas and dark matter could have interacted with each other—sharing their heat and experiencing each other's gravitational pull—only if they were both within a radius of 400,000 light-years of a common point. The clumps in Figure 80.7 are each regions of about this radius.

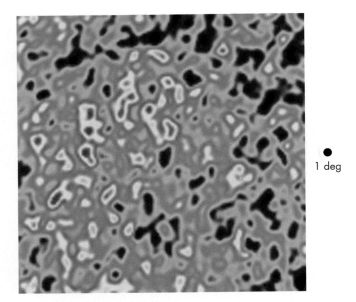

1 deg

FIGURE 80.7

A portion of the cosmic microwave background radiation. The red bumps are slightly denser and hotter regions, while the dark blue regions on the map are lower-density, cooler regions. At this redshift 1° is about 800,000 light-years.

If we could have been present at the location where the Milky Way was going to form, not long after the last scattering epoch, what might we have observed?

In the clumps of dense matter visible in the CMB, we are seeing the distribution of matter in distant regions of space as they were before any galaxies formed. It is possible that those regions today, almost 13.7 billion years later, contain galaxies in which astronomers are looking in our direction and seeing a bright region where the Local Group of galaxies is going to form.

The angular size of these clumps provides another clue about the properties of the universe because the path of the light that we see is affected by the overall gravitational field of the universe. The entire universe can be curved by gravity, and this can profoundly alter our notions about what it means for space to have an edge, as we examine next.

80.4 THE CURVATURE OF THE UNIVERSE

Einstein's general theory of relativity has shown that the mass and energy of the universe curve its space. Such bending can be difficult to picture. In some ways it is easier to picture the bending of space by an object like a black hole (Unit 68.2) or a cluster of galaxies (Unit 78.3) because the strong bending is localized to one region. But in fact, the enormous mass and energy of the entire universe can change how space connects to itself, bending and shaping the universe.

We are a little like an ant that crawls into a rubber hose lying on the ground. If the hose is straight, representing uncurved space, the ant walking in a straight line will reach the other end of the hose and emerge. However, suppose the hose is bent and twisted in large loops. If the curves are gradual, an ant crawling through the hose may still feel like it is traveling in a straight line, even though its direction is changing. And if the other end of the hose was brought around so that the ends met, the ant could crawl forever in a seemingly straight line and would never reach the hose's end.

We experience curved space on the surface of the Earth. Athletic fields and parking lots appear flat, and if we ignore hills, the surface of the Earth from horizon to horizon certainly looks flat. We know, however, that if we could walk in a "straight" line along this "flat" surface, we would return to our starting point. Perhaps the expanding universe is similarly curved.

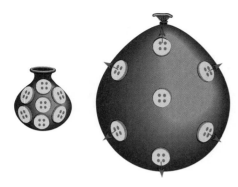

FIGURE 80.8

Representations of the expansion of the universe.

One way of picturing the expansion of space is that it is something like the stretching of a rubber sheet, with buttons on it representing galaxies (Unit 79.2). Imagine, though, that the curvature of space allowed the rubber sheet to curve around so the edges connected to each other, like a balloon. An analogy for the expanding curved universe, then, could be buttons glued to a balloon (Figure 80.8). As the balloon inflates, the space between buttons expands, and the buttons separate. Moreover, no button is near an edge even though space is finite. Space in this analogy is represented by the surface of the balloon, so a photon traveling from one button to the next travels through the curved space of the rubber in the balloon. In our analogy the rubber of the balloon represents only two of the three dimensions of space. The two-dimensional surface of the balloon curves in another dimension so that it can connect back to itself.

General relativity indicates that a similar situation exists for a universe curved around on itself. Such a space looks as if it extends forever in all directions; but if we could travel in a spaceship along an apparently straight line, we would eventually return to our starting position. We could move through this space forever without coming to an end point. A universe with this property is said to have **positive curvature** and is called a **closed universe** because it curves around back on itself. The Earth's surface is also positively curved because if we travel in a straight line on it, we return to our starting point. Positively curved space has other properties as well: Parallel lines meet when extended, and the sum of the interior angles of a triangle drawn on a positively curved surface is greater than 180°, as illustrated in Figure 80.9A.

General relativity holds open the possibility that the universe might have other types of curvature, however. It might be **flat** (that is, have no curvature) or have **negative curvature.** Flat space is what most people picture when they think of space. In it, parallel lines do not meet, and a triangle's interior angles add up to exactly 180°, as illustrated in Figure 80.9B. Negatively curved space is harder to visualize, but you can think of it as being bent into a saddle shape, as shown in Figure 80.9C. Such curvature corresponds with bending space one way along one line and the opposite way along another line. In negatively curved space, the sum of the interior angles of a triangle is less than 180°.

Note that the paths of light through curved space can have surprising properties. If the universe had a very strong positive curvature, we might be able to see all the way around the universe to see ourselves—or perhaps our own galaxy as it looked long ago. Positive curvature, like the curvature of light around a black hole or galaxy cluster, also acts as a lens (Unit 78.3), making distant objects look larger. On the other hand, if space is negatively curved, it will make distant objects look smaller than they would look in flat space. Perhaps like a rearview mirror, a negatively curved universe should have a label saying "galaxies in a negatively curved universe are closer than they appear"!

FIGURE 80.9

Figures illustrating how the shape of a surface determines the geometry of a triangle on it. (A) The interior angles of a triangle on the surface of the Earth add up to more than 180°. (B) The interior angles of a triangle on a flat surface add up to 180°. (C) The interior angles of a triangle on a saddle-shaped surface add up to less than 180°. The top row of figures illustrates that a distant object will appear larger if the curvature is positive (A) and smaller if the curvature is negative (C).

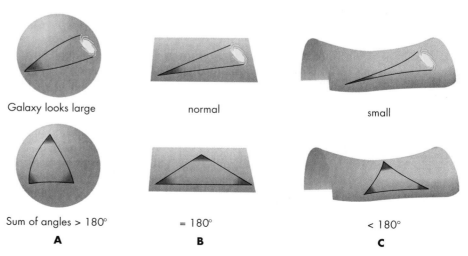

Galaxy looks large   normal   small

Sum of angles > 180°   = 180°   < 180°

A   **B**   **C**

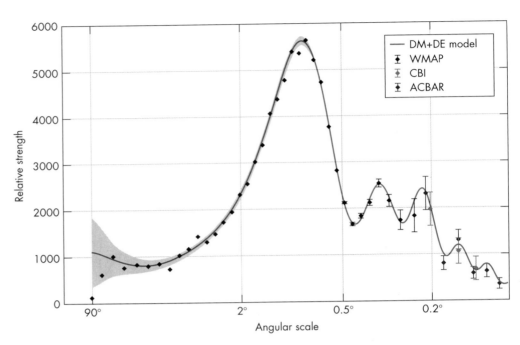

FIGURE 80.10

The strength of fluctuations of different angular sizes measured in the CMB is shown by the points in this plot. The fluctuations in brightness are caused by the clumping of matter at the last scattering epoch. The red line shows a plot of the predicted size of the fluctuations for a particular model containing dark matter and dark energy.

The sizes of clumps on the CMB map give us a way to determine the curvature of the universe. Because the largest clumps should each have a radius of about 400,000 light-years, we can compare this to the size predicted for the different geometries. The measurements of the CMB show that the strongest fluctuations are about 1° across (about twice the apparent size of the full moon). From detailed analyses, this indicates that the universe's geometry must be very nearly flat.

By analyzing the distribution of sizes of these fluctuations, astronomers can estimate H and other cosmological parameters. The relative strengths of different sizes of fluctuations in the CMB (Figure 80.10) depend on the different ways that matter, light, and dark matter interact. Computer simulations can predict the relative strengths of different fluctuations depending on the relative abundance of each component. Astronomers have been able to obtain beautiful matches to the distribution of sizes of these clumps in their simulations, but only if dark matter—about 10 times more than normal matter—is included in the mix. This provides additional evidence that dark matter exists. It also provides hints of an even more mysterious quantity known as "dark energy," which will affect the ultimate fate of the universe, as we will discuss in Unit 82.

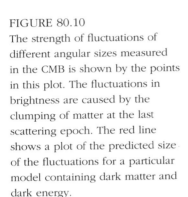

KEY TERMS

closed universe, 663

cosmic horizon, 656

cosmic microwave background (CMB), 660

flat universe, 663

homogeneous, 655

last scattering epoch, 658

negative curvature, 663

Olbers' paradox, 656

positive curvature, 663

visible universe, 657

QUESTIONS FOR REVIEW

1. What is meant by a cosmic horizon?

2. What is Olbers' paradox?

3. What is the cosmic microwave background? What is its origin?

4. What is the significance of fluctuation in the cosmic microwave background?

PROBLEMS

1. When the Solar System formed about 4.5 billion years ago, approximately what would the temperature of the CMB radiation have been? Would the structure in the CMB have looked the same as today? Explain your answer.

2. Imagine that there were so many galaxies that 30 K dust inside their interstellar clouds blocked our view in every direction. What would their redshift have to be for this to have a temperature like the CMB? What problems are there with this model for explaining the observed properties of the CMB?

3. At the time of the last scattering epoch (at a redshift of $z \approx 1000$),
   a. What was the average density of normal matter if today it is about 3×10^{-31} kilograms per liter?
   b. How large a sphere would have contained 10^{11} $M_\odot$, about as much normal matter as a galaxy today? Express your answer in parsecs.

4. On a flat surface, the angles of a triangle always add up to 180°, but on a spherical surface they may add up to a greater number.
   a. Find a triangle on a spherical surface that contains three 90° angles. Illustrate your example.
   b. What's the largest sum of the angles you can find for a triangle on a spherical surface?

TEST YOURSELF

1. The cosmic background radiation comes from a time in the evolution of the universe
   a. when protons and neutrons were first formed.
   b. when the Big Bang first began to expand.
   c. when the first quasars began to shine.
   d. when X-rays had enough energy to penetrate matter throughout the universe.
   e. when electrons began to combine with nuclei to form atoms.

2. In which part of the sky is the cosmic background radiation brightest?
   a. It is equally bright in all parts of the sky.
   b. Toward the Virgo cluster
   c. Toward the center of the universe
   d. Toward the spot where the Big Bang took place
   e. Around the locations of quasars

3. Which of the following is important for explaining why the sky is not bright at night?
   a. Dust in our galaxy blocks the light from distant galaxies.
   b. The universe has a finite age.
   c. There are no stars beyond a distance of about 10 Mpc.
   d. The universe did not expand from a dense hot state, but instead had a temperature close to absolute zero when it exploded.
   e. Galaxies are much farther apart than was believed at the time Olbers proposed the paradox.

81 The Beginnings of the Universe

The early universe was hot and dense. During the first 400,000 years of its existence, the universe was full of radiation that interacted constantly with the matter. This period is hidden from direct view because of this constant interaction—the universe became transparent only after the *last scattering epoch,* when the density had dropped low enough. However, it is possible to reconstruct the history at these early times from a variety of evidence.

Figure 81.1 depicts our current understanding of the history of the universe from its beginning, 13.7 billion years ago. Before the last scattering epoch, radiation was not only constantly interacting with matter, but for the first 50,000 years or so the radiation actually generated most of the gravitational force. Both mass and energy exert gravitational influences—the conversion between the two is based on Einstein's famous formula, $E = m \times c^2$. During this early time light "weighed" more than matter! As the universe cooled, the radiation lost energy, and we left the **radiation era** and entered the **matter era,** where the gravity of matter became the dominant force in the universe.

During the radiation era, there were none of the large gravitational structures we see today. The universe was a sea of subatomic particles and light, initially at extremely high temperatures, but cooling to temperatures comparable to the surface of a star by the end of the era. The behavior of matter at these high temperatures has been explored by physicists with *particle colliders.* These are machines that accelerate subatomic particles to nearly the speed of light, then crash them together, and for an instant create conditions similar to the universe just after the Big Bang. This allows us to predict what happened at these times and to search for confirming evidence.

In this unit we work our way step-by-step from a few minutes after the Big Bang back toward even earlier times—to a tiny fraction of a second after time began as depicted in Figure 81.1. Although scientists are confident about the first few of these steps back in time, in the final steps, when the entire universe visible today was far smaller than an atom, we reach the limits of our current understanding.

81.1 THE ORIGIN OF HELIUM

When astronomers extrapolate their understanding of the universe back to times before the last scattering epoch, their first major prediction is that a significant fraction of the hydrogen in the newborn universe would have fused to form helium. We can see why from the following argument: If we take all the matter in the universe today and extrapolate the amount of compression there was a few minutes after the Big Bang, the temperature and density throughout the universe would have been similar to that in the core of a massive star. Therefore, fusion should have occurred, just as it does in the cores of stars.

Such reasoning led George Gamow to hypothesize in 1948 that helium and heavier elements should have been created in the young universe—if the Big Bang

Background Pathways: Unit 79

666

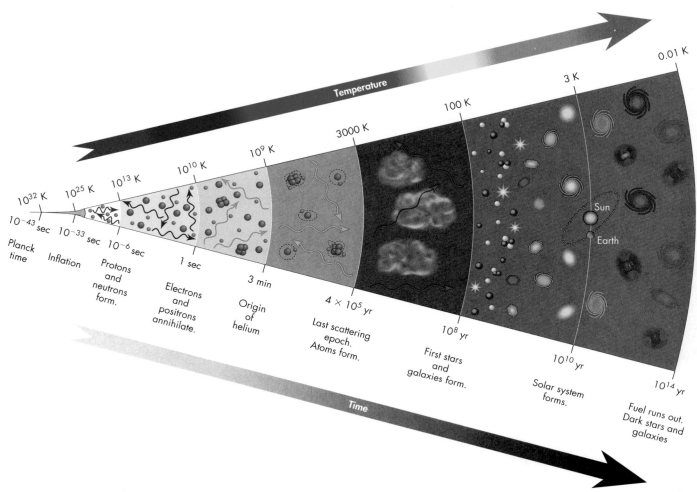

FIGURE 81.1

A sketch depicting the history of matter and radiation in the universe.

model was correct. Gamow's hypothesis is supported by more recent detailed simulations as well as by observations.

We now estimate that about 1 second after the universe began the temperature of the universe would have been 10 billion Kelvin, and its matter would have had a density of about 0.1 kilogram per liter. This may not sound very dense, but it is roughly a thousand-trillion-trillion times denser than the universe is at present. At these densities and temperatures, all of the matter in the universe would have broken apart into its constituent particles: electrons, protons, and neutrons. In gas this hot, nuclear fusion could take place, binding protons and neutrons into deuterium (^{2}H). However, at such a high temperature, the intense radiation would have broken apart the newly fusing particles of deuterium almost as fast as they could form.

As the universe expanded, it steadily cooled, and the radiation became weaker so that the deuterium nuclei—the first step in the proton–proton chain (Unit 50.2)—built up in number. At around 100 seconds, expansion had cooled the universe to a temperature of about 1 billion Kelvin, and particles of deuterium were rapidly fusing into helium ^{4}He. But not only had the universe cooled, it had grown less dense (to a density of about 10^{-3} kg/liter); so the rate of collisions became infrequent, and fusion slowed and then stopped in another few minutes.

The total amount of helium formed depends on how hot and dense the universe was and how long those conditions lasted. These conditions were in turn determined

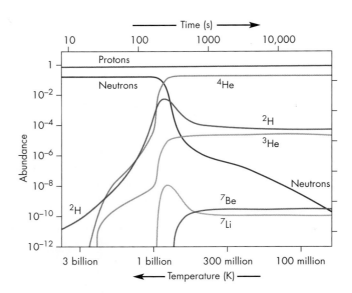

FIGURE 81.2

During the first few minutes after the Big Bang, the universe was hot and dense enough to fuse hydrogen into heavier elements. The predicted amounts of the elements closely match what we observe today.

by how rapidly the universe was expanding. Big Bang models predict that about 24% of the matter should have been fused into helium. This amount agrees remarkably well with the amount of helium observed in the atmospheres of the oldest stars.

Trace amounts of other light elements would also have been produced during these first few minutes after the Big Bang, including ^{3}He and the element lithium (^{7}Li). This is shown in Figure 81.2, where the amount of each element formed as the particles interacted is traced, beginning about 10 seconds after the Big Bang. Depending on precisely how hot the universe was and how rapidly it was expanding, different amounts of deuterium would have completed the fusion process all the way to helium. By determining the amounts of each of these elements in the most primitive surviving stars and gas clouds we detect today, astronomers can set tight constraints on the conditions a few minutes after the universe began. They can estimate the rate of expansion and the total amount of matter in the universe.

For example, if the expansion was rapid or the density low, slightly less ^{4}He would have had a chance to form, leaving significantly more deuterium and ^{3}He unfused. Slow expansion and higher densities would have allowed more complete fusion, resulting in much lower amounts of ^{3}He and deuterium, and providing an opportunity for more of the heavier elements such as lithium to form. Therefore, from the elements that they observe in the oldest stars, astronomers can deduce the properties of the expansion of the early universe; and from the properties of the expansion, they can deduce the total mass of the universe. These results suggest that our universe has a relatively slim amount of matter and has been expanding so rapidly from the start that it will continue to expand forever (see Unit 82).

The excellent consistency between the Big Bang helium production predictions and the percentages of all the light elements in the oldest matter in the universe provides further strong support for the Big Bang theory. It is interesting that in more than 13 billion years since this early epoch, stars have succeeded in fusing only about 3% more of the matter in the universe into helium. All of the fusion in all of the stars in the whole history of the universe has produced little more than a tenth of what occurred during a few minutes after the Big Bang!

 If there had not been any fusion in the early universe, how might the evolution of stars have been altered?

81.2   RADIATION, MATTER, AND ANTIMATTER

Next let us take another step closer to the beginning of the universe. During the first second after the Big Bang, the universe was so hot that matter and radiation mingled without the sharp distinction we see between these two entities today.

Energy into matter

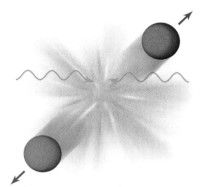

Radiation creates particle and antiparticle.

A

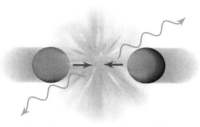

Particle and antiparticle annihilate, creating radiation.

B

FIGURE 81.3

(A) The energy of electromagnetic radiation can be converted into mass with the formation of a particle and its antiparticle. (B) The collision of a particle and its antiparticle leads to their annihilation and conversion of their mass back into energy.

Physicists can generate antimatter in particle colliders. How do you suppose they can keep it from immediately destroying their experimental apparatus?

Astronomers call this period the **early universe.** It was a time when matter and radiation behaved almost as a single entity.

That the distinction between matter and radiation can become blurred is predicted by Einstein's equation $E = m \times c^2$, which indicates the equivalence between an amount of energy, E, and an amount of mass, m. (The term c^2 is the conversion factor between mass and energy—the speed of light, c, squared.) This equation lets us calculate, for example, the amount of energy that is released when hydrogen fuses into a slightly smaller mass of helium in the core of a star (Unit 50). The energy–mass equivalence also permits the reverse: If a photon of energy E is concentrated in one place, it may be turned into particles whose mass equals E/c^2. Scientists observe matter created this way from high-energy gamma rays in laboratory experiments.

In the early universe, extremely high-energy radiation filled all of space. Creation of matter from energy would have occurred everywhere. Such transformations of energy into mass, however, cannot occur in just any way. Laboratory experiments show that the conversion of energy into matter always creates a pair of particles, as illustrated in Figure 81.3A. Moreover, the particle pair has two special properties: The particles must have opposite charge (or no charge), and one of the pair must be made of ordinary matter, whereas the other must be made of what is called **antimatter** (Unit 4). For example, the electron has an antimatter particle, the **positron,** with identical mass but opposite electric charge.

Antimatter has the important property that a particle and its antiparticle are annihilated on contact, leaving only high-energy radiation, as depicted in Figure 81.3B. Thus, radiation (electromagnetic energy) can become matter in the form of a particle–antiparticle pair, and matter can become radiation. In the early universe, this shifting between mass and energy happened continually as particles formed and were annihilated.

Energetic radiation cannot turn into just any pair of particles: The pair's combined mass multiplied by c^2 must be less than the radiation's energy. For example, to become an electron–positron pair, the radiation needs a wavelength shorter than about 0.0012 nanometers. Any excess energy of the photon becomes the kinetic energy of the electron and positron, which race away from each other until they encounter and annihilate another of their antiparticle partners. Radiation with such large energies occurs only at extremely high temperatures. To have enough energy to make an electron–positron pair, the temperature must be more than about 6×10^9 Kelvin, which was true for about the first second of the universe's existence.

More massive particle–antiparticles pairs were created at earlier times. Protons, for example, have a corresponding antiparticle called the antiproton. To become a proton–antiproton pair the radiation needs roughly 1800 times more energy than to create the electron–positron pair. This requires an even shorter-wavelength high-energy gamma ray. To make a proton–antiproton pair, the gas temperature must be more than about 10^{13} Kelvin. Such high temperatures do not occur in ordinary stars, but they were reached at earlier instants after the Big Bang.

81.3   THE ORIGIN OF MATTER

As we move back to very early times, there was another important transition at approximately 1 microsecond (10^{-6} sec) after the Big Bang. At that time, the universe we can presently see was packed into a volume about 10 times bigger than the Solar System. Before this time, the universe had a temperature greater than 10^{13} K, hot enough for its radiation to be converted into not only protons and antiprotons but even into **quarks** and antiquarks, the particles that make up protons and neutrons.

The universe was an inferno, jammed with quarks and antiquarks created from the high-energy radiation. This period is estimated to have begun about 10^{-35} seconds after the Big Bang. The entire universe we see today was then packed into a volume less than a meter across but was expanding at enormous speeds.

As space expanded, this dazzling sea of matter and radiation cooled so that by 10^{-6} seconds, the radiation was no longer energetic enough to create new quark–antiquark pairs. Thereafter quarks annihilated antiquarks, and those that remained combined to make protons and neutrons. However, that any quarks remained at all is a surprise because quarks and antiquarks should have been created from energy in equal numbers. For some reason, though, an imbalance left an excess of quarks over antiquarks. This is fortunate for us because otherwise there would have been no matter in the universe at all. There would have been nothing in the universe but radiation.

Physicists have been trying to understand the origin of this imbalance in high-energy collider experiments. They have discovered that the weak force (Unit 4) has what they describe as an **asymmetry.** An asymmetry can allow a pathway for an antiquark to decay, for example, but no equivalent pathway for its partner quark to decay as well. The result is that at the end of the first microsecond of the universe's history there was a tiny excess of quarks and electrons—literally, one in a billion—surviving with no antiquark or positron to annihilate.

The ordinary matter and radiation continued to expand, and for the first second, the universe was still hot enough to create electron–positron pairs (as described in the previous section). At the end of that brief interval (about the time for one heartbeat), the universe cooled below 10 billion Kelvin, and the positrons annihilated the electrons. In addition to creating the photons that eventually became the cosmic microwave background, these annihilations also would have produced a similar number of neutrinos. These neutrinos would today be redshifted to energies that are currently too low to detect, with energies only about 2° above absolute zero. But better detectors might someday give us a direct glimpse of these neutrinos and the first moments of the universe. If we could detect these primordial neutrinos, it might be as helpful to our understanding of the early universe as the detection of solar neutrinos has proved to be for understanding the workings of the Sun (Unit 50.3).

81.4   THE EPOCH OF INFLATION

The time before 10^{-33} seconds remains at the frontier of our understanding. Particle colliders cannot yet reach the energies of such early times, but a variety of evidence suggests what may have happened in the first instants of the universe.

The very earliest time to which we can apply our current understanding of the forces of nature is known as the **Planck time.** The Planck time is a moment about 10^{-43} seconds after the universe began, when the entire universe we can see today would have been packed into a volume far smaller than a proton, and the temperature would have been greater than 10^{33} Kelvin. At the Planck time we encounter an interface between our understanding of the subatomic nature of matter and the way space is curved by gravity. Physicists today studying the subatomic behavior of matter observe **quantum fluctuations,** where particles can flash into and out of existence and where changes occur according to probabilities

Q If some regions of the universe had been left with a surplus of antimatter instead of matter, how might they appear different? How might they appear similar?

rather than certainties. Before the Planck time, extremely dense bits of matter would have fluctuated in and out of existence, causing space to curve in an unpredictable way that is difficult to reconcile with our current understanding of the nature of gravity.

Experiments that physicists have conducted with particle colliders give us a number of clues about conditions after the Planck time. The universe was almost pure energy, and subatomic particles of all types flashed in and out of existence—their masses almost nothing in comparison to the enormous amount of energy present. It appears possible that the fundamental forces of nature (Unit 4) interacted at these extremely high energies quite differently from the way they do today. The transition of the strong, weak, and electromagnetic forces into their current form may have provided the mechanism that "launched" the rapid expansion of the universe in a process that cosmologists call **inflation.**

Inflationary models are based on hypotheses that attempt to unify the basic forces of nature. For example, in the 1970s it was discovered that the electromagnetic force and the weak force, which is responsible for radioactive decay, are really two aspects of the same force. The primary difference between these forces is in the mass of the carriers of the force. Photons, which carry the electromagnetic force, have no mass and can therefore be generated easily and travel large distances. W and Z particles, which are corresponding carriers of the weak force, have a very large mass. They therefore require high energies to be created and rarely travel more than a subatomic distance before decaying and giving up their energy. However, in the early universe, when there was so much free energy, W and Z particles could be created easily. Physicists have found that the carriers of the electromagnetic and weak force form a closely linked family that becomes almost indistinguishable above about 10^{15} Kelvin. Today physicists generally refer to the two forces in combination as the *electroweak force* to emphasize the concept of unification.

Temperatures in the very early universe were far higher than can be reached with current experiments, but cosmologists have developed hypotheses for how the strong force unifies with the electroweak force at the temperatures found shortly after the Planck time. An analogy to how these forces change at different energy levels is a BB flicked at a pane of glass with your finger. At a low energy the BB bounces off the glass as if repelled. However, if it is shot from a pellet gun (high energy), it pierces the glass—the repelling effect vanishes at high energy. So, too, in the high-energy environment of the early universe, forces may have interacted and behaved differently from the way they do today in our low-energy world, making their unification a possibility.

Currently there are several competing hypotheses of unification, called **grand unified theories** or **GUTs,** for short. According to most versions of GUTs, the strong and electroweak forces would have remained unified until the universe's temperature dropped to about 10^{27} K. This would have happened about 10^{-35} seconds after the Big Bang. At this temperature the strong force began to "freeze out" of the unified mix. The destruction of unification had extremely important consequences. GUT models suggest that the separation released a vast amount of energy. This energy release has been compared to the heat released when water freezes—ice absorbs energy to melt, and it releases the same amount of energy when it freezes.

As the universe began this transition to a separate strong force, it was caught briefly in a peculiar state, where the creation of energy everywhere led to an accelerating expansion of the universe that drove space to expand more and more rapidly. In each 10^{-35} sec, the universe approximately doubled in size, hence growing exponentially. A region the size of a proton at first grew slowly, but after doubling and doubling again perhaps a hundred times over, by 10^{-33} sec it had grown

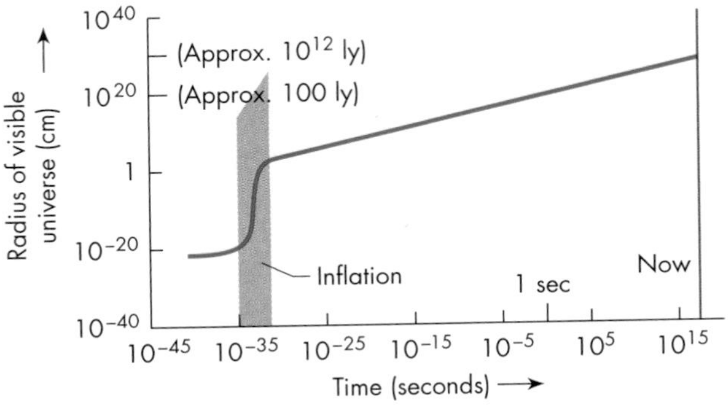

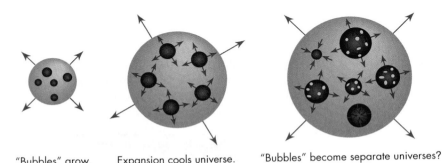

FIGURE 81.4

A sketch showing how the radius of the universe suddenly inflated.

FIGURE 81.5

A sketch depicting bubbles that might become separate universes forming in an inflating universe.

"Bubbles" grow.   Expansion cools universe.   "Bubbles" become separate universes?

to the size of a basketball. The rapid increase in size during the inflationary period is illustrated in Figure 81.4.

Regions of space at the end of this inflationary expansion were moving away from one another at speeds far exceeding the speed of light. This might sound like an impossibility according to Einstein's theory of relativity, but it is not. Motions of different regions of space relative to one another can have any speed, even greater than light speed. No light or matter can travel *through* any part of space at greater than light speed, and no observers will ever measure a photon traveling *by* them through space at any speed other than *c*. But different portions of the universe can travel extremely fast *relative* to one another, carrying along the matter within them.

One of the intriguing possibilities suggested by some inflation models is the existence of other universes forever separated from our own. It appears possible that inflation may not have occurred everywhere at the same time. After the Planck time, different regions of space may have inflated independently, and today they may be completely separated from one another with no possible contact between them. Such isolated regions might well be considered separate universes, independent of ours and completely unobservable by us. Furthermore, according to inflationary theories, these other universes may be either expanding or contracting.

Inflation creates such sections of space much as bubbles form in a pot of boiling water. As water begins to boil, regions of the liquid suddenly change from liquid to gas (steam) and begin expanding. We see that region of steam as a bubble. Some bubbles expand, rise to the surface, and burst, whereas others collapse. So, too, in inflationary theories, entire universes may form and dissolve, entirely unknowable to us (Figure 81.5). It would be amazing indeed if all the wonderful intricacy and beauty of our universe is merely a single bubble in an even vaster sea of space, a space that we can never see.

81.5   COSMOLOGICAL PROBLEMS SOLVED BY INFLATION

The rapid acceleration generated during an inflationary period offers an explanation of how the Big Bang might have been launched. It also explains several features of our universe that cosmologists had previously found difficult to explain.

The first of these is the uniformity of the cosmic microwave background, as depicted in Figure 81.6A. In every direction we look, the radiation we observe has the same temperature, 2.726 Kelvin to within about 1 part in 100,000. This is much smoother than the Big Bang model predicts without inflation. Regions of the CMB we see that are more than about 1° away from each other at the time the CMB was formed were too far apart to reach a common temperature (Unit 80.3). This separation was larger than the distance light could have traveled from any intermediate point within the age of the universe at that time, so these regions could

FIGURE 81.6

A map of the cosmic microwave background, made by the Wilkinson Microwave Anisotropy Probe satellite (WMAP). The map shows the brightness distribution of the cosmic microwave background across the entire sky, much as a map of the Earth depicts its features on a flat surface. (A) The top map depicts the uniformity of the radiation seen—constant everywhere to within 1 part in 100,000. (B) The small deviations from uniformity are brought out in this map. Red regions are brighter (warmer by a few millionths of a degree) and blue regions are fainter (colder). They show irregularities in the temperature of the universe at the time of recombination, when the universe was about 400,000 years old. The satellite that gathered the data for the map was named for David T. Wilkinson, a pioneer in the study of the cosmic microwave background. The term *anisotropy* refers to the lack of uniformity from point to point across the sky.

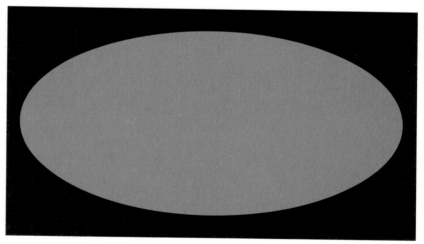

A

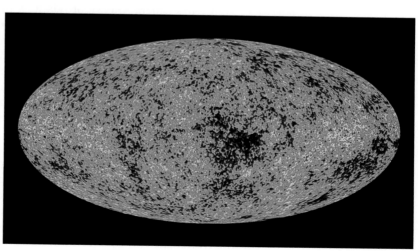

B

never have had an opportunity for their radiation and matter to mix smoothly and reach a common temperature. Thus, without inflation the Big Bang model predicts that the brightness of the cosmic microwave background should be very uneven.

With the inclusion of an inflationary period after the Big Bang, however, the visible universe we see today was originally one-trillionth the size of a proton and was not expanding rapidly at first. Even though only 10^{-35} second had gone by, light could travel thousands of times back and forth across the region that became our present visible universe, so the temperature had an opportunity to even out. That smoothness was preserved during inflation and is retained to this day in the high degree of uniformity of the cosmic microwave background. The faster-than-light expansion speed of inflation carried these regions so far apart in a trillionth-trillionth-trillionth of a second that the regions remained out of sight of each other for billions of years. Cosmic microwave background photons (Unit 80) are only now reaching us from regions that had been closer to us than the distance between two protons in an atomic nucleus before inflation began.

Although inflation makes the universe uniform on large scales, it also leads to the development of smaller irregularities. The cosmic microwave background is the same in every direction except for tiny variations that amount to about 1 part in 100,000. Just as water will spread and level out to a uniform height—sea level— it still contains small fluctuations—waves—that deviate from this uniform level. The map shown in Figure 81.6B shows the whole sky, spanning 360° across its center. There are many cool and warm regions that are far bigger than the 1° limit for communication at the time of the CMB. These large-scale fluctuations must have been produced before inflation began.

The inflation model suggests that microscopic quantum fluctuations from the time before 10^{-35} seconds were inflated tremendously, growing to the size of superclusters of galaxies today (Unit 76.3). Based on the inflationary model, cosmologists have predicted a range and distribution in the sizes of fluctuations generated by quantum fluctuations, and then preserved in the patterns of large-scale structure much later. This closely matches what is seen imprinted on the CMB. The slightly higher-density regions acted as the sites where dark matter began to congregate, and later on normal matter collected. Thus, the first steps toward the formation of large-scale structure in the universe (Unit 76.4) appear to have been taken when the universe was smaller than an atom.

The huge expansion associated with inflation also leads to the prediction that whatever curvature the universe had, due to gravity, was eliminated by inflation. Einstein's general theory of relativity predicts that space may be curved in different ways by gravity (Unit 80.4), depending on the balance between gravitational and kinetic energy. Inflation removes this curvature much as inflating a balloon makes its surface less curved. While the balloon is small, the curvature of its surface is obvious. However, if the balloon could be inflated to the size of the Earth, its surface would appear nearly flat, just as the surface of the Earth appears to a local observer as nearly flat. The rapid inflation of the early universe does the same to space, taking an initially **curved space** and effectively flattening it.

Inflationary models of the first moments of the universe mark the frontier of our understanding of the cosmos. They offer tentative explanations of some of its puzzling features, and they offer some intriguing possibilities. Some models suggest that the universe may have been created literally from nothing. Some suggest that space has more dimensions than the four we are familiar with: the three of space and one of time. In fact, according to some versions, called *string theories*, space has as many as 10 or 11 dimensions.

KEY TERMS

antimatter, 669

asymmetry, 670

curved space, 674

early universe, 669

grand unified theories (GUTs), 671

inflation, 671

matter era, 666

Planck time, 670

positron, 669

quantum fluctuation, 670

quark, 670

radiation era, 666

QUESTIONS FOR REVIEW

1. Why do astronomers believe the early universe was hot and dense?
2. How was helium formed from hydrogen in the early universe?
3. What problems of the Big Bang theory are resolved by inflationary models?
4. What is "unified" in grand unified theories?
5. What are the basic requirements in the creation of matter from energy?

PROBLEMS

1. One second after the Big Bang, the density of the universe was about 0.1 kilogram per liter.
   a. How big a volume of the universe contained as much mass as the Solar System?
   b. Estimate approximately how much bigger the universe would have been after three minutes. How big would the Solar System mass region be now? Compare this to the present size of the Solar System.
2. Examine Figure 81.6B. The map in the figure is 360° wide and 180° tall. An angle of 1° on this map corresponds to about 250 kpc.
   a. Approximately how big are the largest high-density regions you see in degrees? How large is this in kiloparsecs?
   b. How large a region would that be today?
   c. Relate this size to components of the large-scale structure we see today (Unit 76).

3. Cosmologists have commented that the universe expanded more in the 10^{-33} seconds of inflation than in the rest of the history of the universe. What they are referring to is the relative amount of expansion—the ratio of size at the end to the initial size. Compare the amount of expansion that occurred during inflation (doubling in size 100 times over) to the expansion of the present visible universe (13.7 billion light-years in radius) from a region with just a 1-cm radius.
4. Compare the amount of energy released when a proton and antiproton annihilate each other to the energy released when four protons fuse to form helium.

TEST YOURSELF

1. What is meant by *inflation* in the early universe?
   a. The force of gravity suddenly grew stronger in the distant past.
   b. Protons expanded to the size of stars, which was how our Sun formed.
   c. The universe increased dramatically in size in an extremely brief period.
   d. The number of galaxies that we see at large distances is much greater than the number we can see near us.
   e. The diameter of distant galaxies is much greater than the diameter of galaxies near us.
2. Which of the following statements about the first few minutes of the Big Bang is true?
   a. The universe was transparent.
   b. The universe was very cool.
   c. Hydrogen was converted to helium.
   d. Galaxies began to form.
   e. The cores of stars began to form.
3. Which of the following observations is explained by a brief period of extremely rapid inflation?
   a. The high uniformity of the cosmic background radiation
   b. The large number of black holes in the universe
   c. The fact that some globular clusters are older than the universe
   d. The presence of helium in the universe
   e. The existence of more matter than antimatter

82

The Fate of the Universe

The universe is currently expanding. But will it expand forever? This question is fundamental to the nature of the universe because according to the Big Bang theory, all the hydrogen in the universe originated with the Big Bang. If the universe expands forever, one by one its stars will consume their hydrogen and die, eventually leaving the universe a black, cold, empty space.

But what is the alternative? If the universe does not continue expanding, perhaps its gravity will someday force it to stop expanding and fall back together. The overwhelming force of gravity means that all the objects and atoms within it will be compressed to higher and higher densities, crushing everything that has ever existed in what some astronomers have called the **Big Crunch**.

Neither fate is particularly appealing. Are there any alternatives? Cosmologists are exploring the possibilities by tracing the history of the universe's expansion and seeing how well it matches the trajectories predicted for simple expansion or future collapse. The latest results show that there may be a third, more bizarre fate in store for the universe.

Background Pathways: Unit 78

82.1 EXPANSION FOREVER OR COLLAPSE?

To determine which fate awaits the universe, we need to understand more fully what factors affect its expansion or collapse. But before we look at details, a simple analogy may help. Suppose you throw a stone into the air. As the stone moves upward, Earth's gravitational force slows its rise and eventually stops it so that the stone falls back to Earth. On the other hand, if you could throw the stone faster than the Earth's escape velocity, it would rise forever. The stone's behavior depends on the strength of gravity and the upward impulse given to it. So too with the expansion of the universe.

Near the time of its birth, some process began the universe's expansion (the possibilities are explored in Unit 81). Since then the gravitational forces between galaxies and all the other contents of the universe have slowed its expansion. If the universe's gravitational force is weak enough, or if the initial impetus of expansion is strong enough, we would expect the universe to expand forever. On the other hand, if the gravitational force is strong or the initial impetus is weak, we expect that the expansion will stop and the universe will collapse.

To measure the relative strength of these effects for the universe, we can compare the gravitational energy holding the universe together to the energy of its expansion. If gravity dominates, the expansion will stop. If the expansion energy dominates, the expansion will continue forever (Figure 82.1). Therefore, we can learn (at least in principle) whether the universe will expand forever or collapse back on itself by seeing if the universe has a gravitational pull strong enough to stop its expansion.

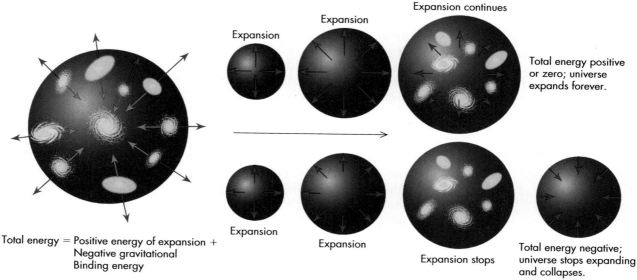

Total energy positive or zero; universe expands forever.

Total energy = Positive energy of expansion +
Negative gravitational
Binding energy

Total energy negative; universe stops expanding and collapses.

FIGURE 82.1
A sketch of how gravity and the energy of expansion determine the behavior of the universe.

Q Do you have a personal preference about the eventual fate of the universe? What is it, and why?

To see how strong the effect of gravity is, astronomers use Newton's law of gravity, modified to account for relativity theory. Essentially, gravity's effect depends on the universe's mass. It turns out, however, that it is far easier to work with the universe's density (the amount of mass contained within a given volume). The reason is simple: Astronomers can measure the density of the universe but not its mass. To measure its mass, we would need to observe all of space. To measure its density, we need only measure the mass within some large, representative volume of the cosmos.

82.2   THE DENSITY OF THE UNIVERSE

To find the density of the universe, astronomers choose a large volume of space and count the galaxies in it. Next we measure the mass of each galaxy, add the masses, and divide by the volume. For example, to measure the local density, astronomers might choose the Local Group, which contains three large galaxies and about two dozen smaller ones. The total mass of gas and stars in the Local Group is estimated to be about 10^{12} solar masses, which we can convert to kilograms if we multiply by the Sun's mass, 2×10^{30} kilograms. This gives a mass for the group of about 2×10^{42} kilograms.

Next that mass is divided by the Local Group's volume, assumed to be a sphere whose radius is the distance from the center of the Local Group to the next nearest galaxy group, which is about 3 Mpc (about 9×10^{22} meters) away. Using the formula for the volume of a sphere $(4/3\pi R^3)$ gives a volume for the Local Group of about 3×10^{69} cubic meters. Dividing this volume into the mass gives a density of about 7×10^{-28} kilograms per cubic meter, or 7×10^{-31} kilograms per liter.

The vicinity of the Local Group is a somewhat richer-than-average environment, however. To get a more representative sample, we must look at a much larger region that includes both clusters and voids, and we must include intergalactic gas, particularly in clusters. A similar calculation for these much larger volumes of the universe gives a slightly smaller value of about 3×10^{-31} kilograms per liter or, on average, roughly 2 hydrogen atoms per 10 cubic meters. This fantastically low density gives some indication of just how sparse the density of the universe is today.

ρ_C = critical density

H = Hubble's constant

G = gravitational constant

To determine whether the universe will expand forever or recollapse, astronomers compare this observed density to a theoretically calculated **critical density,** which is written with the Greek letter rho as ρ_C. If the actual density is greater than the critical density, the universe will recollapse; if it is less, the universe will expand forever. We can calculate the critical density by comparing the gravitational potential energy in a volume to the kinetic energy of expansion in that same volume. At the critical density, the two energies are equal. It turns out that mathematically the critical density ρ_C is

$$\rho_C = \frac{3H^2}{8\pi G}$$

where H is the Hubble constant and G is Newton's gravitational constant. Without carrying out the full derivation, we can see why the equation has this form. The gravitational potential energy depends on the density and Newton's gravitational constant, whereas the kinetic energy depends on the square of the expansion speed, which can be related to Hubble's constant. Thus, we are equating values proportional to $G \times \rho$ and H^2 and then solving for ρ.

Astronomers use a quantity called **Omega (Ω)** to indicate how close the observed density is to the critical density. This is because, as the last letter of the Greek alphabet, omega is sometimes used to suggest the "end of things." For the amount of matter in the universe, then, cosmologists define Ω_M as the value of the actual density of matter divided by the critical density:

$$\Omega_M = \rho/\rho_C.$$

With this notation,

If $\Omega_M > 1$, the universe will recollapse.
If $\Omega_M < 1$, the universe will expand forever.
If $\Omega_M = 1$, the universe is exactly at the critical density.

We specifically note that this is the value of Ω for *matter* because a variety of quantities can contribute to the overall value of Ω, and not all of them have the same effect on the universe's expansion, as we will see later in this unit.

The past and future sizes of the universe are illustrated for various values of Ω_M in Figure 82.2. Each of these models is assumed to match the present size and rate of expansion of the universe. The rate of expansion is known from Hubble's constant, and it is indicated by the slope of the curve—steeper slopes indicating faster expansion. If Ω_M is much larger than 1, the universe expands rapidly after the Big Bang, but the large density of matter causes the universe to slow its expansion quickly. In this case the expansion halts sometime in the future, and the universe then collapses back in on itself, ending in a "Big Crunch."

ANIMATIONS

The future of expansion

INTERACTIVE

Cosmology

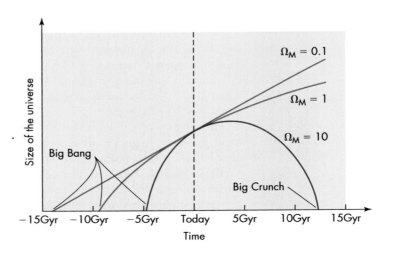

FIGURE 82.2

The size of the universe in the past and future as calculated for different densities. The slope of each curve today is determined by the Hubble constant.

If Ω_M is much less than 1, the expansion of the universe is barely slowed at all, and the rate of expansion remains nearly constant. The time since the Big Bang in this type of universe is close to the Hubble time $1/H = 14.0$ billion years (Unit 79.4) because the speed of expansion has remained nearly constant. Models of the universe with larger values of Ω_M all have shorter times since the Big Bang.

The $\Omega_M = 1$ case is interesting because it implies that the universe's gravitational potential energy exactly cancels its kinetic energy—a condition that some theories of the beginnings of the universe suggest makes the most sense. In the $\Omega_M = 1$ cosmology, the universe's expansion slows, approaching zero speed in the infinite future but never quite stopping.

When we compare the observed and critical densities, what do we find? If we use the value for H of 70 km/sec per Mpc, the critical density is 9.3×10^{-30} kg per liter. This tiny density—less than a billionth-billionth of the density of the best vacuum created in a laboratory on Earth—would be sufficient to halt the universe's expansion. As small as this density is, it is about 30 times larger than the measured density of visible stars and gas. This implies that the gravitational pull of all the visible matter in the universe is too weak to stop the expansion. For the universe to recollapse, it would need about 30 times more mass than this. This can be written as $\Omega_{VM} \approx 0.03$ (where VM stands for visible matter). This value is consistent with estimates made based on the amount of fusion that created helium and other light elements during the first few minutes of the universe (Unit 81.1).

However, this result does not take into account dark matter. Observations of galaxies and clusters imply large quantities of unseen matter, many times more than is directly observed (Unit 78). To explain the rapid rotation of the outer parts of spiral galaxies and motions within galaxy clusters, about 10 times more dark matter is required than is observed in stars and gas, so $\Omega_{DM} \approx 0.3$. Combined with the estimate for visible matter, this is still less than 1; but given the difficulty in estimating the amount of dark matter, a value of $\Omega_M = \Omega_{VM} + \Omega_{DM}$ as large as 1 could not be ruled out conclusively.

This was the state of cosmology at the end of the 20th century. Data on the density of matter in our vicinity, combined with the measured expansion rate of the universe, indicated that the galaxies should continue sailing apart forever. The fate of this universe would be determined by the total amount of dark matter. Even though estimates of the amount of dark matter in galaxies and clusters of galaxies suggested that Ω was still below 1, many cosmologists predicted that more dark matter would be found, pushing Ω_M to a value of 1. One of the cosmologists' arguments for a value of 1 was that inflation makes the universe almost exactly "flat" (Unit 81.5), which can be shown to imply that $\Omega = 1$.

There were problems with models of the universe with so much dark matter, however. As you can see in Figure 82.2, a value of $\Omega_M = 1$ implies that the Big Bang occurred less than 10 billion years ago. This is because a decelerating universe used to be expanding faster, so it can reach a large size more quickly than can a universe that maintains a slower fixed rate of expansion. An age for the universe of under 10 billion years was a major problem, though, because there are globular clusters that are estimated to be 12–13 billion years old—and nothing in the universe can be older than the universe itself.

Things were not adding up right, so it seemed something must have been left out of our calculations.

Q If the universe contained nothing but the visible matter we have observed, what would its expansion look like compared to the models shown in Figure 82.2?

82.3   THE SUPERNOVA TYPE IA FINDINGS

To improve our predictions about the eventual fate of the universe, cosmologists have developed different means of examining how the universe is expanding. Besides measuring the current rate of expansion and density of the universe,

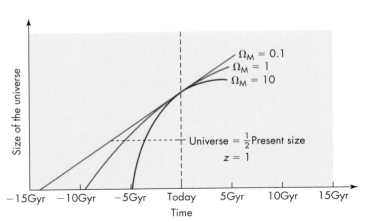

FIGURE 82.3

By studying the rate of expansion at large redshifts, for example at $z = 1$ (when the universe was half its current size), we can determine which cosmological model applies.

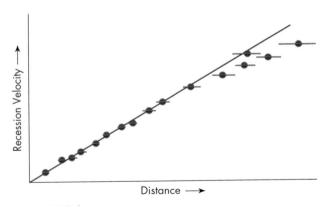

FIGURE 82.4

A graph of the distance and recession velocity of distant galaxies. The red line shows how the speed should change if the universe has expanded at a constant speed. The thin blue lines show the distance uncertainty for each galaxy. The graph is schematic and is based on observations of S. Perlmutter et al. Note that the distant galaxies tend to lie below the red line.

another way is to look at the history of expansion. Figure 82.3 focuses on the history of expansion that was illustrated in Figure 82.2, showing again the same three models of the history of expansion between the Big Bang and the present. Although the three models predict different ages for the universe, this is difficult to measure directly. We can instead look for differences between the models at different redshifts.

When we observe a galaxy at a redshift of $z = 1$, the wavelength of the light coming from it has been stretched by a factor of 2 (Unit 79.2). This means the universe was only half its current size when the light left the galaxy. The horizontal dotted line in Figure 82.3 indicates when the universe was half as big as it is now.

In the low-Ω_M model, the rate of expansion slows down only a little; so when the universe was about half as big, it was about half as old as it is now, and the light has taken about 6.5 billion years to reach us. In the high-Ω_M model, the universe contains much more mass, which makes the universe slow down quickly. In this model the light from an object at $z = 1$ travels for under 4 billion years. In other words, the distance to a galaxy with a particular redshift will be measured to be smaller if Ω_M is larger. In principle, then, we can determine the value of Ω_M if we can measure the redshifts and distances of distant galaxies. This test is straightforward, but doing it well enough to test the expansion theory requires precise distance measurements to faraway galaxies.

Supernova observations allow such measurements, though they are often difficult to obtain for such remote galaxies. In the 1990s, using the Hubble Space Telescope, astronomers were able to observe supernova blasts in some very remote galaxies. In particular, they detected supernovae of Type Ia (Unit 65), which result from the explosion of white dwarfs. These have been found to be excellent standard candles, so we can determine accurate distances based on their known luminosity and observed brightness. This allows astronomers to compare redshifts and distance of galaxies out to very large distances.

The results are shown in Figure 82.4, and they have delivered quite a shock! The line drawn in the figure is the prediction for a low-Omega universe. It was expected that the measured distances would all be smaller than this, lying to the left of the line. Instead, the measurements lie to the right of the line. This implies that the universe was expanding *more slowly* in the past than it is now—that the expansion of the universe is *speeding up*. How can this be?

82.4   DARK ENERGY

Our discussion of the fate of the universe to this point has been based on the assumption that expansion is affected only by the gravity of the matter within it. This originally appeared to be a good description of how the universe expands; but gravity's pull can only slow down the expansion, and this is not what the supernova result shows.

The best current explanation of this strange result comes from Einstein's early efforts to develop the general theory of relativity. Einstein developed equations to describe how matter and energy curve space and create the force of gravity. When he solved the set of equations he had put together to describe gravity, his mathematical solution showed an extra term. This extra term was called the **cosmological constant** because the mathematics of general relativity suggested that it should be the same everywhere and at all times.

A way of describing the cosmological constant is as an energy that fills all of space. This energy is unlike energies we are familiar with. It remains constant everywhere, existing even where there is nothing but space, and it does not grow more dilute as space expands. This is different from how matter and electromagnetic energy behave—they spread out and become more dilute as the universe expands.

No measurements in Einstein's time could determine the value of the cosmological constant. However, if the cosmological constant is not zero, it can counteract the effects of gravity. This constant level of energy everywhere creates a kind of cosmic repulsion, driving space to expand faster. Astronomers have given the cosmological constant the descriptive name **dark energy** because it is a kind of counterpart to dark matter.

Actually, Einstein developed general relativity before Hubble had discovered the expansion of the universe, and at that time the universe was thought to remain static rather than expand or contract. General relativity predicted that the universe should be in motion, so Einstein suggested that the cosmological constant might be quite large and balance gravity's attraction, making the universe static. Then, within a few years of Einstein's suggestion about the cosmological constant, astronomers discovered that the universe was in fact expanding, and Einstein concluded he should have set the cosmological constant to zero all along. He called this his "greatest blunder" because he might have actually predicted the expansion of the universe in addition to his many other remarkable discoveries. Since that time, cosmologists have noted the possibility of a cosmological constant, but for decades most assumed that it was zero.

The recent findings that the universe's expansion may be speeding up have revived interest in the cosmological constant. If dark energy in fact pervades all of space, it has major implications for our understanding of the universe. For example, because dark energy speeds up the universe's expansion, our estimates of the age of the universe must be modified. The result is an age that agrees better with the ages of the oldest known stars. The current best estimate of the universe's age, accounting for both gravity and the cosmological constant, is 13.7 billion years. This model uses a value for Hubble's constant of 71 km/sec per Mpc, and it assumes that there is somewhat more than twice as much dark energy in the universe today as the total of normal and dark matter. The comparison of energy and matter in this way is based on Einstein's formula $E = m \times c^2$, so matter (both normal and dark) represents about 3×10^{-30} kg/liter, while dark energy is about 6×10^{-30} c^2 kg/liter $\approx 5 \times 10^{-13}$ joules/liter.

In the last few years, additional evidence for the existence of dark energy has accumulated from studies of the cosmic microwave background (Unit 80). Detailed observations of the background reveal tiny variations in brightness from place to

Q The discovery that the universe can accelerate as well as decelerate opens up the possibility that the universe could have been "stationary" for some period in the past before resuming its expansion. If there had been such a period, would galaxies seen from that time have zero redshift?

place. Those tiny variations come from slight "lumps" in the universe's density shortly after its birth. The physical sizes of these lumps are determined largely by the density of both visible and dark matter, and the best fit to these results indicates that $\Omega_M \approx 0.3$. However, the *apparent* size of these lumps (their angular size) depends on the curvature of the universe (Unit 80.4), which cosmologists have shown implies that $\Omega = 1$. The difference between these results for Ω and Ω_M shows that something else must be contributing to the curvature of the universe.

One of the intriguing aspects of dark energy is that although it works against gravity to speed the expansion of the universe, the amount of dark energy should be *added* to the density of matter in the universe in determining its curvature. If dark energy contributes the missing fraction of Omega—$\Omega_{DE} \approx 0.7$—then both supernova results and the curvature results can be explained. This suggests there is more than twice as much dark energy as matter. The contributions of the various components of the universe to its overall composition are shown in Figure 82.5.

The pattern of expansion predicted for a universe with the current estimates of dark energy and dark matter is shown in Figure 82.6. In this model, the universe initially expanded rapidly, but the gravitational pull of its matter began to slow it down. As the universe expanded, the matter was thinning out and its gravitational pull was growing weaker. Meanwhile, though, the dark energy would have remained constant, so its repulsive effect began to take over to accelerate the universe. The universe has been picking up speed since then, and if we extrapolate into the future, the universe should expand faster and faster.

If this model is correct—and it *is* the best-supported model at present—it implies that the universe will tear itself apart in the distant future. Space will expand

The inflation of the universe in the first tiny fractions of a second was another episode of exponentially increasing expansion (Unit 81.4).

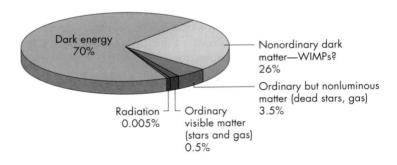

FIGURE 82.5
The makeup of the universe as deduced from observations of the brightness variations in the cosmic microwave background and other data. The percentages have been rounded off, so they do not add up to exactly 100%.

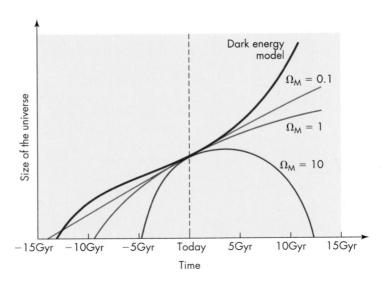

FIGURE 82.6
A universe with dark energy will accelerate more rapidly in the future.

faster and faster until matter that today is close together is overwhelmed by the rapid expansion of space. This is an alternative fate in which the universe does not just expand forever: It expands so fast that every part of the universe is eventually pulled away from every other part at speeds that become so fast that they all disappear from even the sight of each other.

Is this fate a certainty? It is too early to say. Astronomers are in the early stages of studying dark energy and developing tests to detect it more directly. The science of cosmology is beginning, however, to sketch the overall structure of the universe. It is a universe in which the stars and galaxies—all of the kinds of matter that we are familiar with—are just a minor constituent of the universe. We are along for the ride in a universe driven by far larger forces generated by dark matter and dark energy.

KEY TERMS

Big Crunch, 676

cosmological constant, 681

critical density, 678

dark energy, 681

Omega (Ω), 678

QUESTIONS FOR REVIEW

1. Why do astronomers think the universe will expand forever?

2. What is meant by *dark energy*?

3. What role does the cosmological constant play in cosmology? What evidence makes astronomers think it may exist?

PROBLEMS

1. The critical density depends on the value of Hubble's constant H. If H turns out to be 65 km/sec per Mpc, calculate how much the critical density changes compared to $H = 71$ km/sec per Mpc.

2. Dark energy retains a constant density even as the universe expands. Today it has a value of $\Omega_{DE} = 0.7$. Matter, by contrast, grows more dilute as space expands, so its density would have been larger in the past by a factor that depends on the redshift,

as $(z + 1)^3$. At what redshift would dark energy and matter (currently $\Omega_M \approx 0.3$) have had the same density?

3. If dark energy in the universe currently has about twice as large a density as matter, then how many joules of energy fill the Solar System? (Assume the Solar System is spherical with a radius of 40 AU.) If this energy could be converted into matter, how much mass would it be?

TEST YOURSELF

1. Dark energy
   a. is the same as the cosmological constant.
   b. is repulsive.
   c. is distributed evenly through space, never dissipating.
   d. causes space to expand ever faster.
   e. All of the above

2. Which of the following is true of the expansion of the universe if there is no dark energy?
   a. The expansion rate will increase.
   b. The expansion rate will decrease.
   c. The expansion rate will remain constant.
   d. The expansion rate will first increase and then decrease.

3. What causes the universe to curve around onto itself?
   a. Its own mass and energy
   b. The cosmic microwave background
   c. The distribution of galactic superclusters
   d. Black holes suck in all the space around them

83

Astrobiology

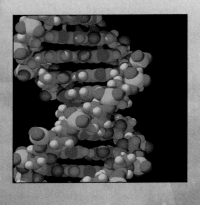

The evolution of the universe from Big Bang to galaxies, stars, and planets has along the way created us. This evolutionary path from raw quarks to thinking beings who seek to understand how they fit into this grand scheme—what eighteenth-century scientists called "the great chain of being"—is one of many marvels of the universe.

To understand the possibilities for life elsewhere in the universe, we turn to life on Earth and how it may have formed and developed. We will see that the existence of life on our own planet suggests the possibility of its existence elsewhere.

We begin our discussion by examining the long history of life on Earth, looking at the factors thought to play a role in its development here. Finally, we will speculate about the role of life in the history of our planet—the so-called Gaia hypothesis—and whether our existence may even say something about the nature of the universe itself.

83.1 LIFE ON EARTH

Life has existed on Earth for nearly the entire history of our planet. For example, fossil algae and bacteria such as those illustrated in Figure 83.1 occur in rocks 3.5 billion years old. In fact, one of the most ancient life forms, algal colonies called **stromatolites,** look identical to the stromatolites that exist today in shallow, warm, salty bays such as those along the coast of western Australia.

Given that the Earth formed about 4.5 billion years ago and that parts of its surface were probably molten rock for almost 1 billion years, life began quite quickly.

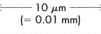

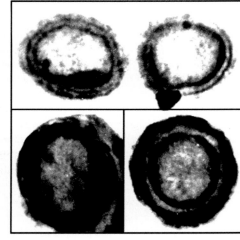

FIGURE 83.1
Photographs of fossil bacteria.
Note: 10 μm $\approx$ 0.004 inch.

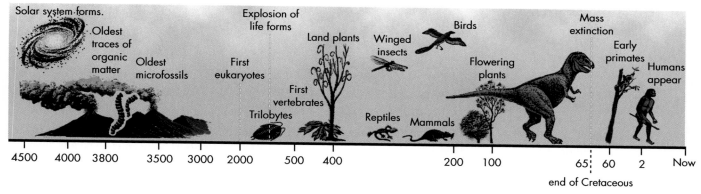

Time (millions of years before present)

FIGURE 83.2

A time line illustrating the history of life on Earth. (Note that the time scale is greatly expanded for more recent times.)

FIGURE 83.3

A few of Earth's many complex life forms.

Not all scientists agree that the microscopic features in these 3.5 billion-year-old rocks are fossil bacteria. There are more widespread fossils with ages about 1 billion years younger, as well as possible isolated examples with intermediate ages.

Perhaps we can appreciate this speed by examining Figure 83.2, which shows a time line for the history of life on Earth. This diagram shows the epoch in which various life forms first appear in the fossil record. Their ages have been measured by radioactive dating of their associated rocks (see Unit 33.1). Along with the ages, the figure also illustrates some of these ancient life forms.

Scientists conclude from such evidence that only extremely simple life forms such as algae and single-celled animals existed for roughly three-fourths of the Earth's history. Then, approximately 600 million years ago, more complex forms such as invertebrates developed, and about 500 million years ago, only a little more than one-tenth of the Earth's lifetime, shells and crustaceans appeared. Mammals and dinosaurs both appeared roughly 250 million years ago, but the latter were wiped out about 65 million years ago in the great Cretaceous extinction, caused, we think, by a totally chance event—an asteroid's collision with the Earth (Unit 48). Hominids, our immediate ancestors, appeared roughly 5.5 million years ago; but our own species, Homo sapiens, evolved only about 500,000 years ago. Earth has therefore existed as a planet roughly 10,000 times longer than we have as a species. If all of Earth's history was compressed into a year, humans would appear only in the last hour, and what we call civilization in the last few minutes.

Life on Earth shows a dazzling variety of complex forms from butterflies to whales, and mushrooms to trees (Figure 83.3). Despite that variety, all living beings on Earth have an amazing underlying unity of structure, reproduction, and metabolism. For example, all living organisms use primarily the same kinds of atoms for their structure and function: hydrogen, oxygen, carbon, and nitrogen. These atoms are not only the most commonly used by Earth's life forms, but they are also some of the most abundant throughout the universe. Even widely differing organisms

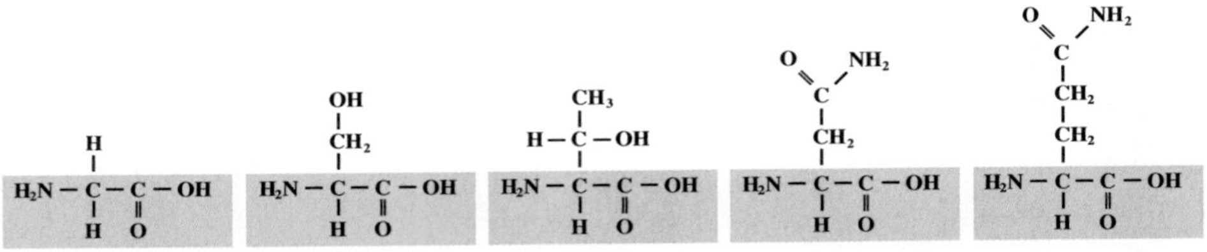

| Glycine | Serine | Threonine | Asparagine | Glutamine |

FIGURE 83.4

Diagrams illustrating the structure of several amino acids. H, C, N, and O represent atoms of hydrogen, carbon, nitrogen, and oxygen. The basic building block (shown in yellow) of amino acids is the same.

An organic molecule is one that contains carbon atoms. Organic molecules do not require biological organisms for their creation.

use the same chemical substances. Life tends to draw on the materials with which our planet is most richly endowed; we are made primarily of the same substances that make up our ocean and atmosphere, the same elements that are the primary products of nucleosynthesis in stars.

The chemical elements that compose living things are linked into long-chain molecules for their structure and function. Many of these chain molecules are built up from various arrangements of some 20 amino acids. **Amino acids** are organic molecules containing atoms of hydrogen, carbon, nitrogen, oxygen, and sometimes sulfur. Their structure is the same except for a "side chain" that makes each one slightly different (Figure 83.4).

Moving up to the next level of complexity, we find that all living things use amino acids as the structural units of more complicated molecules called proteins. Protein molecules give living things their structure. For example, the scales of reptiles and fish are composed of the protein keratin, and the stiff cartilage of our own bodies is composed of the protein collagen. Proteins not only give cells their structure but also supply their energy needs. That energy supply reveals yet another underlying unity among living things: All cells use the same kind of molecule, adenosine triphosphate (abbreviated as ATP) to supply energy for action and growth.

This unity of structure and energy supply is echoed in the reproduction of life forms. All single-celled organisms divide in two to create new cells; all multicelled organisms produce egg cells, which, once fertilized, divide to create new cells. Moreover, in all cases, parents pass on genetic information to their offspring by the same molecule, **DNA.**

DNA stands for *deoxyribonucleic acid.* It is a massive molecule consisting of two long chains of smaller molecules wound around each other and linked by molecules called base pairs. DNA has the appearance of a twisted ladder with the role of the connecting rungs played by the base pairs (Figure 83.5). When a cell reproduces, the DNA separates into two individual chains. Each chain then adds a new base pair at each point along its length to match the one previously present on the other chain. After division and replication, two new DNA molecules exists that are identical to the original one. Nonetheless, copying errors—*mutations*—sometimes occur. Most are harmful and kill the organism, but the nonlethal ones make evolution possible. That all terrestrial organisms use DNA for their reproduction gives life a truly remarkable unity.

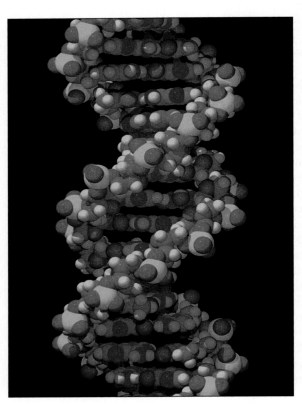

FIGURE 83.5

A diagram illustrating the structure of the DNA molecule. The blue horizontal structures in this model are the base pairs.

Unity can also be seen in how more complex organisms function, such as the chemical processes that move biologically important molecules into and out of cells. They also use similar molecules for different purposes. For example, plants derive their food from carbohydrates that they manufacture during photosynthesis. The chlorophyll they use in this process is structurally similar to the hemoglobin that transports oxygen in the blood of animals. These two molecules differ primarily in that chlorophyll is built around a magnesium atom, whereas hemoglobin is built around an iron atom. Similarly, cellulose, the structural material of plants, has a molecular structure very similar to chitin, the structural material in insect bodies and crustacean shells.

What conclusions can we draw from the chemical similarities of living beings and their ancient history as deduced from the time line? First, such chemical similarity almost inescapably suggests that all life on Earth had a common origin. Second, the antiquity of life suggests that under suitable conditions, life develops rapidly from complex molecules. Therefore, we might infer that wherever such conditions occur, so too will life.

We should treat these conclusions with great caution, however, because they are based on only a single case: life on Earth. For example, if the first time you hit a golf ball you make a lucky hole-in-one, it does not mean that you will do so again anytime soon. Likewise, the fact that life exists on Earth may say next to nothing about its chances elsewhere. Supposing, however, we so choose to speculate: What *can* we say about the origin of life and its likelihood elsewhere?

83.2   THE ORIGIN OF LIFE

Most scientists today think that terrestrial life originated from chemical reactions among complex molecules present on the young Earth. This idea dates back to at least the time of Charles Darwin, but it was strongly bolstered in 1953 by an experiment performed by Stanley Miller—then a graduate student at the University of Chicago—and his professor, Harold Urey. Miller and Urey filled a sterile glass flask with water, hydrogen, methane, and ammonia—gases thought to be present in the Earth's early atmosphere. They then passed an electric spark through this mixture as shown in Figure 83.6. The spark generated both visible and ultraviolet

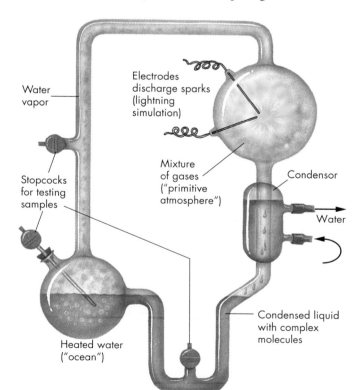

FIGURE 83.6

The Miller–Urey experiment attempted to simulate conditions on the primitive Earth. A mixture of gases received sparks (simulating lightning) and were repeatedly condensed and evaporated as they might have been in the young oceans. A variety of complex organic molecules were generated.

Water vapor

Electrodes discharge sparks (lightning simulation)

Stopcocks for testing samples

Mixture of gases ("primitive atmosphere")

Condensor

Water

Heated water ("ocean")

Condensed liquid with complex molecules

radiation, which triggered reactions in the gas and water mixture. At the end of a week they analyzed the mixture, finding in it a variety of organic molecules as well as five of the amino acids used to make proteins. They therefore concluded that at least some of the ingredients necessary for life could have been generated spontaneously on the early Earth.

The success of the **Miller-Urey experiment** motivated other researchers to perform similar but more complex and realistic experiments. For example, the U.S. chemist Sydney Fox took complex organic molecules, such as those made by Miller and Urey, and by repeatedly heating and cooling them in water was able to create short strands of proteins called **proteinoids.** More interesting, however, Fox discovered that these proteinoids spontaneously formed droplets reminiscent of cells. These droplets had no power to reproduce, nor did they show any of the complex structure commonly found in living cells; but they demonstrated that organic material can spontaneously give itself structure and wall itself off from its environment, a step probably taken by the first living things.

Even the energy needs of living things can be produced in Miller–Urey types of experiments. For example, the Sri Lankan–American biochemist Cyril Ponnamperuma and the U.S. astronomer Carl Sagan showed that ATP can be synthesized from compounds similar to those generated by Miller and Urey, demonstrating one additional step in the creation of life from nonliving material.

So why do we not observe similar processes occurring today on Earth? One reason is that our atmosphere now contains oxygen, which quickly attacks unprotected organic molecules, breaking them down. The absence of oxygen earlier in the Earth's history—an absence born out by analysis of ancient rocks—may have been crucial for the development of life. Another reason was given by Charles Darwin in his speculations on life's origin: Today's living things consume such organic material before it has an opportunity to develop further.

The importance of low oxygen abundance to the origin of life has led some researchers to speculate that life may have originated deep in the ocean near sea floor vents. These environments are rich in minerals and energy, and today they support a variety of life forms. In ancient days, before oxygen began to accumulate in our atmosphere, these same environments might have been especially congenial to the assembly of complex molecules. Even now, hot, oxygen-poor environments harbor life forms, such as the bacteria that live in boiling hot springs at places such as Yellowstone National Park.

While the Miller–Urey experiment showed that organic material could have developed on the young Earth, it is possible that this material may have been present on our planet right from the start. Interstellar matter in fact contains a rich mix of organic molecules (Unit 72.4). These include several that are precursors to amino acids. Because the Solar System and Earth formed from such clouds, complex organic molecules must have been present on or near the Earth from its birth. Even if these molecules were destroyed during the Earth's formation, fresh molecules may have been carried to its surface by planetesimals at the end of the period of heavy bombardment (Unit 33.5). In fact, carbonaceous chondritic meteors—probably surviving fragments of these ancient planetesimals—are rich in organic material, and some even contain amino acids (Unit 48). Finally, some scientists have demonstrated experimentally that the compression and heating of gases resulting from planetesimal bombardment may be as fruitful as the electric discharges used by Miller and Urey for forming complex organic materials. For a variety of reasons, then, astronomers are confident that organic molecules were common on the ancient Earth.

Regardless of the origin of organic matter on the early Earth, whether created by lightning, ultraviolet radiation, or planetesimal bombardment, many steps are needed from amino acid–filled droplets in a primordial ocean to living cells. For example, the structures created must be able to reproduce, a feature not yet seen in any origin-of-life experiments.

Q One objection to the Miller–Urey type of proposal for the origin of life is that chemicals in the ocean are far too dilute to undergo long series of interactions. In what kinds of environments on the young Earth do you imagine chemicals might have become concentrated?

Cells in Earth's present life forms replicate using the double-chained molecule DNA. Some researchers have speculated, however, that the earliest forms of life used the slightly simpler molecule called *ribonucleic acid* (RNA). In fact, bio-chemists have demonstrated that if certain enzymes are present, RNA can replicate in the absence of living cells. Moreover, they have also discovered several other molecules far simpler in structure than RNA that can replicate in a mechanical sense. That is, if such molecules are put in a solution containing the necessary raw materials, the ensuing chemical interactions allow the molecules to spontaneously produce copies of themselves. Furthermore, one particular mix of molecules not only replicates but also can "mutate" into a slightly better replicator. But although experiments can create self-replicating molecules, even a simple cell is far more complex than this, and simple life forms must be able to form cells that work co-operatively.

The first cellular life on Earth was probably far simpler than the cells we see today. For example, the earliest cells probably had no nucleus, a configuration now found only in bacteria and certain algae. **Prokaryotes,** as these cells without nuclei are known, may accidentally have merged with other cells, which benefited both cells by allowing them to grow and reproduce more rapidly. For example, a cell good at storing energy might have combined with a cell good at reproducing, thereby creating a more complex cell. From such simple beginnings, **eukaryotes** (cells with nuclei) probably evolved. Microscopic fossils of eukaryotes have been found that are up to 1.5 billion years old (Figure 83.7). Fossils of multicelled or-ganisms began to appear about 600 million years ago. Today most cells have a complex *membrane* that holds them together and allows food to enter and waste products to leave. Moreover, most cells also have a *nucleus* in which genetic infor-mation is stored. Cells also have *mitochondria,* which are tiny bodies within the cell that convert food into energy that the cell uses.

The greater complexity of multicelled organisms would in some cases allow them to exploit resources unavailable to simpler beings. For example, an organism that can move from nutrient-poor to nutrient-rich environments may be able to reproduce more prolifically than a stationary creature. Such reproductive advan-tages, combined with changing environments, have led to the incredible diversity of life that we see around us today. In fact, reproductive advantage (*natural selec-tion*) and random mutations are crucial to evolution and the development of more complex life forms, as Charles Darwin explained in 1859 in *On the Origin of Species by Means of Natural Selection,* a work of monumental importance.

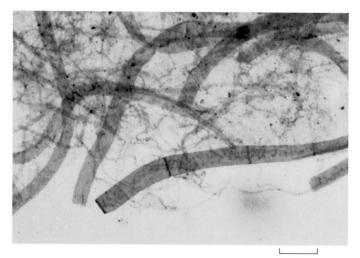

FIGURE 83.7
1 billion-year-old fossil of eukaryote algae.

100 μm

83.3   THE GAIA HYPOTHESIS

In 1974 James Lovelock, a British chemist, and the U.S. microbiologist Lynn Margulis suggested that life creates a single "larger entity" with a planet, a symbiosis of life and planet that they called the **Gaia hypothesis.** They chose this name from the Greek goddess of the Earth, Gaia (pronounced *GUY-uh* in this context but *JEE-uh* elsewhere).

Q Think about your everyday activities. What is the smallest environment that you impact? The largest?

According to the Gaia hypothesis, life does not merely respond to its environment but actually alters its planet's atmosphere and temperature to make it more hospitable. Plant life has almost certainly done just that on Earth. For example, by photosynthesis, plants have created an oxygen-rich atmosphere on Earth that shields them from dangerous ultraviolet radiation. Similarly, through photosynthesis plants have altered our planet's temperature by removing CO_2—a greenhouse gas—and helped to lock it away in soil and ultimately rock. This means that plants can modify the Earth's temperature by adjusting the greenhouse effect (Unit 36). If the planet is too cold, plant metabolism is slowed, and less CO_2 is converted to oxygen. The greater abundance of CO_2 warms the planet and enhances plant growth. Conversely, if the planet gets warmer, plants grow faster and produce lots of oxygen, reducing CO_2 abundance and thereby also reducing greenhouse warming.

Changing levels of CO_2 on Earth have almost certainly prevented a catastrophe for life on our planet. Over the last 4 billion years, it is estimated that the Sun's luminosity has increased by about 40% (Unit 61.4). If greenhouse gases had not been removed steadily from Earth's atmosphere throughout this time, the planet might have suffered a runaway greenhouse effect such as that on Venus (Unit 39). By contrast, the Gaia hypothesis would suggest that Mars never had life because if it had ever taken hold, conditions conducive to life would probably have been maintained.

Q If the runaway greenhouse effect were to take hold, what processes might help to bring it into check?

Lovelock and Margulis have also proposed that many other facets of the environment, such as humidity, salinity of the oceans, sea level, and even plate tectonics may be controlled by living organisms so as to optimize their reproduction. Such an intimate linkage between the living and nonliving is an appealing idea. It almost suggests that the planet is "alive."

But is the Gaia hypothesis correct? Some of the mechanisms described certainly are, but many scientists are skeptical of its more extreme claims (for example, that life can control plate tectonics). Nevertheless, living beings exert a remarkable control over their local environment, and many unsuspected links between life and the physical world surely remain to be found.

83.4   THE ANTHROPIC PRINCIPLE

In 1961 Robert Dicke, a physicist at Princeton University, wrote a paper concerning some remarkable cosmological coincidences. Dicke noted, for example, that the age of the universe is not much different from the lifetime of stars like the Sun, and he went on to argue that this "coincidence" is in fact not remarkable at all. Rather, it is a *necessity* if cosmologists are to exist to note it.

Dicke pointed out that life requires elements such as carbon, silicon, and iron that are made in massive stars. So for life to form, enough time must pass for massive stars to evolve and make the heavy elements and then eject them into space with a supernova explosion. Moreover, additional time is needed for the ejected material to be incorporated back into interstellar clouds and to form new stars. Only with this second generation of stars does life become possible, so intelligent life cannot evolve in a universe only a few billion years old. Finally, he argued that

Q It has been suggested that if the physical constants were slightly different, life in the universe might have been impossible. Can you think of examples where this might be true?

life requires stars, and so it can exist only in a universe young enough that stars have not all died. Therefore, the existence of an intelligent observer in the universe requires that the age of the universe fall within certain limits.

In 1974 the English physicist Brandon Carter took Dicke's idea further and proposed what he called the **anthropic principle.** According to the anthropic principle, "what we can expect to observe must be restricted by the conditions necessary for our presence as observers." Since Carter's work, many scientists have shown how "fine-tuned" the universe must be for life to exist. Thus, no matter how unlikely some aspects of the universe may appear to us, if those aspects are necessary for life, then that is what we will observe. For example, we might argue how truly marvelous it is that conditions on Earth are just right for life. According to the anthropic principle, the rude reply is, "Of course conditions are just right for life. If they were not, there would be no life here to do the marveling."

In some respects the anthropic principle suggests that there are many potential Earths, or even potential universes, in which life might have formed. Among all these possibilities, life arises only where all the necessary conditions are present. Perhaps there are billions of other almost-Earths where life did not succeed, or barren universes that spawned no one to marvel at them. If this is true, then our universe and Earth were the ones that just happened to achieve conditions where life like us would inevitably form.

KEY TERMS

amino acid, 686

anthropic principle, 691

DNA, 686

eukaryote, 689

Gaia hypothesis, 690

Miller-Urey experiment, 688

prokaryote, 689

proteinoid, 688

stromatolite, 684

QUESTIONS FOR REVIEW

1. What is the earliest evidence for life on Earth?

2. How old are the earliest life forms thought to be?

3. What features of life suggest a common origin?

4. What is the Gaia hypothesis?

5. What is the anthropic principle?

PROBLEMS

1. A bacterium has a mass of about 10^{-15} kg. It can replicate itself in one hour. If it had an unlimited food supply, how long would it take the bacterium, doubling in number every hour, to match the mass of the Earth? (Hint: solve for the number of hours N where 2^N gives the number of bacteria you estimate.)

2. A human brain has a mass of about 2 kg. If an individual brain cell has a mass of about 10^{-13} kg, how many cells does a brain contain? How does this number compare to the number of stars in our galaxy? To the number of bits of information in the largest computers today?

TEST YOURSELF

1. Evidence that life on Earth is very ancient comes from
   a. fossil algae in rocks more than 3 billion years old.
   b. fossils buried thousands of feet below the surface.
   c. fossil algae in rocks a few million years old.
   d. the discovery of silicon-based life forms in ancient rocks.
   e. the discovery of silicon-based life forms in a Miller–Urey type of experiment.

2. The Miller–Urey experiment demonstrated that
   a. very simple bacteria can be created from the chemicals present in the atmosphere of the ancient Earth.
   b. Earth's early atmosphere was too harsh for life to have formed until recently.
   c. conditions on the early Earth were suitable for the creation of many of the complex organic molecules found in living things.
   d. if simple life forms landed on the early Earth, they could have survived.
   e. early life forms were silicon-based.

3. According to the anthropic principle,
   a. we are the only civilization in the universe.
   b. without human beings to observe the universe, the universe would not actually exist.
   c. human beings are the highest form of life.
   d. the universe could not be very different, or we would not be here to observe it.
   e. the greenhouse effect is beneficial to us.

84

The Search for Life Elsewhere

Background Pathways: Unit 83

Humans have speculated for thousands of years about life elsewhere in the universe. For example, about 300 B.C.E., the ancient Greek philosopher Epicurus wrote in his "Letter to Herodotus" that "there are infinite worlds both like and unlike this world of ours." Epicurus went on to add that "in all worlds there are living creatures and plants." In a similar vein, the Roman scholar Lucretius wrote about 50 B.C.E. that "it is in the highest degree unlikely that this Earth and sky is the only one to have been created." Such views were not universal, however. For example, Plato and Aristotle argued that the Earth was the only abode for life, a view that prevailed through most of the Middle Ages.

Another possibility that has been debated for centuries is whether terrestrial life descended from organisms created elsewhere in the universe. This is sometimes called **panspermia**. According to this hypothesis, simple life forms (perhaps bacteria) from some other location drifted from their place of origin across space to Earth. However, panspermia does not really simplify the problem of the origin of life: It just shifts it elsewhere. Moreover, it further requires getting the life to Earth, a perilous voyage even for a bacterium.

Today, however, influenced by science fiction movies, television, and books, many people find no difficulty at all in believing that extraterrestrial life exists. In fact, supermarket tabloids make silly claims almost weekly about encounters with aliens. If we look at the facts, however, there is not a shred of evidence that life exists elsewhere. This does not mean there is no extraterrestrial life; it simply means we do not know whether there is. Thus, with no evidence, what can we say meaningfully about life elsewhere in the universe?

Life arose quite soon after Earth's formation, but billions of years passed before a species arose that could wonder if we are alone in the universe. Do other intelligent civilizations exist? Astronomers are strongly divided about such possibilities. Some argue that such life is highly probable, whereas others argue that it is highly improbable. One goal of this unit is to describe the evidence that goes into forming such conclusions.

84.1 THE SEARCH FOR LIFE ON MARS

Scientists have long wondered whether living organisms developed on Mars. Several hundred years ago, after it was realized that the Earth was a planet, many scientists and philosophers assumed that the other planets must have been made as abodes for life.

Mars became the particular focus of attention after it was discovered in the 1700s to have polar caps and an atmosphere and to rotate at about the same rate as the Earth (Unit 40). In 1820 the German mathematician Karl Friedrich Gauss even proposed sending a mathematical message to the Martians by planting wheat fields and pine trees in a huge right triangle. He calculated that the triangle would be visible to the Martians and would show them that intelligent life was present on Earth.

The belief in Martian life took on a new twist based on a linguistic misinterpretation of observations made in 1877 by the Italian astronomer Giovanni Schiaparelli. Schiaparelli saw what he took to be straight-line features on Mars and called them

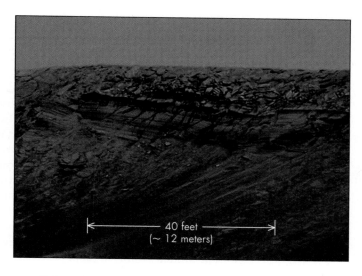

40 feet
(~ 12 meters)

FIGURE 84.1

Cliffs on the inner wall of Endurance Crater, which was visited by *Opportunity*, one of the Mars rovers. The layers (strata) that you can see here suggest that this area of Mars once was covered with water. Because water is vital to life here on Earth, scientists are eager to explore regions on other planets and moons where water is, or was once, present—in the hope of finding traces of present or past life on other worlds.

canali, by which he meant "channels." In English-speaking countries the Martian **canali** became *canals,* with the implication of intelligent beings to build them. Some astronomers thought they could see a complex system of interconnected canals just at the limits of telescope resolution. Most astronomers could see no trace of the alleged canals, but they did note seasonal changes in the shape of dark regions— changes that some interpreted as the spread of plant life in the Martian spring.

Popular belief continued to fuel speculation about life on Mars into the mid-1960s, when the first of a series of *Mariner* spacecraft sent back pictures of a far more desolate world than had been previously imagined. The images of Mars made by the early *Mariner* spacecraft (Unit 40) demonstrated that the straight "canal" features were not real except for the giant Valles Marineras canyon, and the changes in darkness were caused by widespread dust storms. And when the United States landed two *Viking* spacecraft on the planet in 1976, all tests for signs of carbon chemistry in the soil or metabolic activity in soil samples were negative or ambiguous. However, subsequent missions have found geological evidence that there was once fairly abundant water on Mars (Figure 84.1), and it may have been able to support life during its early history.

In 1996 a group of American and English scientists reported possible signs of fossil life in rocks from Mars. These were not samples returned to Earth by spacecraft but meteorites found on Earth. They arrived here after being blasted off the surface of Mars by the impact of a small asteroid. Such impacts are not uncommon, but most fragments are scattered in space or fall back to Mars. Detailed analysis of the composition of these meteorites shows a close match with Martian rock, and they are unlike any other known source.

Microscopic examination of samples from the interior of one such meteorite revealed many tiny, rod-shaped structures (Figure 84.2). These look much like ancient terrestrial bacteria but are much smaller. To some scientists they look like fossilized primitive Martian life. However, over the last few years, scientists have shown that ordinary chemical weathering can form similar structures. As a result, most scientists today are unconvinced that any meteorite yet studied shows evidence of Martian life. Nothing found by the *Spirit* or *Opportunity* rovers provides any evidence of life either; but perhaps if any life exists or once existed on that remote red world, it lies buried well below the hostile surface of the planet.

The meteorite discovery does raise an interesting issue, however. If fossil life is ever found on Mars, we will have to consider the possibility that panspermia occurred. A bacteria-laden rock might have been blasted off Earth's surface in the remote past and reached Mars when it was more habitable. Or perhaps if life formed on Mars first, our own ancestors might have been Martians!

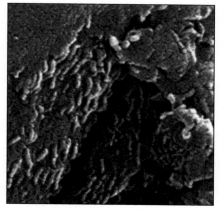

FIGURE 84.2

Fossils of ancient Martian life? The tiny rod-shaped structures in this microscopic image look similar to primitive fossils found in ancient rocks on Earth. However, some scientists think these structures formed chemically.

84.2   LIFE ON OTHER PLANETS?

Is there any evidence for life elsewhere in the universe? Not really. The absence of hard evidence may seem discouraging; but as the British astronomer Sir Martin Rees has said, "Absence of evidence is not the same as evidence of absence."

Even if life is abundant in the universe, we have so far examined only a tiny portion of it: some half dozen spots on the Moon, five on Mars (by robot space landers), and some asteroid fragments picked up on Earth. The Moon's lack of atmosphere and water makes it so inhospitable that astronomers did not expect to find life there, nor did they. Mars seemed more promising because its environment, though harsh, is more like the Earth's than that of any other planet, and laboratory experiments indicated that some terrestrial organisms (bacteria and lichens) could in fact survive Martian conditions, even though they did not reproduce or grow.

The limited sample of planets we have to explore in the Solar System has led scientists to consider other ways to search for signs of life on other planets. One interesting method involves looking for unexpected gases in the atmosphere of a planet—gases that might be produced by living things.

In the late 1960s James Lovelock, a British chemist (see also Unit 83), noted that several of the gases making up the Earth's atmosphere would disappear if Earth had no living beings to replenish them. For example, oxygen is highly reactive and readily combines with surface rock. Lovelock argued that the presence of such gases in a planet's atmosphere is an indication of life there. Rather than sending robot spacecraft to explore a planet's surface, astronomers might find evidence of life by analyzing the light from a planet's atmosphere.

Detecting an Earth-mass **exoplanet** (a planet orbiting another star—Unit 34) remains challenging, let alone detecting oxygen from its atmosphere. One exoplanet (HD 209548b), with a mass close to Jupiter's, is in a tight orbit around its star, passing between us and the star during each orbit. Astronomers have detected oxygen and carbon in its atmosphere, seen in absorption against the star's light. This finding is quite unlikely to signify life, however. The planet orbits so close to its star that the planet's atmosphere is boiling away. The oxygen and carbon were probably the by-products of other molecules being broken down when they were heated by the star and driven away from the planet. What we probably detected was a large cloud of gas streaming off the planet. Nonetheless, this first indication of oxygen in the atmosphere of another planet gives us encouragement that the next generation of space telescopes may eventually lead to the discovery of more hospitable planets.

Finally, we note that life may have taken very different routes in places that would be incompatible with life as we know it. For example, Jupiter's moon Europa, and possibly Ganymede and Callisto, probably have giant oceans of liquid water hidden beneath their frozen crusts. In addition, we detect complex organic molecules in Titan's atmosphere, but temperatures are so cold there that its atmosphere rains liquid methane that cuts rivers into its frozen ice surface—conditions inhospitable to Earth-based life. Could life have formed in these environments? Could intelligent life have evolved? These environments are so alien to us that we have little grasp of the possibilities.

84.3   ARE WE ALONE?

Do other intelligent life forms exist in the universe? Hundreds of light-years from Earth, is there perhaps a "student alien" who is right now reading a book discussing the possibility of life on planets such as Gzbhλx? Scientists do not know, and in fact

Suppose astronomers found clear evidence of another civilization. What effect do you think this would have on our society? What effect would it have on you?

Drake originally wrote his equation in terms of the rate of star formation in our galaxy and estimated the number of civilizations that were currently sending out communications that we might detect.

they are strongly divided into two groups. One group of scientists (let us call them **many-worlders**) thinks that millions of planets with life exist in the Milky Way, and many of these may have advanced civilizations. The other group (let us call them **loners**) argues that we are the only intelligent life in the galaxy. We will discuss below how these two radically different points of view arise.

Many-worlders argue as follows: Earth-like planets are common. Life has formed on Earth. Therefore, it would be surprising if life did not exist elsewhere. In fact, even if only a small fraction of such planets have life, many of them have probably developed advanced civilizations. This may even vastly underestimate the number of planets inhabited by intelligent beings because life may be able to arise in environments intolerable to terrestrial organisms.

To make this argument more quantitative, the U.S. astronomer Frank Drake showed how we can estimate the number of civilizations by multiplying together the probability of each condition necessary for them to exist. He developed a mathematical estimate of the number of communicating civilizations known as the **Drake equation.** We will follow a modified form of Drake's method here. We can estimate the number of planets in our galaxy that contain intelligent life N_I by taking the number of suitable stars $N_\star$ and multiplying that by the fraction of them that have a habitable planet, f_P, and the probability of life forming on such a planet, f_L, and the probability of intelligent life forming given that life had formed, f_I. This can be written in equation form as follows:

$$N_I = N_\star \times f_P \times f_L \times f_I.$$

Unfortunately, the accuracy of each of these terms is not well known, but this approach helps focus our discussion. To see how such numbers are deduced, let us begin by estimating the number of Earth-like planets in the Milky Way.

We start by asking how many Sun-like stars are in the Milky Way. We limit ourselves to such stars because luminous blue ones fuse their fuel so rapidly that they will die and explode as supernovae before life has much chance to evolve, whereas dim red stars are so cool that planets must be very close to them for conditions to be warm enough for a terrestrial type of life. Given these restrictions on star types, we turn to census counts of stars, which reveal that stars like the Sun make up about 10% of our galaxy's hundred billion stars, for a total of about $N_\star = 10^{10}$ such bodies.

Of these 10^{10} stars, we next ask, How many of them have Earth-like planets? Our own system has two planets out of a total of nine that might harbor life as we know it: Earth and Mars. Let us be conservative and say that most other systems are not quite so lucky with respect to having habitable planets. We know that most stars are in multiple-star systems (Unit 56), and of the planetary systems detected so far, most do not resemble the Solar System. We might therefore estimate that 1 in 10 have a habitable planet orbiting them, so $f_P = 0.1$.

How many of these actually have life on them? That depends on how easy it is for life to form, a probability that the Miller–Urey experiments (Unit 83) suggests may be high, but that some astronomers have thought to be quite low. In fact, the British astrophysicist Fred Hoyle compared the likelihood of life arising spontaneously to the likelihood of a box containing several hundred tons of aluminum being shaken and by chance assembling itself into a 747 jet. Although many scientists find such a comparison silly, it illustrates the divergence of opinion about the likelihood of life forming in the universe. The difficulty in assessing that probability is that we have only the single case of life here on Earth as our basis. Many-worlders point to the rapidity with which life developed on Earth. They also argue that the success of the Miller–Urey experiment in making molecular precursors to life in just a few days indicates that the chance of life starting is fairly high: 1 in 100, say, so that $f_L = 0.01$.

We now go to the next step and ask, On how many of these life-bearing planets do intelligent beings form? There is no certain way to know whether the probability of more complex life evolving is high or low. Moreover, even if life succeeds in starting, perhaps it will be annihilated by asteroidal impact. Or if it develops high technology and is careless or thoughtless, life may destroy itself with pollution, nuclear war, or some other self-inflicted disaster. On the other hand, disasters might provide a means of natural selection for intelligence. If an asteroid had not killed the dinosaurs (Unit 48), we might not have gotten a chance to evolve—and we are developing technologies that will protect us from a future asteroid impact. Suppose we are conservative and say that if life forms, it has a 1 in 1000 chance of developing a technological civilization. This sets our final factor, the fraction of life-bearing planets, to $f_I = 0.001$.

The many-worlder carries out this calculation and estimates that the number of intelligent life forms in our galaxy is approximately

$$N_I = 10^{10} \times 0.1 \times 0.01 \times 0.001 = 10{,}000.$$

Therefore, with conservative estimates (from a many-worlder's perspective), there should be many intelligent civilizations within our galaxy with which we might communicate.

By contrast, the loners suggest we are the only advanced life in the Milky Way. The loner argument was first proposed by the Italian physicist Enrico Fermi and is sometime called the **Fermi paradox.** Fermi argued that, based on the rapidity with which our species has developed technology, it is just a matter of time before we spread across the galaxy, or at least fill it with the radio waves of our own transmissions. Even if it takes a technological civilization thousands of years to travel to other stars, once they have mastered space flight, they will need at most a few million years to colonize the entire galaxy, and then their radio communications would fill the galaxy. A few million years is a brief time compared to the billions of years since such civilizations could have arisen, so evidence of their presence should be easy to find if they existed.

Accordingly, loners argue that because no other civilization has been seen, no other technological civilization has already appeared. Therefore, they argue, we are probably alone in the Milky Way, or nearly alone with at most only a handful of other civilizations. However, this argument, like that of the many-worlders, rests on a number of assumptions that are difficult or impossible to substantiate. For example, it assumes (1) that civilizations are driven to colonize, (2) that they seek rather than avoid contact, and (3) that it is possible to master interstellar travel.

The loners' argument is not based solely on our failure to already find such a civilization, though. They point out that the various probabilities for life forming may be significantly lower than the many-worlders suggest. For instance, the Earth and Solar System have many unusual attributes. Perhaps a large moon is critical for stabilizing a planet's axis. Or maybe a large outer planet is necessary to perturb the orbits of water-bearing comets, thereby delivering a critical late shower of water to a planet. In addition, from the extrasolar planetary systems so far detected (Unit 34), it is clear that the nearly circular, regularly spaced orbits of the planets in the Solar System are unusual. What if such orbits are critical for creating stable conditions for the formation of life? It may be that the fraction of planets that are suitable for life f_P is much smaller than the many-worlders suggest.

Perhaps a star's location in the Milky Way is critical for the survival of life. Because the Sun is fairly far from the galactic center, where most of the stars are concentrated, the Earth has not been exposed to as much radiation from nearby nova and supernova explosions. Earth also has a much stronger magnetic field than either Mars or Venus, and maybe that too has been critical for protecting life from high-energy particles. Maybe the great majority of exoplanets are sterile because of the radiation they suffer, and therefore the fraction where life forms f_L is smaller too.

All in all, there could be enough factors such as these to make the probability of life forming far lower. Supposing f_P and f_L were each just 1% of what the many-worlders estimate, we would find

$$N_I = 10^{10} \times 0.001 \times 0.0001 \times 0.001 = 1.$$

By the loners' calculation, there is just one planet with intelligent life in the galaxy—the Earth.

84.4   SETI

On the chance that there may be technologically capable life elsewhere in our galaxy, a few astronomers are searching for radio signals from other civilizations (Figure 84.3). Such signals might be either deliberate broadcasts sent to us or communications directed elsewhere that we would simply overhear. On Earth, radio transmissions have been generated for about a century, so our signals might be picked up by an alien civilization on another planet if there are any within 100 light-years of us.

We can estimate how far away the next nearest civilization might be by assuming that the N_I civilizations we predict are present are spread evenly throughout the disk of the Milky Way, as depicted in Figure 84.4. We can calculate the average distance, d, separating them as follows: Imagine drawing a sphere of radius $d/2$ around each civilized world. Then we ask how big would the average sphere be such that N_I of them would cover the Milky Way's disk. If we make the approximation that the Milky Way's disk is a circle of radius R, then its area is given by the formula πR^2. So we set the area covered by the N_I spheres (each of radius $d/2$) equal to the area of the galaxy's disk:

$$N_I \times \pi \left(\frac{d}{2}\right)^2 = \pi(R)^2.$$

We can cancel π on both sides of the equation, multiply both sides by $(2^2/N_I)$, and take the square root of both sides to give

$$d = \frac{2R}{\sqrt{N_I}}.$$

FIGURE 84.3

One of the radio telescopes used to search for signals from other civilizations.

Suppose we set R equal to the Milky Way's radius, taken to be 40,000 light-years, and $N = 10,000$, the value we deduced from the many-worlder argument. These choices give a value for d of approximately 800 light-years. Therefore, even if such a large number of civilizations exists in our galaxy, they are all likely to be very far from us, and none has likely yet seen any of the transmissions we have generated.

Other civilizations may have begun transmitting a much longer time ago than we did, so there may be detectable signals passing by Earth right now. Astronomers began listening for extraterrestrial radio signals in 1960 with Project Ozma, which monitored radio emissions from several nearby star systems, hoping to detect emissions other than those produced naturally. To date, no signals have been detected, but far more sensitive and sophisticated receiver systems have been designed to continue the *Search for Extra-Terrestrial Intelligence* (**SETI**), mostly based on private funding.

Searching for signals is not easy because even if they exist, they are probably very weak and therefore are buried in cosmic static. To have any hope of success, astronomers use considerable care when choosing wavelengths to monitor. For example, at very long wavelengths, interstellar gas in the galaxy emits strongly, overwhelming all but the

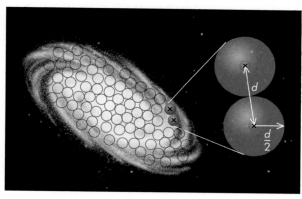

FIGURE 84.4

A sketch illustrating how to estimate the distance between potential civilizations in the Milky Way, if there are N_I civilizations spread evenly within the Milky Way.

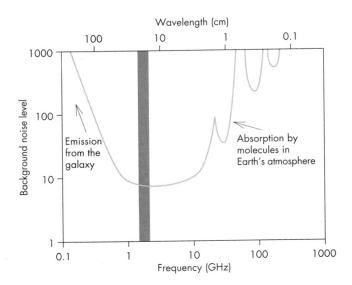

FIGURE 84.5
The *waterhole* is a range of wavelengths between prominent spectral lines of H and OH at 21 and 18 cm. It falls in a clear region of the radio spectrum that is free of interference from natural sources of radio emission and is not blocked by absorption in an Earth-like atmosphere.

Project Ozma was named for the fictional princess in the series of stories of the *Land of Oz*. The princess supposedly sent radio reports about Oz to the author.

Q What if a civilization exhausted its supply of fossil fuels and had to live on limited energy supplies? How important has an abundant energy resource been for human development of technologies?

most powerful signals. On the other hand, at very short wavelengths the molecules in our atmosphere block such signals. It turns out that the optimum wavelengths are near 10 centimeters (Figure 84.5). This clear window for radio signals spans a range that includes the 21-cm line emitted by neutral hydrogen and the 18-cm line emitted by the hydroxyl (OH) molecule in the cold gas clouds of our galaxy. These two wavelengths are distinctive, and any other intelligent civilization would probably have discovered their significance and would have equipment that could receive or broadcast at these wavelengths. Thus, many searches have focused on signals in the range of 18 to 21 cm, which is sometimes called the **waterhole** because it is bracketed by the spectral lines of H and OH—which chemically add up to H_2O.

The most recent SETI projects use receivers that automatically scan billions of radio wavelengths, covering most of the wavelength range where signals from other civilizations might be detected. Many of these programs have been designed to "ride piggyback" on the major radio telescopes around the world—they collect radio signals from whatever direction is being studied by astronomers for other experiments. The SETI projects require huge amounts of computer time to search within the collected data for potential signals. To achieve the computer processing power needed to analyze these data, one project, called SETI@home, has developed a system to transmit data from the central search telescope to a network of millions of volunteers who allow their home computers to process the information when they are not using their machines.

Other than the "loner" hypothesis, there are many possible reasons why SETI has not yet detected any signals. Other civilizations might not be transmitting radio signals. They may not have developed radio communications technology, or they may rely on more carefully targeted transmission systems, such as fiber optics or laser communications. However, if other civilizations are like ours, they inadvertently broadcast a wide range of signals. For example, radio and television stations and radar transmit extremely powerful signals that would be detectable, with sensitive equipment, over most of our galaxy.

Another potential limitation for SETI is that even if communicating civilizations develop elsewhere, they might not survive on their planet for long. Humans have been capable of transmitting communication for less than 100 years. Will we continue to transmit detectable signals for the remaining 5 billion years of the Sun's lifetime, or just another 100 years? And what is likely to happen on other planets? The period during which we could be in contact with them might be so narrow that the opportunity for communication never occurs. For example, if a civilization existed in the nearby Alpha Centauri system, but they somehow destroyed themselves even as recently as 100 years ago, we could not have communicated with them by radio, and the chance has now been lost forever.

The immense distances likely to separate us from any other civilizations also make it unlikely that we will be able to exchange information with other civilizations. To communicate with another civilization within a human lifespan, we would need to be within about 30 light-years from them. At that distance, if you sent a message when you were 10 years old, and if they replied immediately, you would get a response when you were 70! For it to be likely that another civilization was that close, there would have to be millions of other civilizations in our galaxy by our earlier calculation.

Humans *have* sent at least one intentional message to any aliens who might be listening. During a ceremony in 1974 celebrating the renovation of the huge Arecibo radio telescope in Puerto Rico (Unit 27), astronomers broadcast a signal toward the globular cluster M13. This cluster is more than 20,000 light-years away from us, so the message of our existence is still just beginning its voyage to the stars.

Q What message would *you* send to an alien civilization? What dangers might sending a message pose for us on Earth?

KEY TERMS

canali, 693

Drake equation, 695

exoplanet, 694

Fermi paradox, 696

loner, 695

many-worlder, 695

panspermia, 692

SETI, 697

waterhole, 698

QUESTIONS FOR REVIEW

1. What is the significance of the meteorites from Mars?
2. How might we detect a life-sustaining planet?
3. What is the Drake equation?
4. Why are radio wavelengths near 20 cm considered a likely place to find interstellar radio communications?

PROBLEMS

1. Use a method similar to the Drake equation to estimate the number of people getting a haircut at some given moment. That is, identify the different factors involved in deciding to get a haircut, estimate the probabilities of each, then multiply the probabilities times the number of people in, for example, the United States (approximately 260 million).

2. The U.S. Department of Labor estimates that there were 651,000 barbers, hairdressers, and cosmetologists in the United States in 2002. Does this workforce level appear to be consistent with your result from Problem 1?

3. Make your own estimate of the number of habitable planets within the Milky Way. What is the average distance between them?

4. After 20 years, the two *Voyager* spacecraft have reached about twice Pluto's distance from the Sun. How long will it take them to travel 4.2 light-years—the distance to the nearest star?

TEST YOURSELF

1. The Drake equation attempts to determine
   a. the conditions under which life originated on Earth.
   b. the optimum wavelength for communicating with extraterrestrials.
   c. the age of life on Earth.
   d. the number of other technically advanced civilizations.
   e. the lifetime of our own civilization.

2. Current searches for extraterrestrial life use _____ to identify signals from intelligent beings.
   a. X-rays
   b. ultraviolet radiation
   c. visible light
   d. radio waves

3. If you wished to participate in the search for extraterrestrial life, you could
   a. volunteer your computer's downtime for SETI.
   b. set out cruelty-free traps baited with green cheese, a favorite of little green creatures.
   c. lay your speakers outside facing straight up and blast your stereo at its highest volume.
   d. get the best receiver you can afford, and tune it to receive 21- and 18-cm wavelengths.

APPENDIX

SCIENTIFIC NOTATION

Scientific notation is a shorthand method for expressing and working with very large or very small numbers. With this method we express numbers as a few digits times 10 to a power, or exponent. The power indicates the number of times that 10 is multiplied by itself. For example, $100 = 10 \times 10 = 10^2$. Similarly, $1,000,000 = 10 \times 10 \times 10 \times 10 \times 10 \times 10 = 10^6$. Note that we do not always need to write out $10 \times \ldots$. Instead we can simply count the zeros. Thus 10,000 is 1 followed by four zeros, so it is 10^4.

We can write the numbers 1 and 10 in scientific notation as well: $1 = 10^0$ and $10 = 10^1$.

To write a number like 300, we break it into two parts: $3 \times 100 = 3 \times 10^2$. Similarly, we can write $352 = 3.52 \times 100 = 3.52 \times 10^2$. Any number can be expressed in scientific notation as a value between 1 and 10 multiplied by 10 to a power.

We can also write very small numbers (numbers less than 1) using powers of 10. For example, $0.01 = 1/100 = 1 \times 10^{-2}$. We can make this even more concise, however, by writing 1×10^{-2} as 10^{-2}. Similarly, $0.0001 = 10^{-4}$. Note that for numbers less than 1, the power is 1 more than the number of zeros after the decimal point.

We can write a number like 0.00052 as $5.2 \times 0.0001 = 5.2 \times 10^{-4}$.

Suppose we want to multiply numbers expressed in powers of 10. The rule is simple: We add the powers. Thus $10^3 \times 10^2 = 10^{3+2} = 10^5$. Similarly, $2 \times 10^8 \times 3 \times 10^7 = 2 \times 3 \times 10^8 \times 10^7 = 6 \times 10^{15}$. In general, $10^a \times 10^b = 10^{a+b}$.

Division works similarly, except that we subtract the exponents. Thus $10^5/10^3 = 10^{5-3} = 10^2$. In general, $10^a/10^b = 10^{a-b}$.

The last operations we need to consider are raising a power-of-10 number to a power and taking a root. In raising a number to a power, we multiply the powers. Thus "one thousand to the fourth power" is $(10^3)^4 = 10^{3 \times 4} = 10^{12}$. Care must be used if we have a number like $(2 \times 10^4)^3$. Both the 2 and the 10^4 are raised to the third power, so the result is $2^3 \times (10^4)^3 = 8 \times 10^{4 \times 3} = 8 \times 10^{12}$.

Taking a root is equivalent to raising a number to a fractional power. Thus the square root of a number is the number to the $\frac{1}{2}$ power. The cube root is the number to the $\frac{1}{3}$ power, and so forth. For example, $\sqrt{100} = 100^{1/2} = (10^2)^{1/2} = 10^1 = 10$.

SOLVING DISTANCE, VELOCITY, TIME (d, V, t) PROBLEMS

Many problems in this book (and in science in general) involve the motion of something. In such problems, we often know two of the three quantities distance, velocity, and time (d, V, t), and we want to know the third. For example, we have something moving at a speed V and want to know how far it will travel in a time t. Or we know that something travels with a speed V and want to find out how long it takes for the object to travel a distance d. We can usually solve such problems in our heads if the motion involves automobiles. For example, if it is 160 kilometers to a city and we travel at 80 kilometers per hour, how long does it take to get there?

Or how far can we drive in 2 hours if we are traveling at 60 miles per hour? Because we solve such problems routinely, you might find it easier to think of astronomical (d, V, t) problems in terms of cars.

Regardless of your approach, the method of solution is simple. Begin by making a simple sketch of what is happening. Draw some arrows to indicate the motion. Label the known quantities and put question marks beside the things you want to find. Then write out the basic relation $d = V \times t$. If you want to find d and know V and t, just multiply them for the answer. If you want to find the time and are given V and d, solve for t by dividing both sides by V to get $t = d/V$. If you want the velocity, divide d by t: $V = d/t$.

In some problems the motion may be in a circle of radius r. In that case the distance traveled will be related to the circumference of the circle, $2\pi r$. For such cases you may need to use the expression $V = 2\pi r/t$.

In most problems you will find that it is helpful to write the units of the quantities in the equation. For example, suppose you are asked how long it takes to travel 1500 km at a velocity of 30 kilometers per second. Insert the quantities so that

$$t = \frac{d}{V} = \frac{1500 \text{ km}}{30 \text{ km/sec}} = \frac{1500 \text{ km}}{30 \text{ km}} \text{ sec} = 50 \text{ sec}.$$

Note that the units of km cancel out and leave us with units of seconds, as the problem requires.

APPENDIX TABLE I   Physical and Astronomical Constants

Physical Constants

Velocity of light (c)	$= 2.99792458 \times 10^8$ m/s
Gravitational constant (G)	$= 6.67259 \times 10^{-11}$ m³-kg⁻¹-s⁻²
Planck's constant (h)	$= 6.62608 \times 10^{-34}$ joule-s
Mass of hydrogen atom (M_H)	$= 1.6735 \times 10^{-27}$ kg
Mass of electron (M_e)	$= 9.1094 \times 10^{-31}$ kg
Stefan-Boltzmann constant (σ)	$= 5.6705 \times 10^{-8}$ watts-m⁻²-K⁻⁴
Constant in Wien's law ($\lambda_{max}T$)	$= 2.90 \times 10^6$ K·nm

Astronomical Constants

Astronomical unit (AU)	$= 1.495978706 \times 10^{11}$ m
Light-year (ly)	$= 9.4605 \times 10^{15}$ m $= 63,240$ AU
Parsec (pc) $= 3.26$ ly	$= 3.085678 \times 10^{16}$ m $= 206,265$ AU
Sidereal year	$= 365.2564$ days $= 3.1558 \times 10^7$ s
Mass of Earth	$= 5.974 \times 10^{24}$ kg
Mass of Sun ($M_\odot$)	$= 1.989 \times 10^{30}$ kg
Equatorial radius of Earth	$= 6,378.0$ km
Radius of Sun ($R_\odot$)	$= 6.96 \times 10^8$ m
Luminosity of Sun ($L_\odot$)	$= 3.83 \times 10^{26}$ watts

APPENDIX TABLE 2
Metric Prefixes

nano (n)	$= 10^{-9}$	$= 1$ billionth
micro (μ)	$= 10^{-6}$	$= 1$ millionth
milli (m)	$= 10^{-3}$	$= 1$ thousandth
centi (c)	$= 10^{-2}$	$= 1$ hundredth
kilo (k)	$= 10^3$	$= 1$ thousand
mega (M)	$= 10^6$	$= 1$ million
giga (G)	$= 10^9$	$= 1$ billion

APPENDIX TABLE 3   Conversion Between American and Metric Units

Length

1 km	$= 1$ kilometer	$= 1000$ meters	$= 0.6214$ mile
1 m	$= 1$ meter	$= 1.094$ yards	$= 39.37$ inches
1 cm	$= 1$ centimeter	$= 0.01$ meter	$= 0.3937$ inch
1 μm	$= 1$ micrometer	$= 10^{-6}$ meter	$= 3.93 \times 10^{-5}$ inch
1 nm	$= 1$ nanometer	$= 10^{-9}$ meter	$= 3.93 \times 10^{-8}$ inch
1 mile	$= 1.6093$ km		
1 inch	$= 2.5400$ cm		

Mass

1 metric ton	$= 10^6$ grams	$= 1000$ kg	$= 2.2046 \times 10^3$ lb
1 kg	$= 1000$ grams	$= 2.2046$ lb	
1 g	$= 1$ gram	$= 0.0022046$ lb	$= 0.0353$ oz
1 lb	$= 0.4536$ kg		
1 oz	$= 28.3495$ g		

APPENDIX TABLE 4   Some Useful Formulas

Geometry

Circumference of circle $= 2\pi r$

Area of circle $= \pi r^2$

Surface area of sphere $= 4\pi r^2$

Volume of sphere $= \frac{4}{3}\pi r^3$

Distance Relationships

Distance–Velocity–Time: $d = V \times t$

Linear size–Angular size: $L = d \times A/57.3°$

Distance from parallax: $d(\text{in parsecs}) = 1/p(\text{in arcsec})$

Gravity

Kepler's 3rd law—orbits around Sun with semimajor axis a (in AU) and period P (in years): $P^2 = a^3$

Gravitational force between masses M and m: $F_G = G\dfrac{M \times m}{d^2}$

Newton's modified form of Kepler's 3rd law for the total mass of two orbiting bodies: $M = \dfrac{4\pi^2}{G} \times \dfrac{d^3}{P^2}$

Mass of object producing an orbital speed V at distance d:

$$M = \frac{d \times V^2}{G}$$

Escape velocity from a mass M at radius R: $V_{esc} = \sqrt{\dfrac{2GM}{R}}$

Light

Frequency (ν)–Wavelength (λ) relation: $\lambda \times \nu = c$

Energy of a photon: $E = h \times \nu = \dfrac{h \times c}{\lambda}$

Stefan-Boltzmann law—luminosity L of thermal source at temperature T: $L = \sigma T^4 \times (\text{Surface area})$

Wien's law—temperature of thermal source from wavelength of maximum emission: $T = \dfrac{2.9 \times 10^6\ \text{nm} \cdot \text{K}}{\lambda_{max}}$

Brightness B–Luminosity L relation: $B = \dfrac{L}{4\pi d^2}$

Doppler effect: Radial velocity $= V_R = c \times \dfrac{\Delta\lambda}{\lambda}$

Other Physical Relationships

Density $= \dfrac{\text{Mass}}{\text{Volume}}$

Newton's 2nd law—acceleration a produced by force F on mass m: $a = F/m$

Kinetic energy $= \frac{1}{2} mV^2$

Conservation of angular momentum:

(Mass) $\times$ (Circular velocity) $\times$ (Radius) $=$ Constant

Lorentz factor for special relativistic contraction at speed V:

$$\gamma = \frac{1}{\sqrt{1 - V^2/c^2}}$$

APPENDIX TABLE 5   Physical Properties of the Planets

Name	Radius (Eq) (Earth units)	Radius (Eq) (km)	Mass (Earth units)	Mass (kg)	Average Density (g/cm³)
Mercury	0.382	2,439	0.055	3.30×10^{23}	5.43
Venus	0.949	6,051	0.815	4.87×10^{24}	5.25
Earth	1.00	6,378	1.00	5.97×10^{24}	5.52
Mars	0.533	3,397	0.107	6.42×10^{23}	3.93
Jupiter	11.19	71,492	317.9	1.90×10^{27}	1.33
Saturn	9.46	60,268	95.18	5.68×10^{26}	0.69
Uranus	3.98	25,559	14.54	8.68×10^{25}	1.32
Neptune	3.81	24,764	17.13	1.02×10^{26}	1.64
Pluto	0.181	1,160	0.00216	1.29×10^{22}	2.05

APPENDIX TABLE 6   Orbital Properties of the Planets

Name	Distance from Sun* (AU)	Distance from Sun* (10^6 km)	Period Years	Period Days	Orbital Inclination[†]	Orbital Eccentricity
Mercury	0.387	57.9	0.2409	(87.97)	7.00	0.206
Venus	0.723	108.2	0.6152	(224.7)	3.39	0.007
Earth	1.00	149.6	1.0	(365.26)	0.00	0.017
Mars	1.524	227.9	1.8809	(686.98)	1.85	0.093
Jupiter	5.203	778.3	11.8622	(4,332.59)	1.31	0.048
Saturn	9.539	1,427.0	29.4577	(10,759.22)	2.49	0.056
Uranus	19.19	2,869.6	84.014	(30,685.4)	0.77	0.046
Neptune	30.06	4,496.6	164.793	(60,189)	1.77	0.010
Pluto	39.53	5,900	247.7	(90,465)	17.15	0.248

*Semimajor axis of the orbit.   [†]With respect to the ecliptic.

APPENDIX TABLE 7   Satellites of the Solar System*

Planet Satellite	Radius[†] (km)	Distance from Planet (10^3 km)	Orbital Period (days)	Mass (10^{20} kg)	Density (g/cm^3)
Earth					
Moon	1,738	384.4	27.322	734.9	3.34
Mars					
Phobos	$13 \times 11 \times 9$	9.38	0.319	1.3×10^{-4}	2.2
Deimos	$8 \times 6 \times 5$	23.5	1.263	1.8×10^{-5}	1.7
Jupiter[‡]					
Metis	20	127.96	0.295	9×10^{-4}	—
Adrastea	$13 \times 10 \times 8$	128.98	0.298	1×10^{-4}	—
Amalthea	$135 \times 82 \times 75$	181.3	0.498	8×10^{-2}	—
Thebe	$55 \times ? \times 45$	221.9	0.6745	1.4×10^{-3}	—
Io	1,821	421.6	1.769	893.3	3.57
Europa	1,565	670.9	3.551	479.7	2.97
Ganymede	2,634	1,070.0	7.155	1,482	1.94
Callisto	2,403	1,883.0	16.689	1,076	1.86
Leda	~8	11,094	238.7	4×10^{-4}	—
Himalia	92.5	11,480	250.6	8×10^{-2}	—
Lysithea	~18	11,720	259.2	6×10^{-4}	—
Elara	~38	11,737	259.7	6×10^{-3}	—
Ananke	~15	20,200	631R	4×10^{-4}	—
Carme	~20	22,600	692R	9×10^{-4}	—
Pasiphae	~25	23,500	735R	1.6×10^{-3}	—
Sinope	~18	23,700	758R	6×10^{-4}	—
Saturn[‡]					
Pan	10	133.58	0.574	4.2×10^{-6}	—
Atlas	$19 \times ? \times 14$	137.64	0.602	1.6×10^{-4}	—
Prometheus	$74 \times 50 \times 34$	139.35	0.613	1.4×10^{-3}	0.27
Pandora	$55 \times 44 \times 31$	141.7	0.629	1.3×10^{-3}	0.42
Epimetheus	$69 \times 53 \times 53$	151.42	0.694	5.6×10^{-3}	0.63

(*continued*)

APPENDIX TABLE 7   (continued)

Planet Satellite	Radius[†] (km)	Distance from Planet (10^3 km)	Orbital Period (days)	Mass (10^{20} kg)	Density (g/cm^3)
Saturn (cont.)					
Janus	99 × 95 × 76	151.47	0.695	2.0×10^{-2}	0.65
Mimas	210 × 197 × 193	185.52	0.942	0.370	1.12
Enceladus	256 × 247 × 244	238.02	1.370	0.65	1.00
Tethys	536 × 528 × 526	294.66	1.888	6.17	0.98
Calypso	15 × 8 × 8	294.66	1.888	4×10^{-5}	—
Telesto	15 × 13 × 8	294.67	1.888	6×10^{-5}	—
Dione	559	377.4	2.737	10.8	1.49
Helene	18 × ? × 15	377.4	2.737	1.6×10^{-4}	—
Rhea	764	527.04	4.518	23.1	1.24
Titan	2,575	1,221.85	15.945	1,345.5	1.88
Hyperion	185 × 140 × 112	1,481.1	21.277	0.28	—
Iapetus	720	3,561.3	79.331	15.9	1.0
Phoebe	115 × 110 × 105	12,952	550.4R	0.1	—
Uranus[‡]					
Cordelia	13	49.75	0.336	1.7×10^{-4}	—
Ophelia	16	53.77	0.377	2.6×10^{-4}	—
Bianca	22	59.16	0.435	7×10^{-4}	—
Cressida	33	61.77	0.465	2.6×10^{-3}	—
Desdemona	29	62.65	0.476	1.7×10^{-3}	—
Juliet	42	64.63	0.494	4.3×10^{-3}	—
Portia	55	66.1	0.515	1×10^{-2}	—
Rosalind	29	69.93	0.560	1.5×10^{-3}	—
Belinda	33	75.25	0.624	2.5×10^{-3}	—
Puck	77	86.00	0.764	5×10^{-3}	—
Miranda	240 × 234 × 233	129.8	1.413	0.66	1.26
Ariel	581 × 578 × 578	191.2	2.520	13.5	1.65
Umbriel	584.7	266.0	4.144	11.7	1.44
Titania	788.9	435.8	8.706	35.2	1.59
Oberon	761.4	582.6	13.463	30.1	1.50
Neptune[‡]					
Naiad	29	48.2	0.296	1.4×10^{-3}	—
Thalassa	40	50.0	0.312	4×10^{-3}	—
Despina	74	52.5	0.333	2.1×10^{-3}	—
Galatea	79	62.0	0.396	3.1×10^{-2}	—
Larissa	104 × ? × 89	73.6	0.554	6×10^{-2}	—
Proteus	218 × 208 × 201	117.6	1.121	0.6	—
Triton	1,352.6	354.59	5.875R	214	2.0
Nereid	170	5,588.6	360.125	0.31	—
Pluto					
Charon	635	19.6	6.38718	17.7	1.83

*Note: Authorities differ substantially on many of these values. "R" means orbit is retrograde.

[†] $a \times b \times c$ values for the radius are the approximate lengths of the axes for irregular moons.

[‡] Astronomers have recently found many new moons orbiting each of the outer planets. All are small and many are as yet unnamed. Their orbital properties and sizes are uncertain. They may be captured asteroids. A relatively up-to-date list for each planet is available at the website solarsystem.nasa.gov/planets. Click on the planet of interest and then click on the tab labeled "moons."

APPENDIX TABLE 8   Properties of Main-Sequence Stars*

Spectral Type	Luminosity ($L_\odot$)	Temperature (K)	Mass ($M_\odot$)	Radius ($R_\odot$)
O5	790,000	44,500	60	12.0
B0	52,000	30,000	17.5	7.4
B5	830	15,400	5.9	3.9
A0	54	9,520	2.9	2.4
A5	14	8,200	2.0	1.7
F0	6.5	7,200	1.6	1.5
F5	3.2	6,440	1.3	1.3
G0	1.5	6,030	1.05	1.1
G5	0.79	5,770	0.92	0.92
K0	0.42	5,250	0.79	0.85
K5	0.15	4,350	0.67	0.72
M0	0.08	3,850	0.51	0.6
M5	0.01	3,240	0.21	0.27
M8	0.001	2,640	0.06	0.1

*Authorities differ substantially on many of these values, especially at the upper and lower mass values. Note also that the values do not always agree with the Stefan-Boltzmann law.

APPENDIX TABLE 9   The Brightest Stars

Star	Name	Apparent Visual Magnitude	Distance (ly)	Spectral Type	Visual Luminosity ($L_\odot$)
—	Sun	−26.72	0.0000158	G2V	1
α CMa A	Sirius	−1.44	8.6	A1V + DA2	26
α Car	Canopus	−0.74	310	F0II	14,000
α Cen	Rigel Kentaurus*	−0.27	4.4	G2V + K1V	1.6
α Boo	Arcturus	−0.05	36.7	K2III	110
α Lyr	Vega	0.03	25.3	A0V	50
α Aur	Capella*	0.08	42	G5III + G0III	130
β Ori	Rigel	0.10	770	B8Ia	40,000
α CMi	Procyon	0.40	11.4	F5IV/V + DA	7
α Ori	Betelgeuse	0.3–0.6**	430	M1-2 1a-1ab	10,000
α Eri	Achernar	0.45	143	B3V	1,100
β Cen AB	Hadar	0.61	500	B1III	12,000
α Aql	Altair	0.77	16.8	A7V	11
α Cru	Acrux*	0.78	300	B0.5IV + B1V	4,000
α Tau A	Aldebaran	0.78–0.93**	65	K5III + M2V	150
α Vir	Spica	0.92–0.98**	260	B1III-IV + B2V	2,200
α Sco A	Antares	1.02	600	M1Ib + B3V	11,000
β Gem	Pollux	1.16	33.7	K0III	30
α PsA	Fomalhaut	1.17	25	A3V	17
α Cyg	Deneb	1.25	3,000	A2Ia	250,000
β Cru	Mimosa	1.25	350	B0.5IV	3,000

*A number of these stars look like individual stars to the unaided eye, but actually are in multiple-star systems, as indicated by the multiple spectral types listed. For most of these systems, the brightest star contributes the great majority of the light; however, a few would shift location in the table if seen alone: Rigel Kentaurus (Alpha Centauri) is a binary of stars with magnitude −0.01 and 1.34; Capella is a spectroscopic binary of two stars having apparent magnitudes of about 0.6 and 1.1; Acrux is a visual binary of stars having magnitudes of 1.33 and 1.73. In addition to the multiple systems noted in the table, several others have dim companions whose spectroscopic types have not been determined.

**Variable star

APPENDIX TABLE 10   The Nearest Stars

Name	Apparent Visual Magnitude	Distance (ly)	Spectral Type	Visual Luminosity ($L_\odot$)
Sun	−26.72		G2	1
Proxima Centauri	11.09	4.23	M5.5V	0.00005
α Cen A	−0.01	4.35	G2V	1.6
B	1.34	4.35	K0V	0.5
Barnard's Star	9.55	5.98	M4V	0.0004
Wolf 359	13.45	7.80	M6V	0.00002
Lalande 21185	7.47	8.23	M2V	0.005
Luyten 726-8 A	12.41	8.57	M5.5V	0.00006
B	13.2	8.57	M6V	0.00003
Sirius A	−1.44	8.57	A1V	26
B	8.44	8.57	DA2 (white dwarf)	0.003
Ross 154	10.47	9.56	M3.5V	0.0005
Ross 248	12.29	10.33	M5.5V	0.0001
ε Eridani	3.73	10.67	K2V	0.3
Ross 128	11.12	10.83	M4V	0.0003
Luyten 789-6 ABC	12.33	11.08	M5V	0.0001
Groombridge 34 A	8.08	11.27	M1.5V	0.006
B	11.07	11.27	M3.5V	0.0004
ε Indi	4.68	11.29	K5Ve	0.13
61 Cygni A	5.22	11.30	K5V	0.08
B	6.03	11.30	K7V	0.04
BD + 59° 1915 A	8.90	11.40	M3V	0.003
B	9.68	11.40	M3.5V	0.001
τ Ceti	3.50	11.40	G8Vp	0.4
Procyon A	0.40	11.41	F5IV–V	7
B	10.7	11.41	DA (white dwarf)	0.0005
Lacaille 9352	7.34	11.47	M1.5V	0.01
GJ 1111	14.79	11.83	M6.5V	0.00001
GJ 1061	13.03	12.06	M5.5V	0.00007

Source: From "Our Nearest Celestial Neighbors" by Joshua Roth and Roger W. Sinnott, *Sky and Telescope* 92 (October 1996), compiled by Roger W. Sinnott. © 1996 Sky Publishing Corp. Reprinted with permission of the publisher.

Notes: BD stands for Bonner Durchmusterung catalog and GJ for Gliese-Jahreiss catalog; *p* indicates a peculiar spectrum; *e* indicates that the star shows emission lines in its spectrum.

APPENDIX TABLE 11 — Known and Suspected Members of the Local Group of Galaxies

Name of Galaxy	Right Ascension (hours and minutes)	Declination (degrees and minutes)	Galaxy Type	Distance (Mpc)	Diameter (kpc)	Visual Apparent Magnitude	Approximate Luminosity (millions of solar luminosities)
WLM	0 02	−15 28	Irr	1.0	3	11	50
IC10	0 20	+59 18	Irr	0.7	1	11	300
Cetus	0 26	−11 02	E	0.8	—	14	1
NGC 147	0 33	+48 30	E5	0.7	2	10	100
Andromeda III	0 35	+36 30	E5	0.8	1	15	1
NGC 185	0 39	+48 20	E3	0.7	2	9	150
NGC 205	0 40	+41 41	E5	0.8	5	8	300
Andromeda VIII	0 42	+40 37	E	0.8	14	9	150
M32	0 43	+40 52	E2	0.8	2	8	300
M31	0 43	+41 16	Sb	0.8	40	3	25,000
Andromeda I	0 46	+38 02	E3	0.8	0.5	14	4
SMC	0 53	−72 50	Irr	0.06	5	2	600
Sculptor	1 00	−33 42	E3	0.09	1	9	1
Pisces	1 04	+21 53	Irr	0.8	0.5	18	1
IC 1613	1 05	+02 07	Irr	0.7	3	9	100
Andromeda V	1 10	+47 38	E	0.8	—	15	1
Andromeda II	1 16	+33 25	E2	0.7	0.7	13	4
M33	1 34	+30 40	Sc	0.8	16	6	3,000
Phoenix	1 51	−44 27	Irr	0.4	0.6	12	1
Fornax	2 40	−34 27	E3	0.14	0.7	9	15
UGCA 92	4 32	+63 36	Irr	1.44	0.6	14	5
LMC	5 23	−69 45	Irr	0.05	0.6	1	2,000
Carina	6 42	−50 58	E4	0.10	0.7	16	1
Canis Major	7 35	−28 00	Irr	0.08	220	—	—
Leo A	9 59	+30 45	Irr	0.7	1	12	3
Sextans B	10 00	+05 20	Irr	1.41	1.6	12	30
NGC 3109	10 03	−26 09	Irr	1.38	5	10	100
Antlia	10 04	−27 20	E3	1.41	0.6	15	2
Leo I	10 08	+12 18	E3	0.2	0.7	10	4
Sextans A	10 11	−04 43	Irr	1.60	1.8	12	35
Sextans	10 13	−01 37	E	0.09	1	12	1
Leo II	11 13	+22 09	E0	0.2	0.7	12	1
DDO 155	12 59	+14 13	Irr	2.4	0.4	14	8
Ursa Minor	15 09	+67 13	E5	0.06	0.5	11	1
Draco	17 20	+57 55	E3	0.08	0.8	10	1
Milky Way	17 46	−29 00	SBc	0.01	40	—	20,000
Sagittarius	18 55	−30 29	Irr	0.03	—	10	30
SagDIG	19 30	−17 41	Irr	1.2	1.5	15	5
NGC 6822	19 45	−14 48	Irr	0.5	2	9	200
Aquarius	20 47	−12 51	Irr	1.0	0.6	14	2
Tucanae	22 42	−64 25	E4	0.9	0.5	15	1
UGCA 438	23 26	−32 33	Irr	1.4	0.5	14	5
Cassiopeia	23 27	+50 42	E	0.7	0.5	13	5
Pegasus	23 29	+14 45	Irr	0.8	1	13	7
Pegasus II	23 52	+24 35	E	0.8	0.9	14	3

APPENDIX TABLE 12 — The Brightest Galaxies Beyond the Local Group*

Name of Galaxy	Right Ascension (hours and minutes)	Declination (degrees and minutes)	Galaxy Type	Distance (Mpc)	Diameter (kpc)	Visual Apparent Magnitude	Approximate Luminosity (millions of solar luminosities)
NGC0253	00 48	−25 17	Sc	2.6	21	7.3	17,000
NGC0300	00 55	−37 41	Scd	2.2	13	8.2	2,700
NGC1068 (M77)	02 43	−00 01	Sb	13.8	28	8.7	60,000
NGC1291	03 17	−41 06	S0-a	9.2	27	8.7	25,000
IC 342	03 47	+68 06	SBc	2.5	15	8.5	51,000
NGC2403	07 37	+65 36	Sc	3.4	23	8.4	6,800
NGC3031 (M81)	09 56	+69 04	Sab	3.7	23	6.9	33,000
NGC3034 (M82)	09 56	+69 41	Irr	3.7	10	8.3	15,000
NGC4258 (M106)	12 19	+47 18	Sb	7.4	41	8.4	34,000
NGC4472 (M49)	12 30	+08 00	E	16.3	43	8.3	110,000
NGC4486 (M87)	12 31	+12 24	E	15.6	32	8.6	81,000
NGC4594 (M104)	12 40	−11 37	Sa	9.2	24	8.1	58,000
NGC4736 (M94)	12 51	+41 07	Sab	5.2	11	8.1	14,000
NGC4826 (M64)	12 57	+21 41	Sab	7.4	17	8.4	30,000
NGC4945	13 05	−49 28	SBc	4.6	29	8.6	37,000
NGC5055 (M63)	13 16	+42 02	Sbc	7.7	21	8.6	25,000
NGC5128 (Cen A)	13 25	−43 01	S0	3.7	32	6.7	32,000
NGC5194 (M51)	13 30	+47 12	Sbc	8.0	26	8.3	33,000
NGC5236 (M83)	13 37	−29 52	Sc	4.6	15	7.3	28,000
NGC5457 (M101)	14 03	+54 21	Sc	7.4	47	7.8	34,000
NGC6744	19 10	−63 51	Sb	10.7	65	8.4	62,000

*Galaxies in Table 12 were selected from the HyperLeda database (http://leda.univ-lyon1.fr) based on their total visual apparent magnitude.

APPENDIX TABLE 13 — The Messier Catalog*

Messier Number	Popular Name	Object Type	Right Ascension (hours and minutes)	Declination (degrees and minutes)	Visual Apparent Magnitude	Angular Size (arc minutes)	Distance (kpc)
1	Crab Nebula	Supernova remnant	05 35	+22 01	8.4	6 × 4	1.9
2		Globular cluster	21 34	−00 49	6.5	12.9	11
3		Globular cluster	13 42	+28 23	6.2	16.2	10
4		Globular cluster	16 24	−26 32	5.6	26.3	2.2
5		Globular cluster	15 19	+02 05	5.6	17.4	7.5
6	Butterfly Cluster	Open cluster	17 40	−32 13	4.2	25	0.4
7	Ptolemy's Cluster	Open cluster	17 54	−34 49	4.1	80	0.2
8	Lagoon Nebula	Diffuse nebula	18 04	−24 23	6	90 × 40	1.5
9		Globular cluster	17 19	−18 31	7.7	9.3	8.1
10		Globular cluster	16 57	−04 06	6.6	15.1	4.4
11	Wild Duck Cluster	Open cluster	18 51	−06 16	6.3	14	1.8
12		Globular cluster	16 47	−01 57	6.7	14.5	4.9
13	Hercules Cluster	Globular cluster	16 42	+36 28	5.8	16.6	7.6
14		Globular cluster	17 38	−03 15	7.6	11.7	8.8
15		Globular cluster	21 30	+12 10	6.2	12.3	10
16	Eagle Nebula	Open cluster	18 19	−13 47	6.4	7	2.1
17	Omega or Swan Nebula	Diffuse nebula	18 21	−16 11	7	11	1.5
18		Open cluster	18 20	−17 08	7.5	9	1.5
19		Globular cluster	17 03	−26 16	6.8	13.5	8.7
20	Trifid Nebula	Diffuse nebula	18 03	−23 02	9	28	1.5

(continued)

APPENDIX TABLE 13   (continued)

Messier Number	Popular Name	Object Type	Right Ascension (hours and minutes)	Declination (degrees and minutes)	Visual Apparent Magnitude	Angular Size (arc minutes)	Distance (kpc)
21		Open cluster	18 05	−22 30	6.5	13	1.3
22		Globular cluster	18 36	−23 54	5.1	24	3.1
23		Open cluster	17 57	−19 01	6.9	27	0.6
24	Sagittarius Star Cloud	Milky Way patch	18 17	−18 29	4.6	90	3.0
25		Open cluster	18 32	−19 15	4.6	32	0.6
26		Open cluster	18 45	−09 24	8	15	1.5
27	Dumbbell Nebula	Planetary nebula	19 60	+22 43	7.4	8.0 × 5.7	0.3
28		Globular cluster	18 25	−24 52	6.8	11.2	5.7
29		Open cluster	20 24	+38 32	7.1	7	1.2
30		Globular cluster	21 40	−23 11	7.2	11	8
31	Andromeda Galaxy	Spiral galaxy	00 43	+41 16	3.4	178 × 63	800
32		Elliptical galaxy	00 43	+40 52	8.1	8 × 6	800
33	Triangulum Galaxy	Spiral galaxy	01 34	+30 39	5.7	73 × 45	860
34		Open cluster	02 42	+42 47	5.5	35	0.4
35		Open cluster	06 09	+24 20	5.3	28	0.8
36		Open cluster	05 36	+34 08	6.3	12	1.2
37		Open cluster	05 52	+32 33	6.2	24	1.3
38		Open cluster	05 28	+35 50	7.4	21	1.2
39		Open cluster	21 32	+48 26	4.6	32	0.2
40		Binary star	12 22	+58 05	8.4	0.8	0.1
41		Open cluster	06 46	−20 44	4.6	38	0.7
42	The Orion Nebula	Diffuse nebula	05 35	−05 27	4	85 × 60	0.4
43		Diffuse nebula	05 36	−05 16	9	20 × 15	0.4
44	Praesepe or Beehive Nebula	Open cluster	08 40	+19 59	3.7	95	0.1
45	Pleiades or Seven Sisters	Open cluster	03 47	+24 07	1.6	110	0.1
46		Open cluster	07 42	−14 49	6	27	1.6
47		Open cluster	07 37	−14 30	5.2	30	0.4
48		Open cluster	08 14	−05 48	5.5	54	0.4
49		Elliptical galaxy	12 30	+08 00	8.3	9 × 7.5	16,000
50		Open cluster	07 03	−08 20	6.3	16	0.9
51	Whirlpool Galaxy	Spiral galaxy	13 30	+47 12	8.3	11 × 7	8,000
52		Open cluster	23 24	+61 35	7.3	13	1.5
53		Globular cluster	13 13	+18 10	7.6	12.6	18
54		Globular cluster	18 55	−30 29	7.6	9.1	27
55		Globular cluster	19 40	−30 58	6.3	19	5.3
56		Globular cluster	19 17	+30 11	8.3	7.1	10
57	Ring Nebula	Planetary nebula	18 54	+33 02	8.8	1.4 × 1.0	0.7
58		Spiral galaxy	12 38	+11 49	9.7	5.5 × 4.5	17,000
59		Elliptical galaxy	12 42	+11 39	9.6	5 × 3.5	16,000
60		Elliptical galaxy	12 44	+11 33	8.8	7 × 6	16,000
61		Spiral galaxy	12 22	+04 28	9.7	6 × 5.5	15,000
62		Globular cluster	17 01	−30 07	6.5	14.1	6.9
63	Sunflower Galaxy	Spiral galaxy	13 16	+42 02	8.6	10 × 6	7,700
64	Blackeye Galaxy	Spiral galaxy	12 57	+21 41	8.4	9.3 × 5.4	7,400
65		Spiral galaxy	11 19	+13 05	9.3	8 × 1.5	8,000
66		Spiral galaxy	11 20	+12 59	8.9	8 × 2.5	8,000
67		Open cluster	08 50	+11 49	6.1	30	0.8
68		Globular cluster	12 40	−26 45	7.8	12	10
69		Globular cluster	18 31	−32 21	7.6	7.1	8.5
70		Globular cluster	18 43	−32 18	7.9	7.8	9

APPENDIX TABLE 13   (continued)

Messier Number	Popular Name	Object Type	Right Ascension (hours and minutes)	Declination (degrees and minutes)	Visual Apparent Magnitude	Angular Size (arc minutes)	Distance (kpc)
71		Globular cluster	19 54	+18 47	8.2	7.2	3.8
72		Globular cluster	20 54	−12 32	9.3	5.9	16
73		System of four stars	20 59	−12 38	9	2.8	0.6
74		Spiral galaxy	01 37	+15 47	9.4	10.2 × 9.5	7,400
75		Globular cluster	20 06	−21 55	8.5	6	18
76	Little Dumbbell Nebula	Planetary nebula	01 42	+51 34	10.1	2.7 × 1.8	1.0
77		Spiral galaxy	02 43	−00 01	8.7	7 × 6	14,000
78		Diffuse nebula	05 47	+00 03	8.3	8 × 6	0.4
79		Globular cluster	05 25	−24 33	7.7	8.7	12
80		Globular cluster	16 17	−22 59	7.3	8.9	10
81	Bode's Galaxy	Spiral galaxy	09 56	+69 04	6.9	21 × 10	3,700
82	Cigar Galaxy	Irregular galaxy	09 56	+69 41	8.3	9 × 4	3,700
83	Southern Pinwheel Galaxy	Spiral galaxy	13 37	−29 52	7.3	11 × 10	4,600
84		Elliptical galaxy	12 25	+12 53	9.1	5	17,000
85		S0 galaxy	12 25	+18 11	9.1	7.1 × 5.2	16,000
86		Elliptical galaxy	12 26	+12 57	8.9	7.5 × 5.5	17,000
87	Virgo A	Elliptical galaxy	12 31	+12 24	8.6	7	16,000
88		Spiral galaxy	12 32	+14 25	9.6	7 × 4	17,000
89		Elliptical galaxy	12 36	+12 33	9.8	4	16,000
90		Spiral galaxy	12 37	+13 10	9.5	9.5 × 4.5	17,000
91		Spiral galaxy	12 35	+14 30	10.2	5.4 × 4.4	16,000
92		Globular cluster	17 17	+43 08	6.4	11.2	8.1
93		Open cluster	07 45	−23 52	6	22	1.1
94		Spiral galaxy	12 51	+41 07	8.1	7 × 3	5,200
95		Spiral galaxy	10 44	+11 42	9.7	4.4 × 3.3	10,000
96		Spiral galaxy	10 47	+11 49	9.2	6 × 4	10,000
97	Owl Nebula	Planetary nebula	11 15	+55 01	9.9	3.4 × 3.3	0.7
98		Spiral galaxy	12 14	+14 54	10.1	9.5 × 3.2	17,000
99		Spiral galaxy	12 19	+14 25	9.9	5.4 × 4.8	17,000
100		Spiral galaxy	12 23	+15 49	9.3	7 × 6	16,000
101	Pinwheel Galaxy	Spiral galaxy	14 03	+54 21	7.8	22	7,400
102	Spindle Galaxy	S0 galaxy	15 07	+55 46	9.9	5.2 × 2.3	15,000
103		Open cluster	01 33	+60 42	7.4	6	2.6
104	Sombrero Galaxy	Spiral galaxy	12 40	−11 37	8.1	9 × 4	9,200
105		Elliptical galaxy	10 48	+12 35	9.3	2	11,000
106		Spiral galaxy	12 19	+47 18	8.4	19 × 8	7,400
107		Globular cluster	16 33	−13 03	7.9	10	6.4
108		Spiral galaxy	11 12	+55 40	10	8 × 1	14,000
109		Spiral galaxy	11 58	+53 23	9.8	7 × 4	17,000
110		Elliptical galaxy	00 40	+41 41	8.5	17 × 10	830

*Data in Table 13 drawn primarily from the SEDS Messier Catalog Web pages. By Hartmut Frommert, Christine Kronberg, and Guy McArthur. SEDS, University of Arizona Chapter, Tucson, Arizona, 1994–2005. http://www.seds.org/messier/.

The data in the preceding tables come from many sources, including

Lang, Kenneth R. *Astrophysical Data: Planets and Stars.* New York: Springer-Verlag, 1992.

Landolt-Bornstein: *Zahlenwerte und Funktionen aus Naturwissenschaften und Technik/Numerical Data and Functional Relationships in Science and Technology.* Neue Serie/New Series. *Group VII: Astronomy, Astrophysics, and Space Research,* vols. 1–2, Astronomy and Astrophysics. Berlin, New York: Springer-Verlag, 1965–1982.

Donald K. Yeomans of JPL for satellite dimensions and masses.

APPENDIX TABLE 14 — The Periodic Table of Elements

Key (example):
9 — Atomic number
F
Fluorine
19.00 — Atomic mass

1 1A	2 2A	3 3B	4 4B	5 5B	6 6B	7 7B	8 8B	9 8B	10 8B	11 1B	12 2B	13 3A	14 4A	15 5A	16 6A	17 7A	18 8A
1 **H** Hydrogen 1.008																	2 **He** Helium 4.003
3 **Li** Lithium 6.941	4 **Be** Beryllium 9.012											5 **B** Boron 10.81	6 **C** Carbon 12.01	7 **N** Nitrogen 14.01	8 **O** Oxygen 16.00	9 **F** Fluorine 19.00	10 **Ne** Neon 20.18
11 **Na** Sodium 22.99	12 **Mg** Magnesium 24.31											13 **Al** Aluminum 26.98	14 **Si** Silicon 28.09	15 **P** Phosphorus 30.97	16 **S** Sulfur 32.07	17 **Cl** Chlorine 35.45	18 **Ar** Argon 39.95
19 **K** Potassium 39.10	20 **Ca** Calcium 40.08	21 **Sc** Scandium 44.96	22 **Ti** Titanium 47.88	23 **V** Vanadium 50.94	24 **Cr** Chromium 52.00	25 **Mn** Manganese 54.94	26 **Fe** Iron 55.85	27 **Co** Cobalt 58.93	28 **Ni** Nickel 58.69	29 **Cu** Copper 63.55	30 **Zn** Zinc 65.39	31 **Ga** Gallium 69.72	32 **Ge** Germanium 72.59	33 **As** Arsenic 74.92	34 **Se** Selenium 78.96	35 **Br** Bromine 79.90	36 **Kr** Krypton 83.80
37 **Rb** Rubidium 85.47	38 **Sr** Strontium 87.62	39 **Y** Yttrium 88.91	40 **Zr** Zirconium 91.22	41 **Nb** Niobium 92.91	42 **Mo** Molybdenum 95.94	43 **Tc** Technetium (98)	44 **Ru** Ruthenium 101.1	45 **Rh** Rhodium 102.9	46 **Pd** Palladium 106.4	47 **Ag** Silver 107.9	48 **Cd** Cadmium 112.4	49 **In** Indium 114.8	50 **Sn** Tin 118.7	51 **Sb** Antimony 121.8	52 **Te** Tellurium 127.6	53 **I** Iodine 126.9	54 **Xe** Xenon 131.3
55 **Cs** Cesium 132.9	56 **Ba** Barium 137.3	57 **La** Lanthanum 138.9	72 **Hf** Hafnium 178.5	73 **Ta** Tantalum 180.9	74 **W** Tungsten 183.9	75 **Re** Rhenium 186.2	76 **Os** Osmium 190.2	77 **Ir** Iridium 192.2	78 **Pt** Platinum 195.1	79 **Au** Gold 197.0	80 **Hg** Mercury 200.6	81 **Tl** Thallium 204.4	82 **Pb** Lead 207.2	83 **Bi** Bismuth 209.0	84 **Po** Polonium (210)	85 **At** Astatine (210)	86 **Rn** Radon (222)
87 **Fr** Francium (223)	88 **Ra** Radium (226)	89 **Ac** Actinium (227)	104 **Rf** Rutherfordium (257)	105 **Db** Dubnium (260)	106 **Sg** Seaborgium (263)	107 **Bh** Bohrium (262)	108 **Hs** Hassium (265)	109 **Mt** Meitnerium (266)	110 **Ds** Darmstadtium (269)	111 **Rg** Roentgenium (272)	112	113	114	115	116	(117)	(118)

Lanthanides and Actinides:

58 **Ce** Cerium 140.1	59 **Pr** Praseodymium 140.9	60 **Nd** Neodymium 144.2	61 **Pm** Promethium (147)	62 **Sm** Samarium 150.4	63 **Eu** Europium 152.0	64 **Gd** Gadolinium 157.3	65 **Tb** Terbium 158.9	66 **Dy** Dysprosium 162.5	67 **Ho** Holmium 164.9	68 **Er** Erbium 167.3	69 **Tm** Thulium 168.9	70 **Yb** Ytterbium 173.0	71 **Lu** Lutetium 175.0
90 **Th** Thorium 232.0	91 **Pa** Protactinium (231)	92 **U** Uranium 238.0	93 **Np** Neptunium (237)	94 **Pu** Plutonium (242)	95 **Am** Americium (243)	96 **Cm** Curium (247)	97 **Bk** Berkelium (247)	98 **Cf** Californium (249)	99 **Es** Einsteinium (254)	100 **Fm** Fermium (253)	101 **Md** Mendelevium (256)	102 **No** Nobelium (254)	103 **Lr** Lawrencium (257)

Legend:
- Metals
- Metalloids
- Nonmetals

The 1–18 group designation has been recommended by the International Union of Pure and Applied Chemistry (IUPAC) but is not yet in wide use. In this text we use the standard U.S. notation for group numbers (1A–8A and 1B–8B). No names have been assigned for elements 112, 114, and 116. Elements 113, 115, 117, and 118 have not yet been synthesized.

GLOSSARY

This glossary includes definitions of the key terms and other important terms relevant to the study of astronomy. Section numbers (in parentheses) indicate where terms are introduced in the text.

A

aberration of starlight the angular shift in the apparent direction of a star caused by the orbital speed of the Earth. (52.5)

absolute magnitude the apparent magnitude a star would have if it were at a distance of 10 parsecs. (54.4)

absolute zero the lowest possible temperature, at which a material contains no extractable heat energy. Zero on the Kelvin temperature scale. (23.3)

absorption the process in which an atom or molecule intercepts light or other electromagnetic radiation and takes up its energy. For example, ozone in our atmosphere absorbs ultraviolet radiation. (21.5)

absorption lines dark lines superimposed on a continuous spectrum, produced when a gas absorbs specific wavelengths of light from a background source of light. (55.1)

absorption-line spectrum a spectrum in which certain wavelengths are darker than adjacent wavelengths. The missing light is absorbed by atoms or molecules between the source and the observer. (24.2)

acceleration the rate of change in an object's velocity (either its speed or its direction). This can include speeding up as well as slowing down (deceleration). (15.0)

accretion the addition of matter to a body. Examples are gas falling onto a star and asteroids colliding and sticking together. (33.3)

accretion disk a nearly flat disk of gas or other material held in orbit around a body by its gravity. (65.1, 68.3)

active galactic nucleus (AGN) a small central region in a galaxy that emits abnormally large amounts of electromagnetic radiation and sometimes ejects jets of matter. Examples are radio galaxies, Seyfert galaxies, and quasars. (77.0)

A.D. abbreviation of Latin *Anno Domini*, "year of the Lord" in the Christian calendar. Equivalent to the nondenominational designation C.E. or "common era." (9.5)

adaptive optics a technique for adjusting a telescope's mirror or other optical parts to compensate for atmospheric distortions, such as **seeing**, thereby giving a sharper image. (30.2)

æther a substance that was once hypothesized to fill space and through which light waves propagated. Einstein's theory of special relativity overturned this idea. (53.1)

alpha particle a helium nucleus: two protons plus (usually) two neutrons.

alt–az mount a support structure for a telescope that allows it to rotate up and down (in altitude) and parallel to the horizon (in azimuth). (31.3)

altitude an object's angular distance above the horizon. (31.3)

amino acid a carbon-based molecule used by living organisms to build protein molecules. (83.1)

angstrom unit a unit of length used in describing wavelengths of radiation and the sizes of atoms and molecules. One angstrom is 10^{-10} meters.

angular momentum a measure of an object's tendency to keep rotating and to maintain its orientation. Mathematically, it depends on the object's mass, M, radius, R, and rotational velocity, V, and is proportional to the product $M \times V \times R$. (20.2)

angular shift the change in the apparent position of an object. For example, an object will show an angular shift relative to background objects when viewed from different places. (52.2)

angular size a measure of how large an object looks to an observer. It is defined as the angle between lines drawn from the observer to opposite sides of the object. For example, the angular size of the Moon is about 1/2° seen by an observer on Earth. (10.4)

annular eclipse an eclipse in which the Moon passes directly in front of the Sun but does not completely cover the Sun. A bright ring of the Sun's photosphere remains visible around the dark disk of the Moon. We therefore see a ring (annulus) of light around the Moon. (8.2)

Antarctic Circle the latitude line of 66.5° south, marking the position farthest from the South Pole where the Sun does not rise on the date of the summer solstice (the start of winter in the Southern Hemisphere). (6.5)

ante meridian (A.M.) Latin for before noon. (7.1)

anthropic principle the principle that the properties of the universe are limited to those that permit our existence—otherwise we could not be present to observe the universe. (83.4)

antimatter a type of matter that, if brought into contact with ordinary matter, annihilates it, leaving nothing but energy. The positron is the antimatter partner of the electron. The antiproton is the antimatter partner of the proton. Antimatter is observed in cosmic rays and can be created from energy in the laboratory. (81.2)

aperture the opening in a telescope or other optical instrument that determines how much light it collects. (26.1, 27.2)

aphelion the point in an orbit where a body is farthest from the Sun.

Aphrodite a continent-like highland region on the planet Venus. (39.2)

Apollo asteroids asteroids whose orbits cross the Earth's. (41.4)

apparent double star two stars that lie close to each other in the sky but that are at different distances and not physically associated. (56.1)

apparent magnitude a unit for measuring the brightness of a celestial body. Also simply called *magnitude;* the word *apparent* helps to distinguish it from **absolute magnitude.** The smaller the magnitude, the brighter the star. Bright stars have magnitude 1, whereas the faintest stars visible to the unaided eye have magnitude 6. The largest telescopes can detect stars of about magnitude 30; bright objects like the Sun, Moon, and planets can have "negative" magnitudes. (54.4)

apparent noon the time at which the Sun crosses the meridian from any particular location. This usually differs from noon on a clock, which represents an average within a time zone and over the year. (7.1)

arc minute (arcmin) an angular measure equal to 1/60th of 1 degree. (12.1, 52.1)

arc second (arcsec) an angular measure equal to 1/60th of 1 arc minute, or 1/3600th of 1 degree. (52.2)

Arctic Circle the latitude line of 66.5° north, marking the position farthest from the North Pole where the Sun does not rise on the date of the winter solstice (the start of winter in the Northern Hemisphere). (6.5, 7.2)

Aristarchus (about 310–230 B.C.E.) Greek astronomer who first estimated the sizes and distances of the Moon and Sun. (10.2)

Aristotle (384–322 B.C.E.) Greek philosopher who developed many early ideas of physics and astronomy. (10.1)

association see **stellar association.**

asterism an easily identified grouping of stars, sometimes a part of a larger constellation (such as the Big Dipper) or extending across several constellations (such as the Summer Triangle). (5.2)

asteroid a small, generally rocky, solid body orbiting the Sun and ranging in diameter from a few meters to hundreds of kilometers. (32.1, 41.0)

asteroid belt a region between the orbits of Mars and Jupiter in which most of the Solar System's asteroids are located. (32.1)

astronomical unit (AU) a distance unit based on the average distance of the Earth from the Sun. (1.6, 52.1)

astrophysics the branch of science in which physical laws are applied to interpret astronomical observations and derive chemical and physical properties of remote objects. (54.0)

asymmetry a variation in properties where none is expected. For example, the weak force shows an asymmetry in its behavior with some particles and their antiparticles. (81.3)

atmosphere layer of gas held close to a planet or star by the force of gravity. Also a unit of pressure in which 1 atmosphere is the average pressure of Earth's atmosphere at sea level. (36.1)

atmosphere effect an alternative name for the greenhouse effect. (36.3)

atmospheric pressure the pressure produced by the weight of overlying atmosphere. (36.1)

atmospheric window a wavelength range in which our atmosphere absorbs little radiation. For example, our atmosphere allows a range of wavelengths at about 300 to 700 nm to pass through. This is our "visible" atmospheric window. The atmosphere also has a wide window at radio wavelengths and narrow windows at infrared wavelengths. (30.1)

atom a submicroscopic particle consisting of a nucleus and orbiting electrons. The smallest unit of a chemical element. (4.2)

atomic mass the mass of an atom expressed in units approximately equal to the mass of a proton or neutron. The isotope of carbon with 6 protons and 6 neutrons has an atomic mass of 12.0. (21.4)

atomic number the number of protons in the nucleus of an atom. Unless the atom is ionized, the atomic number is also the number of electrons orbiting the nucleus of the atom. (21.4)

aurora the light emitted by atoms and molecules in the upper atmosphere. This light is a result of magnetic disturbances caused by the solar wind, and appears to us as the Northern or Southern Lights. (36.1, 43.4, 51.1)

autumnal equinox the beginning of fall in the Northern Hemisphere, on or near

September 23 when the Sun crosses the celestial equator. (6.5)

averted vision looking slightly to one side of a dim object so that its light strikes slightly off-center in your field of view. This allows you to more easily discern faint objects, although at a sacrifice of sharpness. (31.2)

azimuth a coordinate for indicating the direction of objects on the sky. The azimuth direction is measured from north to the point on the horizon below the object. A star that is directly above the point due east is at azimuth 90°; south is at 180°; west at 270°; and north at 360° or 0°. (31.3)

B

Balmer lines a series of absorption or emission lines of hydrogen seen at visible wavelengths. (55.4)

barred spiral (SB) galaxy a galaxy in which the spiral arms wind out from the ends of a central bar rather than from the nucleus. (75.1)

Barringer crater an impact crater more than 1 km in diameter located in Arizona, probably produced about 50,000 years ago. (48.4)

basalt a dense rock of volcanic origin. (37.1)

baseline the distance between two observing locations used for the purposes of triangulation measurements. The larger the baseline, the better the resolution attainable. (52.1)

B.C. (Before Christ) an abbreviation used in the Christian calendar to indicate dates before the birth of Jesus. Biblical scholars believe the actual birth year was between 4 and 8 years earlier than the date indicated by this calendar system. (9.5)

B.C.E. (Before the Common Era) a nondenominational designation for referring to B.C. dates. (9.5)

belt a dark, low-pressure region in the atmosphere of Jovian planet, where gas flows downward. (43.1)

Big Bang the event that began the universe according to modern cosmological theories. It occurred about 13.7 billion years ago and generated the expanding motion that we observe today. (2.5, 79.0)

Big Crunch point of final collapse of a bound universe. (82.0)

binary star two stars in orbit around each other, held together by their mutual gravity. (56.0, 57.2)

bipolar flow the narrow columns of high-speed gas ejected by a protostar in two opposite directions. (59.2, 60.2)

blackbody an object that is an ideal radiator when hot and a perfect absorber when cool. It absorbs all radiation that falls upon it, reflecting no light; hence it appears black. Stars are approximately blackbodies. The radiation emitted by blackbodies obeys Wien's law and the Stefan-Boltzmann law. (23.1)

black dwarf a white dwarf that has cooled to the point where it emits little or no visible radiation. (64.4)

black hole an object whose gravitational attraction is so strong that its escape velocity equals the speed of light, preventing light or any radiation or material from leaving its "surface." (59.2, 68.0)

blueshift a shift in the wavelength of electromagnetic radiation to a shorter wavelength. For visible light, this implies a shift toward the blue end of the spectrum. The shift can be caused by the motion of a source of radiation toward the observer or by the motion of an observer toward the source. For example, the spectrum lines of a star moving toward the Earth exhibit a blueshift. See also **Doppler shift.** (25.0)

Bode's rule a numerical formula for predicting the approximate distances of most of the planets from the Sun. (41.1)

Bohr model first theory of the hydrogen atom to explain the observed spectral lines. This model rests on three ideas: that there is a state of lowest energy for the electron, that there is a maximum energy beyond which the electron is no longer bound to the nucleus, and that within these two energies the electron can only exist in certain energy levels. (21.3)

Bok globule a small, dark, interstellar cloud, often approximately spherical. Many globules are the early stages of protostars. (60.2)

B.P. (Before Present) a designation used primarily by archaeologists and geologists to indicate a number of years before the present time. (9.5)

Brahe, Tycho (1546–1601) Danish astronomer who made the finest pretelescopic measurements. His observations disproved the idea that comets and novae were atmospheric phenomena and provided the basis for Kepler's laws about the motions of the planets. (12.1)

brightness a measure of the amount of light received from a body, often measured in watts per square meter detected by a telescope or other detector. (21.2, 54.2)

brown dwarf a body intermediate between a star and a planet. A brown dwarf has a mass between about 0.017 and 0.08 solar masses—too low to fuse hydrogen in its core, but massive enough to fuse deuterium as it contracts. (34.2, 60.3)

bulge the dense, central region of a spiral galaxy. (70.4)

C

Callisto the second largest of Jupiter's satellites. (46.1)

Caloris Basin the largest impact crater region yet seen on Mercury. (38.1)

carbonaceous chondrite a type of meteorite containing many tiny spheres (chondrules) of rocky or metallic material stuck together by carbon-rich material. (48.2)

carbon detonation the ignition of carbon fusion for stars in the range of about 4 to 8 solar masses in the late stages of their evolution. Some astronomers hypothesize that this occurs with enough explosive force to destroy the star. (64.1)

Cassegrain telescope a type of reflecting telescope in which incoming light hits the primary mirror and is then reflected upward toward the prime focus, where a secondary mirror reflects the light back down through a small hole in the main mirror into a detector or eyepiece. (28.2)

Cassini's division a conspicuous 1800-kilometer-wide gap between the outermost rings of Saturn. (45.3)

CCD charged-coupled device: an electronic device that records the intensity of light falling on it. CCDs have replaced film in most astronomical applications. (26.2)

C.E. (Common Era) a designation for the year-numbering system of the most widely used calendar system. A nondenominational equivalent to **A.D.** (9.5)

celestial equator an imaginary line on the celestial sphere lying exactly above the Earth's equator. It divides the celestial sphere into northern and southern hemispheres. (5.3)

celestial pole an imaginary point on the sky directly above the Earth's North or South Pole. (5.3)

celestial sphere an imaginary sphere surrounding the Earth. Ancient astronomers pictured the stars as all being attached to it, all at the same distance from the Earth. (5.1)

center of mass the "average" position in space of a collection of massive bodies, weighted by their masses. (16.2, 56.3)

centripetal force a force that causes the path of a moving object to curve. (17.1)

Cepheid variables a class of yellow giant pulsating stars. Their pulsation periods range from about 1 day to about 70 days. Cepheids can be used to determine distances. See also **standard candle.** (63.2)

Ceres the first-discovered and largest of the asteroids, over 900 km in diameter. Often called a planet during the early 1800s. (41.1)

Chandrasekhar limit the maximum mass of a white dwarf stellar remnant. Approximately 1.4 solar masses—above this limit the remnant will collapse. Named for the astronomer who first calculated that such a limit exists. (65.2)

Charon the largest moon of Pluto. (46.4)

chemical energy the energy stored or released when bonds form or break between atoms and molecules. (20.1)

Chicxulub the site of a major impact crater that is the most likely explanation for the extinction of the dinosaurs about 65 million years ago. (48.5)

Chiron a large icy body discovered orbiting between Saturn and Uranus in the 1970s. (47.4)

chondritic meteorite a meteorite containing small spherical granules called chondrules. (48.2)

chondrule a small spherical silicate body embedded in a meteorite. (48.2)

chromosphere the lower part of the Sun's outer atmosphere that lies directly above the Sun's visible surface (photosphere). (49.4)

circumpolar star a body close enough to a celestial pole that always remains above the horizon. Circumpolar stars are closer to the celestial pole than the latitude of the observer. (5.4)

closed universe geometry that the universe as a whole would have if the density of matter is above the critical value. A closed universe is finite in extent and has no edge, like the surface of a sphere. It has enough mass to stop the present expansion and will eventually collapse. (80.4)

cluster a group of objects (for example, stars or galaxies) held together by their mutual gravitational attraction.

CNO cycle/process a reaction involving carbon, nitrogen, and oxygen (C, N, and O) that fuses hydrogen into helium and releases energy. The process begins with a hydrogen nucleus fusing with a carbon nucleus. Subsequent steps involve nitrogen and oxygen. The carbon, nitrogen, and oxygen act as catalysts and are released at the end of the process to start the cycle again. The CNO cycle is the dominant process for generating energy in main-sequence stars that are hotter and more massive than the Sun. (61.2)

coherent a relationship indicating that electromagnetic waves are in synchronization with each other. Waves that are coherent exhibit persistent interference effects. (27.1)

collecting area the total aperture over which a telescope can gather light. (27.2)

coma the gaseous atmosphere surrounding the head of a comet. (47.1)

comet a small body in orbit around the Sun, consisting of a tiny, icy core and a tail of gas and dust. The tail forms only when the comet is near the Sun. (32.1)

compact stars very dense stars whose radii are much smaller than the Sun's. These stars include **white dwarfs, neutron stars,** and **black holes.**

comparative planetology the study of bodies in the Solar System, examining and understanding the similarities and differences among worlds. (42.0)

condensation conversion of free gas atoms or molecules into a liquid or solid. A snowflake forms in our atmosphere when water vapor condenses into ice. (33.3)

cone light-sensing cell in the eye that gives us our color vision. (31.1)

conglomerate a rock made of eroded material and pebbles and later fused together by heat and pressure. (40.3)

conjunction the appearance of two astronomical objects in approximately the same direction on the sky. For example, if Mars and Jupiter happen to appear near each other on the sky, they are said to be in conjunction. **Superior conjunction** refers to when a planet is approximately in line with the Sun as seen from the Earth, but on the far side of the Sun. **Inferior conjunction** refers to when a planet is aligned between the Sun and the Earth. (11.3)

conservation law a theory describing how certain properties, like electric charge, energy and angular momentum, remain constant despite interactions. (20.0)

conservation of angular momentum a principle of physics stating that the angular momentum of a rotating body remains constant unless forces act to speed it up or slow it down. Mathematically, conservation of angular momentum states that $M \times V \times R$ is a constant, where M is the mass of a body moving with a velocity V in a circle of radius R. One extremely important consequence of this principle is that if a rotating body shrinks, its rotational velocity must increase. (20.2, 67.1)

conservation of energy a principle of physics stating that energy is never created or destroyed, although it may change its form. For example, energy of motion may change into energy of heat. (20.1)

constellation an officially recognized grouping of stars on the night sky. Astronomers divide the sky into 88 constellations. (5.2)

continuous spectrum a spectrum with neither dark absorption nor bright emission lines. The intensity of the radiation in such a spectrum changes smoothly from one wavelength to the next. (24.3)

convection the rising and sinking motions in a liquid or gas that carry heat upward through the material. Convection is easily seen in a pan of heated soup on a stove. (35.4)

convection zone the region immediately below the Sun's visible surface in which its heat is carried upward by convection. (49.3)

Copernicus, Nicolas (1473–1543) astronomer who proposed that the Earth was just one of the planets orbiting the Sun. (11.3)

core the innermost, usually hot, region of the interior of the Earth or another planet. (32.1)

Coriolis effect the deflection of an object moving within or on the surface of a rotating body. It is caused by the conservation of angular momentum when the moving object gets closer to or farther from the rotation axis of the rotating body. The Coriolis effect makes storms on Earth spin, generates large-scale wind systems, and creates cloud belts on many planets. (36.4)

cornea the curved transparent layer covering the front part of the eye that does most of the focusing of light that we see. (31.1)

corona the outer, hottest part of the Sun's atmosphere. (49.4)

coronal hole a low-density region in the Sun's corona, probably related to the structure of the Sun's magnetic field. The solar wind may originate in these regions. (49.4)

coronal mass ejections a blast of gas moving outward through the Sun's corona and into interplanetary space following the eruption of a prominence. (51.1)

cosmic horizon the greatest distance it is possible to see out into the universe given its age and rate of expansion. The horizon lies at a distance in light-years approximately equal to the age of the universe in years. (80.1)

cosmic microwave background (CMB) the radiation that was created during the Big Bang and that permeates all space. At this time, the temperature of this radiation is 2.73 K. (80.2)

cosmic rays extremely energetic subatomic particles (such as protons and electrons) traveling at nearly the speed of light. Some rays are emitted by the Sun, but most come from more distant sources, such as exploding supernovae. (50.3)

cosmological constant a term in the equations that Einstein developed to describe the effects of gravity on the entire universe. The term has the effect of a repulsive "force" opposing gravity. (82.4)

cosmological principle the hypothesis that on average, the universe looks the same to all observers, no matter where they are located in it. (79.2)

cosmology the study of the evolution and structure of the universe. (79.0)

Crab Nebula a supernova remnant in the constellation Taurus. Astronomers in ancient China and the Far East saw the supernova explode in A.D. 1054. A pulsar lies in the middle of the nebula. (26.4)

crater a circular pit, generally with a raised rim and sometimes with a central peak. A crater can be formed by the impact of a solid body, such as an asteroid, with the surface of a planet, moon, or another asteroid. Crater diameters on the Moon range from centimeters to several hundred kilometers. (37.1)

crescent moon a lunar phase during which the Moon appears less than half full. (8.1)

Cretaceous period a period of time in the geologic timescale ranging from about 146 million to 65 million years B.P. (48.5)

critical density the density necessary for a closed universe. If the density of the universe exceeds the critical density, the universe will stop expanding and collapse. If the density is less than the critical density, the universe will expand forever. (82.2)

crust the solid surface of a planet, moon, or other solid body. (35.1)

curvature of space the "bending" of space or motion given to space by gravity, as described by Einstein's general theory of relativity. For example, stars curve the space around them so that the path of light traveling by them is bent from a straight line. The universe too may be curved in such a way as to make its volume finite. (68.2)

curved space space that is not flat. See also **curvature of space.** (81.5)

D

dark adaptation the process by which the eye changes, through chemical changes in the retina and by expanding the pupil, to become more sensitive to dim light. (31.2)

dark energy a form of energy detected by its effect on the expansion of the universe, also known as the **cosmological constant.** It causes the expansion to speed up. The nature and properties of dark energy are unknown. (2.5, 82.4)

dark-line spectrum see **absorption-line spectrum.**

dark matter matter that emits no detectable radiation but whose presence can be deduced by its gravitational attraction on other bodies. (2.5, 73.1, 78.0)

dark matter halo the extended "cloud" of dark matter in which galaxies lie. The dark matter comprises most of the mass of a galaxy, but is only detected from the rapid speed of rotation of the outer parts of the galaxy. (78.1)

dark nebula a dense cloud of dust and gas in interstellar space that blocks the light from background stars. (72.1)

daughter atoms the atoms produced by the decay of a radioactive element. For example, uranium decays into daughter atoms of lead. (33.1)

daylight saving time the time kept during summer months by setting the clock ahead one hour. This gives more hours of daylight after the workday. (7.4)

declination (dec) one part of a coordinate system (with **right ascension**) for locating objects in the sky. The declination indicates an object's distance north or south of the celestial equator. Declination is analogous to latitude on the Earth's surface. (5.5, 31.3)

degeneracy pressure the pressure created in a dense gas by the interaction of its electrons. Electrons cannot occupy the same space if they have the same energy, so the gas cannot be compressed further once the

electrons are tightly packed unless some of their energies are raised. (62.3)

degenerate gas an extremely dense gas in which the electrons and nuclei are tightly packed. Unlike a normal gas, the pressure of a degenerate gas does not depend on its temperature. (62.3)

Deimos one of the two small moons of Mars, probably a captured asteroid. (40.5)

density the mass of a body or region divided by its volume. (3.1, 32.4)

density wave a spiral-shaped compression of the gas and dust in a spiral galaxy kept from dispersing by the gravitational attraction between all the matter in the wave. Density waves may account for the spiral arms of galaxies. According to the theory, stars and gas in one region of a galaxy may become packed more closely together (reaching a higher density). As the galaxy spins, these dense regions are pulled apart, but their gravitational influence on neighboring regions may cause them to compact in succession. This can produce "waves" of higher density that travel through the disk of a galaxy, pulling stars and interstellar gas into a spiral pattern. (73.3)

deuterium an isotope of hydrogen containing one proton and one neutron. (50.2)

differential gravitational force the difference between the gravitational forces exerted on an object at two different points. The effect of this force is to stretch the object. Such forces create **tides** and, if strong enough, may break up an astronomical object. See also **Roche limit.**

differential rotation rotation in which the rotation period of a body varies with latitude or radius. Differential rotation occurs for gaseous bodies like the Sun and Jovian planets as well as for the disks of galaxies. (73.1)

differentiation the separation of previously mixed materials inside a planet or other object. An example of differentiation is the separation that occurs when a dense material, such as iron, settles to the planet's core, leaving lighter material on the surface. (33.5)

diffraction bending of the path of light or other electromagnetic waves as they pass through an opening or around an obstacle.

Diffraction limits the ability of a telescope to distinguish fine details. (29.1)

diffraction pattern a pattern that a telescope imparts to light from even something as simple as a single point of light. This is caused by diffraction around various structural elements of a telescope—such as the spikes seen surrounding the image of a stars, which are caused by the support structure for secondary mirrors. (29.1)

dirty snowball a model of the structure of a comet, suggesting it is made primarily of ices with grains of heavy elements scattered within it. (47.1)

disk the flat, round portion of a galaxy. The Sun lies in the disk of the Milky Way. (70.4)

dispersion the spreading of light or other electromagnetic radiation into a spectrum. A rainbow is an example of the dispersion of light caused by raindrops. (28.3)

DNA deoxyribonucleic acid. The complex molecule that encodes genetic information in all organisms here on Earth. (83.1)

Doppler shift the change in the observed wavelength of radiation caused by the motion of the emitting body or of the observer. The shift is an increase in the wavelength if the source and observer move apart and a decrease in the wavelength if the source and observer approach. See also **redshift** and **blueshift.** (25.0)

Doppler shift method a means of detecting planets orbiting other stars by detecting the Doppler shift caused by the star's own reflex motion in response to the orbiting planet. (34.2)

Drake equation expression that gives an estimate of the probability that intelligence exists elsewhere in the galaxy, based on a number of conditions thought to be essential for intelligent life to develop. (84.3)

dust tail a comet tail containing dust that reflects sunlight. The dust in a comet tail is expelled from the nucleus of the comet. (47.3)

dwarf a small dim star.

dwarf galaxy small galaxy containing a few million stars. (75.1, 76.1)

dynamo see **magnetic dynamo.**

E

early universe the first second or so of the universe's history after the Big Bang when the universe was composed of nothing but energy and elementary particles. (81.2)

eclipse the blockage of light from one astronomical body caused by the passage of another between it and the observer. The shadow of one astronomical body falling on another. For example, the passage of the Moon between the Earth and Sun can block the Sun's light and cause a solar eclipse. (8.2)

eclipse seasons the times of year, separated by about six months, when solar and lunar eclipses are possible because the Moon is crossing the **ecliptic** while it is full or new. (8.2)

eclipsing binary a binary star pair in which one star periodically passes in front of the other, totally or partially blocking the background star from view as seen from Earth. (56.1, 57.2)

ecliptic the path followed by the Sun around the celestial sphere. The path gets its name because eclipses can occur only when the Moon crosses the ecliptic. (6.2)

electric charge the electrical property of objects that causes them to attract or repel one another. A charge may be either positive or negative. (4.2)

electric force the force generated by electric charges. It is attractive between unlike charges (+ and −) but repulsive between like charges (+ and + or − and −). (4.3)

electromagnetic force the force arising between electrically charged particles, between charges and magnetic fields, and between magnets. This force holds electrons to the nuclei of atoms, makes moving charges spiral around magnetic field lines, and deflects a compass needle. (4.3)

electromagnetic radiation a general term for any kind of electromagnetic wave. (22.0)

electromagnetic spectrum the assemblage of all wavelengths of electromagnetic radiation. The spectrum includes the following wavelengths, from long to short: radio, infrared, visible light, ultraviolet, X-rays, and gamma rays. (22.0)

electromagnetic wave a wave consisting of alternating electric and magnetic energy. Ordinary visible light is an electromagnetic wave, and its wavelength determines the light's color. (21.1)

electron a low-mass, negatively charged subatomic particle. In an atom, electrons orbit the nucleus, but may at times be torn free if ionized. See also **ionization.** (4.2)

electron degeneracy a condition resisted by electrons because they cannot occupy the same space if they have the same energy. (65.2)

electronic transition a shift in energy level that an atom undergoes when an electron shifts from one energy level to another. (21.3)

electroweak force a combined form (or unification) of the **electromagnetic force** and **weak force** that occurs at high energies.

element a fundamental substance, such as hydrogen, carbon, or oxygen, that cannot be broken down into a simpler chemical substance. Approximately 100 elements occur in nature. (21.4)

ellipse a geometric figure related to a circle but elongated along one axis. (12.2)

elliptical (E) galaxy a galaxy in which the stars smoothly fill an ellipsoidal volume. Abbreviated as E galaxy. The stars in such systems are generally old (Pop II). (75.1)

emission the production of light, or more generally, electromagnetic radiation, by an atom (or other object) when the atom's energy drops to a lower level. (21.5)

emission-line spectrum a spectrum consisting of bright lines at certain wavelengths separated by darker regions in which little or no radiation is emitted. (24.2)

emission nebula a hot gas cloud in interstellar space that emits light. (72.1)

energy a quantity that measures the ability of a system to do work or cause motion. (20.1)

energy level any of the numerous levels that an electron can occupy in an atom, roughly corresponding to an electron orbital. (24.1)

epicycle a fictitious, small, circular orbit superimposed on another circular orbit. Epicycles were proposed by early astronomers to explain the retrograde motion of the planets and to make fine adjustments to the predictions of planets' positions. (11.2)

equation of time a relationship describing the offset of time measured by a sundial from the average 24-hour period from noon to noon. The offsets vary through the year because of the Earth's elliptical orbit and the tilt of its axis. (13.5)

equator the imaginary line that divides the Earth (or any other body) symmetrically into its Northern and Southern Hemispheres. The equator is perpendicular to a body's rotation axis.

equatorial mount a support structure for a telescope that allows it to rotate along celestial coordinates: north and south (in declination) and around the celestial poles (in right ascension). (31.3)

equinox the time of year when the Sun crosses the celestial equator. The number of hours of daylight and night are approximately equal, and the two dates mark the beginning of the spring and fall seasons. (6.5)

Eratosthenes (276–195 B.C.E.) Greek astronomer who first estimated the diameter of the Earth. (10.3)

erosion the wearing down of geological features by wind, water, ice, and other phenomena of planetary weather. (36.2)

escape velocity the speed an object needs to move away from another body in order not to be pulled back by its gravitational attraction. Mathematically, the escape velocity, V, is defined as $\sqrt{2GM/R}$ where M is the body's mass, R is its radius, and G is the gravitational constant. (18.2, 68.1)

Europa the smallest of the four Galilean satellites of Jupiter. (46.1)

eukaryotes cells with nuclei. Most cells in current terrestrial organisms have nuclei and are thus eukaryotes. (83.2)

Evening Star any bright planet, but most often Venus, seen low in the western sky after sunset. (13.4)

event horizon the location of the "boundary" of a black hole. An outside observer cannot see in past the event horizon. (68.2)

excited state the condition in which the electrons of an atom are not in their lowest energy level (lowest orbital). (21.5)

exclusion principle the condition that no more than two electrons may occupy the same energy level in an atom. This limitation leads to **degeneracy pressure.** (62.3)

exoplanet a planet orbiting a star other than our Sun. (34.0, 84.2)

expansion redshift a lengthening of the wavelength of electromagnetic radiation caused when the waves travel through expanding space. The redshift of distant galaxies is an example of this phenomenon. (74.3)

exponent a number placed to the right and above another indicating the power by which to raise the other. For example, in the expression T^4 the exponent is 4. (3.1)

extinction the dimming of starlight due to absorption and scattering by interstellar dust particles. (72.3)

extrasolar planet a planet orbiting a star other than the Sun (also called an exoplanet). (34.0)

F

false-color image a depiction of an astronomical object in which the colors are not the object's real colors. Instead, they are colors arbitrarily chosen to represent other properties of the body, such as the intensity of radiation at other than visible wavelengths. (26.3)

fault in geology, a crack or break in the crust of a planet along which slippage or movement can take place. (39.2, 45.2)

Fermi paradox the question why, if there are many advanced civilizations in the galaxy, none of them are visiting us or have left indications of their existence. (84.3)

filter (for light) a material that transmits only particular wavelengths of light. (27.3)

first-quarter moon a phase of the waxing Moon when Earth-based observers see half of the Moon's illuminated hemisphere. (8.1)

fission the splitting of an atom into two or more smaller atoms.

flare an outburst of energy on the Sun. (51.1)

flat rotation curve a plot showing that the orbital speed in a galaxy remains nearly constant out to large radii. (78.1)

flat universe a universe that extends forever with no curvature. Its total energy is zero. (80.4)

fluorescence the conversion of ultraviolet light (or other short-wavelength radiation) into visible light. This occurs, for example, when an atom is **excited** into a high energy level by an ultraviolet photon, and then descends to the ground state in a series of steps, emitting lower-energy photons. (47.3)

focal length the distance between a mirror or lens and the point at which the lens or mirror brings a distant source of light into focus. (31.3)

focal plane the surface where the lenses and/or mirrors of a telescope form an image of distant object. (28.1)

focus (1) one of two points within an ellipse used to generate the elliptical shape. The Sun lies at the focus of each planet's elliptical orbit. (12.2) (2) A point in an optical system in which light rays are brought together; the location where an image forms in such systems. (26.1, 28.0)

force a push or pull. (14.3)

fovea a central part of the retina, where most of our color vision is concentrated. This region is not very sensitive at low light levels. (31.1)

frequency the number of times per second that a wave vibrates. (22.1)

full moon phase of the Moon in which it appears as a completely illuminated circular disk in the sky. (8.1)

fully convective a condition seen in some stars (such as very low-mass main-sequence stars) in which convection circulates matter between the core and the surface. (61.2)

fundamental forces the four basic forces of nature: gravitation, the electromagnetic force, and the strong and weak forces. According to some modern theories (see also **GUTs**), these forces are different forms of a single, more fundamental, unified force. (4.3)

G

g the acceleration due to gravity on the surface of the Earth—about 9.8 meters/second^2. (16.3)

G Newton's universal gravitational constant, which allows us to determine the force between objects if we know their masses and separation. (16.2)

Gaia hypothesis the hypothesis that life does not merely respond to its environment but actually alters its planet's atmosphere and temperature to make the planet more hospitable. For example, by photosynthesis, plants have created an oxygen-rich atmosphere on Earth, which shields the plants from dangerous ultraviolet radiation. Gaia is pronounced *GUY-uh* in this context. (83.3)

galactic cannibalism the capture and merging of one galaxy into another. (71.2, 75.4)

galactic plane the flat region defined by the disk of the Milky Way. (70.2)

galaxy a massive system of stars, gas, and dark matter held together by their mutual gravity. Typical galaxies have a mass between about 10^7 and 10^{13} solar masses. Our galaxy is the Milky Way. (2.2, 70.0, 74.0)

galaxy cluster a set of hundreds or thousands of galaxies held together by their mutual gravitational attraction. The Milky Way lies on the outskirts of the Virgo cluster. (2.4, 76.0)

galaxy group a system of from two to several dozen galaxies held together by their mutual gravitational attraction. The Milky Way belongs to the Local Group galaxy cluster. (76.1)

Galilean relativity a method for determining the relative speeds of motion

seen by observers who are moving with respect to each other. This method successfully describes motions at slow speeds, but not when speeds grow to an appreciable fraction of the speed of light. (53.1)

Galilean satellites the four moons of Jupiter discovered by Galileo: Io, Europa, Ganymede, and Callisto. (12.3, 45.1, 46.1)

Galilei, Galileo (1564–1642) Italian physicist and astronomer who studied inertia and published the first studies of the sky with a telescope. (12.3)

gamma ray electromagnetic waves that have the shortest wavelength. (22.6)

Ganymede the largest moon of Jupiter and the largest satellite in the Solar System—even larger than Mercury. (46.1)

gas giant a Jovian planet. (32.1)

general theory of relativity Einstein's theory relating how mass and energy can change the structure of space and time, "curving" them. (68.0, 79.2)

geocentric model a hypothesis that held that the Earth is at the center of the universe and all other bodies are in orbit around it. Early astronomers thought that the Solar System was geocentric. (11.2)

giant a star of large radius and large luminosity. (62.0)

gibbous appearance of the Moon (or a planet) when more than half (but not all) of the sunlit hemisphere is visible from Earth. (8.1)

glitches abrupt changes in the pulsation period of a pulsar, perhaps the result of adjustments of its crust. (67.2)

globular cluster a dense grouping of old stars, containing generally about 105 to 106 members. They are often found in the halos of galaxies. (69.1, 70.3)

globule see **Bok globule.**

gluon a particle that carries (or exerts) the strong force between quarks. (4.3)

grains small solid particles that condense from the heavy elements in a gas when it cools. (64.2)

grand unified theory (GUT) a theory that describes and explains the four fundamental

forces as different aspects of a single force. (81.4)

granulation texture seen in the Sun's photosphere. Granulation is created by clumps of hot gas that rise to the Sun's surface. (49.3)

grating a piece of material with many fine, closely spaced parallel lines used to create a spectrum. Light may be reflected off a grating or diffracted as it passes through a transparent grating. (24.0)

gravitational force force exerted on one body by another due to the effect of gravity. The force is directly proportional to the masses of both bodies involved and is inversely proportional to the square of the distance between them. (4.2)

gravitational lens an object whose gravity curves space in a way that can focus the light from an object behind it. The light from the more distant object may be magnified, distorted, or turned into multiple images by the lens. See also **curvature of space.** (78.2)

gravitational potential energy the energy stored in a body subject to the gravitational attraction of another body. As the body falls, its gravitational potential energy decreases and is converted into kinetic energy. (20.1)

gravitational redshift the shift in wavelength of electromagnetic radiation (light) caused by a body's gravitational field as the radiation moves away from the body. Only extremely dense objects, such as white dwarfs or neutron stars, produce a significant gravitational redshift of their radiation. (68.4)

gravitational waves a wavelike disturbance in the curvature of space that is generated by the acceleration of massive bodies. (68.5)

gravity the force of attraction between two bodies that is generated by their masses. (2.0)

Great Attractor a large concentration of mass toward which everything in our part of the universe apparently is being pulled. (76.3)

greatest elongation the position of an inner planet (Mercury or Venus) when it lies farthest from the Sun on the sky. Mercury and Venus are particularly easy to see when they are at greatest elongation. Objects may be at greatest eastern or western elongation according to whether they lie east or west of the Sun. (11.3)

Great Red Spot a reddish elliptical spot about 40,000 km by 15,000 km in size in the southern hemisphere of the atmosphere of Jupiter. The Red Spot has existed for at least 3½ centuries. (43.3)

greenhouse effect the trapping of heat by a planet's atmosphere, making the planet warmer than would otherwise be expected. Generally, the greenhouse effect operates if visible sunlight passes freely through a planet's atmosphere, but the infrared radiation produced by the warm surface cannot escape readily into space. (36.3)

greenhouse gas a molecule (such as carbon dioxide, methane, or water vapor) that efficiently absorbs infrared radiation. (39.1)

Gregorian calendar the calendar devised at the request of Pope Gregory XIII in 1582. It is the common civil calendar used throughout the world today. It improved upon previous calendars by omitting the leap year for century years not divisible evenly by 400. (9.4)

grooved terrain regions of the surface of Ganymede consisting of parallel grooves. Believed to have formed by repeated fracture of the icy crust. (46.1)

ground state the lowest energy level of an atom. (21.3)

H

half-life the time required for half of the atoms of a radioactive substance to disintegrate. (33.1)

Halley's comet a comet with a period of about 76 years famous because it was the first comet whose return was predicted. (47.0)

halo the approximately spherical region surrounding spiral galaxies. The halo contains mainly old stars, as are found, for example, in globular clusters. The halo also appears to contain large amounts of dark matter. (70.4)

Hawking radiation radiation that black holes are hypothesized to emit as a result of quantum effects. This radiation leads to the extremely slow evaporation of black holes. (68.3)

heliocentric models models of the Solar System in which Earth and the other planets orbit the Sun. (11.3)

helium flash the beginning of helium fusion in a low-mass star. The fusion begins explosively and causes a major readjustment of the star's structure. (62.3)

hertz a unit of frequency: one cycle per second. Named for Heinrich Hertz, who first produced radio radiation. (22.1)

HI emission 21-cm wavelength radio waves produced by atoms of un-ionized hydrogen. (72.4)

HII region a region of **ionized** hydrogen. HII regions generally have a luminous pink/red glow and often surround luminous, hot, young stars. (72.2)

highlands the older, most heavily cratered regions on the Moon. (37.1)

high-mass stars stars born with masses above about $8M_\odot$; these stars will end their lives by fusing elements up to iron in their cores and then explode as supernovae. (59.2)

Hipparchus (about 190–120 B.C.E.) Greek astronomer who made one of the first detailed charts of the stars. (52.1)

homogeneous having a consistent and even distribution of matter that is the same everywhere. (80.1)

horizon the line separating the sky from the ground. See also **cosmic horizon** and **event horizon.** (5.1)

H-R diagram (or Hertzsprung-Russell diagram) a graph on which stars are located according to their temperature and luminosity. Most stars on such a plot lie along a diagonal line, called the **main sequence,** which runs from cool, dim stars in the lower right to hot, luminous stars in the upper left. (58.0)

Hubble's constant the multiplying constant H in **Hubble's law:** $V = H \times D$. The reciprocal of Hubble's constant (in appropriate units) is approximately the age of the universe. (74.3)

Hubble's law a relation between a galaxy's distance, D, and its recession velocity, V, which states that more distant galaxies recede faster than nearby ones. Mathematically, $V = H \times D$, where H is Hubble's constant. (74.3)

Hubble Space Telescope a 2.4 meter diameter telescope launched by NASA in 1990. Because it observes from above the blurring effects of the Earth's atmosphere, it is able to make more detailed images than most other telescopes. (74.2)

Hubble time an estimate of the age of the universe obtained by taking the inverse of Hubble's constant. The estimate is valid only if there has been no acceleration or deceleration of the expansion of the universe. (79.4)

hydrosphere refers to the "layer" of water on the Earth consisting of oceans, lakes, rivers, ice caps, and other liquid water and ice. (36.0)

hydrostatic equilibrium the condition in which pressure and gravitational forces in a star or planet are in balance. Without such balance, bodies will either collapse or expand. (49.5, 61.1)

hypernova a term used to describe a hypothesized kind of supernova explosion of a star so massive that its core collapses directly into a black hole. (68.1)

hypothesis an explanation proposed to account for some set of observations or facts. (4.1)

I

Iapetus the third largest moon of Saturn particularly noted because it has one side that is dark black and the other is bright white. (45.2)

ice giants a term sometimes used to distinguish Uranus and Neptune from the gas giants Jupiter and Saturn. Despite the name, these planets have interiors hotter than Earth's, but the name is descriptive of their likely origin from the accretion of icy planetesimals. (44.0)

ideal gas law a law relating the pressure, density, and temperature of a gas. This law states that the pressure is proportional to the density times the temperature. (49.5)

impact the collision of a small body (such as an asteroid or a comet) with a larger object (such as a planet or moon). (37.1)

inclination the angle by which an astronomical object or its orbit is tilted.

inertia the tendency of an object at rest to remain at rest and of a body in motion to continue in motion in a straight line at a constant speed. See also **mass.** (14.1)

inferior conjunction see **conjunction.**

inferior planet a planet whose orbit lies between the Earth's orbit and the Sun. Mercury and Venus are inferior planets.

inflation the enormously rapid expansion of the early universe. Between about 10^{-35} and 10^{-32} seconds after the Big Bang, the universe expanded by a factor of perhaps 10^{100}. (81.4)

infrared a wavelength of electromagnetic radiation longer than visible light but shorter than radio waves. We cannot see these wavelengths with our eyes, but we can feel many of them as heat. The infrared wavelength region runs from about 1000 nm to 1 mm. (22.3)

initial mass function the relative number of stars born onto the main sequence at each mass. (69.3)

inner core the inner portion of Earth's iron–nickel core. Despite its high temperature, the inner part of the core is solid because it is under great pressure. (35.1)

inner planet a planet orbiting in the inner part of the Solar System. Sometimes taken to mean Mercury, Venus, Earth, and Mars. (32.1)

instability strip a region in the H-R diagram containing stars that pulsate. These stars have layers below the surface at a temperature that makes them unstable, so the outer layers expand and contract rhythmically. (63.2)

instrumentation a branch of astronomy dealing with the development of new kinds of detectors. (26.1)

interfere, interference a phenomenon in which electromagnetic waves mix together such that their crests and troughs can alternately reinforce and cancel one another. (21.1, 29.1)

interferometer a device consisting of two or more telescopes connected together to work as a single instrument. Used to obtain a high resolution, with the ability to see small-scale features. Most interferometers currently operate at radio, infrared, or visible wavelengths. (29.3)

interferometry the use of an **interferometer** to measure small angular-size features. (57.1)

international date line an imaginary line extending from the Earth's North Pole to the South Pole, running approximately down the middle of the Pacific Ocean. It marks the location on Earth at which the date changes to the next day as one travels across it from east to west. (7.3)

interstellar cloud a cloud of gas and dust in between the stars. Such clouds may be many light-years in diameter. (33.2, 59.2)

interstellar dust tiny solid grains in interstellar space, thought to consist of a core of rocklike material (silicates) or graphite surrounded by a mantle of ices. Water, methane, and ammonia are probably the most abundant ices. (60.1, 70.3)

interstellar grain microscopic solid dust particles in interstellar space. These grains absorb starlight, making distant stars appear dimmer than they truly are. (33.2)

interstellar medium (ISM) gas and dust between the stars in a galaxy. (72.0)

inverse-square law (1) any law in which some property varies inversely as the square of the distance, d (mathematically, as $1/d^2$). (2) The law stating that the apparent brightness of a body decreases inversely as the square of its distance. (21.2, 54.2)

Io the innermost of the Galilean satellites of Jupiter, noted especially for its active volcanoes. (46.1)

ionize remaining one or more electrons from an atom, leaving the atom with a positive electric charge. Under some circumstances, an extra electron may be attached to an atom, in which case the atom is described as negatively ionized. (21.5)

ionosphere the upper region of the Earth's atmosphere in which many of the atoms are ionized. (36.1)

ion tail a stream of ionized particles evaporated from a comet and then swept away from the Sun by the solar wind. (47.3)

iris the region of the eye surrounding the pupil that can control the amount of light entering the eye. (31.1)

iron core the endpoint of fusion in a massive star. When the iron core grows massive enough, the star undergoes a supernova. (59.4)

irregular (Irr) galaxy a galaxy lacking a symmetrical structure. (75.1)

irregular variables a class of variable star that changes in brightness erratically. (63.1)

Ishtar a continent-like region on Venus. (39.2)

isotopes nuclei with the same number of protons but different numbers of neutrons. (50.2)

J

jet narrow stream of gas ejected from any of several types of astronomical objects. Jets are seen extending from protostars and active galactic nuclei. They may be funneled into their direction of flow by magnetic fields or accretion disks surrounding the central object. (77.1)

joule a unit of energy. One joule per second equals one watt.

Jovian planet a giant gaseous planet such as Jupiter, Saturn, Uranus, or Neptune. The name *Jovian* is based on "Jove," an alternative name for Jupiter. (32.1)

Julian calendar a 12-month calendar devised under the direction of Julius Caesar. It includes a leap year every four years. (9.3)

K

Kapteyn's Universe a model of the known universe in wide usage in the early 1900s. Based on observations of the time, it was limited to stars within the disk of the Milky Way surrounding the Sun. (70.2)

Kelvin the most commonly used temperature scale in science, defined such that absolute zero is 0 K and water freezes at 273.15 K. (23.3)

Kepler's three laws mathematical descriptions of the motion of planets around the

Sun. The first law states that planets move in elliptical orbits with the Sun off-center at a focus of the ellipse. The second law states that a line joining the planet and the Sun sweeps out equal areas in equal times. The third law relates a planet's orbital period, P, to the semimajor axis of its elliptical orbit, a. Mathematically, the law states that $P^2 = a^3$, if P is measured in years and a in astronomical units. (12.2)

kiloparsec a unit of distance, equal to 1000 parsecs (pc), often used to describe distances within the Milky Way or the Local Group of galaxies. (70.2)

kinetic energy energy of motion. Kinetic energy is given by half the product of a body's mass and the square of its speed ($1/2\ mV^2$). (20.1)

Kirchhoff's laws a set of three "rules" that describe how continuous, bright line, and dark line spectra are produced. (24.3)

Kirkwood gaps regions in the asteroid belt with a lower-than-average number of asteroids. Some of the gaps result from the gravitational force of Jupiter removing asteroids from orbits within the gaps. (41.4)

Kuiper belt a region containing many large icy bodies and from which some comets come. The region appears to extend from the orbit of Neptune at about 30 AU, past Pluto, out to approximately 55 AU. (32.1, 47.2)

L

laminated terrain alternating layers of ice and dust seen in the ice caps of Mars. The layers appear to reflect climate changes caused by long-term orbital variations. (40.1)

large-scale structure the structure of the universe on size scales larger than that of clusters of galaxies. (76.4)

last scattering epoch the period of time a few hundred thousand years after the Big Bang when the density and temperature of gas in the universe dropped sufficiently that photons could begin to travel freely through space. The time of the formation of the **cosmic microwave background**. (80.2)

latitude the angular distance of a point north or south of the equator of a body. The equator has a latitude of 0° while the poles are at latitudes of 90° north and south. (5.4)

law in science, generally a theory that can be expressed in a mathematical form. (4.1)

law of gravity a description of the gravitational force exerted by one body on another. The gravitational force is proportional to the product of their masses and the inverse square of their separation. If the masses are M and m and their separation is d, the force between them, F, is $F = G\,Mm/d^2$, where G is a physical constant. (16.2)

leap second a time adjustment added every few years to clocks around the world to adjust them to account for Earth's gradually slowing spin. (7.5)

leap year a year in which there are 366 days implemented under the Julian and Gregorian calendar systems. (9.4)

lens an optical instrument, made of glass or some other transparent material, shaped so that parallel rays of light passing through it are bent to arrive at a single focus. (28.1)

light electromagnetic energy. (21.1)

light curve a plot of the brightness of a body versus time. (57.2, 63.1, 65.3)

light pollution the illumination of the night sky by waste light from cities and outdoor lighting. Light pollution makes it difficult to observe faint objects. (30.1)

light-year a unit of distance equal to the distance light travels in one year. A light-year is roughly 10^{13} km, or about 6 trillion miles. (2.3, 3.3, 54.0)

liquid core see **outer core.**

liquid metallic hydrogen a form of hot, highly compressed hydrogen that is a good electrical conductor, found in the interiors of Jupiter and Saturn. (43.2)

lobe see **radio lobe.**

Local Group the small group of about 30 galaxies to which the **Milky Way** belongs. (2.4, 76.1)

Local Supercluster the cluster of galaxy clusters in which the **Milky Way, Local Group,** and **Virgo cluster** are located. (76.3)

loners those who argue that humans are probably the only intelligent beings within our galaxy. (84.3)

longitude a coordinate indicating the east–west location of a point on the Earth's surface. The line of 0° longitude runs north–south through an observatory in Greenwich, England. (5.4)

Lorentz factor a term that describes by how much time, space, and mass are altered as a result of their motion. The Lorentz factor is very close to 1 (indicating very little change) except at speeds approaching the speed of light. (53.2)

low-mass star star with a mass less than 8 times that of the Sun; progenitor of a white dwarf. (59.2)

low surface brightness galaxy a galaxy with a low density of stars, making it difficult to detect in visible-wavelength surveys. (75.1)

luminosity the amount of energy radiated per second by a body. For example, the wattage of a lightbulb defines its luminosity. Stellar luminosity is usually measured in units of the Sun's luminosity (approximately 4×10^{26} watts). (23.2, 54.1)

lunar eclipse the passage of the Earth between the Sun and the Moon so that the Earth's shadow falls on the Moon. (8.2)

lunar month the time period of approximately 29½ days during which the Moon cycles from the new phase through all of its phases back to new. (8.1)

lunisolar calendar a calendar system that maintains links to both the lunar month and the solar year, such as the Chinese or Jewish calendars. Because the lunar month does not divide evenly into the year, lunisolar calendar systems must insert an extra month every few years. (9.2)

M

Magellanic Clouds two small companion galaxies of the **Milky Way.** (76.1)

magnetar a highly magnetized neutron star that emits bursts of gamma rays. (67.3)

magnetic dynamo a physical process that generates magnetic fields in astronomical bodies. In many cases, the process involves the generation of electric currents from an interaction between rotation and convection.

magnetic field a representation of the means by which magnetic forces are transmitted from one body to another. (35.5)

magnetogram an image of the Sun coded in a way to show the strength and direction of the magnetic field. (51.1)

magnitude see **apparent magnitude.**

main sequence the region in the H-R diagram in which most stars, including the Sun, are located. The main sequence runs diagonally across the H-R diagram from cool, dim stars to hot, luminous ones. See also **H-R diagram.** (58.1)

main-sequence lifetime the time a star remains a main-sequence star, fusing hydrogen into helium in its core. (61.3)

main-sequence star a star, fusing hydrogen to helium in its core, whose surface temperature and luminosity place it on the **main sequence** on the Hertzsprung-Russell diagram. (59.2, 61.0)

mantle the outer part of a rocky planet immediately below the crust and outside the metallic core. (35.1)

many-worlders those who argue that the development of life and intelligence are probably common occurrences within our galaxy. (84.3)

mare a vast, smooth, dark, and congealed lava flow filling basins on the Moon and on some planets. Maria often have roughly circular shapes. (37.1)

maria plural of **mare.**

maser an intense radio source created when excited gas amplifies some background radiation. *Maser* stands for "microwave amplification by stimulated emission of radiation."

mass a measure of the amount of material an object contains. A quantity measuring a body's **inertia.** (14.1, 32.1)

massive compact halo object (MACHO) collective name for bodies of planetary or stellar mass that may contribute a significant amount of mass to our galaxy's halo, but

that generate relatively little detectable light. (78.4)

mass–luminosity relation a relation between the mass and luminosity of stars. Higher-mass stars have higher luminosity, so that $L \approx M^{3.5}$ where L and M are measured in solar units. (58.2)

mass transfer process by which one star in a binary system transfers matter onto the other. (65.1)

matter era the times after the end of the **radiation era,** when radiation from the Big Bang cooled so that matter became the dominant constituent of the universe. (81.0)

Maunder minimum the period from about A.D. 1600 to 1740 during which the Sun was relatively inactive. Few sunspots were observed during this period, and Earth's temperature was cooler than normal. (51.4)

Maxwell Montes the highest mountain range on Venus, one of the first features identified by radar imaging. (39.2)

mean solar day the standard 24-hour day. The mean solar day is based on the average day's length over a year. Using a mean is necessary because the time interval from noon to noon, or sunrise to sunrise, varies slightly throughout the year. (13.5)

megaparsec a distance unit equal to 1 million **parsecs** and abbreviated Mpc. (74.3)

merger the resultant galaxy from when galaxies collide and form a single larger system. (75.3)

meridian the line passing north–south from horizon to horizon and passing through an observer's zenith. The meridian is the dividing line between the eastern and western halves of the sky. (7.1)

metal astronomically, any chemical element more massive than helium. Thus carbon, oxygen, iron, and so forth are metals. (71.3)

meteor the bright trail of light created by small solid particles entering the Earth's atmosphere and burning up; a "shooting star." (47.5, 48.0)

meteor shower an event in which many meteors occur in a short space of time, all from the same general direction in the

sky, typically when the Earth's orbit intersects debris left by a comet. The most famous shower is the Perseids in mid-August. (47.5)

meteorite the solid remains of a meteor that falls to the Earth. (33.1, 48.0)

meteoroid small, solid bodies moving within the Solar System. When a meteoroid enters our atmosphere and heats up, the trail of luminous gas it leaves is called a **meteor.** When the body lands on the ground, it is called a **meteorite.** ("A meteoroid is in the void. A meteor above you soars. A meteorite is in your sight.") (48.0)

method of standard candles see **standard candle.**

metric prefix a term that can be attached to the front of a measurement unit to indicate a power of 10 times the unit. For example, the prefix *kilo-* means a thousand, so a kilometer is a thousand meters. (3.1)

metric system a set of units developed in the late 1800s to replace the confusing systems then used. See also **MKS system.** (1.1, 3.1)

microwaves a range of electromagnetic radiation lying between radio and infrared wavelengths, typically with wavelengths of from about 1 to 100 millimeters. (22.5)

mid-ocean ridge an underwater mountain range on Earth created by plate tectonic motion.

Milky Way the galaxy to which the Sun belongs. Seen from Earth, the galaxy is a pale, milky band in the night sky. (2.2, 70.0)

Miller–Urey experiment an experimental attempt to simulate the conditions under which life might have developed on Earth. Miller and Urey discovered that amino acids and other complex organic compounds could form from the gases that are thought to have been present in the Earth's early atmosphere, if the gases are subjected to an electric spark or ultraviolet radiation. (83.2)

millisecond pulsar a pulsar whose rotation period is about a millisecond. (67.2)

Miranda an fairly small but very unusual moon of Uranus that shows huge cliffs and canyons and dramatic differences in its surface features in neighboring regions. (45.2)

MKS system the form of the **metric system** using meters, kilograms, and seconds to measure length, mass, and time, respectively. (3.1)

model a theoretical representation of some object or system. (4.1)

modified form of Kepler's third law a generalized version of the law ($P^2 = a^3$) relating the periods and semimajor axes of orbits for systems in which the mass is not dominated by the Sun. The modified form can be expressed as $(M_1 + M_2) P^2 = a^3$ where the sum of the masses is measured in solar masses. See also **Kepler's three laws.** (17.2)

molecule two or more atoms bonded into a single particle, such as water (H_2O—two hydrogen atoms bonded to one oxygen) or carbon dioxide (CO_2—one carbon atom bonded to two oxygen atoms).

molecular clouds relatively dense, cool interstellar clouds in which molecules are common. (60.1, 72.4)

month an interval of time loosely resembling the lunar month in some calendar systems, but closely tied to the cycle of moon phases in others. (8.0)

Moon illusion the optical illusion that the Moon appears larger when nearer the horizon than when seen high in the sky. (8.4)

Morning Star any bright planet visible in the eastern sky before dawn, but usually used in reference to Venus. (13.4)

mountain ranges long parallel ridges of mountains formed at the boundaries of plates. See also **plate tectonics.** (35.4)

N

nadir the point on the celestial sphere directly below the observer. The opposite of the zenith. (5.2)

nanometer a unit of length equal to 1 billionth of a meter (10^{-9} meters) and abbreviated nm. Wavelengths of visible light are several hundred nanometers. The diameter of a hydrogen atom is roughly 0.1 nm.

neap tide the weak tides occurring when the Moon is at first or third quarter and the Sun's and Moon's gravitational effects on the ocean partially offset each other. (19.2)

nebula a cloud in interstellar space, usually consisting of gas and dust.

negative curvature a form of curved space sometimes described as being "open" because it has no boundary. Negative curvature is analogous to a saddle shape. (80.4)

Nereid a moderately large satellite of Neptune with an irregular orbit. (46.3)

neutrinos tiny neutral particles with little or no mass and immense penetrating power. These particles are produced by stars when they fuse hydrogen into helium. They are also created during a supernova when a star's core collapses to form a neutron star. (4.4, 48.0, 50.0)

neutron a subatomic particle of nearly the same mass as the proton but with no electric charge. Neutrons and protons comprise the nucleus of the atom. (4.2, 50.1)

neutron star a very dense, compact star composed primarily of neutrons. Protons and electrons combine to form neutrons because of the enormous gravitational pressure on the cores of massive stars with iron cores. (59.2, 67.0)

new moon phase of the moon during which none of the sunlit side is visible. (8.1)

Newtonian telescope a reflecting telescope designed so that the focused light is reflected by a small secondary mirror out to the side of the telescope, where it can be viewed. (28.2)

newtons the standard unit of force in the metric system. (15.2, 16.3)

Newton's first law of motion a body continues in a state of rest or in uniform motion in a straight line unless acted upon by an external force. See also **inertia.** (14.2)

Newton's second law of motion $F = ma$. The amount of acceleration, a, that a force, F, produces depends on the mass, m, of the object being accelerated. (15.2)

Newton's third law of motion when two bodies interact, they exert equal and opposite forces on each other. (15.3)

nonthermal radiation radiation emitted by charged particles moving at high speed in a magnetic field. The radio emission from pulsars and radio galaxies is nonthermal emission. More generally, *nonthermal* means "not due to high temperature."

north celestial pole the point on the **celestial sphere** directly above the Earth's North Pole. The star Polaris happens to lie close to this point, and other objects in the northern sky appear to circle around this point.

North Star any star that happens to lie very close to the north celestial pole. Polaris has been the North Star for about 1000 years, and it will continue as such for about another 1000 years, at which time a star in Cepheus will be nearer the north celestial pole. (6.6)

nova a process in which a surface layer of hydrogen builds up on a white dwarf and then fuses rapidly into helium, causing the star to brighten for several weeks in an explosively luminous burst. Nova explosions may be recurrent. (63.1, 65.1)

nuclear fusion the binding of two lighter nuclei to form a heavier nucleus, with some nuclear mass converted to energy—for example, the fusion of hydrogen into helium. This process supplies the energy of most stars and is commonly called "burning," but burning is a chemical process that does not alter atoms' nuclei. (50.1)

nucleosynthesis the formation of elements, generally by the fusion of lighter elements into heavier ones—for example, the formation of carbon by the fusion of three helium nuclei. (66.1)

nucleus the core of an atom around which its electrons orbit. The nucleus has a positive electric charge and comprises most of an atom's mass. (4.2)

nucleus of a comet the core, typically a few kilometers across, of frozen gases and dust that make up the solid part of a comet. (47.1)

nucleus of a galaxy the central region of a galaxy. (70.4)

O

O-B-A-F-G-K-M-L-T the sequence of stellar spectral classifications from hottest to coolest stars. (55.3)

Occam's razor the principle of choosing the simplest scientific hypothesis that correctly explains any phenomenon. (11.2)

Olbers' paradox an argument that the sky should be bright at night because of the combined light from all the distant stars and galaxies. (80.1)

Olympus Mons a volcano on Mars, the largest volcanic peak in the Solar System. (40.1)

Omega (Ω) the ratio of the average density of matter or energy in the universe to the **critical density**. In a **flat universe**, $\Omega = 1$. (82.2)

Oort cloud a vast region in which comet nuclei orbit. This cloud lies far beyond the orbit of Pluto and may extend halfway to the next nearest star. (32.1, 47.2)

opacity the blockage of light or other electromagnetic radiation by matter.

open cluster a loose cluster of stars, generally containing a few hundred members. (69.1)

opposition the configuration of a planet when it is opposite the Sun in the sky. If a planet is in opposition, it rises when the Sun sets and sets when the Sun rises. (11.3)

orbit the path in space traveled by a celestial body.

orbital an energy state in which an electron can exist in an atom. (21.3)

orbital plane the flat plane defined by the path of an orbiting body. In the Solar System, most of the bodies orbiting the Sun have similar orbital planes. (32.2)

orbital velocity the speed needed to maintain a (generally) circular orbit. (18.1)

Orcus a recently discovered icy world orbiting the Sun at about the same distance as Pluto, with about two-thirds Pluto's diameter. (46.4)

organic compounds a compound containing carbon, especially a complex carbon compound; not necessarily produced by life. (48.2)

outer core the molten portion of the Earth's iron-nickel core surrounding the solid inner core. (35.1)

outer planet a planet whose orbit lies in the outer part of the Solar System. Jupiter, Saturn, Uranus, Neptune, and Pluto are outer planets. (32.1)

ozone a molecule consisting of three oxygen atoms bonded together. Its chemical symbol is O_3. Because it absorbs ultraviolet radiation, ozone in our planet's stratosphere shields us from the Sun's harmful ultraviolet radiation. Too much ozone in the lower atmosphere, however, is harmful to human respiration. (36.3)

P

pancake dome an unusual surface feature on Venus that may be produced by volcanic activity. (39.2)

panspermia a theory that life originated elsewhere than on Earth and came here across interstellar space either accidentally or deliberately. (84.0)

parallax the shift in an object's perceived position caused by the observer's motion. Also a method for finding distance based on that shift. (10.2, 52.1)

parsec a unit of distance equal to about 3.26 light-years (3.09×10^{13} km). It is the distance at which an object would have a parallax of one arc second. (3.3, 52.3)

partial eclipse an alignment of two celestial bodies during which only a part of the light from one body is blocked by the other. (8.2)

period–luminosity relation the relationship between the period of brightness variation and the luminosity of a variable star. Cepheid variables with longer periods are more luminous. (63.3)

perfect gas law see **ideal gas law.**

penumbra the outer part of the shadow of a body where sunlight is only partially blocked by the body. (8.2)

penumbral eclipse a lunar eclipse in which the Moon passes only through the Earth's **penumbra.** (8.2)

perihelion the point in a planet's orbit where it is closest to the Sun. (38.2)

period the time required for a repetitive process to repeat. For example, *orbital period* is the time it takes a planet or star to complete an orbit. *Pulsation period* is the time it takes a star to expand and then contract back to its original radius. (12.2, 63.1)

phases the changing illumination of the Moon or other body that causes its apparent shape to change. The following is the cycle of lunar phases: new, crescent, first quarter, gibbous, full, gibbous, third quarter, crescent, new. (8.1)

Phobos one of the two small moons of Mars, thought to be a captured asteroid. Its close orbit will cause it eventually to spiral into the planet. (40.5)

photo dissociation the breaking apart of a molecule by intense radiation.

photoelectric effect emission of an electrons from a material when light strikes it. Regardless of the brightness of the light, no electrons are emitted unless the photons' energy is greater than a value that depends on the material. (26.2)

photometer a device that measures the total amount of light received from a celestial object over a chosen range of wavelengths. (54.2)

photon a particle of visible light or other electromagnetic radiation. (4.2, 21.1)

photopigment a chemical that undergoes a chemical or physical change when light shines on it. Vision is possible because of cells in the eye that have photopigments, and the chlorophyll in plants is a photopigment. (31.1)

photosphere the visible surface of the Sun. When we look at the Sun in the sky, we are seeing its photosphere. (49.2, 55.1)

pixel a "picture element," consisting of an individual detector in an array of detectors used to collect light to construct an image. (26.2)

Planck time the brief interval of time, about 10^{-43} second immediately after the Big Bang, when quantum fluctuations are so large that current theories of gravity can no longer describe space and time. (81.4)

planet a body in orbit around a star. (1.1, 11.0)

planetary nebula a shell of gas ejected by a low-mass star late in its evolutionary lifetime. Planetary nebulae typically appear as a glowing gas ring around a central star. (59.2, 64.0)

planetesimal one of the numerous small, solid bodies that, when gathered together by gravity, form a planet. (33.3)

plasma a fully or partially ionized gas. (49.2)

plate tectonics a model of the movement of the Earth's crust, caused by slow circulation in the mantle. The Earth's surface is divided into large regions (plates) that move very slowly over the planet's surface. Interaction between plates at their boundaries creates mountains and activity such as volcanoes and earthquakes. (35.4)

Pleiades a nearby star cluster in the constellation Taurus. (69.1)

plutino a trans-Neptunian object whose orbital period (like that of Pluto) is in a 3:2 resonance with the orbit of Neptune. (46.4)

Pluto a small icy "planet" discovered in 1930. Recent discoveries of similar and larger bodies in the outer Solar System have ignited a debate about whether it should be categorized as a planet. (46.4)

Polaris a moderately bright star in the constellation Ursa Minor. Polaris is also known as the "North Star" because it currently lies near the north celestial pole. (5.3)

polarity the property of a magnet that causes it to have north and south poles.

poor cluster a galaxy cluster with few members. The galaxy cluster to which the Milky Way belongs, the Local Group, is a poor cluster. (76.2)

Population (Pop) I the younger stars, some of which are blue, that populate a galaxy's disk, especially its spiral arms. (71.1)

Population (Pop) II the older, redder stars that populate a galaxy's halo and bulge. (71.1)

Population (Pop) III a hypothetical stellar population consisting of the first stars that formed, composed of only hydrogen and helium. (71.3)

positive curvature bending of space leading to a finite volume. A **closed universe** with positive curvature is analogous to the surface of a sphere. (80.4)

positron a subatomic particle of antimatter with the same mass as an electron but a positive electric charge. It is the electron's antiparticle. (50.2, 81.2)

post meridian (P.M.) Latin for after noon— after the Sun has crossed the **meridian.** (7.1)

potential energy the energy stored in an object as a result of its position or arrangement. For example, a mass lifted to a greater height gains gravitational potential energy. (20.1)

powers-of-10 notation see **scientific notation.**

precession the slow change in the direction of the pole (rotation axis) of a spinning body. (6.6)

pressure the force exerted by a substance such as a gas on an area, divided by that area. That is, pressure = force/area. (49.1)

primary mirror the large, concave, light-gathering mirror in a reflecting telescope. (28.2)

prism a wedge-shaped piece of glass that is used to disperse white light into a spectrum. (28.3)

prokaryotes cells without nuclei. The first life forms on Earth were probably prokaryotes. (83.2)

prominence a cloud of hot gas in the Sun's outer atmosphere. This cloud is often shaped like an arch and is supported by the Sun's magnetic field. (51.1)

proper motion the position shift of a star on the celestial sphere caused by the star's own motion relative to the Sun. (52.4)

proper motion method a technique for detecting a planet orbiting another star by observing the side-to-side "wobble" of the star on the sky caused by the pull of the orbiting planet. (34.4)

proteinoids strings of amino acids that can form through chemical and physical interactions. A possible precursor to the origin of life. (83.2)

proton a positively charged subatomic particle. The constituent of an atom's nucleus that determines the type of the atom—any atom with six protons is a carbon atom, for example. (4.2, 50.1)

proton–proton chain the nuclear fusion process that converts hydrogen into helium in stars like the Sun and thereby generates their energy. This is the dominant energy generation mechanism in main-sequence stars with masses smaller than the Sun's. (50.2, 61.2)

protoplanetary disk a disk of material encircling a protostar or a newborn star. (34.1)

protostar a star still in its formation stage. (59.1, 60.1)

Proxima Centauri the nearest star to the Solar System at a distance of 1.30 parsecs. Despite its proximity, this M-type main-sequence star is too dim to be seen without a small telescope. (52.3)

Ptolemy, Claudius (about 100–170 C.E.) Roman astronomer who developed a geocentric model of the Solar System that could predict most of the observable motions of the planets. (11.2)

pupil the aperture of the eye, which can be adjusted in size to allow more or less light into the eye. (31.1)

pulsar a spinning **neutron star** that emits beams of radiation each time the star spins. These beams flow out along the directions of the poles of the strong magnetic fields of these stars. If the beam happens to point in the Earth's direction, we observe the radiation as regularly spaced pulses as the star spins. (67.1)

pulsating variable a variable star that expands and contracts in size, producing a repetitive pattern of variations in its luminosity and surface temperature. (63.1)

P wave a "primary" or "pressure" wave. A type of seismic wave produced in Earth by the compression of the material. (35.1)

Pythagoras (about 560–480 B.C.E.) ancient Greek mathematician who developed early ideas about the nature of the Earth and sky and geometrical techniques for studying them. (10.1)

Q

quadrature the points in the orbit of an outer planet when it appears (from Earth) to be at a 90° angle with respect to the Sun. From the outer planet at the same times, the Earth would appear to be at **greatest elongation.** (11.3)

quantized the property of a system that allows it to have only discrete values. Subatomic matter is generally quantized in nearly all of its properties. (21.1)

quantum fluctuation a temporary random change in the amount of energy at a point in space. These tiny subatomic variations are constantly occurring everywhere in space. (81.4)

quark a fundamental particle of matter that interacts via the strong force; basic constituent of protons and neutrons. (4.4, 81.3)

quarter moon a phase of the Moon when it is located 90° from the Sun in the sky. (8.1)

quasar a highly energetic source in the nucleus of a galaxy generally seen at a large redshift. Quasars are among the most luminous and most distant objects known to astronomers. (76.5, 77.2)

R

radial velocity the velocity of a body along one's line of sight. That is, the part of a body's motion directly toward or away from the observer. (25.0, 52.4)

radiant the point in the sky from which meteors in showers appear to originate. See also **meteor shower.** (47.5)

radiation era the period of time up until about 50,000 years after the Big Bang, when radiation rather than matter was the dominant constituent of the universe. (81.0)

radiation pressure the force exerted by photons when they strike matter. (47.3)

radiative zone the region inside a star where its energy is carried outward by radiation (that is, photons). In the Sun, this region extends about two-thirds of the way out from the core. (49.3)

radioactive decay the spontaneous breakdown of an atomic nucleus into smaller fragments, often involving the emission of other subatomic particles as well. (33.1, 35.3)

radioactive element an element that undergoes **radioactive decay** and breaks down into a lighter element. (35.3)

radio galaxy a galaxy, generally an elliptical system, that emits abnormally large amounts of radio energy. See also **radio lobe.** (77.1)

radio lobe a region lying outside the body of a radio galaxy where much of its radio emission comes from. Radio lobes are usually located on opposite sides of the galaxy and contain hot gas ejected from an **active galactic nucleus** into intergalactic space through **jets.** (77.1)

radio waves long-wavelength electromagnetic radiation. (22.5)

rays long, narrow, light-colored markings on the Moon or other bodies that radiate from young craters. Rays are debris "splashed" out of the crater by the impact that formed it. (37.1)

recession velocity the apparent speed of a galaxy (or other object) away from us, caused by the expansion of the universe. (74.3)

recurrent nova a white dwarf in a binary system which undergoes repeated **nova** outbursts. (65.1)

reddening the alteration in a star's color seen from Earth as the star's light passes through an intervening interstellar dust cloud. The dust preferentially scatters the star's blue light, leaving the remaining light redder. (72.3)

red giant a cool, luminous star with a radius much larger than the Sun's. Red giants are found in the upper right portion of the H-R diagram. (58.1, 59.2)

red giant branch the section of the evolutionary track of a star corresponding to intense hydrogen shell burning, which drives a steady expansion and cooling of the outer envelope of the star. As the star gets larger in radius and its surface temperature cools, it becomes a red giant. (62.1)

redshift a shift in the wavelength of electromagnetic radiation to a longer wavelength. For visible light, this implies a shift toward the red end of the spectrum. The shift can be caused by a source of radiation moving away from the observer or by the observer moving away from the source (see **Doppler shift**). Light traveling through expanding space undergoes an **expansion redshift,** and light traveling away from a strong gravitational source undergoes a **gravitational redshift.** (25.0, 74.3)

reflection nebula an interstellar cloud in which the dust particles reflect starlight, making the cloud visible. (72.1)

reflector a telescope that uses a mirror to collect and focus light. (28.2)

refraction the bending of light when it passes through one substance and enters another. (28.1)

refractor a telescope that uses a lens to collect and focus light. (28.1)

regolith the surface rubble of broken rock on the Moon or other solid body. (37.2)

regular orbit an orbit of a satellite that is nearly circular, in the same direction as the planet spins, and close to the plane of the planet's equator. (45.1)

resolution the ability of a telescope or other optical instrument to discern fine details of an image. (26.1, 29.0)

resonance a condition in which the repetitive motion of one body interacts with the repetitive motion of another to reinforce the motion. Sliding back and forth in a bathtub to make a big splash is an example. (41.4, 45.3)

rest frame a system of coordinates that appear to be at rest with respect to the observer. (53.1)

retina the back interior surface of the eye onto which light is focused. (31.1)

retrograde motion the westward shift of a planet against the background stars. Planets usually shift eastward because of their orbital motion, but they appear to reverse direction from our perspective when the Earth overtakes and passes them (or when an inferior planet overtakes and passes the Earth). (11.1)

retrograde spin a spin backward from the usual orbital direction. For example, seen from above their north poles, most of the planets orbit and spin in a counterclockwise direction; however, a few have a retrograde spin. (39.3)

revolve to orbit around another body. (6.1)

rich cluster a galaxy cluster containing hundreds to thousands of member galaxies. (76.2)

rifting the pulling apart of a geological plate by currents in the mantle beneath it. (35.4)

right ascension (RA) a coordinate for locating objects on the sky, analogous to longitude on the Earth's surface. Usually measured in hours and minutes of time. Together with **declination,** this defines the position of an object on the celestial sphere. (5.5, 31.3)

rille narrow canyon on the Moon or other body. (37.1)

ring (planetary) consists of numerous small particles orbiting a planet within its Roche limit. (45.0)

ring galaxy a galaxy in which the central region has an abnormally small number of stars, causing the galaxy to look like a ring. Caused by the collision of two galaxies. (75.4)

ringlets any one of numerous, closely spaced, thin bands of particles in Saturn's ring system. (45.3)

ring system material organized into thin, flat rings encircling a giant planet, such as Saturn. (45.3)

Roche limit the distance from an astronomical body at which its gravitational force breaks up another astronomical body. (45.4)

Roche lobe the region around a star in a binary system in which the gravity of that star dominates. Matter pushed outside the Roche lobe (which may occur during a star's red giant phase) may be captured by the companion star. (65.1)

rods light-sensing cell in the eye that gives us our black-and-white vision. (31.1)

rotate to spin on an axis. The distinction between revolving and rotating can be confusing—a galaxy may rotate, but the stars in it revolve around the center. (6.3)

rotation axis an imaginary line through the center of a body about which the body spins. Analogous to the handle and point of a spinning top. (6.3)

rotation curve a plot of the orbital velocity of the stars or gas in a galaxy at different distances from its center. (73.1, 78.1)

round off to express a number with fewer digits by replacing the trailing digits with the nearest decimal value. For example, 12.34567 can be rounded off to 12.346, to 12.3, or to 12 depending on the precision needed. (3.4)

RR Lyrae variables a type of pulsating variable star with a period of about one day or less. They are named for their prototype star in the constellation Lyra, RR Lyrae. (63.2)

runaway greenhouse effect an uncontrollable process in which the heating of a planet leads to an increase in its atmospheric greenhouse effect and thus to further heating. The process significantly alters the composition of the planet's atmosphere and the temperature of its surface. (39.1)

S

S0 galaxy galaxy that shows evidence of a disk and a bulge, but that has no spiral arms and contains little or no gas. (75.1)

Sagittarius A* (Sgr A*) the powerful radio source located at the core of the Milky Way Galaxy. (73.2)

satellite a body orbiting a planet. (1.2, 32.1)

scarp a cliff produced by vertical movement of a section of the crust of a planet or satellite. (38.1)

scattering the random redirection of a light wave or photon as it interacts with atoms or dust particles in its path. (72.3)

Schwarzschild radius the distance from the center of a black hole to its **event horizon.** No light (or anything else) can escape from within the Schwarzschild radius. (68.1)

scientific method the process of observing phenomena, proposing hypotheses to explain them, and testing the hypotheses. Scientifically valid hypotheses must offer a means of testing and rejecting them. (4.1)

scientific notation a shorthand way to write numbers using 10 to a power. For example, $1,000,000 = 10^6$. Numbers that are not simple powers of 10 are written as a number between 1 and 10 and 10 to a power—for example, $123.45 = 1.2345 \times 10^2$. (3.2)

scintillation the twinkling of stars, caused as their light passes through moving regions of different density in the Earth's atmosphere. (30.2)

secondary mirror in a reflecting telescope, a mirror that directs the light from the primary mirror to a focal position. (28.2)

Sedna a large body (about two-thirds the diameter of Pluto) orbiting far beyond Pluto's orbit. (46.4)

seeing a measure of the steadiness of the atmosphere during astronomical observations. Bad seeing results from **scintillation** and makes fine details difficult to observe. (30.2)

seismic waves waves generated in the Earth's interior by earthquakes. Similar waves occur in other bodies. Two of the more important varieties are S and P waves. S waves can travel only through solid material; P waves can travel through either solid or liquid material. (35.1)

selection effect an unintentional selection process that omits some set of the objects being studied and leads to invalid conclusions about the objects. For example, selecting the brightest stars in the night sky does not provide a representative sample of all star types. (69.3)

self-propagating star formation a model that explains a galaxy's spiral arms as arising from the explosion of massive stars, triggering the birth of other stars around them. The resulting pattern is then drawn out into a spiral by the galaxy's differential rotation. (73.4)

semimajor axis half the long dimension of an ellipse. (12.2)

SETI Search for Extra Terrestrial Intelligence. Some SETI searches involve automated "listening" to millions of radio frequencies for signals that might be from other civilizations. (84.4)

Seyfert galaxy a type of galaxy with a small, abnormally bright **active galactic nucleus.** Named for the astronomer Carl Seyfert, who first drew attention to these objects. (77.1)

shell fusion nuclear energy generation in a region surrounding the core of a star rather than in the core itself. This occurs during the red giant phase of stellar evolution. (62.1)

shepherding satellite a satellite that by its gravitational attraction prevent particles in a planet's rings from spreading out and dispersing. Saturn's F-ring is held together by shepherding satellites. (45.3)

shielding the process in which dust particles in an interstellar cloud block light from entering the interior of the cloud, thereby allowing it to cool. (72.2)

short-period comet a comet whose orbital period is shorter than 200 years. For example, Halley's comet has a period of 76 years. (47.4)

sidereal day the length of time from the rising of a star until it next rises. The length of the Earth's sidereal day is 23 hours 56 minutes. (7.1)

sidereal month the length of time (about 27.3 days) required for the Moon to return to the same apparent position among the stars. Compare with **lunar month.** (8.1)

sidereal period the time it takes a body to turn once on its rotation axis or to revolve once around a central body, as measured with respect to the stars.

sidereal time a system of time measurement referenced to the motion of stars across the sky rather than of the Sun. (7.1)

significant digits all the leftmost digits in a number (excluding any leading zeros) whose values are well determined by measurement or calculation. For example, someone's height might be measured as 1.93524 meters, but if the accuracy of the measurement is only about a millimeter (0.001 meters), the significant digits are 1.935. (3.4)

silicate mineral composed primarily of silicon and oxygen. Most ordinary rocks are silicates. For example, quartz is silicon dioxide, and a wide variety of minerals are made of silicates containing an additional element. (35.1)

singularity a theoretical point of zero volume and infinite density to which any object that becomes a black hole must collapse, according to the **general theory of relativity.** (68.1)

solar activity phenomena that cause changes in the appearance or energy output of the solar atmosphere, such as sunspots and flares. (51.0)

solar cycle the cyclic change in solar activity, such as that of sunspots and of solar flares, that increases and decreases every 11 years. (51.3)

solar day the time interval from one sunrise to the next sunrise or from one noon to the next noon. That interval is not always exactly 24 hours but varies throughout the year. For that reason, we use the **mean solar day** (which, by definition, is 24 hours) to keep time. (7.1)

solar eclipse the passage of the Moon between the Earth and the Sun so that our view of the Sun's photosphere is partially or totally blocked. During a **total eclipse** of the Sun we can see the Sun's chromosphere and corona. See also **total eclipse.** (8.2)

solar nebula the rotating disk of gas and dust from which the Sun and planets formed. (33.2)

solar nebula theory the theory that the Sun and planets formed all at approximately the same time from a rotating cloud of gas and dust. (33.2)

solar seismology the study of pulsations or oscillations of the Sun to determine the characteristics of the solar interior. (49.6)

Solar System the Sun, planets, their moons, and other bodies, such as meteors and comets, that orbit the Sun. (1.5, 32.0)

solar wind the outflow of low-density, hot gas from the Sun's upper atmosphere. It is partially this wind that creates the tail of a comet by pushing a comet's gases away from the Sun. (47.3, 49.4)

solid core see **inner core.**

solstice (winter and summer) the beginning of winter and summer. The solstice occurs when the Sun is at its greatest distance north (about June 21) or south (about December 21) of the celestial equator. (6.5)

south celestial pole the imaginary point on the **celestial sphere** directly over the Earth's South Pole. Objects on the sky appear to circle around this point as viewed from Earth's Southern Hemisphere.

special relativity a theory developed by Einstein to explain why light is always measured to travel through space at the same speed—regardless of the motion of the source of light or the observer. This very well-established theory shows that the measurements of lengths, times, masses, and other quantities change depending on the speed of the observer relative to the object being measured. (53.0)

speckle interferometry a technique to overcome the limitations imposed by the effects of scintillation in our atmosphere by analyzing a set of images collected over very short time intervals. (57.1)

spectral type an indicator of a star's temperature. A star's spectral type is based on the appearance of its spectral lines, with different strengths and weaknesses indicating stellar composition and temperature. The fundamental types are, from hot to cool, O, B, A, F, G, K, M, L and T. (55.3)

spectrograph a device for making a spectrum by spreading light out into its component wavelengths.

spectroscopic binary a type of binary star in which the spectral lines exhibit alternating red and blue Doppler shifts. This is caused by the orbital motion of one star around the other, causing the stars to alternately move away from us and back toward us. (56.1)

spectroscopy the study and analysis of spectra. (24.0)

spectrum electromagnetic radiation (for example, visible light) spread into its component wavelengths or colors. A rainbow is a spectrum produced naturally by water droplets in our atmosphere. (55.0)

spicule a thin column of hot gas in the Sun's chromosphere. (49.4)

spindown the slowing down of a body's rotation, often referring to pulsars. (67.2)

spiral arm a long, narrow region containing young stars and interstellar matter that winds outward in the disk of spiral galaxies. (70.4)

spiral density wave a mechanism for the generation of spiral structure in galaxies: A density wave interacts with interstellar matter and triggers the formation of stars. Spiral density waves are also seen in the rings of Saturn. (45.3)

spiral (S) galaxy a galaxy with a disk in which its brighter stars form a spiral pattern of two or more **spiral arms.** (75.1)

spring tide the strong tides that occur at new and full moon, when the Moon's and Sun's tidal forces work in the same direction to make the tides more extreme. (19.2)

standard candle a type of star or other astronomical light source in which the luminosity has a known value, allowing its distance to be determined by measuring its apparent brightness and applying the **inverse-square law.** Good examples include Cepheid variable stars and Type Ia supernovae. (54.3)

standard model the current theoretical model that describes the fundamental particles and forces in nature. (4.4)

standard time an agreed common time kept within a given region so that all clocks in that region agree. (7.3)

star a massive, gaseous body held together by gravity and generally emitting light. Stars generate energy by nuclear reactions in their interiors. (1.4)

starburst galaxies galaxies in which a very large number of stars have formed recently. (75.4)

star cluster a group of stars numbering from hundreds to millions held together by their mutual gravity. (69.0)

steady-state theory a model of the universe in which the average properties of the universe do not change with time. An early competitor of the **Big Bang** model.

Stefan-Boltzmann constant constant that appears in the laws of thermal radiation (23.5)

Stefan-Boltzmann law the amount of energy radiated by one square meter in one second by a hot, dense material depends on

the temperature T raised to the fourth power (σT^4). (23.5, 57.3)

stellar association an extended, loose grouping of young stars and interstellar matter. (69.1)

stellar evolution the gravity-driven changes in stars as they are born, age, and finally run out of fuel. (2.1, 59.0)

stellar models the result of a theoretical calculation of the physical conditions in the different layers of a star's interior. (59.1)

stellar remnant the body remaining after a star ceases supporting itself with the heat and pressure generated by nuclear fusion in its interior, such as a white dwarf, neutron star, or black hole. (59.1)

stellar spectroscopy the study of the properties of stars based on information that can be learned from their spectra. (55.3)

stellar wind an outflow of gas from a star, sometimes at speeds reaching hundreds of kilometers per second. (59.2, 64.2)

stratosphere the region of the Earth's atmosphere extending from about 12 to 50 km above the surface. A protective layer of **ozone** is located in the stratosphere. (36.1)

stromatolites layered fossil formations caused by ancient mats of algae or bacteria, which build up mineral deposits season after season. (83.1)

strong force the force that binds quarks together and holds protons and neutrons in an atomic nucleus. Sometimes called the *nuclear force.* (4.3, 50.1)

subatomic particles particles making up an atom, such as electrons, neutrons, and protons, as well as other particles of similar small size.

subduction the sinking of one crustal plate under another where they are driven together by plate tectonics. (35.4)

sublimate to change directly from a solid into a gas without passing through a liquid phase. (47.2)

sunspot a dark, cooler region on the Sun's visible surface created by intense magnetic fields. (12.3, 51.1)

supercluster a cluster of galaxy groups and clusters. Our Milky Way belongs to the Local Group, which is but one of many galaxy

groups making up the Local Supercluster. (2.4, 76.0)

supergiant a very large-diameter and luminous star, typically at least 10,000 times the Sun's luminosity. (58.3, 64.1)

superior conjunction see **conjunction.**

superior planet a planet orbiting farther from the Sun than the Earth. Mars, Jupiter, Saturn, Uranus, Neptune, and Pluto are superior planets. (32.1)

supernova an explosion marking the end of most massive stars' evolution. Type Ia supernovae occur in a binary system in which one star is a white dwarf. The explosion is triggered when mass from a companion star falls onto the white dwarf, raising its mass above the **Chandrasekhar limit** and causing the star to collapse. Collapse heats the white dwarf so that its carbon and oxygen fuse explosively, destroying the star and leaving no remnant. Type II (and Types Ib and Ic) supernovae probably occur when a massive star's iron core collapses. A Type II supernova leaves either a neutron star or, if the collapsing core is massive enough, a black hole. (59.2, 63.1, 66.2)

supernova remnant the debris ejected from a star when it explodes as a supernova. Typically this material is hot gas, expanding away from the explosion at thousands of kilometers per second. (65.3, 66.4)

surface brightness the amount of light emitted by a source divided by the area from which the light is emitted. (31.5, 74.1)

surface gravity the acceleration an object will experience near the surface of a planet (or other body) because of the planet's gravitational pull. (16.3)

S wave a "secondary" or "shear" wave. A type of seismic wave produced by side-to-side motion. S waves can travel only through solid material and move more slowly than **P waves.** (35.1)

synchronous rotation the condition that a body's rotation period is the same as its orbital period. The Moon rotates synchronously. (37.4)

synchrotron radiation a form of nonthermal radiation emitted by charged particles spiraling at nearly the speed of light in a magnetic field. Pulsars and radio galaxies emit synchrotron radiation. The radiation gets its name because it was first seen in synchrotrons, a type of atomic accelerator. (67.2, 77.1)

synodic period the time between successive configurations of a planet or moon. For example, the time between oppositions of a planet or between full moons. (13.4)

T

tail the plumes of gas and dust from a comet. These are produced by the solar wind and radiation pressure acting on gas and dust evaporated from the comet's nucleus. The **dust tail** and **ion tail** point away from the Sun and get longer as the comet approaches perihelion.

telescope an instrument for gathering and focusing light and magnifying the resulting image of remote objects. (26.1)

terrestrial planet a rocky planet similar to the Earth in size and structure. The terrestrial planets are Mercury, Venus, Earth, and Mars. (32.1)

Tharsis bulge a large volcanic region on Mars rising about 10 km above surrounding regions. (40.1)

theory a hypothesis or set of hypotheses that have become well-established through repeated and diverse testing. (4.1)

thermal energy the kinetic energy associated with the motions of the molecules or atoms in a substance. (20.1, 23.3)

thermal radiation the continuous spectrum of electromagnetic radiation emitted as a result of the thermal energy in any relatively dense material. Also see **blackbody**. (23.0)

thermonuclear runaway a condition in which a nuclear reaction generates enough heat to cause an increasing rate of reaction, rapidly accelerating into an explosion. (65.1)

third-quarter moon a phase of the waning Moon when Earth-based observers see half of the Moon's illuminated hemisphere. This occurs three-quarters of the way through the lunar month. (8.1)

tidal braking the slowing of one body's rotation as a result of gravitational forces exerted on it by another body. (37.4)

tidal bulge a bulge on one body created by the gravitational attraction on it by another. Two tidal bulges form, one on the side near the attracting body and one on the opposite side. (19.1)

tidal force relative forces arising between the parts of a body as a result of differences in the strength of the gravitational attraction by another body. (19.1)

tidal friction friction within an object that is caused by a tidal force. (7.5)

tidal lock circumstance in which tidal forces have caused a moon to rotate at exactly the same rate at which it revolved around its parent planet, so that the moon always keeps the same face turned toward the planet. See also **synchronous rotation.** (38.4)

tides the rise and fall of the Earth's oceans created by the gravitational attraction of the Moon. Tides also occur in the solid crust of a body and its atmosphere. (19.0)

time zone one of 24 divisions of the globe at every 15 degrees of longitude. In each zone, a single standard time is kept. Most zones have irregular boundaries, usually because they follow geopolitical borders. See also **universal time.** (7.3)

Titan the largest moon of Saturn, even larger than Mercury. Titan has a substantial atmosphere and "weather," but its surface temperature is far colder than the terrestrial planets. (46.2)

torque a twisting force that can change an object's angular momentum. (20.2)

total eclipse an eclipse in which the eclipsing body completely covers the other body. (8.2)

transit the passage of a planet directly between the observer and the Sun. At a transit, we see the planet as a dark spot against the Sun's bright disk. From Earth, only Mercury and Venus can transit the Sun. (52.1)

transit method a method for detecting planets orbiting other stars by detecting the slight dimming if the planet's orbit causes it to cross in front of the star. (34.2)

trans-Neptunian objects (TNOs) objects orbiting in the **Kuiper belt** and outer Solar System beyond Neptune's orbit. This includes Pluto and many other objects ranging up to sizes even larger than Pluto. (32.1, 46.4)

triangulation a method for measuring distances. This method is based on constructing a triangle, one side of which is the distance to be determined. That side is then calculated by measuring another side (the baseline) and the two angles at either end of the baseline. (52.1)

trigonometry a mathematical method for relating the sizes of the sides and angles of a triangle. (52.1)

triple alpha process the fusion of three helium nuclei (**alpha particles**) into a carbon nucleus. This process is sometimes called *helium burning*, and it occurs in many old stars. (62.2)

Triton the largest moon of Neptune, somewhat larger than and probably fairly similar to Pluto. Triton shows evidence of a thin atmosphere and some volcanic activity. (46.3)

Tropic of Cancer the latitude line of 23.5° north, marking the distance farthest north where the Sun can pass directly overhead (on the summer solstice). (6.5)

Tropic of Capricorn the latitude line of 23.5° south, marking the distance farthest south where the Sun can pass directly overhead (on the winter solstice). (6.5)

troposphere the lowest layer of the Earth's atmosphere, extending up to an elevation of about 12 km, within which convection produces weather. (36.1)

true binary a pair of stars that actually orbit each other as opposed to just appearing to be close to each other on the sky (see **apparent double star**). (56.1)

T Tauri star a type of extremely young star or protostar that varies erratically in its light output as it settles toward the main-sequence phase. (60.2)

Tunguska event a large aerial explosion that occurred over Siberia in 1908, probably when a large meteor entered the atmosphere and heated up and exploded before reaching the ground. (48.4)

turnoff point the location on the main sequence where a star's evolution causes it to move away from the main sequence toward the red giant region. The location of the turnoff point can be used to deduce the age of a star cluster. (69.2)

21-centimeter radiation radio emission from a hydrogen atom caused by the flip of the electron's spin orientation. (72.4)

twin paradox a supposed paradox in special relativity arising from the differences in time measurement for two observers moving at speeds relative to each other. The paradox is usually phrased in terms of the relative aging of one twin taking a trip in a spacecraft while the other remains on Earth. Special relativity (and experiment) shows that there is no paradox—the twin on the spacecraft ages less. (53.5)

2003 UB313 the temporary designation of a body discovered in 2005 and orbiting beyond Pluto's orbit. This body appears to be larger than Pluto, but there is controversy over whether it should be called a planet. It had not been named as of late 2005. (46.4)

Type Ia supernova an extremely energetic explosion produced by the abrupt fusion of carbon and oxygen in the interior of a collapsing white dwarf star. See **supernova.** (65.3)

Type II supernova see **supernova.**

U

ultraviolet a portion of the electromagnetic spectrum with wavelengths shorter than those of visible light but longer than those of X-rays. By convention, the ultraviolet region extends from about 10 to 300 nm. (22.3)

umbra The inner portion of the shadow of a body, within which sunlight is completely blocked. (8.2)

uncompressed density the density a planet would have if its gravity did not compress it. (42.3)

unit a quantity used for reporting measurements. (1.6, 3.0)

universality the assumption that the physical laws observed on Earth apply everywhere in the universe. (4.2)

universal time the time kept at Greenwich, England. Universal time is the same as Greenwich mean time. This is the starting point from which the Earth's 24 **time zones** are calculated. (7.3)

universe the largest astronomical structure we know of. The universe contains all matter and radiation and encompasses all known space. (2.4)

Ursa Major group the nearest star cluster to the Solar System, including the stars of the Big Dipper and many other bright stars over a large part of our sky. (69.1)

V

Valles Marineris a huge canyon feature on Mars stretching thousands of kilometers. (40.1)

valve mechanism a process occurring in some variable stars in which conditions in the atmosphere can cause radiation to be trapped (like a closed valve), making the atmosphere heat and expand. The radiation is no longer trapped when the atmosphere is expanded enough (open valve), and the star shrinks back down in size. (63.2)

Van Allen radiation belts doughnut-shaped regions surrounding the Earth containing charged particles trapped by the Earth's magnetic field.

variable star a star whose luminosity changes over time. (31.2, 59.2, 63.0)

velocity a physical quantity that indicates both the speed of a body and the direction in which it is moving. (14.2)

vernal equinox or spring **equinox.** Spring in the Northern Hemisphere begins on the vernal equinox, which is on or near March 21. (6.5)

Virgo Cluster the nearest large galaxy cluster. The gravity of the thousands of galaxies in this cluster is predicted to eventually pull in the Milky Way and Local Group. (2.4)

virtual particle one of a particle or its antiparticle, which are created simultaneously in pairs and which quickly disappear. Virtual particles are created from **quantum fluctuations.** (68.3)

visible spectrum the part of the electromagnetic spectrum that we can see with our eyes. It consists of the familiar colors red, orange, yellow, green, blue, and violet. (22.1)

visible universe the portion of the universe in which light has had time to reach us within the history of the universe. This region is limited, therefore, to distances for which the travel time of the light is less than about 13.7 billion years. (80.1)

visual binary star a pair of stars, held together by their mutual gravity and in orbit about each other, that can be seen with a telescope as separate objects. (56.1)

void a region between clusters and superclusters of galaxies that is relatively empty of galaxies. (76.4)

volatile element or compound that vaporizes at low temperature. Water and carbon dioxide are examples of volatiles. (32.3)

vortex a strong spinning flow within a gas or liquid, such as the Great Red Spot on Jupiter. (43.3)

W

wane to gradually decrease, as in the "waning crescent Moon" or the "waning gibbous Moon." (8.1)

waterhole the interval of the radio spectrum between the 21-cm hydrogen radiation and the 18-cm OH radiation. Proposed as likely wavelengths to use in the search for extraterrestrial life. (84.4)

water line the distance from the Sun in the solar nebula beyond which it was cold enough for water to condense and therefore become a major component of planetesimals forming there. (42.4)

wavelength the distance between wave crests. Wavelength determines the color of visible light and is generally denoted by the Greek letter lambda (λ). (22.1)

wave–particle duality the theory that electromagnetic radiation may be treated as either a particle (photon) or an electromagnetic wave. All subatomic particles appear to have both wavelike and particlelike properties. (21.1)

wax to gradually increase, as in the "waxing crescent Moon" or the "waxing gibbous Moon." (8.1)

weak force the force responsible for radioactive decay of atoms. Now known to be linked to electric and magnetic forces in what is called the **electroweak force.** (4.3, 50.4)

weakly interacting massive particle (WIMP) a hypothetical class of subatomic particles that interacts only through the weak force, making it difficult to detect except by gravitational interactions and a possible dark matter candidate. (78.4)

weight the gravitational force exerted on a body by the Earth (or another astronomical object). Sometimes more broadly used to indicate the net force on a body after other forces (centripetal forces, buoyancy, and so on) and effects of the motion of the surrounding rest frame are accounted for. Thus, astronauts in an orbiting spacecraft are "weightless" because their spacecraft is in orbit; however the gravitational force on them is only slightly less than on the Earth's surface. (14.1)

white dwarf a dense star whose radius is approximately the same as the Earth's but whose mass is comparable to the Sun's. White dwarfs burn no nuclear fuel and shine by residual heat. They are the end stage of stellar evolution for low-mass stars like the Sun. (58.1, 59.2, 64.0, 65.0)

white light visible light exhibiting no color of its own but composed of a mix of all colors. Many artificial light sources appear "white"; however, our eyes also adapt to color differences. (22.2)

Wien's law a relation between a body's temperature and the wavelength at which it emits radiation most intensely. Hotter bodies radiate more intensely at shorter wavelengths. Mathematically, the law states that $\lambda_{max} = 2.9 \times 10^6 / T$, where λ_{max} is the wavelength of maximum emission in nanometers, and T is the body's temperature on the Kelvin scale. (23.2)

X

X-ray the part of the electromagnetic spectrum with wavelengths longer than gamma rays but shorter than ultraviolet. (22.6)

X-ray binaries a binary star system in which one of the stars, or the gas associated with a star, emits X-rays intensely. Such systems generally contain a collapsed object such as a neutron star or a black hole. See also **X-ray pulsars.** (68.3)

X-ray pulsars neutron stars from which periodic bursts of X-rays are observed. These are thought to consist of a neutron star and a normal star in a close binary system. Mass from the normal star spills onto the neutron star, where it slowly accumulates and heats. Eventually, the temperature of the neutron star becomes large enough to initiate nuclear fusion. The released energy heats the material further, causing more fusion and leading to an explosion that we observe as the X-ray burst. (67.3)

Y

year the time that it takes the Earth to complete its orbit around the Sun; that is, the **period** of the Earth's orbit.

yellow giant a phase stars pass through after moving off the main sequence, and sometimes again after starting helium fusion in their cores. Many yellow giants become unstable and pulsate as they cross the **instability strip.** (63.2)

yellow supergiant a yellow giant star with a luminosity greater than about 10,000 solar luminosities. (63.2)

Z

Zeeman effect the splitting of a single spectral line into two or three lines by a magnetic field. The Zeeman effect allows astronomers to detect magnetic fields in objects from the appearance of their spectral lines. (51.1)

zenith the point on the celestial sphere that lies directly overhead from the observer's location. (5.2)

zero-age main sequence main sequence on the H-R diagram for a system of stars that have completed their contraction from interstellar matter and are just beginning to derive all their energy from hydrogen-to-helium fusion in their core. (61.4)

zodiac a set of 12 constellations in a band running around the celestial sphere around the ecliptic. The planets move through the constellations of the zodiac as they orbit the Sun, as does the Moon each month and the Sun over the year. (6.2)

zone a bright, high-pressure region in the atmosphere of a Jovian planet, where gas flows upward. (43.1)

zone of avoidance a band running around the sky in which few galaxies are visible. It coincides with the band of the Milky Way and is caused by dust that is within our galaxy. This dust blocks the light from distant galaxies. (74.1)

ANSWERS TO TEST YOURSELF

Part One

Unit 1: 1–e; 2–e; 3–d.
Unit 2: 1–c; 2–e; 3–b.
Unit 3: 1–a; 2–d; 3–b.
Unit 4: 1–b; 2–d; 3–a.
Unit 5: 1–b; 2–c; 3–d.
Unit 6: 1–e; 2–a; 3–c.
Unit 7: 1–d; 2–a; 3–c.
Unit 8: 1–b; 2–b; 3–b.
Unit 9: 1–b; 2–a; 3–a.
Unit 10: 1–a; 2–d; 3–a.
Unit 11: 1–b; 2–a; 3–a.
Unit 12: 1–e; 2–a; 3–b.
Unit 13: None

Part Two

Unit 14: 1–e; 2–d; 3–d.
Unit 15: 1–c; 2–d; 3–b.
Unit 16: 1–c; 2–d; 3–d.
Unit 17: 1–e; 2–c; 3–d.
Unit 18: 1–b; 2–c; 3–e.
Unit 19: 1–a; 2–c; 3–c.
Unit 20: 1–b; 2–c; 3–b.
Unit 21: 1–b; 2–c; 3–b.
Unit 22: 1–e; 2–c; 3–a.
Unit 23: 1–c; 2–d; 3–e.
Unit 24: 1–d; 2–d; 3–e.
Unit 25: 1–a; 2–d; 3–b.
Unit 26: 1–d; 2–b; 3–c.
Unit 27: 1–b 2–a; 3–c.
Unit 28: 1–b; 2–c; 3–b.

Unit 29: 1–c; 2–d; 3–a.
Unit 30: 1–a; 2–c; 3–d.
Unit 31: 1–b; 2–c; 3–a.

Part Three

Unit 32: 1–c; 2–a; 3–b.
Unit 33: 1–c; 2–e; 3–d.
Unit 34: 1–b; 2–d; 3–a.
Unit 35: 1–a; 2–e; 3–d; 4–d.
Unit 36: 1–b; 2–a; 3–b.
Unit 37: 1–b; 2–a; 3–d.
Unit 38: 1–d; 2–a; 3–c.
Unit 39: 1–c; 2–a; 3–b.
Unit 40: 1–c; 2–b; 3–a.
Unit 41: 1–d; 2–d; 3–a.
Unit 42: 1–b; 2–d; 3–d.
Unit 43: 1–d; 2–c; 3–a; 4–b
Unit 44: 1–b; 2–c; 3–c.
Unit 45: 1–a; 2–a; 3–d.
Unit 46: 1–c; 2–a; 3–d.
Unit 47: 1–e; 2–b; 3–d.
Unit 48: 1–d; 2–c; 3–c.

Part Four

Unit 49: 1–d; 2–a; 3–b.
Unit 50: 1–d; 2–d; 3–e.
Unit 51: 1–c; 2–a; 3–b.
Unit 52: 1–d; 2–a; 3–d.
Unit 53: 1–e; 2–c; 3–b.
Unit 54: 1–a; 2–e; 3–d.

Unit 55: 1–e; 2–e; 3–a.
Unit 56: 1–b; 2–e; 3–a.
Unit 57: 1–e; 2–c; 3–c.
Unit 58: 1–a; 2–b; 3–e.
Unit 59: 1–e; 2–a; 3–c.
Unit 60: 1–e; 2–b; 3–d.
Unit 61: 1–d; 2–a; 3–d.
Unit 62: 1–b; 2–d; 3–e.
Unit 63: 1–d; 2–e; 3–e.
Unit 64: 1–e; 2–c; 3–a.
Unit 65: 1–a; 2–b; 3–c.
Unit 66: 1–b; 2–c; 3–c.
Unit 67: 1–d; 2–b; 3–c.
Unit 68: 1–c; 2–e; 3–d.
Unit 69: 1–a; 2–d; 3–b.

Part Five

Unit 70: 1–b; 2–b; 3–b.
Unit 71: 1–a; 2–e; 3–b.
Unit 72: 1–e; 2–b; 3–c.
Unit 73: 1–e; 2–a; 3–d.
Unit 74: 1–b; 2–c; 3–c.
Unit 75: 1–d; 2–b; 3–d.
Unit 76: 1–e; 2–e; 3–a.
Unit 77: 1–b; 2–a; 3–e.
Unit 78: 1–e; 2–d; 3–e.
Unit 79: 1–b; 2–b; 3–a.
Unit 80: 1–e; 2–a; 3–b.
Unit 81: 1–c; 2–c; 3–a.
Unit 82: 1–e; 2–b; 3–a.
Unit 83: 1–a; 2–c; 3–d.
Unit 84: 1–d; 2–d; 3–a.

CREDITS

Photographs

Unit 1 Figure 1.1: Courtesy of NASA; 1.2a: Courtesy of Stephen E. Strom; 1.2b: Dr. F. A. Ringwald; 1.3(all): Pictures of planets courtesy of NASA/JPL; 1.4 (Jupiter): Courtesy of NASA/JPL/University of Arizona; (Earth): Courtesy of NASA; (Sun): Courtesy of SOHO, ESA & NASA.

Unit 2 Figure 2.1a: Atlas Image courtesy of 2MASS/UMass/IPAC - Caltech/NASA/NSF; 2.1b: Image adapted from NASA; 2.2a: Courtesy of Anglo-Australian Observatory, photograph by David Malin; 2.2b: Courtesy of Anglo-Australian Telescope Board; 2.3: Courtesy of George Greany; 2.5: Color image creation by Jean-Charles Cuillandre, The Anada-France-Hawaii Telescope Corp. Copyright © 2001 CFHT; 2.6: Courtesy of NASA.

Unit 3 Figure 3.1: © The Bridgeman Art Library/Getty.

Unit 5 Figure 5.2a(left): From Roger Ressmeyer, digitally enhanced by Jon Alpert; 5.2a(right): Courtesy of Eugene Lauria; 5.2b(both): Courtesy of Eugene Lauria.

Unit 6 Figure 6.1a: Courtesy of Tony Stone/Rob Talbot; 6.1b, 6.9b: © English Heritage Library.

Unit 8 Figure 8.1: Courtesy of Antonio Cidadao; 8.3b: © CNES/Jean-Pierre Haignere', 1999; 8.3c: © Roger Ressmeyer/CORBIS; 8.4b: John Walker.

Unit 9 Figure 9.2: Steve Schneider; 9.3: Chinese Rare Book Collection/Asian Division/Library of Congress; 9.4: © Araldo de Luca/CORBIS; 9.5: © Bettmann/CORBIS.

Unit 10 Figure 10.1b: Rod Somerville.

Unit 11 Figure 11.6: © Art Resource/Erich Lessing.

Unit 12 Figure 12.1: © Bettmann/CORBIS; 12.2: © Stapleton Collection/CORBIS; 12.3: © Art Resource/Erich Lessing; 12.6: Courtesy of the Bridgeman Art Library/Getty.

Unit 13 Figure 13.5: © Hermann Eisenbaiss/Photo Researchers, Inc.

Unit 14 14.2: © Bettmann/CORBIS.

Unit 15 Figure 15.4: © Stone/Getty Images.

Unit 16 Figure16.3: NASA Johnson Space Center (NASA-JSC).

Unit 19 Figure 19.3b(both): Tom Arny.

Unit 22 Figure 22.1: © Ewing Galloway, Inc.; 22.4(Pulsar): Courtesy of NASA/CXC/SAO; 22.4(Sun): SOHO (ESA & NASA); 22.4(Other Stars): Courtesy of NOAO and Anglo-Australian Observatory, photograph by David Malin; 22.4(Interstellar cloud): Courtesy of Anglo-Australian Observatory; photograph by David Malin; 22.4(Active galaxy): NRAO/AUI/NSF.

Unit 23 Figure 23.4: Spitzer Science Center and the Infrared Processing and Analysis Center; 23.5: Courtesy of SOHO/MDI consortium. SOHO is a project of international cooperation between ESA and NASA.

Unit 24 Figure 24.7a(top): © Courtesy of Stephen M. Larson, University of Arizona; 24.7a (bottom): © Courtesy of Mees Solar Observatory, University of Hawaii.

Unit 26 Figure 26.1: © Roger Ressmeyer/CORBIS; 26.4a: Courtesy of NRAO/AUI; 26.4b: Courtesy of NASA/CXC/SAO/Rutgers/J. Hughes; 26.5b: Courtesy of Richard Wainscoat; 26.5c: Courtesy of NRAO; 26.5d(X-ray): NASA/CXC/ASU/J. Hester et al.; 26.5d(Optical): NASA/HST/ASU/J. Hester et al.

Unit 27 Figure 27.1: Courtesy of California Association for Research in Astronomy; 27.2: © European Southern Observatory; 27.3: Courtesy of the NAIC - Arecibo Observatory, a facility of the NSF.

Unit 28 Figure 28.2: Tom Arny; 28.4: Photo courtesy of Yerkes Observatory; 28.8a: © Roger Ressmeyer/CORBIS; 28.8b: US Gemini Project/AURA/NOAO/NSF; 28.9:

The Video Encyclopedia of Physics Demonstrations © The Education Group.

Unit 29 Figure 29.1a: © E.R. Degginger; 29.1b: R. Thompson, (University of Arizona) and NASA; 29.2: California Institute of Technology; 29.3: Courtesy of Steve Criswell; 29.4a-b: Courtesy of Andrea Ghez, UCLA.

Unit 30 Figure 30.2a: Courtesy of Kitt Peak National Observatory; 30.2b: Courtesy Astronomical Society of the Pacific; 30.2c: W.T. Sullivan, © 1993; 30.4: Courtesy of USAF; 30.5c: Photo courtesy of Patrick Watson; 30.6(HST): Courtesy of NASA/JPL; 30.6(EUVE): Courtesy of NASA, ESA, and Max Planck Institute for Extraterrestrial Physics; 30.6(Chandra): CXC/TRW; 30.7(all): Courtesy of NASA and The Hubble Heritage Team.

Unit 31 Figure 31.3: Courtesy of Alan Dyer; 31.4: Joe Orman; 31.5a: © Roger Ressmeyer/CORBIS; 31.5b: © Carol B. Ivers and Gary Oleski.

Unit 32 Figure 32.1(Planets): Courtesy of NASA/JPL; 32.1(Sun): Courtesy SOHO/ESA/NASA.

Unit 33 Figure 33.2: Courtesy of Anglo-Australian Observatory; photograph by David Malin; 33.4: © Fundamental Photographs/Diane Schiumo; 33.7a: Courtesy of NASA.

Unit 34 Figure 34.1a: Courtesy of A.M. Lagrange, D. Mouillet, and J.L. Beuzit, Grenoble Observatory; 34.1b: Smith (University of Hawaii), G. Schneider (University of Arizona), E. Becklin and A. Weinberger (UCLA) and NASA; 34.5: © ESO.

Unit 35 Figure 35.1: Copyright © 2005 Planetary Visions Limited; 35.2: © Taxi/Getty Images; 35.5: Photo courtesy of NASA; 35.12: Tom Arny.

Unit 36 Figure 36.1: Courtesy of NASA; 36.2: Steve Schneider; 36.5a: Courtesy Paul Schneider; 36.5b: Courtesy of NASA; 36.6: Sylvain Grandadam/Photo Researchers, Inc.; 36.7(top): © Maso Hayashi/Photo

Researchers; 36.7(bottom): © Douglas/Photo Researchers; 36.8: SVS/TOMS/NASA; 36.11: Courtesy of NOAA.

Unit 37 Figure 37.1, 37.2(both): Photo © UC Regents/Lick Observatory. Unauthorized use prohibited; 37.3(both): Courtesy of NASA/National Geographic Society; 37.4(both), 37.5a: Courtesy of NASA; 37.5b: © John Gillmoure/CORBIS; 37.6: © UC Regents, UCO/Lick Observatory image; 37.9: Courtesy of NASA; 37.12b: Courtesy of Robin Conup, Southwest Research Institute.

Unit 38 Figure 38.1: USGS and NASA; 38.2: Courtesy of JPL/California Institute of Technology/NASA; 38.3, 38.4, 38.5: Courtesy of NASA.

Unit 39 Figure 39.1: Courtesy of NASA; 39.2: National Space Science Data Center; 39.3, 39.4, 39.5(all): Courtesy of NASA/JPL.

Unit 40 Figure 40.1: Courtesy of HST; 40.2: Courtesy of A.S. McEwen, USGS; 40.3: MOLA Science Team, MGS, NASA; 40.4a: Courtesy of A.S. McEwen, USGS; 40.5: Courtesy of NASA; 40.6a: Courtesy of A.S. McEwen, USGS; 40.6b-c: Courtesy of NASA/JPL; 40.7a: Courtesy of NASA; 40.7b: NASA/Calvin J. Hamilton; 40.8a: Courtesy of NASA/JPL; 40.8b: Courtesy of NASA/JPL/Malin Space Science Systems; 40.8c: ESA/DLR/FU Berlin (G. Neukum); 40.9a: Courtesy of NASA; 40.9b: NASA/JPL/Malin Space Science Systems; 40.10: Courtesy of NASA/JPL; 40.11: Courtesy of Mars Exploration River Mission, JPL, NASA; 40.12: Courtesy of NASA/JPL/Cornell; 40.13: Courtesy of NASA/JPL/Cornell/ARC.; 40.14a, 40.15: Courtesy of NASA/JPL.

Unit 41 Figure 41.2(Veta): NASA Goddard Space Flight Center (NASA-GSFC); 41.2(Eros): NEAR Shoemaker (JHU/APL); 41.2(Ceres): NASA, ESA, J. Parker (Southwest Research Institute), P. Thomas (Cornell University), and L. McFadden (University of Maryland, College Park); 41.3: NASA Jet Propulsion Laboratory (NASA-JPL); 41.4: Courtesy of NASA/JPL.

Unit 42 Figure 42.1a(Mercury): NASA Headquarters - Greatest Images of NASA (NASA-HQ-GRIN); 42.1b-j: Courtesy of NASA/JPL/USGS; 42.1k(swirling clouds): NASA Johnson Space Center - Earth Sciences and Image Analysis (NASA-JSC-ES&IA); 42.1l-v: Courtesy of

NASA/JPL/USGS; 42.3(Jupiter, Saturn, Neptune): Courtesy of NASA/JPL; 42.3(Earth): RF PhotoLink/Getty Images; 42.3(Uranus): Lawrence Sromovsky, University of Wisconsin-Madison/W.M. Keck Observatory; 42.5(all): Courtesy of NASA.

Unit 43 Figure 43.1(Jupiter): Courtesy of NASA/JPL; 43.1(Saturn): Courtesy of Cassini Imaging Team/SSI/JPL/ESA/NASA; 43.1(Earth): Courtesy of NASA; 43.2: Courtesy of NASA; 43.3: Bjorn Jonsson; 43.6a: Courtesy of NASA/JPL; 43.7: NASA/JPL/Space Science Institute; 43.8a,c: Courtesy of NASA/JPL; 43.8b: J.T. Trauger (Jet Propulsion Laboratory) and NASA.

Unit 44 Figure 44.1a-b: Courtesy of NASA/JPL; 44.4: Courtesy NASA/JPL; 44.5: Courtesy of STScI; 44.7: Lawrence Sromovsky, University of Wisconsin-Madison/W. M. Keck Observatory.

Unit 45 Figure 45.1: Courtesy Don Wheeler and Brien Dunn/Louisiana Delta Community College; 45.3: NASA Jet Propulsion Laboratory (NASA-JPL); 45.3(Io - Titan): NASA Jet Propulsion Laboratory (NASA-JPL); 45.3(Triton): National Space Science Data Center; 45.4(Amalthea): © Calvin J. Hamilton; 45.4(Enceladus - Oberon): NASA Jet Propulsion Laboratory (NASA-JPL); 45.4(Larissa & Proteus): NASA Goddard Space Flight Center; 45.4(Nereid): NASA Jet Propulsion Laboratory (NASA-JPL); 45.5(left): NASA, Voyager 1, © Calvin J. Hamilton; 45.5(right): NASA Jet Propulsion Laboratory (NASA-JPL); 45.6(both): NASA Jet Propulsion Laboratory (NASA-JPL); 45.7(both), 45.8,45.9: Courtesy of NASA/JPL; 45.10(Uranus): NASA and Erich Karkoschka, University of Arizona; 45.10(Jupiter & Neptune): NASA Jet Propulsion Laboratory (NASA-JPL); 45.11: Courtesy of NASA/JPL; 45.12: Bjorn Jonsson; 45.13: Courtesy of STSci/HST.

Unit 46 Figure 46.1(Io, Europa, Ganymede, Callisto, Titan, 2003UB313, Orcus, Sedna): NASA Jet Propulsion Laboratory (NASA-JPL); 46.1(Triton): National Space Science Data Center; 46.1(Pluto & Charon): Images courtesy of Marc W. Buie/Lowell Observatory; 46/1(Moon): © Digital Vision RF/Punchstock; 46.1(Earth): © Stocktrek/Getty Images RF; 46.2(all) & 46.3: NASA Jet Propulsion Laboratory (NASA-JPL); 46.4a(top): NASA Jet Propulsion Laboratory (NASA-JPL);

46.4a(bottom): NASA Goddard Space Flight Center; 46.4b-c(all): NASA Jet Propulsion Laboratory (NASA-JPL); 46.5a-b: Courtesy of NASA/JPL; 46.5c: Courtesy of Cassini Imaging Team/SSI/JPL/ESA/NASA; 46.6(all), 46.7: NASA Jet Propulsion Laboratory (NASA-JPL); 46.8a: © UC Regents; UCO/Lick Observatory image; 46.8b: Courtesy of NASA; 46.9: Courtesy of STScI.

Unit 47 Figure 47.1: James R. Westlake, Jr.; 47.3a: Copyright 1986, Max Planck Institute for Aeronomy; courtesy of Harold Reisema, Ball Aerospace; 47.3b-c: NASA Jet Propulsion Laboratory (NASA-JPL); 47.4: NASA/JPL-Caltech/UMD; 47.7: Courtesy of Mike Skrutskie, University of Virginia; 47.10: Courtesy Juan Carlos Casado and Isabel Graboleda.

Unit 48 Figure 48.1: Courtesy of Ronald A. Oriti, Santa Rosa Junior College, Santa Rosa, Calif.; 48.3a: Tom Arny; 48.3b: Courtesy of John A. Wood, Harvard-Smithsonian Center for Astrophysics; 48.4: NASA Jet Propulsion Laboratory (NASA-JPL); 48.5a: Courtesy of John A. Wood, Harvard-Smithsonian Center for Astrophysics; 48.7: Courtesy of Landsat/EOS; 48.8: Courtesy of Walter Alvare/University of California, Berkeley; 48.9b: A. Hildebrand, M. Pilkington, and M. Connors.

Unit 49 Figure 49.1: SOHO/MDI project; 49.2: Courtesy of SOHO/Extreme Ultraviolet Imaging Telescope (EIT) consortium; 49.5: Prof. Goran Scharmer/Dr. Mats G. Löfdahl/Institute for Solar Physics of the Royal Swedish Academy of Sciences; Courtesy of Jacques Guertin, Ph.D.; 49.7: Courtesy of NOAO/AURA/NSF; 49.8: 1988 eclipse image courtesy High Altitude Observatory (HAO), University Corporation for Atmospheric Research (UCAR), Boulder, Colorado. UCAR is sponsored by the National Science Foundation; 49.9: Courtesy of LMSAL and NASA; 49.12: Courtesy of NOAO/AURA/ NSF.

Unit 50 Figure 50.4: Courtesy of Ernest Orlando, Lawrence Berkeley National Laboratory.

Unit 51 Page 412: Courtesy of NOAO/AURA/NSF; Figure 51.1: Courtesy of Royal Swedish Academy of Sciences; 51.2a-b: Courtesy of NOAO/AURA/NSF; 51.4: NASA/ESA; 51.5b-d: Courtesy of NOAO/AURA/NSF; 51.6a: Courtesy of SOHO - EIT Consortium, ESA, NASA;

of Chicago) and Anatoly Klypin (New Mexico State University).

Unit 77 Figure 77.1: AURA/STSci/NASA; 77.2(all): Image courtesy of NRAO/AUI; 77.3: Courtesy of NOAO; 77.4: Courtesy of John Bahcall, Institute for Advanced Study; M. Disney, University of Wales; and NASA; 77.7(left): Courtesy of W. Jaffe, Leiden Observatory, and H. Ford, Johns Hopkins University, Space Telescope Science Institute; and NASA; 77.7(right): Courtesy of NRAO and California Institute of Technology.

Unit 78 Figure 78.5a: Courtesy of NASA; 78.6: Courtesy of STScI.

Unit 80 Figure 80.1a: Courtesy of Seldner, M., Siebers. B: Groth, E.J., and Peebles, P.J.E., A.J. 82, 4, "New Reduction of the Lick Catalog of Galaxies;" 80.6: Courtesy S. Beckwith and the HUDF Working Group (STScI), HST, ESA, NASA; 80.7: Courtesy of NASA/WMAP Science Team.

Unit 81 Figure 81.6: Courtesy NASA/WMAP Science Team.

Unit 83 Figure 83.1: Courtesy of Andrew H. Knoll; 83.3(left): © Brandon Cole/Visuals Unlimited; 83.3(middle): © CORBIS RF; 83.3(right): © DV RF/Getty; 83.5: Courtesy of Ken Eward/BioGrafx; 83.7: Andrew H. Knoll, Harvard University.

Unit 84 Figure 84.1: Courtesy of NASA/JPL/Cornell; 84.2: Courtesy of NASA; 84.3: Courtesy of Seth Shostak.

Looking Up Features

Looking Up at Northern Circumpolar Constellations
Background(both): © Akira Fujii/DMI; **M52:** Hartmut Frommelt; **M81 and M82:** © Robert Gendler; **M101:** Adam Block/NOAO/AURA/NSF.

Looking Up at Ursa Major
Background: © Akira Fujii/DMI; **M97:** Gary White and Verlenne Monroe/Adam Blcok/NOAO/AURA/NSF; **Mizar and Alcor:** Courtesy of DSS/Processing by Coelum (www.coelum.com); **M51:** © Tony and Daphne Hallas.

Looking Up at M31 & Perseus
All photos: © Akira Fujii/DMI.

Looking Up at Summer Triangle
Background: © Akira Fujii/DMI; **M57:** Courtesy of H. Bond et al., Hubble Heritage Team (STScI/AURA), NASA, **M27:** © IAC/RGO/Malin; **Alberio:** Courtesy of Randy Brewer.

Looking Up at Taurus
Background: © Akira Fujii/DMI; **M1:** Courtesy of R. Wainscoat; **M45:** Courtesy of Anglo-Australian Observatory; photographs by David Malin; **Hyades:** © Akira Fujii/DMI.

Looking Up at Orion
Background: © Akira Fujii/DMI; **Close-up of Horsehead Nebula:** Courtesy NOAO/AURA/NSF; **Horsehead Nebula:** Courtesy of Anglo-Australian Observatory;

photograph by David Malin; **Betelgeuse:** Courtesy A. Dupree (CFA), NASA, ESA; **Pink Orion Nebula:** Courtesy of Carol B. Ivers and Gary Oleski; **Orion Nebula with Dust and Gas:** Courtesy Gary Bernstein (U. Pennsylvania), copyright U. Michigan, Lucent; **Protoplanetary Disk;** Courtesy of A.M. Lagrange, D. Mouillet, and J.L. Beuzit, Grenoble Observatory.

Looking Up at Sagittarius
Background: Courtesy of Till Credner, AlltheSky.com; **M16 Close-up:** Courtesy of NASA, HST, J. Hester & P. Scowen (ASU); **M16 with Gas Cloud:** Courtesy of Bill Schoening/NOAO/AURA/NSF; **M20:** Courtesy of Jason Ware; **M22:** Courtesy of N.A. Sharp, REU program/NOAO/AURA/NSF.

Looking Up at Centaurus and Crux, The Southern Cross
Background: Anglo-Australian Observatory/David Malin Images; **Centaurus A:** Courtesy Peter Ward, 2004; **Omega Centauri:** Al Kelly; **Eta Carinae:** J. Hester/Arizona State University NASA; **The Jewel Box:** Research School of Astronomy and Astrophysics, the Australian National University.

INDEX

Using the Star Charts➤

These star charts were produced at Stephen F. Austin State University by Professor Dan Bruton and his students, with some additions by the publisher. The charts cover the entire sky and can be used to identify stars and constellations from any location on Earth and at any time of year.

The large rectangular chart covers a wide region around the celestial equator, and the two circular charts cover large portions of the northern and southern hemispheres. The charts are labeled with Right Ascension and Declination coordinates, which are like longitude and latitude on the celestial sphere (see Unit 5). Stars visible to the naked eye have sizes on the charts that depend on their brightness (or apparent magnitude—see Unit 54). Lines connecting the stars help to identify the pattern and primary stars in each constellation. The pale blue strip shows the approximate location of the Milky Way on the chart.

For viewers in the northern hemisphere, to observe the northern half of the sky, use the **Northern Region** chart. If you are looking at the sky around 8 P.M., you should locate the closest date along the outer perimeter of the northern region chart. Rotate the chart so that the correct date is at the top. For each hour earlier/later that you are outside, you should rotate the map so that a Right Ascension an hour earlier/later is at the top. For example, if it is 8 P.M. on September 18, you would rotate the map so that "Sep 20" (Right Ascension of 20^h) is at the top. On the same date at 9 P.M., you would rotate the map so that a Right Ascension of 21^h is at the top. Held with this orientation, the chart should match what you see over the northern sky. The stars that are due north will match what you see at the center of the chart. Stars at the bottom of the chart will be below the horizon and stars near the top of the chart will be overhead.

Looking southward from the northern hemisphere, use the **Equatorial Region** chart. Find the date along the top of the chart, and adjust to an earlier or later Right Ascension according to how much earlier or later than 8 P.M. you are observing the sky as explained for the Northern Region chart. Stars located at the Right Ascension you have identified will lie straight up from the point on the horizon that is due south of you, along the "meridian" (see Unit 7). For observers north of latitude 30°, some stars along the bottom of the chart will never rise above the horizon.

The curving "sine wave" line running through the middle of the Equatorial Region chart is the ecliptic (Unit 6), which marks the path of the Sun among the stars. The dates along this line indicate where the Sun is located throughout the year. The Moon and planets also remain close to this line as they move around the celestial sphere.

The star charts can be used from the southern hemisphere using the same instructions but swapping the words "north" and "south" wherever they occur. Looking northward from the southern hemisphere, you will also need to hold the Equatorial Region chart upside down.

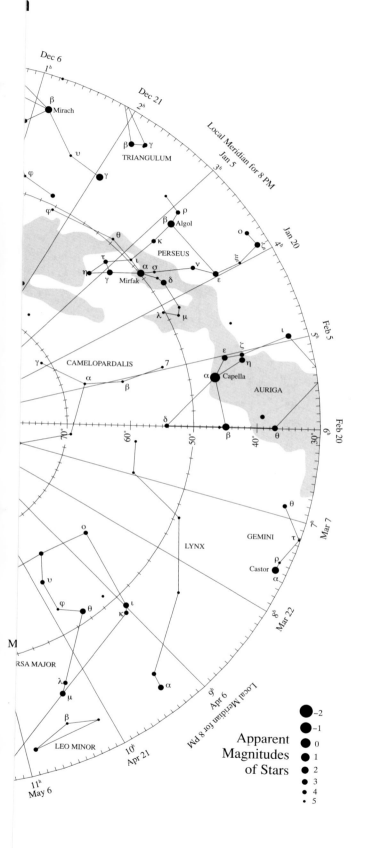

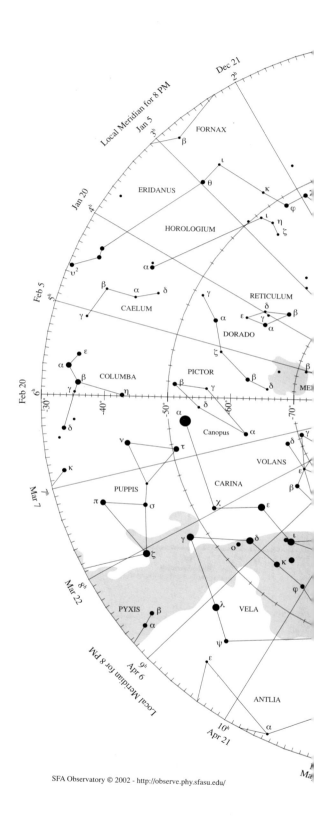

Southern Region

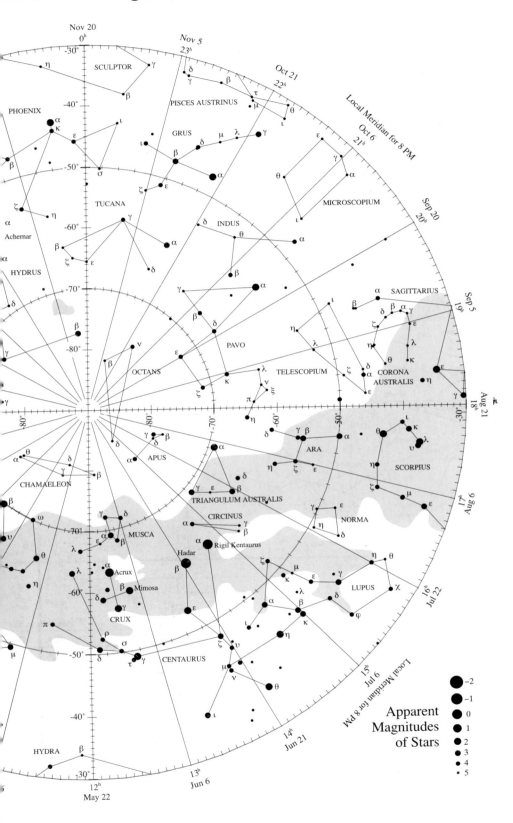

"So, what do you want to do with your Life?

You should be a Lawyer, a doctor, an accountant a consultant," BLAH BLAH, BLAH...

Everywhere you turn, people try to tell you who to be and what to do with your Life.

WE call THAT THE "NOISE."

Block it. Shed it. Leave it for the Conformists.

As a Generation, we need to get back to focusing on INDIVIDUALITY.

Self-Construction rather than MASS Production.

Define your own Road in Life instead of traveling down somebody else's.

Listen to yourself. Your Road is THE Open Road. Find it,

FIND THE OPEN ROAD.

finding
THE OPEN ROAD

a guide to self-construction rather than mass production

find the open road @ www.roadtripnation.com

roadtrip nation

Mike Marriner, Brian McAllister, and Nathan Gebhard

a **roadtrip nation** book

TEN SPEED PRESS
Berkeley | Toronto

Ten Speed Press
Box 7123
Berkeley, California 94707
www.tenspeed.com

Distributed in Australia by Simon & Schuster Australia, in Canada by Ten Speed Press Canada, in New Zealand by Southern Publishers Group, in South Africa by Real Books, and in the United Kingdom and Europe by Airlift Book Company.

Creative Directors: Lori Werstein, Dan Marriner
www.unofficialcharacter.com

Library of Congress Cataloging-in-Publication Data
Marriner, Mike.
Finding the open road : a guide to self-construction rather than mass production : a road-trip nation book / Mike Marriner, Brian McAllister, and Nathan Gebhard.
p. cm.
Summary:"A compilation of the wisdom gleaned from Roadtrip Nation's informational interviews and experiences on the road, including a how-to-roadtrip guide"—Provided by publisher.
Includes index.
ISBN-10: 1-58008-721-3 (pbk.)
ISBN-13: 978-1-58008-721-6 (pbk.)
1. Career development. 2. Vocational guidance. 3. Counseling. I. Gebhard, Nathan. II. McAllister, Brian. III. Title.
HF5381.M35 2005
650.1—dc22   2005004868

Printed in China
First printing, 2005

contents

The essence of
discovery is not in
seeing
new landscapes,
but in having
new eyes.

—Marcel Proust

find the open road @ www.roadtripnation.com

roadtrip nation

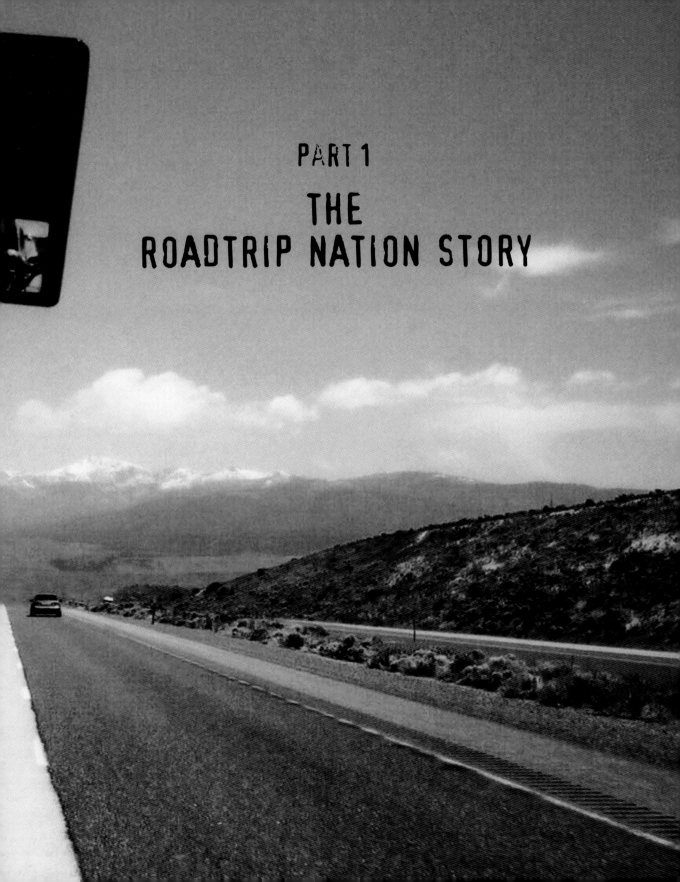

PART 1

THE
ROADTRIP NATION STORY

The Roadtrip Nation Story

Imagine yourself as a truck driver. You've been driving the same route from Seattle to Chicago for the past thirty-five years, and whenever you go through Montana you see an exit that always catches your eye. At the bottom of the exit is a beautiful valley with a winding river that has towering rock formations on either side. You've always thought about pulling off the freeway to go check it out and skip a few rocks . . . but you never have. Today is just another normal day along your route, and that exit is coming up. Do you power through to Chicago as usual?

Or do you slow down, take a closer look, and exit the freeway?

Why don't people exit the freeway? Why do they stick with majors they dislike, stay in jobs they hate, or travel roads they have no passion for? We know about this, because we were on that path.

Seniors in college and overall dutiful patrons of society, we had been fulfilling someone else's expectations for virtually our entire lives.

"Get good grades. Get your SAT scores up. Get into a rated college. Get focused on your major. Get your resume built up. Get a high-paying job with benefits. Get the mortgage and 2.5 kids. Get a solid investment portfolio so one day you can retire and *then* start to live your life."

But one night, sitting around our college apartment, we started questioning things for the first time. "Whoa, wait a second, what am I doing on this freeway? Does it have anything to do with what I'm passionate about? And how can I really be sure if this is the right path if I haven't ever seen anything else? If I haven't explored, experimented, or been exposed to all the different roads out there?"

We began to realize that all those expectations had been slowly shaping the individuals we were becoming and the major life decisions we were making. And now, with graduation just over the horizon, we stood waiting on the loading docks. Packaged in our newly minted college degrees, we were headed into the world, ready to fill whatever space it needed us to take up.

One of us was on his way to medical school (Mike) and was getting set to take the MCATs after four years of pre-med study. Another one of us (Brian) was a communications major who was the third generation to be funneled into the family business. And the third one of us (Nathan) was studying business administration; he was really interested in art and design but had no way of expressing it through that major.

All of us had passions outside of these boxes, but we had no idea how we could link what excited us to what we could do with our lives. We didn't even have a clue as to what our life's work could be. At twenty-three years old, we were already settling in, with no idea of what we were settling into.

The Idea

The brainstorming continued into the night, when the idea first hit: "What if, for one summer, we could take a roadtrip across the entire country and talk to people about how they found their roads in life?"

We wanted to meet with people who had learned how to shed the noise. How did they do it? Where were they when they were our age? How did they resist pressures to conform? How did they identify what they were passionate about? How did they create roads around their individuality? We basically just wanted to hear their stories. Maybe they could help us shed the constrictions we were feeling, and define our own roads.

In school, we had been taught the template for building traditional paths like medicine, law, or accounting. Those paths might be great for some, but not for us. Our passions didn't lie in those boxes. We hoped that the road would give us a new kind of education—a broader education that would expose us to stories from every eclectic area of life.

We wanted to learn how a lobsterman on the coast of Maine got where he was. How the conductor of the Boston Philharmonic discovered his passion for music. Or how the president of the Sierra Club built a life around an issue he stood for. We were empty cups, anxious to experience these stories firsthand. Maybe they would reveal things about ourselves, and the world, that we didn't know were there.

The idea of hitting the road stirred in our heads for a few days until it became something we couldn't ignore. After a series of conversations addressing "How would we pay the issues and passions they stood up for became the seeds that would one day create their own roads.

Before Tom First cofounded Nantucket Nectars, he was a Brown University alumnus humbly working as a fisherman on Nantucket Island with his friend Tom Scott. "I thought to myself, 'What the hell am I doing? I went to Brown. My parents are pissed at me. I should be doing something different.'" But he didn't buckle under the pressure, and continued to follow his passion for building his own business.

Before Beth McCarthy was the director of *Saturday Night Live,* she worked for nearly nothing as a secretary at MTV and had to work at the Gap on weekends to pay the bills. Rather than conforming and getting a more glamorous job at CNN, she continued on her own path.

The breadth of stories we encountered on the road was eclectic and individualistic. But all the people we met shared a common ideology that guided their decision making. They challenged society's view of success, stood for ideals that drove their life's work, rarely followed a road map, had the courage to change roads when needed, and realized that we're living in a new world that demands a more creative and individualistic approach to life.

These commonalities started shaping and defining how we began to look at the world. The get-to-the-next-mile-marker perspective we had before our trip was tossed out the window somewhere in the middle of Iowa. For the first time in our lives, our roads were being shaped by what we stood for, what we were passionate about, and what we believed in.

Coming Home

Coming home from the trip was more of a challenge than we could have ever expected. But we were filled with the stories of eighty-two people who had faced those very same pressures and overcome them.

School loans, an impounded car, massive credit card debt, nagging family members—you name it, we faced it. The most common phrase we heard was, "Okay, now that your little roadtrip is over, what are you really going to do with your life?" Some form of graduate school was a very popular solution, but we knew in our hearts that it would only prolong the inevitable.

One of the hardest parts of coming home was seeing how "successful" our peers had become since we had last seen them. A lot of the kids we had graduated from high school or college with were now working as real estate agents, accountants, or investment bankers. We would see them on the street and share what we were up to, and at times it was downright embarrassing. We had been living out of a green motor home for three months, while they were launching off into successful careers, leasing fancy cars, and preparing for their life of ultimate coosh-ness. But the trip had given us a sense of confidence that those were not the important things in life. They represented a view of "success" that society had predetermined for us. Besides, we knew it was okay to not have it all figured out at the beginning of our roads.

Over the next few months, our own convictions and passions slowly started to emerge. The experiences we went through on the road inspired us to look at life through an entirely different lens, and the more this new perspective sank in, the more we wanted to share it with the people around us.

Our friends would come over from time to time and watch some of the interviews from our trip. We couldn't believe the impact the footage had on them. We could see their eyes being opened to how big the world is, to all the ways they could live their lives, and to new ways of thinking. For the first time, we could see them questioning whether the freeway they were on was really the right road for them.

As more and more friends started asking about our travels, we began to define the specific impact we took away from the trip. There was so much energy around what we had done, amongst ourselves and friends, that we started to think about broader ways of sharing these experiences.

First, we wrote a short piece that put some definition to our thinking and later became the ideology for our project. We titled it the "Roadtrip Nation Manifesto."

"So, what do you want to do with your life? You should be a lawyer, a doctor, an accountant, a consultant . . ." Blah, blah, blah.

Everywhere you turn, people try to tell you who to be and what to do with your life. We call that the "noise."

Block it. Shed it. Leave it for the conformists.

As a generation, we need to get back to focusing on individuality. Self-construction rather than mass production.

Define your own road in life instead of traveling down somebody else's. Listen to yourself. Your road is the open road. Find it.

Find the open road.

The manifesto then sparked a larger idea. We thought to ourselves, "What if there was a way we could create a documentary from the footage we collected on our trip?" We knew the footage was extremely raw and we had no editing experience whatsoever, but we were driven to share what we had learned. Maybe we could independently release it online, enter it into film festivals, or try to show it on college campuses.

Around this time, a national magazine heard about our trip and was interested in learning how we had booked the interviews in such a rogue fashion. They decided to run a tiny four-inch article on our journey.

That small article opened up new doors for our project. It led to a book deal with a major New York publisher and financial support from a small group of organizations that would help fund the production of our documentary and get it broadcast on public television stations across the country. Suddenly, the exit we took off the freeway had led us onto an entirely different road. Where that road was taking us we weren't actually sure, but we were committed to riding it out.

Once all of these new opportunities were in front of us, we really had to make it happen. We spent an entire year editing the documentary, writing the book, and launching a website that would share the experience we had on the road. None of us were film majors, writers, or website designers, so we approached the production and writing process in the same spirit that got us around the country in that old RV: making it up as we went, and not being afraid to fall on our faces.

When the book came out later in the spring and the documentary began airing on public television, things really started to take off. All of a sudden we were on the *Today* show, CNN, CNBC, CBS News, and NBC's *Carson Daly* show. And publications like *Anthem Magazine, Teen People, Newsweek, LA Times, New York Post, Chicago Tribune,* and several others were writing stories about our roadtrip.

The Movement

As the project progressed and people became aware of our travels, the question of "What is Roadtrip Nation?" was also changing and evolving in our heads. Deep down, we could feel that people wanted and needed to hit the road themselves.

That's when we began to envision Roadtrip Nation as something bigger than just our first trip. "What if Roadtrip Nation was less about our first trip, and more about a movement of people who hit the road to discover their paths in life?"

We started knocking on the doors of campus career centers to see if they would run Roadtrip Nation programs that would put students on roadtrips to interview people themselves. A few of them were really excited about it and promoted the opportunity to their student body, while others declined due to our "renegade" style and unorthodox approach to career development. However, the response on some campuses was amazing. Applications poured in from teams of students. We ended up selecting three students from Georgetown and NYU to hit the road in the green RV for the summer and interview their own list of nonconformists. Ryan, Randy, and Mike would be the first Roadtrip Nation student team, starting an annual cycle that would put students on the road in green RVs every summer.

With the money we had left over from our first documentary, we filmed their trip in the same way we filmed our own—totally spontaneously and making it up as we went.

We had already broadcast our first roadtrip documentary on public television, so we went back to the same programmers and approached them with a new idea—"What if we shot an annual documentary series about these student roadtrips that are happening every summer and then aired it on public television? We already have the first one shot. Check it out." They were excited about it, which led to financial support from another group of organizations that funded the production of the series and fueled our new roadtrips year after year.

To make it happen, we created our own grassroots production company called Roadtrip Productions to produce the series, started partnering with more college campuses across the country to run these programs, bought three more RVs, hired our friends to help film the trips, and rented a small house in California to edit the footage.

Lastly, we found a new publisher that was excited about releasing another book that would be based less on our first trip and more on the broader movement to help other people get on the road themselves. We spent another full year editing a new documentary series and writing this book, which we hope will express how everyone has the freedom to hit the road, explore the world with new eyes, and define their own path in life.

Ideology

At its core, Roadtrip Nation is an ideology—a way to define our own roads in life. But the way we individually mobilize that ideology is what we hope to encourage. The road is open to everyone. And if you take the space to temporarily exit off your own freeway and experience it, the road will reveal things to you that you never knew existed—about the world, and yourself.

The movement to hit the road is not limited to people in green RVs or students on college campuses. The movement is open to all of our minds, and can be actualized in all of our lives.

Hitting the road is less about logging miles on your odometer, and more about changing the way you look at the world. It's about conditioning yourself to look left if you always look right. It's about talking to people whom you wouldn't ordinarily talk to. Learning from experiences that you wouldn't ordinarily learn from. Questioning things that you wouldn't ordinarily question. And being open to opportunities that you wouldn't normally let yourself be open to.

At the end of the day, hitting the road is simply about opening your eyes. Physically hitting the road is a great way to do that, but changing the way you look at the world is even better. As Marcel Proust, the renowned French novelist, said: "The essence of discovery is not in seeing new landscapes, but in having new eyes."

Yes, we hope that this book inspires you to create your own trip and get out there to experience the road for yourself. But many people travel to the far ends of the earth and still don't see anything at all. We hope that by absorbing the stories and experiences in this book, you'll be able to approach the world with "new eyes"—eyes that will help you stay true to yourself and define your own road in life.

See ya out there,

Mike, Brian, and Nathan

PART 2

THEMES FROM THE ROAD

Themes from the Road

During the Copernican Revolution, there was a huge earthquake in the way people saw themselves in relation to the cosmos. The assumption that everyone had made was that the sun went around the earth. But the scientists kept seeing errors in their calculations. Then Copernicus came along and said, "Hang on, what if the earth goes around the sun?" When they tested that out, all the calculations worked. But people were outraged because of the view that we were the center of God's creation. So Copernicus was persecuted for even suggesting it. This wasn't just a new theory on the cosmos, it was a complete paradigm shift.

—*Sir Ken Robinson,*
senior advisor on education policy at the Getty Museum

Out on the road, one thing that became apparent was how different the world is today. Change is everywhere—in culture, technology, society, and science. Even the world of work has completely changed. Dan Regis, a venture capitalist we met in Seattle, paints a picture of what it used to be like: "In the sixties, Boeing gave me an offer as an engineering aide. I can remember to this day walking into the engineering room and there was a football field–long series of gunmetal gray desks and guys with pocket protectors and white shirts rolled up at the sleeves. You could see your life in twenty, twenty-five, thirty, forty, fifty years—then dead. That's not the world we're living in anymore."

The question is just begging to be asked: Since we're living in a completely different world, why are we still approaching our lives in the same linear manner as past generations? Shouldn't new times call for new ways of thinking?

Sir Ken Robinson, a former professor at Warwick University in England and current senior advisor on educational policy at the Getty Museum in Los Angeles, shared with us how the world today is filled with more choice and diversity. "When I was growing up, there weren't very many magazines and we only had three television channels—we were told what to think about without many other options. Now we have 400 TV channels. So increasingly, we are living in a world driven by choice, individuality, and diversity. And it's being interlaced with the growth of technology that is becoming more and more pervasive. Like never before, we need to break the mindset that we get to be what we are by following a linear path and moving sequentially from one piece to the next. There are more choices and more possibilities in the world today."

The people we met with on our trip all embraced the diversity and creativity this new world has to offer. None of them has worked in the same job for their entire life. In fact, most reinvented themselves every few years. Part of that was out of necessity. As Alastair Paulin, managing editor of *Mother Jones* magazine, shared with us, people don't even have the option to stay in one job for life anymore: "The definition of work has changed since when I was growing up—you had one job and that was it—now most people change careers something like five times. The contract between the employer and the employee has been somewhat broken down by the economic changes of the past decade. It's not like you commit to us and we'll commit to you and we'll see you after sixty-five for that nice pension. Those days are gone. In a way it's a terrible thing for workers, but in a way it's very liberating."

In this new world that is driven by choice, individuality, and diversity, how can we all build lives around things we believe in and are excited about? On the road, the stories we collected didn't give us a definitive road map to follow; instead, they gave us a set of guiding principles that could help us navigate the world for ourselves.

The information we gathered and the experiences we absorbed were easy to take in with an open mind because we had gone straight to the source, not a book on the shelf, a resource in the career center, or a motivational speaker. People from all walks of life sat down with us, shared their stories, and expressed what worked for them in their own lives.

We've broken the experience down into a dozen themes, backed up by the interviews we conducted, that express the impact the road had on our lives. When we came home from the trip, these ideals resonated in our hearts and gave us the courage to stay on our own paths. We hope they do the same for you.

I honestly hope that none of you ever say, "Well, gosh, I'm twenty-four now, I have a degree, and now it's time to grow up and be sensible." That's the worst advice you could ever give yourself. Life has to be discovered all the time. You can never say, "I have done everything I can up to the age of twenty-five, and now I have to settle down and be a responsible, boring adult in a responsible, boring world." You don't have to do that.

—Ric Birch,
executive producer of the Sydney
and Barcelona Olympic Games ceremonies

Don't Fall Asleep at the Wheel

If it's not working for you—the company is bad or the people are bad—get out of there. Move. Every day is a wasted day after that. Life's just too short.

—*Pat O'Donnell,*
CEO of Aspen Skiing Company

During our roadtrips, we had a few extremely late-night drives that tested our ability to stay awake at the wheel. Sometimes no matter how much coffee we had or how loud we turned the music up, our heads kept nodding toward the floor. We had to pull over, find a place to sleep, and be ready to start the next day fresh. On those nights, we didn't cover as much ground as we had wanted, but we got some much-needed rest and woke up the next morning with new eyes.

We began to realize that prior to our trip, we had been going through our lives asleep at the wheel—stuck in a groove that provided no passion or meaning. We had wasted valuable years of our lives that could have been spent being aware of new possibilities.

Many of the people we met on our roadtrip fell asleep at the wheel at one point in their lives, but they knew when to wake up and make a move. Pat O'Donnell, CEO of Aspen Skiing Company, majored in engineering so he could go to work at his father's company when he graduated. "I hated every living minute of it. After doing engineering for a couple of years I thought to myself that life's too short, I gotta make a move. At the time, my real interest and passion was rock climbing, so I moved to Yosemite and took a job as a bellman. The pay was ninety cents an hour, but it included a tent and free meals. I worked from 7 a.m. to 3 p.m., but it stays light until 9 p.m., so you could climb six hours a day."

This was the first of many moves throughout Pat's life, but each change allowed him to satisfy his passion for the outdoors. He built ski resorts, directed nonprofits to help conserve the environment, lived on a boat for a year, and at one point became the CEO of Patagonia.

"Every time I made a change I was petrified. When I moved to Yosemite I was petrified. When I did the sailboat thing I was petrified. When I walked into the CEO position at Patagonia, I knew nothing about making clothing. But you survive these changes, and they're never as scary as they seem. I used to have a quote posted up on the wall of my tent in Yosemite. It was by Thoreau—'Most men lead lives of quiet desperation.' Then it's gone. You wake up one day, you're sick and old, and you go what if, what if, what if. I never wanted to be on my deathbed thinking, 'What if?'"

How do you know when it's time to wake up and make a move? Mike Lazzo, senior vice president of programming at the Cartoon Network, has a system in place to make sure he doesn't fall asleep at the wheel. "I have two litmus tests for my work. If I drive by a construction site and want to start digging because manual labor seems better than my current job, then that's a sign it's time to switch jobs. The other test I do is this: Every single day I ask myself if I'm happy. The day I say no is the day I leave."

In your life, as at the wheel, it's easy not to notice when you fall asleep. You become sedated by the goal of reaching a destination and fade out during the process. Luckily, waking up is within all of our means. Some people may need an experience like a road-trip to snap out of it and revive consciousness. Others might be able to wake themselves up through less dramatic means. Either way, if you've nodded off, be honest with yourself and do something about it. Wake up! Life is too short to go through it asleep at the wheel.

MORE ON "DON'T FALL ASLEEP AT THE WHEEL"

Your life is now. Your life is not going to start when you graduate college. Your life is not going to start when you get married. Your life is now. You have to enjoy your life today . . . otherwise you're going to be miserable.

—Katherine Cohen,
founder of IvyWise college counseling

It's amazing how precious life is. Time is so precious. If you're doing a job you don't like because someone else wants you to do it, get out. Don't waste the spark of life.

—Peter Seligmann,
chairman of the board and CEO of Conservation International

The only advice I could give is—"Poof, you're fifty." So what have you done? Did you have fun? Did you have adventures? Did you do stuff that you really loved? Did you live a brave life? Were you not afraid? Because mostly I think it's fear of "am I going to make a mistake?" that holds people back from living their lives. Well, you know what, make a mistake.

—Geoffrey Frost,
senior vice president and chief marketing officer of Motorola

Ten to fifteen years from now, if you look back on your life and ask, "I wonder if I should have done that differently?" I don't think you want to be in a position to say, "Yeah, I should have done that differently." Because you can't get that time back.

—David Jacobs,
CEO and founder of Spyder Active Sports

I see so many students who are so directed. They're in high school and they already want to get into advertising, but there's no life there. No embracing of the world, or the people around them. I mean, how do you know if you want to be an art director if you never looked at other things? If you've never gotten out of your comfort zone and tried anything else?

—Laurie Coots,
chief marketing officer of
Chiat\Day advertising agency

Exit the Freeway

If you don't slow down, you're going to get caught halfway through life and realize all the things you missed when you were capable of doing them.

—Chuck McGrady,
president of the Sierra Club

On one of our roadtrips from Seattle to Chicago, we pulled off I-90, a major truck route connecting the ports in Seattle to the Midwest, at the Clark Fork River in western Montana. We got off to explore a beautiful valley that caught our eye from the freeway. A dark blue river curved around a massive rock formation that had a perfect strip of sand at its base. It was a beautiful sunny day, so we hiked down to the river to play a few games of Over the Line.

While we were at the river we ran into a trucker, all geared up in overalls, greasy cap, dark shades, and boots. He was skipping stones in the river. This was odd; truckers are known for not stopping. The only place you ever see them is at truck stops when they're fueling up.

He started opening up to us about how his ex-wife had passed away the day before from cancer, and for the first time he felt how impermanent life is. "For thirty-five years I've driven by this exit a million times, and I've never stopped." The look on his face was part enlightenment, but part regret. He was an older man who didn't have many years left, and you could sense the sadness he felt from knowing he could have skipped a few more rocks in his day.

Many of the people we interviewed on the road exited the freeway in one form or another throughout their lives. When he was thirty years old, Alastair Paulin, managing editor of *Mother Jones* magazine, left his career as an investment banker to travel the world with his wife for eighteen months. Howard Schultz, chairman of Starbucks Coffee Company, traveled throughout Italy where he discovered the romance of the Italian espresso bar. Pat O'Donnell, CEO of Aspen Skiing Company, cashed out everything he had and lived on a boat for a year.

We thought about that a lot on the road—why don't more people exit the freeway? Not just on a roadtrip, but in life. There are so many beautiful stops, and it only takes a few moments. Sure, you'll get to your destination a little bit behind schedule. But when will you ever have the chance to stop off at a small town in Nebraska, pursue your passion for film, travel to Thailand with your wife, write a book about your experiences, climb a random hill in Vermont, start a juice company with your friend, or stroll down to that river in Montana?

Meeting that trucker and listening to all the stories of the people we met made us wonder, what is life really about anyway? Is it a series of connected events that serve the sole purpose of arriving at a predetermined destination? Or is it about the roadtrip along the way?

MORE ON "EXIT THE FREEWAY"

Then, in January 1984, I got a call from a guy who I'd interned with at the Clean Air Coalition. He was in Iowa organizing the presidential campaign for Walter Mondale and asked if I'd come work for him. He said he needed me in two days. I wasn't sure, but I believed then, as I do now, that you should take an opportunity when it's presented to you. You never know where it might lead. So in the next two days I dropped out of school, quit my job, left my boyfriend, and went to Iowa for what I thought would be a month.

—Deb Callahan,
president of the League of Conservation Voters

I applied to Yale and was accepted, but I decided to take a year off in 1968. There was a lot of pressure to go, and people said things like "You're crazy. What are you doing?" I spent the year bumming around Europe and participating in the revolution. It was amazing. I ended up never going to college.

—Geoffrey Frost,
senior vice president and chief marketing officer of Motorola

Right after my senior year at Brown, while my friends went to Wall Street or law school, I moved to Nantucket and started a business with my friend.

—Tom First,
cofounder of Nantucket Nectars

I didn't know what I wanted to do. I was totally displaced after college. But when I was traveling in Europe, I discovered the romance of the Italian espresso bar in Italy. It was an epiphany for me.

—Howard Schultz,
chairman of Starbucks Coffee Company

My parents thought I had flipped out when I didn't take the banking job. But at some point you just have to follow your gut and what feels right.

—Mike Egeck,
president of The North Face

You can't map out a very long route for yourself.
What you can do is sort of know where you're going. Be half sure,
jump in, and hope you can swim the rest of the way.
It's like Eco-Challenge. The race is five hundred
miles and thirty checkpoints.
The people who don't finish are full of ego and only focus on their finish placement. They do that even when they're leaving the start line.

People who are successful focus on the immediate checkpoint.

When they reach checkpoint one, then they worry about checkpoint two, and so on.

That's the philosophy you need to keep to.

—Mark Burnett,
creator and producer of *Eco-Challenge, Survivor,* and *The Apprentice*

There Is No Road Map

There is no road map. No black or white answers. People really like black and white—go to school, boom, boom, boom, and you could be this. No one should tell you, "This leads to that" or "You need to do this," because there are a lot of options out there.

—Gary Erickson,
founder and CEO of Clif Bar

What if the trip doesn't end up like you expected? What happens if the road map you're following takes you to a destination that isn't even there anymore? Road maps can't possibly keep up with the rate of change in society today, and over time they can close your life off to pivotal opportunities.

One person we met who kept his life open to opportunity by throwing out the road map was Gary Erickson, founder of Clif Bar. "I had grown up with evangelical parents in suburbia, where things seemed pretty black and white. But after college I traveled the world, and when I came back to the United States I realized that there is no black and white. You have to be comfortable living in the gray." Gary went on to become a professional cyclist, selling Greek pastries on the side. One day, he had an epiphany. Combining his experience in cycling and cooking, he invented Clif Bar, an energy bar that is perfect for athletes. Rather than following a specific road map, Gary found his path by seeing the intersection between two finite aspects of his life.

Even the scientist who decoded the human genome, Dr. J. Craig Venter, didn't have a road map. "School was boring so I never paid much attention. It was an unrewarding experience. In high school I only took two science classes, got a D-minus in physics, and came within half a grade of graduating. I left home at seventeen. To support myself, I worked as a night clerk at Sears, putting labels on things, so I could surf during the day." He would later find science through a very circuitous route, working in the medical corps during the Vietnam War. "I loved it."

We're led to think that absolute structure is demanded for all roads in life. But just because it works for someone who *wants* to become an accountant doesn't mean it works for someone who is passionate about writing, art, sports, music, or virtually anything else. Life is not a predictable experience. Very few of the people we met on the road went directly from point A to point B.

In fact, it was clear that if the people we met had followed an exact road map, they would have missed opportunities that played a huge part in defining their roads. If Gary Erickson had gone on to grad school instead of selling Greek pastries to support his cycling career, he never would have started Clif Bar. The world is full of possibilities and potential, but we're the ones who choose to be open or closed to them.

MORE ON "THERE IS NO ROAD MAP"

It's not like when I was your age I had this plan that led me here. Things just go along as a progression of your interests and what you like to do.

—David Jacobs, CEO and founder of Spyder Active Sports

You can't just script your life.

—Jerry Colangelo, chairman and CEO of the Phoenix Suns

In terms of a career path, mine was totally unpredictable. It wasn't anything I ever had in mind at the beginning. It was a series of connected events, but it's only with hindsight that you realize they're connected.

—Ric Birch, executive producer of the
Sydney and Barcelona Olympic Games ceremonies

I had no clue about what I'd be doing. What I thought I was going to do is not even close to what I am actually doing. When I first started college in Seattle, I was an engineering major, and I spent the first part of my college life figuring out that I didn't want to be an engineer.

—Dan Regis, partner at Digital Partners Venture Capital

Don't have a plan now. Throw the plan out. Be determined, but don't plan. A lot of people would come up to me when I was working at the New York Times and they would say, "How do I become a journalist?" I think especially when you're young, it has to be about the ideas. It can't be about the career stuff. Your thirties are for your career stuff. Try to not even think about that shit until then.

—Ann Powers,
rock journalist and senior curator of Experience Music Project

You never quite know where one path is going to take you. How one experience parlays itself into other things. So don't be afraid to take an opportunity that comes your way even if it's not the one you think you've been looking for. Through that opportunity you meet new people, which changes the way you look at the world. It could shoot you out into a totally different direction.

—Doug Ross, television producer

Don't Fall Asleep at the Wheel

Exit the Freeway

There Is No Road Map

Follow What Excites You

Risk?

Success?

Shed the Noise of Conformity

Harness Your Individuality

Work versus Your Life's Work

Don't Get Trapped

Unplug Your Education

Hit the Road

I don't think you can be successful without an unbridled enthusiasm for something you really really love.

You can't fake it

for too long. People will see through it.

—Howard Schultz,
chairman of Starbucks Coffee Company

Follow What Excites You

You have to be passionate about something that makes you wake up in the middle of the night. That's all you need. Then you feel like you're in the flow and doing what you're supposed to be doing

—*Jill Soloway,*
writer and coproducer of Six Feet Under

It seems so simple: just follow what excites you. But many people don't even know they're permitted to do that. Tom First, cofounder of Nantucket Nectars, told us, "I wasn't even sure if I was allowed to do what I wanted to do."

Mike Lazzo didn't have any problem following what excited him. "I had no direction when I was in high school. I was expected to go to college or work in the textile industry like my father had done. But I just couldn't do it because it was too boring. At night, I did what I really wanted to do: watch television, think about it, and bore people with my opinions. It slowly dawned on me that maybe I'd just get a job in the mailroom at a television network. I thought it was important to be close to my natural interest. They could have told me I'd be digging holes; as long as there were TV monitors in the vicinity, I would have been happy."

So Mike went to work in the mailroom at Turner Broadcasting in his hometown, Atlanta. He started by delivering mail. "Delivering mail turned out to be the best job in the world because I got to know everyone at the company." Over time, he became the guy who decided which movies would play on Tuesday nights. "I could actually get paid to watch TV and have an opinion." Today, Mike is the senior vice president of programming for the Cartoon Network, a subsidiary of Turner Broadcasting, and is known as one of the leaders in his field. "There's always something—that thing you would do for no money. Identify that and you've found your perfect job, if you're willing to work at it. My ability to work hard definitely helped me, but natural interest is what keeps people getting out of bed."

We also interviewed a renowned professor at the Harvard Business School in Boston, Dr. William Sahlman. He has watched thousands of students go through his classes and evaluated what they've done right, and what they've done wrong. "I've had four thousand students here at Harvard over the years, so I've had opportunities to watch them go out and evolve. I would say the biggest mistake people make is following other people down a path rather than doing what they're really passionate about."

After hearing so many stories that echoed this same perspective, we started giving a lot of thought to what excited us in our own lives. We started to study ourselves, and when we did, we didn't have to look far to find out what we had a natural interest in. This project was what lit us up. Doing exactly what we were doing. And, sure, it didn't seem like the most realistic of "careers," but three years later we're still doing it. Just like Mike Lazzo is still watching television.

MORE ON "FOLLOW WHAT EXCITES YOU"

For me it was about falling in love with the outdoors. For someone else it could be New York City or it could be driving a bus. You see a million people like this—they're driving the bus or being waiters, and they're the happiest people in the world. That's what it's all about.

—*Pat O'Donnell,*
CEO of Aspen Skiing Company

My only compass was to try and have a passion for what I was doing. Regardless of whether it's going to be great or not, if what you do is your passion, then it never really seems like "work."

—*Rob Bollinger,*
artistic director of Cirque du Soleil's O

How do you recognize a passion? I think it all comes down to curiosity. It's not about the answers that come up, but the questions that come up. So if it's something that you have endless questions about, whether it's a career or whatever, you're probably interested in it. And if you're interested and you're curious, you probably have passion for it. It's just like falling in love I guess.

—*Gary Rydstrom,*
sound director with Skywalker Sound

The people I know who didn't follow their own desires ended up sort of lost and confused. Even if they were successful. They were left not just wondering whether they enjoyed their lives, but wondering who they were. That is a very difficult way to spend your life.

—*Jane Smiley,*
Pulitzer Prize–winning novelist

There is no excuse

for someone in his twenties not to pursue a **dream.**
If it doesn't work out or if you don't like it, who cares? You'll have done
something that others don't have the

courage to try. And you'll be young enough to get on the track

of a so-called "mainstream" career.
That will still be there. So big deal, you start school five years later than
somebody else. But fifteen years later it doesn't make any difference.
That's my only message.

Take a chance.

—Charlie Trotter,
chef and owner of Charlie Trotter's restaurant

Risk?

People said, "Gee you took a lot of risk," but the bigger risk would have been staying at a job that wasn't fulfilling and was wasting my life. Quitting to do something that I really loved wasn't a risk.

—Jim Koch,
founder and brewmaster of Samuel Adams Brewery

The people we met along our journey didn't see the notion of risk the same way that we had before we hit the road. To us, risk meant jeopardizing something that we'd gained or were trying to gain—our "security blanket." But what if the object of our pursuit was created from what we'd been programmed to want out of our lives? In that case, the idea of risk can hold us back from refocusing our paths onto something more meaningful. Something we're truly passionate about.

Jim Koch, founder of Samuel Adams Brewery, left a high-paying job as a consultant to start his own beer company. Jim used to be an Outward Bound instructor, and he gained a new perspective on the notion of risk from that experience. "Something that climbers talk about is the difference between perceived risk and actual risk. A lot of times we're driven and limited by perceived risk. But perceived risk is very unrelated to the actual risk. Those kids in my groups who would rappel down the cliff backward were scared. Especially in the first couple of steps when they had to lean all the way back. There's nothing underneath you but thin air and sharp rocks a couple of hundred feet down at the bottom. That is very high perceived risk, but the actual risk is negligible. That rope would probably hold a car."

When we were in Seattle, we sat down with Jonathan Poneman, cofounder of Sub Pop Records (Nirvana's original record label), who had his own take on risk. "If everything in your life is characterized as 'risk versus safety,' the human instinct is to choose safety. But what if you use a whole different standard for reevaluating your life, such as 'necessary versus unnecessary.' Things like happiness, passion, and love are all necessary." From that perspective, the real risk would be not having happiness in our lives. And the idea of refocusing our lives in order to feel more fulfillment becomes less of a risk and more of a necessity.

When people told us, "It was something I *had* to do," what they were really saying was that the traditional risk associated with taking action in that direction was nothing compared to the internal pull they felt to bring passion and meaning into their lives. That was the "necessary" part of the equation.

This is not to say that if you take risks and try to build a meaningful life, you won't have financial instability. Most likely, you'll be scraping by in the beginning. Before Jonathan started Sub Pop, he was a janitor at the Seattle Westin Hotel and a copy manager at Kinko's. He did anything to pay the bills so he could keep working in the Seattle music scene. During that part of his life, he traded a security blanket for happiness, passion, and love. By not focusing on the perceived risk associated with that lifestyle, he ended up defining a whole new road for himself based on what mattered to him.

Our decision to hit the road in the first place definitely felt like a risk. But looking back on it with some perspective, the greater risk would have been to stay on the freeway, continuing to live lives without passion or meaning. We were too young to start settling, but it wasn't until after the trip that we realized that.

MORE ON "RISK?"

My parents and I went through a time when we talked about my dropping out of college. They thought it was a very bad idea. They said stuff like "You've got to get your priorities straight" and "What's going to become of you?" So I dropped out, but I knew that if the company didn't take off I could go back to school. It wasn't like college wasn't going to be there in another six months or a year.

—Michael Dell,
founder and CEO of Dell Computers

Take that leap of faith. If you stay sheltered and in the shadows, you're not going to live life. If you hold back and never do anything, you just kind of float.

—Rob Bollinger,
artistic director of Cirque du Soleil's O

There is never a perfect situation, but you have to be willing to take a risk. Not a blind risk, but a calculated risk. I think that people in general are afraid to make decisions because they're afraid of the downside, but you can't ever be afraid to fail.

—Jerry Colangelo,
chairman and CEO of the Phoenix Suns

You have to risk all that you have. If you just risk a little of it, that's not a risk. If you're going to do it, you have to go all the way. I feel very strongly that we must all go through something, if we are to get something. If you don't lose your life, you can't gain it. Lose something, so you can live.

—Reverend Cecil Williams,
CEO and minister of Glide Memorial United Methodist Church

Success on whose terms?
Are you going on society's terms, your parents' terms, or your terms?

—Rob Bollinger,
artistic director of Cirque du Soleil's *O*

Success?

Success is a funny word. Don't buy that one. You have to figure out what your own definition is—and everybody has their own definition. You shouldn't buy that "being a doctor is successful." Why is that exactly?

—*Alan Webber,*
founding editor of Fast Company *magazine*

When we were driving through Maine, we bumped into a lobsterman who had a lasting impact on us. We would interview him later on his boat at dusk, right after he brought in the day's catch. He was one of the most successful people we met on our trip.

During our conversation, Manny shared how he went from working in a secure job that he hated to becoming a lobsterman. "For a while I did the nine-to-five, yes sir, no sir, can I have a raise sir. But the whole time I was doing it, I wasn't happy. I couldn't wait to get back out here on the water. So I went into lobstering. Now I wouldn't do anything else if given the choice."

For Manny, success wasn't evaluated by traditional standards. He loved smelling the salt in the air and feeling the satisfaction of bringing in the day's catch. Success meant being happy. That was the way Manny measured his life, even though there were people pressuring him to become successful in a more traditional sense. "My father put a lot of pressure on me. I was going to live my father's dream, my father's life."

But even the most traditionally successful people we met on our trip backed Manny's version of success. Michael Dell, CEO of Dell Computers, shared with us his perspective: "Everyone has his or her own definition of success. For me, it's happiness. That's the most important thing. Do I enjoy what I'm doing? Do I enjoy the people I'm doing it with? Do I have time to do things with my family and do things I like? That's what it's all about."

The day after we met with Manny on the coast of Maine, we roadtripped down to Boston to interview Ben Zander, conductor of the Boston Philharmonic. Ben had his own take on living a successful life: "I don't measure life by success. I measure it by how much opportunity I have to contribute. I've always had an absolutely clear notion that I could be a teacher, and in that sense my life is about contribution."

The common thread between Manny the lobsterman, Michael Dell, and Ben Zander is that they all have their own definitions of success. They took the space to identify what it was they valued in life, then built their lives around it. They don't base their lives on external pressures that condition them to conform. For them, success is a life that directly reflects what they believe in and who they are.

MORE ON "SUCCESS?"

Forget about what sells. Forget about what is going to make you famous. Do what you like, what makes you happy, and success will follow.

—*Jill Soloway,*
writer and coproducer of Six Feet Under

It's not about the money. It's not about what you're supposed to do or what somebody tells you to do. Just do what you love and it will be awesome.

—*Kim Walker,*
owner of Outdoor Divas

Can you achieve your dream if it's financially driven? Perhaps in the short-term, but it won't be sustainable because it's shallow. It will not fill your heart. It will not fill your life.

—*Howard Schultz,*
chairman of Starbucks Coffee Company

It's all about sincerity—not thinking about money or how much you could sell something for, just thinking of something that's pretty cool. Hell, I felt successful just for being the guy you could come to for a wallet.

—*Paul Frank,*
cofounder of Paul Frank Industries

Most of the noise that people give you is a reflection of their own sense of inadequacy about what they did. So what you have to find is the peace and the silence inside yourself to know what you want. It's your life, not anybody else's.

—Peter Seligmann,
chairman of the board and
CEO of Conservation International

Shed the Noise of Conformity

Even if people say "No, you don't have any talent" or "No, you won't earn any money," I say go ahead and do it. Do it with your whole commitment. Do it knowing that you're the one who wants to do it—it's your choice and your responsibility.

—*Jane Smiley,*
Pulitzer Prize–winning novelist

On the road, we got away from all that noise back home, telling us who to be and what to do with our lives. We experienced firsthand the importance of casting off that interference so we could see ourselves more clearly. If we hadn't done that, what we thought excited us could have been just a reflection of what society had conditioned us to be excited about.

Randy Komisar, a start-up guru in San Francisco, used to be a product of this noise. He went to Harvard Law School and "didn't like it from the day I got there." He went to work in the legal profession until he determined that he had "no affinity or passion for being a lawyer and needed to deal more with ideas and creativity." So he extricated himself from the pressures holding him back from leaving a "very successful" career, and left. "When you speak to people who have all the trappings of success but are really unhappy, there's a common syndrome: They've crossed a lot of hurdles, but they weren't their own hurdles. They were someone else's hurdles."

So who sets up these hurdles for us? And why do we continue to jump over them? Randy is the perfect example. "There I was on my path, completely unhappy but very successful. I was making good money, but I wasn't one iota closer to happiness."

Determined to bring happiness back into his life, Randy started the process of shedding the noise. "My transformation was not immediate. First I stripped away layer after layer of social bias. I had to deal with my fear that, by not being a lawyer, I would lose something that I had gained."

After living this process, he has an enlightened perspective on what life is really about. "Engage in what truly motivates you now. Don't defer it by wearing a suit and going to work on Wall Street in the hope that you'll put away enough money to figure out what you really want to do. You will never get there that way. I can spout off about this because I made that mistake."

Listening to Randy and others share their stories about how they freed themselves may have saved us from making that very same mistake. We were on the brink of going into medical school, consulting, and the family business. The roadtrip and the stories we heard got to us right in the nick of time. Without that experience, we might never have realized that the noise was in fact just noise—and that we could shed it from our lives and define our own roads. As Howard White from Nike told us, "Everybody tries to define who you are. So are you going to let them define you, or are you going to define yourself?"

MORE ON "SHED THE NOISE OF CONFORMITY"

I remember in school when I was taking film production classes I had a professor telling me, "It's a one in a million chance to get into visual effects" and "It will never happen to you." Well, you know, sometimes I just want to call him up and say, "Here ya go."

—Jim Mitchell,
visual effects supervisor with Industrial Light and Magic

We put expectations on people that don't have to exist—you need time to figure out what you want to do and what you're good at. I think that's why the twenties are so difficult because you have this expectation that you have to figure it out when you really don't have to.

—Phil Marineau,
president and CEO of Levi Strauss

When I was in college, I thought the whole purpose of the game was to convince other people that I was good. I spent all my time and energy compelling others to believe that I was good at what I was doing. And after college I think I had this realization that it wasn't about that, it was about convincing *myself* that I was good at what I was doing. It sounds really simple, but it really changes the way you look at the whole game. Once you've convinced yourself, all the rest falls in line.

—Chris Flink,
product designer with IDEO

I talk to a lot of guys from my generation who went into university, got a degree, and are unsettled and unhappy because they did what their parents wanted them to do—but they know in their hearts it's not what they wanted to do.

—Ric Birch, executive producer of the
Sydney and Barcelona Olympic Games ceremonies

You've got to listen to your inner voice. What do you like to do? What do you really get excited about? Everything you do in life should revolve around those things.

—Katherine Cohen,
founder of IvyWise college counseling

Harness Your Individuality

Distinction is everything. If you look at any example of people who have been successful at what they are doing—not monetarily successful, successful in their life—they are generally pretty distinctive individuals. I think that can be a farmer in the boonies of Vermont, and it can be the most progressive artist in New York City.

—Michael Jager,
founder and creative director of JDK Design

Most of the people we interviewed out on the road focused on harnessing their individuality rather than stifling it. They identified what made them distinct and then built their lives around it.

When we roadtripped through Vermont, we sat down with Michael Jager, founder of the design firm that does all the creative work for Burton Snowboards. "I realized I sucked at math early on, so I focused on artwork. I always had a feel for how to draw, so I channeled myself in that direction." Building on what made him distinct, Michael went to design school in Montreal, and later he went on to start his own small design firm out of his basement in Burlington. When Jake Burton first started his snowboard company in the late 1970s, Michael did all of his design work. Over time, as Burton Snowboards exploded into an entire industry, Michael's design firm grew. Today JDK Design operates out of a four-story brick building in Burlington Vermont, where it works with several clients, including Swatch and IBM.

After dedicating much of his life to allowing himself to be his own person, Michael had this to say: "I think it was Simon Woodruff who said, 'The world will conspire to support you if you really magnify what it is you believe in.' I think you'll find that if you're going to be a sculptor or if you're going to be a kick-ass accountant. If you really believe it, wear it on your shoulder, and do it with distinction, the world will conspire to support you."

When we roadtripped through San Francisco, we met with Beatrice Santiccioli, a color designer who works for Apple Computer. Growing up in Italy, her passion was always the colors she saw around her—in the fields and on the trees. "Since I was a child, I loved to work with colors." Similar to Michael, she harnessed her individuality and went to art school in Italy to further develop her talents. Today, she creates the color schemes for Apple's computers. She works out of her own small design studio in San Francisco and has built a life around what makes her distinct—her passion for color.

When we got home from our trip, to a certain extent we harnessed our own individuality. What we went through on the road differentiated us, so we started to build our lives around sharing that adventure. Similar to Beatrice and Michael, what made us unique would not launch us onto traditional career paths. Instead of letting that deter us, we harnessed our distinctive experience, and the world slowly conspired to support us.

MORE ON "HARNESS YOUR INDIVIDUALITY"

For me it started with a vision of knowing who I was and what I wanted to do. In college I rode bikes, I raced bikes, I worked on bikes—I totally loved bikes. So I thought that there must be somebody somewhere who made decisions on how bikes came to be. Someone who decided what color it was going to be and what components are on it. I thought that's something I'd like to do, so I went from there.

—Steve Gluckman,
bike designer with REI

In Sunday school, I put out the Sunday school newspaper. When I was in high school, I put out the high school newspaper. And when I was in college, I put out the college newspaper. I realized that there was a pattern there.

—Alan Webber,
founding editor of Fast Company *magazine*

The first step is self-knowledge, and self-knowledge is a complicated process because it's something that should be happening all the time. It's really a frame of mind. You can go through every day in your life gaining absolutely zero self-knowledge, or with the exact same circumstances you can be very introspective and you can learn a great deal.

—Randy Komisar,
virtual CEO and start-up guru

I would do what I do
even if nobody was willing to pay me for it.
It's the one thing I'm good at. It's the one thing I love doing all day every day.

—Stan Richards,
founder of The Richards Group

Work versus Your Life's Work

I love to cook. I wanted to learn about cooking. Forget the career. If the career happens as a byproduct of my pursuit of learning more about food and cooking, so be it.

—*Charlie Trotter,*
chef and owner of Charlie Trotter's restaurant

Before we hit the road, we had been programmed with the notion that "work" was the be-all-end-all of human existence. There was no higher calling. But on the road, we found a clear distinction between work and your life's work: *work* meaning something you do to pay the bills, and *life's work* meaning something you do to bring passion and fulfillment into your life.

We learned that work and your life's work can be separate, or they can be integrated—and they can be separate at some points in your life and integrated at other points. For most people we met, work and their life's work only became integrated later in their lives; very few people got paid to do their life's work right off the bat. But the important thing was that their life's work always had priority over their work. They figured out how to pay the bills, but not at the expense of what meant most to them.

Ann Powers, a writer and former music editor for the *New York Times,* talked about the idea of "dedicated poverty": "You have to be willing to live on ramen noodles for five months straight. That's how I did it when I was making no money." At any time, Ann could have gotten a sales job to bring in more money for nicer meals, but that wasn't her priority. Writing was.

Ben Younger, a filmmaker we met in New York, had a similar story. He was a successful young campaign manager in the political world, but he left it all to follow his passion. "Who cares if I lived on nothing? I ate rice and beans for the next three years, but I was happier than when I had more money and a car." For years Ben worked as a waiter and did several odd jobs, all the while making short films on the side. He would go on to write and direct movies such as *The Boiler Room* with Ben Affleck and *Prime* starring Uma Thurman and Meryl Streep.

The people we met were driven by a higher set of ideals than simply receiving a paycheck. They got us thinking about what our life's work could be. Of course, we'd be facing student loans and debt when we returned from the trip and would have to deal with that, but what about our life's work? What mattered to us? What were we passionate about? What could we contribute to the world? The answers didn't come quickly, but our thinking had been jump-started by all of these stories, and over time our own roads would become clear.

MORE ON "WORK VERSUS YOUR LIFE'S WORK"

I think the biggest sense of fulfillment I've had career wise is going to see the film on the big screen, totally anonymously, in the dark, listening to peoples' reactions. These little tiny decisions that you made at like five o'clock in the morning, over fifty cups of coffee, and you had no idea whether they were good decisions or really dumb decisions. You watched it a few times and you think in your own head that it's going to work, but you don't really know until you sit with an audience and you can feel a reaction and feel the laughter or you feel the tension around you, and you're like, wow, it worked, and nobody knows who I am.

—Jehane Noujaim,
filmmaker

My first day in a kitchen they paid me $3.10 an hour. Since that day, nineteen years ago, almost to the week, I don't think I've ever had a job. I feel guilty at times because I do something that I can't believe that I can make a living at because I would do this gladly for a lot less.

—Charlie Trotter,
chef and owner of Charlie Trotter's restaurant

The very first time I ever did sound, there was something magical about it. A film comes to life in an interesting way as you add sound to it: it becomes three-dimensional. You take a flat movie that's not really working, and you add some simple sounds, and it comes to life.

—Gary Rydstrom,
sound director with Skywalker Sound

No amount of money can make you feel happy and productive. There are people that go into certain professions because they think they'll make a lot of money.
I have friends in those professions, and they hate it, but they can't get out because they need to make that money to preserve their lifestyle.
To me, that is a disaster.

—Tracy Westin,
professor at USC Law School

Don't Get Trapped

I have friends who are truly unhappy now. They got married, had a couple of kids, and are locked down. It's not that they don't love their families, but they just got into things so fast before they figured out what they wanted to be as individuals.

—Ben Younger,
filmmaker

Jehane Noujaim, a thirty-three-year-old filmmaker from New York City, went to Harvard and double majored in art and political science. Art was her passion, and political science was her way to make a living and have a successful career. When she graduated she had to make a decision—would she go to law school and buckle down? Or would she follow her passion? She chose the latter. "The first year out of college was probably the hardest year of my life. A lot of my friends went into consulting or banking because they didn't know what else to do. They all had assistants and were making $60,000 a year while I was traipsing around with a little video camera. I had no idea where I was going."

Jehane spent her twenties struggling financially, but now in her thirties, she is a noted documentarian and has carved out her own niche around something she's passionate about. Meanwhile, her friends who trapped themselves in the banking world are burned out and starting over in their thirties. "The people I know who went into consulting and banking had a really hard time. The analyst programs are hellish; they burn people out after two years. There's no way you can work that hard if your heart isn't in it."

Getting stuck on a road that doesn't resonate with what excites you closes you off to opportunities. If Jehane had gone on to law school and not pursued her passion for film, she would have never met Kaleil, the CEO of a start-up company that would be the subject of her first hit documentary film.

In Colorado, we met with Gary Neptune, a mountain climber who opened up a local outdoors shop in downtown Boulder called Neptune Mountaineering. Gary has been on numerous mountain climbing expeditions, including to Mt. Everest, and has lived one of the most untrapped lives we've seen. "Don't get yourself locked into something you can't get out of. I know people who have gone to med school, and they end up in unbelievable debt. Sure they have a way bigger house than I do and drive a fancy car—but they are so far in debt, they can't get out of their job and some of them hate it. They're being worked to death, and almost exploited. And they're barely making ends meet—they can't go on a climbing trip with us. For all that stuff they seem to have, they're actually nearly broke. They don't have any more than I do."

Once you're passionate about something and it's a path that's an honest reflection of what you stand for as an individual, go for it. Everyone echoed that message. At the same time, the people we met urged us to take a second look at the decisions we make and question whether those are our decisions or someone else's. Michael Jager, founder and creative director of JDK Design, told us, "The thing that saddens me most is when you see an individual go down a path and burn about ten years of their life on a channel they now realize was the wrong one." Don't be that person. Don't get trapped.

MORE ON "DON'T GET TRAPPED"

I'm in contact with people I went to college with who have these jobs where they work forty hours a week and it goes toward paying off their mortgage, buying that car, or saving up for that one weekend in Bermuda. Then, here I am touring Europe for two months with a band. Tell me, who is living a better life?

—*Mike Thorne,*
coordinator of Maximumrocknroll *magazine*

So ask yourself, "What can't I do when I'm encumbered with family and home ownership?" It's key that you listen to yourself and know what you really want. You can't ever go back in time, and some choices are harder to make when you're older.

—*Deb Callahan,*
president of the League of Conservation Voters

Unfortunately a lot of people conform and go through life being unhappy. If you find yourself in that situation, do everything you can to get out of it. If you're in a place where creative thinking isn't appreciated, maybe you should go to a different place.

—*Michael Dell,*
founder and CEO of Dell Computers

There is no class for working at Burton Snowboards.

—Dave Schriber,
vice president of marketing of Burton Snowboards

Unplug Your Education

I always tell my students there's an inverse correlation between the grade they get in my class and how well they're likely to do when they graduate. In a lot of structured environments, you grade along a dimension. I don't necessarily measure creativity. I don't measure attitudinal characteristics about certain kinds of things. I think you have to set standards for yourself that are very personal, not standards that other people dictate.

—William Sahlman,
professor at Harvard Business School

On the road, we discovered that most of the people we met with had not built their lives around what they studied in school. Ben Younger, a filmmaker in NYC, was a political science major. Pat O'Donnell, CEO of Aspen Skiing Company, studied engineering. Michael Dell was pre-med.

The idea that we have to choose a major in college and then use that major to define our lives isn't accurate anymore. While academia is a valuable part of your life that teaches you how to think and look at the world, the boxes it puts you in can isolate you from discovering your life's work.

We ran into Sir Ken Robinson at the Getty Museum in Los Angeles, where he works as a senior advisor on educational policy. It turned out that Sir Ken was a former professor at Warwick University in England, which gave him a unique perspective on how academia can shape how people look at our lives.

"The whole education system has a hierarchy of subjects built into it that is very skewed and partial. At the top is languages and mathematics, then the humanities, and at the bottom are the arts. Doing art is not thought to be as worthwhile as doing mathematics and sciences. At fourteen or fifteen years of age this hierarchy starts to kick in, and it overrides the real talents that most people have. It's why a lot of prominent artists, musicians, and dancers didn't do well in school—because what they could do well wasn't valued. They felt alienated. One of the consequences of this model is that most people go through education never discovering what they're good at because schools aren't looking for what you're good at. They're looking for skills that they can sell off. People think that education is about following the natural grain of your ability. But it really isn't about that. Education is intended to be a system of social engineering. It was designed, for the most part, in the eighteenth and nineteenth centuries to be a compulsory system that everyone was required to go through. And the reason it came into being was the industrial revolution. If you look at the industrial economy, it was 80 percent manual work and 20 percent professional work. But the world now is so unlike the world during the industrial revolution. There is an economic and cultural revolution happening, and the education system hasn't kept pace with it. Our education system is still based on conformity. The industrial model runs right through it. But because of this new revolution, we can better expose what we're capable of. And by doing that, we can literally create our own realities."

Sir Ken helped us to understand the frustrations we were feeling with academia before we hit the road. It didn't mean that we should sit back and blame our universities for our confusion, but it did mean that we needed to hit the road to broaden our educational experiences and not rely solely on what we learned in the classroom.

Prior to hitting the road, we were planning on defining our lives based on an academic perspective. But on the road, we found reasons not to. We encountered stories from people like Michael Dell, Ben Younger, and Pat O'Donnell, whose lives didn't follow that template. And we were exposed to viewpoints from people like Sir Ken and Dr. Sahlman, who questioned the current academic model. For the first time in our lives, we started to think for ourselves and define our roads based on what mattered most to us as individuals, rather than a box we had lived in during school.

CASE STUDY #1: "Unplug Your Education"

Early in the morning on November 22, our "Mini Roadtrip Nation" from The Future Is Mine: Student Project at Brownsville Area High School hit the road in our big yellow school bus. We set out for the Laurel Highlands mountains for three interviews. First on our agenda was Mr. Joseph Hardy, owner and founder of 84 Lumber and Nemacolin Woodlands Resort and Spa.

It was a foggy morning, and the kids and I were nervous wrecks. We made several stops on the way up the mountains to film little pieces for our video. We would stop the bus when we saw a location that grabbed our attention, and out we would jump with our film crew borrowed from the school's drama class. On our way to the interviews, we still couldn't believe that we had gotten through to a man recently listed as one of the four hundred richest men in America. The kids had simply made one cold call, and before we knew it we were being contacted by Mr. Hardy's personal assistant and were being "penciled in."

When we arrived at Nemacolin, we were greeted royally and felt so important. We were lead through one of the most beautiful resorts we had ever seen to a grand ballroom that had been set up especially for us. When Mr. Hardy joined us, we were surprised at how down to earth he was. He spoke to the kids for hours and was so candid and frank with them, answering any questions they had. He talked to them about how dreams can come true, but he showed them how you have to get out there in the world and see what it has to offer to make things happen. We all left there wanting to be entrepreneurs.

After leaving Mr. Hardy, our day was just starting; we still had two exciting interviews set up, one with the executive chef at the Golden Trout at Nemacolin and the other with a supervising park ranger at Ohio Pyle State Park. It was so much fun going to three such different environments.

This whole process was an awesome experience for all of us; it really built the ego of the entire team. The kids saw firsthand that sometimes all it takes is a simple call and a little initiative to make things happen. I think we all started to realize what a big world it is and how much more we wanted to see. We loved being on the road that day, and we are looking forward to our bus pulling out again.

—Lynn Jellots,
instructor at Brownsville Area High School, western Pennsylvania

CASE STUDY #2: "Sacramento Roadtrip Experience"

On January 25, 2005, with little more guidance than an assignment sheet and the enthusiasm generated by watching the Roadtrip Nation documentary, forty freshmen from Sacramento State University ventured into California's capitol city "Roadtrip style." Equipped with a video camera, a digital camera, or simply a tape recorder and notebook, they interviewed people from various career fields to learn how they got there. The kids traveled by car, bus, or on foot, and the results were amazing.

The students are enrolled in the Educational Opportunity Program (EOP), a system-wide program of the California State University. They take a freshman seminar course designed to help them understand the college environment and how a student's personal and educational growth relates to higher education and to selecting a college major. The Roadtrip Experience was introduced as part of the course to help students learn firsthand what the career exploration process is all about. The Sacramento Roadtrip project was awarded a certificate of recognition for unique programming by EOP leadership and was one of the projects featured at the semester-closing event attended by all EOP freshman seminar students and faculty.

The Sacramento Roadtrip Experience is one of four major assignments in the freshman seminar course. This is a brief outline:

1. Discuss Roadtrip Nation's "finding the open road" experience. Show the Roadtrip Nation documentary film.

2. Students form teams of three to four members and receive a written copy of the assignment.

3. Instruct teams to interview five of the most interesting individuals in the community.

4. Students are workshopped on how to cold call individuals, book the interviews, and videotape or record the conversations.

5. Students prepare a presentation for class and write a summary paper for the instructor. Class presentations can consist of oral reports, video clips, PowerPoint slide shows, posters, and photo binders. Project must be completed within six weeks.

Students surprised themselves with their accomplishments. One student said, "Last year when I was a senior in high school, I wouldn't have believed I'd be out in the community asking leaders questions about their jobs and their personal lives." Another said, "I got over a lot of my shyness. I'm not so afraid of talking to strangers at college or any place now."

The Roadtrip assignment helped accomplish many of the course objectives and goals related to managing time, finding and using resources, building self-confidence, and understanding the process of learning. Students also had to work cooperatively and develop priorities within time limitations. In general, students felt empowered. Some students actually did not want to stop their interviews. They wanted to "go next door" to find out what the next person had to say. Overall, the program was a powerful way for students to become engaged with their learning experience and exposed them to some of the vital links between the university and the community.

—Al Striplen,
counselor and instructor at California State University, Sacramento

The more experiences you have, the easier it is to **connect the dots** later in life.

—Tinker Hatfield,
vice president of Innovation at Nike

Hit the Road

Today, the number of things you can do with your life is so diverse and so interesting that you will never guess what they are. You have to go explore the world and experience them for yourself.

—Tracy Westin,
professor at USC Law School

Since the new world we're living in is so broad, and since the educational system is not set up to expose all the new possibilities, we need to get out there and explore for ourselves. The road can be a new style of education, a process rooted in exploration and experiential learning. Ben Franklin said it best: "All education is self-education." By engaging in this process of self-education, you'll expose yourself to new roads that you never thought existed—roads that could lead to your life's work.

As Alastair Paulin, managing editor of *Mother Jones* magazine, told us, "Experience the world. When you get that time off, take that opportunity. Take that roadtrip across Laos. Follow that band for the summer. Yeah, you could spend your summer studying for the LSAT, but if you can create some space, if you can create some kinks and creases to get off the path and do other stuff, go for it. Those experiences will have amazing repercussions in ways you don't even know about. In ways you can't explain to your parents."

Those words perfectly capture the impact the road had on our lives—completely unpredictable, but totally life changing. The experience challenged the roads we were on, clarified what was important to us, and gave us the confidence to redirect our lives.

The road gives you that clarity and confidence by exposing you to vantage points that let you see your life in a new light. As Tom First, cofounder of Nantucket Nectars, told us, "It's like hiking. You get to a spot and see a vista that you never knew existed. And then you come to another place and see another unexpected view. Out of college, you're right at the base of the hill. You wonder what the hike will be like, which path you should take. But there are a million different paths and many of them connect. After twelve years out of college, I felt like I had gotten to another place, and all of a sudden I saw things that I had never seen before."

And here's the most interesting part: To see those vistas that you never knew existed, you don't even have to physically leave your part of the country—because hitting the road is less about putting miles on the odometer and more about changing the way you look at the world.

If you open yourself to experience, hitting the road can happen every day of your life. Try starting up a conversation with that person you're sitting next to in the coffee shop, whom you wouldn't normally speak with. Try driving to work through a neighborhood you've never seen, instead of on the freeway. Try picking that independent documentary you wouldn't normally rent at the video store. Hitting the road is simply about opening your eyes to the world around you. Doing that on a daily basis will condition you to encounter new experiences, revealing potential and possibilities that you didn't know were there.

Since we've returned from our first trip, we've been on many others. What started as a temporary exit off the freeway turned into a totally new way of looking at the world. We've thrown out the road map, opened our lives up to possibility, followed what excited us, shed that noise around us, and redefined how we look at success. Now we're in the middle of finding our life's work—which happens to be helping others find theirs.

Did we need to hit the road in the first place to figure this out? Maybe. But it doesn't mean that you have to. Everybody's set point is different—maybe you're Clark Kent hiding your true self in a suit and tie, or maybe you're an avant-garde artist shaking things up in New York City. Either way, we all have room in our lives to hit the road—and the day we think we don't is the day we start moving backward.

We all have room in our lives to hit the road.

"So, what d(

watch road

PART 3
A GUIDE TO CREATING YOUR OWN ROADTRIP

Do You Need to Hit the Road?

Why did I need to hit the road? Well, it was spring of my junior year of college, and I was at a point where my life was standing still. Everywhere around me, people were moving, pushing, asking, "What's next?" I felt like I was being pulled in different directions, yet I was merely a passive observer, watching as my peers on their rigidly planned tracks assessed my uncertainty with smug smiles. The phrase "waking up in someone else's dream" was starting to become more than lyrics I would sing along with—it became my reality—waking up in the middle of the night, short of breath, thinking, "What am I doing? What next?" The "itch" was more than just restlessness; it was a gnawing fear telling me I had to do something. There was more for me to see, and I hoped my fears would be put to rest somewhere along the road.

—*Candace Elliott,*
age 22, Chicago

Do you need to slow down for a moment, take that next exit, look around, and reassess the road you're traveling on? Maybe you'll become more confident that you're traveling on the right path. Or maybe the exit will reveal something new to you—a road you never knew existed, something inside yourself that you didn't know was there, or a new opportunity that just happened to be over the horizon. In any case, all you lose is a few moments off the freeway—but what you gain may be a significant life experience that launches you down an entirely new path.

Whether you're a college student anxious about the next step, a twenty-something still waiting to find that spark, or a thirty-something locked into a track but searching for something new, a roadtrip is a great way to stir things up. Out on the open road, far from home and everything associated with it, things are clearer. Our lives, what we value, what we stand for, what we're passionate about—all start to emerge in a new light. That noise that surrounded you at home—society telling you who to be and what to do with your life—is reduced to a subtle buzz, letting answers emerge for the first time.

And it doesn't take long. If you're open to it, sometimes you can feel the impact of the road in the first few weeks, maybe even the first few days. It doesn't take a three-month cross-country expedition to create some space in your life. Launching away from your comfort zone and experiencing new events that open your mind is all you need.

Our first roadtrips were not of the three-month variety in green RVs. They were ten-day trips up the coasts of Northern California and Oregon to find new waves to surf. That was our itch, and through it, we changed. There was nothing like packing up all the boards and pointing our old truck north toward Oregon. Along the way we started talking to people about their stories, and a new dimension was added to our journey. But in the beginning, it was about hitting the road—getting away from Southern California and everything associated with it—and creating fresh experiences, seeing new places, and meeting different people.

A ten-day roadtrip may work for some people, but others may need to really open their lives up by taking an extended trip. When Alastair Paulin was thirty years old, he was living in San Francisco and working as an investment banker. After years of "faking it," he quit his job and hit the road for eighteen months to figure out who he was and what he wanted out of life.

An Extended Trip

So I ended up at this investment bank. It was the last place in the world that I should have been, but they offered me more money than I ever imagined I would make. I hated it. You know, I was just faking it. So I quit that job, took the money that they paid me, and my wife and I went backpacking around the world for eighteen months. We sold everything. We gave up our apartment in the city, had one little backpack each, and just hit the road. Neither one of us was doing what we wanted to be doing with our lives, and we hoped that a big part of that trip would be to find a sense of who we really were. Who would we be when we didn't have daily expectations? What choices would we make if we didn't have to be somewhere the next day—when we didn't have to go to a meeting or be on that conference call? We thought that maybe the freedom would help us work out what we really wanted to do with our lives. So in a way, it was our own Roadtrip Nation; we were definitely searching for those answers. Maybe it wasn't a great idea financially, but I would say that it was an investment in ourselves and in our life experience—an investment that would help us figure out what we wanted out of life.

—*Alastair Paulin, managing editor of* Mother Jones *magazine*

When Alastair got home from his trip, he had to start completely over, which ended up being a good thing. One morning he was going through the *San Francisco Chronicle* and found a help-wanted ad for assistant editor of *Mother Jones,* an independent magazine whose roots lie in a commitment to social justice, funded primarily by the Foundation for National Progress. When he saw that ad, he thought to himself, "That's the exact place that I could see myself. That is the right fit for me." So he applied, got the job, and over time worked his way up to being the magazine's managing editor.

Alastair needed to go through a significant life experience that showed him who he was and what he wanted out of life. Not everyone needs an eighteen-month trip to find that clarity, but some of us may. At the end of the day, you're the only one who knows your appetite for the road and how long you need to be out there.

Everyone has a different starting point, but all of us are programmed to a certain extent. We all have pressures and expectations circling around us that influence many of our actions. The beautiful thing about hitting the road is that the experience can deprogram our perspective on the world so our lives can start to be an honest reflection of the individuals we really are.

Andrew Smith hit the road with two of his friends from Pratt Institute in Brooklyn, New York. He had just graduated from college and was looking for an experience that would bring him some clarity. Out on the road, he felt this deprogramming happening in his life.

REASONS WHY YOU MAY NEED TO HIT THE ROAD:

- You've never seen what's outside your comfort zone.
- You're curious about what other roads are out there.
- You're feeling frustrated about where your life is at.
- You need to get away from the noise of conformity.
- You're still looking for what truly excites you.
- You're not sure what to do after college.
- You're stuck in a job but don't know what else to do.
- You feel like you've been living someone else's life.

Finding Clarity on the Open Road

On the roadtrip, I learned so much about different regions all across the country. But more importantly, I learned about who I am as an individual. There is no way to fully explain the personal changes I went through last summer on the roadtrip. Going someplace new every day allows you this strange freedom of self. Away from all the pressures you face back home, you're more open and receptive to the world and to the person you could be. I haven't felt that anywhere else but on the road.

—Andrew Smith, age 22, Brooklyn

Once the road begins to add this clarity to your life, listening to people's stories will significantly increase the impact on you. Their experiences will disprove many of the "truths" you've been taught, allowing you to see your life through a completely new lens. The world begins to look like a much bigger place, and you start to see the space in which you can carve out your own road.

Candace Elliott hit the road with her two friends from the University of Chicago in the summer of 2004. She shares how one of the people she met along the way changed the way she looks at her life.

Meeting People along the Way

When we began the trip, I was hoping I would be hit with a bolt of inspiration. However, I braced myself, pretty sure that I would probably just chalk up the summer as an amazing experience and then head off to law school in a year. Our interview with Keith Stegall, an award-winning producer and songwriter in Nashville, served as my bolt of inspiration and sent me from the interview reeling. I was able to draw striking connections between his story and mine. What resonated most was the way he took a huge risk for something he was really passionate about and how that risk led to success after success. After our interview with him, my tentative dream to try my hand in the music industry became more than just a thought in the back of my head—it became a tangible reality. I was overwhelmed to the point of tears to think that, yes, I could go to Nashville and try to break into the business I adore. As of today, this is what I plan to do.

—Candace Elliott, age 22, Chicago

Whether you're a college student in Chicago searching for your road, or a thirty-year-old investment banker whose life is not reflecting what you think it could be, one thing is certain: in this new world there is space for all of us to define our own roads. What will that road be? Creating your own roadtrip may not provide all the answers, but it will give you some of them. And those answers will lead to others. But in the beginning, it starts with finding the answer to one simple question: do you need to hit the road?

WHY IT'S IMPORTANT TO MEET WITH PEOPLE ALONG YOUR TRIP:

- Their stories reveal paths that you didn't know were there.
- Your life's work could be on one of those paths.
- People love to talk about their story.
- You could find a mentor to help you along your path.

Building Your Roadtrip

Create Your Route

Make an Interview Target List

Book the Interviews

Get Yourself Organized

Create Your Route

How do you create this experience for yourself? As Mike Lazzo, senior vice president of programming at the Cartoon Network, said to us, so much of it is "just about getting behind the wheel and going." But with some forethought, building a bit of structure into the trip can heighten the experience. That said, leaning too much on "the plan" could inhibit you from discovering those spontaneous experiences that lie just off the main highways.

The first step is to craft your route: Where will your journey take you, and what positions will you put yourself in to have the experiences you want to have? More specifically, what kinds of people do you want to meet? If you're looking to meet more of the boat-builder type, Maine or the Pacific Northwest might be your best bets. If you're into the art scene, maybe a roadtrip to New York is the perfect experience for you. Music? Nashville and Austin are definitely a must.

Whatever your reasons for hitting the road, make sure you create a route that puts you in places you've never seen before. Make the experiences fresh. Make them new. So much of the road is about getting out there and putting yourself in foreign environments that open up parts of yourself that you never knew existed. If you're traveling to

DIFFERENT METHODS OF TRAVELING:

- Car (a truck with a shell or a minivan is best).
- Motor home (borrow or rent an old one).
- Train (sleeper cars are a real bonus).
- Bike (for the brave).
- Foot (for the very brave).

DIFFERENT LOW-BUDGET PLACES TO SLEEP ALONG YOUR TRIP:

- Friends' and family's houses.
- Hostels.
- Campsites.
- Kinko's (some are open 24/7).
- Your car or RV in a hotel parking lot (our favorite—try to find one with a pool for bathing).
- The local baseball field dugout (last resort).

places you've already been, at least make sure you're doing things that you've never done there.

Throughout the journey, you'll get to know yourself a little better each day, coming a few steps closer to finding a road that connects with your individuality. But that all starts with what types of experiences you're putting yourself into, which may depend entirely on the route you choose for your journey.

Planning Your Route

We knew from the beginning that we wanted to go through the South. Not only do we all hold a soft spot in our hearts for country music, wide open spaces, and people who say "darlin'" and "y'all," we were excited by the incredible diversity the South has to offer. By simply crossing a state border, we would see various landscapes, taste distinct cuisines, and experience different ways of life. There were a few cities that we agreed upon as "must stops": Austin, New Orleans, Nashville, Key West, and Charleston. Also, a few interviews dictated specific stops. We then planned around these based on time constraints and logistics.

—Diana Dravis, age 21, Chicago

Once you've broadly defined the route around your interests, do some general research on what other types of things you could do along the way. With a little planning, you can route some extra experiences into your journey. And if you take it a step further, you can even set up some time to talk with the folks at the place you stop.

Meeting People at Your Travel Stops

We planned our route over dinner one night at our favorite Mexican restaurant in Brooklyn. Then we spent the next week gathering the specifics: What days did we want to be in what cities? How many miles is it from city to city? How long would it take us to drive? It's not easy to chart a five-week cross-country trip, but it sure is exciting. We planned to see places we had never seen before. I had never seen Devils Tower, never even been to Wyoming, so that was at the top of my list. Then I thought, "Why don't we interview someone at Devils Tower?" So we called up the woman who runs the park and sat down with her a few weeks later. That is one of the greatest things about a roadtrip. Sometimes the ideas are being made up as you go, and you catch yourself truly living in the spirit of the road.

—*Samantha Weiss, age 23, Detroit*

Make an Interview Target List

Later on, when you look back on your roadtrip, what will stick with you the most are the people you met along the way. To enhance that part of the journey and put yourself in situations that allow you to really connect with the people you meet, we suggest you dedicate some serious thought and time to creating a target list of individuals to interview.

So how do you figure out who should be on that list? First, the sky's the limit. If you approach people in an honest manner, they are more open and accessible than you might think. So don't limit yourself because "That person is too big-time" or "There is no way they would fit me into their schedule." A group of students from the University of Chicago really wanted to interview Hugh Hefner, the founder of *Playboy* magazine. Thousands of people try to interview Hugh every year, but that didn't stop these roadtrippers from giving it a go. After months of cold calling and emailing, they booked an interview with him. Sure, many others rejected the students' requests, but some agreed to meet with them. And Hugh was one of those people.

The point is: Swing for the fence. Don't let someone's title or what they've achieved in life stop you from trying to have a conversation with them. The worst thing that could happen is that they say no.

Have you ever read a book that piqued your interest? Put the person who wrote it on your list. How about your favorite magazine that you religiously read every month? Try booking a meeting with the managing editor and learn how they got to where they are. You can keep it practical if you want; if your passion is writing and you want to break into being a freelance writer, put someone who has done it on your list.

Everything that interests you represents a potential interview. When you're done watching the basketball game on TV, read the credits to see who the producers, writers, and directors are. The bookstore that blew your mind in Austin, figure out who started it. Your mom's high school friend who now works for the United Nations—why not learn how she got to where she is? Interesting people are all around you; all you need to do is take that next step and go listen to their story.

Where were they when they were your age? Did they ever face pressures to conform rather than explore what meant most to them? How did they turn their passion into their life's work? These people didn't always have it figured out. Most people were lost at some point and can relate with someone trying to define their own road. If your motives are pure (meaning, you're not trying to get a job out of it), you'll be surprised to see how many people are excited to share their stories with you and pass on what worked for them.

Carefully reading magazines is a great way to find interesting people to approach for an interview. By reading an article on someone, you should get enough information to know whether you want to dive in deeper or not. We've used this technique on all of our roadtrips—combing through magazine racks to learn about people we'd never heard of before. Erica Cerulo used the same tactic when she was trying to find people to interview for her own trip.

Use Magazines as a Resource

I am a magazine junkie, so flipping through the pages of issue after issue seemed like a reasonable and painless way to start our hunt. I was sold on Julie Gilhart from a piece I had already read in *Vogue*, Diana became enamored with Walt Mossberg from a profile in *Wired*, and Candace was impressed with an article on Neal Stewart in *Fast Company*. Sometimes we were particularly interested in a specific field and then went on a quest for someone prominent within that sphere. For instance, we searched for Pulitzer Prize–winning cartoonists to find Ann Telnaes. Friends and family are also good resources. My dad saw Lally Brennan and Ti Martin on a Food Network special, and one of Candace's professors hooked us up with Joel Klein. Ultimately, talking to interesting people about your adventure will lead you to other interesting people.

—*Erica Cerulo, age 23, Chicago*

When you're deciding who to put on that list, keep in mind that some of the people you choose should have nothing to do with what you're directly interested in—they can broaden your horizons to what the world holds. For example, Bernardo, Cristina, and Gloria, from California State University, Sacramento, interviewed Al Merrick, one of the world's leading surfboard shapers in Santa Barbara, California. The students didn't even know how to surf, let alone have a career interest in shaping surfboards, but the interview had a huge impact on their lives. Al challenged them to define their own versions of success and encouraged them to focus not on financial goals, but rather on what made them happy. They didn't walk away with new career goals, but they did come away with something much greater—a broader perspective on the world and a new way to measure their lives.

By meeting with people unrelated to your direct interests, you might veer onto new roads that unexpectedly inspire you. When Cristina was on the road, she interviewed someone at the FBI headquarters in Washington, D.C. She had no previous interest in the FBI, but then again, she had never experienced it—so how could she have known if there was an attraction if she didn't venture outside of her direct interests?

Finding New Roads

Prior to the roadtrip, my master plan was to get my Ph.D. in psychology, become a clinical psychologist, and open my own practice. I was set on this idea because I had declared psychology as my major. Even though I didn't think it would make me happy, once I had this plan I didn't want to question it because I didn't want to feel lost again. So I stayed with that major, and over time it became my safety net.

Once we decided to take the roadtrip, I was determined to book an interview with a psychologist. When the day came to interview the psychologist, I thought it would light me up and I would be at the edge of my seat and have all my questions answered. But I didn't. I walked out of the interview questioning, even more so than before, if becoming a psychologist was the road for me.

So the roadtrip continued to Washington, D.C., where we had an interview at the FBI headquarters. I had no special expectations for that day. We sat down with My Harrison and began our conversation. I'm not sure I know exactly why, but I couldn't wipe away this huge smile I had on my face. I was smiling so much throughout the interview that My Harrison even commented on it. Once the

»

interview was over, we headed to her office where she had a carpet depicting the FBI seal, which I wanted to make sure I stood on. She also had an FBI flag, which I made sure to stand next to. My smile continued throughout our stay in the offices.

Once we left the FBI headquarters, I took a major step back to evaluate what I was experiencing. I was petrified because I had never had this feeling before and wasn't sure what to do with it or what it was. This feeling and interest was not in my master plan. I started thinking back to the interview with Hardy Garrison, the storyteller. He had told us that we would know what our road was when we felt a spark. When he said that, I wanted to know when and where the spark would come, but he had just said, "You will know." So I started analyzing my feelings and realized that I had found my spark at the FBI headquarters. There was no way I could ignore this, not after speaking to all the people on our trip who followed their sparks and are happy. While I was afraid of this possible detour in my life, the numerous stories we had heard throughout the trip reassured me that I would be okay.

Since this experience, I am open to the possibilities that present themselves in my life, and I am currently following this call to someday work for the FBI. I wonder what my road would be like if I hadn't had the opportunity to find that spark . . . I'm happy to not know what that would be like.

—*Cristina Barajas, age 21, Sacramento*

At the same time, you want to put people on your target list who are directly involved with what you're naturally interested in. If you love animation, meet with someone at the Cartoon Network headquarters in Atlanta. If you love beer, meet with Paul Shipman in Seattle, who started Redhook Brewing Company. Are you into the art scene? Of course New York City is full of amazing artists, but you can uncover some of the most inspiring and creative people in Burlington, Boulder, Jackson Hole, or even Kansas City.

Do you love documentaries? There are independent filmmakers working on projects everywhere. Do some research, see who is working on what, and call them up to see if they'll sit down with you and share their story. Meeting with someone who is living the things that you're passionate about may show you how you can build your life around those interests.

Or, as with Cristina's experience in relation to psychology, meeting with people in your interest area may show you that your passion for that road is not as strong as you think, saving you years of time on the wrong path and giving you the chance to redirect your life. As Laurie Coots, chief marketing officer at the ad agency Chiat\Day, told us, "Sometimes learning what you don't like is just as important as learning what you do like."

Bottom line: When you're making your list of people that you want to meet on your trip, keep it balanced. Have a few people who are more practical, and have a few others who are out in left field—people who could blow your mind and show you possibilities and viewpoints that you never knew existed.

Book the Interviews

Once you've created a list of prospective interviewees, then it's time to take that next step and fill your calendar. All it takes is a little work, a lot of creativity, and a willingness to be flat-out rejected by one assistant after another. That said, you'll be amazed when you find how many people will rally to support your effort. Most of the time, you'll be dealing with the assistant to the person you want to meet, and these people are inundated with phone calls all day long from people who are trying to sell them this or pitch them that. They rarely get calls from people who are coming from such a genuine perspective: simply trying to listen to their story, no strings attached.

How do you find the contact info for these people? The answer is painfully simple—just consult the phone directory to find their office or home phone number. That's the best way to start. If they're not listed, try going online. If the organization they work for has a website, it will most likely have a "contact us" section that will have the general phone line. If a phone number isn't listed, don't let that stop you. You usually can find a general email address for the organization, such as info@organization.com. Also, you can just guess what your target's email address is; there are usually only six possibilities:

raysmith@organization.com
ray.smith@organization.com
ray_smith@organization.com
rays@organization.com
rsmith@organization.com
ray@organization.com

Don't feel awkward about digging around a little bit. If people are in hiding, it's not from people like you, it's from folks who are trying to solicit them for money or a job.

Remember, you're simply asking them to share their story with you. It's not a very demanding or threatening thing to ask of someone.

Cold calling is when the games begin. Call up the organization's general phone number and ask to be transferred to your prospective interviewee. Once you're connected, you will most likely talk to their voice mail or to their assistant. Sometimes you may talk with them directly, but that's pretty rare. Once you get through to a live person or to their voice mail, you've got to have your pitch down perfect. Here's an example of what Cristina Barajas did to book interviews along her trip.

Cold Calling

When I first started cold calling to book the interviews for our trip, I would wait for the person on the other end to answer and start on my minute-long speech that covered all the details of why I was calling:

"Hello my name is Cristina, and I'm traveling around the country interviewing people to learn how they got to where they are today. I was wondering if I could speak to Chris Wink and sit down with him to learn from his story. Maybe his insight could help me find my path. My friends and I will be in town on June 2 and would love to take him out for a cup of coffee."

Some people would direct me to someone else; others asked me to explain myself. To this day, I wonder if the person listening to my little speech understood what I was saying. Through the process of cold calling, my minute-long speech turned into the statement: "I would like to speak to someone who loves what they do." This phrase was the attention grabber. The people who answered the phone would start laughing and were curious and willing to help me find the person that was passionate about their job. Who would have thought this would do the trick?

—*Cristina Barajas, age 21, Sacramento*

Be as genuine and authentic as possible with your pitch. Be honest; you're someone who is simply in search of your own road in life, and you think that listening to their story may give you a spark. If you make the commitment easy for them—"I would love thirty minutes of your time; I'll buy the coffee"—who wouldn't want to pass on what they've learned in life?

Everyone gets nervous when they first start doing the cold calls. It definitely can be an awkward experience: here you are, calling completely out of the blue to ask someone to share their personal story with you. Even though the person on the other end of the line is often delighted to offer you their insight, that doesn't make the experience any less awkward.

Once you get it down, it's like riding a bike, but getting over the beginning hump can be a challenge. Here's Gloria's account of how she conquered her nerves.

Getting Over Your Cold-Call Nerves

When it came time to start making our actual calls to the people we wanted to interview, I was terrified. I had spent all this time and effort to find the individuals I wanted to meet, but the fear of picking up the phone and speaking to them was difficult to overcome.

Each week would begin with telling myself, "I need to call three people three times each this week." At nine in the morning I'd say, "I'll call at ten," and then at ten I would say, "I'll call at eleven." And this would go on until about five, when I knew that they were about to leave for the day. Then I would hurriedly call and feel so relieved about getting it over and done with.

Then someone recommended that I could work through this fear by calling information. Information? What did this have to do with anything? I wondered. The task was simple: find out how to cook a potato. Easy, you say, but this mission impossible still seemed too much for me. Petrified, I made the call.

The first thing the operator said was "What city please?" Then I'd say, "Well, actually, I want to know if you can tell me how to cook a potato." The operator quickly let me know that he had no clue on how to do this. This is crazy, I thought. Even I know how to cook a potato. The second time I called and got the same question, "What city please?" I let that operator know that I needed to cook a potato for my brother's birthday. This time, success—no I didn't get my answer, but I got transferred! So I was gaining confidence and started to elaborate on my story to the second operator. I told her that I needed to cook this potato for my brother's birthday because it is his favorite food. She kindly let me know that she didn't know how to cook a potato, but went around to the nearby operators to ask for help. Politely, I said thanks and hung up the phone. Now, not even thinking about being nervous, I called again. This time I was determined to get my answer. After speaking to three operators, the last one shared with me how to bake a potato. All I had to do was ask and I received. I didn't give up with the first *no* that came my way, a lesson that would help me go on to book the interviews we did on the roadtrip.

—*Gloria Pantojas, age 22, Sacramento*

Whether you follow Gloria's tactics or develop your own, at the end of the day just ask yourself, "What's the worst thing that could happen?" You'll find that the repercussions are not even real; they're just simple fears that have sprouted in your head. When you overcome those fears, you'll find that all things are possible and that any interview is open for you to try to book. Ignore the nerves. Just go for it.

Once you've nailed your pitch, the person you're talking to will most likely ask you to follow up with an email detailing when you'll be roadtripping through the area and exactly what you want to talk about. Sometimes they'll ask for a letter, but try to push for an email, because most physical letters get tossed. Most people would rather correspond by email anyway. Once you've sent them the dates and other info, check back in from time to time to ask if they need any more information from you. Especially as the date of the meeting gets close, reminding them about who you are and why they agreed to meet with you is a good idea. They probably have a million things going on, and by the time you arrive in town, your phone conversation may be a distant memory, even though you're scheduled on their calendar.

Get Yourself Organized

Unfortunately, this is a lesson we had to learn the hard way. Don't wait until the drive there to get directions to the interview location. And never give yourself just enough time to get there. Always try to get there twenty minutes early, because if you're anything like we can be, twenty minutes early always ends up being just barely on time. These people have given you a slice of time out of their crazy schedules, and being punctual is one of the best ways to show how much you appreciate their openness. Besides, if you actually do end up getting there twenty minutes early, you can go grab a cup of coffee or otherwise clear your head for a potentially beautiful conversation.

Before arriving at the interview, try to do some general research on the person you're going to meet with. Don't educate yourself too much on their story, or the conversation won't be fresh and organic. Just have some basic information on what they're known for and what they've done recently. For example, when we interviewed Lee Clow, creative director at Chiat\Day ad agency, we didn't know his complete story going into the interview, but it helped to know that he was the creative mastermind behind Apple Computer's "Think Different" and iPod marketing campaigns.

On the Road

Living the Open Road

While you're on the road, your guiding philosophy should be carpe diem, seize the day. Those little ideas that are in the back of your head, make them happen. Take that exit that strikes your imagination. Jump in that body of water that's calling your name. Go talk to that musician on the street who captures your curiosity. See the world with new eyes—without filters, inhibitions, or boundaries.

You can proactively build a sense of carpe diem into your travels by creating road-trip rituals. Maybe you want to have a cup of coffee at the local diner in every small town you hit, hike to the top of the tallest peak in every state you visit, or watch the sun go down every evening of your trip. These rituals will help you look at your trip with fresh, excited eyes, heightening the experience and helping you live outside your comfort zone.

On the road, Bernardo and his friends made a ritual of diving into every body of water they saw. Talk about getting out of your comfort zone—Bernardo didn't even know how to swim.

A Roadtrip Ritual

At the beginning of the road-trip, we had agreed to jump into every major body of water from the West to the East Coast. I willingly agreed to this, even though I can't swim. I thought to myself, "When will I ever be in any of these places again?" I felt it was a great opportunity to experience something outside of my normal life. And if there was one piece of advice I kept hearing on the road, it was "Take advantage of every opportunity." Coincidentally, when we interviewed Rita Simo, founder of the People's Music School in Chicago, she used the metaphor of the only way to learn how to swim is to jump in. There are two options: I can sink or swim— but if I want it bad enough I won't let myself sink. The Snake River was a little intimidating, but I wasn't going to learn how to swim by just standing on the shore looking in.

—*Bernardo Pantojas,*
age 20, Sacramento

TIPS FOR LIVING THE OPEN ROAD:

- Carpe diem.
- Create roadtrip rituals.
- Focus on absorbing the experience.
- Ask the people you interview what there is to do in the area.
- Share your travels with people to make local friends.
- Travel the side roads as much as possible.
- Stay away from guaranteed experiences.
- Leave your prejudices at home—on the road, you can learn from anyone.

Being in the moment and absorbing everything around you is essential during a roadtrip. After crossing the country for three weeks on our first big roadtrip, we arrived at the Atlantic Ocean in Bar Harbor, Maine. We had come so far, and finally seeing the sun going down with our backs to the ocean was enough to make us forget to grab the camera. We don't have a photo of that moment, but we'll always have the experience imprinted in a much deeper place.

On the road, you'll find that the majority of the people out there are open and welcoming to outsiders—if you approach them with a light and respectful attitude. Of course, there are some close-minded areas of the country, but even in those parts there are kind and open people who will point you in the right direction.

Most places you go to will have a thriving local culture that may be unlike anything you've ever encountered. If you can tap into it, your experience of that region could be that much deeper. When you're out around town, share your story with the people you meet. Let your waitress know that you're on this roadtrip, and ask her what there is to do around town. Most people will be intrigued to learn about your travels and may give you a heads-up about local stuff you can do or even interesting people you can meet with in the area.

One thing you can do is try to hang out at the non-generic spots around town. Instead of getting lunch at McDonald's, try going to the local diner on Main Street. Instead of staying in and renting a movie, try walking the streets if there's a downtown area. That little diner on the corner could serve up some sketchy food, but it could also serve up the best waffle you've ever had. Main Street could be completely dead, but you could also meet some locals who lead you to an entirely new experience. At home, you can get that Big Mac or rent that movie, but when will you ever be in that small town in the middle of Iowa again?

Once you start doing your interviews, be open to letting the experience lead to something more. For example, after we interviewed chef Charlie Trotter at his five-star restaurant in Chicago, we got to eat the meal of our lives in the kitchen with his entire staff. It costs about $300 a person to dine at this restaurant, so the thought of helping ourselves to anything we wanted was definitely an unexpected bonus—especially in the middle of a roadtrip devoid of decent food.

At the same time, don't try to force something like this to happen. Your instincts will tell you whether it's right or not. If the person seems to be in a hurry, just be appreciative of the time they've given you and let them run off.

The point is to be open if those opportunities and experiences present themselves. When Andrew, Samantha, and Jeff were on their roadtrip, they interviewed the park superintendent at Devils Tower. She ended up offering them a camping spot to park the RV in for the night. After interviewing the park superintendent, hiking up to the monument at sunset, and barbecuing at the campsite, the conversation definitely turned into a larger experience than expected.

When Bernardo, Gloria, and Cristina were on their roadtrip, they interviewed Chris Wink, cofounder of the Blue Man Group, a music, art, and dance performance group. They were offered tickets to the show, which allowed them to see what the person they met with had created. Seeing the show helped them understand how amazing Chris was, and interviewing him helped them understand what he had to go through to bring his vision to life.

More Than an Interview

Fortunately for us, Lauren at Blue Man Group was really cool and got us into the Blue Man show the night before we interviewed Chris Wink. None of us had seen the show, so it gave us an opportunity to see into Chris's personality and understand what he really enjoyed doing. It was an incredible show and one of the most memorable events on the trip. The show was music, comedy, and philosophy all rolled together. Chris and his friends had created this art from many different aspects of life that they enjoyed. The idea of mixing and matching what *I* like finally seemed so real.

—*Bernardo Pantojas, age 20, Sacramento*

Bernardo's experience with the Blue Man Group echoes the basic theme for living the open road: Make yourself open to experiences. Try things you've never tried before, look at things in a way you've never looked at them before, and engage in things that you've never experienced before. By following that basic philosophy along your journey, the road will influence you on a deeper level than you ever thought possible.

Interview Conversation Topics

So you've spent a few weeks booking the meeting, or you just ran into someone playing drums outside Wrigley Field. Either way, the day has come and it's time to sit down and listen to this person's story. What do you talk about? How do you start the conversation? How should you handle yourself?

STAPLE INTERVIEW TOPICS:

- Where were you when you were my age?
- Did you feel pressure from people around you?
- Did you always know that you wanted to do this?
- Were you ever lost?
- How did you get from there to where you are today?
- Did anyone help point you in the right direction?
- What kind of risks did you have to take?
- What would you consider your life's work to be?
- Do you have any final words of wisdom for people trying to figure out what they want to do with their lives?

Don't worry about dressing up; this should be a casual conversation, and you want your presence to reflect that. Right when you meet the person, be respectful, look them in the eye, shake their hand, and show them how much you appreciate their taking time with you. It's a big deal that they were open enough to share themselves with a stranger, so help them feel comfortable about making that decision. Bottom line: You want to be yourself with a good dose of respect mixed in.

Once you sit down and start getting into it, open up the conversation by sharing a little bit about yourself. Tell some of your own story, where you're at in life, and some of the issues you're facing. Be passionate and real. It's kind of like physics—every action has an equal and opposite reaction. If you express a certain level of authenticity and enthusiasm, the person you're interviewing will be comfortable enough to do the same.

We don't think it's a good idea to have a list of questions prepared—that will seem too much like a formal interview. Just have a few talking points in your head that you want to discuss: Where was this person when they were your age? Did they ever have pressure from family to conform? How did they deal with that? How did they figure out what they were passionate about and make it their life's work? How do you build a life around something that you believe in while still paying the bills?

During the conversation, you want to strike a balance between discussing current events and examining the person's personal story. If you're in Chicago and the Cubs just beat the Yankees, definitely indulge the person if they want to talk baseball, but don't let it go too far or you'll miss the meat of the conversation. You'll know when it's right to quit the small talk and take it to a deeper level. Sometimes the person may go off on a rogue tangent—let them go for it to a certain extent, but just as with the small talk, know when to rope them back in. This conversation is about their personal story, not the Cubs game or that dog they had in junior high school.

One thing you may encounter is that some people start to talk at length about the company they work for. This can be a big red flag. Their public relations departments usually train these people to talk about the latest earnings report or their plans for 2010. Your job is to get them off PR autopilot and back to their personal story. If you're interested, you can get all that other information from their website or a press release. Gently get the conversation back on track to learn from their story because that's something you can't get anywhere else.

During the interview, take notes if you want, but don't get so focused on writing that you miss the point of the experience. This is definitely not an assignment for school. Above all, make sure you're present and absorbing the conversation. Since we have taped all our interviews, we have an amazing library of audio recordings that have been great to go back to when we needed some inspiration or were feeling stuck. Also, if you use an audio recorder, you won't be so pressed to take notes, so you can really experience the conversation.

Lastly, if your conversation is beginning to take on a life of its own, let it. Sometimes a meeting scheduled for thirty-minutes will become a three-hour cup of coffee that deeply affects both you and the person you're meeting with. Be flexible enough to let the conversation take you where it may. If your interviewee has a friend come in during your talk, that new person may become interested in speaking with you too, and off you go to your next conversation.

A Wrigley Field Experience

When we interviewed Gary Pressy, the organist at Wrigley Field in Chicago, I was beyond excited. I thought the interview would be great on top of the thrill of just going to Wrigley Field for the first time. During the interview, we met the head of marketing for the Cubs, who was a really great guy. We asked him if we could talk with him after we were done with Gary, and he obliged. And on top of that, we got to have the conversation on the grass at Wrigley Field! I stood at home plate on Wrigley Field. I am a fan of baseball, but you don't need to be a big baseball fan to appreciate standing at home plate at one of the oldest stadiums in the country. After that, the public relations woman who helped us book the interview with Gary got us a great deal on tickets to the game that afternoon. Our seats were fantastic, and in the third inning we (each one of us touched it) caught a fly ball. Unbelievable! We had planned to just interview the organist, but as the experience evolved right in front of us, it ended up becoming something much better.

—*Andrew Smith, age 22, Brooklyn*

After the interview is over, be appreciative of the time the person just gave you. If you felt as though a bond was created during the conversation, reach out and ask if it would be okay to correspond from time to time. Usually email is best.

Finding Spontaneous Interviews

It's great to have a list of scheduled meetings for your roadtrip, but don't end your search there. While you're on the road, you'll spontaneously come across people you never would have discovered in magazines, online, or through friends and family. These people are diamonds in the rough. Society may not consider them to be leaders or cutting-edge thinkers, but they do hold many of the insights needed to define your own road in life—simply because they've done it themselves.

If someone is an independent filmmaker in New York City, she's not doing it for the paycheck. She's doing it because she loves that life, and she's overcome a ton of obstacles to make that path work for her. Imagine telling your parents, "Sorry, I'm not going to law school, I'm going to make documentaries." The wisdom such people have gained from going through their process is as good as any bit of advice you would get from the CEO of National Geographic, Michael Dell, or the director of *Saturday Night Live.* Be sure to leave your prejudices at home. On the road, you can learn from anyone.

The process of finding these people can be ambiguous. There's no specific formula, no research source, and no network for locating them. Often they're in their own worlds—their own niches. So to find them, you simply have to change how you look at the world.

On one of our trips, we had a flat tire on the Texas-Louisiana state line. We maneuvered the RV into an old truck stop at the next exit, and out came a small fireball of a man with a patchy beard, a black trucker hat, and a huge belt buckle holding up his grease-soaked jeans. Of course, he was scratching his head at our big green motor home, but it was clear that he ran one of the tightest truck stops we had ever seen. Within about ten minutes, his team had a new tire on our rig, and we were ready to get back on the road.

Rather than immediately pushing off for Austin, we asked him if we could buy him a cup of coffee and listen to his story. He obliged and told us to follow him. From behind the shop, he pulled out one of the most beautiful Harley-Davidson motorcycles we'd ever seen, fired it up, and rode off down the road. We tried unsuccessfully to keep up, but eventually found him a few exits down the freeway at the nearest diner. There, we had about five cups of coffee over a span of three hours. It was one of the most mind-blowing conversations of our journey. He was a helicopter pilot in the Vietnam War who was shot down and spent eighteen months in a POW camp. When he got home, the government paid for his education at MIT, where he got his Ph.D. in electrical engineering. He worked as a U.S. marshal for a few years, but his real passion was handling trucks, so he started up his own shop, which is how we serendipitously ran into him.

On another roadtrip, Ryan, Randy, and Mike were walking through the streets of downtown Boulder when they came across an awesome outdoors shop. They cruised in, checked the place out, and met the woman who started it. She seemed really easygoing, so they mentioned that they would be in town for a few days and would love to sit down with her and hear how she got to where she is today. Learning how she went from being a ski bum in Aspen to starting her shop in Boulder was inspiring, to say the least.

Gloria, Cristina, and Bernardo also spontaneously found a person to talk with in Chicago. The students were simply looking for people who love what they do, and they found someone who fit that criterion right on the side of the road.

The Bucket Man

While in Chicago, we did a spontaneous interview with Mike, a.k.a the Bucket Man. By coincidence, all of our interviews were close to Wrigley Field. We spent a lot of time circling the stadium, and after each of the Cubs games there was this man playing buckets as drums. He put his all into it and got the crowd dancing and singing. He really knew how to work the crowd. We realized that the Bucket Man seemed to really enjoy what he was doing, so we wanted to hear his story. Once he was done playing, we walked up to him and asked if he would be willing to tell us how he got to be where he is today. He agreed, and we ended up having an amazing conversation with him, learning how he created a life for himself around his passion for music.

—*Gloria Pantojas,
age 22, Sacramento*

While you're on the road, you'll find that you start collecting a valuable list of referrals for other potential conversations. When we left for our three-month trip, we had about thirty interviews booked. We came home with more than eighty completed. Many of these were referrals from people we met while we were on the road.

When we were in Washington, D.C., we met someone studying environmental law at George Washington University. She mentioned that we should meet with Deb Callahan, a local environmental hero and president of the League of Conservation Voters. We called her office and mentioned we were in town, and she agreed to meet with us. A similar thing happened in Seattle to Randy, Ryan, and Mike—a person they randomly met introduced them to one of the local music journalists. It turned out that she had been the music editor for the *New York Times* and had an amazing story to tell. If they hadn't been flexible enough with their minds and schedule to allow space for spontaneous meetings like that one, their journey would not have been half as rich.

Stepping Back from the Road

Reflecting on the Experience
Coming Home

Reflecting on the Experience

When an interview concludes and it's time to get back on the road, take some time to reflect on the experience. Stir through what you learned from the person you just interviewed: how they discovered things that excited them, how they overcame pressure to conform, and how they built lives around issues that mattered to them.

On one of our roadtrips, Jim Collins, an author in Boulder, Colorado, challenged us to "Never slip into a life of gray mediocrity." He asked us to agree or disagree with that challenge before we exited the conversation. We all agreed, and he asked us to email him in one year to let him know where we were in our lives. It was a way of holding us accountable and making sure that we were staying true to ourselves.

After that interview, we reflected on what Jim had said for a long time. We wrote about it in our journals, talked about it on late-night drives, and really let his words sink in. His challenge was for us to stay true to ourselves—not to make a lot of money, buy big houses, or drive fancy cars, but to find our own roads based on a set of ideals we would have to define for ourselves. To Jim, mediocrity meant slipping away from our own ideals and not "painting a masterpiece" with our lives. Jim's speaking fee is about $20,000 an hour, but he sat with us for more than three hours and didn't charge us anything except that email due back to him in one year. The interview had a lasting impact on our lives, but it wouldn't have been so significant if we hadn't reflected on how his words could be applied to our own roads.

Coming Home

Coming home can be the most defining part of your trip. While it reminds you of how loud the noise of conformity is, it also reveals how much you've grown. Your comfort zone has expanded. You've changed: You're more open and miles closer to the person you really are.

The noise doesn't seem as loud as it used to. It's now a distant buzz that you're aware of but not affected by. This is the point where you break free—where you distill what you've learned from the inspiring folks you've met and apply it to your life.

On the one hand, you hear all those constricting voices: "How can you make a living doing that? How can you pay off your school loans? What kind of an example are you setting? Okay, your roadtrip is over. Now what?"

But on the other hand, you're filled with the stories of people who have broken free of this noise, defined their own roads, and come out the other end enlightened and passionate about their lives. You learned how a filmmaker worked three side jobs instead of going to law school so he could finish writing his first screenplay. You listened to a political cartoonist talk about pursuing her life's work even if it meant breaking away from family career traditions. You've seen firsthand that it is possible to follow what you have passion for, regardless of what other people think.

You know it can be done, because you've seen it, and you've gleaned insight about how you can do the same. And the funny part is that it's not rocket science. It's basic. And it's open to everyone, although few people actually follow it. It's about defining a set of values for your life, having the courage to let those ideals steer you, and then not letting anyone push you off track.

Coming home from the road isn't the end of your travels. All it means is that you're beginning an entirely new journey.

PART 4

INTERVIEWS FROM THE ROAD

WEST
ALTERNATE
SOUTH
58
NORTH
11
75
EAST
TO
ALTERNATE
SOUTH
58
75

Interviews from the Road

It's an important thing to expose people to others who have taken different paths and who are emotionally enthusiastic and attached to what they're doing—who love getting up in the morning.

—William Sahlman,
professor at Harvard Business School

In the past three years, Roadtrip Nation has interviewed hundreds of individuals who have defined their own roads in life. How did these people get to where they are? Where were they when they were our age? Did they face the same pressures of conformity? How did they shed that noise and harness their individuality to find the open road?

Through more than three hundred interviews, we've found that the same themes of individuality and passion have guided how all these people discovered their life's work. But the one thing that is different and distinct in each interview is the personal story. The diversity of roads that people have taken to get to where they are mirrors the breadth of the new world we live in.

At the end of the day, advice is simply words on paper, but people's stories are much more. Their paths offer insight that isn't based on empty promises; it's grounded on years of experiences that expose the failures, success, hesitations, and realizations that have shaped their lives.

If someone tells you to "shed the noise," but has never done it themselves, why should you listen to them? But if someone shares their story with you and expresses how they resisted these pressures in their own lives, that's something worth learning from.

Listening to one person's story can give you a great spark, but exploring several can give you a selection of insights to pick and choose among, deciding which ones resonate with your own life and discarding those that don't. Absorbing a wide collection of stories will allow you to develop your own set of ideals that can guide your journey and keep you on track to define a road in life that reflects the person you really are.

Interest is not enough.
You must be passionate
about what you do.

ROBERT MONDAVI

Founder of
Robert Mondavi Winery

NAPA VALLEY, CALIFORNIA

- Father comes to America as a hard-working Italian peasant.
- Robert grows up in a loving Italian family.
- Attends Stanford University for under-graduate work.
- Decides to go into wine because "it has an affinity for family life."
- Convinces father to buy a struggling winery.
- Starts to bottle his own wine and makes it a successful business.
- At eighty-nine years old, shares, "You must have confidence and faith in yourself."

I'm eighty-nine years old. The most important thing that I've learned over all those years is that you have to believe in yourself. Many people will tell you that you can't do something. Well, all of my life people told me I couldn't do it, but I didn't agree with them. I just kept working at it and, little by little, things worked out. My parents helped me to believe in myself from a very early age. They were Italian and they were very family oriented, so there was a lot of love in my family.

Ever since I was a child I wanted to excel. My father, who came to this country as a hardworking Italian peasant, instilled that in us at an early age. I also realized that I was not that smart. I was average, but I had faith in myself and always put my whole heart and soul into everything I did, regardless of what other people said. I was only 140 pounds when I went to college, but I went out for the rugby team and became the most valuable player.

One day my father came to me when I was a junior in college, and he asked me what I wanted to do upon graduation. I said, "Dad, I want to be a businessman and go into wine." I felt that the wine business had an affinity for family life. Being an Italian family we were very emotional—we had the highs and the lows. It was a good way of life. I heard Napa Valley was an outstanding wine-growing region for both red and white table wine, so I thought, "Why not start there?"

So I took chemistry in my senior year to learn more about the chemical process, and my brother, who was a year behind me, said that he would be interested in helping me. My brother and I then convinced my dad to buy the controlling interests of a bulk winery in Napa. I was going to go to business school, but instead I decided to tutor under a professor who taught ecology, grape growing, and wine making at the University of California at Berkeley. I spent two months of my summer vacation with him.

Then I went to work in the bulk winery, but I realized that we couldn't stay in the bulk wine business in Napa Valley because our price of grapes was much higher than in the San Joaquin Valley. They had a higher production rate there. We had made about a million gallons of wine at the bulk winery, and I realized that if we bottled it ourselves we could make a dollar a gallon instead of twenty-eight cents a gallon.

That was during the World War II years, so companies everywhere were going up for sale. I heard that the Charles Winery was going to be up for sale and knew that we could use it to bottle our wine and make a great business from the wine we had already made. So I went to my father and shared with him these facts. But he didn't want to do it. He said, "Bobby, I don't want to run a winery. I'm happy with what I have." He then went up to bed. Mother was there, and I said, "Mother, we've got to do something! This is too good of a deal!"

At seven o'clock the next morning I was trying to figure out what I could tell my father so he would go see the winery, but I didn't have to say a word. Dad came in and said, "Bobby, when do we go see the winery?" Well, you know what took place in that bedroom the night before with my mother [laughs].

So we went to the winery. It was a beautiful piece of property in Napa Valley. It was about 150 acres, and about half of it was planted. There was a stable and an older home on it. The price was seventy-five thousand dollars. He realized that it was a damned good deal, turned to my brother and me, and said, "If the two of you are willing to work through this thing and see that it becomes a reality, I'll consider buying it."

So we went to San Francisco and sat down in the banker's office. Right as we sat down, the telephone rang. The banker picked up the phone, paused, and said, "Oh, I'm sorry. I just sold the property to Mr. Mondavi." I looked at my father, looked at the banker, and breathed a sigh of relief. We took out a loan, signed the papers, and bought the property right there. I couldn't sleep for two months, I was so excited.

Building this thing was not easy, but anything good takes hard work. People always ask me what the lessons are that have helped me build my life. Well, there are some basic tenets that I believe in.

To succeed in business or in life, I don't think you need fancy schooling or a highly technical expertise. What you need is common sense, a commitment to hard work, and the courage to go your own way. This is the foundation, and on top of this, there are fourteen other qualities that have served me well.

First and foremost, you must have confidence and faith in yourself.

Second, whatever you choose to do, make a commitment to excel. Pour your entire heart and soul into it with complete dedication.

Third, interest is not enough. You must be passionate about what you do. Find a job you love, and you'll never have to work a day in your life.

Fourth, establish a goal beyond what you think you can do. When you achieve that, establish another. This will teach you to embrace risk.

Fifth, be completely honest and open. I never had secrets. I would always share my knowledge and experience with others, because I had confidence that there was enough room for all of us.

Sixth, generosity pays, so learn to initiate giving. What you give will enrich your life and come back to you many times over.

Seventh, only make promises and commitments you know you can keep. A broken promise can damage your credibility and reputation beyond repair.

Eighth, you must understand you cannot change people. You might be able to improve someone a little bit, but you can't change anyone but yourself, so accept them the way they are. Accept their differences. I learned this late in life, and it's amazing what peace of mind I found when I finally understood it.

Ninth, live and work in harmony with others. Don't be judgmental. Instead, cultivate tolerance, empathy, and compassion. This one wasn't easy for me.

Tenth, it is very important that we understand one another. We need to learn how to bridge those spaces of misunderstanding. To do this, listen carefully, and when you talk, be sure people understand you. On important issues, have people repeat back to you what you've said to make sure there are no errors of confusion or conflict.

Eleventh, rarely will you find complete harmony between two human beings, but if you find it, maintaining this harmony requires individuals to have complete confidence in one another. Make time to be alone, share experiences, and appreciate the beauty of life. Open all of yourself to that person, emotionally, physically, spiritually, and intellectually. And always, always make time for playfulness and laughter. There is no better way to keep love alive and vibrant than through laughter and good cheer. Remember that.

Twelfth, in both life and work, stay flexible. Whether in a country, a company, or a family, dictatorship and rigidity rarely work. Freedom and elasticity do.

Thirteenth, always stay positive. Use your common sense. Common sense is not common. There are many more intelligent people than people with common sense.

And fourteenth. Out of all the rigidities and mistakes of my past, I have learned one final lesson that I'd like to see engraved on the desk of every businessman, teacher, and parent in America. The greatest leaders don't rule. They inspire.

The greatest leaders don't rule. They inspire.

I never set out to do this. I always did it because it was fun.

PAUL FRANK
Cofounder of
Paul Frank Industries
HUNTINGTON BEACH, CALIFORNIA

- Not social in high school—spends time doing music.
- Spends ten years at a junior college so he could say, "I'm in school."
- Plays in a band during that time and meets Ryan.
- Starts making wallets to sell as band merchandise.
- Ryan and Paul start selling wallets at trade shows.
- They run orders for UPS out of their garage.
- John joins the team, and they get more notice at bigger trade shows.
- Move into first office; "just grew from that."

I was in community college for about ten years. I didn't know what the hell to do. I thought if I was in school, I wasn't such a big loser and I could at least say, "I'm in school." I went to Orange Coast Community College just to do art, and that was cool. Art classes are fun because you just make whatever you want. I did a lot of racy things in art classes, and the teacher loved it.

I was always in a band, which really affected my life. When you're in a band, you meet other kids in bands and you have a whole different social group than someone who plays football. I was a nerd, played guitar every night, and didn't go to parties or anything too social. But it was cool. I stayed in and learned music, which paid off later when I could actually play for a real band and meet lots of other creative people.

That's how I met Ryan, my business partner. He was friends with another band we knew. He was really cool. I liked the way he dressed; he liked the way I dressed. We'd actually get together and talk about clothes. We always talked about starting something but didn't do it at first 'cause his dad wasn't into it.

A few years went by, and I started to do a few custom things, like sew patches on my girlfriend's sweaters. I got a sewing machine when I was twenty-five, but I didn't really use it at first 'cause it was hard to thread. I just put it away. But I got it out eventually 'cause my girlfriend needed me to sew things on her sweaters. I once made Ryan a hat out of a sweater—I cut the arm off and made a beanie out of it.

Then I made a wallet. It was like a toy wallet, the kind you get at the Grand Canyon or Las Vegas—a cheapo one with the lacing on it. I noticed a lot of my friends had Hello Kitty wallets and Sesame Street wallets, so I thought, "I'm gonna make something cooler than that."

I made a few more wallets, and we sold the first ones at our shows as band merchandise. We got together with Eric Stefani, Gwen from No Doubt's brother, and he did a cool logo for me. We put that on the wallets and hand colored them. I used to have these coloring parties where people would come over to my house and help make wallets for our next show.

That's what was fun about life—making stuff for our next show. We would make a papier-mâché planet to hoist up behind the band, or we would paint the drum kit with all different things on it. We weren't artists, we were just creative kids who would do stuff like paint our guitars different colors and make cool flyers.

After I made the first wallet, everybody wanted one. I couldn't stop making them. Kids would come up to me at the newsstand I worked at and say, "Are you that guy who makes the wallets?" It was pretty cool. So I kept busy with that for a long time.

At the time, Ryan worked in the PR department at Mossimo. We decided to start our own company, so I started making all of the stuff in his garage. We had to sneak into the first trade show we ever went to. We shared a booth with this lady—she let us have a little table in her booth for free that had all of my samples on it.

The things I made started to sell. I sold three or four hundred wallets to some Japanese guys. After that I thought, "Whoa, this is the big time." Then some people I knew who owned stores let me put my wallets on their racks.

It just kept evolving like that—a couple more in this store, a couple more in that store. Ryan's stepmom had a store in Belmont Shore. She took some for us, sold out, and ordered some more. Stores were timid at first, but then we met John Oswald—he was Ryan's roommate's boyfriend. He was a venture capitalist and saw the momentum growing; eventually he wanted to get in on it. I don't know why he would risk all that money on us, but thank god he did. So John brought in some money and became our other partner.

That's when we went to the Action Sports Retail Show in San Diego and had our own booth. That was awesome. We made a little house that was like our own cottage. We had a cool coffee table. I think it impressed people. For little guys, we sure got a lot of attention.

For our first office, we rented part of a space from a guy who was a contractor. We rented the front office and he got the warehouse. We just grew from that. We moved within the same complex, but this time we got our own building. Then we rented the one next to it and knocked out the wall between them. Then we moved into the place we're in now. And today we're moving into a new place. The new building is five times the size of this one, at least—it's huge. We need to get those razor scooters to get around.

I never set out to do this. I always just did it because it was fun. People shouldn't have too much of an agenda to do something that doesn't come naturally to them. If you love art and being creative, don't wait for school to give you the go-ahead. Do it after school or before school. It's all about how much you want to learn. It's like learning how to use a computer: you just do it. I taught myself how to sew, so you could teach yourself how to build a car if that's what you want. It's just a matter of how much passion you have for it.

But it can't be fake. It won't work if it's not sincere. It's all about sincerity—not thinking about money or how much you could sell something for, just thinking of something that's pretty cool. Hell, I felt successful just being the guy you could come to for a wallet. When that's all you've got, that's pretty good.

I wanted to be a part of the world effort
to go out there and understand
the universe.

KIM WEAVER
Astrophysicist with NASA

GREENBELT, MARYLAND

- Grows up in West Virginia where "for a woman to do anything with her life was shocking."
- As a child, gets hit by a car on the day of the first moon landing—finds passion.
- Grandfather takes her on drives to the "big radio telescope."
- Is one of only a few female physics students in college.
- Has two mentors who "made me know I could do it."
- Starts work at NASA—pursues passion for figuring out "What's the universe all about?"

We landed on the moon when I was five years old. My family was taking a trip and I wanted to get home to watch the moon landing. I was so excited! Then we stopped to get gas, and my mother wanted to cross the road to see something on the other side. While we were crossing the road, a car came by and hit me. In the ambulance I remember thinking, "Oh no, I'm not gonna get to see them land on the moon." It was very devastating for me as a five-year-old, but it made me understand how much I cared about the space program and astronauts. I really thought I wanted to be an astronaut.

That childhood event made me realize that everyone needs to figure out what they loved when they were a little kid, because that's the thing that is fundamentally part of your personality—that love you had for something before the world influenced you.

As I grew older, I learned that I didn't like to fly. So I knew that being an astronaut was probably not for me, but I still liked stars and using telescopes. Where I grew up in West Virginia, there's a big radio telescope at Green Bank. My grandfather would take us out there on Sunday drives. I'd see this big radio dish and say, "Wow that's really cool. I'd love to use that some day." So as a kid, even though I didn't know I was going to be an astrophysicist, I thought astronomy and space were really neat. I always wanted to know what else is out there.

When I went to school, that ended up being my direction. I think I subconsciously picked classes that took me there. This was before anyone told me, "You should take math and physics to be an astrophysicist." And then when I got to college, I learned I had to take math and physics. For a while I thought it was too much work, and I waffled. I tried some geology classes; I tried psychology classes; I tried biology classes. I thought I liked science, but I wasn't sure because the physics seemed a little scary.

But I always felt that I might want to work for NASA, so I started systematically going through the program, and it took a long time—four years in college and then six years as a graduate student. That's a lot of your life. But if you find something you're passionate about, then it doesn't seem like a lot of your life.

Even though I thought I might want to work at NASA, I still wondered all the time. Even though I knew I loved science, I didn't know if it was really going to be the thing for me. I think those are healthy questions that people should ask themselves.

I tried to keep all of my doors open. When you're young, all of your doors are open and you can do anything you want. Then as you go through life, you start shutting doors. You want to choose to shut them yourself, rather than letting other people shut them for you. But you really want to shut as few as you can. For example, in college I looked at being a geology major. I realized I didn't want to do that, so I shut that door. I looked at being a psychology major, but realized I didn't want to do that either, so I shut that door. But there are some other things, like archaeology, that I loved. So someday I may change and become an archaeologist. I haven't ruled that out.

I don't think anybody should close certain doors just because they're getting older. What if you don't find your real passion until you're forty-five? You want to be able to change and realize it if you don't like something anymore. The key is to know yourself, so you gotta keep asking all those questions. The answers are there as long as you know who you are and what you want out of life.

I think one of the biggest problems today is when adults counsel kids too much and tell them what they should do with their life. Then the kids never figure out who they are.

A lot of parents have had their life doors shut on them. Especially my parents' generation, where women didn't have much career choice. So a lot of parents try to live through their children's lives so they can be happier. They don't mean to be hurtful, but they reach out and project themselves onto their children.

I grew up in West Virginia, and in my culture for a woman to do anything with her life was shocking. Most of the women I knew were secretaries or worked at a grocery store. They didn't go to college. I was the first woman in my family to go to college. I'm not knocking on my family at all; they just didn't really understand what it meant to get a professional degree and to become a scientist. They said things like, "Wouldn't you rather get married and have kids?"

I ended up getting engaged when I was twenty. I wasn't sure I wanted to get married yet, but I felt the pressure. The day after I got engaged, I went to class with my engagement ring on and my physics professor saw it. In front of the whole class he said, "Well, I can see you're just going to waste our time and your time, because you're just going to get married. What's the point of having you in this class to become a physicist? Because you're never going to use this education."

I took my ring off and put it in my pocket because I thought, "If he thinks I'm not worthy as a scientist, I can't show anybody my engagement ring because they're gonna think I'm no good." But then I realized that I didn't really want to get married anyway, so I broke the engagement off.

Even my high school and college guidance counselors said, "You're an intelligent woman, but you may not want to be doing science." So everyone had their opinion on what I should do with my life, and it took me a long time to realize that I'm not here to please them. I'm really here because I want to do what I love to do.

It helps to have good mentors. Someone who takes you under their wing and encourages you to do a good job. I had a couple of really good bosses. I had two senior male scientists who were very supportive of me. Having them trust me and tell me that I could do a good job made me know that I could do it.

Whether it's in college or out of college, I would encourage anybody to find specific role models and mentors who can show you the ropes and help build your self-confidence. The world is just waiting and trying to take away your self-confidence, so you need people around you who can help build it up.

What was driving me was more than just doing science. The most important question you can ask yourself is, "Why am I here, where did I come from, and what's the universe all about?" So, not only did I want to do astronomy, I really wanted to find out how to answer those questions. My path was to get involved in some agency or scientific endeavor that was going to be answering those big questions. I wanted to be a part of the world effort to go out there and understand the universe.

I know a lot of people who are trying to be professional singers. Sometimes they may not be that good at a certain type of singing, so they give up. Well I would say to them, you could still be a professional singer, but you might have to do it slightly differently. You may have to do it in a different place. You may need to teach for a while to make some money. It might not be the exact way you envisioned it, but you can find a way to do what you love.

you

can

find

a

way

to

do

what

you

love.

I knew if I was still doing that in ten to fifteen years, I wasn't going to be happy with how I lived my life. So I decided to leave and follow something I was passionate about—beer.

JIM KOCH

Founder and brewmaster of Samuel Adams Brewery

BOSTON, MASSACHUSETTS

- Goes to law school, but drops out after one year.
- Spends three years as an Outward Bound instructor—finds a new definition for risk.
- Goes back to school to finish his law degree.
- Gets a job at a prestigious, high-paying consulting firm.
- Not happy being a consultant; leaves to start his own brewery.
- Realizes that "Quitting to do something that I really loved wasn't a risk."

When I was twenty-four years old, it dawned on me that I would need to decide what to do with my life, so I went to law school and business school. But one year into it, I dropped out and was gone for three and a half years. I spent most of that time outdoors as an Outward Bound instructor. It was the type of experience that you can only have in your twenties. I knew I wanted to do mountaineering and kayaking. If I had waited until now to do it, I probably would have never done it.

When you're in your twenties, that's the time to have the experiences that you look back on later and say, "I'm glad I took some chances and did that." Maybe those experiences don't lead to any big career achievements, but that's okay.

I could only work for Outward Bound three to four months a year, so I knew it wasn't something that I could live on year-round. I got to be about twenty-seven and realized I couldn't do it for the rest of my life, so I went back to school and finished my J.D. and M.B.A.

When I graduated, I went to a firm here in Boston called Boston Consulting Group. I was there for almost seven years, made a lot of money, and flew all around the country first-class. But after five or six years, I started asking myself, "Is this what I want to do with the rest of my life? Is this it?" That was a scary thought. I knew if I was still doing that in ten to fifteen years, I wasn't going to be happy with how I lived my life. So I decided to leave and follow something I was passionate about—beer.

A unique thread in my background is that I am the only sixth-generation brewmaster in the United States. I'd always enjoyed beer, and in retrospect there is some irony that I got degrees from Harvard and yet my biggest job qualification is that I love beer.

But my parents didn't want me to go into the beer business. When my dad got into brewmaster school in 1948, there were about 1,200 breweries in the U.S. and for the most part they were all doing well. When I started Sam Adams twenty years ago, that number was down to thirty. So all of those jobs for brewmasters went away. Being a brewmaster was not a good career decision. My dad said, "Jim it took us 150 years to get the smell of a brewery out of our clothes, I don't want you going back into that."

At that time I was in my mid-thirties, had gained a lot of experience in business, and, as an Outward Bound instructor, had learned a lot about leadership, self-reliance, and challenging yourself. The Outward Bound paradigm is probably encapsulated in instructing someone to rappel off a cliff backward. As an instructor, you know that the rope would probably hold a car. Everything is safe, solid, and secure, but to the student, they're still walking off a cliff backward.

A lot of times we're driven and limited by perceived risk. But perceived risk is unrelated to the actual risk. Those kids who would rappel down the cliff backward were scared. Especially in the first couple of steps, when they had to lean all the way back. There's nothing underneath you but thin air and sharp rocks a couple of hundred feet down at the bottom. That is very high perceived risk, but the actual risk is negligible.

When I left the consulting firm to start Sam Adams, it was a similar situation—the perceived risk was high. At the consulting company I had a high-paying job and a great office on the thirty-third floor that looked out over Boston. Then one day it all disappeared and I was a brewer. People said, "Gee, you took a lot of risk." But the bigger risk would have been staying at a job that wasn't fulfilling and was wasting my life. Quitting to do something that I really loved wasn't a risk.

Sam Adams started with nothing. We didn't have a computer, a phone, or an office. I made the first batch of Sam Adams in my kitchen, and I didn't have huge expectations. I didn't think that it would become what Sam Adams has become. I thought it would be this nice little local company in Boston. I didn't expect much in the financial and ordinary form of success.

But I'd learned how to live on no money from Outward Bound. I came out of my twenties with no real need to have a lot of money and all this fancy stuff. It wasn't until I was twenty-nine that I could afford real furniture. I thought that getting rich was life's big booby prize. People who aren't happy want to be rich.

When I started my own company, I wanted to start something that would make me happy in the big scheme of things. I think that most people would rather be happy than rich. Most people who start businesses aren't going to be rich; the odds are not in your favor. But if you pick the right thing that will make you happy, you get the real prize.

We built Sam Adams step-by-step, and we were able to change the world of American beer. We started by really believing in what we were doing. Then we got bar owners, bartenders, and store managers to believe in our vision that America can brew great beer.

It's good business to focus on the things that you really enjoy, because those will be the things that you're good at. I've learned that if I don't really enjoy doing something, then I probably shouldn't do it because there is someone else out there who can do it a whole lot better than I can. Today, the two things that I focus on are the quality of the beer and the culture of the company. Those are the things that I can't hire someone else to take care of. I still taste a bottle of every batch of Sam Adams. I know—it's work, work, work.

I took a train down there with fourteen dollars in my pocket and stayed at my uncle's apartment. I would take bus trips around town to knock on doors of radio stations.

LARRY KING
Host of *Larry King Live*
LOS ANGELES, CALIFORNIA

- Father dies when he is nine years old.
- Mother has to beg the dean to let Larry graduate from high school.
- Does many odd jobs—UPS, department store, mail boy, etc.
- Takes train to Miami with only fourteen dollars to pursue dream of radio.
- Knocks on radio station doors across Miami and finds an opening.
- Echoes Branch Rickey's advice, "Luck is the residue of design."

My father died when I was very young. I was only nine. I wasn't a very good student, and my mother had to beg the dean so I could graduate from high school. I never went to college, so after high school I did a bunch of odd jobs—I delivered packages for UPS, worked at Hearn's department store, and was a mail boy—but in the back of my mind I always wanted to be a broadcaster. I never wanted to do anything else.

One day, when I was a little over twenty-two years old, I met a man named James Sermons, who worked at CBS. I asked him where was a good place to break into radio, and he said Miami. So I went down to Miami and that's where I ended up breaking in to radio.

I took a train down there with fourteen dollars in my pocket and stayed at my uncle's apartment. I would take bus trips around town to knock on doors of radio stations. One small radio station said that the next opening they had, I would be hired. So I hung around that station, and on May 1, 1957, I started. I was a disc jockey.

My birth name was Larry Zeiger. So it's my first day on the air, and I'm finally going to get my big shot. I've got my record all cued, and my manager says to me, "What name are you going to use?" I said, "What's wrong with Larry Zeiger?" He told me it was too ethnic, opened the newspaper, and there was an ad for King's Wholesale Liquors. He said, "Why not Larry King?"

Then I sat down. I'm ready to go on to the air, and I turn the mic on, but nothing comes out. I fade the record down and I fade the record up, and in that thirty seconds I said to myself that I didn't have the guts to do this. All my life I'd wanted it, but I was too scared. The general manager, the late Marshall Simmons, kicked open the door to the control room and said, "This is a communications business, dammit. Communicate!"

He slammed the door shut. I turned on the microphone and did what he said. I said, "Good morning. This is my first day on the air. My name is Larry King. They just gave me that name, and the general manager just kicked open the door and told me to communicate!"

I was never nervous again. Never. I learned something that day that I've kept with me for forty-six years: never lie to the audience. If a situation develops, tell them. Be authentic. What are they going to do to you? It ain't the end of the world. They ain't going to shoot you. I also learned the only secret in my business is that there is no secret. Just be yourself, and if it's good enough, it's good enough.

Once I got past those first thirty seconds, I never had a fear of failure. I knew that I could do it. I knew that I could communicate. I knew that I could tell a story well and ask good questions. I knew I was curious and had a sense of pace, which is important.

I don't believe there's any barrier to getting what you want if you really want it, especially in creative positions. If you want to be a writer, you'll be a writer, but you've got to really want it. It can't be half-assed. You can't just think you want to be a broadcaster, an artist, or an engineer, because then you're not going to make it. You'll be ordinary and you don't want to be ordinary.

Branch Rickey, the great general manager of the old Brooklyn Dodgers, said, "Luck is the residue of design. You make your luck." I asked a great lawyer if he ever got lucky as a lawyer. He said, "Yes, at 4 a.m. in a law library." So it's a combination of fortitude and perseverance.

In my business it's so competitive, and everybody really loves it. So there isn't one person on the air in America today who doesn't love it. Whether you're a disc jockey on the air in Biloxi, Mississippi, or Peter Jennings, you've got to love it. And Peter Jennings might be making a lot more money than the disc jockey in Biloxi, but when that disc jockey goes on the air he's just as happy as Peter Jennings.

I've followed my heart, so I haven't worked in forty-six years. I don't go to work; I go to the network and I do a show. The last time I worked was probably when I was on the UPS truck. That was work. It's such a break in life to do what you absolutely love to do. Sure, there are days that are better than others, but there isn't a show that I don't want to do.

In any creative business it's the same. If you're in the chorus line of a Broadway show dancing, you love it as much as Julia Roberts loves doing a movie. I never met a successful person who didn't love what they do. That's why I believe it's hard to be born with money, because a person's drive is taken away from them.

My drive was never financial. The financial end was always a byproduct of what I wanted to do. Bill Gates told me that his drive in school was gadgetry. He was the kid who wanted to be in the "build the thing" club. His money was a byproduct of that. If his goal had been money, he wouldn't have made it. I can't emphasize that enough: If your goal is money, you're not going to make it.

It had such a powerful effect on me
that I dropped everything and did
an editorial cartoon. I had no intention
of publishing it or anything.

ANN TELNAES
Pulitzer Prize–winning political cartoonist
WASHINGTON, D.C.

- Goes to ASU for college where she "was floundering."
- Drops out of ASU and applies to art schools.
- Does freelance design work for Disney.
- Watches Tiananmen Square events on TV and draws first political cartoon in response.
- The event awakens her passion for editorial cartoons.
- Starts drawing a series of political cartoons and sending to newspapers.
- After several years of "plugging away," wins the Pulitzer Prize.

When I was twenty-one I was floundering. I was going to Arizona State University to study art, but I wasn't being very serious about it. I made a deal with my dad that I would minor in art and major in journalism so I could have something to fall back on. For some reason that went absolutely nowhere. I don't even remember taking a journalism class.

I went to college in the early eighties, which was when everyone was majoring in business and going to Wall Street. But I could never see myself doing something that I wasn't passionate about.

I ended up dropping out of ASU so I could apply to art school and start studying animation. Art school was very expensive, so I worked in a casino for almost a year to make enough money for tuition.

That was a big deal for me. I had finished two and a half years at Arizona State, so I felt like I was throwing it away. But now I'm a firm believer that people shouldn't jump right out of high school into college. I really think that young people need to travel, see the world, and open up their minds before they get so locked in.

When I first got out of college I was a freelance animator. I didn't like it at all that much because I wanted the regular paycheck. One night I was up late doing a job for Disney, and I watched the Tiananmen Square events unfold on television. It had such a powerful effect on me that I dropped everything and did an editorial cartoon. I had no intention of publishing it or anything. But that awakened my passion for it.

A light went off in my head, and I realized that I had a talent. But there's a big difference between people who have talent and people who apply themselves. God gives people a certain amount of talent, but it's what you do with it that makes the difference.

Then, when the Anita Hill hearing happened in 1991, I was outraged. Having worked in private industry for several years and knowing about sexual harassment, to watch Senate hearings where a bunch of guys are sitting up there saying, "There's no such thing as sexual harassment anymore because we passed laws for that," made me realize I had to do something. That's when I decided to draw a series of editorial cartoons and send them out to various newspapers. That's how it started.

It was sort of hit-and-miss. I had a few papers that were interested and would publish me as a freelancer. But I had to really plug away. That's where a lot of people fall off. It takes several years. I started in 1991 and I wasn't nationally syndicated until 1994. I still had to work other jobs on the side, so it was really about perseverance. You have to be willing to do that for something you really enjoy. I was doing freelance work almost until I won the Pulitzer.

When I was doing freelance work at Disney, I would get up at 3 a.m., do my own drawings for a few hours, and then drive to work. Sometimes I would fall asleep in the car. I did that for two years. But that's what coffee's for. And when you're passionate about something, it just all seems to work.

At the time, I was really trying to get an on-staff job at Disney because then I would at least have a daily platform and a solid paycheck. But that brings a whole different set of limitations and restrictions. Looking back on it, even though it was painful to not get the full-time job at Disney, I don't think it would have been right for me.

Make sure you're willing to take other doors. I think the hardest thing was when I was training to be an animator and all I wanted to do was work at Disney. It's so easy to get tunnel vision and not look at all the other doors. What if there is something else that you are better suited for? I had a lot of doors opened to me that I never thought about when I was a student. I took them, and it worked out really well. But if I had been narrow-minded and just focused on a particular job, I wouldn't have been open to what else was available.

You can do anything you want with your life, but I didn't believe that when I was your age. I think it's part of being scared and frightened of the unknown. Why not just try it? Even if you end up failing—big deal! If you're passionate about it, just keep trying.

I had six hundred bucks and no return ticket, and I only knew one person in Los Angeles.

MARK BURNETT

Creator and producer of *Eco-Challenge, Survivor,* and *The Apprentice*

LOS ANGELES, CALIFORNIA

- Grows up in England and becomes a British paratrooper.
- Comes to America planning to travel to Nicaragua and fight in war.
- Mother convinces him to not go to Nicaragua at last minute.
- Stuck in America with no money, becomes a nanny in Beverly Hills.
- Starts doing adventure racing.
- Creates his own adventure race and TV show, *Eco-Challenge.*
- Creates the TV show *Survivor*, then *The Apprentice.*

I didn't know anything about TV. There were only three channels when I was growing up in England. I never could have predicted that I would be in television.

At the age of nineteen I was leading a team of men in Northern Ireland for the British Parachute Regiment, hunting terrorists. Then, when I was twenty-two, I was in the Falklands War. We were fighting Argentina because they invaded the Malvinas, which is the Falkland Islands. The transition from that to what I'm doing today was huge. The biggest transition was probably coming to America.

I came to America because of the conflict in Nicaragua. By international law, the United States could only have fifty-five Americans down there, so there were jobs for foreign soldiers, working under America, to train rebels.

But I ended up not going down to Nicaragua. As I was leaving the London airport, my mom grabbed me with tears in her eyes and told me not to go. She had a really bad feeling about it. She never felt bad about any of the other military stuff, but this was different. So I sat on a Pan Am flight, going to America for the first time, wondering if I should take that job in Central America. I realized it would be totally screwed up to do that to my parents, so I decided to find something else.

I had six hundred bucks and no return ticket, and I only knew one person in Los Angeles. I called the guy I knew, who offered me a job and gave me a place to sleep for a couple of nights. Then I found a job taking care of children. I was basically a nanny [laughs]. It was the weirdest experience in the world—parachute regiment guy from the war applies to be a nanny in Beverly Hills. I got the job because I spoke English and because they wanted to have some security around the house.

I tried to start various businesses: a clothing company where I sold t-shirts on Venice Beach, a marketing company for credit cards, and a few others. Then I decided to race in some adventure races in France and New Zealand. I loved the racing because it was like being back in the special forces. I ended up racing three years in a row and actually made a really good living.

I think I made a quarter of a million dollars a year from sponsorships. I went to clothing companies and shoe companies and pieced together these different sponsorship deals at fifty grand each a year. I sold little patches on my uniform, but I had one problem: the exposure wasn't big enough for the sponsors to get a good return on their investment. I realized that television was the only solution, so I decided to make TV shows of my racing to fulfill the sponsors' investment. I got channel 9 in Los Angeles to send an anchor and a camera crew. I convinced the French organizer of the race to provide me all the footage from their fifteen camera crews. I then gave the show to ESPN for free, but kept half the commercials. I parlayed the commercials to my sponsors, and that's how I ended up getting the money. It was a different approach to TV financing, but it worked.

I used that same approach to start *Eco-Challenge* and bring expedition racing to America. *Eco-Challenge* grew from its early days on MTV to being on the Discovery Channel with 145 million households throughout the world.

In the sixth year of *Eco-Challenge,* I decided that the social psychology study that was *Eco-Challenge* could be better if the subject matter of the documentary wasn't being chased over five hundred miles. My friend Charlie Parsons had an idea to put a few people on an island and have a mad billionaire kick them off one by one. I met with Charlie and we did a deal. I changed the idea a little bit and made it into the drama that is now *Survivor.*

Looking back on my path, I realize that you can't map out a very long route for yourself. What you can do is *sort of* know where you're going. Be half sure, jump in, and hope you can swim the rest of the way. It's like *Eco-Challenge.* The race is five hundred miles and thirty checkpoints. The people who don't finish are full of ego and only focus on their finish placement. They do that even when they're leaving the start line. People who are successful focus on the immediate checkpoint. When they reach checkpoint one, then they worry about checkpoint two, and so on. That's the philosophy you need to keep to.

Around the time I was marketing credit cards, my mother died from cancer. It made me realize how transient life really is. That's when I started to dream big and write down what things I'd like to do in my life. So when I read about those races in New Zealand and France, I knew it was something I had to do.

People around me told me I was insane. I was making great money in credit card marketing, but I was bored to death. I wasn't inspired. It's not about making the money without the fun. You want money so you have choices, but you better have some fun along the way; otherwise, when you get into a rocking chair when you're eighty, you'll be bummed that your dreams became regrets.

To build that into your life, you have to have the courage to not need all the answers. People who need 100 percent assuredness tend to do nothing—they don't get married, they don't change jobs, they don't do anything risky—because nothing is 100 percent sure.

Especially in your early twenties, don't worry about knowing exactly where you're going. You just need to have a general idea of what you want out of life. Keep yourself open to opportunity. Realize that while you're busy on one project, other things will come your way. Positive energy attracts positive energy. The most important thing is to recognize those opportunities. Then have the courage when nine thousand people tell you it's stupid. If you know it's good, do it anyway.

One of
my first
teachers
told my
mom to
encourage
my artistic
ability.
Luckily for
me, my
mom
listened
to him.

1998 PRIMETIME EMMY AWARDS
OUTSTANDING COMMERCIAL
APPLE COMPUTER - THINK DIFFERENT
TBWA/CHIAT/DAY

LEE CLOW

Chairman and chief creative director of Chiat\Day advertising agency

MARINA DEL REY, CALIFORNIA

- Works in a bowling alley, but "always left the window open for experiences."
- Directs most of his emotional energy into surfing.
- Discovers a passion for art.
- Meets a girl he likes who motivates him to get serious.
- Can't afford to enroll in the Art Center, so teaches himself advertising.
- Gets a job at Chiat\Day, a "really creative" advertising agency.
- Proves himself and gets to lead accounts like Apple Computers.

You hear that stupid saying, "You're only young once." But you really are only given that ten or fifteen years when you're old enough to be on your own and young enough to not have too much responsibility. During those years of my life, I worked in a bowling alley. I cleaned offices in Beverly Hills. I always found ways to feed myself, but I always left the window open for experiences and discovery.

It's very tough to know exactly what you want to do when you're young. When I was a kid I could draw pretty well, and one of my first teachers told my mom to encourage my artistic ability. Lucky for me, my mom listened to him. She always pushed me in that direction, as opposed to being a doctor or a lawyer.

I started surfing when I was really young, so most of my emotional energy went into that. But taking art classes wasn't bad. I liked it. Then in the seventh grade, they made us do a term paper on what you wanted to be when you grow up. They asked you to go talk to companies in related fields. It was easy for me. I just said art, even though I didn't know what exactly you could do with it.

So I looked into it and discovered an animation company called Playhouse Pictures, which was making animated commercials at that time—the late fifties, sixties. I saw these animations that I liked, so I ended up writing my term paper on wanting to be a commercial artist. I didn't say advertising because I wasn't quite sure what that was yet. I did a storyboard for an animated commercial as part of the assignment.

By virtue of that class I was forced to look into what I wanted to do with this art thing. I began to see that you can actually make a living with it. After high school I went to Santa Monica Junior College and kept taking art classes, except on days when there were waves. This was during the Vietnam War, and I had a flaky semester where I was surfing more than I was going to class, so I had incompletes in a bunch of classes. My draft notice came through when my college situation wasn't in place, and I ended up getting drafted into the army. Around that time I also met a lady I really wanted to be with.

So, between the army and this new girl that I had met, I started trying to figure out how I could get a bit more serious with my life and this art thing. I spent my time in the army but got out early to go to Long Beach State. I studied design, built an advertising portfolio, and tried to start finding work. I discovered advertising again. It seemed fascinating because it was about ideas and about creating succinct messages. Then there was the artistic dimension of how you execute it visually, whether it's film, animation, print ads, or outdoor billboards. So I found my way in the side door of this business, and the more I learned from looking at industry journals, magazines, and annuals that are put out to celebrate good work, the more I connected emotionally with what's exciting about this business.

Art Center actually had a great advertising curriculum, but I couldn't afford to go there. So I taught myself. I think the way I learned was kind of a neat way, because I had to emotionally seek out what was the soul of this business. The most important message from that whole learning curve was that I found something that I loved to do. I really enjoyed it, and I had people encourage me to pursue it.

That's the thing that blows me away every day. I found something I love to do, and I do it well enough to make a good living and be part of a pretty amazing company and its history. I get up in the morning thinking about it and wanting to do it. And I married the lady that I met. I've been married to her for thirty-five years.

I don't know what gave me that compass, but I think you'll find that a lot of really successful people never had the goal to make a lot of money. I mean, I wanted some independence, and I wanted to have my own place, but I didn't get into this to make a lot of money.

To get into Chiat\Day—a really creative agency in Los Angeles—was one of my big ambitions early on. I was in a couple of clunky little agencies where I was learning stuff, but I was also learning what I didn't want to do because a lot of ad agencies compromise themselves dramatically—as you can see from all the crappy advertising out there.

I never felt like I was roughing it or having to compromise because I was pursuing what I really wanted to do. When we were in our first little apartment, driving used cars, it wasn't like, "Oh, shit, I'm really frustrated. I gotta get the house and the car." I was totally focused on being good at what I liked to do. And I was very lucky to find a woman who supported my efforts and wasn't sitting there saying, "This dining room's kinda small." My wife was always really cool. That's a very lucky thing—finding something you like to do and finding somebody who wants to do it with you.

So when I finally got to Chiat\Day, I wanted to prove that I could be great at advertising. I was amongst some of my heroes, so I did more than was expected on any assignment that was given to me. I eventually worked my way up through some exciting projects and proved that I was the guy who should lead the accounts.

I think creative people are an interesting combo of ego and insecurity. The ego part is wanting to be great, and the insecurity part is not believing you're good, so you keep working your ass off. Those two things pushed me all the time.

I'm a really passionate person, and I think my desire was more important than my talent. I've met a lot of people in this business who are more talented than I am, but I just wanted it really bad. I got so in love with it that everything worked out.

They took a huge risk on me. I was a twenty-six-year-old kid launching a magazine.

ATOOSA RUBENSTEIN

Editor of
Seventeen magazine

NEW YORK, NEW YORK

- Father dies when she is growing up; mother works to put her through college.
- Growing up, she is "just a kid who loves magazines."
- Goes to college in New York and does magazine internships.
- Gets a job offer at *Cosmo* on graduation day.
- Starts at *Cosmo* as an editorial assistant; becomes a fashion editor.
- *Cosmo* starts a teen magazine to compete with *Teen People*.
- Is chosen to be editor-in-chief of *Cosmo Girl* at age twenty-six.
- Moves over to *Seventeen* magazine.
- Aspires to help families in need similar to her own when she was growing up.

People want to think that your first sign of ambition is going to be very traditional. But it's not always like that—I was just a kid who loved magazines. I wasn't the smartest girl in high school. Even in college I didn't care a great deal about my classes. And I didn't necessarily have gigantic plans for myself. But I always loved fashion. I was always a kooky dresser because that's how I expressed myself. My brother went to the University of Chicago to get his Ph.D., so when I decided to go into magazines my family was like, "Oh my god! You'll be starving!"

When I was twenty-one, I had two different internships, one at *Sassy Magazine* and one at *American Health and Fitness. American Health* said I could work as an editorial assistant for them when I graduated, which was nice, but it just wasn't right for me. Still, I was relieved to at least have a job offer.

So on graduation day, I was packing up my room in my cap and gown when the phone rang. I assumed it was one of my friends, but it was the fashion director at *Cosmopolitan*. She had heard about me, was looking for an assistant, and asked if I could come in for an interview. I played it cool, but when I hung up the phone I was screaming my head off. I'm not a religious person at all, but when that happened I thought, "Wow." It was either god, or the stars, or whatever you're into.

So in the beginning of June, I started as a fashion assistant at *Cosmo.* Over time I worked my way up to being a fashion editor, and one day the editor-in-chief of *Cosmo* told me they were thinking about starting a teen magazine called *Cosmo Girl.* She asked me if I would like to work at the new magazine and be a fashion editor, which was basically the same thing I was already doing, so I said no. But I spoke my mind about what I thought the magazine could be and said that I thought it was going to be a great thing. The next thing I know, I'm meeting with the president of Hearst, the company that owns *Cosmo.* The president, Cathy Black, is known as the first lady of magazine publishing and just wanted to pick my brain since I had interned at *Sassy.*

I thought it was no big deal, so I showed up wearing pants slit up to my hips, something covering my breasts, and stiletto boots. I used to dress really crazy. When I walked into her office, the first thing she said to me was, literally, "Nice pants." So again, I just spoke my mind and didn't censor anything. At the end of the meeting, she said, "Will you do a project for me?" I said, "Sure," but really had no idea what she meant. So I went back to the editor-in-chief at *Cosmo* and asked what it meant to do a project for Cathy Black, and she told me that I was up for the editor-in-chief job at the new teen magazine, and that I was to present my vision for *Cosmo Girl.*

I worked really hard for two days, preparing my presentation for what I thought the magazine could be. I didn't second-guess anything I believed, didn't ask anybody for their opinion, and just did what I thought was right. I handed it in but figured that I wasn't going to get it. But I got it. Cathy Black looked at me and said, "It looks like we have ourselves a magazine."

They took a huge risk on me. I was a twenty-six-year-old kid launching a magazine. There were some people who were much more senior who didn't get the job and were very bitter. To try to smooth things out, I sent an email to those people and said, "You have such great experience. This is new for me, and if you know anybody we should hire, please let me know." One of them meant to reply to another one, but copied me on the email by accident. She wrote, "Oh look, the fashion girl needs a grammarian." Once she realized she had copied me on the email, she of course came bolting into my office to apologize. It was a rough start.

One year after I started *Cosmo Girl* I was doing an interview with the *New York Post,* and the first thing the reporter said to me was, "Did it feel like no one thought you could do it?" But I never thought that for a second. I knew that I could do it—and I did.

If someone was to look at my life and try to learn something, it would have to be that I always had tremendous faith in myself. I think it came from my parents, who were very loving. My dad died when I was growing up, so my mom, who barely spoke English, worked really hard to put my brother and me through college. Amidst all that, my mom was always telling me how beautiful I was even though I was the ugliest duckling on the face of the earth.

Today, all of the what-ifs don't mean anything to me. I don't listen to anybody's pressure. I do exactly what I want. If I want a Big Mac, as I did for lunch today, I go and I have that Big Mac. I don't care if someone says, "What if you gain weight?"

After I launched *Cosmo Girl,* I was ready for bigger things. So when the *Seventeen* thing came up, it seemed like a new project that was with an age group that I adore. I wanted to create a magazine for girls who had matured and maybe passed *Cosmo Girl.* I wanted to grow up with them, and that's sort of happened.

An old cliché is "Listen to your inner voice." But it's a cliché because it's true. That fire in my heart that always told me I could do things—well, that fire was right. All of the people who rejected me, be it the National Honor Society, different colleges, or different internship programs like the one at *Rolling Stone*—none of that mattered. It was all just people's opinions, and people's opinions are just distractions.

I guess I'm living proof that if you actually listen to that voice, have courage in yourself, and follow that path, powerful things can happen. Have you seen *The Wizard of Oz*? They keep walking down the yellow brick road even though there are all those bad people trying to get to them, and they finally get to Oz. Don't let anybody take you off that path. Oz is defined by you.

I figured out what my Oz was when I was twenty years old. I was interning at *American Health,* and Oscar de la Renta had created a scarf where the proceeds were donated to curing cancer. Before my father died, he lost his leg, so we had organizations like Meals On Wheels bringing him food. When I heard that Oscar de la Renta was doing that campaign for cancer, I hoped that one day I could have enough success that I could go to a big producer and say, "I need twenty thousand dollars a table to throw some kind of charity event that could help organizations like the ones that helped my family." That's my Oz.

Lakes in my hometown were being filled with concrete, and all my friends could talk about was how well they did on Wall Street or how high the value was on their house. Something was missing.

JOHN PASSACANTANDO

Executive director of
Greenpeace USA

WASHINGTON, D.C.

- Studies economics in college.
- Goes to work selling right-wing econ books on Wall Street.
- Reads environmental books—reconnects with a love for nature.
- Realizes that right-wing economics is an "environmental cannon."
- Starts his own environmental group called Ozone Action.
- Gets hired as executive director of Greenpeace USA.

Normally we orient our life questions toward "What should I do?" We look at what majors are available, what are other people doing, how much you get paid for doing something. But when you orient it that way, you don't analyze your life based on what you like to do. In fact, you're probably not even asking that question. And if you're not even asking the question, it's unlikely that you're going to find what you love to do.

When I went to college, I had no idea what I wanted to be. I used to tell everybody that I wanted to go to law school so older people wouldn't pin me down. I did it so much that I almost convinced myself that I wanted to go to law school! It was the eighties, so economics was the story of the day. There was all this recession going on and inflation was high. I was fascinated by that.

After college I ended up selling medical insurance door-to-door. Then I ended up selling computer systems to auto parts stores and warehouses in the New York metropolitan area. Some of these places looked right out of *The Sopranos.* I even had someone pull a gun on me once.

It was the school of hard knocks. I learned a lot, but I was still fascinated by economics. There were these guys who were supply economists—basically Reaganomics—right by where I lived in Morristown, New Jersey. So I just popped in there a few times and they gave me all their literature for free, which they were selling on Wall Street. After popping in enough times, they asked me if I wanted to work for them. So I ended up selling their research to people on Wall Street for four or five years. I was fascinated by economics because I thought, "Well, maybe this is how the world works."

And so I was learning. I've always wanted to be on a steep learning curve. If you're really fascinated by what you're doing, you'll tend to do well at it. And then success will follow.

Here's the contradictory part of my story. I grew up in northern New Jersey, hiking, fishing, kayaking, and camping up in Maine during the summers. I loved the environment. And so while I'm studying and supporting this brand of very conservative economics, a friend of mine is dropping off used books to my house every couple of weeks that start to open my eyes to environmentalism. I learned that the brand of economics that I was supporting was actually an environmental cannon.

It really hit home, because I was seeing the development boom across New Jersey. Lakes in my hometown were being filled with concrete, and all my friends could talk about was how well they did on Wall Street or how high the value was on their house. Something was missing. And when I started reading those books, the world became much more textured and fabulous than I could have imagined from the narrow perspective of economics. I found what was missing. I found something that I really believed in. I began to realize that my generation's great fight was ecological protection. I saw it as a challenge that I had to engage in.

Around this time, an old friend asked me if I would help him run a foundation that was funding a lot of environmental groups. My salary would go way down, but I jumped on board anyway. That experience allowed me to see an inch deep and a mile wide. It really gave me a clear view of the environmental movement, which made me want to dive in even deeper.

That foundation helped me start an environmental group to work on ozone depletion and global warming. We made nothing but mistakes in the first year— ridiculous mistakes because we didn't know what we were doing. But that was better than a graduate school program in how to create and run an environmental group. The only way you're going to learn is by doing everything wrong. From that, you'll figure out what not to do, which will lead you to what to do. We ran Ozone Action for about eight years. We organized scientists, students, mayors— all in favor of supporting an international treaty to stop global warming. On that journey I learned many things.

Then I heard Greenpeace was looking for a new executive director. I survived the interview process, and they offered me the job. I said I would only do it if I could merge these two groups together. So I ended up tucking Ozone Action into Greenpeace. That's how I got to where I am today.

So I've had a bizarre path. I was doing very conservative economic research and then I found my calling to do environmental work. I love doing the work I do. If I was independently wealthy, I would still do this work—which is a really good test.

When you're young, you simply haven't been exposed to enough possibilities. You really don't have enough experience to figure out what you want to do with your life. You can't analyze what it's like to run Greenpeace or to run an environmental group. You just don't have enough information.

So what that means is you have to try different things. Why should you have to pick a path for life when you're young? On average, people have three to five different careers in their life. It's not like you're an apprentice to the shoe cobbler and you're going to be a shoe cobbler for the rest of your life. It doesn't work that way anymore. It makes things a little more confusing, but it's also a great gift.

Society used to be much more rigid and concrete. It was much more of a caste system. Where you could go with your life was based on race, ethnicity, or your income class. That has broken down, which is a very good thing. But one of the things we lost with that was a mentoring system. If someone wanted to be a silversmith, he would go and apprentice under a silversmith. What happens now is you have to do the work on your own. You have to go seek out that ancient connection and find your own mentor.

Offer to fetch coffee for someone in the field that you're interested in. Tell them that you just want to be their wingman, do their errands, or answer their phones. Say that you'll do it free for two weeks. Most people will say no. But a handful will take you up on it. You'll be able to see into their situation, learn, and absorb. Before you know it, you could have an entirely different path. Not because you saw a classified ad, but because you experienced it firsthand.

Deciding what's good or bad for your life should be based on how it feels in your heart, not on someone else's rule book. So here I am running Greenpeace, but if you feel passionately about doing conservative political work, you should do that. If you feel passionately about working for the companies that I think are pillaging the earth, you should do that. You should do the best job you can for them, and you shouldn't let somebody else judge you. Other people's judgments can get you off course. But your own heart will never lead you astray.

Even though we didn't know what we were doing . . .

. . . we were doing it.

RICHARD WOOLCOTT

Cofounder and president/CEO of Volcom Stone clothing company

NEWPORT BEACH, CALIFORNIA

- Grows up skateboarding and surfing.
- Pursues being a pro surfer.
- Breaks neck in five spots: "My dream was taken away from me in a split second."
- Refocuses his road on the business side of surfing.
- Goes to work at Quiksilver making surf videos.
- After a snowboarding trip, decides to leave Quiksilver and start Volcom.
- Realizes that people should gravitate "toward things that make them excited."

I was very active in skateboarding and surfing when I was growing up. I always had business sense in my teenage years, but I was trying to build my professional surfing career. That's really what I was going after, but I had an accident my first year in college. I broke my neck in five different places. I should have died, but I didn't.

Everything in my life shifted after that. It was a traumatic time not just because of the injury, but because I had worked for five years trying to build my competitive portfolio, which is what you need when you turn professional, and it was all taken away from me in a split second. But now that I look back on it, I see that it was necessary. It helped me to focus on college full-time, get a degree, and go from there. While I was in college studying business, I started working in the action sports industry at Quiksilver.

Quiksilver gave me the opportunity to work on a few surfing films with some really talented cinematographers. I did a couple of surf films and then got to make one on my own with Kelly Slater, who's now the six-time world champion. At the time, he was just starting out, and the film we made did really well.

I started to learn that I had a passion for filmmaking. I guess I should have studied film in college, but I just had no idea at the time that I would be excited about it. I happened to bump into it later on. Even today we have a very active film department at Volcom that does skate, snow, and surf videos.

Around the time I was working at Quiksilver, I started getting into snowboarding. My friend Tucker Hall had just lost his job, so we went on this snowboarding trip. Tucker started talking about doing an outerwear company, and we started getting really excited about the idea of doing our own thing.

I also had the bug to do more at Quiksilver, but it just wasn't happening.

When I got home I approached the Quiksilver guys and asked if I could do more at the company. They basically told me that I had reached my ceiling, so I said, "I gotta go." They were cool with it, so we went out and started our own company.

We started Volcom out of my bedroom. It was a little scary, but not too much because I didn't have a family or anything else to support. I didn't need a lot of money to live on, and in my head I just knew that Volcom was going to be the greatest thing ever. Even though we didn't know what we were doing, we were doing it.

In your twenties, you've got a lot more freedom in your thinking and in your mindset. You're more open to things. And hey, if it fails, so what? It doesn't matter because you can just move on to the next thing. You'll find something that works.

One good thing was that we didn't have a lot of bills. It was never like, "Oh my God, the money's running out!" I suggest that to anybody who wants to start their own business—keep your overhead low. I watch a lot of people who get all these investors and bring in a ton of money to start their business. They end up getting huge, grand offices and company cars—but they go out of business before you know it. Everything came too easily for them, so it didn't last.

It's a good thing to take it slow and just go from one door to the next. If one shuts on you, then move on. After a while, something will spark. And if you don't have a lot of overhead, you can maintain that spark for a little while. That's how it worked for Volcom, and for me.

You're going to spend the majority of your waking life working, so make sure you're doing something that you like. That's the best advice I could give—go with what your heart's telling you. You may find yourself moving around, but make sure that you keep gravitating toward things that make you excited. If you're really passionate about what you're doing, money and all that other stuff doesn't matter.

I got to
Seattle and
had no idea
what I
wanted to
do. I didn't
know I
wanted to
be on the
radio or in
the music
business; I
just knew I
loved
music.

JOHN RICHARDS
DJ at KEXP radio
SEATTLE, WASHINGTON

- Grows up in eastern Washington.
- Follows passion for music to Seattle.
- Enrolls at the University of Washington.
- Does grunt work at a local radio station.
- Gets his own late-night radio show.
- Radio station hires him full time—drops out of college.
- Becomes one of the top DJs in Seattle.

I was always good at making mix tapes. Who knew you could make a living at it? I grew up in Spokane, Washington, about three hundred miles east of Seattle. It's not a place for someone who's kinda forward thinking about music, art, culture, politics, or life in general. It's a nice place for kids and retirees, but not for a guy who loves music.

I had seen all these great bands touring through town. I would sneak into the club and see Screaming Trees, Love Battery, Alice in Chains, and I had no idea that they were Seattle bands. This was in the eighties.

The only reason I stayed in Spokane was because of my girlfriend. But she ended up breaking up with me, so I said, "I'm outta here," and headed to Seattle. My brother lived over there and was sending me demos. I was getting these tapes of Nirvana and Pearl Jam and couldn't believe this was going on just three hundred miles from Spokane.

I got to Seattle and had no idea what I wanted to do. I didn't know I wanted to be on the radio or in the music business; I just knew I loved music and I wanted to be in Seattle. I fell in love with the city right away.

So I'm at shows, seeing good music, and could care less about what I was doing with my future. But it hit me eventually—I didn't really want to work in a restaurant anymore. I decided to go to the University of Washington and thought communications would be a good place to start if I wanted to get into music of some sort.

I started listening to Marco Collins, who was a big DJ here in Seattle and broke Beck's *Loser*. I would listen to him break bands and get excited about things he was playing and thought that had to be the coolest job on earth.

I soon realized that I wanted to get into radio and my communications degree wasn't going to be enough. So I decided to check out KCMU, which was the public radio station on campus. I walked over there and knocked on the door, and this guy named Patrick Cohn answered. I'll never forget it. He said, "What can I do for you?" And I said, "What can I do for you?" I would do anything—forty hours a week, no pay. I didn't care. I learned the magic words in radio: "I'll work for free" and "I'll work as many hours as you want."

Patrick told me to come on in and asked if I knew how to do production. I said yes, but I really had no idea. He put me in a room with a reel, a razor blade, and a piece of chalk to start making produced spots for the station. The pieces I did ended up being pretty good, but what really got me in was that I was willing to do anything at any hour. I was just so enthusiastic about it.

That's what got me into radio. It wasn't school, it wasn't anything else but being really enthusiastic about the place I was at. A lot of people would come in and say, "I want to be on the air right away. I want a show." But they weren't passionate about the station.

Eventually a late-night on-air shift opened up. The program director at the time called me in and said, "Hey, we've got a spot for you." I took it and didn't even care about the hours. It ended up being Thursday mornings from 1 a.m. to 6 a.m. I loved it. I had these crazy freight crew dudes calling me in the middle of the night. They loved my show.

So I continued to do that and filled in shifts when other people wouldn't show up. I was still going to the University of Washington, but going to school and being in radio didn't really gel because it took a lot of time to be on the air. So I had to drop out of school.

It's tough when kids come through the radio station today and ask for advice. You hate to say, "Don't go to school." That's not good advice; you should be well-rounded. But for something like radio, you just have to dive in and don't look back. I think that goes for a lot of things: starting a record label, writing a column, or working at recording companies.

For me it was about finding something I was passionate about. I knew I was passionate about radio when I didn't mind getting to the office at 6 a.m. and not being able to go out the night before. I was willing to work for free and be totally enthusiastic about it.

But even though I was passionate about radio, I still got frustrated. The most well-known DJ in Seattle is DJ Riz. He was on every night before me. I was doing the late-night shift, and sometimes I would just feel like nobody was listening. I thought I was no good and that I couldn't keep coming in. I talked to Riz about it, and he told me to keep coming in. He told me to believe in myself and that I was a good DJ. He said I was only going to get better.

Then the next thing I know, right when I was feeling that frustration, they decided to go five days a week with DJs and make it a paying gig. So right when I was ready to give up, I got those five days a week to be on the air and play music. That launched me onto my path.

I ended up achieving what I wanted to do with my life at twenty-five years old. I was on the radio, met a nice girl, had a nice place to live. I had simple goals and I achieved those, which is great! If I fall flat on my face now, that's okay; I've already done a lot of the things I want to do.

I think you guys are doing the best thing you can be doing right now—asking people how they got to where they are. I wish I'd done that—I wish I'd have gone and talked to someone in radio. People are willing to talk to you, and you're gonna get the plusses and minuses of what it's like to do it.

So I got sucked into this world of poetry, not by pursuing it, but . . .

. . . by being **open** to it.

DEVORAH MAJOR

Poet laureate of San Francisco

SAN FRANCISCO, CALIFORNIA

- Goes to San Francisco State University.
- Drops out to travel through Asia for a year.
- Re-enrolls, but changes major to black studies.
- Reads poetry on television by chance.
- Gets invited into a poetry program.
- Starts doing poetry readings and teaching.
- Becomes the San Francisco poet laureate.

I went to San Francisco State University and was committed to becoming a dancer. I wrote poetry, but never thought about being a poet. When I was nineteen, I dropped out of college for a year and traveled throughout Asia. I went to Hong Kong, Bangkok, and all of those fabulous places. It was my second time out of the country. The first time I had lived in a tent in Europe for a year with my father. Especially being a black person in the United States, those experiences really gave me a sense of freedom. I became a citizen of the world—not a San Franciscan or an American.

By the time I got to be twenty-one years old, I was all about just putting one foot in front of the other in the direction of least resistance. So I got back into college and switched my major to black studies because I wanted to be a part of the revolution that was going on. I was in the TV department doing a show for dance, and they asked if I could come and read a poem. I said that I wasn't a poet, but I would be happy to read some things that I'd written.

The next thing I knew, I was asked to be a part of a poetry program by Carol Lee Sanchez, a fabulous Laguna Pueblo Native American woman. I said, "Okay, but I'm a fraud." She said not to worry about it, and told me to write ten poems for her. She really took me under her wing.

I was still acting and dancing in the Black Light Explosion Company and the Black West Repertory, so I was just writing poetry on the side. Then I started to get gigs making money for reading my poetry. I couldn't believe I was getting money for this! I think my theater training allowed me to present it in a unique way. Then I was asked to start reading and teaching poetry to little kids. I helped to coordinate the third-world poetry reading series.

So I got sucked into this world of poetry, not by pursuing it, but by being open to it. I was writing letters to my mother, who at the time was out of the country, and I remember telling her, "I think my dancing was a conduit to my writing."

I was doing poetry in the schools, but I was also working at the hospital as a ward clerk on the weekends because I thought that would be my career. I never really thought that being a poet would be my career. That seemed kind of absurd— I thought there was no way to make a living doing that.

Being in the hospital wasn't working for me anymore, and I kept getting more work doing poetry in school systems, so I started committing myself more toward poetry. Then, a woman who had heard about my poetry wanted to use it in a show. So I filled out the application for the grant, turned it in, and ended up getting a lot of money every month.

I started to take myself seriously as a poet: I had my work published, I was teaching in the schools, and I was having open readings a few times a week. I became the director of poets for two years, where we trained poets to teach in the classroom. But that job was a little straight for me, so I quit. People said, "But what are you going to do?" I said that I was going to follow my dreams.

At the time I was splitting up with my kids' father, so we were dirt-poor. But jobs came to me. Someone called from a local magazine and asked me to be their editor. I did some teaching. There was a lot of pressure to get a straight job, but my mother said, "A straight job will be too seductive with all the money. You're supposed to write."

So I stayed on my path, and over the years I've done several poetry books, written a handful of novels, and was selected to be the San Francisco poet laureate.

Today I'm getting into writing screenplays for movies and doing more books. Before I came here for the interview, I was on the phone about this gig on Friday—we're doing poetry for peace, and I'm getting paid for it. What can be better than to get a little bit of money to promote what you believe in? I love my life.

Life isn't a straight road; it's curvy. And sometimes there are dead ends, so you have to be able to backtrack. Other times you have to know when to knock over mountains. But at every turn you need to take counsel with yourself and ask the hard questions.

And there's not always an immediate payoff. Life is full of hard work. But it's like I used to always tell my kids, it's either hard-good or hard-bad. And when you get to the end of your life, you want to make sure that it was hard-good.

I was writing because I loved to write. There was a fire in my belly for it. It would be 3 a.m. in the morning, and I would still have that fire. I loved it enough to do it even when I was not recognized and was making very little money. When you find something that excites you and it grows tangentially, involve yourself in it. Don't allow yourself to be paralyzed. Come alive. It's like Einstein said, "Energy needs to move."

Growing up in the Depression era of the 1930s, I looked back at the Roaring Twenties as the party that I had missed as a kid.

HUGH HEFNER

Founder of *Playboy* magazine

HOLLYWOOD, CALIFORNIA

- Born in 1926 in a "Methodist home with not a lot of affection."
- Escapes that way of life by "diving into fantasy and dreams."
- Creates small magazines as a child.
- Majors in psychology at the University of Illinois.
- Works for *Esquire* magazine; leaves to start *Playboy*.
- Scrounges up $8,000 to produce the first issue.
- Doesn't put an address on first issue because "I didn't know if there would be a second."

I was born in 1926 and was raised in a very typical Midwestern Methodist home with not a lot of affection. My parents were farm people from Nebraska and their parents had not shown much affection for them, so they didn't show much to us. That's just part of the American heritage.

My parents were Puritans. I'm an eleventh-generation descendant of William Bradford, who came over on the Mayflower. Being raised in an emotionally oppressive manner, I was fascinated by the hypocrisy and the inability to love one another in a more open and kind way.

I escaped that way of life very early on by diving into fantasy and dreams. At five years old, I wanted to be a cartoonist. I started doing comic books to entertain my friends. Then I started my first penny newspaper when I was ten years old and did a school paper called the *Pepper* when I was in sixth grade. Later, in my teens, I did *Shutter Magazine,* about horror. I loved mystery stories and horror.

When I was in high school I had a crush on a girl, so I put her in a little movie I created called *Return from the Dead.* That didn't work—she ended up rejecting me. After that I reinvented myself: I changed my wardrobe and started referring to myself as Heff instead of Hugh. That was the only name for myself that I ever cared about.

After I served in World War II, I got my degree from the University of Illinois. I took writing and art courses, but my major was not journalism, it was psychology. I majored in psychology because I wanted to understand why we are the way we are.

When I graduated from the University of Illinois, I did one semester of postgraduate work at Northwestern, in sociology. I had a course in criminology in which I did a paper comparing the first Kinsey Report, on the sexual behavior of the human male, and our United States sex laws at the time. I got an A for the research but got marked down to a B-plus because I concluded that our laws were wrong.

I went to work on a couple of magazines in Chicago prior to beginning *Playboy.* I worked for *Esquire,* and then I started thinking about creating a men's magazine. I had worked for a publisher as a circulation manager, so I knew the names of newsstands and wholesale dealers around the country. But the problem was that I didn't have any money. So I formed a corporation, borrowed six hundred dollars from a locally owned company, put my furniture up for collateral, and got a little money from the bank. Then I went to anybody who would listen and got a few hundred here and few hundred there—a total of eight thousand dollars. I used that to produce the first issue. I didn't even put the date or an address on the first issue because I didn't know if there would be a second.

From the moment I started the magazine, it was almost like Clark Kent going into a phone booth and coming out Superman. I felt as if I was doing what I was born to do. Suddenly I had tremendous confidence.

In that first summer the staff was simply me and an art director I hired part-time to design the magazine. I created the magazine on a card table in my living room. For the first year the magazine was made out of a combination of things. I had created a little coterie of talented writers and artists in Chicago who did much of the work. I ran a couple of Sherlock Holmes stories in the first few issues. And then I came upon *Fahrenheit 451,* by Ray Bradbury, which had never been run in a magazine. I made a deal to publish that, and a whole lot of young people started reading the magazine.

We think of the fifties as being a very conservative time, and it was. The skirt lengths were down to the ankles. I found that very depressing. During the Roaring Twenties, after World War I, the skirt lengths went up, and that was the time frame that I identified with. In other words, growing up in the Depression era of the 1930s, I looked back at the Roaring Twenties as the party that I had missed as a kid.

So I think that after World War II, when everything went so very conservative—politically, socially, sexually—I was waiting for the party to begin again, but it wasn't there. I think that was part of the inspiration for the magazine. I think the magazine was an attempt to create that mystique. What I didn't know was that a whole new generation of young people was now growing older and identified with the magazine immediately. It was successful from the very outset.

We started the magazine in Chicago but eventually moved it to Hollywood. My dream was always to come to Hollywood, so when I moved out here it was kind of like coming home to where my dreams came from. When we're young we dream impossible dreams, but I think that holding on to a certain part of that impossible dream is what keeps you alive. It makes life worthwhile.

I think my life has been a quest for a world where the words in songs are true. That's an impossible dream, but it has worked for me. As fascinating as my life appears to a great many people on the outside, it is even better on the inside.

I think there's **more** of those **expectations** today. When I grew up, it wasn't like my parents had this grand plan to make me a genius. People today are programmed to be successful. They think, "How do I get into the right nursery school? How do I get into the right grade school? What college am I going to?"

MICKEY DREXLER
CEO of J.Crew
NEW YORK, NEW YORK

- Grows up in the Bronx; father works in the garment business.
- During a college summer, works at a department store—"I loved it."
- After college, becomes a buyer in another department store; climbs the ladder.
- Becomes president and CEO of Ann Taylor; turns the business around.
- Becomes president, then CEO of the Gap.
- Grows the Gap from $200 million to $15 billion over twenty years.
- Leaves the Gap and becomes CEO of J.Crew.

Growing up in the Bronx in the fifties, it seemed like everyone's father worked in the garment business. My dad worked for a junior coat manufacturer buying buttons and piece goods; over the years, I absorbed a lot of that. I would go to work with him on Saturdays starting in my teen years. He worked really hard— six days a week, which a lot of parents did in the Bronx.

So I went to college in Boston, and during my first summer break, I was looking for a job in New York. The first thing that came along was at Abraham & Straus, which was a department store. They were going to pay me $125 a week, so I took the job.

I ended up in the jeans department working for a guy named Ken Hirsch. I did that for the summer of 1967, and I really loved what I did. It was a respectable $125 a week job, and I was learning a lot.

When I went back to school I got a job working part time at Gilchrist's department store. I did that for about four months, but it was the quietest place ever and I ended up getting fired because I used to throw away the sales figures.

So I graduated from college and applied to all the New York–based retail companies because I really liked my experience that one summer. I interviewed at JCPenney and I remember very clearly what the HR guy said to me: "Follow your gut, and never take a job that isn't right for you." I did, and passed on the job and went to Bloomingdale's.

It was a really exciting place to be. My first day there, they put me in a department where the buyer was on maternity leave, and I ended up being the temporary buyer of junior sportswear. For the first three months there, I bought the goods that I liked for the New York store, and no one told me anything. I loved it, and I learned a lot about using my gut.

I then got moved to a different store and started moving up the ladder, but it got less fun as I moved up. I stayed for six years, but left in 1974 because I was bored out of my mind. I took a job with Macy's New York, which was going through a turnaround and I was excited by the new leader. I soon realized that I had made a mistake. It was too much of a men's club rather than a creatively driven company. I always liked creative organizations where it was all about the product. So I left to go back to Abraham & Straus where I had originally worked on summer break. But it also ended up being a very bureaucratic department store environment. There was no room for being creative. There was no room for buying the goods without layers of approval.

Around that time the guys who owned Ann Taylor called me and asked me if I would be interested in running that company, which was then relatively small. I said no at first because I would be working under the founder. But I was unhappy at Abraham & Straus and was trying to figure out what I wanted to do with my life. I was still a young guy at thirty-four years old.

Then I had dinner with an older, wiser friend of mine. I told him that Ann Taylor wanted me to be president, but I didn't want to take it. He said to me, "What are you, crazy? Do it. You can be president of a small company, and you won't have to deal with lots of people." I called the next morning and said, "I changed my mind. I want the job." That's kind of where my life began.

I ended up really turning the business around. It was a great experience. I hired good creative people and followed my gut, and it became my learning process for what I still do today.

So I started to explore again, and I ended up going to the Gap, which had the same attributes as Ann Taylor when I first went to work there—a well-known name and good real estate, but it was a lousy business. So I took a shot, moved to San Francisco, and ended up staying for eighteen years, during which time we built the company from being worth $200 million to being worth $15 billion. Then, I was pushed out by the board of directors—I was shocked.

One of the greatest things I learned from that experience is that you should treat people the way you want to be treated. Whether they're good at what they do or not, treat them with respect, understanding, and sensitivity.

After the Gap, I had to figure out what I wanted to do with my life again. I had always admired what Emily Woods, the founder's daughter, had done with J.Crew, so I invested in this company, came aboard as CEO, and here I am two years later.

I never said, "I wanna be there." I remember one of my first days on the job at Bloomingdale's, one of the young guys was asked, "What do you want to be doing in five years?" The guy answered, "I want to be vice president of this joint." I cringed when I heard that. I never said that I wanted to be anything in my life. I didn't have those expectations initially, and I'm glad that I didn't.

I think there's more of those expectations today. When I grew up, it wasn't like my parents had this grand plan to make me a genius. People today are programmed to be successful. They think, "How do I get into the right nursery school? How do I get into the right grade school? What college am I going to?" and so on. A friend of mine last week was looking at different grade schools for his kid. But have you ever heard of anyone whose life was dramatically changed by what grade school they went to?

When you're young, you never know what you know when you're older. Yet there is all this pressure on young people to figure out what they want to do with their lives. I never felt that I wanted anything more than simply being able to do what I felt I needed to do.

I remember him telling me,
"You're never going to make a living
drawing monsters."

MARK FOSTER
Founder of Heroin Skateboards and skateboard filmmaker

LONDON, ENGLAND

- Grows up in Manchester, a blue-collar city in northern England, where everyone "got jobs at the factory."
- Discovers skateboarding and art at a young age.
- Goes to a design school in London.
- Starts his own t-shirt company, which takes off.
- Starts a skateboard company that creates videos and boards.

Ever since I was five years old, I wanted a skateboard. All the kids on my block had bikes, and then there was one kid with a skateboard. I just wanted to have a go at it. Skateboarding became a really big part of my life. I think it saved me from a lot of things. It distracted me from drugs and drinking. I always just wanted to go skateboarding so I never got into any of that stuff.

Even though I live in London now, I didn't grow up there. I'm from Manchester, and the kids I grew up with finished school and then got jobs at the factory. If they did really well, they became the manager of an electric shop. I didn't see my future there.

When I was around sixteen my dad started asking me what I wanted to do with my life. He told me that I should be a plumber, an electrician, a builder, or something like that, but the only thing I was interested in was art and creativity. My mom ended up convincing him to let me start taking some art classes, but the whole time I took them my dad told me that I was wasting my time. I remember him telling me once, "You're never going to make a living drawing monsters."

At the time, all of my friends were letting their parents control their lives. If their dad told them to give up skateboarding, they gave up skateboarding. But that never made any sense to me. I always had to do what I felt I needed to do.

I applied to a college in Manchester to do illustration, but it fell through. That really gave me second thoughts about pursuing art. Then I heard there were spaces available at a college in London to study design. I didn't want to bother going to the interview, but my mom encouraged me to go. So we took a train to London and saw the college, and it was amazing. It really felt like it was meant to be.

So I moved there and went to college. I was always known as the kid who skated. I was always going into town, skating, hanging out, drinking coffee, coming back, painting, and listening to music. It was really a good time.

Around that time I started a little t-shirt company with my friend Hiro. We started doing shirt graphics that looked like sixties horror comics. I was really into it, but we had no money. It cost me everything I had to get it going. Student loans just rinsed out of my bank account so we could make those t-shirts. But it was so much fun and so spontaneous.

We would go around to shops to see if anybody wanted to buy our shirts, and we got a really good response. We'd see an Oasis video on TV, and one of the guys in the crowd would be rocking out in one of our t-shirts. We couldn't believe people were so into it.

After a while we stopped doing the t-shirt thing, and I started thinking about what else I could do. Around that time I was skating, painting, and filming skateboarding a lot. Then I broke my wrist, and in the hospital I realized that I could start a skateboard company. I was working at a big skateboard distributor, so I had all the contacts. And I could use my artwork as the basis for the designs on the boards.

The first video we made was really rough. I just went to different spots and filmed what looked good. The next thing I knew, we had loads of footage that I had to crash edit on two VCRs. It was really sketchy, but that's how we made our first video. It was called *Good Shit,* and that's what kicked off our company. That's how I started Heroin Skateboards.

It's kind of funny, because now I *am* making money drawing monsters. One day my landlady said that she was going to sell our house, and I asked her why she didn't ask me to buy it. And she actually said, "But you're a skateboarder." Unbelievable! My wife and I bought the house, and we're still living in it today.

It seems like people want to put you in a box because it's simpler for them to understand. It's like, "What do you do?" "I'm an accountant." "Okay, that guy's an accountant. That's simple enough." But if you run a skateboard company, that's a little more strange and harder to understand.

You need to listen to yourself more than anybody else. You can take advice from people, but you've got to be selective about what you listen to. I had an interview with a skateboarding magazine, and they were asking me how all of my skateboarding stuff worked out. I basically told them that I just didn't listen to other people who tried to tell me how to live my life. Learn from other people's mistakes, but you have to make sure that you're living your own life.

It was an **open** country. The reservation fence
was behind my house, but in the other direction was
land as far as you could see.

GERARD BAKER
Park superintendent of Mount Rushmore
KEYSTONE, SOUTH DAKOTA

- Government takes his family's tribal land away.
- Father moves to a new land and builds a log cabin.
- Gerard is born in 1953, and grows up in that log cabin.
- Goes to college, but drops out because "I was partying pretty crazy."
- Finds passion for the Park Service—goes to school in Northern California.
- Becomes park superintendent at Custer Battlefield.
- In a controversial move, works to change park's name to Little Bighorn Battlefield.
- Has "seven confirmed death threats" after changing the name.
- Becomes park superintendent of Mount Rushmore.

I grew up on a reservation outside of a little town called Mandaree. Our reservation had two different tribes—the Ricaree and the Minnetaree. And traditionally the Ricaree were Pawnee. They joined us around 1845 to 1850.

Our people were the hub of the trading network. We would trade with our enemy tribes: the Sioux. We would trade with the Blackfoot and the Cree. We would trade hides with the white man. Everyone talks about the Mall of America in Minnesota, but we were the biggest mall in the world at the time, because the river was our highway.

We had two epidemics of smallpox—one was in the 1700s, and the other was in 1837. The most devastating one was in 1837, which killed 80 percent of our population.

Our people used to live down in the valley. We were farmers and ranchers, and life was really good. We had a 1 percent unemployment rate and no welfare. But in the early 1950s, the government flooded our land, and we had to move to the top of the valley.

We moved into Indian towns, but there were no jobs. It was a tough transition. My mom went to school as a first-grader when she was sixteen years of age. She didn't know a word of English. She wasn't allowed to speak Indian in school, and when she was caught they would put her in the corner and make her sit there all night long.

My dad was a rancher, so when the government moved us they gave him a homestead plan. He left the valley with some horses, a plow, some cows, and that was it. He got some land from the government, cut all the trees down, and built a log house. I was born in that log house in 1953 and raised there.

My first thought of what I wanted to do with my life was to be a rancher. It was an open country. The reservation fence was behind my house, but in the other direction was land as far as you could see. We ran cows, we had a tractor, and we cut grass. That was a really good life.

I learned about the Park Service when I was in fourth grade. We lived about thirty miles from the nearest national park, but we didn't have a car so we couldn't go there. When we did get a car, we went down there, and I thought it was a different world.

When I was a freshman in high school, I got sent to boarding school because my parents didn't want me to speak broken English. My parents believed in two kinds of education—traditional ceremony education and academic education.

When I graduated from high school, I went to college near South Dakota and played basketball, but flunked out because I was partying pretty crazy back then. One of my clan grandpas gave me a hell of a time when I flunked out of college. He said it really hurt because I was not being a warrior. I didn't understand, but he said that today there is a new way of being a warrior—to do something with your life.

I started to think more about what I could do. I liked being outdoors and I didn't like working at a desk, so I started thinking about how I could get into the Park Service. I talked to some of the rangers to find out what the best way to do that was, and they said that I should get a law enforcement degree. They said to go out to the College of the Redlands in Northern California where the college is free, and enroll in the police science courses. I lived in a tent in Eureka for a week, but it was right on the coast so it was really pretty. We eventually got a place to live, and I paid my rent by breaking horses.

My first job in the Park Service was cleaning toilets. I picked up trash and registered people for camping. I started at the bottom rung, and by the time I got through it, I realized that was what I wanted to do for a career.

The Park Service is the closest I can get to the Indian philosophy. We're here to preserve and protect the land, and to tell stories. I eventually got a job with the U.S. Forest Service. I went to Beartooth Mountain in Montana and was an assistant ranger for three years. It was okay, but I missed the Park Service, and I got this call that there was a job opening at the Custer Battlefield in Montana. So I applied and got the job to be the park superintendent.

It was a really controversial area. The Custer Battlefield was the only place in the National Park System where we didn't honor the land—we honored General Custer. I wanted to change that, so I brought seven bands of Indians to the park. Their people had also fought in the battle, and we needed to hear their side of the story to create an accurate depiction of what had really happened. According to Custer, Indian people had no soul. He thought they were animals. There were women and kids involved in that fight, and it was not a warrior village. When I brought the Indians to the park, all of a sudden about twenty thousand Custer buffs, some of whom were terrible, came out to show their side of the story. To a lot of people, Custer was a god and a hero. They got really mad because we changed the name of the national monument from Custer Battlefield to Little Bighorn Battlefield, which gave us the right to change the prospectus. Then we started to hire Indian people to work with the park.

When all this was happening, I had a guy come up to me and say, "I wish this was two hundred years ago and I could throw your babies in the air and catch them with my bayonet." I started getting threatening calls in the middle of the night. By the time it was over, I had seven confirmed death threats.

I like challenges, which made the work at Little Bighorn good for me. But I wanted to come closer to home, so my boss offered me this job at Mount Rushmore.

Now we're dealing with how to interpret this monument. This was somebody's land at one point, and it's still a sacred site. I'm the first American Indian to be the superintendent of this place, and I'm very proud of that. But it's a heck of a challenge because a lot of American Indians won't come up here.

Especially right now, this place is extremely important. In a time of war, it's even more important to know who we are as American people. By the time I leave here, I want this park to express and preserve the philosophy of the Indian people. A respect for the environment, culture, self-sufficiency, and conquering the enemy. The damn government has gotten our people, but we can come out fighting again on many fronts. We can be warriors on a new battlefield.

When I was a senior in high school, a nun told me that the best thing I could do with my life was pump gas. She told me that just two days before I was to graduate. The highlight of my life was when I was asked to return to my old school and speak to the students. When I returned, I told her, "No, I'm not working at a gas station. I'm the park superintendent at Mount Rushmore National Memorial."

When you have a
dream, you're going to
work for that dream no
matter what.

RITA SIMO
Founder of
People's Music School
CHICAGO, ILLINOIS

- Born in the Dominican Republic—family was all musicians.
- Wins a piano competition, gets a scholarship to Julliard School of Music in New York.
- Becomes a concert pianist, but finds that the work is "very detached."
- Decides to start her own tuition-free classical music school.
- Needs to get a green card, so she becomes a nun—experience was "very negative."
- Moves to Chicago to enroll in a Jesuit theology course—finds support.
- The Jesuit priests help her to open the People's Music School.
- She gets a job on the side to pay the bills.
- Has a street in Chicago named after her—"Rita Simo Way."

I was born in the Dominican Republic many years ago. We lived in a nice town, but it was the third world—we had no running water. My family was all musicians. Some of us were professional and some played for fun.

When I was sixteen, I was in a piano competition and I won. The prize was a full scholarship to the Julliard School of Music in New York. Everyone realized that it was an amazing opportunity, so my mother put me on a plane.

After I graduated from Julliard, I was sent to play in concerts for public schools across the country. I loved talking to the kids after the concerts. That's when I knew that I wanted to teach. When you're giving concerts all over the place, you never get to know any of the people who ask for your autograph. I wanted to get to know those people.

I saw that the life of the concert pianist is very detached from reality, meaning that there is not a sense of belonging—and I wanted to belong to something. I wanted to start my own music school and make the classes free for the students. In the Dominican Republic, the music classes were free—it was part of the National Conservatory. But in America that's not the case. If you want to learn, you have to pay.

Before I started my own music school, I needed to get my green card. To do that I needed to be a part of an organization, and the only organization I knew of was the Catholic Church, so I became a nun.

I got a directory of all the nuns in New York City and found that there were at least two groups who seemed to be very involved in their schools. So I went and talked to the groups and ended up landing a job teaching in Sinsinawa, which is somewhere in the Midwest. It wasn't even on the map. My experience at that convent was very negative. The students were very good, but the more I stayed the more I realized it wasn't for me, so I left.

I came to Chicago and took a course by a Jesuit priest called "The Theology of Hope." I told them about my idea to start the music school, and that was the first time that nobody said to me, "You're crazy." They said that if I went for it, they would help me. So in November 1975 we went downtown and incorporated the People's Music School. I had to put my house address on the forms because we didn't have anything else. We decided to have a board of directors, and we had our first board meeting on Beethoven's birthday. In the process, I wrote a letter to everyone I knew. Almost everyone wrote me back and said, "It's about time you started your own music school. You've been talking about it for too long. We'll send you money monthly." I ended up with six hundred and twenty-five dollars but that wasn't enough, so I went looking for another job.

I got myself a job at the local Episcopal church as the activities director. I directed the lunch program for the senior citizens, which was great because I could eat with them and save my food money.

I'm very good at conning people into doing something. I got two pianos from a priest who had them in his basement. I called storage facilities to see if people had left any instruments behind. One of the senior citizens in my activities program was a piano tuner, so he fixed up the instruments. When you're desperate, all kinds of strange ideas come into your head.

Around that time, one of the priests helped me to sign a lease on a space for our school. It was a former beauty salon that had been closed down for two years. When we walked in there you could peel the dirt off the walls, so I put up a sign that said, "Cleaning party—beer provided." Ask and you shall receive, right? I had a bunch of people come in and help us clean the place up and get it ready for the students. We opened the space on February 22, 1976, and it's been downhill ever since. It's amazing to look at our building today—it cost two million dollars to build, and it's all paid off. We've come a long way.

When you have a dream, you're going to work for that dream no matter what. And if you're a realist, as I am, you know that things are not going to come on a silver platter. Anything worthwhile will never be handed to you. You have to work for it.

I never liked Saint Paul much because he was an asshole—women weren't important to him—but he did say something good. He said that God never gives us troubles that we can't handle. That's been very true in my life.

There's that poem by Robert Frost, "The Road Not Taken." It talks about a road diverging in the wood, and when I look back on my life, I encountered that many, many times. I always took the road less traveled.

You have to believe that you can do what you love.

We were burning it at both ends—working twenty hours a day, seven days a week, to **make snowboarding happen.**

MICHAEL JAGER

Founder and creative director of JDK Design

BURLINGTON, VERMONT

- Grows up in Vermont, twenty miles from Canadian border.
- "Sucked at math" early on, so he focuses on art.
- Goes to design school in Montreal—gets heavily into punk music.
- Moves back to Burlington; "I thought it would just be a pit stop."
- Designs for "IBM during the day and . . . local punk bands at night."
- Gets work designing for a young Burton Snowboards.
- As Burton "invented a culture," JDK Design grows with them.

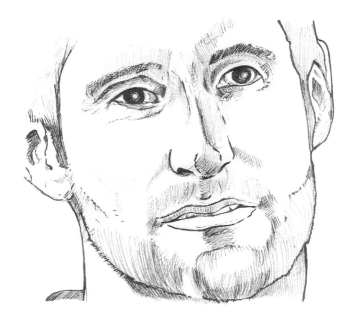

As you guys are trying to prove here, you have to go with what you believe in and what you really want to do. The thing that saddens me most is when you see an individual go down a path and burn about ten years of their life on a channel they now realize was the wrong one.

I realized I sucked at math early on, so I focused on artwork. I always had a feel for how to draw, so I channeled myself in that direction. I didn't have a calculated path. My parents didn't have a clue about that kind of stuff. My father worked in a factory making cottage cheese, and my mother was a hairdresser from Maine.

I had lived in Vermont my whole life and grew up in a small town called St. Albans, which is about fifteen minutes from the Canadian border. I did funky portraits and paintings, so I decided

I wanted to go to art school. I ended up going to Dawson College in Montreal. I had never even been to Montreal before, but somehow I convinced my father that it was a good idea.

I was seventeen years old and was right in the heart of Montreal, which was a really cool experience. I remember one eye-opener was when I saw about twenty guys get into a full-on brawl right in the middle of the street. They were real leather-jacket, platform shoe–wearing guys, and it was just a full-on war. People were running over each other with cars. It was pretty hard-core for a seventeen-year-old. I had never seen anything like that so close-up. It really affected my artwork.

Around that time punk music was starting up. It was 1978, and my first date with my wife was a Clash concert in Montreal. She was Italian, and I was some whacked-out punk American guy. We got married when we were eighteen.

After design school I came back to Burlington. I thought it would just be a pit stop; I wanted to move to New Orleans because I had never been there before and I really like the architecture down there. I ended up working at a small design shop. We would design for IBM during the day, and do stuff for local punk bands at night. It was a weird collision of those worlds—it really blurred the lines between fine art and communication.

The job gave me the opportunity to do a lot of traveling, which is when I realized how cool Burlington was. I didn't want to do the standard thing for a designer and go to New York, Chicago, or L.A. I didn't feel like I needed to prove anything by doing that. If you're dedicated and have integrity, the world will hear about it.

I also think the cities have a more potent influence on you when you don't live there. You can step in and see things that people who live there don't see. I would rather go to Tokyo, London, and L.A to visit than to live. Plus, here I've got the mountains and the lake, people's guard is down, and the vibe is right.

Getting the Burton Snowboards account was a big deal for us. At the time, there were only three of us working out of my basement. Burton was also tiny at the time—they only did four snowboards. It turned out that they were looking for someone to help do the graphics for the boards, so a woman named Basey asked me to come in for a meeting. She had heard about my work, and we connected immediately.

When I talk to her today, she says that she gave us the job because we dressed really differently. Distinction is everything. If you look at any example of people who have been successful at what they're doing—and not monetarily successful, but successful in their lives—they are generally pretty distinctive individuals. That can be a farmer in the boonies of Vermont or it can be the most progressive artist in New York City.

As soon as we got a few projects, we started delivering really good work, and then it all started to move quickly. Business got a lot busier as the snowboard industry launched. That Burton thing was a once-in-a-lifetime deal. Somehow I happened to hook up with someone who invented a culture, not just a sport. We were burning it at both ends—working twenty hours a day, seven days a week, to make snowboarding happen.

Snowboarding is more prevalent now, but in the beginning it was a cause. In the early days, if you rode a snowboard on a mountain you were an exception. In our design work, we tried to make the mainstream feel like baboons—like they had no imagination. In fact, one catalog cover was the face of a baboon to represent this. Even today, there is still a level of freedom that snowboarding represents that's linked to skateboarding and surfing. There's something about standing sideways on a board that is still a cause. Ten years ago that cause was unstoppable.

The idea of designing a life is interesting. In design, nothing stays still. There is a constant evolution. You can't ever have a house style; you need to be willing to change it up. If you apply that perspective to someone's life, you have a new format as compared to the traditional view. I love seeing resumes of people who have done fifteen different things in five years. They were in the Peace Corps for a while and then came home and did some sales. If you look at our design firm, we've been around for fifteen years, and it's been fifteen different businesses. You have to change as culture changes, which keeps things interesting. If that wasn't the case I would have left a long time ago.

You have to believe that you can do what you love. Follow what you love is kind of a cliché answer, but not a lot of people have the backbone to do that. I think it was Simon Woodruff who said, "The world will conspire to support you if you really magnify what it is you believe in." I think that's true whether you're going to be a sculptor or a kick-ass accountant. If you really believe it and do it with distinction, the world will conspire to support you.

At six o'clock in the morning I was standing in the Bombay airport wondering what the hell I had done.

WALTER HOLMES
Couture designer turned yoga guru

CHICAGO, ILLINOIS

- Grows up poor in England, youngest of eleven kids.
- Is released from the British army at age twenty.
- Starts doing sketches for friends in London.
- Is introduced to Julie Andrews and Audrey Hepburn.
- Starts sketching and designing for famous actresses.
- Moves to America; becomes a famous designer.
- Goes on a trip to India; finds himself.
- Becomes a yoga teacher.

I always get asked, "How did you get to where you are?" And my answer is, "I don't know." I guess it's not very inspiring, but on the other hand, it's very inspiring.

I grew up in London and was the youngest of eleven children. We were very poor, so I only had two years of schooling, from twelve to fourteen years old. When I was twenty, I got released from the British Army and came back to London, where I started doing sketches for some friends who were actresses. Those friends introduced me to Julie Andrews and Audrey Hepburn, who then asked me to start sketching and designing for them.

So one thing leads to another, and the next thing I know I'm being called a "designer." I actually was designing, but I didn't know that I was. I was having a lot of fun doing it and figured I would just do it for as long as I could.

I kept evolving and found myself working as a gofer for top fashion photographers in London. That led to me coming to America and meeting the people of Marshall Fields, whom I soon started to design for. I liked the idea of coming to America, because I figured I could make mistakes and nobody from back home would catch me. But I didn't make any mistakes. I was just expressing myself and enjoying life.

I have never believed that there is any formula for life. Your life is a process, and when you accept that, you permit these things to unfold, allowing many different opportunities to be presented to you. The only limitations people have are the ones they put on themselves. And most of those limitations are based on a fear of being unsuccessful. But how do you know if you'll be unsuccessful if you don't try?

So I became a very famous couture designer in America, and I won many awards. I designed for some major celebrities and television shows. That led to me doing what we call "couture ready-to-wear" for some of the largest couture elegance stores in America.

My work was then brought to the attention of Hugh Hefner, and I was selected to create design concepts for the flight attendants on his first plane—the big black one with the bunny on the tail. That led me to doing other design concepts for him, which led me to consulting for the magazine, which led me to becoming the men's fashion editor for *Playboy* magazine. I never once thought that any of this was strange or bizarre. I just thought how remarkable it all was. If something didn't turn out, so what?

At that time, I had a private restaurant, I was living in a mansion across the street from Hefner, and I started practicing hatha yoga. I don't know why I started doing yoga; I was just drawn to it. After three months, I was asked by my yoga teacher to go to India with her. I had never thought about going to India, but three months later my whole life was in storage and at six o'clock in the morning I was standing in the Bombay airport wondering what the hell I had done. I went from living in a mansion and sleeping on a king-sized bed covered in fur to sleeping on a concrete floor. But the interesting part was that it didn't bother me. That was when I started to realize that life is not about all the "stuff."

Most people do not have a good relationship with themselves. In our society, we're not encouraged to do that. We're encouraged to be greedy and look for something else to keep us happy, which keeps us consuming, which in turn keeps the economy going.

I was supposed to be in India for three months, but ended up staying for six months and practicing hatha yoga with Iyengar, who was one of the highest people in India. I had the ultimate experience of mind, body, and spirit. Afterward, I decided not to do anything frivolous anymore, so I decided to stop design and start teaching yoga. I didn't have those cravings for ego or attention anymore, and I began to look at life completely differently. Instead of looking for what was missing from my life, I looked for what could be celebrated in my life.

I can't even begin to express some of the rubbish I hear from people. Why not give me the good news? You got up this morning, you're healthy, you have a great place to live, and so on. It's important to be grateful and live your life with an attitude of gratitude. Your life is the way you think it is: your mind, thoughts, actions, and words literally create your life. All any of us can do is be the best version of ourselves as human beings. Unfortunately, I think many of us become what we relate to: physical appearance, financial status, labels, skin color—the list is endless. If you're not careful, you become a series of labels that other people have placed on you, and you get lost. That's why we live in a world where people are so discontent and dissatisfied.

Today I absolutely will never own anything with any designer's name on it. I don't need to do that anymore because I have my own identity. If anything, I'll put my own name on a shirt.

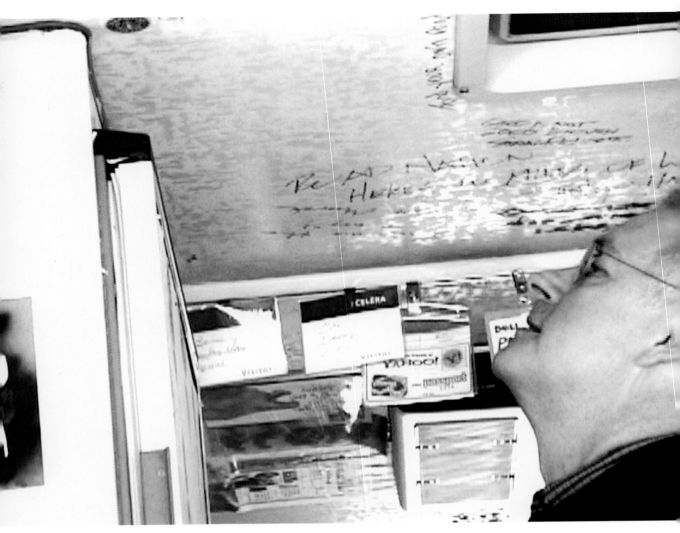

Education takes place in brief moments in time.

Inspirational connections that people make end up having a profound impact on the path you take.

WILLIAM SAHLMAN
Professor at Harvard Business School
BOSTON, MASSACHUSETTS

- Grows up in Florida where "the University of Florida was the center of the known universe."
- Has an influential teacher in high school who went to Princeton.
- Attends Princeton for undergrad; has an influential teacher who went to Harvard.
- Attends Harvard Business School for his Ph.D. and starts teaching there.
- Shifts his career to topic of "entrepreneurship" and becomes an expert.

I've had four thousand students here at Harvard over the years, so I've had opportunities to watch them go out and evolve. I would say the biggest mistake people make is following other people down a path rather than doing what they're really passionate about.

People often behave like lemmings. And I think it's become more true of late. For example, here at Harvard Business School, there was a dramatic shift in where kids went to work. They started going to work for McKinsey or Goldman Sachs, which wasn't hard to understand. McKinsey was paying over one hundred thousand dollars to twenty-six-year-olds with relatively little business experience. So was Goldman. You could make three or four hundred thousand dollars in a few years in that kind of career, so it became very popular. All of a sudden, if you didn't get a job offer from a consulting firm or an investment bank, you really weren't cool.

But many of those kids who went into those consulting jobs hated them. And most didn't really know what the other choices were, because those choices don't recruit on college campuses. McKinsey and Goldman Sachs have certain kinds of corporate recruiters who show up on campus and walk away with a bevy of people. But people who run small businesses and the like don't come in and try to grab college graduates.

We don't live in a world that's always clear-cut. I often think we're in these pachinko machines, where you tumble from experience to experience and often you meet someone who has a profound impact on the path you take.

In my case, I had a few of those teachers or professors. There was one guy who had gone to Princeton and always described it as the center of the known universe. Well, I grew up in Florida, where the University of Florida was the center of the known universe, so the idea that I would go to Princeton was something I hadn't really thought about. Because of that teacher, I ended up going to Princeton, which turned out to be a critical decision. Then I had a professor at Princeton who had a Harvard M.B.A. I never really focused on going to Harvard Business School, but this guy was the best professor I'd ever had. So I ended up getting to know him and applying to the business school and came here. So it cascaded into a career, which wasn't necessarily the way I thought it would all pan out when I was eighteen.

Beginning in the early 1980s, I shifted my career to the topic of entrepreneurship. So for the past twenty-two years I've been able to meet lots of entrepreneurs, written something like 130 case studies on individual entrepreneurial firms, and been involved in lots of different kinds of projects. When you do that, you become aware of the great diversity of backgrounds that people have.

For example, take the history of Intuit—the company that created Quicken software. There was a point at which Scott Cook, the founder of Intuit, would gladly have given you 100 percent of the stock if you would just take the financial obligations off his hands. He went out and tried to raise money, but he couldn't raise a nickel. He scraped together credit card money and a variety of things, and that company's worth five or six billion dollars today. Scott Cook is an important player in the business scene and an incredibly charismatic, capable guy. But it wasn't a straight path.

So I think it's an important thing to expose people to others who have taken different paths and who are emotionally enthusiastic and attached to what they're doing—who love getting up in the morning and going to work.

Being optimistic is a fantastic way to approach life. Lots of kids end up on the cynical side, and they miss opportunities to learn from people because they simply shut themselves off. People too often have switches that they turn on or off very quickly. It's the common problem of stereotyping. If I meet someone who is different from me, I automatically assume I'm not going to learn anything from them because they're not interesting or whatever. But stereotyping in any dimension turns out to be extraordinarily dangerous because most of the opportunities to learn are from differences.

I always tell my students there's an inverse correlation between the grade they get in my class and how well they're likely to do when they graduate. In a lot of structured environments, you grade along a dimension. I don't necessarily measure creativity. I don't measure attitudinal characteristics about certain kinds of things. I think you have to set standards for yourself that are very personal, not standards that other people dictate.

I have this view of both high school education and college education: We worry about the overall quality of education, but in fact, education takes place in brief moments in time. Inspirational connections that people make end up having a profound impact on the path you take.

PART 5

Day 1
and so we
started . . .

Days 2–4
Las Vegas
to Arizona

Days 5–9
Colorado to
Washington

A ROADTRIP JOURNAL

Days 10–11
Seattle

Days 12–16
San Francisco

Days 17–19
Los Angeles

Day 20
and so we started . . .

Day 396
New York—1 year later

A Roadtrip Journal

So we could in fact do a roadtrip of our own. Turns out Roadtrip Nation, the group who drove the RV through campus, was looking for the "next generation" of kids to take the wheel and head out on their own. For three twenty-one-year-old kids dealing with schoolwork, internships, and the summer job hunt, a free ride across the country to film a documentary seemed like a bit of a pipe dream.

—*Ryan Duffy,*
age 21, New York

In the spring of 2003, Roadtrip Nation started a program on college campuses called Behind the Wheel. The point was to provide opportunities for students to hit the road, explore the world around them, and broaden their understanding of what they could do with their lives.

Ryan Duffy, Randy D'Amico, and Mike Sussman, friends from New Jersey, were among the applicants that first year. They filled out the application form, booked an interview with a local leader, filmed the conversation, sent us the footage, and waited.

Their application wasn't polished or professional, but it was genuine and sincere. There was an authentic friendship between these three guys—a connection that we knew would lead to a rich roadtrip experience.

Once they were selected, the trio went to work building their trip. They chose the route, decided whom to meet, and cold called the interviewees to book the conversations. After months of prep, Ryan, Randy, and Mike flew to California to pick up the RV and start their journey.

This final section of the book shares Ryan's personal roadtrip journal. It gives a first-hand glimpse into life on the road and the interviews they conducted along the way. We hope you will experience a bit of the road through his journal, but more than that, we hope his reflections inspire you to hit the road yourself.

Since this first Behind the Wheel roadtrip, the program has grown, giving many other students the opportunity to get out on the road. But what we're most excited about is how these student roadtrips have inspired teachers to offer experiential roadtrip learning curricula, career centers to support career exploration programs, and people everywhere to hit the road independent of green RVs.

Why do we care so much about helping people find their open roads? Because if everyone found that path, just think of the impact it could have on the world. One person's search to find their open road, linked to the next person's search, linked to the next person's search, linked to the next person's search, leads to global solutions.

A collective chain of personal discovery, harnessed to mobilize potential in ourselves, and in our world.

DAY 1: and so we started . . .

On Thursday, May 29, Mike, Randy, and I boarded a plane out of JFK and headed for Laguna Beach, California. The Roadtrip Nation crew met us at the airport, and we headed back to the RTN HQ for a few days of prepping: getting acclimated to the RV (cleaning out the mounds of junk left in it from the last trip), learning how to use the cameras, and doing some pretrip interviewing. The time flew by, and when it finally registered with us that we were actually in California, it was Saturday morning and time to hit the road.

And so we started.

DAY 2: Las Vegas

Our first destination was Las Vegas (great way to ring in the trip, eh?). We piled into the RV and reluctantly pulled out of beautiful Laguna Beach, heading straight into the desert. We were five total: the three of us and two guys from Roadtrip Nation acting as film crew. That desert is no joke out there. New York and D.C. are pretty devoid of cacti, so we pulled the *City Slickers* routine for the first mile or two, along with many a "Wow, look at that." It was a rough initiation into the roadtrip lifestyle, a baptism by fire of sorts. There was no air-conditioning or refrigeration unit to speak of. They had been ripped out along with the toilet and shower to make room for more storage. The heat, mixed with our excitement, anxiety, and whatever else we were feeling, made for an exhilarating ride into Vegas.

Pulling into Vegas was completely surreal. You are just driving through desert and then . . . *Vegas.* Even during the day it hits you like some kind of neon seizure, coming out of nowhere and completely overwhelming your senses. We pulled our bright green RV into town and it barely stood out. Our first interview, with Rob Bollinger, the artistic director for Cirque du Soleil, wasn't scheduled until later that afternoon, so we were left to our own devices. We did the tourist bit, driving down the strip, reciting lines from *Swingers,* pretending the RV was a late-model Cadillac—the obligatory first-time idiots-in-Vegas routine.

Before we knew it, it was time to head over to the Bellagio and meet up with Rob. He was our first interview of the trip, so we were definitely apprehensive. However, interviewing him in the RV was cool because he was on our turf, so to speak.

We parked the RV around back of the Bellagio. When Rob came out to the RV, we were a little shaky at the beginning and didn't know who should speak first, when to ask questions, when to listen, what to do with the microphones, and so on. Fortunately, Rob was able to put us at ease and we fell into comfortable conversation.

ROB BOLLINGER

Artistic director of Cirque du Soleil's *O*

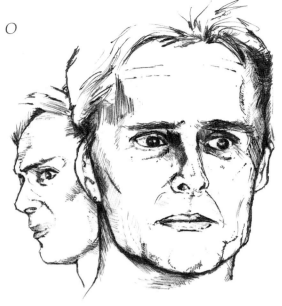

ROADTRIP NATION: This is day one of our roadtrip. We're interviewing people all across the country to learn how they got to where they are today. Basically, the goal is to get a sense of your story and how you got here. So, how did you get here?

ROB: My publicist told me that I had an interview, and she walked me out the back door [laughs].

RTN: Gotcha [laughs]. But where were you when you were our age? We are all finishing college. Put yourself in our shoes a little bit and give us an idea of your story.

ROB: I went to Indiana University in Bloomington during the Bobby Knight era with Isaiah Thomas and chair-throwing incidents. I was a competitive springboard diver and was on a scholarship. My whole world at that time was really focused on the Olympics. I qualified in 1979 to try out for the Olympics in 1980, which I did, but the U.S. didn't end up going because we boycotted those games. I didn't make the team, but I competed in the trials. I graduated in 1982 but stayed in Indiana to train for the 1984 Olympics. So at that point in my life I really didn't have a lot of thought beyond that, because when you are training for something like that, it's pretty intense and that's your world. I majored in marketing and business management because I knew I would have to do something beyond diving, but what it was going to be, I had no idea. I really didn't have a clue.

RTN: How was it for you once that ended?

ROB: It's tough, because your mind and your whole world were focused on something and then, in my case, I didn't make the Olympic team. There was a certain emotional letdown from that. So here's a goal I had set for myself; I worked for four years or more to achieve it, and I didn't reach it. I think that's been my drive—because I didn't reach that, I channeled it into something else. It takes a while though. You go through that period of time where you really don't know. You need some time to chill and relax and figure out what it is that you really want to do and then be open to other opportunities, because when you're training and you're doing something so focused you're not necessarily open. It took me a couple of years. I did a couple of different jobs coming out of college. I worked for a defense contractor for a while on a B-1 bomber project. It was a classified project that should have been very cool, but it was boring for me. Something inside of me was itching to leave. So I got out and I did insurance for a little while, but then the entertainment thing came around, and that's when I started to come out of the lull and started going up.

RTN: Coming from B-1 bombers and insurance, how did you know that you wanted to move on? What was it, a passion for entertainment?

ROB: I just knew at the time. You know, when you're into something and you're giving it an effort but it just doesn't feel right. You're just not there. I didn't know where I was going to end up, but I knew that it wasn't there. So I went into insurance. It was just a step, some interim time to think. With insurance, there was a little bit more freedom to think on your own. During that time I started to look at entertainment opportunities. Then all of a sudden, the door opened a little wider and I started to go in that direction.

RTN: It sounds so seamless when you say it like that, but obviously you didn't walk right into the Bellagio and become the artistic director for Cirque du Soleil. What happened?

ROB: I started off as a diver, and then I got into doing diving shows—small companies that would do theme parks in different places, Sea World and things like that, which led to contracts over in Europe. It was great. I had a lot of fun and I met my wife. Then, because of my trampoline and diving background, I thought that I could get into the stunts business—

I thought it could be something I could really excel in. We saw an ad in the paper for auditions in Orlando for the Indiana Jones stunt show. So one thing led to another. We ended up in Orlando, Florida. That gig didn't work out, but they were filming and doing episodic television and motion pictures, so my wife and I got into that business. So I left the performing end and focused on the motion picture business. I worked on some big movies and had a great time. While my wife was working on *Jurassic Park*—she was doubling the little girl—Cirque du Soleil was in Santa Monica. They had an audition going on at the pier. She went and they wanted her. And they were looking for a guy with trampoline background, so I sent them a videotape of some of the shows that I had done and they liked it. We were invited to join Cirque du Soleil. I stayed on the performance side for about five years. Then they wanted somebody to come over and coach and put together the acts for this new show, *O,* so I ended up leaving performing to be on the other side of the stage in production. Then I went from coach to artistic coordinator and ultimately artistic director.

RTN: So, going through all of this, from the insurance to the stunt stuff, were you ever scared?

ROB: Every step of the way. Absolutely. A little fear is good because it keeps you alert and attentive. Especially in this kind of businesses: You don't want to be lackadaisical when you are doing hundred-foot high diving or when you're burning yourself on a stunt. So fear is a good thing under control, and a little bit of nervousness or uncertainty always keeps you questioning. My only compass was to try and have a passion for what I was doing. Regardless of whether it's going to be great or not, if what you do is your passion, then it never really seems like work. Follow your dream in that sense, whatever it is. If you guys love roadtripping and asking questions, then that's a passion.

RTN: I think that's what everyone wants to hear: If you follow your dream, it's going to work out. But I know personally and from friends, sometimes there's resistance. Going down to Orlando, was there any resistance? Did you ever say, I've got to do this for myself?

ROB: I was pretty fortunate. I met my soul mate, and we knew that we wanted to do this together and we still are. She's still performing in the shows, and we've always worked together. My parents were somewhat skeptical but supportive. You're gonna get some of that—your parents or peers may give you support, others may not, and that's when you have to say, "This is my dream. This is what I need to do." Yeah, it's scary and you have to take risks, but, you know, no risk, no reward. It's like gambling, if the chips aren't on the table, buddy, you're not going to win the jackpot.

RTN: You've talked about so many different things that you've done. Where does the road go now? What is there left to achieve? What keeps you going?

ROB: I don't want to give the impression that I'm always looking ahead and that I'm not really living in the moment, because this is great, and I'm having an awesome time working on this show. But I'm always going to look ahead and see what's the

next thing, and I think you would be foolish not to think like that. I can't answer the question "Where am I going to be in five years or even two years?" I'm keeping an open mind and an open ear to see what it could be. I think people need to do that unless you have a specific plan and that's your thing for the rest of your life. That's not necessarily the case for me. I'm going to have a little different approach to it and take some risks, and the next opportunity will present itself.

RTN: Is there a point where you may have had to compromise to make ends meet?

ROB: No, because I made the decision with my wife that we would raise a family. So that changes your whole perspective and dynamic. It was a choice we made, so you can't look at it as a compromise; you look at it as "That was my choice and I wouldn't change that for the world." It's an awesome thing to be a dad twice over and have these two guys ripping up the house—it's a lot of fun. It's not a compromise in the negative sense, it's an adjustment—a change of perspective.

RTN: What kind of advice would you give to kids like us, people just trying to find their roads in life?

ROB: Not to sound too cliché, but I guess it's again going back to the passion. Do what you enjoy, what you love. If you're going to do that, then you're going to give it the extra effort that is going to make you successful with it. If you're into something you don't like, you're not going to give it everything you have, so you're not going to get anything out of it. Whatever it is, whatever you do, there are a plethora of opportunities out there. It might not be apparent at the moment, so you need to go out there and find the path. I think what you're doing is an awesome start. It's a great experience and you guys have a passion. It's simple; it's not a revelation. Love what you do and do what you love.

RTN: How do you handle people trying to tell you what success is and what you should do to get there?

ROB: Well, success on whose terms? Are you going on society's terms, your parents' terms, or your terms? I know people who are successful and don't have a lot of money, but they're doing what they love and they're living the life and they're successful. So you have to figure out in your own mind your own definition of success. Is it having a lot of money, is it a lot of friends, is it doing what you love, or is it all those things? Balance it out. That's part of the journey—everyone has to find it out for himself. That's the beauty of life. I'm still figuring it out.

RTN: You're a person who embraces change. What would you say to someone our age to get them over the hump to make a change?

ROB: Change always happens—I think if you accept that and go with it, you'll be open to opportunity. In our show, we have a performance twice a night, five times a week, and the show is never the same. It's difficult, because there's the technical group that wants everything to be the same so they can plan and forecast. Then there's the artistic side that wants to blow that out of the window and see change, because that is how you grow and how you develop new artistic concepts for the

show. To answer your question, you have to embrace change. You have to go back and say, "There's some risk and some fear with it," but you take that leap of faith. And if you've developed yourself up to that point and you have the courage to make the leap, then you're going to have the courage to deal with what is on the other side. That's the fun of it. If you stay sheltered and in the shadow of things, you're not going to really live the life that is out there. If you hold back and you never do anything, you just kind of float.

RTN: Are you ever scared that you jumped too many times?

ROB: They have this saying: "Look before you leap, but he who hesitates is lost." When you're in a position where you may have to jump, something has to give. Something has to cause a change in the path of least resistance. At some point you may wonder and regret if you did something wrong, but you can't once you make that choice. You have to live with it and move on.

RTN: To sum it up, what advice do you have for our generation and people in general who are trying to find their roads in life?

ROB: I would say to be socially responsible—environmentally responsible. I find a lot of people are so wrapped up in their own thing that they don't really care all that much about their neighbors or people around them. If we could be a little more helpful and sensitive and caring with each other, that would go a long way. You have to take care of yourself and achieve your own level of success, but don't do so at the expense of other people. ■

Before we began the trip, we decided that we would carry on a tradition that the original roadtrippers had started. **Every person they met had signed the ceiling of the RV after their interview.** It was a cool ritual and gave us a way to visually track our progress. There was a small room in the back that had been renovated just for our trip. It had been decided that it would serve as our piece of the Roadtrip Nation ceiling. So after we wrapped up, we asked Rob for our first signature, and he readily obliged. It was exciting to watch what was sure to be the first of many messages get tacked onto the ceiling.

When he was done, Rob headed back to rehearsal, and his assistant gave us tickets to that evening's performance. That evening was one of many crazy experiences we had on our roadtrip. We had no business being at that show, unshowered and in our smelly clothes, but we walked right in like VIPs and were treated to one of the most exhilarating performances I've ever seen in my life.

As we headed for bed after the show, we decided to leave the RV in the parking lot of the Bellagio where it had sat all day. Guessing that the security guards would not have been down with the plan, we kept a low profile and had the lights off early. Unfortunately, it was 110 degrees during the day and it didn't cool off much at night.

DAY 3: Las Vegas

We woke up early the next morning, probably because it was three thousand degrees by 8 a.m. I felt like I had slept in fire. We brushed our teeth in the parking lot, applied seventeen layers of deodorant, and headed out for the day. Our first interview was scheduled for 3:30 p.m. with Julian Serrano, a renowned chef at Picasso in the Bellagio. Mike had booked the interview through a friend of his. We were interested in talking to Julian, as our combined culinary skills include burning Pop-Tarts and microwaving veggie dogs. It was the most extravagant restaurant I had ever been in—there were original Picasso paintings on the wall. Julian was clearly taking time out of a very busy schedule to talk to us, and we sat down at a corner table next to a beautiful patio and got started.

JULIAN SERRANO
Head chef of Picasso restaurant at the Bellagio Hotel and Casino

RTN: How did you get into cooking?

JULIAN: I went to culinary school in Spain when I was sixteen years old. At the beginning I was not planning on being a chef, I simply wanted to travel to different countries.

RTN: You mentioned that once you found your path, your family and friends were kind of pulling you back. Would you say it is important to stay on your path?

JULIAN: Absolutely. It's very important. If you do not follow your path, then it's tough to be happy and successful. You have to follow your path—whatever it is. It's also very important to listen to people as well. But whatever they say, you still have to follow your own path. ■

Julian's passion for cooking was a marketable skill that allowed him to make money all over the globe, which fulfilled his passion for travel. He decided cooking was something that transcended the language barrier. He made the decision to focus on becoming a chef and put all his energy into it. He started working on cruise ships out of Miami as a way to get into the United States. Eventually, he got fed up with the poor working conditions and left the cruise ships, moving to Nashville. Julian went on to receive his green card through cooking and continued to hone his skills while following his passion for traveling. Ultimately, his hard work paid off when he became the head chef at Picasso in the Bellagio Hotel.

After a great interview, we left Julian to his genius and headed off to meet up with Pavel Brun, the associate director of Celine Dion's show. Pavel turned out to be one of the most intelligent, well-spoken people we had ever come in contact with. Beyond that, he was totally psyched on our project. At one point the dancers were rehearsing below us and the sound was filtering into our mics and causing a little disruption. Without hesitation he called down and had them tone it down. We had an awesome discussion with Pavel about his story and a whole lot more, including the politics of performance art and buying Beatles records on the black market in Soviet Russia.

PAVEL BRUN
Artistic director of Celine Dion's
A New Day

RTN: We're honored to be speaking with you. Would you take us back to where you were at when you were our age? How did you get to where you are today?

PAVEL: There was not any logic behind my life's journey, in which the Celine Dion show *A New Day* is a very important stop, but not the final destination. I want you to accept the reality that you are talking to a forty-six-year-old person who is still thinking, "What am I going to do when I grow up?" I was raised behind a huge iron curtain in the Soviet Union. My parents were architects, and they really wanted me to follow their footsteps. In the meantime, I hated their guts and I hated what they did—even though they were educating me. I am thankful for them now, because they helped me think visually. It helps me immensely, but at the time I wanted to be rebellious. I wanted to do something else. I wanted to explore something I did not know. They took it very painfully. So I asked my parents to leave me alone, although not that politely, and left. I went in search of ultimate fusion in life. I started doing underground theater, listening to the music that was not legal to listen to, and putting together certain things that are seemingly incompatible.

RTN: You have been able to draw from all of these different experiences that wouldn't naturally be associated with each other. Do you think they are still helpful to you?

PAVEL: Absolutely. I know one thing for sure—well, maybe I know a couple of things—number one is that I don't know enough and the other thing is that everything is diffusion. As long as you acknowledge the existence of diffusion and are driven to find your own fusion, your own alchemy of incompatible elements, you'll be fine.

RTN: You mentioned you left home and went into the underground theater. Was schooling involved, or was it a complete passion that you pursued on your own?

PAVEL: Schooling was secondary—I mean, it was almost forceful. The desire and drive to be a performer, a director, a choreographer came first. The necessity to obtain my master's degree became supportive to this. It worked and it's still working right now, and I'm okay with that.

RTN: Were there any pressures that were working against you—thoughts in your mind or noise around you?

PAVEL: There is a constant state of disappointment, but I would consider disappointment a very positive thing because you cannot achieve anything better until you are disappointed first. When I was making steps farther, broader in my life, it was always dictated by disappointment. Okay, I was working under parameters of modern art, "I can do this, I can do that better," but what is the point? How can I expand the horizons? When I'm disappointed, there must be something else for

me to explore. So from disappointment to fascination, that's a very interesting dynamic that I was trying to follow.

RTN: You said that Celine Dion's show is not the final destination. How do you know when it's time to move on?

PAVEL: When the passion is going down, it means that I am done. When I have a recipe, it means that I am done. When I feel that I can do something well and there is still a lot left inside, then I feel that I am done. That's pretty much it. I always keep myself under a big magnifying glass—I'm trying to question myself all the time. It's a constant dialogue within myself. When I stop recognizing the person inside of me or if I fold under peer pressure, then I'm screwed. ■

After the interview, Pavel came down to sign the RV, and we once again got hooked up with outlandishly good tickets to the performance that night. After the show, we hopped right into the RV—we had an interview in Phoenix the next morning and needed to start traveling through the desert again.

We tried to get as far as we could that night before we passed out; we couldn't be late for our interview at 11 a.m. in Phoenix. Mike was beat, so Randy took the wheel and I rode shotgun for a good stretch. We drove through the middle of nowhere, played music as loud as we wanted, and just talked—that's everything I love about long roadtrips, right there in one night. Pavel's interview was still fresh in our minds, and we talked about the aspects of his story we found compelling.

Toward the end of the interview, Pavel had mentioned that he was thinking of just stepping away from it all and scaling down, working in a small independent theater on a production he could really throw himself into. After hearing the passion with which he spoke about performance and creation throughout the interview, we believed every word. It wasn't about the accolades or the success of the show; he honestly didn't mention that even once. Instead he talked about the creative common ground among cast and crew and how excited he was to work with such talented people. It was awe inspiring to talk to a person doing exactly what he wants and loving every second of it, not giving any thought to elements that could distract him (success, fame, and so on). Randy and I took the RV all the way into northern Arizona, just talking and singing out of tune.

DAY 4: Phoenix

Mike got an early start and we were already coming up on Phoenix by the time Randy and I woke up. Our interview was with Jerry Colangelo, Chairman of the Diamondbacks and CEO of the Suns and all-around rich businessman. I knew who Jerry was from ESPN, so it was my first "whoa, *that* guy" experience on the road. Our interview was at America West Arena, which was cool in itself. Thankfully, we showed up on time and, after a brief wait in the lobby, headed into Jerry's office to do the interview.

JERRY COLANGELO
Chairman and CEO of the Phoenix Suns

RTN: We're on a roadtrip talking with people from all walks of life—would you tell us your story? What was your road?

JERRY: Well, what you have to understand is that you can't just script your life. I can't tell you that at twenty or twenty-one I had it all figured out. I feel very fortunate and blessed that I grew up the way I did, in a kind of poor ethnic community in Chicago, an Italian American community. We didn't have much, but I did the best I could with every opportunity I had. In my case, there was no financial help coming from home because no one had anything. When I accepted the job in Phoenix, I was the youngest general manager in professional sports history at twenty-eight years of age. So when we left Chicago to come here, it was like starting a whole new life. I came with my wife, three kids, nine suitcases, and three hundred bucks in my pocket.

RTN: Any words of wisdom for someone who may be in the same shoes now?

JERRY: If you have a dream, pursue it. Something that I learned through competition and growing up in that neighborhood is that you can't ever be afraid to fail. When someone says to me, "You got a lucky break," I say, "Those who sit around waiting for their ship to come in don't recognize opportunity when it goes right by them." There is never a perfect situation, but you have to be willing to take risk—not blind risk, but calculated risk. I think that often young people in particular and people in general are afraid to make decisions because they are afraid of the downside. They are afraid to fail. If it can happen to me it can happen to anyone; it absolutely can. You can't do anything about yesterday, you certainly can't do anything about tomorrow, but you can control what you are doing today. ■

After the interview, we gave ourselves a quick tour of the arena, which was really cool. We saw all the areas that you usually get yelled at for trying to see. The best part by far was heading onto the basketball court.

Afterward, we decided to get out of town. Phoenix seemed all right, but we were ready to be done with hundred-plus-degree temperatures. We were on our way to the mountains of Colorado (and a thirty-degree temperature drop). Our next interview wasn't until Wednesday in Boulder, and we had a lot of driving ahead of us. In a way, it was our first chance to really breathe and take a moment to sit back and figure out what we were doing.

With the speed of things and all the new experiences, the realization that we

were on the road and meeting all these folks came slower than it normally would. The interviews in Vegas and Phoenix had gone really well and our confidence was building. We realized what we wanted to learn from the interviews, and as a result we looked forward to them. I got excited listening to what came out of my mouth when I called home to talk to my friends and parents.

We drove all the way through Arizona with the windows down. Outside of Flagstaff we saw an absolutely beautiful sunset. We were making good time, so we pulled over to the side of the road to check it out. Our hectic schedule had made it seem like we'd never do the proverbial stop and smell the roses. **Climbing on top of the RV to watch the sun go down, take some photographs, and just relax was awesome. When it got dark, we hopped back on the road and kept on charging.**

DAY 5: Colorado

Once again, I woke to the sounds of gravel speeding by my head. I rolled out of my makeshift bed in the back of the RV to find we were well into Colorado and jamming (okay, plodding along at fifty, but I was still psyched) toward Boulder. We each took a turn at the wheel while the other two alternated between navigating and leaning out the window to take blurry pictures that would be indistinguishable to our friends but mean everything to us.

Southern Colorado was absolutely beautiful and, coming from the cramped confines of Brooklyn and D.C., foreign to us. There were vast expanses of absolute nothingness. It was the first true "roadtrip" drive—we had shot through a bunch of desert the first day, and most of our other drives had been at night. Now we took it all in: the sun-scarred sand and cacti had given way to huge fields and then to tall evergreens and small streams. It was great to roll down the windows and drive on through, slowing down partly because of narrow, windy roads around waterfalls and partly because who on earth would want to rush through this?

Closer to Boulder, we hit more of what is called "civilization" back home. We stopped at a Kinko's and transformed it into an all-purpose Roadtrip Nation pit stop: we charged phones, checked email, bathed in the Comfort Inn pool next door, sent faxes, made follow-up calls for upcoming interviews, printed out directions, skateboarded, and shaved. It was

nice to catch up on these things and talk to friends and family, but it was also weird. We spent a few hours there, and while the hygiene maintenance and air-conditioning were nice, we couldn't stay too long: we were already hooked on the road.

We piled back in the RV and got moving. We actually pulled into Boulder around dinnertime and parked on a side street that looked as if it could serve as the Roadtrip Nation hotel that night (it did). We had all heard good things about the town, so we were anxious to check it out. Randy and I were excited: we figured we had a good chance of finding some decent vegan food, as it was a liberal college town.

We popped our heads into stores up and down Pearl Street. Getting a feel for a town and its life was always a fun part of the trip. Glimpsing into lifestyles, comparing new places to home, and imagining living there was interesting. In spite of our East Coast arrogance, it was cool to see these other spots and realize how great they were in their own respect.

The pace and scheduling of the trip had put us on abnormally normal sleeping schedules. Usually, we were proponents of the typical college kid "sleep all day, up all night" philosophy, but once midnight rolled around on the road, we were done. For as much fun as we were having, we were also putting work into getting to our meetings and touring new cities—it was exhausting.

DAY 6: Boulder

In keeping with the roadtrip lifestyle, we woke up early the next day. We had an interview with David Jacobs, the founder of Spyder Active Sports, which outfits the U.S. ski team in that flashy ski apparel. We set up the camera equipment in a private room and David was right on time.

DAVID JACOBS
CEO and founder of
Spyder Active Sports

RTN: As you can see out the window, we're traveling the country in that huge green thing to talk with people about how they got to where they are today.

DAVE: It's not as if when I was your age I had a plan that just led me to here. Usually things just go along as a progression of your interests and what you like to do. I started out ski racing when I was eighteen or so and headed to college. My father wanted me to be an engineer, so I went to St. Lawrence University in New York. The reason I went there was because they had a great ski coach and a great ski team. So my dad said, "Well, okay, I will send you there, but promise me you'll go to engineering school." So I said, "Great," but I was only interested in the coach.

I went to St. Lawrence and raced and became the best in the country. I was also on the Canadian ski team that won the national championships. I came back and transferred to MIT for a year, but I realized that engineering was not for me. I flunked out and went back to St. Lawrence. In my twenties I was an instructor at a ski school and helped design and build a ski area for a company in Ontario. My personal rule was that every position that I took should be a progression from the one before it. So if I was a ski school director one year, I would not go back to being a ski teacher the next year.

Later, when I was coaching the national team, I met Bob Lange and ended up working with him at his factory. We designed a race boot in 1966, which became the best race boot in the world at that time. I went on to work in the international department at Lange, and while I was in Europe I saw all of these little kids with great-looking ski clothes that were made in France. It was one of those epiphanies; in the middle of the night I woke up and said, "I've got to do this." So I told Bob, "Look, I am going to leave." This was 1972; I just felt that I had to move on. Now Spyder is in forty countries—it's the biggest in the world in the high-end ski apparel market. At the same time, it was never about how much money I made, it was always about the skiing.

RTN: A lot of people today are not told to follow their passion—what would you say to people in that situation?

DAVE: You have to ask yourself—ten to fifteen years from now, will I look back on my life and say, "I wonder if I should have done that differently?" I don't think you want to be in a position to say, "Yeah, I should have done that differently," because you can't get that time back. ■

M.Z., who works with David, offered to show us around, which was really nice. Spyder HQ seemed like an awesome place to work; everyone was having a good time because they were all involved with something they loved. As she took us around, M.Z. mentioned that a lot of the workers would use their lunch breaks to head up to the mountains and get in a few quick runs. The culture of the place was totally dedicated to the core of Spyder—skiing.

After the interview we headed a bit out of town for an amazing hike to some waterfalls. Then we needed to get moving to our next interview. While walking around Boulder the day before, we had met a woman named Kim. Turns out Kim actually owned the store Outdoor Divas, right on Pearl Street, so we decided to set up an interview with her. We were excited because it seemed like this was exactly what the Roadtrip Nation project was about: We met an intriguing person with a worthwhile story while on the road and sat down and talked to her. Kim's story would no doubt be just as helpful and interesting as that of Jerry Colangelo or the CEO of Levi's or anyone else we spent weeks booking. We were excited that Kim made some time for us, and we interviewed her in the back of her store.

KIM WALKER
Owner of Outdoor Divas

RTN: Was skiing always your passion?

KIM: It actually wasn't. I started coming to Colorado in high school and I loved it. I fell in love with the state and then I picked up skiing. I had actually skied a few times in Michigan, but I really took up skiing because I love living here.

RTN: What ideas were in your head when you went to college? What led you to declare an English major?

KIM: When I went to college, it was the first time I hadn't lived with my parents and it was like, "Sweet!" I was excited to live somewhere else for a while. For me, college was more about being out on my own and seeing what was out there. It was an opportunity to discover "What is life about?"

RTN: How did you start this outdoor shop? Did you have anyone's blueprint to follow?

KIM: My boyfriend and I talked about doing our own thing for years. Whether it was owning a coffee shop or doing something else, we really wanted to be our own bosses. My parents were supportive of that, and they had some entrepreneurial friends who told us, "You don't need money. You don't need anything. You just need the idea." They told us, "Be the idea and follow it through."

RTN: Is there any particular piece of advice that you could share with us?

KIM: Do as much as you can. Experience as much as you can. It's not about money; it's not about what you are supposed to do or what somebody tells you to do. Just do what you love and it will be awesome. ■

We were impressed with what Kim had to say. Her story was simple but, sadly, not common. She had a passion and wanted to keep it in her life, so she took a risk and is doing what she loves.

Sure, she worked hard at it—planning for the store, writing proposals, and doing research—but it was worth it because it allowed her to do what she wanted with her life. After the interview, Kim walked to the RV with us and added her signature to what was becoming a formidable collection of names.

DAY 7: Colorado

Our interview with Gary from Neptune Mountaineering was early the next morning—our last day in Boulder. We headed over to Gary's gigantic outdoors shop, sans coffee, and tried to get into interview mode. This was a bit of a struggle: Gary was great, one of the nicest folks we had met on the trip so far, but it took a few minutes to wake up and get the interview going.

GARY NEPTUNE
Founder of Neptune Mountaineering

RTN: We're sitting in your store and it's amazing, but at some point you were twenty-one years old, just like us—so how did you get here?

GARY: I went to college at Rice University down in Houston—I was kind of a restless soul, I guess. I changed majors from biology to geology and eventually came to the realization that although these things interested me, I did not really want to earn a living that way. By that time I had been bitten by the climbing bug. That goes way back, but at that point it was a pretty strong deal. I read a book called *Our Everest Adventure,* by John Hunt. The mountain [Mt. Everest] had just been climbed, and I became fascinated with it. So I dropped out of college and moved to Colorado. I didn't really know what I was going to do, but I knew that I had to somehow work in the mountains. That's all I ever really cared about at that time. My parents couldn't understand that at all. They were like, "Why didn't you just finish the thing and have the degree to fall back on?" I just didn't want to—I had to get out of there.

So I came up here and got a job at an outdoor store that was similar to this one. I worked there for about two years. When the business climate changed, I saw a little niche open up—so I thought about opening a tiny climbing shop. A guy who I climbed with knew something about boot repair, so we took a very small amount of money and bought some boot repair machinery and a little bit of retail goods and opened a store. Gradually, the whole thing grew over thirty years, and here we are. During that time, I continued to climb a lot. Early on I ended up going out to Yosemite and doing several routes on El Capitan. I've always been interested in big mountains—I climbed McKinley twice and then started going over to the Himalayas. Ultimately, I got a permit for Everest in 1990 and climbed it.

RTN: Based on your life and your experiences, do you have any words of wisdom for our generation?

GARY: Don't get locked into something you can't get out of. I know people who have gone to med school and ended up

in unbelievable debt. Sure, they live in a bigger house than I do and drive a fancy car—but they are so far in debt, they can't get out of their job and some of them hate it. They're being worked to death and almost exploited and they're barely making ends meet. They can't go on a trip with us, and with all that stuff they seem to have, they're actually almost broke. ■

Gary came out to sign the RV, and then it was time to head toward Aspen. We hopped onto I-70 and headed west. Our next interview was later in the afternoon with Pat O'Donnell, the CEO of Aspen Skiing Company.

The stretch we drove through next had to be some of the most beautiful scenery in the world. Honestly, it was straight out of postcards and Ansel Adams pictures. Huge rolling hills covered with evergreens gave way to ice-capped giants in the background. For a few miles, the road ran parallel to a rushing stream, with several waterfalls right along the road. It was breathtaking. We may have set records for the amount of film used in such a short time.

After we got off the freeway, we drove through Glenwood Springs, Colorado. It was easily the happiest place we'd ever been in our lives. The town was full of happy folks and pretty scenery, and I think the sun was actually smiling. By the time we got to Aspen, we were feeling pretty good. We had heard great things about Pat, so we went into the interview with high hopes. He did not disappoint.

In the sixteen days that I walked out, primarily by myself, I was really reflecting back on how short life was and why I didn't go—
why I was spared.

PAT O'DONNELL
CEO of Aspen Skiing Company

ROADTRIP NATION: You say that you have a strange story. What has your path been like? Did you always have it figured out? Were you ever confused?

PAT: I think it's normal to be confused about where to go. I can relate because I was beside myself at twenty-seven years old. I didn't know what I wanted to do. Now, reflecting back, I see that it's normal. Things will smooth out. Right now it's all fuzzy, but it will come into focus.

I went to school for engineering and hated every living minute of it. I chose the major for one reason: that's what my father did, and he had a company and said, "If you don't do that, then you're not going to get financial support. And secondly, if you want to enter the family business, the only way you're going to survive is to be an engineer." So I finished it and was working in San Francisco and really was a member of TGIF. I couldn't wait for Friday to hit so I could go and live my real life and do what I wanted to do. I was really floundering around, probably not happy in my heart. At the time, my real interest and passion was rock climbing in Yosemite Valley. So I would drive all night on Friday, climb all weekend, drive back on Monday, and begrudgingly go back to work. Finally, after doing engineering for a couple of years, I thought to myself that life's too short, I gotta make a move. I was actually climbing on El Capitan, and afterward I walked into the local Yosemite corporate offices, not to really seek a job, but just kind of find out how you could live in the beautiful Yosemite National Park.

I found out that you had to work for the park and become a ranger, or you had to work for the concessionaire—they have all the hotels, horses, and ski areas. So three weeks later I went back up and said, "I want to come up here and work for the coming summer and then go back to the real world after that." They said, "We'd really love to put you into management training," and I said, "I don't think that's for me. I'll be straight up with you: I'm really here to climb." So I took a job as a bellman. The pay was ninety cents an hour, but it included a tent and free meals. I worked from 7 a.m. until 3 p.m., but it stays light until 9 p.m., so you could climb six hours a day. That was the beginning of how I got into the hospitality business. I worked all summer, and then at the end of the summer they said, "We'd like you to stay on in management."

So I stayed for the winter and became maître d' for the world-famous Ahwahnee Hotel dining room. I couldn't spell maître d' and then became it. In the middle of winter, the ski area manager became deathly ill, and my boss said, "Do you want to see skis in your tent?" I said, "Yeah." He said, "Great, you're the new ski area manager in Yosemite." So in a very short period of time I did a multitude of jobs and basically followed my heart and my value system. But I knew that you could wake up in this place one day like Rip van Winkle, thirty years later, and never really progress. So I left and became the opening vice president and general manager at Kirkwood ski resort in Tahoe. I was very green behind the ears, and there was nothing there—it was basically wilderness. The guy that saved my rear was the mountain manager, who was very experienced from Squaw Valley. We spent the next four months putting up a day lodge, putting up four chairlifts, and cutting the runs. We opened on time.

The next thing I know, I landed in Colorado in the early seventies as the chief operating officer in charge of all ski operations at Keystone. It was a big place and I spent ten years there. At the same time, they were kind enough to let me pursue my passion, which had graduated from rock climbing to Himalayan climbing. They let me go on numerous Himalayan expeditions until 1980, when a bunch of us went to climb a peak called Annapurna, the eighth-highest mountain in the world, just north of Everest. We were on that mountain for two and a half months, and on the last day the entire north face just let loose. Three of my best friends were swept away and killed. In the sixteen days that I walked out, primarily by myself, I was really reflecting back on how short life was and why I didn't go—why I was spared.

When I came back to Keystone to work, I couldn't get away from it. I was interacting with their families and their kids. I was going to work every day and just going through the motions, so I said, "I'm going to resign." I wanted to get out of there, but I really had no place to go. They said, "Where are you going to go?" I said, "I'm going to buy a sailboat and sail away." People said, "You know nothing about sailing. What do you mean, you're going to do that?" I said it was a fantasy of mine that I always wanted to fulfill. So I cashed out everything I had in my little house and went to San Diego with a friend who knew all about sailing, bought a boat, paid cash, and moved on board. Then I finally sailed out of the bay, hung a left, and didn't come back for a year and a quarter.

I learned a lot about people and cultures, and a lot about being frightened. I remember sailing by San Diego and feeling like I was really copping out. I went by the freeway and all these people were going to work, making a positive contribution to society and doing something meaningful with their lives, and here I was living on a boat. I felt guilty about that for the first month.

When I came back about a year later, somebody said, "When are you going to go back to work?" I said, "You've got to be kidding—work!" I'd been diving for lobsters, cleaning boats, climbing masts for twenty-five dollars, and surviving with no money whatsoever. I had found another way to live and really be happy, but I got an offer I couldn't refuse. They wanted me to run an institute based out of San Francisco. It was a nonprofit organization that taught adolescents about environmental awareness. Robert

Redford, Cap Weinberger, and a whole bunch of other prominent active people made up the board of directors. One day I was working there under the Golden Gate Bridge in my little office when I got a phone call from Yvon Chouinard, founder of Patagonia. I had known him from my climbing history, but not all that well. He said, "I'm looking for someone to run Patagonia." I said, "You must have the wrong number. You know I'm a ski guy, and now I'm running this outdoor environmental education organization." He said, "No, as you know, we give a large part of Patagonia's profits away to support the environment, and I know your outdoor background, and you're doing the environmental thing, which is exactly my thing, but I need a CEO who believes it. I can teach you how to make fuzzy jackets."

So next thing I know, I'm the CEO of Patagonia. I did that for about six years, and then I just got worn out from worldwide travel. I found myself consistently on the road, but most importantly, I wasn't living my own value system anymore. I wasn't getting to the outdoors and doing what I wanted to do. So I went to Yvon and resigned, and he was great—he understood it. I was paid a lot of money, but what good is the money if you're not living the dream?

To make a long story short, I got hired as the CEO of Whistler. We lived in British Columbia for three years. Things were going well, and then this Aspen job came open.

In the summertime, my legs are turning on my road bike or mountain bike at 5:30 a.m. seven days a week. I can't wait for the weekend to have fun; I have to be doing something every day. I have to be snowboarding. I have to be riding my bike. If I can't do that, I'll just be a school bus driver. Money doesn't matter to me. Whatever it takes, the real happiness is in the heart. I found that out way back in the Yosemite days, and it's steered every decision I've made. Titles mean nothing to me, still don't to this day. The main thing is really being happy.

Young people ask me all the time, "How did you get here? How did you do this?" I think it's hard to establish what those values are in your twenties. There's a lot of pressure from everybody—peer pressure, societal pressure. I think that if you stay true to your values, you are going to be a better person.

Whatever my values are, life is too short, because I'm going to be sixty-five next month. I look at life now and say life is just precious.

RTN: So it seems like you have a good handle on what makes you happy and what you're passionate about. You mentioned the TGIF thing back in San Francisco. When did the lightbulb go off that showed you that you could make a change?

PAT: At the time I was petrified. I had my security blanket working for my father in his little company and I knew my check was coming. It wasn't much but it was there. When I made the move to Yosemite, I was absolutely petrified. Every time I made a change I was petrified. When I walked into the CEO position at Patagonia, I knew nothing about making clothing. I knew nothing about that stuff. But you survive these changes, and they're never as scary as they seem.

I used to have a quote posted up on the wall of my tent in Yosemite. It was by Thoreau—"Most men lead lives of quiet desperation." Then it's gone. You wake up one day, you're sick and old, and you go what if, what if, what if. I never wanted to be on my deathbed thinking, "What if?"

RTN: You've been talking a lot about Yosemite. I think that's amazing how you just picked up and moved there. What did those years in Yosemite mean to you?

PAT: Yosemite is just a metaphor for whatever you want to do. For me, I know that I loved the outdoors. I was kind of a loner out there, and that worked for me, but you can't expect that to work for everyone. For someone else it could be New York City or it could be driving a bus. You know, you've seen a million people like this, where they're driving the bus or waiting tables—and they're the happiest people you know. They're the happiest people I've ever seen in my life. That's what it's all about.

RTN: Any last words of wisdom? Any thoughts for how people like us can find our roads in life?

PAT: It's not always going to be upbeat, and it's okay to be frightened. You are going to be scared. You can't see the light at the end of the tunnel, and you want to know what the outcome is going to be. You want that security blanket. I say, go without it every now and then; it will work out. The real security blanket is your values and your visions, and that is what you come back to all the time. You keep hanging on to that. If it's not working for you—the company is bad or the people are bad—get out of there. Move. Every day is a wasted day after that. Life's just too short. ◼

There were some long silences during Pat's interview, not because we didn't have a question to ask, but because we were totally in awe. Some of the things he had experienced and learned from were so far removed from our lives that it was hard to process what he was telling us. The most tangible thing we got from Pat came through not in his words (which were great), but in how excited he was to speak them. You could tell he squeezed every minute out of every day and couldn't imagine living his life any differently.

After the interview, Pat showed us around the grounds and signed the RV. After talking with Pat and seeing Aspen, it was interesting to think about all the different lifestyles that existed. Pat lived in a manner far distant from my daily routine back at NYU. I'm not sure that Mike, Randy, or I could necessarily live in the mountains, but it was good for us to get out of our comfort zones and think about what it would be like to live in a town where it only took five minutes to go skiing, mountain biking, or hiking.

Our next destination was Grand Junction, Colorado, to bunk with a friend of Mike's for the night before getting on the road to Seattle. It felt like a pause in the trip. The deserts of Vegas and Phoenix had given way to the mountains of Colorado in only five days. In some ways we'd accomplished a lot, and in others we'd barely started. Either way, there was no doubt that something was happening. We'd shed our apprehensions as soon as we got out of the desert, and we'd begun to adopt the Roadtrip Nation philosophy as our own.

A lot was going on in our heads as we tried to synthesize the impact of these interviews and travels while still flying through them. We'd heard many stories about the paths people traveled and how they figured out their own roads and what made them happy. Simultaneously, we were traveling on our own path, and it really made us think about the extended metaphor in concrete terms. The roadtrip put us in a cocoon, away from school, jobs, and money problems, and we had to figure out how to absorb what all these folks were telling us. Could we take what was in this temporary world back to our daily lives? It was a daunting but exciting task, and we went to bed ready to start again.

DAY 8: Oregon

Mike and Randy had covered a lot of road the night before, which was great because we were able take it easy rather than drive straight through the next day. In the morning we got off the freeway and parked the RV down by a river in a state park in northeastern Oregon. It was time well spent and felt like our last deep breath before heading toward Seattle and then down the coast. A lot of interviews were lined up over the next two weeks, and the schedule was going to get much more intense, especially as we got into San Francisco and L.A.

Overall it was nice to enjoy a quieter day with no rushing or cities to get lost in. We'd finished up one week of our trip, and this was a chance to take stock and enjoy being on the other side of the country. We felt, I don't want to say "proud" and sound boastful because it wasn't that; maybe it was more "lucky"? I remember thinking and talking about what we would've been doing back home: working all day, sleeping in on the week-ends, seeing the same faces we saw every day. Of course, there's absolutely nothing wrong with those places or faces—we love them. But there was something about being in a new place, trying new things, messing up, and figuring things out. It was invigorating, and it felt right.

DAY 9: Wellpinit

As usual on traveling days, I woke up after we'd already covered considerable ground and moved into Washington. Before heading into Seattle, we had an interview scheduled with Bob Brisbois, the director of the Spokane Tribe, in Wellpinit. It was farther out of the way than we bargained for and the problem was exacerbated by our spotty directions (*spotty* and *non-existent* are interchangeable terms, right?). It wasn't easy to find the place and the RV didn't take too kindly to the narrow, unpaved roads we were steering it down. Considering the driving conditions, we were in pretty high spirits. This was an interview we had been looking forward to. Bob was going to tell us about experiences that were entirely different from ours, and we were anxious to hear them. By some stroke of luck, we managed to end up in the right place, and Bob hopped in the RV to get the interview started.

BOB BRISBOIS
Council member of the
Spokane Tribe of Indians

RTN: This is our first time on a reservation—would you tell us a bit about it?

BOB: When you crossed the bridge at Little Falls, you came on the reservation. It was established on January 18, 1881, by the executive order of President Hayes.

RTN: What about you, what's your road?

BOB: I made some bad choices when I was younger. Maybe not bad choices, but they weren't the right ones for me. I had to redirect my whole life, but I had a strong foundation of self-discipline—from there, I could take charge of things.

RTN: In your study of mental health, you mentioned the concept of self-actualization. That doesn't sound too far off from what we're doing. We're finding the open road, finding our niche—like you said, what makes us happy and what makes us who we are. This is a really general question, but what's the key to that?

BOB: I suppose for some people, they have it. I didn't have it all together, and it took me a while to get it together. When I found what I wanted to do, I got up to go to work and I felt good. When I went home I felt good about what I'd done during the day and I didn't have to go home and dump it on my wife and kids. On the inside I just found balance between here and here [points to his head and his heart]. You've always got to keep evaluating yourself to make sure you feel good between your head and your heart, and that your spirit feels good and your psyche feels good about what you're doing. Hopefully you guys will have it—it will come to you. But if it doesn't, don't be afraid to keep trying different things. ■

The interview was mind opening and we were glad we had made the out-of-the-way stop to speak with Bob. He was candid about some of the hardships of growing up on a reservation, which doubled our admiration of his dedication and drive. Psyched from the interview and with legitimate directions in hand, we made our way back onto the major roads for Seattle.

It was about three hundred miles from Wellpinit to Seattle, but the time flew by. We alternated turns at the wheel and took advantage of the time to make calls, read, and talk about the upcoming West Coast swing. I was looking forward to stopping in Portland, Mike was excited about interviewing Mason Gordon in L.A., and Randy talked about San Francisco and the interview at *Maximumrocknroll.* As we moved toward Seattle, we were too excited, too talkative, to even notice the land passing us by. Before we knew it, we were rolling the RV over the sloping streets of Seattle. We found something to eat, then pulled into a hotel parking lot on the fringe of the city (this by now had become routine) and slept.

DAY 10: Seattle

In the morning, we woke up feeling eager for clichés, so we headed into Seattle to get a cup of coffee before our interview with Jonathan Poneman, cofounder of Sub Pop Records (Nirvana's original record label). There was a coffee shop right around the corner from the Sub Pop office, so we loaded up on caffeine and talked about the upcoming interview. As 11 a.m. rolled around, we finished our coffee and walked next door to sit down with Jonathan.

JONATHAN PONEMAN

Cofounder of Sub Pop Records

ROADTRIP NATION: We're taking a roadtrip across the country to interview different people to get their story. We're all twenty-one right now, at that "what do we do now" point in our lives. Where were you at our age?

JONATHAN: It's kind of a weird story. I'm now forty-three years old. I graduated from high school in Phoenix, Arizona. I went there after getting kicked out of a boarding school in Detroit, Michigan. I had to move in with my folks in Arizona. At the time they were trying to retire and live the good life, and then they had their messed-up son show up and bring a little bit of hell back into their lives. That was for only six months—then I moved back to Michigan, where my girlfriend was living, and one day she came back from college and said, "What do you think about the idea of moving to Washington State?"

I had never given a thought to living in Washington State, so I said, "What the hell." It was the day that Elvis Presley died. We loaded up her car and drove to Bellingham, Washington. We spent a couple of years there—pumped a lot of gas—then my girlfriend, who was a little older than me, turned twenty-one and said, "You know what? I want to get out of here and go to bars." So I tried to move to Vancouver with her, but you know, if you are an illegal immigrant in Vancouver, it's easier if you're a woman. There was really no way I could live in Vancouver, so I decided to go down to Seattle and go to college.

I went to the University of Washington and Central Community College and played in a bunch of crummy rock bands—you know, they seemed like new wave but were really a bunch of nonsensical, awful rock bands. I later got involved in the local radio station, KCMU, which is now KEXP here in Seattle. I became a disc jockey and inherited the local music program. I started becoming familiar with a lot of the local artists and started booking shows at this place in the university district called the Rainbow Tavern, which gave me a further awareness of what was going on in the Seattle music scene.

Then somewhere along the line I learned—well I didn't really learn it—I came to accept that I would never become a pop star [big grin], and I thought, "If I can't be a pop star, then I might as well be their pimp." So me and this guy Bruce Pavitt got involved in doing this on a full-time basis on April Fools' Day, 1988.

RTN: Well, not to sound rude, but when you first moved to Seattle you didn't have much going for you. What was keeping you here?

JONATHAN: Going back home was basically not an option. My parents, rest in peace, they are both gone now, but they had their issues. I loved them dearly, but they were older parents and they were very interested in their own lives. My mother did say from time to time, "Why don't you come back and live in Scottsdale? You can go to ASU and it would be great." But the whole idea of going home and pursuing that "by the numbers existence" just wasn't for me.

Around that time I was nineteen. I quit smoking and cleaned up my life a bit, so I was really happy to come to Seattle and just go through a series of menial jobs and hang out. I didn't have the fire in the belly, like you gentleman have, to drive up and down the coast and do these interviews. I was just trying to get real with living and being by myself, and then I took steps in trying to go to college because, you know, college is a great thing, but I don't know if you can relate or not when you are younger. You don't really know what you want to do. It's like a smorgasbord, and the theory is that if you try enough options, you're going to find something that tastes good. Well, the only thing that ever tasted good to me was pop stardom, and that wasn't going to happen. So I got into radio in 1982. From then on it was this slowly building momentum as I got more and more involved in the business side.

RTN: I'm about to graduate, and that idea that you sample this and that and then eventually something is going to click—I'm not sure if it's going to happen to me.

JONATHAN: Well, it doesn't happen that way, and I don't think that it should happen that way. I think that at the end of the day you are going to live and die, succeed or fail, by your wits. And I think the important thing is to have the courage of your convictions, and if you fail it's just a chapter in your story; it's not the story itself. It took me a long time to face that, because even though I paint a happy-go-lucky picture, there were plenty of times when I was working at a yarn store or as a janitor at the Westin Hotel. You know, I would call my parents and say, "What am I doing with my life?" Well, I wouldn't exactly put it that way because I had a little more pride, but in the way that parents do they were able to extract that from me.

RTN: I am totally interested in the formation of Sub Pop. How did your work in radio turn into a record label?

JONATHAN: Sub Pop started as "Subterranean Pop," a column that became known in *Option Magazine.* From that, Bruce [Sub Pop cofounder] started a fanzine to focus on music being made out of non-major media centers. You would always hear about music coming out of New York and L.A.—but the truth was that the vital music in this country was being made in Athens or Minneapolis or Lawrence or Seattle, for that matter.

So Bruce liked to review and pay attention to bands like the Oil Tasters from Milwaukee, who were actually doing a Morphine-type thing twenty years before it happened. Anyway, one of the bands that I put on at the Rainbow Tavern was Soundgarden. I actually had a pivotal experience at this point: I knew nothing about Soundgarden. I had booked them based on a recommendation, and I dropped in at the beginning of their set to settle some money. I saw Soundgarden for the first time that night, and they totally blew my mind.

Frankly, at that time in my life I was like, "I have to get serious." We started formulating the record label in 1987, and I was having conversations with Soundgarden probably a year before the record came out, so I was like twenty-six years old. So here I was floundering, my non-career in ruins, working at Kinko's, and I just said, "You know, I want to do one successful thing in music, one f—ing successful thing before I go back and be an accountant." I, who can't do math [laughs]. I told Soundgarden that I wanted to put out their record. I knew that they were going to make a great record, so I cashed in my savings bonds, around $15,000, and we borrowed another $5,000 or $6,000. And we were in business. That was April 1, 1988. We rented a space right down here in downtown Seattle, and I think we blew that money in three weeks.

RTN: At the point where you were floundering, throwing $15,000 into the Seattle music scene sounds good, but did it seem that way at the time?

JONATHAN: That investment to us seemed like a colossal amount of money, but you have to go with your gut, and for me there is nothing better in life than to go, "I f—ing *know* it; I know this is going to hit."

The idea of making it huge was so different then. We were thinking that

Nirvana was going to sell one hundred thousand or maybe two hundred thousand records—and coming from the underground that was colossal. To think that they would sell ten million records—that was completely mind-blowing.

RTN: You've taken a lot of risks in your life, like moving across the country when you were seventeen and throwing all this money that had been a safety blanket into this one venture that you believed in. What led you to that? Why not take the easy way?

JONATHAN: I didn't see those as being risks, I saw those as being necessary. If everything in your life is characterized as "risk versus safety," the human instinct is to choose safety. But what if you use a whole different standard for reevaluating your life, such as "necessary versus unnecessary." Things like happiness, passion, and love are all necessary. So for me it was just changing the way one evaluates one's life and taking the whole idea of security out of the equation. ∎

We were mega psyched. Jonathan was super relaxed and friendly, and I think Randy in particular got a lot out of it. At one point he asked Jonathon what advice he would give to someone looking to start up a label; his suggestion, in short, was, "Do it." All in all, we'd had a very cool interview focused on his story about a passion for music that was eventually fulfilled through an alternate route. When we were done, the Sub Pop guys hooked us up with about fifteen of their newest and upcoming releases, as if they needed to give us anything else.

Our next interview, at the Bill & Melinda Gates Foundation, was on the outskirts of the city. On the way there, we ended up behind a school bus full of little kids who were unbelievably fascinated by our big green ride. If only the little guys had been inside, we said, surrounded by Mike's laundry bag and the rotting mysteries inside the fridge, they might not have been so into it. Complain as we might, I can't imagine that we would've traded our decrepit RV for anything else. Our interview that afternoon was with Sylvia Mathews, chief operating officer at the foundation.

SYLVIA MATHEWS
Chief operating officer and executive director of the Bill & Melinda Gates Foundation

RTN: Thanks so much for your time. We're coming to the end of our college careers and are searching for what is next—where were you at this point?

SYLVIA: I'm entering my senior year of college and basically coming to two paths: I'm going to apply for a job as well as scholarships, because I'm excited to move on and get a job, but if that doesn't work I'll go back to school. So I start the process—recruiters, meetings, and so on, but fortunately I get a scholarship to keep going in school. I go on to study for

three years in England, and at this point I realize, "This has to come to an end." So after seven years of university, I realize it's time to get a job. So I start applying for jobs in different areas, and I end up getting a job at McKinsey & Company and go to New York City for two years. Then I get a call from Gene Sperling, who is running the Bill Clinton campaign for the presidency, and he says, "Do you want to come to Little Rock or not? I have one slot and they have agreed to hire you, so you have forty-eight hours to decide." So I go to Little Rock and work in the war room. Clinton is elected, and I work for his administration for eight years. After that, I decide I'm going to climb Mt. Kilimanjaro, be an institute of policy fellow, and work. During that time I'm going to decide what I'm going to do for the next phase of my life. Then, as often happens with life, you can't define your path. I receive a call from the Bill and Melinda Gates Foundation, and we start conversations. I said, "No, no, no; I am going to the private sector." But this opportunity was so unique that here I am.

RTN: That's an amazing road. What thoughts can you offer to people out there who are trying to figure out what they want to do with their lives?

SYLVIA: I think you should have a set of guiding principles that steer your decision making. Also, make sure that you are actively seeking those things that are making you happy. Everyone is different, and you shouldn't be embarrassed about it. Whatever is important to you, act on it. For me, it's been more about principles than a place or a path. I never would have predicted that I would be sitting here. ∎

Sylvia's story was amazing—and to think that she works with one of the largest philanthropic budgets in the world is crazy. Think of all the people and issues she's helped over the years. It definitely inspired us to find a path that enriches other peoples lives, not just our own.

DAY 11: Seattle

We woke up early the next morning to the notorious Seattle gloom and rain we had avoided up to that point. At 8 a.m., we headed to the first interview of the day: Dan Regis, a venture capitalist at Digital Partners Venture Capital. Mike had booked the interview through a contact back at Georgetown, and I think he was looking forward to the opportunity to sit down with someone as accomplished as Dan. Randy and I didn't have the faintest idea what venture capitalism was, so it was good for us to be there and expand our horizons. We arrived at Dan's office but, in an ironic twist, we ended up a few minutes late because I momentarily lost my wallet. So I can't even be trusted to keep twenty dollars in my pocket, yet every day Dan is trusted with sums of other people's money that would boggle my mind. Clearly, I had a lot to learn.

DAN REGIS
Partner at Digital Partners Venture Capital

RTN: We're roadtripping around to talk with all sorts of interesting folks to learn how they wound up where they are today, what roads they followed, and how they stumbled along the way.

DAN: I had no clue about what I'd be doing. What I thought that I was going to do is not even close to what I am actually doing. When I first started college in Seattle, I was an engineering major, and I spent the first part of my college life figuring out that I didn't want to be an engineer. When I graduated from high school, I waited a year to go to college because I did not have the money and I also didn't have the strongest desire.

Anyhow, I worked at Boeing and got a great first job during college working at the plant. My second year of college, I had upgraded my skills and Boeing gave me an offer as an engineering aide. But I knew it wasn't me. When I returned to college after the summer, I switched majors to accounting. After college I joined Price Waterhouse in the mid-sixties, which at the time was the largest and most prestigious of the accounting firms. I learned so much—I really just loved my job, absolutely loved it. I was with the firm for thirty years until I left to join a small venture capital company. I was in front of the bubble and invested at a time when we thought we were picking good companies, but the market rewarded us way beyond our wildest imagination. At one time I put two million dollars from our little fund into a company, and it shot to being worth 440 million in street value. It was quite a ride.

RTN: You obviously enjoy what you're doing. If it's based on your enjoyment, what about it appeals to you? What is it about crunching numbers?

DAN: Because it's not crunching numbers. You see, that is what I thought as I started into business as well. But what I love is that I'm always involved in all aspects of business. I was hiring people, and there is nothing more rewarding, incidentally, than bringing in somebody along behind you. The concept of how to make businesses succeed is an interesting art, and I use the word *art* purposely. ■

After some discussion about Dan's job and some half-brained proposals for venture capitalism schemes of our own, we were off to our second interview of the day with Ann Powers, a rock journalist. If talking to Dan had been an experience that correlated with Mike's interests,

speaking to Ann would certainly have the same implications for me.

Ann used to live in New York City and worked at the *Village Voice*. She still wrote for a bunch of music magazines—in fact, as we walked into her office, she was finishing up a piece for *Blender*. Ann moved to Seattle to become a curator for the Experience Music Project, an insanely cool rock 'n' roll museum that opened in the summer of 2000. We met in her office above the museum and sat down to talk.

ANN POWERS
Rock journalist and senior curator of Experience Music Project

---

RTN: Why writing?

ANN: I sort of believe in the whole Freudian cycle thing—what made you feel like a whole person when you were five years old is what you'll want to do. So for me, I was a nerdy kid, had no friends, and read a lot. That was the world that accepted me—the world of magazines.

RTN: You said you were an English major in college and that you always wanted to write. Was there ever any pressure from family or society to do something else?

ANN: I never really listened to my parents [laughs]. I guess I shouldn't say that, but my dad would always say, "You should be an accountant; oh, you should get your teaching certificate," which isn't a bad idea. The thing you have to ask yourself is, "Are you willing to be poor?" Are you willing to accept this idea that they call "dedicated poverty," which is that maybe you weren't born poor, but you're willing to eat nothing but ramen noodles for about five months solid? Are you willing to get all your clothes from bags on the street? That's how I did it when I was making no money.

RTN: You spoke about how you had anxiety when you were younger, but now you're confident. How do you overcome that?

ANN: I think one thing is being able to take failure. That's really important. Once you live through a couple of very deep, painful rejections—that helps. It's good to fail, it's good to be rejected, and it's good to have that morning-after feeling. It's like after breaking up with someone; you say, "I will never love again!" But you do.

RTN: Right now, when I go home for a holiday I'm inevitably asked, "What happens when you graduate?" To which I respond, "Something—probably, hopefully—something will fall into place." What you're saying sounds great right now because we can sit back in your office and talk about it, but it's scary, right?

ANN: A lot of kids used to come up to me when I was working at the *New York Times*, and they would say, "How do I become a journalist? How do I get into *Spin* magazine?" I think especially when you're young, it's got to be about the ideas; it can't be

for the career stuff. Your thirties are for your career. Try not to think about it until then. Don't have a plan now. Throw the plan out. Be determined, but don't plan. So be focused to the extent that you're focused on your own love, believe in yourself, but don't have the focus be external— that focus has to be from within. ■

The interview went even better than I could've hoped; it remains one of the highlights of the trip for me. As Ann was walking down to sign the RV and we were talking about what I was going to do when I graduated, it was one of those moments when I just stepped back in total "What the f—k am I doing? This is the coolest thing ever" mode. From the minute we walked in, it was more like a conversation—about writing, passion, and perseverance—than a formal interview. This type of interview always yields better results. Ann was uber-friendly, super knowledgeable, and what's more, her story was really encouraging. As a young writer looking to do a lot of the same things she had done, I found it an unbelievably fortunate experience.

Seattle was a fresh experience for us, with tons of friendly folks and a handful of great interviews. We were more or less at the halfway point on our roadtrip; the first week and a half had flown by and we knew things wouldn't slow down anytime soon. Our trip was going to finish with a flurry: San Francisco and L.A., with lots of interviews packed into each. Our exhaustion was overridden by excitement and restlessness as we left Seattle and started driving toward California.

DAY 12: San Francisco

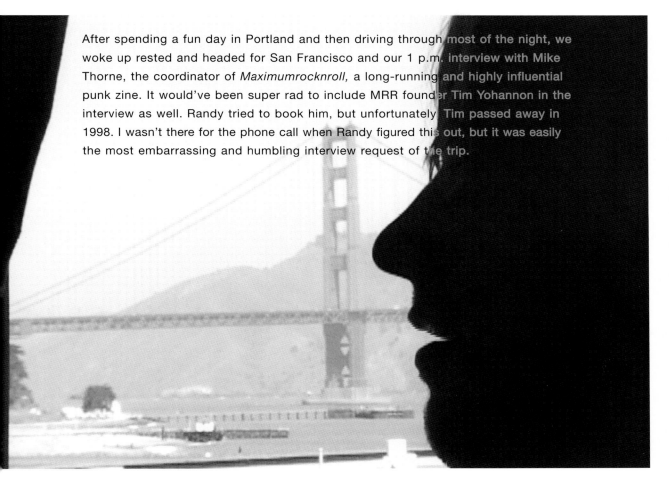

After spending a fun day in Portland and then driving through most of the night, we woke up rested and headed for San Francisco and our 1 p.m. interview with Mike Thorne, the coordinator of *Maximumrocknroll,* a long-running and highly influential punk zine. It would've been super rad to include MRR founder Tim Yohannon in the interview as well. Randy tried to book him, but unfortunately Tim passed away in 1998. I wasn't there for the phone call when Randy figured this out, but it was easily the most embarrassing and humbling interview request of the trip.

Randy and I were really stoked—we'd been talking about this interview all trip. We lived by this zine when we were younger, so we couldn't wait to check out MRR headquarters, meet Mike, and hear what he had to say. We pulled up to Mike's apartment, which doubles as *Maximum*'s home base, and were promptly dumbfounded by a library of records that seemed to extend for miles. It was punk rock heaven, except we were alive and in a pretty nice apartment in the middle of S.F. I was in a haze somewhere between admiration and trying to figure out the nearest escape route and exactly how many LPs would fit under my shirt. At some point Mike pulled Randy and me away from the vinyl, and we sat down around the kitchen table and got started.

MIKE THORNE
Coordinator of *Maximumrocknroll* magazine

RTN: This place is sweet! Can you tell us a little bit about the magazine?

MIKE: *Maximum* covers completely independent punk rock bands and bands that support and are part of the underground network of friends from across the world. We don't support or cover anything that is on a major label. We're primarily concerned with covering independent punk bands that aren't going to get any coverage anywhere else.

RTN: Right on. A lot of that philosophy is similar to what we're doing with Roadtrip Nation; it's just more geared to how people have lived their lives. Can you tell us a little bit about your own road?

MIKE: I never gave a shit about making a career. What I care most about is creating a life for myself where I'm sane. That's what's important to me. I don't make a lot of money, but I don't care. I've never understood the whole American psyche of when you're eighteen years old you have to decide what you are going to do for the rest of your life.

RTN: *Maximum* is a big magazine, but when you look at the opportunities you might have had with *Spin* and *Rolling Stone,* you could have done something more mainstream with your music background. Did you ever feel pressure from parents or society to go in other directions?

MIKE: My parents didn't and still don't get it. My dad is waiting for my punk phase to end. My parents still want me to get a real job—although I feel like I have a real job because I work about sixty hours a week. They would definitely like it if I were doing something different. You know, I'm doing this and I like doing this. It's my life and the only person I have to please is myself. I'm in contact with people I went to college with who have these jobs where they work forty hours a week and it goes toward paying off their mortgage, buying that car, or saving up for that one weekend in Bermuda, and here I am touring Europe for two months with a band. Tell me, who is living a better life? ■

After we finished the interview, we went back to the RV to regroup. We made a few more calls, got some directions to our remaining S.F. interviews, and then set out to explore the city.

S.F. was a very cool place. Maybe it was just because we didn't have the faintest clue where we were going, but it seemed really expansive. While riding in our cab, I realized that all three of us were plastered to the back seat in fear. The cabbie wasn't necessarily speeding, but after being in the gigantic, lumbering RV for a couple of weeks, the combination of speed and closeness to the ground made the taxi feel like a deathtrap.

DAY 13: San Francisco

We woke up early for our interview with Alastair Paulin, the managing editor of *Mother Jones* magazine. In my brief experience with magazine publishing, it seemed like a tough industry to break into. The opportunity to sit down and learn how Alastair did exactly that was exciting.

ALASTAIR PAULIN
Managing editor of *Mother Jones* magazine

ROADTRIP NATION: Would you put yourself back in our shoes—where were you when you were our age and how did you get to where you are today?

ALASTAIR: I went to university in my own country, New Zealand, and when I had one year left I took a year off to come work in the States. I got an exchange visa and basically spent six months being a ski bum in Utah.

I was looking for a way to extend my visa, so I hooked up with an organization that puts foreigners in touch with summer camps to work as counselors. I did that and met this girl there. I thought that it would be a summer fling and nothing would come of it—now thirteen years later, I'll be married to her for ten years. So I ended up going back to New Zealand to finish up my final year of college, and then I was faced with a choice. I really wanted to come back to San Francisco to be with my girlfriend, but at the same time I was working in the drama department, and the chairman of the department came to me and said, "We are creating this new position; it's going to be like a lecturing fellow kind of thing. I think you would be really good for it." I was like,

"Wow, here's a chance to waltz into academia, not do any postgrad work, and have this cushy job teaching theater." So I did apply, not really thinking anything would come from it because I didn't think I was qualified. At the same time, I was ambivalent because I wanted to get back to San Francisco.

So summer rolls around and I want to leave, but they come and tell me, "We shouldn't be telling you this, but you are number two on the short list. Hang around because it may still happen." So I stick around all summer and the day before school starts this other guy takes the job.

I went right to the travel agent and got my ticket.

I ended up painting houses for a year, which was a good experience. Once you do hard manual labor for a year, it makes you realize that you really don't want to do it for the rest of your life. Then I got a temp job in an office, but I couldn't type and didn't know anything about computers.

So I ended up at this investment bank. It was the kind of place that if you worked hard and were kind of smart you could go far. I'm not an investment banker and it was the last place in the world that I should have been. It was so not the right thing for me to be doing, but they offered me more money than I ever imagined I would make. So I stayed there for a couple of years, and by the time I left I managed a portfolio of basically bad loans on shopping malls and apartment complexes for about twenty million bucks. I didn't know what the hell I was doing. You know, I was just faking it. I was an actor, so I could get through that. I hated it and it wasn't the right thing for me, so I quit that job, took the money that they paid me, and my wife and I went backpacking around the world for eighteen months.

We sold everything. We gave up our apartment in the city, had one little backpack each, and just hit the road. We spent most of our time in Asia, but we had to come home suddenly because my wife's father died. So we come back, I'm thirty, we're broke, all my credit cards are maxed out, and we're living at my mother-in-law's house.

You know, my wife is grieving and it's just not a good time, so I tell myself, "It's time I figure out what I want to do for the rest of my life." I really thought about

it. I had always been the literature type, so I really wanted to get into publishing. A friend in publishing said, "You should really sell books," so I got a job at a big independent bookstore down on the peninsula. They made me take a test to see what I knew about books, and I guess I did well because they said, "No, we don't want you to sell books, we've got this great job for you in the big office." So I'm there for about a week and I see an ad in the *San Francisco Chronicle* to be the assistant to the editor of *Mother Jones* magazine. I thought, "That's the exact place that I could see myself. That is the right fit for me."

I basically talked the publisher into hiring me because I had a business background. I was like, "Look, I'm not one of those flaky liberal arts grads who is gonna tell you anything to get in the door and then they really just want to be the editor. That's not me; I'm really interested. I've got a business background. I've always been interested in a wide range of things. I want to know the whole business. I want to help this business grow and make things happen around here." So I ended up being the assistant to the editor for a while, and that was my foot in the door on the editorial side. Then I slowly worked my way up to the managing editor position.

It took me five and a half years to find where I belong, and this is the perfect fit for me—I don't think I knew this at the time. Growing up, I don't even think I knew what a magazine editor did.

I don't think I would've ended up here if I hadn't had all that life experience I gained before I was thirty—and the fact that by the age of thirty I was prepared to say, "Yeah, I'll be somebody's assistant, I'll book lunch, I'll make twenty grand a year in order to get to where I want to be."

RTN: You said that when you were about twenty years old and in New Zealand, you didn't have a great knowledge of magazines. What were you studying?

ALASTAIR: I studied ancient literature. The school I went to was based on the English model—old-fashioned, very conservative. It wasn't really what I wanted and I spent nearly all my time hanging out in the theater. It becomes all-consuming; it becomes your social life. I directed and produced a few shows; I did a little of this and a little of that. So maybe I didn't know what to do for a job from that, but I realized that I am one of those people who likes to know a little bit about everything, and I like to be the one connecting people and making sure stuff happens.

RTN: Right. I'm actually a twenty-one-year-old student about to enter my senior year, and I do have an internship in publishing. But all of a sudden I got this feeling of "I don't really want to enter my senior year; I'd like to go to Portland and waiter and bum around and enjoy Portland." This just happened a few days ago when we were up there, so it's pretty interesting that you are saying this now—that you took a year off and went to the States.

ALASTAIR: Well, I had always been a fan of that. I had actually been to the States before. My senior year in high school I got an international scholarship. They sent me off to New Mexico. I really wanted to go to Italy, but whatever. It turned out to be a great experience that put me behind a little bit in school, but the time away just opens up your mind so much.

So I went to college with this international experience behind me, when most kids go to college straight out of high school and have never been anywhere. Their idea of fun is going down to the pub on a Friday night and getting pissed. That is part of going to college, but I felt like I had seen so much more.

You know, people get hung up on, "You're going to get behind; you're not going to achieve this by the time that you are twenty." But in the grand scheme of things, that is not going to count for much. The opportunity to get to see things before you get caught up with too many commitments and get too locked into something is great.

RTN: You said that you waited until you were thirty, when you came back from that trip, to decide what you wanted to do with your life. How did having that traveling experience behind you help you to figure stuff out?

ALASTAIR: Well, I think I had a lot more clarity about what I wanted from my life because I had tried a few things. I had seen a lot of stuff and I knew what I wanted nothing to do with. I think that people get paralyzed by indecision: I could do this, or I could do that, or I could go to med school or law school. Trying some things out is the only way to find out.

I was one of those students in university who people thought, "Oh, he's an academic; he's going to graduate and then go on to postgraduate." Although that appealed to me, when it came down to it I thought to myself, "Do I love something enough to study it for however many years to get my Ph.D.?" I just couldn't face the idea of being in the library for another five years.

I think the definition of work has changed since when I was growing up—you had one job and that was it—now most people change careers something like five times. The contract between the employer and the employee has been somewhat broken down by the economic changes of the past decade. It's not like you commit to us and we'll commit to you and we'll see you after sixty-five for that nice pension. Those days are gone. In a way it's a terrible thing for workers, but in a way it's very liberating. You realize that whatever you're going to do, you don't have to lock into doing it for the next forty years.

RTN: When you quit the investment banking stuff and you and your wife blew through all your money traveling the world for eighteen months, how did you handle coming back to nothing?

ALASTAIR: We really didn't think it through very clearly. That was a really challenging time. It was May in northern Europe and we were sitting in a cheap hotel room with the money running out and just freaking out, realizing that it was time to face up to some responsibilities. Maybe it wasn't a great idea financially, but it was like, "Hey, we may never have this opportunity again," and my wife had never been to Europe. Also, neither one of us was doing what we wanted to be doing with our lives. We hoped that a big part of that trip would be to find a sense of who we really were. Who would we be when we didn't have daily expectations?

What choices would we make if we didn't have to be somewhere the next day—when we didn't have to go to a meeting or be on that conference call?

We thought that maybe the freedom would help us work out what we really wanted to do with our lives. So in a way, it was our own Roadtrip Nation; we were definitely searching for those answers. We wanted to go see how the rest of the world lived because we had a feeling that living in a small apartment in San Francisco and working downtown is not everything there is to life. We spent three months living with a family in India. We saw how those people live, and we realized that we didn't need so much stuff to make a good life and be happy—to have a fulfilling family and a rich life. I would say that it was an investment in ourselves and in our life experience—an investment that would help us figure out what we wanted out of life.

RTN: You talked about the theater and how that ties into the publishing, but what is it that you are passionate about?

ALASTAIR: Well, as it turned out, I'm kind of passionate about magazines. I think it's because I'm one of those people who is interested in a lot of stuff. Magazines and the cycle by which you work on a magazine give you a sort of manageable time frame for you to delve into a subject. You get to know a lot about it, you develop a really good relationship with the writer, you work through a story, and a couple of months later you move on. You're doing that with about four or five stories at a time.

I had a few epiphany experiences with magazines. They made me realize there is someone who does this, who takes and blends subjects and writers and captions and photos and all this stuff and packages it together to make this wholly satisfying experience for a reader—and wow, wouldn't it be great to do that?

I don't think of myself as creative. I can't draw to save my life, and if I paint my bedroom three colors, they'll look horrible together. I enjoy writing, but I don't have a novel bursting inside me. It's drawing from many different sources and creating something.

RTN: If you could give us a package of insight to take away—not just us, but people everywhere who are trying to figure out what to do with their lives—what would that be?

ALASTAIR: Take vacation. No, I'm serious. This culture is puritanical and work obsessed. People take two weeks' vacation in a year. In Europe you take six weeks minimum, and a whole bunch of people save it up to have three months vacation. And they don't feel guilty about taking it.

Experience the world. Work is not the be-all or end-all. It's not the measure of who you are as a person. Be prepared to experiment. When you get that time off, take that opportunity. Take that roadtrip across Laos. Follow that band for the summer. You're not always going to have that opportunity, so do it.

The way we've set up society is there are these breaks, these times that you get to use them. Yeah, you could spend your summer studying for the LSAT to get into grad school, but if you can create some space, if you can create some kinks and creases to get off the path and do other stuff, go for it. Those experiences will have amazing repercussions in ways you don't even know about. In ways you can't explain to your parents. ■

Not only was the interview a great experience, but *Mother Jones* is an awesome magazine that I somehow failed to notice up until that point. We were all struck by the outsider perspective Alastair brought to the conversation. Having been born in New Zealand, he had some interesting observations on the American work ethic and its place in our culture. What he said about the place of "work"—work to live versus live to work—confirmed why we were on the road.

After Alastair's interview, we took advantage of our brief downtime to walk the streets of San Francisco on a gorgeous day before driving to the outskirts of town to meet with Joe Haller and Ian Hannula, the cofounders of Nice Collective clothing. We had a chance to look at some of their line and saw how creative they were within what I had imagined must be a confined medium. They were definitely dealing with shirts in ways that I never would've come up with, let alone had the ability or guts to execute.

IAN HANNULA and JOE HALLER
Cofounders of Nice Collective clothing

RTN: We're going around and inter-viewing people from all walks of life, listening to their stories of how they got to where they are today. So if you could, take us back and tell us the story of how you got here.

IAN: I came out here from Atlanta. I was sick of the music scene there—or the nonexistent music scene there [laughs]. I was a DJ; I sold everything I had and said, "What the hell, San Francisco is where I am going."

JOE: I graduated with a degree in med-ical biology and radiology technology. I was kind of surprised when I graduated and nothing amazing happened to me [laughs]. I was done with school and started working in the hospital, and, you know, it was interesting, but I hated get-ting up every day and driving in to work—you know, the whole day looking at the clock. I said, "There has to be something where I don't do this all day long." So I met Ian, and we thought we would start a company. We thought it was going to be a record label, and so we started doing t-shirts to promote it. I had been friends with Moby from New York, so he came out here to visit and said, "God, everyone I know that has moved to San Francisco became a drug addict." He was impressed that we were trying to start something. He had just gotten signed to Electra Records. He took a bunch of our t-shirts back with him and said he would wear them at press shoots and everything. He didn't care what style shirts—he wore everything.

IAN: This is when we had one bedroom and one rack of t-shirts, and we had put everything on our credit cards—all the wrong ways of doing it. We grabbed the yellow pages and looked under t-shirts.

JOE: Then all of a sudden it was like, "Did you see the cover of *Mix* magazine? The shirt that Moby is wearing is ours!" It added a lot of legitimacy and we started to grow. We didn't think about overhead or anything like that; we just started it out of our passion. It took us a while to find out that there was no right way—but you just have to have a passion. If we didn't have the passion, we would have quit a million times. ■

It was clear that Joe and Ian had worked long and hard to make it in a notoriously fickle industry. Although it certainly wasn't the easy solution, it was what they absolutely loved doing every day. They were the embodiment of "Wow, we're really doing this!" Listening to them talk—both the words they were saying and maybe even more how they said them—was an encouragement to follow our passions.

After saying our good-byes, we headed to Santa Cruz. We still had plenty of time to spend in San Francisco, and we were anxious to see as much of California as possible during our brief stay.

We slept right off the main drag that night, with the RV parked on a relatively quiet residential street. An interview with Jane Smiley in Carmel Valley was scheduled for the next day, and then a full day of driving through the fabled Big Sur region, which runs along the coast just south of where Jane lives.

DAY 14: Big Sur

I woke up on our way to Carmel. The drive was apparently really nice, and if I could've kept my eyes open for any of it I'd tell you all about it. I guess it was the warm weather or all the days of waking up early and running around like maniacs, but I was beat. If the view from Jane's house was any indicator, I probably should've gotten out of bed. Once we maneuvered the RV up a steep, highly impractical, slalom-type driveway, Jane and her husband invited us into their living room, where a huge picture window overlooked the surrounding hills of Carmel Valley.

JANE SMILEY
Pulitzer Prize–winning novelist

RTN: Writing is not the most well-paid industry. Were there times when you were scared?

JANE: No, because I was an idiot [laughs]. First of all, it was very cheap to live in those days. We were content to not have running water, to not have wood heat—we thought it was fun. Not everybody lived like that in the 1970s, but we did. So I lived in Iowa City all through the seventies, until I finished my graduate work in creative writing and medieval English literature. Then I was going along being a moderately successful literary novelist, and boom—in 1992 I won the Pulitzer Prize for a book called *A Thousand Acres,* which was a rewriting of *King Lear.*

RTN: So do you think that becoming a novelist is probably more due to hard work than talent in some respects?

JANE: It has to be something that you really, really want to do. Do you guys ride skateboards? [A collective "yes."] When you were nine or ten, when you were first starting out to learn to ride your skateboard, you weren't humiliated by first putting your two feet on the skateboard and just going ten feet. You were thrilled by it, and all those times you fell off, you weren't shamed and humiliated. You didn't bother to be shamed and humiliated because you just wanted to do it so much. The ones who were shamed by the embarrassing aspects did not get anywhere because they were focused on the wrong thing. They were not focused on going forward; they were focused on, "What do I look like?" Everything is like that.

RTN: You said that those who get too much praise sometimes wind up stopping, and the ones who get just enough are the ones who keep going. Why do you think that is?

JANE: I think people like a challenge, and it's the act of rising to the challenge itself that convinces you that this is something that you really want to do. I think you actually have to exercise your convictions that "This is what I want to do" many times in order to get anywhere.

RTN: Do you have any final words of wisdom for people like us, who are trying to find their roads in life?

JANE: The people I know who didn't follow their own desires ended up sort of lost and confused. Even if they were successful, they ended up lost and confused in their late forties and early fifties. They were left wondering who they were—not just wondering whether they enjoyed their lives, but also wondering who they were. That is a very difficult way to spend your life, especially your later life—not knowing who you are and wondering if you wasted your time and energy. So even if people say, "No, you don't have any talent" or "No, you should not do that because you won't earn any money," I would say go ahead and do it. Do it with your whole commitment. Do it knowing that you're the one who wants to do it—it's your choice and your responsibility. ∎

It was striking to hear Jane talk about the dilapidated shack she used to live in while we sat in the living room of her absolutely spectacular home. After the interview, Jane and her son came down to sign the RV; by this point, the back room hosted quite the impressive collection of signatures. We maneuvered the RV back down the driveway and headed down Highway 101 and into Big Sur.

The drive along the coast was unbelievable. It was like driving along a huge wall plastered with gigantic postcards, one right after the other. We pulled over for a while, relaxing on rocks by the ocean. It was cool to reflect at that point and take a closer look at the past couple of weeks: we had looped through Arizona, up into Colorado, up to the northwest corner of the country, and right down the coast. The roadtrip was almost done. This was the homestretch and we were spending the final days bouncing around S.F. and L.A. That night we made a concentrated effort to get to sleep a little early; the next morning we'd be back in S.F.

DAY 15: San Francisco

Glide Memorial is in downtown San Francisco, and it's not just any church. Located in the heart of the troubled Tenderloin district, Glide is half church and half community outreach organization, with more than fifty programs, from free meals to resume counseling services. We were lucky to have a meeting scheduled with the chief executive officer of Glide Memorial Church, the Reverend Cecil Williams. But first, we were going to Sunday morning services.

The church was absolutely packed, and the service turned out to be really eye-opening. It was super musical and, consequently, it felt much more engaging—less preaching at, more talking with. After the service, it was apparent that we were lucky to be getting some of the reverend's time. I don't know that I've ever met a busier person in my life—every thirty seconds someone needed something, asked a question, requested approval, and so on. We were thrilled to get whatever time we could.

CECIL WILLIAMS
CEO and minister of Glide Memorial United Methodist Church

RTN: We're really interested in your story. We're all twenty-one and wondering what we should be doing with our lives, not sure which decisions to make. How did you figure it out?

THE REVEREND: I went to Southern Methodist University. I was one of the first African Americans at SMU. I was there for three years and studied theology. It was quite an ordeal to be there because here is this white university just admitting five African Americans. At that particular time, it was not easy, to be honest with you. So I graduated, and the Methodist Church assigned me to a church in Hobbs, New Mexico. It was an oil boom town, what they call a "wildcat" town. I was there for just one year, and I found out what the Methodist Church had done was send me out there to create a segregated church. I said, "There ain't no way in the world, man, that I am going to go somewhere to create segregation." So I left and moved to the West Coast to study at Berkeley for two years. I finished my work there and was assigned to a church in Kansas City. It was an awful church, so I decided I was going to stay there for a while and do something. In a short span of time, what I found was common in Kansas City was that there were a lot of poor folks there—both whites and blacks. And so I realized that there was something in common to them both—they are poor. So I began to work with them. The first thing you know, whites and blacks were coming to that church together. Boy, it hit me then: What we need to do is work with people where they hurt most of all. It came to me—what we must do is empower especially the poor. So I created my own theology based on this.

RTN: What about risk? The way you speak about risk, it seems like it's been worth it for you every time. But it also has to be tough.

THE REVEREND: Yes. Yes. But the bottom line to me is that love means living dangerously. And to live dangerously, you have to risk all that you have. If you just risk half of it or you risk a little of it, that's not risk. If you're going to do it, you have to go all the way. I feel very strongly that we must all go through something if we are going to get something. So I put it this way: If you don't lose your life, you can't gain it. You have to lose something. Lose something so you can live. Life comes after death. ■

Afterward, we gave a quick call back to our respective homes since it was Father's Day, and then we had the rest of the day to ourselves. We took an hour going over the Golden Gate Bridge a couple of times for some video footage and checking out the impressive views on all sides of the bridge. Then we did a little shopping, hunted down some records, and continued to explore the city. This was the last free time we'd have in San Francisco; we had one more meeting scheduled in the morning with the CEO of Levi's, and then it was off to a couple more interviews out of town. By the end of the night, we'd be on our way to L.A., and after three days of interviews there we'd be on our way back to JFK. We realized we had covered a ton of ground both physically and metaphorically, yet we were baffled by the idea that it was rapidly coming to an end. This was definitely an odd feeling, because in many ways we felt as if the roadtrip had been going on for months. Sitting at a coffee shop in San Francisco that afternoon, we talked about the beginning of the trip, our first interview with Rob, and the Cirque du Soleil show. It all felt so distant—we had crammed an unbelievable amount of travel, learning, and excitement into the last two weeks. But at the same time, it seemed impossible that it could be ending so soon, that this thing we had spent so much time and energy booking and preparing for was soon to be over, and all of us would be back on the East Coast wondering if it really had happened.

We all resolved to squeeze as much as we could out of the remaining days on our trip. Rather than trying to synthesize what had happened already or what could happen at the end, we were content to live it now and save the analysis for later. With three interviews on the next day's schedule and eight in the three days in L.A., it didn't look like we'd have a hard time staying in the moment.

DAY 16: San Francisco

We woke up early and made the quick drive over to the Levi's HQ, where we were set to meet with Phil Marineau, president and CEO of Levi's. We were given a quick tour of the place first—including an insider's overview of the Levi's timeline and a few pairs of jeans dating back to medieval times—and then headed to Phil's office upstairs. I was impressed by how genuinely nice Phil seemed—he was so friendly and interested in our project. Mike had booked this interview through a school connection; Phil had attended Georgetown as well, so they talked about that for a bit and then we got down to business.

PHIL MARINEAU
President and CEO of Levi Strauss

RTN: Where are you from?

PHIL: I was originally from Chicago, and I wanted a job in the packaged-goods marketing business, so I went to Quaker Oats. They were the place to go. I went there never expecting to stay that long, but I spent twenty-three years there. I marketed everything from Captain Crunch to Kibbles 'n Bits dog food [sings the Kibbles 'n Bits theme song with perfect pitch].

RTN: Marketing sounds like a late-blooming interest for you, since you were a history major. We're struggling with the idea that when you're eighteen you have to know your major, get on the right track, get into grad school, med school, law school, what have you. I don't know what I want to do with the rest of my life at twenty-one, and apparently you didn't either, but it worked out for you.

PHIL: I think the twenties are a period for discovery. This notion that you have to commit to a lifetime career that you've discovered at twenty-one, when the average life expectancy is eighty-plus years, is absolutely crazy. I think for a lot of people that's why the twenties are so difficult, because you have this expectation that you've got to figure it out, when you really don't have to. I sort of had a track that I was on because I went to grad school and then I went to the army, but I really didn't get serious about anything until I was twenty-seven or twenty-eight.

We put expectations on people that don't have to exist; you need the time to figure out what you want to do, and what you're good at. That's part of the problem: What do I like to do and what am I good at, and how do I marry the two? ■

Talking with Phil was great, and I think it turned out to be one of Mike's favorite stops of the trip. Phil was a perfect example of doing what you love, doing it well, and becoming very successful as a result.

Our next interview was at Industrial Light and Magic. ILM is a Lucasfilm company that basically makes every movie you've ever seen look really cool: *Star Wars*, *Raiders of the Lost Ark*, *Poltergeist*,

you get the picture. Despite getting lost twice on the way, we made it in time to get a brief tour and see all kinds of amazing stuff, like Darth Vader costumes. Jim Mitchell, who works on visual effects, came out to the RV, and we got the ball rolling.

JIM MITCHELL
Visual effects supervisor with Industrial Light and Magic

RTN: You mentioned that you graduated from school and knew that you wanted to be in film and animation. What made you say, "This is what I want to do"?

JIM: There was not just one thing. I just remember, as a kid, my dad was always watching Westerns on TV. I wasn't very interested in the stories. I was more interested in "How did they do that?" I wanted to figure it out. Back when I was going to school, there were no classes that taught you to become an animator or how to do computer graphics. It was all doctoral degrees and things like that. So when I got out of school, I moved to D.C. and started working for a video postproduction house called Interface Video. But again, film was always the goal. The whole time I was there, I was always trying to send my resume to ILM. I got turned down a few times, but you just have to be persistent. So in the summer of 1990, as they were just starting to get ramped up for *Terminator 2,* I eventually got the call to come out here. I started as a videotape operator—that was thirteen years ago. It was just a golden time—everything started happening. That's when computer graphics were really just growing. We had *Jurassic Park* in here and I became part of the team that did that. Then I went on to be the visual effects supervisor on five or six big pictures, and I haven't looked back ever since.

RTN: Film seems like a hard thing to break into. Was there anyone who pressured you to do something else?

JIM: Well, I remember that when I was not doing electrical engineering in school, I was trying to take film production classes. I remember having a television production professor telling me "it's a one-in-a-million chance to get into visual effects" and "it will never happen to you." Well, you know, sometimes I just want to call him up and say, "Here you go."

RTN: Maybe you just did. ■

Jim's story was great because it was so simple: he'd loved visual effects since he was a kid, and through a lot of hard work and a ton of dedication, he was able to make it his livelihood. He was a truly friendly guy on top of it all, and his job is mind-bogglingly cool. After we finished, Jim signed the back of the beast and we met up with Suzie, a friendly Lucasfilm lady who helped us book these interviews. She was to act as our guide out to the fabled Skywalker Ranch, which apparently

they keep pretty much under wraps. We weren't told where exactly it was; we had to follow Suzie's car all the way out there. The ranch is out in the middle of nowhere as far as I could tell, and I'd love to tell you more about the ride if I didn't think someone would cut off my hand with a lightsaber.

The ranch itself was gorgeous. We didn't get much of it on film because the guards there said no filming, and believe me—they were not messing around. They chased us down to inform us that taking a shot of some cows was not allowed. The ranch was seemingly endless, full of rolling hills, fields, lakes, and other ranch stuff. We were surprised that it's actually a working ranch. We always figured that

was just a cute little moniker, and that it really consisted of studios and big ugly buildings. But there were plenty of grazing animals and farmhands. We were scheduled to meet with Gary Rydstrom, the sound designer from Skywalker Sound. He has seven Academy Awards, so you might say he's big-time. When we walked into his office, he was tooling around with ideas for the new Mac start-up sound—the sound your computer makes when you turn it on. He gets to do stuff like that all the time, just experimenting with a lot of insane and expensive equipment. He showed us a few tricks and told us a few secrets—the alien voices in *A.I.* are just a slowed-down recording of a kid reciting the Torah at his bar mitzvah. Afterward, we headed over to the cafeteria with Gary for some food and the interview.

GARY RYDSTROM
Sound director with
Skywalker Sound

RTN: What do you love about sound?

GARY: The very first time I ever did sound, there was something magical about it. A film comes to life in an interesting way as you add sound to it—it becomes three-dimensional. You take a flat movie that is not really working, and you add some simple sounds and it comes to life.

RTN: You decided really early on that you wanted to get into film. You mentioned that you were only twelve years old. Was it really that easy for you to say, "I'm doing film"?

GARY: I was lucky enough to have support, at least to my face. My parents and other people did not laugh when I said I wanted to work in film. Everyone told me, "You're not going to get a job." But I did it because I loved film and I wanted to study it.

RTN: You mention confidence a lot. I'm someone who doesn't take risks because I'm not so confident in what the outcome will be. So how do you build the confidence?

GARY: If you go into something, as you start doing it you're never as good as you are later. And you don't know how good you are at things until you try them. So that's why you have to be led by your passion. If it's something that really intrigues you, you will probably put your energy into being good at it down the road. To be naturally good at things right away is impossible.

RTN: This might be a silly question, but how do you recognize a passion?

GARY: I think it all comes down to curiosity. It's not about the answers that come up, but the questions that come up. So if it's something you have endless questions about, whether it's a career or whatever, you're probably interested in it. And if you're interested and you're curious, you probably have passion. It's just like falling in love, I guess. ■

Gary ended up being a consensus favorite at the end. He was so laid-back, friendly, and humble, especially considering how much unbelievable stuff he's been a part of. He dropped that seven Academy Awards thing into the conversation like he was mentioning that he took his dog for a walk that morning. This was one of the interviews that turned into a regular conversation almost immediately. It was almost as if the cameras weren't there and the four of us were all just having an interesting talk over lunch. Gary had a ton of great stuff to say and came off as someone who totally understood this idea about following a passion and couldn't imagine it any other way.

After the interview, we made our way back to civilization and headed toward L.A., still talking about Gary's insight on

taking risks and identifying what we're passionate about. It's safe to say this was one of the high points of a trip that was pretty much unbelievably great the whole time. Instead of letting ourselves get bummed out that we were nearing the end, we were psyched on how the day had gone. Three great and diverse interviews, genuinely nice folks, and simply a great time overall. It felt as if we were finishing up with a bang—taking everything we had gained from the roadtrip so far and already learning from it and applying it. We all took turns at the wheel on what was, for all intents and purposes, the last extensive drive of our trip. We got close to L.A. and pulled into a hotel parking lot around 1 a.m. We went to bed still feeling great from our day and ready to take on our three-interviews-a-day schedule for the final leg.

DAY 17: Los Angeles

I woke up outside Jeremy Weiss's house in Silver Lake. This was slightly problematic because we were supposed to be interviewing him, not applying deodorant in his front yard. Fortunately, Jeremy was no more of a morning person than I and let us raid his kitchen for a much-needed cup of coffee before we got started. He also took us out back to show us his toolshed-turned-darkroom, which made me insanely envious. Eventually we settled into his living room and got started.

JEREMY WEISS
Independent photographer

RTN: Sorry for parking our RV in front of your house.

JEREMY: No worries.

RTN: Anyway, as we mentioned, we're traveling the country in that thing out there, interviewing people to learn how they got to where they are today. What's your story?

JEREMY: I don't really know [laughs]. I always read a lot of magazines as a kid, and I loved shooting photos of my friends with my mom's camera. I would take photos of them when they were skating or playing music or whatever. I liked to see if I could make them look cool. I never realized that it was something I could make a profession out of. I never even thought about it; I didn't put two and two together. Once I figured out that people were making a living by shooting photos for magazines, I was like, "That's what I want to do." I went to the only photography school that I could afford, which was in Boston. I graduated in 2000 and I've been doing photography ever since. I've done a lot of freelance stuff, been broke sometimes, and had a lot of money at other times.

RTN: You do a lot with bands, such as the White Stripes and some others. Was music something you were passionate about when you were younger?

JEREMY: I've always been involved with music. Most of my friends are in bands, so I can't avoid it.

RTN: Did anyone ever pressure you about going down a more traditional path?

JEREMY: No. My mom and dad got divorced when I was two. My dad is the most laid-back guy ever, and I'd say he's an artist in a way. He's a carpenter, but he's always inventing. He's always in the garage making crazy stuff. He definitely saw where I was coming from. He probably wished he hadn't worked a nine-to-five job his whole life. He waited until he was fifty to leave his job and become a carpenter. There was no pressure from him because that's what he always wanted to do.

RTN: You said that sometimes you have money and sometimes you don't. Have you ever doubted that you should be doing this? Ever thought about switching?

JEREMY: No. Never. If it were for the money, I would definitely be doing something else. I went in knowing that I wasn't going to make a lot of money. I don't think I could ever do anything else. I'm obsessed with taking photos. I couldn't see myself doing anything else. ■

The interview with Jeremy was different than most of the others. On the trip, we had talked with a lot of older people who were looking back on lives and careers that had been successful and fulfilling. Jeremy was awesome because, really, he wasn't that far removed from us. He's a twenty-six-year-old kid trying to make it doing what he loves, and he's kind of still in the midst of it all. So it was a different perspective on the whole thing. Rather than looking back and saying, "Hey, I loved taking pictures, I followed that passion, and it turned out great," Jeremy's perspective was more along the lines of "I love photography and I'm trying; it's tough at times, but I can't really imagine stopping." I think that was really good for us to hear.

After we wrapped up with Jeremy, we got right back on the road. When we booked the L.A. interviews, our limited knowledge of the geography and poor communication amongst us had led to an interview schedule that had us darting back and forth across town through the notoriously painful traffic. Just now, we were on our way to meet up with Mason Gordon, the creator of *Slamball*. Mike and I were looking forward to this one. A couple of months earlier, we had been sitting in Mike's living room compiling a list of prospective interviews for this trip. *Slamball* came on TV, and on a whim we threw it down on the list—up at the top, in fact. We were in love with the idea of huge athletes jumping off trampolines and dunking on each other's heads, and fortunately for us, the interview had worked out. On our way to Mason's office, I was driving and Mike was giving directions. Unfortunately, Mike either hates being on time or can't read directions, because he got us very lost. By the time we got to Mason's, we were a solid forty minutes behind schedule. Fortunately, everyone at the office understood the hellish L.A. roads and traffic.

MASON GORDON
Creator and producer of *Slamball*

RTN: What was life like for you in your early twenties—before you started *Slamball*?

MASON: I came out of college and promised myself one year in the entertainment industry. I had only two goals: one was to date a wannabe actress and the other was to go to a movie premier. I took a job as a bottom-feeder at a little production company. I took out the trash, made coffee and photocopies, and answered the phone. I ended up dating the wannabe actress, and I even went to a movie premier—although it was for Nickelodeon's *Good Burger* [laughs]. My life goals were pretty much getting checked off the list.

Then I struck upon this idea for *Slamball*. I wrote up a three-page prospectus and took it to this guy that I worked for at an entertainment company. Somehow he got it; he could see what I was seeing in my head. He said, "You can't launch a new sport nowadays, but what you can do is get something on TV. And if you can get something on TV, then you can back your way into the traditional sports model." That was really the genius; it was the eureka moment. After a lot of hard work, I went from just this goofy idea on a napkin to a national broadcast deal eighteen months later.

RTN: What was going through your head when you had to take more than a year off to really pursue this?

MASON: When I got the urge to follow my idea, it was really like putting blinders on. You just run forward. You're not interested in anything else; you're just interested in going forward. I think young people who have ideas need to be encouraged to run with those ideas. This is your time, this is your platform, this is your stage, and the only question is whether you're going to go after it or not. ■

Mason was so excited about *Slamball* that it was impossible not to get wrapped up in it. He showed us tons of clips from the season and pointed out his favorite players and specific dunks that were, as he said, "Bananas—totally bananas." And really, how "find the open road" is it to make up a new sport? Nobody does that. That's like inventing a new facial expression or flavor of meatloaf. But Mason came up with this insane idea that he totally loved, which is more or less a video game come to life, and took it from crappy trampolines in old warehouses to a national stage.

Next we were on our way to meet with Larry Weintraub, the cofounder of Fanscape, an artist management agency based in L.A. We had hooked this one up through a friend's brother whose band worked with Larry and said that he'd be an awesome person to talk to. We arrived at the Fanscape offices, not quite as horrifyingly late as at our previous interview,

and set up our stuff in the upstairs conference room. Within a few minutes Larry joined us and we were ready to go.

LARRY WEINTRAUB
CEO and cofounder of Fanscape
Artist Management Agency

RTN: So this interview isn't necessarily going to be, "Larry, tell us about Fanscape." We're more interested in your story—how you came to be.

LARRY: My family had nothing to do with the entertainment business, but I remember my father always playing music. I remember him playing Elvis in the morning when he was shaving. That's my earliest memory of music, and I just fell in love with it. Then I got to high school here at Santa Monica High School. I could never play an instrument, but I just wanted to be in a band somehow. So I started managing bands in high school. They played shows, and I would send tapes out to clubs in L.A. and try to get them booked.

RTN: In Phoenix we interviewed Jerry Colangelo, who is CEO for a couple of sports teams. He had a passion for sports, but he ended up throwing out his arm. The most interesting part of his story was that he still didn't give up his dream of sports—he just found an alternate route to it. With you and music, did you always see that wraparound route so clearly?

LARRY: Yes. I always knew, because I loved music and I knew I wanted to be involved in it, but I just couldn't play an instrument. I came from a pretty poor family—lower middle class. We weren't starving, but my dad said, "You aren't going to Harvard, so just forget about it." So I applied to the University of California schools, and I ended up going to UCSD, where I joined the radio station and became a DJ. I also ran the concert committee and was still managing bands on the side. I went to work at A&M Records after UCSD. I was in the artist development department and it was awesome. This was before Nirvana and those bands had broken, and A&M had just signed bands like Soundgarden, Blues Traveler, and the Gin Blossoms. I got to work with them all. I was still managing bands on the side; one of them was Offspring, and they ended up breaking huge, so my little secret was out. Then something completely unexpected happened—A&M Records folded. At that time I debated between going to another record label or taking a stab and doing my own label.

RTN: It's my dream to start a record label, but in all honesty, I'm petrified. I have all these school loans and I have all these people saying, "You can't do it." How do you get over that? How do you take the jump?

LARRY: I had loans when I graduated, and I had to pay those off too. I would say, if it's something you're thinking about, do it. At the same time, you're gonna have to get another job to pay your bills and your loans. But it's your life, and if it's something you want to see through, you need to find a way to get in and do it. If you think you're struggling now, it's gonna be harder, but you can do it. ■

The interview with Larry really resonated with Randy, since he has toyed with the idea of getting involved with the music industry in some way. Larry's story had a different twist because he had this love for music but, like 99.9 percent of the population, wasn't really in the position to make it as a professional musician. Instead, he was resourceful and carved out a different path into the music industry. Talking with him was great, too, because he was so honest, and that translated into some really practical advice. He didn't speak in generalities; he told us exactly what he had to do and how hard he had to work to get to where he was.

Not wanting to break tradition, I woke up once we had already parked the RV at our first interview. This time I picked the crusties out of my eyes just in time to meet Jehmu Greene, the executive director of Rock the Vote, those amazing people who actually spend time worrying about the fact that more people vote for *American Idol* than for the "leader of the free world." Talking with Jehmu, we figured she had to be pretty damn passionate about what she did, because if there is a more frustrating entity than the American political system, I have yet to see it. And yet when she walked in, she was full of energy and positiveness. We got the interview rolling to find out if she knew something we didn't.

JEHMU GREENE
Executive director of Rock the Vote

RTN: Tracing back to when you were our age, were you feeling some of the same confusion or did you know exactly what you wanted to do?

JEHMU: I was in Austin and I was working on political campaigns. I wanted to do it as a career and I was trying to figure out how to get into the process. Rock the Vote has been a lifelong connection for me. I was registered to vote by Rock the Vote at a concert at Austin, and then I became a Rock the Vote volunteer. They recruited me to come out here and help them with partnerships, and three years later, I guess I'm now the boss.

RTN: Why politics? Was that something you were always interested in from a really young age?

JEHMU: Absolutely; it came from my parents at a really, really early age. When I was five, I wanted to be the first black senator from Texas. So it definitely has always been a part of my core. When people asked me what I wanted to be when I grew up, the answer would be that I wanted to make a difference. It was always to make a difference and to help people—and I really think that public service and politics is the most direct way to have an impact on a very large number of people.

RTN: So now what's still driving you? What's still there for you to keep getting up in the morning and work this hard?

JEHMU: It's the moment that drives me. Thinking back to the Vietnam War and major social movements in this country, the Civil Rights Movement, I always thought if I was eighteen, twenty-five, thirty, or whatever during those times, what would I have been doing? And in my head I think I would have been out there, getting sprayed with fire hoses and attacked by dogs and protesting. That's what's always been in my head—thinking that's the type of person I am. So we have a moment with this political climate right now, and I have to be active; I have to do everything I can because otherwise I'm just a poser. I have to be about it, because we're in one of those moments in this country.

RTN: I have a question on a more personal level. A while back a friend of mine and I were talking about voting, and I asked him, "Why don't you vote?" He said, "Specifically because when you vote, you acknowledge that the system exists and that it works." I kind of thought about it, and I was like, "Wow, that makes a lot of sense." What are your thoughts on that?

JEHMU: It's a bunch of BS. Because when you're driving down the street, when you turn your water on, when you do anything that you need to do in your daily life, you are proving that the system exists. Your life is surrounded by a system, and if you say, "I'm not going to participate in this," but you see something wrong and you're not even gonna to try and fix it—that's when I look at people and I have to call them on it. Because the system is a part of their lives at all times. Absolutely the system is broken, but there is no reason why it can't be fixed. Get up off your ass and do something, because if you're not participating in it, you're contributing to it. ∎

While Jehmu didn't necessarily know any secrets about the coming revolution, she was completely committed to what she believed in. The interview was really rad as a result; we had an interesting and honest discussion not only about her story—how she got there, her goals, her ambitions—but also about politics, and in each topic, her passion for her job was clear. There's really no way to say this without sounding like a Motley Crue song, but she was honestly really inspiring. Awesome woman.

With new Rock the Vote t-shirts in hand, we had to rush away to meet with Ric Birch, the producer of several Olympic ceremonies. We picked Ric up in the RV and pulled over on a side street to set up for the interview right there. Speaking to Ric in the motor home on our second-to-last day of the trip got us thinking back to that first interview with Rob, also in the RV. We had come full circle. With Rob, we had been stumbling all over each other, awkward, not sure when to ask questions or even what questions to ask. By the time we sat down with Ric, we were a well-oiled interview machine. We gave Ric a brief RV tour and the obligatory "sorry for the inside-of-a-used-sweatsock smell" speech, and then started things off.

RIC BIRCH

Executive producer of the Sydney and Barcelona Olympic Games ceremonies

RTN: Did you have any influences early in life that shaped who you are today?

RIC: Well, I grew up in Australia, and I think one of the important influences in my life was that my father was in the air force. It meant that we moved a lot, and that has been a continuing influence and trend right through my entire life. It has also meant that I have never been scared to move. I'm always ready to change countries, change jobs, change states, whatever. I think for what I do, it's been real important.

RTN: You mentioned that you didn't know you would end up here at all. You had no idea. What appealed to you? How did you realize that this was what you wanted to do?

RIC: Yeah, that's the tough question. It's a bit like the one, "How do you know when you are in love?" Ya know, it's one of those things: no matter how you describe it, until you experience it personally you can't understand what everyone is talking about. It's very hard to think of it abstractly. It was just feeling totally fulfilled and challenged and able to deal with the challenge while still enjoying it. I mean, beyond anything,

enjoying it. I talk to lots of people from my generation who went into university. They got a degree, and they are unsettled and unhappy because they did what their parents wanted them to do, but they know in their hearts it's not what they wanted to do.

So in terms of a career path, mine was totally unpredictable. It wasn't anything I ever had in mind at the beginning. It was a series of connected events, but it's only with hindsight that you realize they're connected. But the starting point was always back in student theater during the university days. I didn't know what I was going to do when I went into university, but the moment I got involved with student theater, particularly backstage, I knew that this is what I enjoy doing.

So I ended up working in television for about sixteen years as a producer and director. I became an executive producer at a network. That led into the Commonwealth Games, and I did the ceremonies for that. They were received well, and that led to an invitation to come to Los Angeles, and I ended up being director of production for the L.A. Olympic Games and ceremonies in 1984. Then in 1992, I was the executive producer for the Barcelona Olympics, and ultimately the Sydney Olympics in 2000. I've returned to America to live, so I spent a lot of time shuffling backward and forward.

RTN: With so much responsibility now, have you matured very much? Ya know, what has changed for you?

RIC: I'd like to think that I haven't matured all that much. I honestly hope that none of you or any of the rest of the people who are involved with Roadtrip

Nation would ever say, "Well, gosh, I'm twenty-four now, and I have a degree and now it's time to grow up and be sensible." That's really the worst advice you could ever give yourself. Life has to be discovered all the time. You can never rule the line under life and say, "Okay, well, I have done everything I can up to the age of twenty-five and now I have to settle down and be a responsible, boring adult in a responsible, boring world." You don't have to. In the end, the world out there is going to be pretty much what you make of it—and what your generation makes of it. You've got an interesting time ahead. ■

I think we were all in awe after talking to this guy who is in charge of so much stuff and sitting him right down in the RV with our shabby-looking crew of fools. Ric seemed to love it too, which was awesome. I think that's one of the things that never stopped surprising us: how into our project so many of the people we interviewed were. They were willing to take time out of some of the busiest schedules I'd ever seen and come hang out in our RV just because they understood exactly where we were coming from. It makes sense that everyone would understand this question of "Hey, what next?" but at the same time we were blown away by the reactions each time.

In a totally baffling failure to learn from our mistakes, we ended up late to our final interview of the day, with Jill Soloway, a writer and coproducer for *Six Feet Under*. As it turned out, she was one of the most genuinely nice and hilarious people ever, so it worked out well. We got a brief tour of her house, she offered us some food, and we met her adorable six-year-old son, who is writing a book about a runaway lemon. We set up the cameras in her living room and finally got the way-late interview started.

JILL SOLOWAY
Writer and coproducer of *Six Feet Under*

We are talking with an eclectic bunch of people to get their stories from when they were in their early twenties. Where has your road taken you from then until now?

JILL: I hope that I can make this interesting. I went to college at Madison, and I wanted to go into advertising. That was the plan. So I graduated and then started working at an advertising agency doing production. I didn't really know what I wanted to do. I thought I was going to work my way up through advertising and production. Then I decided I didn't want to do advertising anymore and wanted to do something in TV and movies. I was working at this documentary company in Chicago, and on the side my sister and I were doing something that was totally from our hearts. We loved the TV show *The Brady Bunch* so much. At one point I was talking to a friend, saying, "I don't know what to do with my life." He asked, "What do you care about?" I said, "*The Brady Bunch.*" That's all I could think of.

It was the only thing that was interesting to me. So we did this stage play where we took all of our friends who were actors and gave them costumes and wigs and did an episode of the Brady Bunch. It was the "Oh, My Nose" episode, when Marsha gets hit in the face with a football. We played this episode out on a stage, basically because we wanted to see our friends in those wigs, reading those lines, and watch it like a play in a theater. It was hilarious. We weren't doing it for the money or anything. It just made us laugh.

I was still working—trying to work my way up at the production company—and then our play became this huge thing, like this cult thing. Tons of people started coming. Then hundreds and thousands of people came to the theater to see it. It was on Tuesday nights and we charged three dollars. One night I was driving down the street to the theater four weeks after we started the show, and we were stuck in traffic. I was like, "F—k, what is all this traffic? We're not going to be able to do the show tonight." It turned out that they were all there to see the play. It was just this insane moment of "What happened, what did we create?"

The fact that we created it not for money but to satisfy ourselves by laughing at our friends was a huge lesson that I would have to learn over and over again as a writer. Forget about what sells, forget about what is going to make you famous; write what you like, what makes you feel good and what makes you happy, and success will follow.

So we did the Brady Bunch thing, and we went to New York and went to Broadway and then we came out here on a national tour. Then my sister and I got an agent. My sister left the business and went to Boston. She now writes and produces rock operas in Boston.

Even though the Brady Bunch play was a success, I was writing sitcoms but getting down the path of "I hate this; I am only doing this for money." So I wrote a short story called "Courtney Cox's Asshole." I was writing it for my friend Becky, the same girl who played Marsha in the Brady Bunch play. We were out hiking one day and she said, "Oh, I hate my job." I said, "I hate my job too. Let's do a show. I'll write a monologue and you can perform it."

So I wrote a monologue for her that was about being in L.A. and being Courtney Cox's assistant. It was totally made up, totally insane, just to see my friend get on stage. That was my motivation: I wanted to see Becky get up on stage and say these things out loud. For some reason I sent it to a literary journal. They said, "This is the funniest thing I've ever read," and they printed it. My agent sent it to Alan Ball, who runs HBO's *Six Feet Under,* and I got hired on that show.

So I went from the most horrible TV show to the best TV show just by doing something from my heart. I had to learn that lesson one more time in my life: don't try to do anything just because you think it will sell; do things that satisfy you and your own senses.

RTN: You said that you went to college for advertising. Where did this writing bug happen?

JILL: I really didn't have a bug to write; it was really just about *The Brady Bunch* and loving the show so much. Then, once the play was a success, agents from Creative Artists Agency took us out to lunch. They were like, "You and your sister could be a writing team. You need to write spec scripts." I did write little stories and stuff when I was in high school, you know, stories about Matt Dillon that teenage girls write to entertain their friends. Anyway, I had a vague idea that I wanted to be a writer, but only recently have I really felt like a writer. Now I am a writer. I have to be a writer. I have to write these things to make myself feel like a human being. But I didn't start off that way.

RTN: You seem like you've always been around entertainment of some sort. Isn't it hard to break into an industry that fickle?

JILL: Yes and no. Now people come up to me and say, "I want to be a TV writer. What should I do?" I sort of encourage everyone to first write a short story, an essay, or a short film and then make it. Instead of trying to write a spec script, write a play; there are actors all over the place. Have them read your lines. Decide if you've written something good or not. If you haven't, change it. As soon as it's good enough, put it in front of an audience. Instead of schmoozing and trying to convince someone that you've got something hot, just write something. Try to get it published. Then take the magazine clipping to an agency and say, "*Esquire* published my short story." Have something besides the knowledge that you're talented.

RTN: You've gotten into producing too, haven't you?

JILL: Well, they call us all "producers" on *Six Feet Under.* But it's really just writing. For *Six Feet Under,* each writer produces their own episodes because Alan Ball is the coolest boss in the world. He lets us cast our own episodes or at least be a part of the casting. We're on the set and have a chair next to the director and get to say things like, "I don't think they really got that take; can you do another take?" We get to take a pass on the cut after the director.

RTN: We spoke with Jeremy Weiss, who is a photographer. He said he doesn't even worry about the selling side of things; he just does what makes him happy—similar to what you've said.

JILL: Yeah, same thing. With *Six Feet Under* I try to write scenes that ring true, that people can watch and say, "Yeah, that's the way I feel." I like something that represents honesty. When you say, "He's funny," and you laugh, then you feel more alive for that moment. You shared a moment. I feel the same way about writing. If I can write a scene that makes people feel, "Oh, I've felt that too" or "I've said that too; I've been there too, but I've never seen that on TV," that's what motivates me—to try and express the truth. I especially try to do that for women and girls, because girls are so defined by what guys think of them. Look at this. [She flips through an *L.A. Weekly*.] These girls are defined by this. Just being a woman and telling the truth about my experiences is more than enough because no one ever does it.

RTN: When you did the Brady Bunch play and you had your other job, was there any doubt about leaving the job and just focusing on the play?

JILL: No, because when the Brady Bunch thing happened, a promoter came and paid to take the show to New York. He paid us all to move the show there, and I was going to get to live in the Village.

RTN: So there was no doubt about the future?

JILL: We were getting paid about a thousand dollars a week, which was more money than we could ever want, and we were getting housing in the Village. There was no question about going; I was twenty-four years old and had nothing to lose.

RTN: We talked to Ann Powers, who wrote for the *New York Times*, and she said that her most difficult thing to do was to find her own voice. Did that come naturally to you?

JILL: No, but I think how you find your voice is to write about something you care about and love. You have to be passionate about something—where you wake up in the middle of the night to think about what characters should be doing. That's all you need; then you feel like you're in the flow and doing what you are supposed to be doing. ■

Jill was great, especially in light of our tardiness. We were instantly in the midst of great conversation rather than formal interview, and Jill's honesty about how she got to where she is now made it easy for us to relate to her. She was definitely one of the people who, although we didn't really know much about her story before we met with her, just seemed perfect for the project. In addition, she even invited us to a comedy show she was hosting the next night, which was to be our last night in town. We took a photograph of her son's lemon on the back of the RV (getaway car?) and made plans to meet up with Jill the next night.

Back in the RV, we decided to splurge a bit to celebrate. The next night we'd be en route back to Laguna Beach to pack up our stuff and head home. So after we left Jill's, we drove into Venice and got ourselves a table at this ridiculously good Italian place the camera guys knew about. The food was terrific and we all had a great time together. Five guys living in such cramped quarters for three weeks is a precarious arrangement, and it was awesome that, despite a few punches and an occasional battle with plastic utensils, all of us could sit around this dinner table at the end of it all, knowing damn well we were having the time of our lives. As we had all resolved by that point, we didn't waste too much time that night looking back on the trip or ahead to its conclusion, but it was impossible to ignore the surreal end-of-the-line feel that sometimes intruded, and I think we lingered at the already-cleared table a bit longer than normal.

DAY 19: Los Angeles

And so we started our final day. We woke up in the morning, all set for our final showdown with L.A. traffic. Our only interview that day was with Patrick Park, a singer/songwriter. Feeling like seasoned pros on the road, we made it over to Patrick's apartment on time, and he came out to the RV for our last interview.

PATRICK PARK
Musician

RTN: Thanks for hopping in the RV to do the interview. Is that shotgun seat working out okay for you?

PATRICK: Sure.

RTN: Great. As we mentioned, we're traveling the country in this thing collecting people's stories. How did you pave your own road around your passion for music?

PATRICK: I grew up in Morrison, Colorado. Music was always in my house, growing up. My dad played guitar, my mom was a poet—so the two were always around writing songs. It came really naturally to me, but initially I don't think they were too excited about me choosing to do it. At a really early age, I pretty much decided that that was what I wanted to do. I barely got through high school and didn't go to college. I then moved out here to sort of try and make it happen. It's a total long shot—there are about nine million kids moving here every year to try to be an actor or a musician.

RTN: Your CD is right there on the table, so you've obviously been successful, but what about when you first graduated? You said you took a job as a janitor at a nightclub. Was there ever a point where you felt it wasn't worth it?

PATRICK: I never really looked at it like that. Feeling like this is what I am—I'm not sure if I could do other things—this is what I want to do and this is what I want to be. It's just a matter of putting it out there. ■

As we spoke to Patrick, I realized that had this interview taken place three weeks earlier, we all would've been rattled early on. It's not that Patrick wasn't talkative or didn't have a great story to tell, but I think that as someone who has to deal with the media and interviews more than most, he was a bit weary of the whole thing, and it took him a little while to get a feel for the project. But I was proud of the way we handled it and even more proud that we stayed on course. It ended up being a great, honest talk about how and why Patrick had stuck it out in such a difficult industry.

By the time we finished up with Patrick, grabbed a quick bite at a burrito place, and made it back across town, it was time for the show Jill had invited us to. We snuck into our seats just as the lights were going down. I had the feeling every single person there was a waiter, by which I mean an out-of-work writer. The show was kind of a half comedy, half monologue—definitely very Hollywood. A lot of it was really funny, and we all enjoyed ourselves. We appreciated the tickets and probably would've been thrilled with anything that delayed the inevitable at that point. But the show finished up and we thanked Jill again; it was time to be on our way.

The ride back to Laguna Beach was surreal. Randy, Mike, and I were all fairly quiet, with the exception of some closing comments we did in front of the camera that were most likely incoherent. Everything was rushing at us at once; though we'd been trying to hold off on analysis, as the experience came to an end there was a lot on our minds. I felt like I needed to try and synthesize or crystallize the experience in a way that I could take back across the country with me so I wouldn't lose sight of it in New York, where I'd immediately be back in the routine of my job, internship, and everything else.

DAY 20: Roadtrip Nation HQ

Back at the Roadtrip Nation headquarters in Laguna Beach, we spent some time looking at the signatures we had collected in the back room. It was impressive. We talked about a few of the interviews, particularly things that stuck out in our minds and that we were reminded of while looking at the names. We packed our clothes and added our own signatures and messages to the back wall. I know what has stayed with us, which parts of the trip still resonate. I remember guiding the RV, just hanging on for the ride, and, most of all, having the time of our lives. With our bags in tow, we locked up the door for the last time, for our last day on the road . . . for the time being. The trip had ended, but that only meant we were starting out on other roads.

And so we started . . .

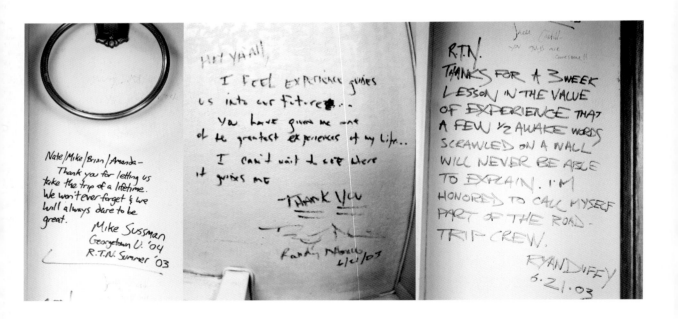

DAY 396: New York—1 year later

More than a year removed from our roadtrip, sometimes it seems as though nothing has changed. We're still kids and we're still trying to negotiate our way through each day without safety nets.

But what has changed is that we aren't afraid of making mistakes anymore. The time we spent on the road, talking to folks who shared similar concerns and amazing stories, really changed the way we think about the roads ahead of us. Nothing is etched in stone when you're in the driver's seat, when you are calling the shots—on the road and in life, you're determining your path as you go.

And in that way, our roadtrip experience has actually changed *everything:* trips like these don't alter some cosmetic aspect of your life that you can point to, one year down the road, and say, "Oh look, X used to be Y. I'm different now." Instead, such a trip dramatically alters points of view; entire methods of thinking and understanding strike fundamental changes in the way you live your life. And if this sounds a bit too much, like I'm grossly exaggerating the impact of a few weeks spent living out of a backpack alongside a bunch of other guys in a shitty old RV smelling of armpits and peanut butter and potato chip sandwiches, then I can't blame you. Before I tracked down that RV and got on board, I would've said the same thing. But having experienced the days, cities, and people I met on that trip, I can't believe I ever thought it would be anything but this, anything but all of this.

At the end of the trip, I had a feeling that we all would remember the important stuff. And now, further removed from that summer, those significant memories are even clearer. I'm able to realize what each person and experience meant by simply looking at my life now and seeing how they've affected me in everything I've done since.

Special Thanks

Thank you to everyone who put their energy into creating this book: Cecily Olson. Lori Werstein. Dan Marriner. Holly Taines White. Jasmine Star. Betsy Stromberg. Tom Southern.

Thanks to our student roadtrippers who contributed to this book and believed in a green RV. Ryan Duffy. Mike Sussman. Randy D'Amico. Gloria Pantojas. Bernardo Pantojas. Cristina Barajas. Candace Elliott. Diana Dravis. Erica Cerulo. Andrew Smith. Samantha Weiss. Jeffrey Klaperich.

We would like to thank our families and friends for supporting us.

And a special thanks to our partners—State Farm, Microsoft, and Xbox—for believing in our vision.

College friends MIKE MARRINER, BRIAN McALLISTER, and NATHAN GEBHARD are the founders and directors of Roadtrip Nation. Mike heads up all publishing projects, including the book and the magazine. Brian oversees all educational programs and manages Roadtrip Nation's partnerships with a hundred college career centers across the country. Nathan takes care of the production side of things, including the annual series on public television. When they're not on the road, they operate out of an old house in Laguna Beach, California.

Index